This is a systematic presentation of quantum f
emphasizing both theoretical concepts and experimental applications.

Starting from introductory quantum and classical mechanics, this book develops the quantum field theories that make up the 'standard model' of elementary processes. It derives the basic techniques and theorems that underly theory and experiment, including those that are the subject of theoretical development. Special attention is also given to the derivation of cross sections relevant to current high-energy experiments and to perturbative quantum chromodynamics, with examples drawn from electron–positron annihilation, deeply inelastic scattering and hadron–hadron scattering.

The first half of the book introduces the basic ideas of field theory. The discussion of mathematical issues is everywhere pedagogical and self-contained. Topics include the role of internal symmetry and relativistic invariance, the path integral, gauge theories and spontaneous symmetry breaking, and cross sections in the standard model and in the parton model. The material of this half is sufficient for an understanding of the standard model and its basic experimental consequences.

The second half of the book deals with perturbative field theory beyond the lowest-order approximation. The issues of renormalization and unitarity, the renormalization group and asymptotic freedom, infrared divergences in quantum electrodynamics and infrared safety in quantum chromodynamics, jets, the perturbative basis of factorization at high energy and the operator product expansion are discussed.

Exercises are included for each chapter, and several appendices complement the text.

An Introduction to Quantum Field Theory

AN INTRODUCTION TO QUANTUM FIELD THEORY

GEORGE STERMAN

Institute for Theoretical Physics,
State University of New York at Stony Brook

Published by the Press Syndicate of the University of Cambridge
The Pitt Building, Trumpington Street, Cambridge CB2 1RP
40 West 20th Street, New York NY 10011-4211, USA
10 Stamford Road, Oakleigh, Melbourne 3166, Australia

© Cambridge University Press 1993

First published 1993

Printed in Great Britain at the University Press, Cambridge

A catalogue record of this book is available from the British Library

Library of Congress cataloguing in publication data

Sterman, George.
An introduction to quantum field theory / George Sterman.
p. cm.
Includes bibliographical references and index.
ISBN 0-521-32258-8. – ISBN 0-521-31132-2 (pbk)
1. Quantum field theory. I. Title.
QC174.45.S78 1993
530.1'43–dc20 92-29556 CIP

ISBN 0 521 32258 8 hardback
ISBN 0 521 31132 2 paperback

KT

For Elise, Adrienne
and my parents

Contents

Preface

The search for the underlying structure of physical reality is as old as speculative thought. Our deepest experimental insights to date are expressed in the language of quantum field theory, in terms of particles that interact at points in space–time, subject to the constraints of special relativity. The theoretical developments that lead to this portrait are the subject of this book. Its aim is to provide a self-contained introduction to relativistic quantum field theory and its applications to high-energy scattering. Some of the methods described predate quantum theory, while others are quite recent. What makes them vital is not only their considerable success thus far, but also the very limitations of that success.

There is every reason to believe that quantum field theory is not a closed chapter. A great deal of freedom remains in the choice of particles and their interactions within the field-theoretic description of fundamental processes. The 'standard model', which describes elementary processes as they are known at this time, is a grab bag of matter and forces, in which breathtaking theoretical elegance coexists with seemingly senseless arbitrariness. Whatever the next step in our understanding of elementary processes, however, the elements of quantum field theory will remain relevant to their description.

Quantum field theory is a vast subject, and an introductory presentation necessarily involves choices of emphasis and of omission. My approach begins with the fundamental considerations of space–time and internal symmetry. These issues are at the heart of gauge invariance, which plays a dominant role in modern field theory. I have emphasized group theory, as a description of the symmetry and invariance properties that are required of any field. In addition, I have chosen to concentrate on the perturbative description of scattering as the best window into the underlying structure of the relevant fields and as the source of our most direct knowledge of the corresponding theories. These topics occupy the first half of the book.

The central formal issues in perturbative field theory concern the self-consistency of the quantum-mechanical expansion, which is the subject of the third quarter of the book. The history of quantum field theory has been driven, in large measure, by a creative tension between the demands of renormalizability and unitarity and the aesthetics of symmetry. Finally, in the fourth part, I discuss the structure of the perturbative expansion at higher orders, by which we may confront quantum field theories with precision experiments. This is a process which, despite some signal successes, is still in its infancy.

This book is intended as both text and reference. Its introductory and intermediate chapters (through Chapter 11) are directed at a reader familiar with classical and quantum mechanics, as taught at the advanced undergraduate or beginning graduate level. Familiarity at a similar level with electromagnetism and complex variable analysis are also assumed, as well as an aquaintance with the basic taxonomy of elementary particles such as electrons, neutrinos and quarks. The more advanced chapters can be used by both those familiar with the first half of the book, or with another text in field theory. I have tried to keep the discussion pedagogical and self-contained, with an emphasis on calculation. Without attempting mathematical rigor, I have tried to indicate where it may be found, as well as the nature of more advanced and formal arguments.

The material outlined below is probably more than can be conveniently presented in a year's course. As a result, it has been organized so that certain more advanced topics, although presented in their natural place in a logical progression, may be bypassed without loss of coherence. An introductory course might extend up to Section 12.4, with the omission of Sections 3.2, 3.3, 8.6, 9.5, 9.6, 10.4, 11.2–11.4 and 12.3. No essential cross references to these sections are made until Chapter 13.

The discussion is divided broadly into four parts. The first develops the methods of field theory through scalar fields. This somewhat simplified context is used to introduce the fundamental applications of group theory (Chapters 1 and 2), canonical quantization and the S-matrix (Chapter 2), the path integral and Feynman rules for diagrams and integrals (Chapter 3), and cross sections (Chapter 4). Many readers may already be familiar with some of the material in Chapter 1, especially group theory and Lorentz transformations. These topics are so central to what follows, however, that I considered it necessary to include them.

In the second part, realistic theories, with intrinsic angular momentum, are introduced and quantized, from the point of view of their space–time symmetries (Chapters 5, 6). Nonabelian fields and spontaneous symmetry breaking are introduced at the outset, on an equal footing with abelian fields. Chapter 6 includes an elementary introduction to unitary representations of the Poincaré group, and their relation to field quantization. Chapters 7 and 8 develop the Feynman rules for the components of the standard model, and give lowest-order applications to experimentally relevant cross sections. Here, representative examples are chosen from quantum electrodynamics, low-energy weak interactions and quantum chromodynamics, including a discussion of the role of ghost fields. Chapter 8 concludes with a brief introduction to the parton model, viewed as a way of interpreting cross sections in quantum chromodynamics.

As mentioned above, the order of chapters within the first two parts follows a presentation in which spin is introduced only after a relatively extensive discussion of scalar fields, up to the computation of cross sections at tree level. This approach is a matter of taste, however, and I have organized the material so that the first eight chapters may also be read in the order 1, 5, 2, 6, 3, 7, 4, 8. Realistic field theories are then discussed earlier, but it takes a little longer to get to the first cross section.

The third part deals with questions of renormalization and unitarity. The method of dimensional regularization is explained, and employed throughout. Chapter 9 also includes an introduction to various general features of Feynman integrals, including Wick rotation, time-ordered perturbation theory and perturbative unitarity. Renormalization is discussed in Chapters 10 and 11, including renormalization schemes and scales, and the renormalization group. For gauge theories, I have concentrated on the issue of unitarity. At the end of Chapter 11, the axial anomaly is used to illustrate the crucial issue of the consistency of classical symmetries at the quantum level.

Finally, in Part IV I undertake a more extensive discussion of perturbative cross sections, with an emphasis on results that extend to all orders in perturbation theory. Loop corrections in quantum electrodynamics are discussed, as well as the infrared problem at all orders. For quantum chromodynamics, I emphasize the concept of infrared safety, and its connection with renormalization, jets and the determination of the strong coupling constant. Chapter 13 treats the analytic structure of Feynman diagrams, dispersion relations and collinear divergences and presents a proof of the Kinoshita–Lee–Nauenberg theorem. Chapter 14 deals with the basis of our understanding of high-energy scattering, including factorization, evolution and the operator product expansion. Although the first two of these results are often identified with perturbative QCD, they are introduced here as general results in field theory. Finally, Chapter 15 briefly presents two issues that go beyond fixed orders in perturbation theory: bound states, and the likely asymptotic nature of the perturbative series.

In the appendices several important topics that do not fit naturally into any chapter are treated. Included are a review of the interaction picture, a derivation of symmetry factors and generating functionals and descriptions of the full standard model Lagrangian, the discrete symmetries of time reversal and charge conjugation, the Goldstone theorem and the role of the chiral anomaly in neutral pion decay. Finally, two appendices consist of *de rigeur* summaries of some useful formulas and Feynman rules.

There are two sets of references within the text. The first set is included for purposes of attribution, and often for historical interest. The second consists of references to reviews and other texts, where more discussion on a relevant topic may be found. These are marked with an 'r' before the date, as for example (r1980). The 'r' is not included in the reference list at the conclusion. In the list of references, the location of each reference is given by section in parenthesis (an 'i' indicates that the reference occurs in the introductary comments in a chapter, an 'e' that it occurs in the exercises).

The list of topics I have omitted would be longer than the list of those included. Generally, however, topics that go under the rubric of 'nonperturbative' have been slighted, including instantons and other types of vacuum structure in QCD, as well as lattice gauge theory. Loop corrections in the weak interactions have also been omitted. I have left out any discussion of supersymmetry, and other extensions of the known symmetries of the standard model. Beyond this, the great questions of gravity and of the nature of space–time remain unaddressed here, and in the standard model. In time, one may expect new theoretical structures to emerge, involving new, more unified field theories, perhaps with supersymmetry, and/or a substructure underlying the fields themselves, perhaps string theory. Whether they fall into the categories already known at this time or not, such theoretical structures will be interpreted in terms of, and their success will be measured by, the quantum field theories of today. I therefore hope that many readers will find what is included here useful in their research and their understanding.

Acknowledgements and thanks

I would like to express my appreciation to the John Simon Guggenheim Foundation for its support in the initial stages of this book, and to the Institute for Advanced Study, Princeton, for its hospitality during that period. I also owe an ongoing debt of gratitude to the Institue for Theoretical Physics at the State University of New York at Stony Brook, to its Director Chen Ning Yang, and to my other colleagues there, for making it possible to undertake this project. Their observations, questions and remarks in passing have continually given me occasion for thought, and have found echos in the pages that follow.

It is commonplace, but none the less true, that by teaching we learn, and I have been fortunate in the excellent group of students with whom I have interacted at Stony Brook. Many have helped me by reading portions of the manuscript, and by giving suggestions. Others graduated even before the book was begun, but all have in one way or another helped in its development. In this regard, I would like to thank in particular Lyndon Alvero, David Appel, Rahul Basu, James Botts, Claudio Coriano, Ghanashyam Date, Vittorio Delduca, Fabian Essler, Keith Kastella, Eric Laenen, Jose Labastida, Hsiang-nan Li, Ma Luo, Lorenzo Magnea, Joseph Milana, Sunil Mukhi, Annibal Ramalho, Ashoke Sen and Michael Sotiropoulos.

Another group of physicists has made this book possible through their encouragement at various crucial stages in my career. In this connection, I would like particularly to recognize Alex Dragt, Joseph Sucher and Ching-Hung Woo, who helped me, as a graduate student, develop a style of research, Shau-Jin Chang and Jeremiah Sullivan, who were so helpful in the early stages of my independent research and Thomas Applequist and Toshiro Kinoshita for generously recognizing that research at a time when it might easily have been forgotten. In addition, thanks are due especially to my collaborators in perturbative QCD, John Collins, Stephen Libby, Robert Ore, Jr, Jianwei Qiu, Dave Soper and Steven Weinberg. Both directly and indirectly, their insights have helped me find my way through the mazes that had to be traversed in the course of writing this book. Thanks also go to Jianwei Qiu for his invaluable and generous help in preparing the figures.

A book in preparation is not always left at the desk, and I thank my wife, Elise, for her patience and constant encouragement, and my daughter, Adrienne, for distracting me from it.

George Sterman

Stony Brook, NY
February 1993

PART I

SCALAR FIELDS

1
Classical fields and symmetries

Our discussion begins with the action principle for classical fields, from which may be derived the field equations of motion. The symmetry properties of a field theory profoundly constrain its time development, through conservation laws. We introduce the Klein–Gordon Lagrangian to illustrate classical space–time and internal symmetries.

1.1 Action principle

Hamilton's principle in point mechanics

A system in classical mechanics is described by a set of generalized coordinates $\{q_i\}$, along with a Lagrangian $L(\{q_i, \dot{q}_i\})$, which depends on the q_i and their associated velocities $\{\dot{q}_i \equiv \mathrm{d}q_i/\mathrm{d}t\}$. The equations of motion for the system are the Lagrange equations,

$$\partial L/\partial q_i - (\mathrm{d}/\mathrm{d}t)(\partial L/\partial \dot{q}_i) = 0. \tag{1.1}$$

Equation (1.1) may be derived from Hamilton's principle that the motion of the system extremizes the action:

$$\delta S \equiv \delta \int_{t_1}^{t_2} \mathrm{d}t\, L(q_i(t), \dot{q}_i(t)) = 0, \tag{1.2}$$

where the variation is taken over paths $\{q_i(t)\}$ between any fixed boundary values, $\{q_i(t_1)\}$ and $\{q_i(t_2)\}$. (The derivation of eq. (1.1) from eq. (1.2) closely follows the field theory argument to be given below.)

Local field theory

In field theory, the analogue of the generalized coordinates, $\{q_i(t)\}$, is a field $\phi(\mathbf{x}, t)$, in which the discrete index i has been replaced by the continuous position vector $\mathbf{x}$. The position $\mathbf{x}$ is *not* a coordinate, but rather a parameter that labels the field coordinate ϕ at point $\mathbf{x}$ at a particular time t. There may be more than one field at each point in space, in which case the fields may carry a distinguishing subscript, as in $\phi_a(\mathbf{x}, t)$. To qualify as a mechanical system, the fields must be associated with a Lagrangian, which determines their time development. Each field describes an infinite number of coordinates, however, and we must make specific assumptions about the Lagrangian to make the system manageable.

We shall be interested in *local* field theories, in which the dynamics does not link different points in space instantaneously. It is then natural to assume that the Lagrangian may be written as an integral over another function, called the *Lagrange density*, $\mathscr{L}$,

$$L(t) = \int d^3\mathbf{x}\, \mathscr{L}(\mathbf{x}, t), \tag{1.3}$$

which depends on the set of fields and their first derivatives,

$$\mathscr{L}(\mathbf{x}, t) = \mathscr{L}(\phi_a(\mathbf{x}, t), \partial\phi_a(\mathbf{x}, t)/\partial x^\mu). \tag{1.4}$$

Conventions

In eq. (1.4) and the following, we employ the conventions

$$v^\mu = (v^0, \mathbf{v}) = g^{\mu\nu} v_\nu, \quad v_\mu = (v^0, -\mathbf{v}) = g_{\mu\nu} v^\nu, \tag{1.5a}$$

for any vector v^μ, where the metric tensor $g_{\mu\nu}(=g^{\mu\nu})$ is the diagonal matrix with nonzero elements $(1, -1, -1, -1)$. For derivatives with respect to the coordinate vector $x^\mu = (x^0, \mathbf{x})$ we use the notation (see Section 1.5),

$$\partial_\mu \phi_a \equiv \partial\phi_a/\partial x^\mu, \quad \partial^\mu \phi_a \equiv \partial\phi_a/\partial x_\mu. \tag{1.5b}$$

Here $x^0 \equiv ct$, where c is the speed of light, so that all the x^μ have dimensions of length. Generally, we shall use lower case Greek letters $(\alpha, \beta, \ldots \mu, \nu, \ldots)$ for space–time vector indices (0, 1, 2, 3), and lower case italic letters $(i, j, \ldots)$ for purely spatial vector indices (1, 2, 3). Finally, except where explicitly indicated, we shall use the convention that repeated indices are summed.

Lagrange equations

Since we are interested in local field theories, the fields in any region, R, of space communicate with the rest of space only through their behavior at the surface σ of R. Thus, if we specify the values of the fields everywhere in R at times t_1 and t_2, and on the surface σ for all times $t_1 < t < t_2$, we ought to have enough information to determine the fields everywhere in R for all times between t_1 and t_2. And indeed, we can derive equations of motion for any Lagrangian of the form (1.3), by demanding that the action S within R be extremal:

$$\delta S = \delta \int_{t_1}^{t_2} \mathrm{d}t \int_R \mathrm{d}^3 \mathbf{x}\, \mathscr{L}(\mathbf{x}, t) = 0. \tag{1.6}$$

This variation is over all possible fields $\phi_a(\mathbf{x}, t)$ inside R,

$$\phi_a(\mathbf{x}, t) \to \phi_a(\mathbf{x}, t) + \epsilon \zeta_a(\mathbf{x}, t), \tag{1.7}$$

where ϵ is an infinitesimal parameter and where the function $\zeta_a(\mathbf{x}, t)$ satisfies

$$\zeta_a(\mathbf{x}, t_1) = \zeta_a(\mathbf{x}, t_2) = 0, \quad \zeta_a(\mathbf{y}, t) = 0, \quad \mathbf{y} \text{ on } \sigma, \tag{1.8}$$

but is otherwise arbitrary. For a fixed $\zeta_a(\mathbf{x}, t)$, the variation in the action is given by

$$\delta S = \epsilon \frac{\partial S}{\partial \epsilon} = \epsilon \int \left[\frac{\partial \mathscr{L}}{\partial \phi_a} \zeta_a + \frac{\partial \mathscr{L}}{\partial(\partial^\mu \phi_a)} \partial^\mu \zeta_a \right] \mathrm{d}^3\mathbf{x}\, \mathrm{d}t. \tag{1.9}$$

Next, we integrate by parts, using eq. (1.8) to eliminate end-point contributions:

$$\delta S = \epsilon \int \left\{ \frac{\partial \mathscr{L}}{\partial \phi_a} - \frac{\partial}{\partial x_\mu} \left[\frac{\partial \mathscr{L}}{\partial(\partial^\mu \phi_a)} \right] \right\} \zeta_a\, \mathrm{d}^3\mathbf{x} \mathrm{d}t = 0. \tag{1.10}$$

This result must be true for *every* choice of ϵ and ζ_a, so we conclude that

$$\frac{\partial \mathscr{L}}{\partial \phi_a} - \frac{\partial}{\partial x_\mu} \left[\frac{\partial \mathscr{L}}{\partial(\partial^\mu \phi_a)} \right] = 0 \tag{1.11}$$

at *every* point inside R for every time between t_1 and t_2. But eq. (1.11) must then hold at every point in space–time, since t_1, t_2 and R were chosen arbitrarily. These are the Lagrange equations for any fields $\phi_a(\mathbf{x}, t)$ that satisfy the assumptions embodied in eqs (1.3) and (1.4).

In point mechanics, the Lagrange equations are total differential equations in time, one for each coordinate. In contrast, eq. (1.11) gives one partial differential equation, involving both spatial and time derivatives, for each field ϕ_a.

1.2 Relativistic scalar fields

There are many examples in which an infinite set of total differential Lagrange equations, eq. (1.1), has a limit in a single partial differential equation of the type (1.11) (see exercise 1.1). The corresponding field theory may then be thought of as the continuum limit of a discrete system. We shall not generally take this viewpoint, however, and instead accept a continuum description as given. The inspiration for particular field theories as elementary processes has most often been found in underlying invariance principles (see Section 1.3). Thus, to begin with, we are interested in theories with definite behavior under the transformations of special relativity. The simplest of these involve relativistic scalar fields.

A scalar field $\phi(x)$ is one whose value at any given space–time point is the same in all inertial frames. Suppose we have two inertial coordinate systems x^μ and $\bar{x}^\mu$ connected by the Lorentz transformation $\Lambda^\mu{}_\nu$:

$$\bar{x}^\mu = \Lambda^\mu{}_\nu x^\nu. \tag{1.12}$$

Let us denote by $\phi(x) = \phi(\Lambda^{-1}\bar{x})$ the field as observed in the x^μ coordinate system and by $\bar{\phi}(\bar{x})$ the same field as observed at the same physical point, but in the $\bar{x}^\mu$ coordinate system. Then if $\phi(x)$ is a scalar field, these two values are equal:

$$\bar{\phi}(\bar{x}) = \phi(\Lambda^{-1}\bar{x}). \tag{1.13}$$

This relation defines the function $\bar{\phi}(\bar{x})$ in terms of the function $\phi(x)$. Here and below we use *passive transformations*, in which the coordinate system changes, while 'physical' objects (fields, points in space, etc.) remain fixed.

Klein–Gordon Lagrangian

The basic Lagrange density for a single relativistic scalar field is the scalar function

$$\mathscr{L} = \tfrac{1}{2}\partial_\mu\phi(x)\partial^\mu\phi(x) - \tfrac{1}{2}(Mc/\hbar)^2\phi^2(x) - V(\phi). \tag{1.14}$$

$\mathscr{L}$ is known as a Klein–Gordon Lagrange density, and the field $\phi(x)$ as a Klein–Gordon field (Schrödinger 1926, Gordon 1927, Klein 1927). $V(\phi)$ is a 'potential' that is a function of the field. To be specific, we may think of it as a polynomial in $\phi(x)$.

Units

In eq. (1.14), c is the speed of light, and we take the constant $\hbar$ to have dimensions of action. As the choice of notation implies, $\hbar$ will be identified with Planck's constant divided by 2π, when we quantize the theory. At present, however, it is just a dimensional quantity. With this choice of units for $\hbar$, M should have units of mass, to make the overall dimensions of the first and second terms in eq. (1.14) the same. To simplify the notation, however, it is useful to define a new parameter, m, which has dimensions of inverse length, by

$$m = Mc/\hbar. \tag{1.15}$$

The Lagrange density (1.14) now takes on the simpler form

$$\mathscr{L} = \tfrac{1}{2}\partial_\mu\phi(x)\partial^\mu\phi(x) - \tfrac{1}{2}m^2\phi^2(x) - V(\phi), \tag{1.16}$$

which we shall use below. In this form, we must require the field $\phi(x)$ to be real, so that the action may be real. $\phi(x)$ has dimensions $(\text{energy/length})^{1/2} = [\text{action}/(\text{length} \times \text{time})]^{1/2}$.

Complex field

A natural generalization to a complex field is

$$\mathscr{L} = \partial_\mu \phi^*(x) \partial^\mu \phi(x) - m^2 |\phi(x)|^2 - U(|\phi(x)|^2), \tag{1.17}$$

where the potential U is a function of the combination $|\phi(x)|^2 = \phi^*(x)\phi(x)$.

Equations of motion

The equations of motion are found from (1.11):

$$\begin{aligned} [(\partial_\mu \partial^\mu) + m^2]\phi(x) &= -\partial V/\partial \phi. \\ [(\partial_\mu \partial^\mu) + m^2]\phi(\mathrm{x}) &= -\partial U/\partial \phi, \end{aligned} \tag{1.18}$$

where, for the complex field, $\phi^*(x)$ is treated as a constant in the derivative of U. In deriving Lagrange equations for the complex field, it is convenient to parameterize ϕ by two real fields according to

$$\phi = 2^{-1/2}(\phi_1 + \mathrm{i}\phi_2), \quad \phi^* = 2^{-1/2}(\phi_1 - \mathrm{i}\phi_2). \tag{1.19}$$

Their equations of motion are $[(\partial_\mu \partial^\mu) + m^2]\phi_i = -\partial \mathrm{U}/\partial \phi_i$.

Free field

Of special interest is the 'free' case, with $U = 0$ or $V = 0$; then

$$[(\partial_\mu \partial^\mu) + m^2]\phi = 0. \tag{1.20}$$

Equation (1.20), of course, implies the equivalent relation for the complex conjugate field. This generalization of the wave equation is known as the Klein–Gordon equation, and its simplicity is part of the reason for considering the Lagrange density eq. (1.16). Its solutions obey the superposition principle, and are of the plane-wave variety:

$$\phi_{\mathbf{k}}(x) = \exp[\pm \mathrm{i}\omega(\mathbf{k})x^0 \mp \mathrm{i}\mathbf{k} \cdot \mathbf{x}], \tag{1.21}$$

where

$$\omega(\mathbf{k}) \equiv \omega_k \equiv (\mathbf{k}^2 + m^2)^{1/2}. \tag{1.22}$$

We shall use these two notations interchangeably below. Note that ω_k has dimensions of $(\text{length})^{-1}$, although it plays the role of a frequency.

Particle analogy

The full particle content of the Klein–Gordon field will only emerge later. Nevertheless, we may gain some insight into this issue, by postulating quantum mechanical relations between frequency and energy, and between wave number and momentum. Suppose we identify the quantity $c\omega(\mathbf{k})$ (which has units of frequency) with an energy, by $\hbar c\omega(\mathbf{k}) \equiv E$, and the wave vector $\mathbf{k}$ with a momentum by $\hbar\mathbf{k} \equiv \mathbf{p}$. Here we define $\hbar \equiv h/2\pi$, h being Planck's constant. If this is the same h as above, then eq. (1.22) is equivalent to the 'mass-shell' equation, $E^2 = p^2c^2 + M^2c^4$, where M is given by eq. (1.15). This suggests a relation between solutions $\phi_{\mathbf{k}}(x)$ to the Klein–Gordon equation with parameter $m = Mc/\hbar$ and particles with momentum $\hbar\mathbf{k}$ and mass M. We make this solution–particle relationship more explicit (and review some elements of quantization) by studying a free relativistic point particle.

First quantization

The action for a classical relativistic point particle may be taken as

$$S = -Mc\int \mathrm{d}s = -Mc^2\int \mathrm{d}t(1 - \mathbf{v}^2)^{1/2}. \tag{1.23}$$

In the first form, $\mathrm{d}s = (\mathrm{d}x_0{}^2 - \mathrm{d}\mathbf{x}^2)^{1/2}$ is the differential invariant length of the particle's world-line. In the second form, $\mathbf{v} = \mathrm{d}\mathbf{x}/\mathrm{d}x^0$ is the velocity (in units of c) and $t = x^0/c$ is the time in any given Lorentz frame. The Lagrangian in this frame is just the integrand of the time integral:

$$L = -Mc^2(1 - \mathbf{v}^2)^{1/2}. \tag{1.24}$$

Now let us consider the quantized version of this particle. To do so, we construct the momenta conjugate to the position coordinates x^i,

$$p_i = \partial L/\partial \dot{x}^i = Mcv_i(1 - \mathbf{v}^2)^{-1/2}, \tag{1.25}$$

where $\dot{x}^i = \mathrm{d}x^i/\mathrm{d}t \equiv cv_i$. The Hamiltonian is

$$H = p_i\dot{x}^i - L = Mc^2(1 - \mathbf{v}^2)^{-1/2} = c(M^2c^2 + \mathbf{p}^2)^{1/2}, \tag{1.26}$$

where we have solved eq. (1.25) to get $\mathbf{v}^2 = \mathbf{p}^2/(M^2c^2 + \mathbf{p}^2)$. Then the Schrödinger equation, $H\phi = \mathrm{i}\hbar\partial\phi/\partial t$ is

$$c(M^2c^2 + \hat{\mathbf{p}}^2)^{1/2}\phi(\mathbf{x}, t) = \mathrm{i}\hbar(\partial/\partial t)\phi(\mathbf{x}, t), \tag{1.27}$$

where $\hat{\mathbf{p}}_i$ is the momentum operator. Proceeding as usual, we represent $\hat{\mathbf{p}}_i$ by the differential operator $-\mathrm{i}\hbar\nabla_i = -\mathrm{i}\hbar\partial/\partial x^i$. Equation (1.27), involving the square root of a derivative, is not so easy to interpret, but its modulus squared, $H^2\phi = -\hbar^2\partial^2\phi/\partial t^2$, is just the free-field Klein–Gordon equation

(1.20). Therefore, the free Klein–Gordon equation is indeed related to the quantum mechanics of a free particle: it follows directly from the Schrödinger equation.

This relationship may be deepened by looking at the nonrelativistic limit. We note that the solutions proportional to $\exp(-i\omega t)$ have time dependence reminiscent of solution of the free-particle Schrödinger equation. Indeed, if we define

$$\psi(\mathbf{x}, t) \equiv e^{imx_0}\phi(x), \tag{1.28}$$

where $\phi(x)$ is such a solution, then, to lowest order in $\mathbf{k}/m$, $\psi(\mathbf{x}, t)$ satisfies the Schödinger equation for a free nonrelativistic particle,

$$-(\hbar^2/2M)\nabla^2\psi = i\hbar(\partial\psi/\partial t). \tag{1.29}$$

The Klein–Gordon equation may thus be thought of as a relativistic generalization of the Schrödinger equation. The identification of $\phi(x)$ with a single-particle wave function is sometimes referred to as the *first quantization* of the Klein–Gordon field. It clearly requires a complex field, at least if it is to describe momentum eigenstates. To anticipate, in Chapter 2 we shall quantize $\phi(x)$ itself, a process known as *second quantization* in view of the interpretation of $\phi(x)$ as a wave function.

Even in the nonrelativistic limit, however, the 'first quantized' interpretation of $\phi(x)$ has an unexpected feature. The general solution to eq. (1.20) is a superposition of plane waves with time dependences of $\exp(-i\omega x_0)$ or $\exp(+i\omega x_0)$ which may be written as (for the normalization, see Chapter 2),

$$\phi(x) = \int \frac{d^3\mathbf{k}}{(2\pi)^{3/2}2\omega(\mathbf{k})} \{a(\mathbf{k})\exp[-i\omega(\mathbf{k})x_0 + i\mathbf{k}\cdot\mathbf{x}] + b^*(\mathbf{k})\exp[i\omega(\mathbf{k})x_0 - i\mathbf{k}\cdot\mathbf{x}]\}. \tag{1.30}$$

$a(\mathbf{k})$ and $b^*(\mathbf{k})$ are arbitrary complex amplitudes, describing the modes with wave vector $\mathbf{k}$. Terms proportional to $a(\mathbf{k})$ have time dependence $\exp[-i\omega(\mathbf{k})x_0]$, like solutions to the Schrödinger equation (1.29). They are referred to as 'positive-energy solutions', and $a(\mathbf{k})$ has a natural interpretation as the amplitude for finding a nonrelativistic particle of momentum $\hbar\mathbf{k}$. Terms proportional to $b^*(\mathbf{k})$, however, have the opposite sign in the exponent, and would seem to correspond to solutions to the Schrödinger equation with negative energy, which have no obvious nonrelativistic interpretation. But in $\phi^*(x)$ the roles of the negative and positive energy solutions are reversed. Since both ϕ and ϕ^* have positive-energy components, we should interpret $b(\mathbf{k})$ as the amplitude for finding another nonrelativistic particle of momentum $\hbar\mathbf{k}$. Thus, even in the nonrelativistic limit, the complex Klein–Gordon field describes not one, but two kinds of particle. This is our first hint of a characteristic feature of relativistic quantum field theory, the inevitability of antiparticles. For now, however, we return to classical field theory, and discuss the role of invariance and symmetry transformations.

1.3 Invariance and conservation

Equations (1.12, 1.13) are one example of a transformation of coordinates and fields. In the following, we consider a general infinitesimal transformation of this type, involving an arbitrary number of fields, which need not be scalars:

$$\begin{aligned} \bar{x}^\mu &= x^\mu + \delta x^\mu(x) = x^\mu + \sum_{a=1}^{N} [\partial x^\mu / \partial(\delta\beta_a)] \delta\beta_a, \\ \bar{\phi}_i(\bar{x}) &= \phi_i(x) + \delta\phi_i(x, \phi_i) = \phi_i(x) + \sum_{a=1}^{N} [\partial\phi_i/\partial(\delta\beta_a)]\delta\beta_a; \end{aligned} \tag{1.31}$$

δx^μ is a function of coordinates, in general, and $\delta\phi_i$ is a function of both the coordinates and the fields. In the second form, we assume that the transformations depend on N infinitesimal parameters $\{\delta\beta_a\}$. Note that $\bar{\phi}_i$ in eq. (1.31) is evaluated at the transformed coordinate value $\bar{x}^\mu$. Equation (1.31) *defines* the transformed fields.

As an example, for an infinitesimal Lorentz transformation and an infinitesimal translation acting on a scalar field, we have

$$\begin{aligned} \delta x^\mu = \delta\lambda^\mu{}_\nu x^\nu + \delta a^\mu, \quad \delta x_\mu = \delta\lambda_{\mu\nu} x^\nu + \delta a_\mu, \\ \delta\phi_i(x) = 0, \end{aligned} \tag{1.32}$$

where $\Lambda^\mu{}_\nu = \delta_{\mu\nu} + \delta\lambda^\mu{}_\nu$ is the infinitesimal Lorentz transformation, and where we define $\delta\lambda_{\mu\nu} = g_{\mu\alpha}\delta\lambda^\alpha{}_\nu$. In this case, the $\delta\beta_a$ may be taken as the δa^μ (or δa_μ) and any independent set of components of the infinitesimal tensor $\delta\lambda$.

Transformed density

Consider a Lagrange density $\mathcal{L}(\phi_i, \partial\phi_i/\partial x_\mu)$, expressed in terms of the untransformed system of coordinates and fields. Let us ask for the Lagrangian in the transformed system. Any density $\mathcal{L}'(\bar{\phi}_i, \partial\bar{\phi}_i/\partial\bar{x}_\mu)$ specifies a set of Lagrange equations in the transformed system. What we require is that every solution $\bar{\phi}_i(\bar{x})$ to these equations should be the transformation of a solution $\phi_i(x)$ to the Lagrange equations in the original system, and that the solutions in the original and transformed systems should be in one-to-one correspondence. In this way, the physical motions generated by the transformed Lagrange density $\mathscr{L}'$ will be in one-to-one correspondence with the physical motions generated by the untransformed density $\mathscr{L}$.

A function that satisfies these criteria is given by

$$\mathscr{L}'(\bar{\phi}_i, \partial\bar{\phi}_i/\partial\bar{x}_\mu) \equiv \mathscr{L}(\phi_i, \partial\phi_i/\partial x_\mu)(\mathrm{d}^4x/\mathrm{d}^4\bar{x}), \tag{1.33a}$$

where $\mathrm{d}^4x \equiv \mathrm{d}x_0\mathrm{d}^3\mathbf{x} = c\mathrm{d}t\mathrm{d}^3\mathbf{x}$. The ratio of differentials on the right ensures that $\mathscr{L}'(\bar{\phi}_i, \partial\bar{\phi}_i/\partial\bar{x}_\mu)$ is a density in the new coordinate system. Let us see why this definition works. Suppose that in the original system we have fields $\phi_i(x)$ in a region R of space–time. In the barred system, the fields are $\bar{\phi}_i(\bar{x})$, and the region is $\bar{R}$. By (1.33a), the actions as seen in the different systems are equal:

$$\frac{1}{c}\int_{\bar{R}} \mathrm{d}^4\bar{x}\, \mathscr{L}'\left(\bar{\phi}_i, \frac{\partial\bar{\phi}_i}{\partial\bar{x}_\mu}\right) = \frac{1}{c}\int_R \mathrm{d}^4x\, \mathscr{L}\left(\phi_i, \frac{\partial\phi_i}{\partial x_\mu}\right). \tag{1.33b}$$

Since the action associated with the transformed fields is invariant, physical fields that are extrema of the action in the original system will give extrema in the transformed system, and will hence satisfy the transformed equations of motion.

Form invariance and symmetry transformations

Again consider the example given by eqs. (1.12) and (1.13), for which $\mathrm{d}^4x = \mathrm{d}^4\bar{x}$. This construction gives the (real) transformed Klein–Gordon Lagrange density as

$$\begin{aligned}\mathscr{L}'(\bar{\phi}, \bar{\partial}^\mu\bar{\phi}) &= \tfrac{1}{2}\{[\partial\bar{\phi}(\bar{x})/\partial x^\mu][\partial\bar{\phi}(\bar{x})/\partial x_\mu] - m^2\bar{\phi}^2(\bar{x})\} - V(\bar{\phi}(\bar{x}))\\ &= \mathscr{L}(\bar{\phi}, \bar{\partial}^\mu\bar{\phi}).\end{aligned} \tag{1.34}$$

The second equality follows after rewriting the derivatives in terms of barred coordinates. $\mathscr{L}'(\bar{\phi}, \bar{\partial}\bar{\phi})$ then has the same functional form as $\mathscr{L}(\phi, \partial\phi)$, and we say the Lagrange density is *form invariant* under the transformation. This means, in particular, that the equations of motion are the same in the barred and unbarred systems. Generally, any transformation that leaves the equations of motion the same is called a *symmetry transformation* of the Lagrange density.

In point mechanics, an invariance of the Lagrangian implies conservation laws. For instance, if the Lagrangian is not explicitly a function of the time, the energy of the system is conserved. Similarly, in a field theory, form invariance under translation and Lorentz transformation leads to a set of invariance principles that include energy, momentum and angular momentum conservation. The relation of invariance to conservation is quite general, and is called Noether's theorem (Noether 1918, Hill r1951).

Noether's theorem

To formulate Noether's theorem, we add to eq. (1.33a) the requirement of form invariance,

$$\mathcal{L}'(\bar{\phi}_i, \partial\bar{\phi}_i/\partial\bar{x}_\mu) = \mathcal{L}(\bar{\phi}_i, \partial\bar{\phi}_i/\partial\bar{x}_\mu), \tag{1.35}$$

to obtain

$$\int_{\bar{R}} \mathrm{d}^4\bar{x}\, \mathcal{L}(\bar{\phi}_i, \partial\bar{\phi}_i/\partial\bar{x}_\mu) - \int_R \mathrm{d}^4x\, \mathcal{L}(\phi_i, \partial\phi_i/\partial x_\mu) = 0. \tag{1.36}$$

The next step is to change variables from barred back to unbarred coordinates in the first integral, using eq. (1.31). This change of variables involves a Jacobian, which is given to first order by $1 + \partial(\delta x_\mu)/\partial x_\mu$. The full first-order term in eq. (1.36) comes from the Jacobian, plus the changes due to the variations of the fields and their derivatives,

$$\int_R \mathrm{d}^4x\{\mathcal{L}[\partial(\delta x_\mu)/\partial x_\mu] + [\partial\mathcal{L}/\partial\phi_i]\delta\phi_i + [\partial\mathcal{L}/\partial(\partial^\mu\phi_i)]\delta(\partial^\mu\phi_i)\} = 0. \tag{1.37}$$

It is now convenient to distinguish that part of the variation of the fields that is due to the difference between $\bar{x}_\mu$ and x_μ alone,

$$\delta\phi_i = \bar{\phi}_i(\bar{x}) - \phi_i(x) = [\partial\phi_i(x)/\partial x_\mu]\delta x_\mu + \delta_*\phi_i(x) + \mathrm{O}(\delta\beta_a{}^2), \tag{1.38a}$$

where we define the 'variation at a point' by

$$\delta_*\phi_i(x) \equiv \bar{\phi}_i(x) - \phi_i(x). \tag{1.38b}$$

(Note that for the scalar field under a Lorentz transformation, $\delta\phi_i = 0$, but $\delta_*\phi_i = -[\partial\phi_i(x)/\partial x_\mu]\delta x_\mu$.) Similarly, we derive from eq. (1.38a) that

$$\delta(\partial^\nu\phi_i) = (\partial^2\phi_i/\partial x_\nu\partial x_\mu)\delta x_\mu + (\partial/\partial x_\nu)(\delta_*\phi_i) + \mathrm{O}(\delta\beta_a{}^2). \tag{1.39}$$

Substituting eqs. (1.38a) and (1.39) into eq. (1.37), we find

$$\int_R \mathrm{d}^4x\{\mathcal{L}[\partial(\delta x_\mu)/\partial x_\mu] + [\partial\mathcal{L}/\partial x_\mu]\delta x_\mu + [\partial\mathcal{L}/\partial\phi_i]\delta_*\phi_i + [\partial\mathcal{L}/\partial(\partial^\mu\phi_i)]\partial^\mu(\delta_*\phi_i)\} = 0, \tag{1.40}$$

where

$$\partial\mathcal{L}/\partial x_\mu = [\partial\mathcal{L}/\partial\phi_i][\partial\phi_i/\partial x_\mu] + [\partial\mathcal{L}/\partial(\partial^\nu\phi_i)][\partial^2\phi_i/\partial x_\mu\partial x_\nu]. \tag{1.41}$$

The Lagrange equations, applied to the third term in eq. (1.40), transform the integrand into the form of a divergence,

$$\int_R \mathrm{d}^4x(\partial/\partial x_\mu)[\mathcal{L}\delta x_\mu + (\partial\mathcal{L}/\partial(\partial^\mu\phi_i))\delta_*\phi_i] = 0. \tag{1.42}$$

Since R is completely arbitrary, it may be taken as small as we like, and eq. (1.42) is then equivalent to the existence of a conserved current,

$$\partial(\delta J^\mu)/\partial x^\mu = 0, \quad \delta J^\mu = -\mathcal{L}\delta x^\mu - \sum_i [\partial\mathcal{L}/\partial(\partial_\mu\phi_i)]\delta_*\phi_i, \tag{1.43a}$$

where the sign is a matter of convention. Once a set of independent parameters for the transformation eq. (1.31) has been identified, we may set their coefficients to zero separately, and derive an independent conserved current for each parameter:

$$\partial J^\mu{}_a/\partial x^\mu = 0, \quad J^\mu{}_a = -\mathcal{L}[\partial(\delta x^\mu)/\partial\beta_a] - \sum_i [\partial\mathcal{L}/\partial(\partial_\mu\phi_i)][\partial(\delta_*\phi_i)/\partial\beta_a]. \tag{1.43b}$$

These results are the content of Noether's theorem. Let us see how it works for the Lorentz transformations and translations of eq. (1.32) acting on a real scalar field with Lagrange density $\mathcal{L}$.

Application to Lorentz transformations

Using eqs. (1.32) and (1.38) in eqs. (1.43), the complete conserved current for the Lorentz transformation on scalar fields, eq. (1.13), may be written (using $\delta x^\mu = g^{\mu\alpha}\delta x_\alpha$) as

$$\delta J^\mu = \left[-\mathcal{L}g^{\mu\alpha} + \sum_i \frac{\partial\mathcal{L}}{\partial(\partial_\mu\phi_i)}(\partial^\alpha\phi_i)\right](\delta\lambda_{\alpha\beta}x^\beta + \delta a_\alpha). \tag{1.44}$$

Since each of the δa_β and $\delta\lambda_{\alpha\beta}$ in eq. (1.44) is independent, their coefficients must also vanish independently.

Consider first the coefficients of δa_ν, the parameters describing translation. We find that

$$\frac{\partial T^{\mu\nu}}{\partial x^\mu} = 0, \quad T^{\mu\nu} = -\mathcal{L}g^{\mu\nu} + \sum_i \frac{\partial\mathcal{L}}{\partial(\partial_\mu\phi_i)}(\partial^\nu\phi_i). \tag{1.45}$$

The divergence of each of the vectors $T^{\mu\nu}$, with ν fixed, vanishes. This implies the existence of four conserved quantities P^ν, where

$$P^\nu = \int d^3\mathbf{x}\, T^{0\nu}, \tag{1.46}$$

and the integral is over space at a fixed time, since

$$\partial P^\nu/\partial x^0 = -\sum_{i=1}^{3}\int d^3\mathbf{x}\,(\partial/\partial x^i)T^{i\nu} = 0. \tag{1.47}$$

The vanishing of the right-hand side of (1.47) follows from Gauss's law and from the assumption that the field is confined to a finite region of space. We can interpret the vector P^ν by looking at its zero component:

$$P^0 = \int d^3\mathbf{x}\, T^{00}, \quad T^{00} = -\mathcal{L} + \sum_i (\partial\mathcal{L}/\partial\dot{\phi}_i)\dot{\phi}_i, \tag{1.48}$$

where we define $\dot{\phi}_i \equiv \partial\phi_i/\partial x^0 = \partial\phi_i/\partial x_0 = (1/c)\partial\phi/\partial t$. Recall that in point mechanics the conjugate momentum for a coordinate q_i is $p_i = \partial L/\partial\dot{q}$, and that the Hamiltonian is given by

$$H = \sum_i p_i\dot{q}_i - L(q_i, \dot{q}_i). \tag{1.49}$$

The quantity

$$\pi_i(\mathbf{x}, x^0) = \partial\mathcal{L}/\partial\dot{\phi}_i(\mathbf{x}, x^0) \tag{1.50}$$

may thus be thought of as a conjugate momentum density, and T^{00} as a Hamiltonian density, so that the complete Hamiltonian of the field is

$$H = \sum_i \int d^3\mathbf{x}\, \pi_i(\mathbf{x}, x^0)\dot{\phi}_i(\mathbf{x}, x^0) - L(x^0). \tag{1.51}$$

In the case of a real Klein–Gordon field, which is form invariant under translations, as we saw in eq. (1.34),

$$\begin{gathered}\pi(\mathbf{x}, x^0) = \dot{\phi}(\mathbf{x}, x^0), \\ H = \frac{1}{2}\int d^3\mathbf{x}\, \{\pi^2(\mathbf{x}, x^0) + [\nabla\phi(\mathbf{x}, x^0)]^2 + m^2\phi^2(\mathbf{x}, x^0)\} + V(\phi).\end{gathered} \tag{1.52}$$

Since the components of P^ν transform as a four-vector, with the total energy as its zero component, it is natural to identify the P^i with the spatial momenta (with dimensions of energy) associated with the field:

$$P^i = -\int d^3\mathbf{x}\, \pi(\mathbf{x}, x^0)[\partial\phi(x^\mu)/\partial x^i]. \tag{1.53}$$

The form invariance of the Lagrange density under translations has thus led to the conservation of four-momentum for the field as a whole.

Very similar reasoning applies to the parameters of the infinitesimal Lorentz transformation $\delta\lambda_{\nu\sigma}$. Since the current, eq. (1.44), is conserved for arbitrary $\delta\lambda_{\nu\sigma}$, and because $\delta\lambda_{\nu\sigma}$ is an antisymmetric matrix (see eq. (1.89) below), we derive, for a single real Klein–Gordon field,

$$\partial_\mu M^{\mu\nu\sigma} = 0, \tag{1.54}$$

$$\begin{aligned}M^{\mu\nu\sigma} &= (x^\sigma g^{\mu\nu} - x^\nu g^{\mu\sigma})\mathcal{L} - [\partial\mathcal{L}/\partial(\partial_\mu\phi)](x^\sigma\partial^\nu\phi - x^\nu\partial^\sigma\phi) \\ &= x^\nu T^{\mu\sigma} - x^\sigma T^{\mu\nu}.\end{aligned} \tag{1.55}$$

We thus find an additional set of conservation laws:

$$\partial J^{\nu\sigma}/\partial x^0 = 0, \ J^{\nu\sigma} \equiv \int \mathrm{d}^3\mathbf{x}\, M^{0\nu\sigma}. \tag{1.56}$$

Examination of eq. (1.55) shows that the J^{ij}, $i, j = 1, 2, 3$, have a natural interpretation as total angular momenta. For, if we identify $p^i \equiv T^{0i}$ as the density of the ith spatial component P^i, eq. (1.53), then we have

$$J^{ij} = \int \mathrm{d}^3\mathbf{x}(x^i p^j - x^j p^i). \tag{1.57}$$

Thus, form invariance of the Lagrange density under Lorentz transformations has led to conservation of angular momentum.

Phase invariance and global symmetry

The energy–momentum and angular momentum tensors for the complex scalar field may be derived in much the same way as for the real scalar field. The Lagrange density of eq. (1.17) is, however, form invariant under yet another symmetry transformation. This is the phase transformation

$$\bar{\phi}(x) = \mathrm{e}^{\mathrm{i}\theta}\phi(x), \quad \bar{\phi}^*(x) = \mathrm{e}^{-\mathrm{i}\theta}\phi^*(x), \quad \bar{x}^\mu = x^\mu, \tag{1.58}$$

whose infinitesimal form is specified by

$$\delta\phi(x) = \mathrm{i}\delta\theta\,\phi(x), \quad \delta\phi^*(x) = -\mathrm{i}\delta\theta\,\phi^*(x), \quad \delta x^\mu = 0, \tag{1.59a}$$

or, in terms of the real fields of eq. (1.19),

$$\delta\phi_1 = -\,\delta\theta\,\phi_2, \quad \delta\phi_2 = \delta\theta\,\phi_1. \tag{1.59b}$$

Such a coordinate-independent transformation is called *global*, because it is the same at every point in space–time. For a global transformation, $\delta_*\phi_i = \delta\phi_i$. When the equations of motion are form invariant under such a transformation, it is known as an *internal symmetry* of the theory. Equations (1.59) are of the general form

$$\delta\phi_i(x) = \mathrm{i}\sum_{a,j}\delta\beta_a(t_a)_{ij}\phi_j, \tag{1.60}$$

where the parameters $\delta\beta_a$ are small and the t_a are fixed matrices. Using eqs. (1.43), the conserved current for such a symmetry transformation is

$$J^\mu{}_a = -\mathrm{i}\sum_{i,j}\frac{\partial\mathcal{L}}{\partial(\partial_\mu\phi_i)}(t_a)_{ij}\phi_j. \tag{1.61}$$

In particular, treating ϕ and ϕ^* as independent components, the infinitesimal transformation (1.58) gives the current (exercise 1.6)

$$J^{\mu} = \mathrm{i}(\phi^{*}\partial^{\mu}\phi - \phi\partial^{\mu}\phi^{*}). \tag{1.62}$$

The conserved quantity

$$Q = \mathrm{i}\int \mathrm{d}^{3}\mathbf{x}(\phi^{*}\partial^{0}\phi - \phi\partial^{0}\phi^{*}) \tag{1.63}$$

is often referred to simply as the total charge associated with the field. More generally, Noether's theorem shows that any internal symmetry results in a conserved current.

1.4 Lie groups and internal symmetries

Group theory is the language of symmetry transformations, and it is now appropriate to discuss the simplest generalizations of the charged scalar field, in order to introduce some relevant concepts and terminology.

Multicomponent fields

A generalization of the charged Klein–Gordon Lagrange density, eq. (1.17), is found by introducing n complex fields, written in the form of a vector, $\boldsymbol{\phi}(x)$,

$$\mathcal{L} = (\partial_{\mu}\boldsymbol{\phi}^{*})\cdot(\partial^{\mu}\boldsymbol{\phi}) - m^{2}\boldsymbol{\phi}^{*}\cdot\boldsymbol{\phi} - V(\boldsymbol{\phi}^{*}\cdot\boldsymbol{\phi}). \tag{1.64}$$

$\mathcal{L}$ is form invariant under the set of global transformations

$$\phi_{i}' = U_{ij}\phi_{j}, \quad U^{\dagger}U = I, \tag{1.65}$$

where U is an $n \times n$ unitary matrix. The set of $n \times n$ unitary matrices defines a Lie group. Let us briefly review what this means. (For more details see, for example, Hamermesh (1962), Wybourne (1974) or Georgi (1982).)

Groups and representations

A group G is a set of abstract elements $\{g_i\}$, along with a rule for multiplication, denoted '·', which together have the following properties.

(i) Closure under '·': if g_i and g_j are in G, then so is $g_i \cdot g_j$.
(ii) Associativity: $g_i\cdot(g_j\cdot g_k) = (g_i\cdot g_j)\cdot g_k$.
(iii) Identity: there is an element g_0 in G such that $g_0 \cdot g_i = g_i$ for every g_i in G.
(iv) Inverse: for every g_i in G there is an inverse element g_i^{-1} such that $g_i^{-1}\cdot g_i = g_0$.

We can easily verify that the $n \times n$ unitary matrices define a group, called $U(n)$, when '·' is matrix multiplication. The $n \times n$ unitary matrices themselves are considered as a matrix *representation* of the group $U(n)$. A matrix representation consists of a set of matrices and a rule that associates with each element g_i of the abstract group a matrix $D(g_i)$ in such a way that if $g_3 = g_2 \cdot g_1$ in the abstract group, then

$$D(g_3) = D(g_2)D(g_1) \tag{1.66}$$

in the sense of matrix multiplication.

The $n \times n$ unitary matrices are, in fact, only one of an infinite set of matrix representations of $U(n)$, called the *defining representation*. Fields obeying eq. (1.65) are said to 'transform according to the defining representation of $U(n)$'. Of particular interest for quantization are *unitary* representations, in which the $D(g_i)$ are unitary matrices. This, of course, is the case for the defining representation of $U(n)$. In addition, for most purposes, we consider *irreducible* representations, which cannot be brought by a similarity transformation into the block diagonal form

$$D^{(R)}(g) = \begin{pmatrix} D^{(R')}(g) & 0 \\ 0 & D^{(R'')}(g) \end{pmatrix} \tag{1.67}$$

for every element $D^{(R)}(g)$ in representation R. Fields that transform according to irreducible representations cannot be broken up into disjoint subspaces.

Lie groups and algebras

$U(n)$ is a *continuous* group. Each element g_i of a continuous group G may be considered as a function of a set of parameters $\{\beta_a\}$, with $a = 1, \ldots, d(G)$, where $d(G)$ is the *dimension* of the group. A *Lie group* has the following additional property. If three elements of the group are related by multiplication,

$$g_3(\gamma_c) = g_2(\beta_b) \cdot g_1(\alpha_a), \tag{1.68}$$

then the $\{\gamma_c\}$ are analytic functions of the $\{\beta_b\}$ and $\{\alpha_a\}$.

Let R denote some representation of a Lie group, and suppose we parameterize the group so that the parameters of the identity element g_0 are $\{\beta_a = 0\}$. Now consider elements that are close to the identity in the sense that their parameters are all infinitesimal. Any such element of representation R may be written in terms of matrices $D^{(R)}(\beta_a)$ known as the *generators* of the group $\{t_a{}^{(R)}\}$ in that representation:

$$D^{(R)}(\beta_a) = 1 + \mathrm{i} \sum_{a=1}^{d(G)} \delta\beta_a \, t_a{}^{(R)}, \tag{1.69}$$

where $d(G)$ is the dimension of the group, and the $\delta\beta_a$ are independent real infinitesimal parameters.

The generators of the $n \times n$ unitary matrices may be taken as n^2 independent hermitian $n \times n$ matrices. Equation (1.69) may be extended to elements with finite parameters β_a by using the identity

$$D^{(R)}(\beta_a) = \lim_{N\to\infty} [I + (\mathrm{i}/N)\sum_a \beta_a t_a{}^{(R)}]^N = \exp\left[\mathrm{i}\sum_a \beta_a t_a{}^{(R)}\right]. \quad (1.70)$$

This exponential form holds as long as the limit exists. We should note that the explicit factor of i in eq. (1.69) is often absorbed into the definition of the $t_a{}^{(R)}$, which then become antihermitian.

For every choice of representation R, the generators $\{t_a{}^{(R)}\}$ obey the matrix commutation relations

$$[t_a{}^{(R)}, t_b{}^{(R)}] = \mathrm{i}C_{ab}{}^c t_c{}^{(R)}, \quad (1.71)$$

where the numbers $C_{ab}{}^c$, called the *structure constants* of the group, are real and independent of R. They are obviously antisymmetric in the first two indices. (Again, if a factor of i is absorbed into the $t_a{}^{(R)}$, the i disappears in eq. (1.71).) The $t_a{}^{(R)}$ specify the representation R of the *Lie algebra* of the group, which is the algebra of the generators of the abstract group. A Lie group may be reconstructed from a knowledge of its structure constants, through eq. (1.70).

The nonabelian groups U(2) and SU(2)

The simplest example of an invariance group has already been encountered for the case when $\boldsymbol{\phi}$ in eq. (1.64) is a vector with one complex component only, as in eq. (1.17). Then the Lagrange density is form invariant under the phase transformations of eq. (1.58). The set of all phases $\exp(\mathrm{i}\theta)$ is the defining representation of the group $U(1)$, the one-dimensional unitary group. There is a single generator for this group, which may be taken simply as the number one – the unit matrix in one dimension.

A more representative example is given when $\boldsymbol{\phi}$ has two complex components. The invariance group is then $U(2)$, defined by the set of two-by-two unitary matrices. The determinant of a unitary matrix is a phase and, for an element u of $U(2)$, we may write

$$u = \mathrm{e}^{\mathrm{i}\theta_u} s, \quad \det s = 1. \quad (1.72)$$

An arbitrary matrix in $U(2)$ is thus the product of an element of $U(1)$ times a unitary matrix of unit determinant. These matrices themselves form a group, called $SU(2)$ (*special unitary* group of 2×2 matrices). More formally, we can write $U(2) = U(1) \times SU(2)$. That is, $U(2)$ is the direct product of $U(1)$ and $SU(2)$.

The generators of $SU(2)$ may be identified from the following general form for any 2×2 unitary matrix of unit determinant (exercise 1.8):

$$s = \exp\left(\mathrm{i}\sum_{a=1}^{3} \beta_a \tfrac{1}{2}\sigma_a\right), \tag{1.73}$$

where the σ_a are the usual Pauli matrices

$$\sigma_1 = \begin{pmatrix} 0 & 1 \\ 1 & 0 \end{pmatrix}, \quad \sigma_2 = \begin{pmatrix} 0 & -\mathrm{i} \\ \mathrm{i} & 0 \end{pmatrix}, \quad \sigma_3 = \begin{pmatrix} 1 & 0 \\ 0 & -1 \end{pmatrix}. \tag{1.74}$$

Thus, the generators of $SU(2)$ are conventionally chosen as $\frac{1}{2}\sigma_a$, and satisfy the Lie algebra

$$[\tfrac{1}{2}\sigma_a, \tfrac{1}{2}\sigma_b] = \mathrm{i}\epsilon_{abc}\tfrac{1}{2}\sigma_c, \tag{1.75}$$

in which the structure constants are $C_{ab}{}^c = \epsilon_{abc}$. Note that $\det s = 1$ follows immediately from the tracelessness of the Pauli matrices and the relation (exercise 1.9),

$$\det(\exp A) = \exp(\operatorname{tr} A), \tag{1.76}$$

valid for any matrix A.

Since the generators of $SU(2)$ do not mutually commute, neither do the elements of the group itself. Such a group is called *nonabelian*, while a group like $U(1)$, in which multiplication is commutative, is *abelian*. In addition, there is no subset of generators of $SU(2)$ that commute with all the other generators. $SU(2)$, or any other Lie algebra with this negative property, is called *semisimple*. All the Pauli matrices, of course, commute with the generator of the overall phase in $U(2)$, which is the unit matrix. The Lie algebra of $U(2)$ is thus not semisimple. This reasoning generalizes to $U(n)$, which may be regarded as $U(1) \times SU(n)$, with $SU(n)$ the $n \times n$ unitary matrices of unit determinant. As we shall see below, the two invariance groups $U(1)$ and $SU(n)$ in $U(n)$ are typically associated with different physics in quantum field theories.

Conserved currents

The parametrization of infinitesimal group elements, eq. (1.69), makes the treatment of internal symmetries according to Noether's theorem particularly straightforward. Suppose that the Lagrange density $\mathscr{L}$ is form invariant when its fields transform according to representation R of the group G:

$$\delta\phi_i(x) = \mathrm{i}\sum_{a,j} \beta_a[t_a{}^{(R)}]_{ij}\phi_j(x), \quad \delta\phi_i{}^*(x) = -\mathrm{i}\sum_{a,j} \beta_a[t_a{}^{(R)}]_{ij}{}^*\phi_j{}^*(x). \tag{1.77}$$

Then eq. (1.61) gives the set of independent conserved currents

$$J_a{}^\mu = \mathrm{i}\sum_{i,j}\left\{\frac{\partial\mathscr{L}}{\partial(\partial_\mu\phi_i{}^*)}\,[t_a{}^{(R)}]_{ij}{}^*\phi_j{}^* - \frac{\partial\mathscr{L}}{\partial(\partial_\mu\phi_i)}\,[t_a{}^{(R)}]_{ij}\phi_j\right\}, \tag{1.78}$$

where the derivative with respect to ϕ^* is carried out with ϕ treated as a constant, and vice versa (exercise 1.6).

For a two-component complex field in the Lagrange density eq. (1.64) the conserved currents associated with $SU(2)$ invariance are

$$J_a{}^\mu = \mathrm{i}\sum_{i,j}[(\partial^\mu\phi_i)(\tfrac{1}{2}\sigma_a{}^*)_{ij}\phi_j{}^* - (\partial^\mu\phi_i{}^*)(\tfrac{1}{2}\sigma_a)_{ij}\phi_j], \tag{1.79}$$

in direct generalization of the $U(1)$ current, eq. (1.62).

1.5 The Poincaré group and its generators

The applications of group theory are not limited to internal symmetries, and coordinate transformations may be regarded in the same light. To close this chapter, we review some relevant features of coordinate transformations, first showing how they define a Lie group.

Lorentz transformations

Let $x^\mu \equiv (x^0, \mathbf{x})$ denote a position vector with a particular choice of coordinates. As above, a Lorentz transformation is a change to another set of coordinates $\bar{x}^\mu$, related to the old set x^ν by

$$\bar{x}^\mu = \Lambda^\mu{}_\nu x^\nu \tag{1.80}$$

that preserves the length of the coordinate vector:

$$x^2 = x^\mu g_{\mu\nu} x^\nu = \bar{x}^\alpha g_{\alpha\beta}\bar{x}^\beta = \bar{x}^2. \tag{1.81}$$

$g_{\mu\nu}$ is the metric tensor, defined in our convention as

$$g_{\mu\nu} = \delta_{\mu0}\delta_{\nu0} - \delta_{\mu\nu}(1 - \delta_{\mu0}\delta_{\nu0}) = (g^{-1})_{\mu\nu} \equiv g^{\mu\nu}. \tag{1.82}$$

Then, for any vector v^μ, $v^2 = (v^0)^2 - (v^1)^2 - (v^2)^2 - (v^3)^2$. The matrix denoted $g^{\mu\nu}$ is defined to be the inverse of $g_{\mu\nu}$, which in this case happens to equal $g_{\mu\nu}$ itself.

We can easily verify that if eq. (1.81) is to hold for every choice of x^μ, then

$$g^{\mu\gamma}\Lambda^\alpha{}_\gamma g_{\alpha\nu} = (\Lambda^{-1})^\mu{}_\nu \equiv \Lambda_\nu{}^\mu, \tag{1.83}$$

where the final form defines a common notation in which the inverse transformation matrix is identified solely by the placement of its indices.

Covariant and contravariant tensors

This is perhaps a useful time to review a few facts about 'up' and 'down' indices. The conventions we have chosen to describe Minkowski space are really necessary only in the context of general relativity; indeed, by using the trick of making the zero component of every vector imaginary, upper and lower indices may be dispensed with in the 'flat' Minkowski space in which we work (see, for example, de Wit & Smith (r1986)). For our purposes, however, upper (contravariant) and lower (covariant) indices will be useful, and representative of much of the literature.

A *contravariant* vector – denoted by a superscript index – is any vector that transforms under the Lorentz group in the same manner as the coordinates x^μ, eq. (1.80). Not every vector transforms in this way, however. In particular, the chain rule and the linearity of Lorentz transformations can be used to show that the gradient vector $\nabla_\mu \equiv \partial/\partial x^\mu$ transforms according to

$$\bar{\nabla}_\mu = (\partial x^\alpha/\partial \bar{x}^\mu)\partial/\partial x^\alpha = \Lambda_\mu{}^\alpha \nabla_\alpha, \tag{1.84}$$

that is, according to the inverse of Λ (see eq. (1.83)). Vectors transforming according to the inverse of Λ are called *covariant*. Covariant and contravariant vectors are in one-to-one correspondence, since if v^μ is any contravariant vector, transforming according to $\Lambda^\mu{}_\nu$, then eq. (1.83) may be used to show that the vector v_σ, defined by

$$v_\sigma = g_{\sigma\mu} v^\mu \tag{1.85}$$

transforms according to the inverse matrix, $\Lambda_\lambda{}^\sigma$. Similarly, if v'_σ is a covariant vector, then

$$v'^\lambda = g^{\lambda\sigma} v'_\sigma \tag{1.86}$$

is contravariant. A tensor $T^\mu{}_\nu$ with one covariant and one contravariant index is identified by its transformation properties

$$\bar{T}^\lambda{}_\sigma = \Lambda^\lambda{}_\alpha \Lambda_\sigma{}^\beta T^\alpha{}_\beta, \tag{1.87}$$

with the generalization to more indices being straightforward. As for a vector, the indices of tensors may be raised or lowered with the metric tensor.

The Lorentz group

Using eq. (1.83), we recognize the following four properties of the matrices Λ, which together show that $L = \{\Lambda\}$ is the (defining) representation of a group called the Lorentz group.

(i) If $\Lambda^{\mu}{}_{\nu}$ and $\Lambda'^{\nu}{}_{\sigma}$ are Lorentz transformations, then so is $(\Lambda\Lambda')^{\mu}{}_{\sigma} \equiv \Lambda^{\mu}{}_{\nu}\Lambda'^{\nu}{}_{\sigma}$. That is, L is closed under matrix multiplication.
(ii) Multiplication of the Λ's is associative, since matrix multiplication is associative.
(iii) The four-by-four identity matrix $I^{\mu}{}_{\nu} = \delta^{\mu}{}_{\nu}$, is in L, and serves as the identity under matrix multiplication. That is, $I\Lambda = \Lambda I = \Lambda$, for any Λ in L.
(iv) Each matrix Λ has an inverse, given by eq. (1.83).

Note that because the matrix $g_{\alpha\beta}$ is not the identity, a typical Λ is neither orthogonal nor unitary. The defining representation of the Lorentz group is therefore not a unitary representation.

Lie algebra

That L is a Lie group is also rather obvious, since the elements of the matrix $\Lambda\Lambda'$ are algebraic, and therefore analytic, functions of the elements of Λ and Λ'. As a result, we may derive a Lie algebra for the Lorentz group. It is found, as usual, from the infinitesimal transformations $\Lambda^{\alpha}{}_{\beta} = \delta^{\alpha}{}_{\beta} + \delta\lambda^{\alpha}{}_{\beta}$, where $\delta^{\alpha}{}_{\beta} \equiv g^{\alpha}{}_{\beta} = \delta_{\alpha\beta}$. Let us first identify an independent set of parameters for the group.

Multiplying eq. (1.83), for an infinitesimal transformation, from the left by $g_{\beta\mu}$, we have

$$(g^{\alpha}{}_{\beta} + \delta\lambda^{\alpha}{}_{\beta})g_{\alpha\nu} = g_{\beta\nu} - \delta\lambda_{\beta\nu}, \tag{1.88}$$

where we have lowered one index of $\delta\lambda$, according to the rules just outlined. We now use the symmetry of $g_{\alpha\nu}$ to lower another index, and obtain

$$\delta\lambda_{\nu\beta} = -\delta\lambda_{\beta\nu}. \tag{1.89}$$

Thus, as claimed in connection with eq. (1.54) above, an infinitesimal Lorentz transformation is generated by six independent variables, the number of independent components in a four-by-four antisymmetric matrix. In terms of the infinitesimal transformations $(1 + \delta\lambda)^{\mu}{}_{\nu}$, with one upper and one lower index, the relation is slightly different:

$$\delta\lambda^{0}{}_{i} = \delta\lambda^{i}{}_{0}, \quad \delta\lambda^{i}{}_{j} = -\delta\lambda^{j}{}_{i}, \quad \delta\lambda^{\mu}{}_{\mu} = 0 \text{ (no sum)}. \tag{1.90}$$

An arbitrary $\delta\lambda^{\mu}{}_{\nu}$ may thus be expanded in terms of the six independent matrices

$$(J_i)^{\alpha}{}_{\beta} = -\mathrm{i}\epsilon_{0i\alpha\beta}, \quad (K_i)^{\alpha}{}_{\beta} = -\mathrm{i}(\delta_{0\alpha}\delta_{i\beta} + \delta_{i\alpha}\delta_{0\beta}), \tag{1.91}$$

where the four-dimensional ϵ-symbol is defined by $\epsilon_{\alpha\beta\gamma\theta} = +1$ when $(\alpha, \beta, \gamma, \theta)$ is an even permutation of $(0, 1, 2, 3)$, -1 for an odd permutation, and 0 otherwise. Then, for instance, we have

$$K_1 = -\mathrm{i}\begin{pmatrix} 0 & 1 & 0 & 0 \\ 1 & 0 & 0 & 0 \\ 0 & 0 & 0 & 0 \\ 0 & 0 & 0 & 0 \end{pmatrix}, \quad J_1 = -\mathrm{i}\begin{pmatrix} 0 & 0 & 0 & 0 \\ 0 & 0 & 0 & 0 \\ 0 & 0 & 0 & 1 \\ 0 & 0 & -1 & 0 \end{pmatrix}, \tag{1.92}$$

and so on. The J's generate infinitesimal rotations, and the K's generate infinitesimal boosts (velocity transformations). In terms of these generators, we parameterize an arbitrary infinitesimal transformation as

$$(I + \delta\lambda)^\alpha{}_\beta = (I + \mathrm{i}\sum_{i=1}^{3}\delta\omega_i K_i - \mathrm{i}\sum_{i=1}^{3}\delta\theta_i J_i)^\alpha{}_\beta \tag{1.93}$$

(note the minus sign on the J's), where we define

$$\delta\theta_i \equiv \tfrac{1}{2}\epsilon_{0i\rho\sigma}\delta\lambda^{\rho\sigma}, \quad \delta\omega_i \equiv -\delta\lambda^{0i}. \tag{1.94}$$

The matrices generated in this way are not unitary, because the K_i are antihermitian. It is easily verified that the J's and K's satisfy the Lie algebra

$$[J_i, J_j] = \mathrm{i}\epsilon_{ijk}J_k, \quad [J_i, K_j] = \mathrm{i}\epsilon_{ijk}K_k, \quad [K_i, K_j] = -\mathrm{i}\epsilon_{ijk}J_k. \tag{1.95}$$

We note, in particular, that the J_i, which generate rotations alone, satisfy among themselves the Lie algebra, eq. (1.75) of $SU(2)$. Thus, the rotation matrices in three dimensions supply a representation of the group $SU(2)$. (For extensive discussions, see Rose (r1957) or Edmonds (r1960).)

The three relations, eq. (1.95), can be summarized in covariant form by defining generators $m_{\mu\nu}$ with antisymmetric tensor indices:

$$\begin{gathered} m_{0i} = -m_{i0} = K_i, \quad m_{ij} = -m_{ji} = \epsilon_{ijk}J_k, \\ (m_{\mu\nu})^\alpha{}_\beta = \mathrm{i}(g_\mu{}^\alpha g_{\nu\beta} - g_{\mu\beta}g_\nu{}^\alpha). \end{gathered} \tag{1.96}$$

Using eqs. (1.94) and (1.96), the infinitesimal transformation, eq. (1.93), may be rewritten in covariant notation as

$$(I + \delta\lambda)^\alpha{}_\beta = (I - \tfrac{1}{2}\mathrm{i}\delta\lambda^{\mu\nu}m_{\mu\nu})^\alpha{}_\beta. \tag{1.97}$$

The Lie algebra eq. (1.95) then becomes

$$[m_{\mu\nu}, m_{\lambda\sigma}] = -\mathrm{i}g_{\mu\lambda}m_{\nu\sigma} + \mathrm{i}g_{\mu\sigma}m_{\nu\lambda} + \mathrm{i}g_{\nu\lambda}m_{\mu\sigma} - \mathrm{i}g_{\nu\sigma}m_{\mu\lambda}. \tag{1.98}$$

We have derived this algebra in the defining representation, but any matrices or operators that satisfy the same algebra generate a representation of the Lorentz group.

Finite and discrete transformations

Consider a finite Lorentz transformation Λ parameterized as the product of a pure rotation (generated by the J's) followed by a pure boost (generated by the K's). Then, following eq. (1.70), we can represent such a Λ as

$$\Lambda = \exp(\mathrm{i}\boldsymbol{\omega}\cdot\mathbf{K})\exp(-\mathrm{i}\boldsymbol{\theta}\cdot\mathbf{J}). \qquad (1.99)$$

It is easy to show (exercise 1.10) that $\exp(-\mathrm{i}\boldsymbol{\theta}\cdot\mathbf{J})$ acts to rotate any spatial vector by angle θ about the unit vector $\boldsymbol{\theta}/\theta$ according to the right-hand rule, and $\exp(\mathrm{i}\boldsymbol{\omega}\cdot\mathbf{K})$ to boost the vector 'at rest', $(1, \mathbf{0})$, into the vector $(\cosh\omega, \hat{\boldsymbol{\omega}}\sinh\omega)$ with 'velocity' $v = \tanh(\omega)$. Note that the action of Λ on $(1, \mathbf{0})$ alone is enough to determine the boost uniquely, so that the decomposition of Λ into a rotation followed by a boost is unique. Of course, we could have chosen other parameterizations for Λ, but it will be useful to have an explicit form at hand.

Because of eq. (1.76) and the tracelessness of the K's and J's, any Λ parameterized according to eq. (1.99) has $\det\Lambda = +1$. On the other hand, the defining condition eq. (1.83), along with the matrix identity

$$\det AB = \det A \det B$$

shows that in fact we may have $\det\Lambda = -1$ as well. The full set of Lorentz transformations, with $\det\Lambda = \pm 1$, makes up what is sometimes called the *extended Lorentz group*; those of the form (1.99) make up a subgroup called the *restricted Lorentz group*. We can parameterize any element of the extended Lorentz group by introducing two diagonal matrices of determinant -1, the discrete *parity* and *time-reversal* transformation matrices, defined by their diagonal elements

$$\operatorname{diag} \mathrm{P}^{\alpha}{}_{\beta} = (1, -1, -1, -1), \quad \operatorname{diag} \mathrm{T}^{\alpha}{}_{\beta} = (-1, 1, 1, 1).$$

Any element in the extended Lorentz group is in one of the sets $\{\Lambda^{(+)}\}$, $\{T\Lambda^{(+)}\}$, $\{\mathrm{P}\Lambda^{(+)}\}$ or $\{\mathrm{PT}\Lambda^{(+)}\}$, where $\Lambda^{(+)}$ denotes an element of the restricted group. These sets are said to be *disconnected*, because transformations with determinant -1 cannot be reached from the identity by a series of infinitesimal transformations.

Finally, it is interesting to note the action of parity on the elements of the Lie algebra:

$$\mathrm{P}^{\alpha}{}_{\alpha'}\mathrm{P}_{\beta}{}^{\beta'}(J_i)^{\alpha'}{}_{\beta'} = (J_i)^{\alpha}{}_{\beta}, \quad \mathrm{P}^{\alpha}{}_{\alpha'}\mathrm{P}_{\beta}{}^{\beta'}(K_i)^{\alpha'}{}_{\beta'} = -(K_i)^{\alpha}{}_{\beta}. \qquad (1.100)$$

Thus $\mathbf{J}$ behaves as a *pseudovector*, which does not change sign under parity, and $\mathbf{K}$ as a *polar vector*, which does. Like the Lie algebra, these relations are independent of the representation.

The Poincaré group and its generators

To interpret the conservation laws and their relation to solutions of the field equations, it is useful to introduce the Poincaré group, which consists of combined translations and Lorentz transformations. An arbitrary element is represented by the pair (a, Λ), where a^μ is the translation vector, and $\Lambda^\mu{}_\nu$ the Lorentz transformation. The multiplication rule for the group in this form is

$$(a_2, \Lambda_2)\cdot(a_1, \Lambda_1) = (a_2 + \Lambda_2 a_1, \Lambda_2\Lambda_1). \tag{1.101}$$

It is obvious that the parameters of the product transformation on the right are analytic functions of the parameters of the two factor transformations on the left, so the Poincaré group is a Lie group. Let $D(a, \Lambda)$ stand for a representation of the group. Because the Poincaré group is a Lie group, the representations of elements close to the identity, $D(\delta a, 1 + \delta\lambda)$, may be expressed in terms of the generators associated with that representation, just as in eq. (1.69). From this specific set of generators we can, in turn, determine the structure constants of the group (its Lie algebra), and study the group as a whole.

Let us look, then, at a particular set of fields $\phi_a(x)$. We shall consider a generalization of eq. (1.13) given by

$$\bar{\phi}_a(x) = S_{ab}(\Lambda)\phi_b(\Lambda^{-1}\, x - a), \tag{1.102}$$

where for an infinitesimal transformation, we write

$$S_{ab}(1 + \delta\lambda) = \delta_{ab} + \tfrac{1}{2}\mathrm{i}\delta\lambda^{\mu\nu}\,(\Sigma_{\mu\nu})_{ab}. \tag{1.103}$$

The $\Sigma_{\mu\nu}$ are matrices, whose exact form depends on the field in question, but which satisfy the Lorentz algebra eq. (1.98), so that S_{ab} is a representation of the Lorentz group. Of course, for a scalar field $S = 1$ and $\Sigma_{\mu\nu} = 0$. Nontrivial examples are given by contravariant and covariant vector fields, for which $S^\alpha{}_\beta = \Lambda^\alpha{}_\beta$ and $\Lambda_\alpha{}^\beta$, respectively:

$$\bar{A}^\alpha(\Lambda x) = \Lambda^\alpha{}_\beta A^\beta(x), \quad \bar{A}_\alpha(\Lambda x) = \Lambda_\alpha{}^\beta A_\beta(x). \tag{1.104}$$

Then, using the representation, eq. (1.97) for Λ, we have, for the contravariant vector,

$$(\Sigma_{\mu\nu})^\alpha{}_\beta = -\,(m_{\mu\nu})^\alpha{}_\beta = -\,\mathrm{i}(g_\mu{}^\alpha g_{\nu\beta} - g_{\mu\beta}g_\nu{}^\alpha), \tag{1.105}$$

where we have used the definitions given in eqs. (1.96). The analogous matrix for the covariant vector is found simply by lowering α and raising β.

Equation (1.102) does not yet exhibit a representation of the full Poincaré group; this can be derived from the expression for $\bar{\phi}_b(x)$ directly in terms of $\phi_a(x)$, evaluated at the same argument. To this end, we rewrite it for an infinitesimal transformation as

$$\begin{aligned}\bar{\phi}_a(x^\mu) &= [\delta_{ab} + \tfrac{1}{2}\mathrm{i}\delta\lambda^{\mu\nu}(\Sigma_{\mu\nu})_{ab}]\phi(x^\mu - \delta\lambda^{\mu\nu}x_\nu - \delta a^\mu)\\ &\equiv D_{ab}(\delta a, 1 + \delta\lambda)\ \phi_b(x^\mu),\end{aligned} \tag{1.106}$$

where D_{ab} is an operator on the fields, expressed to first order in the parameters of the transformation as

$$D_{ab}(\delta a, 1 + \delta\lambda) = \delta_{ab} - \mathrm{i}\delta a^\mu\ \hat{p}_{\mu,ab} - \tfrac{1}{2}\mathrm{i}\delta\lambda^{\mu\nu}\hat{m}_{\mu\nu,ab}, \tag{1.107}$$

where $\hat{p}_{\mu,ab}$ and $\hat{m}_{\mu\nu,ab}$ are their associated generators. Comparing this with eq. (1.69) for an arbitrary Lie group, we identify the infinitesimal parameters as δa^μ and $\delta\lambda^{\mu\nu}$. Their corresponding generators are

$$\begin{aligned}(\hat{p}_\mu)_{ab} &= -\mathrm{i}\partial_\mu\delta_{ab},\\ (\hat{m}_{\mu\nu})_{ab} &= \mathrm{i}(x_\mu\partial_\nu - x_\nu\partial_\mu)\delta_{ab} - (\Sigma_{\mu\nu})_{ab}.\end{aligned} \tag{1.108}$$

The $\hat{p}$'s are referred to as generators of translations, and the $\hat{m}$'s as the generators of Lorentz transformations. These generators are in part differential operators and in part matrices. In the latter aspect they act on the field indices and in the former on the field arguments. Such representations may seem somewhat more complex than the matrix representations of Section 1.4. The principles, however, are the same. A straightforward calculation using eqs. (1.108) gives the following algebra:

$$\begin{aligned}[][\hat{p}_\mu, \hat{p}_\nu] &= 0,\\ [\hat{p}_\mu, \hat{m}_{\lambda\sigma}] &= \mathrm{i}g_{\mu\lambda}\hat{p}_\sigma - \mathrm{i}g_{\mu\sigma}\hat{p}_\lambda,\\ [\hat{m}_{\mu\nu}, \hat{m}_{\lambda\sigma}] &= -\mathrm{i}g_{\mu\lambda}\hat{m}_{\nu\sigma} + \mathrm{i}g_{\mu\sigma}\hat{m}_{\nu\lambda} + \mathrm{i}g_{\nu\lambda}\hat{m}_{\mu\sigma} - \mathrm{i}g_{\nu\sigma}\hat{m}_{\mu\lambda}.\end{aligned} \tag{1.109}$$

Here matrix products are to be understood, so we have suppressed the field indices (a and b). We shall encounter this algebra again in the next chapter where it is represented by quantum mechanical operators.

Exercises

1.1 It is often possible to derive a field theory as the limit of a discrete system. Perhaps the simplest example is an infinite system of point masses, m, separated by springs of spring constant k and equilibrium length a. Let η_i be the displacement from equilibrium of the ith point mass. Derive the exact Lagrangian and Lagrange equations for this system. Then consider the limit

$$m, a \to 0,\ k \to \infty,\ \mu = m/a \text{ and } Y = ka \text{ fixed.}$$

Replacing η_i by a smooth function $\eta(x, t)$, show that in this limit the Lagrangian may be written in the density form

$$L = \int \mathrm{d}x\ \tfrac{1}{2}[\mu(\partial\eta/\partial t)^2 - Y(\partial\eta/\partial x)^2],$$

and write down the corresponding (partial differential) Lagrange equations. (See Soper r1976, Chapter 2, and Goldstein r1980, Chapter 12).

1.2 The electromagnetic field may be specified by a vector $A^\mu(\mathbf{x}, t)$, in terms of which the Lagrange density of the field is

$$\mathcal{L}(x) = -\tfrac{1}{4} F^{\mu\nu} F_{\mu\nu},$$

where

$$F_{\mu\nu} = \partial_\mu A_\nu - \partial_\nu A_\mu.$$

Derive the Lagrange equations for this system, and express them in terms of the free-space field strengths $\mathbf{E} = -\partial_0 \mathbf{A} - \nabla A_0$ and $\mathbf{B} = \nabla \times \mathbf{A}$. How many of Maxwell's equations does this give, and why are the others also satisfied? (See Jackson r1975, Chapter 6.)

1.3 (a) Show that the following modification of the Lagrange density $\mathcal{L}$,

$$\mathcal{L} \to \mathcal{L} + \partial_\mu G^\mu(\phi, \partial_\nu \phi)$$

where G^μ is any vector functional of the fields, leaves the equations of motion unchanged. Show that the definition of form invariance, eq. (1.35), may be extended to include the case where a Lagrange density transforms as above, and derive the form of the resulting Noether currents.

1.4 Suppose a field $\phi(x)$ satisfies the Klein–Gordon equation, and that $\mathrm{d}\phi(\mathbf{x}, 0)/\mathrm{d}x^0 = 0$. What is the relationship between the positive and negative frequency amplitudes in eq. (1.30)? Give a physical interpretation for this result in terms of the particle and antiparticle content of the corresponding state.

1.5 Assume that $\mathrm{tr}\,[(t_a{}^{(R)})^2] \neq 0$ for every generator in some representation R of a Lie algebra with structure constants C_{abc}. Use this to prove the *Jacobi identity*

$$C_{a_1a_2d}C_{a_3de} + C_{a_3a_1d}C_{a_2de} + C_{a_2a_3d}C_{a_1de} = 0.$$

1.6 Use the parameterization given in eq. (1.19) to verify eqs. (1.62) and (1.77).

1.7 Suppose that a global transformation of the form of eq. (1.60) does not leave $\mathcal{L}$ form invariant, but may be written as

$$\mathcal{L}'(\bar{\phi}, \partial_\mu \bar{\phi}) = \mathcal{L}(\bar{\phi}, \partial_\mu \bar{\phi}) + (\partial\mathcal{L}/\partial\beta_a)\delta\beta_a.$$

Show that

$$\partial_\mu J^\mu{}_a = \partial\mathcal{L}/\partial\beta_a,$$

where $J^\mu{}_a$ is the Noether current, defined in eq. (1.61).

1.8 Show that eq. (1.73) for an element in $SU(2)$ may be expanded as

$$\exp\left(\mathrm{i}\sum_{a=1}^{3} \boldsymbol{\theta}\cdot\tfrac{1}{2}\boldsymbol{\sigma}\right) = \cos\tfrac{1}{2}\theta + \mathrm{i}\boldsymbol{\sigma}\cdot\hat{\mathbf{n}}\sin\tfrac{1}{2}\theta,$$

where $\hat{\mathbf{n}}$ is a unit vector in the $\boldsymbol{\beta}$-direction, and show that any 2×2 unitary matrix of unit determinant may be expressed this way.

1.9 Prove eq. (1.76). Hint: derive an equation for $(\mathrm{d}/\mathrm{d}\lambda)\det(\exp\lambda A)$, and integrate it from 0 to 1.

1.10 Let $(j_i)_{jk} = -\mathrm{i}\epsilon_{ijk}$ be the spatial part of the matrices J_i of eq. (1.91). Show that $(\hat{\mathbf{n}} \cdot \mathbf{j})^3 = (\hat{\mathbf{n}} \cdot \mathbf{j})$, and hence that

$$\exp(-\mathrm{i}\theta\hat{\mathbf{n}} \cdot \mathbf{j}) = 1 - \mathrm{i}(\hat{\mathbf{n}} \cdot \mathbf{j}) \sin\theta - (\hat{\mathbf{n}} \cdot \mathbf{j})^2(1 - \cos\theta).$$

Check for the particular case $\hat{\mathbf{n}} = \mathbf{e}_3$ that $\exp(-\mathrm{i}\theta j_3)$ acts on any vector as a rotation, according to the right-hand rule, by angle θ about the $\mathbf{e}_3$ axis. Similarly, show that

$$\exp(\mathrm{i}\omega\hat{\mathbf{n}} \cdot \mathbf{K}) = 1 + \mathrm{i}\hat{\mathbf{n}} \cdot \mathbf{K} \sinh\omega - (\hat{\mathbf{n}} \cdot \mathbf{K})^2(\cosh\omega - 1),$$

and that this matrix boosts the vector $(1, \mathbf{0})$ into $(\cosh\omega, -\hat{\mathbf{n}} \sinh\omega)$.

2
Canonical quantization

We are now ready to explore what it means to quantize a field theory. The scalar field illustrates features typically encountered in quantized relativistic systems with infinite numbers of degrees of freedom. As in the classical case, conservation laws follow from symmetries of the Lagrange density. The free scalar field can be reduced to a system of noninteracting harmonic oscillators, starting from which we define particle states and Green functions. Scattering experiments make it possible to bridge the gap between free and interacting field theories.

2.1 Canonical quantization of the scalar field

Canonical quantization

In a quantum mechanical system we may identify a set of coordinate operators $\{Q_i\}$, acting on a Hilbert space whose vectors are identified with states of the system. The eigenstates of the coordinate operators satisfy

$$Q_i|q_1, q_2, \ldots, q_i, \ldots\rangle = q_i|q_1, q_2, \ldots, q_i, \ldots\rangle, \tag{2.1}$$

and are states in which coordinate i has value q_i. Operators P_i corresponding to classical conjugate momenta $p_i = \partial L/\partial \dot{q}_i$ obey *canonical commutation relations* with the Q_j,

$$[Q_i, P_j] = i\hbar\delta_{ij}, \quad [Q_i, Q_j] = [P_i, P_j] = 0. \tag{2.2}$$

The commutation relations are closely related to the classical Poisson bracket,

$$\{f(\{q_i\}, \{p_i\}), g(\{q_i\}, \{p_i\})\}_{\text{PB}} = \sum_i [(\partial f/\partial q_i)(\partial g/\partial p_i) - (\partial f/\partial p_i)(\partial g/\partial q_i)], \tag{2.3}$$

defined for any functions $f(\{q_i\}, \{p_i\})$ and $g(\{q_i\}, \{p_i\})$ of the coordinates and momenta. The 'fundamental' brackets between the coordinates and momenta are of particular interest:

$$\{q_i, p_j\}_{PB} = \delta_{ij}, \quad \{q_i, q_j\}_{PB} = \{p_i, p_j\}_{PB} = 0. \tag{2.4}$$

In general, for any classical pair that satisfies eq. (2.4), we shall postulate canonical commutation relations of the form (2.2), simply multiplying the right-hand side of the classical relation by $i\hbar$. Other operators, $O(P_i, Q_j)$, are constructed out of ordered products of coordinates and momenta.

In field theory, it is convenient to work in the Heisenberg picture (Appendix A), where operators carry all the time dependence. To be specific, for any operator $O(t)$, we have

$$O(t_2) = \exp[(i/\hbar)H(t_2 - t_1)]\, O(t_1) \exp[(-i/\hbar)H(t_2 - t_1)]. \tag{2.5}$$

Poisson brackets for the scalar field

As we have emphasized, the coordinates of a classical field theory are the field values $\phi(\mathbf{x}, x^0)$ themselves, for each $\mathbf{x}$. To quantize the field, we construct an extension of the Poisson bracket (2.3) to the continuous system.

Imagine a very fine partition of space into cells R_i, each with a very small volume ΔV_i. This partition is supposed to be so fine that $\phi(x)$ and $\pi(x)$ are effectively constant within each cell. Their values in region R_i may then be denoted as $\phi_i(x^0)$, $\pi_i(x^0)$. Using this 'discretized' system (Heisenberg & Pauli 1929), we define new coordinates and conjugate momenta by

$$q_i(x^0) \equiv (\Delta V_i)^{1/2}\phi_i(x^0), \quad p_i(x^0) \equiv (\Delta V_i)^{1/2}\pi_i(x^0), \tag{2.6a}$$

in terms of which the Hamiltonian becomes a sum

$$H = \sum_i (\Delta V_i) T^{00}(q_i(\Delta V_i)^{-1/2}, p_i(\Delta V_i)^{-1/2}). \tag{2.6b}$$

The purpose of the change of variables, eq. (2.6a), is to construct coordinate–momentum pairs whose products have (the canonical) dimensions of action. For such pairs, we may define a standard classical Poisson bracket, eq. (2.3). This Poisson bracket will serve to normalize our quantization procedure in the continuum limit.

Now consider an arbitrary function of the discrete fields and momenta at position i, $f_i(x^0) \equiv f(\phi_i(x^0), \pi_i(x^0))$. The derivative with respect to the field $\phi_j(x^0)$ is nonzero only if $i = j$, and we can write

$$\begin{aligned}\frac{\partial f_i(x^0)}{\partial \phi_i(x^0)} &= \sum_j \frac{\partial f_i(x^0)}{\partial \phi_j(x^0)} \\ &= \sum_j \Delta V_j \left[\frac{1}{\Delta V_j}\frac{\partial f_i(x^0)}{\partial \phi_j(x^0)}\right],\end{aligned} \tag{2.7a}$$

where in the second equality we have simply multiplied and divided by ΔV_j. In the continuum limit, the sum over all positions, weighted by ΔV_j, is replaced by an integral: $\sum_j \Delta V_j \to \int \mathrm{d}^3\mathbf{x}$. On the other hand, the factor on the left-hand side of (2.7a) becomes the derivative $\partial f(\phi, \pi)/\partial\phi$, evaluated at $\phi = \phi(\mathbf{x}, x^0)$. The factor in brackets in the second equality therefore acts as a spatial delta function, times a derivative in the continuum limit,

$$\frac{1}{\Delta V_j}\frac{\partial f_i(x^0)}{\partial \phi_j(x^0)} \to \delta^3(\mathbf{x}_i - \mathbf{x}_j)\left.\frac{\partial f(\phi, \pi)}{\partial \phi}\right|_{\phi=\phi(\mathbf{x}_i x^0)} \equiv \frac{\delta f(\phi(\mathbf{x}_i, x^0), \pi(\mathbf{x}_j, x^0))}{\delta\phi(\mathbf{x}_j, x^0)}. \tag{2.7b}$$

This relation serves as a definition of the *variational derivative* $\delta f/\delta\phi$ of $f(\phi, \pi)$, considered as a function of the functions $\phi(\mathbf{x}, x^0)$ and $\pi(\mathbf{x}, x^0)$, with respect to $\phi(\mathbf{x}, x^0)$, at equal times.

A function of functions is often called a *functional*, and the variational derivative is also referred to as a *functional derivative*. The definition of a functional derivative depends on which parameters are considered as fixed in the functions (in our case x^0), and which are considered as variable (in our case $\mathbf{x}$). The variational derivative produces a delta function in the variable parameters only.

The continuum limit, eq. (2.7b), applies also to variations with respect to the conjugate momentum $\pi(x)$. Using this correspondence and the definitions (2.6a) in the discrete Poisson bracket (2.3), we arrive at the following continuum limit for the bracket of any two functionals $f(\phi, \pi)$ and $g(\phi, \pi)$:

$$\{f, g\}_{\mathrm{PB}} = \int \mathrm{d}^3\mathbf{x}\left[\frac{\delta f}{\delta\phi(\mathbf{x}, x^0)}\frac{\delta g}{\delta\pi(\mathbf{x}, x^0)} - \frac{\delta f}{\delta\pi(\mathbf{x}, x^0)}\frac{\delta g}{\delta\phi(\mathbf{x}, x^0)}\right]. \tag{2.8a}$$

The fundamental brackets in terms of fields are then

$$\begin{gathered}\{\phi(\mathbf{x}, x^0), \pi(\mathbf{x}', x^0)\}_{\mathrm{PB}} = \delta^3(\mathbf{x} - \mathbf{x}'),\\ \{\phi(\mathbf{x}, x^0), \phi(\mathbf{x}', x^0)\}_{\mathrm{PB}} = \{\pi(\mathbf{x}, x^0), \pi(\mathbf{x}', x^0)\}_{\mathrm{PB}} = 0.\end{gathered} \tag{2.8b}$$

Quantization of the scalar field

Proceeding as usual, we use the fundamental Poisson brackets to postulate canonical commutation relations for the real scalar field:

$$\begin{gathered}[\phi(\mathbf{x}, x^0), \pi(\mathbf{x}', x^0)] = \mathrm{i}\hbar\delta^3(\mathbf{x} - \mathbf{x}'),\\ [\pi(\mathbf{x}, x^0), \pi(\mathbf{x}', x^0)] = [\phi(\mathbf{x}, x^0), \phi(\mathbf{x}', x^0)] = 0.\end{gathered} \tag{2.9}$$

In view of the interpretation of $\phi(x)$ as a relativistic wave function (Section 1.2), canonical quantization of the field is sometimes referred to as a *second quantization*, in which the wave function itself is quantized. The following standard developments follow quite closely the principles laid down by Jordan & Klein (1927).

Alternate coordinates

It is useful to re-express the canonical commutation relations in terms of momentum expansion coefficients for a real field, in an analogous way to eq. (1.30). We begin by taking spatial Fourier transforms of the classical scalar field and its conjugate momentum:

$$\begin{aligned}\tilde{\phi}(\mathbf{k}, x^0) &= \int \mathrm{d}^3\mathbf{x} \exp(-\mathrm{i}\mathbf{k}\cdot\mathbf{x})\, \phi(\mathbf{x}, x^0),\\ \tilde{\pi}(\mathbf{k}, x^0) &= \int \mathrm{d}^3\mathbf{x} \exp(-\mathrm{i}\mathbf{k}\cdot\mathbf{x})\, \pi(\mathbf{x}, x^0).\end{aligned} \tag{2.10}$$

These are used to define new variables

$$\begin{aligned}a(\mathbf{k}, x^0) &= (2\pi)^{-3/2}[\omega_k\tilde{\phi}(\mathbf{k}, x^0) + \mathrm{i}\tilde{\pi}(\mathbf{k}, x^0)],\\ a^\dagger(\mathbf{k}, x^0) &= (2\pi)^{-3/2}[\omega_k\tilde{\phi}^\dagger(\mathbf{k}, x^0) - \mathrm{i}\tilde{\pi}^\dagger(\mathbf{k}, x^0)],\end{aligned} \tag{2.11}$$

where $a^\dagger(\mathbf{k}, x^0)$ is the hermitian conjugate of $a(\mathbf{k}, x^0)$. From eqs. (2.11) and (2.9), or by quantizing the relevant Poisson brackets, we find the commutation relations

$$\begin{aligned}[a(\mathbf{k}, x^0), a^\dagger(\mathbf{k}', x^0)] &= 2\hbar\omega_k\delta^3(\mathbf{k} - \mathbf{k}'),\\ [a(\mathbf{k}, x^0), a(\mathbf{k}', x^0)] &= [a^\dagger(\mathbf{k}, x^0), a^\dagger(\mathbf{k}', x^0)] = 0.\end{aligned} \tag{2.12}$$

The inverse forms of eq. (2.11) are

$$\begin{aligned}\phi(x) &= \int \frac{\mathrm{d}^3\mathbf{k}}{(2\pi)^{3/2}2\omega_k}\, [a(\mathbf{k}, x^0)\exp(\mathrm{i}\mathbf{k}\cdot\mathbf{x}) + a^\dagger(\mathbf{k}, x^0)\exp(-\mathrm{i}\mathbf{k}\cdot\mathbf{x})],\\ \pi(x) &= -\mathrm{i}\int \frac{\mathrm{d}^3\mathbf{k}}{(2\pi)^{3/2}2}\, [a(\mathbf{k}, x^0)\exp(\mathrm{i}\mathbf{k}\cdot\mathbf{x}) - a^\dagger(\mathbf{k}, x^0)\exp(-\mathrm{i}\mathbf{k}\cdot\mathbf{x})].\end{aligned} \tag{2.13}$$

In this form, we see that a real classical field corresponds to a hermitian quantum field. We shall continue to refer to this field as 'real', however, even in the quantum theory.

2.2 Quantum symmetries

The operator content of our theory is constrained by the equal-time commutators, eq. (2.9), and it is natural to wonder how a theory quantized in this way can be relativistically invariant. To address this and related questions, we study the action of Poincaré group transformations $\bar{x}^\mu = \Lambda^\mu{}_\nu x^\nu + a^\mu$ in the quantum theory.

Relativistic invariance

The physical information in a quantum theory is contained in the absolute values of inner products $\langle \psi' | \psi \rangle$ between vectors in its (Hilbert) space of states. Given states $\{|\psi\rangle\}$, suppose we change the coordinate system. In general, the states are now represented by a new set of vectors, $\{|\bar{\psi}\rangle\}$. To study the relation between the two sets, we need a rule for identifying physical states with vectors in the original and transformed Hilbert spaces. In addition, if the quantum theory in the transformed frame is to describe the same physics, the absolute values of inner products between vectors representing the same *physical* states should remain unchanged:

$$|\langle \psi' | \psi \rangle| = |\langle \bar{\psi}' | \bar{\psi} \rangle|. \tag{2.14}$$

This suggests the following relation between vectors in the original and transformed spaces:

$$|\bar{\psi}\rangle = U(a, \Lambda)|\psi\rangle, \tag{2.15}$$

where $U(a, \Lambda)$ is a unitary operator. (The classic analysis of Wigner (1939) also allows 'antiunitary' operators, see Appendix D below; we shall not consider this possibility now.) For products of transformations to be meaningful, $U(a, \Lambda)$ must itself be a representation of the Poincaré group, with a product rule

$$U(a', \Lambda')U(a, \Lambda) = U(a' + \Lambda' a, \Lambda' \Lambda). \tag{2.16}$$

Once transformations are represented by unitary operators in the space of state vectors, we may as well identify the original and transformed vector spaces, and eqs. (2.14) and (2.15) become relations between states in the same space.

In summary, a theory is relativistically invariant if we can find: (i) a rule for associating vectors in the Hilbert space with physical states and (ii) an explicit form for a unitary operator satisfying eq. (2.16).

States and the Poincaré algebra

To classify physical states associated with the hermitian scalar field, it is natural to turn to the conserved charges of the classical theory, which are found from Noether's theorem. These are the P^μ, eq. (1.46), and the $J^{\mu\nu}$, eq. (1.56). Correspondence between the classical and quantum theories suggests that they be identified as operators representing linear and angular momentum respectively. Their eigenvectors are then states with a physical interpretation. In the quantum theory, however, we may diagonalize only a maximally commuting subset of these operators. To find such a subset, we compute their mutual commutators, using eq. (2.9):

$$[P_\mu, P_\nu] = 0,$$
$$[P_\mu, J_{\lambda\sigma}] = -i\hbar(g_{\mu\sigma}P_\lambda - g_{\mu\lambda}P_\sigma), \quad (2.17)$$
$$[J_{\mu\nu}, J_{\lambda\sigma}] = -i\hbar(g_{\mu\lambda}J_{\nu\sigma} - g_{\mu\sigma}J_{\nu\lambda} - g_{\nu\lambda}J_{\mu\sigma} + g_{\nu\sigma}J_{\mu\lambda}).$$

Equation (2.17) has two remarkable properties, related to the two requirements of Poincaré invariance identified above.

(i) The components of linear momenta mutually commute, and may thus be diagonalized simultaneously. As a result, a basis for the Hilbert space may be given in terms of state vectors $|q^\mu\rangle$ that are eigenstates of the four operators P^μ and may be identified according to their momentum content. The state represented by vector $|q^\mu\rangle$ in one frame is represented, up to a phase factor, by the state $|\Lambda^\mu{}_\alpha q^\alpha\rangle$ in the frame resulting from the action of Λ on the first frame. In the same way, translations may change momentum eigenstates by at most a phase. Because the P_μ are Hermitian operators, their eigenvectors with different eigenvalues are orthogonal. So, assuming that the range of momenta q^μ is continuous, the space of states in our field theory is infinite dimensional.

(ii) Up to a factor $-i\hbar$, eq. (2.17) is the Lie algebra of the Poincaré group, eq. (1.109). The operators P_μ and $-J_{\mu\nu}$ generate a representation of the Poincaré group via eqs. (1.69) and (1.70),

$$U(\delta a, I + \delta\lambda) = I + (i/\hbar)(\delta a^\mu P_\mu - \tfrac{1}{2}\delta\lambda^{\mu\nu}J_{\mu\nu}). \quad (2.18)$$

Again, because P^μ and $J^{\mu\nu}$ are hermitian, this infinite-dimensional representation is unitary. As such, it is quite different from the familiar four-by-four nonunitary representation of the Lorentz group $\{\Lambda^\mu{}_\nu\}$, which acts on four-vectors. Let us check how momentum eigenstates transform under eq. (2.18).

Transformations of states

An arbitrary (finite or infinitesimal) translation may be put into exponential form using (1.70), so that

$$U(a, 0)|q^\mu\rangle = \exp[(i/\hbar)P_\mu a^\mu]\,|q^\mu\rangle = \exp[i(q_\mu a^\mu/\hbar)]|q^\mu\rangle \quad (2.19)$$

specifies a finite translation. As anticipated, the state changes only by a phase.

For infinitesimal Lorentz transformations we can use

$$[P^\eta, (1 - (i/\hbar)\tfrac{1}{2}\delta\lambda^{\mu\nu}J_{\mu\nu})] = (\delta^\eta{}_\alpha + \delta\lambda^\eta{}_\alpha)P^\alpha \quad (2.20)$$

to show that

$$(1 - (i/\hbar)\tfrac{1}{2}\delta\lambda^{\mu\nu}J_{\mu\nu})|q^\mu\rangle = |(1 + \delta\lambda)^\mu{}_\alpha q^\alpha\rangle. \quad (2.21)$$

Momentum eigenstates of the scalar field thus transform appropriately under the action of infinitesimal Lorentz transformations. Representations of finite Lorentz transformations, under which the state transforms as

$$U(0, \Lambda)|q^\mu\rangle = |\Lambda^\mu{}_\alpha q^\alpha\rangle, \tag{2.22}$$

can be constructed from products of infinitesimal transformations. It is now easy to see that the combined operator

$$U(a, \Lambda) = U(a, 0)U(0, \Lambda) \tag{2.23}$$

obeys eq. (2.16), so that the theory is indeed relativistically invariant.

Transformation of fields

In the classical limit, the expectation values $\langle \psi|\phi_i(x)|\psi\rangle$ are identified with classical fields $\phi_i{}^{(\mathrm{cl})}(x)$ in a system with coordinates x^μ (the expectation value is real for a hermitian field). Similarly, in a transformed system with coordinates $\bar{x} = \Lambda x + a$, the matrix element $\langle \bar{\psi}|\phi_i(\bar{x})|\bar{\psi}\rangle$ should be identified with $\bar{\phi}_i{}^{(\mathrm{cl})}(\bar{x})$. Note the lack of a bar on the field operator $\phi_i(\bar{x})$, which is defined once and for all as an operator in the space of states. Comparison with eq. (1.102) shows that the expectation values must transform as

$$\langle \psi_i|\phi_i(x)|\psi_i\rangle = S_{ij}(\Lambda^{-1})\langle \bar{\psi}|\phi_j(\bar{x})|\bar{\psi}\rangle, \tag{2.24}$$

which, with (2.15), suggests that the field operators transform as

$$U(a, \Lambda)\phi_i(x)U^{-1}(a, \Lambda) = S_{ij}(\Lambda^{-1})\phi_j(\Lambda x + a). \tag{2.25}$$

Equation (2.25), along with (2.15), gives the behavior of matrix elements under translations:

$$\langle q_2|\textstyle\prod_{i=1}^n\phi(x_i)|q_1\rangle = \exp[-\mathrm{i}(q_1 - q_2)\cdot a/\hbar]\,\langle q_2|\textstyle\prod_{i=1}^n\phi(x_i - a)|q_1\rangle. \tag{2.26}$$

We shall make extensive use of this relation.

Causality

Any two spacelike-separated points ($(x - y)^2 < 0$) may be transformed into equal-time points by a Lorentz transformation. Using this fact with the canonical commutation relations, eq. (2.9), we find

$$[\phi(x), \phi(y)] = 0, (x - y)^2 = 0. \tag{2.27}$$

If we invoke the standard complementarity interpretation of commutators, we conclude that spacelike-separated field measurements are independent. This is consistent with relativistic causality, which requires that no signals propagate between spacelike-separated points. Our observations here only scratch the surface of the issues involved in field measurement, which were extensively discussed for electrodynamics by Bohr & Rosenfeld (1933, 1950) (see Corinaldesi r1953).

Internal symmetries

Analogous considerations apply to internal symmetry groups. To illustrate their role in the quantum theory, we may apply canonical quantization to a multicomponent Lagrange density of the form (1.64). The conjugate momenta are

$$\pi_i{}^* = \partial\mathcal{L}/\partial(\partial_0\phi_i{}^*) = \partial_0\phi_i, \quad \pi_i = \partial\mathcal{L}/\partial(\partial_0\phi_i) = \partial_0\phi_i{}^*, \tag{2.28}$$

with canonical commutation relations

$$\begin{gathered}[\phi_i(\mathbf{x}, x^0), \pi_j(\mathbf{y}, x^0)] = [\phi_i{}^*(\mathbf{x}, x^0), \pi_j{}^*(\mathbf{y}, x^0)] = i\hbar\delta_{ij}\delta^3(\mathbf{x}-\mathbf{y}),\\ [\phi_i(\mathbf{x}, x^0), \pi_j{}^*(\mathbf{y}, x^0)] = [\phi_i{}^*(\mathbf{x}, x^0), \pi_j(\mathbf{y}, x^0)] = 0.\end{gathered} \tag{2.29}$$

The commutators of other combinations vanish. The energy and momentum density operators are

$$\begin{aligned} T^{00} &= \sum_i(\pi_i{}^*\pi_i + \nabla\phi_i{}^*\cdot\nabla\phi_i + m^2\phi_i{}^*\phi_i) + V(|\phi|),\\ T^{0k} &= \sum_i(\pi_i{}^*\partial^k\phi_i{}^* + \pi_i\partial^k\phi_i).\end{aligned} \tag{2.30}$$

From these forms we may verify invariance under the Poincaré group, since the momentum and angular momentum operators satisfy the same algebra, eq. (2.17), as the real scalar field. Thus states may again be characterized by their overall momentum. A new feature comes about by considering operator versions of the conserved charges associated with the currents, eq. (1.78):

$$Q_a = -\mathrm{i}\int \mathrm{d}^3\mathbf{x}\{\partial_0\phi^*{}_i[\mathrm{t}_a{}^{(n)}]_{ij}\phi_j - \partial_0\phi_i[t_a{}^{(n)}]_{ij}\phi^*{}_j\}, \tag{2.31}$$

where the $\mathrm{t}_a{}^{(n)}$ are the matrix generators of the defining representation of $U(n)$, i.e., of $U(1)$ (proportional to the identity) and $SU(n)$. From eq. (2.31), the canonical commutation relations and the hermiticity of the $t_a{}^{(n)}$, the Q_a commute with generators of the Lorentz group:

$$[Q_a, P^\mu] = 0, \quad [Q_a, J^{\mu\nu}] = 0. \tag{2.32}$$

In addition, because the $t_a{}^{(n)}$ satisfy the Lie algebra eq. (1.71), we find that the charges satisfy the same algebra:

$$[Q_a, Q_b] = \mathrm{i}C_{ab}{}^c Q_c. \tag{2.33}$$

As for the Poincaré algebra, eq. (2.17), eqs. (2.32) and (2.33) have consequences for the states of the space in which the Q_a act. Equation (2.32) ensures that any of the Q_a may be diagonalized simultaneously with the components of linear momenta, while eq. (2.33) means that not all the Q_a may be diagonalized at the same time. Suppose, however, we find a set of generators $\{D_i, i = 1, \ldots, R\}$, with $R < n$, which commute among themselves,

$$[D_i, D_j] = 0. \tag{2.34}$$

Then we may label basis states of the system, $|q^\mu, \{d_i\}\rangle$ by eigenvalues q^μ of the momentum operator and eigenvalues d_i of the diagonal symmetry generators D_i,

$$P^\mu|q^\mu, \{d_i\}\rangle = q^\mu|q^\mu, \{d_i\}\rangle, \quad D_i|q^\mu, \{d_i\}\rangle = d_i|q^\mu, \{d_i\}\rangle. \tag{2.35}$$

In general, sets of states with the same q^μ transform among themselves under the internal symmetry group according to a specific irreducible representation of the Lie algebra. A basis for any such set is said to form a *degenerate multiplet*, since the invariant mass $\sqrt{q^2}$ for each such state is the same. The simplest example, of course, is the complex scalar field $(n = 1)$, for which there is a single generator Q, called the total charge. The states are then labelled by the momentum and the value of that charge and

$$Q|q^\mu, e\rangle = e|q^\mu, e\rangle. \tag{2.36}$$

In this case, the transformation changes only the phase, and the multiplets are one-dimensional. Another example is given by $SU(2)$, with three charges, I_i, $i = 1, 2, 3$, satisfying the algebra of eq. (1.75). In this case, basis states are labelled by the familiar quantum numbers d_i given by the eigenvalues of the operators $I^2 = I_1{}^2 + I_2{}^2 + I_3{}^2$ and I_3. Degenerate multiplets with eigenvalue $I(I + 1)$ of I^2 are $(2I + 1)$-dimensional, where I is an integer or half-integer. An example is the pair (p, n) (proton and neutron), considered as an $I = \frac{1}{2}$ multiplet of the $SU(2)$ 'isospin' symmetry of the strong interactions. In the absence of isospin-violating forces, the proton and neutron would be degenerate in mass.

Finally, as in the case of Poincaré transformations, the charges generate transformations of the fields, given in the infinitesimal case by

$$\left[1 + \frac{\mathrm{i}}{\hbar}\sum_a \delta\beta_a Q_a\right] \phi_i(x) \left[1 - \frac{\mathrm{i}}{\hbar}\sum_a \delta\beta_a Q_a\right] = \left[\delta_{ij} + \frac{\mathrm{i}}{\hbar}\sum_a \delta\beta_a [t_a{}^{(n)}]_{ij}\right] \phi_j(x) + \mathrm{O}(\delta\beta^2). \tag{2.37}$$

In summary, global symmetries manifest themselves by the existence of extra quantum numbers for states in the field theory.

Having reviewed some of the formal properties of a quantized scalar field theory, it is time to give a concrete example, the free scalar field.

2.3 The free scalar field as a system of harmonic oscillators

It is possible to solve the free scalar theory completely: we can construct a basis for its Hilbert space, and compute the expectation values of any operator constructed from the field and its conjugate momentum.

Natural units

The formalism can be simplified by generalizing the rescaling of time and mass that we introduced in Section 1.2. The units of any dimensional quantity A can be expressed in a unique way as a combination of action (mass × length^2 × (time)$^{-1}$), velocity (length × (time)$^{-1}$) and length. Therefore, for any A there is a unique factor $\hbar^a c^b$ such that $A' = A/(\hbar^a c^b)$ has dimensions of length only. For instance, for a scalar field ϕ, $\phi' = \phi/(\hbar^{1/2} c^{1/2})$ has units of (length)$^{-1}$ while, for the action, $S' = S/\hbar$ is dimensionless. Neither Planck's constant nor the speed of light will appear explicitly in any relation involving only quantities rescaled in this manner to 'natural units'. In the following, except where noted we shall assume that this rescaling has been carried out.

The Hamiltonian in momentum space

The free scalar Hamiltonian is given as a spatial integral by eq. (1.52), and to find its eigenstates we use Fourier transform variables. From eqs. (1.52) and (2.13), we have

$$H = \int d^3\mathbf{k} \tfrac{1}{4}[a(\mathbf{k}, x^0)\, a^\dagger(\mathbf{k}, x^0) + a^\dagger(\mathbf{k}, x^0) a(\mathbf{k}, x^0)], \tag{2.38}$$

where because a and $a^\dagger$ are operators we retain their relative order. In this form, H is an integral over separate operators, one for each wave vector $\mathbf{k}$. By eq. (2.12), these operators commute with each other, and can be diagonalized simultaneously. Indeed, at fixed $\mathbf{k}$, the integrand in eq. (2.38) is just the Hamiltonian for a linear harmonic oscillator, $p^2/2m + m\omega_k{}^2 q^2/2$, in terms of the variable $a_\mathbf{k} = (p + \mathrm{i}\omega_k q)/(2m)^{1/2}$ and its complex conjugate.

This reduction of the free scalar field to a system of independent harmonic oscillators is the paradigm for the treatment of all free fields. It is the basis upon which everything else is built, and will serve as the starting point for the perturbative approach to interacting fields. In fact, in the very first paper on quantum field theory, Dirac (1927) used much the same approach and notation to discuss the electromagnetic field.

In the Heisenberg picture, the time dependence of the operators $a(\mathbf{k}, x^0)$ and $a^\dagger(\mathbf{k}, x^0)$ is derived from eqs. (2.12) and (2.38):

$$\begin{aligned} \mathrm{d}a(\mathbf{k}, x^0)/\mathrm{d}x^0 &= \mathrm{i}[H, a(\mathbf{k}, x^0)] = -\mathrm{i}\omega_k a(\mathbf{k}, x^0), \\ \mathrm{d}a^\dagger(\mathbf{k}, x^0)/\mathrm{d}x^0 &= \mathrm{i}[H, a^\dagger(\mathbf{k}, x^0)] = \mathrm{i}\omega_k a^\dagger(\mathbf{k}, x^0). \end{aligned} \tag{2.39}$$

We have used the time-independence of H to choose x^0 in eq. (2.38) so that the commutators in eqs. (2.39) are equal-time. Equations (2.39) are easily solved to give

$$a(\mathbf{k}, x^0) = a(\mathbf{k}, 0)\exp(-\mathrm{i}\omega_k x_0),\ a^\dagger(\mathbf{k}, x^0) = a^\dagger(\mathbf{k}, 0)\exp(\mathrm{i}\omega_k x_0). \tag{2.40}$$

Substituting eqs. (2.40) back into (2.38), H becomes explicitly time-independent. In the following, therefore, we shall drop the time argument for the operators a and $a^\dagger$ in the free theory. In this notation, the free field and its conjugate momenta become

$$\begin{aligned} \phi(x) &= \int \frac{\mathrm{d}^3\mathbf{k}}{(2\pi)^{3/2} 2\omega_k} [a(\mathbf{k})\exp(-\mathrm{i}\bar{k}\cdot x) + a^\dagger(\mathbf{k})\exp(\mathrm{i}\bar{k}\cdot x)], \\ \pi(x) &= -\mathrm{i}\int \frac{\mathrm{d}^3\mathbf{k}}{(2\pi)^{3/2} 2} [a(\mathbf{k})\exp(-\mathrm{i}\bar{k}\cdot x) - a^\dagger(\mathbf{k})\exp(\mathrm{i}\bar{k}\cdot x)]. \end{aligned} \tag{2.41}$$

Here and below we define $\bar{k}^\mu = (\omega_k, \mathbf{k})$, where $\bar{k}^2 = m^2$, as expected for a relativistic particle of mass m. The free-field operator may therefore be thought of as a superposition of plane-wave solutions to the Klein–Gordon equation. $a(\mathbf{k})$ and $a^\dagger(\mathbf{k})$ are quantized coefficients corresponding to positive- and negative-energy solutions, respectively.

Discrete system

To diagonalize the Hamiltonian (2.38), it is convenient to work with a system in which $\mathbf{k}$ also takes on a discrete, rather than a continuous, range of values. In practice, this may be achieved by imposing finite-volume boundary conditions on the field, often referred to as 'putting the system in a box'. To formulate the system in this way, we introduce a scale V, proportional to the total volume of space, and make the following replacements in eq. (2.38):

$$a(\mathbf{k}) \to V^{1/2}\alpha_{\mathbf{k}},\quad \int \mathrm{d}^3\mathbf{k} \to \frac{1}{V}\sum_{\mathbf{k}},\quad \delta^3(\mathbf{k}-\mathbf{k}') \to V\delta_{\mathbf{k}\mathbf{k}'}. \tag{2.42}$$

Then the $\alpha_{\mathbf{k}}$ obey the canonical commutation relations

$$[\alpha_{\mathbf{k}}, \alpha^\dagger{}_{\mathbf{k}'}] = 2\omega_k\delta_{\mathbf{k}\mathbf{k}'}, \tag{2.43}$$

and

$$H = \sum_{\mathbf{k}} h_{\mathbf{k}},\ h_{\mathbf{k}} = \tfrac{1}{4}(\alpha_{\mathbf{k}}\alpha^{\dagger}{}_{\mathbf{k}} + \alpha^{\dagger}{}_{\mathbf{k}}\alpha_{\mathbf{k}}). \tag{2.44}$$

Let us now review the diagonalization of $h_{\mathbf{k}}$ in the Heisenberg representation.

The one-dimensional harmonic oscillator

Using the commutation relation (2.43), we can rewrite $h_{\mathbf{k}}$ as

$$h_{\mathbf{k}} = \omega_k(N_{\mathbf{k}} + \tfrac{1}{2}),\ \ N_{\mathbf{k}} = (2\omega_k)^{-1}\alpha^{\dagger}{}_{\mathbf{k}}\alpha_{\mathbf{k}}. \tag{2.45}$$

$N_{\mathbf{k}}$ is the *number operator* for wave vector $\mathbf{k}$. Its eigenstates are clearly also eigenstates of the Hamiltonian $h_{\mathbf{k}}$.

Suppose $|m_{\mathbf{k}}\rangle$ is an arbitrary eigenstate of $N_{\mathbf{k}}$, with eigenvalue $m_{\mathbf{k}}$; thus

$$N_{\mathbf{k}}|m_{\mathbf{k}}\rangle = m_{\mathbf{k}}|m_{\mathbf{k}}\rangle. \tag{2.46}$$

We can easily verify from eq. (2.43) that the states

$$|m\rangle = (m!)^{-1/2}(\alpha_{\mathbf{k}}{}^{\dagger})^m|0\rangle \tag{2.47}$$

satisfy eq. (2.46). They are normalized by $\langle m|n\rangle = (2\omega_k)^m\,\delta_{mn}$. The terms *raising* or *creation* operator are often applied to $\alpha_{\mathbf{k}}{}^{\dagger}$, and *lowering* or *annihilation* operator to α_k, referring to their action on states from the left:

$$\alpha_{\mathbf{k}}|m_{\mathbf{k}}\rangle = (2\omega_k)m_{\mathbf{k}}{}^{1/2}|m_{\mathbf{k}} - 1\rangle,\ \ \alpha_{\mathbf{k}}{}^{\dagger}|m_{\mathbf{k}}\rangle = (m_{\mathbf{k}} + 1)^{1/2}|m_{\mathbf{k}} + 1\rangle. \tag{2.48}$$

We can now return to $h_{\mathbf{k}}$, eq. (2.45), and immediately conclude that its eigenstates are labelled by the same number, thus

$$h_{N,\mathbf{k}}|m\rangle = \omega_k(m + \tfrac{1}{2})|m\rangle. \tag{2.49}$$

Note that eq. (2.49), along with the first of eqs. (2.48), shows that the system has a ground state $|0\rangle$ only if the number $m_{\mathbf{k}}$ is a positive integer, called the *occupation number*. The eigenvalue of the lowest state, $\frac{1}{2}\omega_k$, is the zero-point energy of the oscillator.

Fock space as a solution

Since the degrees of freedom associated with different wave numbers are not linked in the Hamiltonian, the space of states $\mathcal{H}$ may be thought of as the direct product of the spaces $\mathcal{H}_{\mathbf{k}}$ associated with each wave vector $\mathbf{k}$:

$$\mathcal{H} = \otimes_{\mathbf{k}}\mathcal{H}_{\mathbf{k}}, \tag{2.50}$$

where the different $\mathcal{H}_{\mathbf{k}}$ are orthogonal. Basis states are labelled by the set of occupation numbers of the $h_{\mathbf{k}}$'s, and denoted $|\{n_{\mathbf{k}}\}\rangle$. Each $h_{\mathbf{k}}$ acts only on its own subspace $\mathcal{H}_{\mathbf{k}}$, so that

$$H|\{n_{\mathbf{k}}\}\rangle = \sum_{\mathbf{k}} \omega_k(n_{\mathbf{k}} + \tfrac{1}{2})|\{n_{\mathbf{k}}\}\rangle. \tag{2.51}$$

Thus the total energy is the sum of the energies in each mode. Similarly, the normalization factor for the eigenstates is

$$\langle\{n_{\mathbf{k}}\}|\{n'_{\mathbf{k}}\}\rangle = \prod_{\mathbf{k}} \langle n_{\mathbf{k}}|n'_{\mathbf{k}}\rangle = \prod_{\mathbf{k}} (2\omega_k)^{n_{\mathbf{k}}} \delta_{n_{\mathbf{k}},n'_{\mathbf{k}}}. \tag{2.52}$$

This is a quantum analogue of the classical principle of superposition.

By analogy with the Hamiltonian, we may also define a total number operator, which commutes with H and whose eigenvalues are the total occupation number of each energy eigenstate:

$$N = \sum_{\mathbf{k}} N_{\mathbf{k}}, \quad N|\{n_{\mathbf{k}}\}\rangle = \sum_{\mathbf{k}} n_{\mathbf{k}}|\{n_{\mathbf{k}}\}\rangle. \tag{2.53}$$

This description of a Hilbert space, in which the basis vectors are chosen as eigenstates of a number operator, is called a *Fock-space* representation. $\mathcal{H}$ may also be thought of as a direct sum of subspaces $\mathcal{H}_N$, each made up of states with total occupation number $N = \sum_{\mathbf{k}} m_{\mathbf{k}}$:

$$\mathcal{H} = \mathcal{H}_0 + \mathcal{H}_1 + \mathcal{H}_2 \cdots. \tag{2.54}$$

From eqs. (2.43) and (2.48) we can compute any matrix element of operators constructed from the α's and $\alpha^\dagger$'s, and hence ϕ and π. In this sense, the Fock-space construction solves the theory.

States of the continuum Fock space

When we return to the continuum, we must specify an occupation number for the continuously infinite set of modes $\mathbf{k}$. A set of states with a convenient continuum limit is related to states of the discrete system by

$$\prod_{\mathbf{k}} (n_{\mathbf{k}}!)^{-1/2}(V^{1/2}\alpha_{\mathbf{k}})^{n_{\mathbf{k}}}|0\rangle \to \prod_{\mathbf{k}} (n_{\mathbf{k}}!)^{-1/2}[a^\dagger(\mathbf{k})]^{n_{\mathbf{k}}}|0\rangle \equiv |\{\mathbf{k}_i\}\rangle. \tag{2.55}$$

The notation for continuum states consists of listing the momentum of each nonzero excitation. For simplicity, we assume that in the continuum limit no two excitations have exactly the same wave vector. Then we have

$$\langle\{\mathbf{k}_i\}|\{\mathbf{q}_j\}\rangle = \sum_{\pi(j)} \prod_i 2\omega_{k_i} \delta^3(\mathbf{k}_i - \mathbf{q}_{\pi(j)}), \tag{2.56}$$

where $\pi(j)$ represents a permutation of the set $\{j\}$.

Although the operators N and H are diagonal in the Fock space, $a(\mathbf{k})$ and $a^\dagger(\mathbf{k})$ connect states in $\mathcal{H}_i$ and $\mathcal{H}_{i\pm1}$ as follows:

$$a^\dagger(\mathbf{q})|\{\mathbf{k}_i\}\rangle = |\{\mathbf{k}_i, \mathbf{q}\}\rangle,$$

$$a(\mathbf{q})|\{n_\mathbf{k}\}\rangle = \sum_j 2\omega_q \delta^3(\mathbf{q} - \mathbf{k}_j)|\{\mathbf{k}_i \neq \mathbf{k}_j\}\rangle, \tag{2.57}$$

$$a(\mathbf{q})|0\rangle = 0,$$

which is the continuum analogue of eq. (2.48). The normalization of states is a product of delta functions for each nonzero occupation number. Note, however, that the vacuum state, for which all occupation numbers are zero, retains the same normalization in the continuum as in the discrete system,

$$\langle 0|0\rangle = 1. \tag{2.58}$$

The completeness relation for the continuum states is then

$$I = |0\rangle\langle 0| + \sum_{n=1}^{\infty} \frac{1}{n!} \prod_{i=1}^{n} \int \frac{\mathrm{d}^3\mathbf{k}_i}{2\omega_{k_i}} |\mathbf{k}_1, \ldots, \mathbf{k}_n\rangle\langle \mathbf{k}_n, \ldots, \mathbf{k}_1|. \tag{2.59}$$

This may be verified by operating on an arbitrary energy eigenstate, and observing that the same state is retrieved. With this normalization, an arbitrary state may be written in the Fock space basis as

$$|\psi\rangle = |0\rangle\langle 0|\psi\rangle + \sum_{n=1}^{\infty} \frac{1}{n!} \prod_{i=1}^{n} \int \frac{\mathrm{d}^3\mathbf{k}_i}{2\omega_{k_i}} f_\psi{}^{(n)}(\mathbf{k}_1, \ldots, \mathbf{k}_n)|\mathbf{k}_1, \ldots, \mathbf{k}_n\rangle,$$

$$f_\psi{}^{(n)}(\mathbf{k}_1, \ldots, \mathbf{k}_n) = \langle \mathbf{k}_1, \ldots, \mathbf{k}_n|\psi\rangle. \tag{2.60}$$

In this way we may build continuum states with unit, rather than delta function, normalizations. Let us now discuss some of the general properties of Fock-space states.

Bose–Einstein statistics

The statistics of an arbitrary Fock-space state of the form of eq. (2.60) follow directly from the canonical commutation relations of the fields. All states in the Hilbert space of the free fields are symmetric under interchange of any pair $(\mathbf{k}_i, \mathbf{k}_j)$. That is, states that differ only by such an interchange are really the same state in the Hilbert space, and should be counted only once in the sum over states. This means that the system is described by Bose–Einstein statistics.

To exhibit the Bose symmetry, consider an arbitrary function of n momenta $f_\psi(\{\mathbf{k}_i\})$, which has no special properties when its arguments are interchanged. We define its symmetric form by

$$f_{\psi,S}(\{\mathbf{k}_i\}) = \frac{1}{n!}\sum_{\pi(i)} f_\psi(\{\mathbf{k}_{\pi(i)}\}), \tag{2.61}$$

where again $\pi(i)$ denotes the permutation $(\pi(1), \ldots, \pi(n))$ of $(1, 2, \ldots, n)$. We claim that using either $f_{\psi,S}$ or f_ψ in eq. (2.60) gives exactly the same state.

Consider, for instance, a state with $n = 2$:

$$|\psi\rangle = \int \frac{d^3\mathbf{k}_1\, d^3\mathbf{k}_2}{(2\omega_{k_1})(2\omega_{k_2})} f_{\psi,S}(\mathbf{k}_1, \mathbf{k}_2) a^\dagger(\mathbf{k}_1) a^\dagger(\mathbf{k}_2)|0\rangle. \tag{2.62}$$

By commuting the creation operators, and interchanging the labels $\mathbf{k}_1$ and $\mathbf{k}_2$, we find

$$|\psi\rangle = \int \frac{d^3\mathbf{k}_1\, d^3\mathbf{k}_2}{(2\omega_{\mathbf{k}_1})(2\omega_{\mathbf{k}_2})} f_{\psi,S}(\mathbf{k}_2, \mathbf{k}_1) a^\dagger(\mathbf{k}_1) a^\dagger(\mathbf{k}_2)|0\rangle. \tag{2.63}$$

These two equal forms may be added and multiplied by one-half to get the symmetric form:

$$|\psi\rangle = \int \frac{d^3\mathbf{k}_1\, d^3\mathbf{k}_2}{(2\omega_{\mathbf{k}_1})(2\omega_{\mathbf{k}_2})} f_{\psi,S}(\mathbf{k}_1, \mathbf{k}_2) a^\dagger(\mathbf{k}_1) a^\dagger(\mathbf{k}_2)|0\rangle. \tag{2.64}$$

For states with $n > 2$ or a superposition of states with different n, the analogous result follows in exactly the same way.

Relativistic invariance

Relativistic invariance requires that states obey eqs. (2.19) and (2.22) under the action of the Poincaré group. We can check these properties for the free field by explicitly constructing the conserved charges that generate transformation. For instance, the total spatial momentum operators may be calculated directly from eqs. (1.53) and (2.41) as

$$P^i = \int d^3\mathbf{k}\, \mathbf{k}^i N_\mathbf{k}, \quad P^i|\{n_\mathbf{k}\}\rangle = \sum_\mathbf{k} \mathbf{k}^i |\{n_\mathbf{k}\}\rangle, \tag{2.65}$$

so we have simultaneously diagonalized the Hamiltonian and spatial momenta. From this, the transformation rule, eq. (2.19), for translations follows immediately. It is not difficult to show that the generators of Lorentz transformations $J_{\mu\nu}$, eqs. (1.56), may also be written as sums over the independent modes. As a result, eq. (2.22) specializes for the free field to

$$U(0, \Lambda)|\mathbf{k}_1, \ldots, \mathbf{k}_n\rangle = |\mathbf{k}'_1, \ldots, \mathbf{k}'_n\rangle, \quad \bar{k}'^\mu \equiv \Lambda^\mu{}_\alpha \bar{k}^\alpha, \tag{2.66}$$

with $\bar{k}^\mu$ the mass-shell vector defined after eq. (2.41).

Normal ordering

Our solution for the free field requires an additional comment. Even in the discrete system described above there are an infinite number of normal modes, so that, from eq. (2.44), the ground state energy is infinite, being the sum of the zero-point energies of all the normal modes. To deal with this problem, we assume that only energy differences, not their absolute values, are observable. We ignore here the question of whether this is consistent with a theory of gravity. This reservation notwithstanding, we define a new form of H by

$$:H: = H - \frac{1}{4}\int \mathrm{d}^3\mathbf{k}[a(\mathbf{k}), a^\dagger(\mathbf{k})] = \frac{1}{2}\int \mathrm{d}^3\mathbf{k}\, a^\dagger(\mathbf{k})a(\mathbf{k}),$$

$$:H:|\{\mathbf{k}_i\}\rangle = \sum_{\mathbf{k}_i} \omega_{k_i}|\{\mathbf{k}_i\}\rangle. \tag{2.67}$$

This subtraction effectively moves raising operators to the left of lowering operators. The notation (: :) is conventional for this reordering, which is known as *normal ordering*. Normal ordering on any operator O produces another operator $:O:$, whose 'vacuum expectation value' – its expectation value in the ground state – vanishes, i.e. $\langle 0|:O:|0\rangle = 0$.

Charged scalar field

Essentially the same procedure may be used to quantize the free charged scalar field, whose Lagrangian is given by $U(|\phi|) = 0$ in eq. (1.17). The Hamiltonian is best expressed in terms of a momentum space expansion of fields analogous to eq. (1.30):

$$\phi(x) = \int \frac{\mathrm{d}^3\mathbf{k}}{(2\pi)^{3/2}2\omega_k}\,[a_+(\mathbf{k}, x^0)\exp(\mathrm{i}\mathbf{k}\cdot\mathbf{x}) + a_-{}^\dagger(\mathbf{k}, x^0)\exp(-\mathrm{i}\mathbf{k}\cdot\mathbf{x})],$$

$$\pi(x) = -\mathrm{i}\int \frac{\mathrm{d}^3\mathbf{k}}{(2\pi)^{3/2}2}\,[a_-(\mathbf{k}, x^0)\exp(\mathrm{i}\mathbf{k}\cdot\mathbf{x}) - a_+{}^\dagger(\mathbf{k}, x^0)\exp(-\mathrm{i}\mathbf{k}\cdot\mathbf{x})], \tag{2.68}$$

and similarly for $\phi^\dagger(x)$ and $\pi^\dagger(x)$. The equal-time commutation relations

$$[a_\pm(\mathbf{k}, x^0), a_\pm{}^\dagger(\mathbf{k}', x^0)] = 2\omega_k\delta^3(\mathbf{k} - \mathbf{k}'),$$
$$[a_+, a_+] = [a_-, a_-] = [a_+, a_-] = [a_+, a_-{}^\dagger] = 0, \tag{2.69}$$

reproduce the canonical commutation relations for the complex field, eq. (2.29), up to a factor $\hbar$, which we have cancelled by rescaling the fields by $(\hbar c)^{-1/2}$. Then the Hamiltonian, found from the energy density in eq. (2.30), is

$$\begin{aligned} H &= \int d^3x[\pi^*(x)\pi(x) + \nabla\phi^*(x)\cdot\nabla\phi(x)] \\ &= \frac{1}{4}\int d^3\mathbf{k}[a_+(\mathbf{k}, x^0)a_+{}^\dagger(\mathbf{k}, x^0) + a_+{}^\dagger(\mathbf{k}, x^0)a_+(\mathbf{k}, x^0) \\ &\quad + a_-(\mathbf{k}, x^0)a_-{}^\dagger(\mathbf{k}, x^0) + a_-{}^\dagger(\mathbf{k}, x^0)a_-(\mathbf{k}, x^0)], \\ :H: &= \frac{1}{2}\int d^3\mathbf{k}[a_+{}^\dagger(\mathbf{k}, x^0)a_+(\mathbf{k}, x^0) + a_-{}^\dagger(\mathbf{k}, x^0)a_-(\mathbf{k}, x^0)], \end{aligned} \tag{2.70}$$

where, as for the real scalar field, normal ordering removes the infinite zero-point energy. Because of eq. (2.69), the terms in the Hamiltonian involving a_+ commute with those involving a_-, and may be treated separately. But each of these parts is of exactly the form of eq. (2.38). Thus, everything that we did for the real scalar field operator $a(\mathbf{k}, x^0)$ above, we can do again for $a_+(\mathbf{k}, x^0)$ and $a_-(\mathbf{k}, x^0)$ independently. The resulting eigenstates of H are labelled by independent occupation numbers for excitations created by $a_\pm{}^\dagger(\mathbf{k}, 0) \equiv a_\pm{}^\dagger(\mathbf{k})$. We label these states as

$$\begin{aligned} |\{\mathbf{k}_i{}^{(-)}\}, \{\mathbf{k}_j{}^{(+)}\}\rangle &\equiv \prod_i a_+{}^\dagger(\mathbf{k}_i)\prod_j a_-{}^\dagger(\mathbf{k}_j)|0\rangle, \\ a_\pm(\mathbf{q})|0\rangle &= 0, \\ :H:|\{\mathbf{k}_i{}^{(-)}\}, \{\mathbf{k}_j{}^{(+)}\}\rangle &= \left[\sum_i \omega_{k_i{}^{(-)}} + \sum_j \omega_{k_j{}^{(+)}}\right]|\{\mathbf{k}_i{}^{(-)}\}, \{\mathbf{k}_j{}^{(+)}\}\rangle. \end{aligned} \tag{2.71}$$

Momentum operators may be treated similarly. The Hilbert space is still a direct sum, but now with two labels for each subspace, which specify the number of excitations of each type. In place of eqs. (2.54) and (2.59), we have

$$\begin{aligned} \mathcal{H} &= \mathcal{H}_{0^{(-)},0^{(+)}} + \mathcal{H}_{0^{(-)},1^{(+)}} + \mathcal{H}_{1^{(-)},0^{(+)}} + \mathcal{H}_{2^{(-)},0^{(+)}} + \mathcal{H}_{1^{(-)},1^{(+)}} \\ &\quad + \mathcal{H}_{0^{(-)},2^{(+)}} + \cdots, \\ I &= |0\rangle\langle 0| + \sum_{n,m=1}^{\infty} \frac{1}{n!m!}\prod_{i=1}^{n}\prod_{j=1}^{m}\int\frac{d^3\mathbf{k}_i\, d^3\mathbf{k}_j}{(2\omega_{k_i})(2\omega_{k_j})} \\ &\quad \times |\{\mathbf{k}_i{}^{(-)}\}\{\mathbf{k}_j{}^{(+)}\}\rangle\langle\{\mathbf{k}_i{}^{(-)}\}\{\mathbf{k}_j{}^{(+)}\}|. \end{aligned} \tag{2.72}$$

Of particular interest in interpreting states of this Hilbert space is the operator corresponding to the conserved charge, eq. (1.63), which is given in terms of creation and annihilation operators by

$$Q = \int\frac{d^3\mathbf{k}}{2\omega_k}[a_+{}^\dagger(\mathbf{k})a_+(\mathbf{k}) - a_-{}^\dagger(\mathbf{k})a_-(\mathbf{k})]. \tag{2.73}$$

This is the difference of the number operators corresponding to $a_\pm(\mathbf{k})$. We may easily verify that Q commutes with the Hamiltonian, and that every state in subspace $\mathcal{H}_{i(-),j(+)}$ has eigenvalue, i.e. charge, $j-i$. We thus interpret the operators $a_{\pm}{}^{\dagger}(\mathbf{k})$ as creation operators for positively and negatively charged particles, and the subspace $\mathcal{H}_{i(-),j(+)}$ as the set of all states with i negatively charged and j positively charged particles.

Antiparticles

The free field $\phi(x)$ in eq. (2.68) creates particles of negative charge and annihilates particles of positive charge. This is what becomes of the negative-frequency and positive-frequency amplitudes in eq. (1.30) in the second-quantized theory. Otherwise indistinguishable states of opposite charge are said to represent particles and antiparticles (which states represent particles is purely a matter of definition). From eq. (2.72), the free Hilbert space is labelled by particle and antiparticle occupation numbers. This result depends only on the existence of a classical symmetry, such as phase invariance under eq. (1.58), that distinguishes the field from its hermitian conjugate, and on the unavoidable occurrence of both positive-energy and negative-energy solutions to the classical field equation. The pairing of positive- and negative-energy solutions is a direct consequence of relativistic invariance of the wave equation, so that particle–antiparticle combinations are characteristic of any nonhermitian relativistic field.

2.4 Particles and Green functions

The solution of the free scalar field allows us to introduce two of the central tools of quantum field theory, particle states and Green functions. Both will be necessary in our study of interacting fields.

Wave packets and particle interpretation

Consider the states for a real scalar field with total occupation number one (elements of subspace $\mathcal{H}_1$),

$$|\chi\rangle = \int \frac{\mathrm{d}^3\mathbf{k}}{2\omega_k} f_\chi(\mathbf{k}) a^\dagger(\mathbf{k})|0\rangle. \tag{2.74}$$

It is from such states that we expect to build particles, as in nonrelativistic quantum mechanics. To merit the term ‘particle’, however, such excitations should be localizable. So, consider the special states

$$|\mathbf{x}, x^0\rangle = \int \frac{\mathrm{d}^3\mathbf{k}}{[(2\pi)^3 2\omega_k]^{1/2}} \exp(\mathrm{i}\bar{k}\cdot x)|\mathbf{k}\rangle, \tag{2.75}$$

which possess the (noncovariant) orthogonality relations

$$\langle \mathbf{x}, x^0 | \mathbf{x}', x^0 \rangle = \delta^3(\mathbf{x} - \mathbf{x}'). \tag{2.76}$$

In the following, these will be considered as position eigenstates, corresponding to the presence of a particle at point $(\mathbf{x}, x^0)$. The formal justification of this intuitively plausible assumption is a bit technical, and may be found in Newton & Wigner (1949) and Schweber (r1961, Sections 3c and 7c). Accepting this interpretation, we introduce a position amplitude for states in $\mathcal{H}_1$ by

$$\psi_\chi(\mathbf{x}, x^0) = \langle \mathbf{x}, x^0 | \chi \rangle = \int \frac{d^3\mathbf{k}}{[(2\pi)^3 2\omega_k]^{1/2}} \exp(-i\bar{k}\cdot x) f_\chi(\mathbf{k}). \tag{2.77}$$

We easily check that if $\langle \chi | \chi \rangle = 1$,

$$\int d^3\mathbf{x} |\psi_\chi(\mathbf{x}, x^0)|^2 = \int \frac{d^3\mathbf{k}}{2\omega_k} |f_\chi(\mathbf{k})|^2 = 1, \tag{2.78}$$

for every x^0, so that $|\psi_\chi(\mathbf{x}, x^0)|^2$ admits interpretation as the probability density for finding a particle at point $\mathbf{x}$ at time x^0. A wave packet has a function $f_\chi(\mathbf{k})$ in eq. (2.77) that is peaked about some value $\mathbf{k}_0$ in such a way that $\psi_\chi(\mathbf{x}, x^0)$ is still localized.

Multiparticle states may be constructed as direct products of single particle states. According to the discussion of statistics in the previous section, the particles will automatically be bosons.

Field matrix elements

The operator structure of a field theory is very rich, since there is a different field operator at every point. The most important matrix elements for our purposes, however, will be vacuum expectation values of products of free fields, such as

$$\langle 0 | \phi^\dagger(y) \phi(x) | 0 \rangle. \tag{2.79}$$

Suppose $y^0 > x^0$. Then this matrix element may be thought of as the amplitude that a measurement of the field ϕ at point x^μ, followed by a measurement of $\phi^\dagger$ at y^μ, will leave the vacuum state undisturbed.

To interpret eq. (2.79), we insert the complex-field completeness relation (2.72) between the two free fields, and evaluate the resulting matrix elements using eqs. (2.69) and (2.71),

$$\begin{aligned}\langle 0 | \phi^\dagger(y) \phi(x) | 0 \rangle &= \int \frac{d^3\mathbf{k}}{2\omega_k} \langle 0 | \phi^\dagger(y) | \mathbf{k}^{(-)} \rangle \langle \mathbf{k}^{(-)} | \phi(x) | 0 \rangle, \\ &= \int \frac{d^3\mathbf{k}}{(2\pi)^3 2\omega_k} \exp[-i\bar{k}\cdot(y - x)]. \end{aligned} \tag{2.80}$$

We can understand this result as follows. The first field operator excites the vacuum, producing a superposition of states with one negatively charged particle. To form the vacuum expectation value, this same superposition of states is absorbed by the second field. Because the particle is produced before it is absorbed, the process is described as *causal*.

Stueckelberg–Feynman Green function

If y^μ is before x^μ, our causal interpretation of eq. (2.79) fails. We may, however, *define* the product of the fields so that an interpretation in terms of the causal propagation of particles is always possible. This requires reordering the field for different relative times. So, when $x^0 > y^0$, rather than (2.79) we choose the opposite order, $\langle 0|\phi(x)\phi^\dagger(y)|0\rangle$. In this case, only positively charged particles are excited. Together, the two choices give the Stueckelberg–Feynman (Stueckelberg 1942, Feynman 1948a, Rivier 1949) *causal Green function*, or *propagator*:

$$\begin{aligned} \mathrm{i}\Delta_\mathrm{F}(x-y) &= \langle 0|\phi(x)\phi^\dagger(y)|0\rangle\,\theta(x_0-y_0) + \langle 0|\phi^\dagger(y)\phi(x)|0\rangle\,\theta(y_0-x_0) \\ &\equiv \langle 0|T[\phi(x)\phi^\dagger(y)]|0\rangle. \end{aligned} \tag{2.81}$$

Note the factor of i on the left-hand side. This is a convention, although not a universal one. T is defined to order any number of operators by putting earlier operators to the right:

$$\begin{aligned} T[\textstyle\prod O_i(x_i)] \equiv O_n(x_n) \times \cdots \times O_2(x_2)O_1(x_1)\theta(x_n{}^0 - x_{n-1}{}^0) \\ \times \cdots \times \theta(x_2{}^0 - x_1{}^0) + \text{permutations}. \end{aligned} \tag{2.82}$$

Adding eq. (2.80) to the contribution from $x^0 < y^0$, we find a useful expression for the causal Green function as a four-dimensional integral:

$$\begin{aligned} \mathrm{i}\Delta_\mathrm{F}(z) &= \frac{1}{(2\pi)^3}\int \frac{\mathrm{d}^3\mathbf{k}}{2\omega_k}\,[\theta(z_0)\exp(-\mathrm{i}\bar{k}\cdot z) + \theta(-z_0)\exp(\mathrm{i}\bar{k}\cdot z)] \\ &= \frac{\mathrm{i}}{(2\pi)^4}\int \mathrm{d}^4k\,\frac{\exp(-\mathrm{i}k\cdot z)}{k_0{}^2 - \mathbf{k}^2 - m^2 + \mathrm{i}\epsilon}. \end{aligned} \tag{2.83}$$

The integral in the second form is defined by introducing the positive infinitesimal imaginary term '$+\mathrm{i}\epsilon$' in the denominator. It is understood that the k_0 integral is to be done first in this case. For $z_0 > 0$ we complete the k_0 contour by adding the semicircle C' of Fig. 2.1, along which the exponential decays to zero at infinity, and close the contour in the lower half-plane, thus picking up the k_0 pole at $k_0 = (\mathbf{k}^2 + m^2)^{1/2} - \mathrm{i}\epsilon = \omega_k - \mathrm{i}\epsilon$. Similarly, for $z_0 < 0$, we close the contour along C'', in the upper half-plane, and pick up the pole at $\mathrm{k}_0 = -\omega_k + \mathrm{i}\epsilon$. In momentum space, the

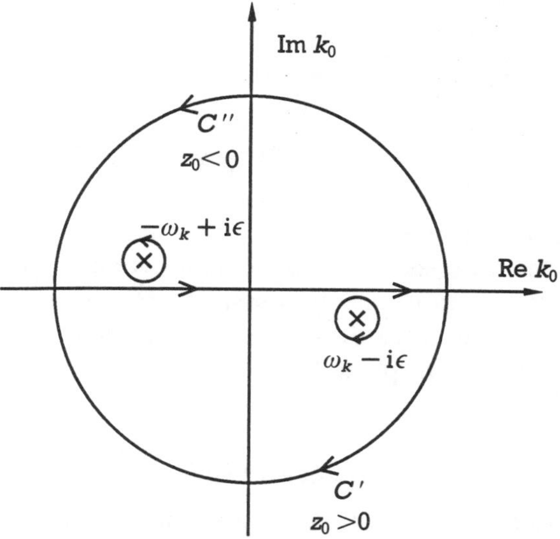

Fig. 2.1 Contour integrals for Feynman propagator.

propagator is a pure pole term:

$$\tilde{\Delta}_F(q) = \int d^4z \exp(iq\cdot z)\, \Delta_F(z) = (q^2 - m^2 + i\epsilon)^{-1}, \qquad (2.84)$$

a form that will occur again and again below. Following common usage, we will generally refer to the causal propagator as the 'Feynman propagator'.

Using (2.83), we find that Δ_F satisfies

$$(\partial_x{}^\mu \partial_{x,\mu} + m^2)\Delta_F(x-y) = -\delta^4(x-y), \qquad (2.85)$$

the equation for a Green function of the Klein–Gordon equation. We say *a* Green function, because many such functions satisfy (2.85), with differing boundary conditions. The time-ordering prescription has fixed the boundary conditions for both positive and negative times. If we take the iϵ prescription, $\omega_k - i\epsilon$, in eq. (2.83) literally, then Δ_F vanishes for $t = \pm\infty$.

2.5 Interacting fields and scattering

We are now ready to introduce some features of scalar theories with nontrivial potentials. An explicit solution, however, will no longer be possible. To see why, let us denote the free-field Hamiltonian (2.38) by H_0, and add a potential of the form $(g/n!)\int d^3\mathbf{x}\, \phi^n(\mathbf{x}, x^0)$. The resulting Hamiltonian may be written in terms of the Fourier-transformed field

$\phi(\mathbf{k}, x^0)$ as

$$H = H_0 + \frac{g}{n!}\int \mathrm{d}^3\mathbf{x}\,\phi^n(\mathbf{x}, x^0)$$

$$= H_0 + \frac{g}{n!}\prod_{i=1}^{n}\int \frac{\mathrm{d}^3\mathbf{k}_i}{(2\pi)^3}\,\phi(\mathbf{k}_i, x^0)\delta^3(\textstyle\sum_{i=1}^{n}\mathbf{k}_i). \qquad (2.86)$$

Comparing this form with eq. (2.38), we see that the energy of the system can no longer be expressed as a simple sum over normal modes $\mathbf{k}$ of the free system. The modes of the free system are said to 'interact'. In the corresponding classical case, the equation of motion becomes nonlinear, and the superposition principle fails.

Spectral assumptions

In the absence of an exact solution, we must make some assumptions about the Hilbert space of the interacting theory. This we do in a basis in which the momentum operators P^μ are diagonal. We present these assumptions heuristically, with no attempt at building a rigorous structure; they will appear automatically in the perturbative expansion of the following chapters. (Such assumptions can, however, be made part of *axiomatic field theory* (Streater & Wightman r1964, Jost r1965).) Our initial assumptions are as follows.

(i) There is a ground state $|0\rangle$, whose total momentum we may choose to be $q_{\mathrm{vac}}{}^\mu = 0$. That is, we assume that the system is stable.

(ii) There is a continuum of 'first-excited' states $|p^\mu\rangle$, with $p^\mu p_\mu = m^2 \geqslant 0$. These states make up the analogue of the subspace $\mathcal{H}_1$ of the free-particle Fock space, and correspond intuitively to single-particle states. The assumption $m^2 \geqslant 0$ means there are no tachyons in the theory, particles that travel faster than light. For simplicity, we limit the present discussion to a single variety of elementary excitation with $m > 0$.

(iii) The spectra of additional states $|q^\mu\rangle$ have $q^2 \geqslant 4m^2$. These states are expected to form a continuum, corresponding to multiparticle excitations.

We shall also require that

$$\langle\phi\rangle_0 \equiv \langle 0|\phi(x)|0\rangle = 0, \qquad (2.87)$$

that is, that the field has zero vacuum expectation value. Translation invariance requires that $\langle\phi\rangle_0$, whatever it is, is constant. Thus we can, if we wish, always redefine the field as $\phi' = \phi - \langle\phi\rangle_0$, $\langle\phi'\rangle_0 = 0$, so eq. (2.87) is not an extra assumption. For fields with nontrivial internal symmetries these considerations become more subtle and interesting (see Section 5.4).

In- and out-states

Consider, for a moment, a system of photons and electrons. We can (and we shall, later) construct free-field theories for both particles. States in these free theories form a Fock space, just as for free scalar theories. The eigenstates of the interacting Hamiltonian – the one that describes the real world – are presumably different. Nevertheless, we do see photons and electrons in the real world behaving as (essentially) free particles when they are sufficiently isolated. By the uncertainty principle, of course, they cannot then be in pure momentum eigenstates; they must be described by wave packets.

Two important classes of wave packet states may be identified. First, consider wave packets that are isolated for $t \to -\infty$, but may overlap at finite times. We call the Heisenberg picture states with this property *in-states*, and denote them as $|\{\mathbf{p}_i\}\text{ in}\rangle$. Second, there are states whose wave packets are isolated for $t \to +\infty$, but may overlap at finite times. These are *out-states*, denoted $|\{\mathbf{p}_i\}\text{ out}\rangle$. Recall (Appendix A) that in the Heisenberg picture states are time-independent, but that each Heisenberg state still represents one complete quantum mechanical evolution of the interacting field, from $t = -\infty$ to $t = +\infty$, with given boundary conditions (in this case, at $t = -\infty$ or $t = +\infty$). Thus, the isolated wave packets described by a Heisenberg in-state at $t = -\infty$ may overlap and scatter at finite t, producing a much more complicated system of waves at finite times. As $t \to +\infty$, the fields described by an in-state are in general no longer a single set of isolated wave packets, but a superposition of many such sets – each of which corresponds to an out-state. This 'new' field configuration, however, is still described by the original in-state.

For the special case of a single particle, there is nothing else with which to overlap or from which to scatter. The single-particle in- and out-states must then be the same, and be identical with the single-particle part of the spectrum identified above. Hence, we write

$$|\mathbf{p}\text{ in}\rangle = |\mathbf{p}\text{ out}\rangle = |\mathbf{p}\rangle. \tag{2.88}$$

At this point we may ask how many in-states and out-states there are. We shall make the assumption that in-states and out-states are independently complete in the Hilbert space,

$$\Sigma|\{\mathbf{p}_i\}\text{ in}\rangle\langle\{\mathbf{p}_i\}\text{ in}| = \Sigma|\{\mathbf{p}_i\}\text{ out}\rangle\langle\{\mathbf{p}_i\}\text{ out}| = I. \tag{2.89}$$

This states that if we wait long enough, *any* system becomes a superposition of states with isolated outgoing particles, and, analogously, if we go far enough back in time, *any* system can be described in terms of states with isolated incoming particles. Is this a reasonable assumption? It would require modification in the presence of bound states of elementary excitations. For the interacting scalar field theory we shall, for simplicity, ignore this possibility.

The S-matrix

By construction, in- and out-states represent field configurations that evolve into isolated particles in the distant past and future, respectively. (Remember, the states are time-independent, but the fields they describe are not.) If we look far into the past of an in-state, the particles get more and more isolated, and analogously for the future of an out-state. As a result, *a matrix element between an in-state and an out-state gives the amplitude for a field configuration that was simple far in the past to evolve into a field configuration that will be simple far in the future.* The matrix elements between Heisenberg in- and out-states define the *S-matrix*, or *scattering matrix*,

$$S_{\beta\alpha} \equiv \langle \beta\,\mathrm{out}|\alpha\,\mathrm{in}\rangle. \tag{2.90}$$

The name is suggestive of the physical process that this matrix element describes. For orthonormal bases $\{|\alpha\,\mathrm{in}\rangle\}$ and $\{|\beta\,\mathrm{out}\rangle\}$, the completeness assumptions eqs. (2.89) imply that

$$S_{\beta\gamma}{S^\dagger}_{\gamma\alpha} = \delta_{\beta\alpha} = {S^\dagger}_{\beta\gamma}S_{\gamma\alpha}. \tag{2.91}$$

That is, S is unitary:

$$S^\dagger S = SS^\dagger = I. \tag{2.92}$$

Let us emphasize that the S-matrix has a standard quantum mechanical interpretation. It is analogous, for instance, to the amplitude for a spin one-half system, prepared with spin up along the z-axis at one time, to be measured up or down along another axis at a later time. As usual

$$|S_{\beta\alpha}|^2 = |\langle \beta\,\mathrm{out}|\alpha\,\mathrm{in}\rangle|^2 \tag{2.93}$$

is the probability for the observation of out-state β after in-state α has been prepared. S thus precisely describes scattering experiments.

Finally, we note that the scattering matrix includes terms where the incoming and outgoing states describe the same sets of particles, with the same momenta. It will be convenient to separate forward scattering by defining

$$S = I + \mathrm{i}T \tag{2.94}$$

where T is the *transition* matrix, which contributes whenever some set of particles (although not necessary *all*) undergo scattering. The computation of the S-matrix in various theories is the essential goal of the rest of the book. The story begins with a specific set of matrix elements, the Green functions.

Green functions

Green functions are vacuum expectation values of time-ordered products of arbitrary numbers of fields:

$$G(x_1, \ldots, x_n) = \langle 0|T[\textstyle\prod_{i=1}^{n}\phi(x_i)]|0\rangle. \tag{2.95}$$

They are the generalization of the two-point function eq. (2.81) in both free and interacting theories, using eq. (2.82) for the time-ordered product.

The most obvious interpretation of Green functions is in terms of field measurements. Our interest, however, will concentrate on their relation to the S-matrix, through *reduction formulas* (Lehmann, Symanzik & Zimmermann 1955).

Reduction formulas

The derivation of reduction formulas involves some rather subtle assumptions, which, however, will turn out to be justified in the perturbative expansion developed in the next two chapters. Our (frankly nonrigorous) arguments may be easier to understand if the conclusion is stated first.

For the sake of completeness, consider a slight generalization of an S-matrix element in a scalar theory, $\langle\{\mathbf{q}_j\}_n \,\text{out}|A(0)|\{\mathbf{k}_i\}_m \,\text{in}\rangle$, $i = 1, \ldots, m$, $j = 1, \ldots, n$, where $A(x)$ is some operator. (We may take A as the field ϕ, but it also may be a composite operator like $\phi^*\,\partial\phi - \partial\phi^*\,\phi$, or a product of operators at different, but fixed, points.) Consider the following specific Green function in momentum space,

$$\bar{G}^{(A)}(q_j{}^\mu, k_i{}^\mu) \equiv \prod_{j=1}^{n} \int d^4y_i \exp(iq_j\cdot y_j) \prod_{i=1}^{m} \int d^4x_i \exp(-ik_i\cdot x_i)$$
$$\times \langle 0|T[\textstyle\prod_{j=1}^{n}\phi(y_j)A(0)\prod_{i=1}^{m}\phi(x_i)]|0\rangle. \tag{2.96}$$

The reduction theorem states that the matrix element above, between in- and out-states, is related to this Green function by

$$\langle\{\mathbf{q}_j\}_n \,\text{out}|A(0)|\{\mathbf{k}_i\}_m \,\text{in}\rangle = [i(2\pi)^{3/2}R^{1/2}]^{-m-n}\prod_{j=1}^{n}(q_j{}^2 - m^2)$$
$$\times \prod_{i=1}^{m}(k_i{}^2 - m^2)\bar{G}^{(A)}(q_j{}^\mu, k_i{}^\mu)|_{k_i^2=m^2, q_j^2=m^2}, \tag{2.97}$$

where the (Lorentz invariant) quantity R is defined by the matrix element

$$R(p_i^2) \equiv (2\pi)^3 |\langle 0|\phi(0)|\mathbf{p}_i\rangle|^2. \tag{2.98}$$

The reduction formula eq. (2.97) implies that $\bar{G}^{(A)}(q_j{}^\mu, k_i{}^\mu)$ has a separate (simple) pole in every variable $(k_i{}^2 - m^2)$ and $(q_j{}^2 - m^2)$, and that the residue of this $(m+n)$th-order pole is the matrix element $\langle \{\mathbf{q}_j\}_n \text{ out}|A(0)|\{\mathbf{k}_i\}_m \text{ in}\rangle$, up to an overall factor given by eq. (2.98). Note that any contribution to $\bar{G}^{(A)}$ that lacks a pole in any one of these variables vanishes in the reduction formula.

To show how this result arises, let us consider the contribution to $\bar{G}^{(A)}$, eq. (2.96), from large negative times $x_i{}^0$ and large positive times $y_j{}^0$. To be specific, we choose to separate the region $\{x_i{}^0 < -\tau\}, \{y_j{}^0 > \tau\}$, for fixed τ, from the remainder of the integral. Then, using the definition of time ordering, and inserting complete sets of in- and out-states, we have

$$\begin{aligned}\bar{G}^{(A)}(q_j{}^\mu, k_i{}^\mu) = {} & \prod_{j=1}^{n} \int_\tau^\infty \mathrm{d}y_{j,0} \exp(\mathrm{i}q_{j0}y_{j0}) \int \mathrm{d}^3\mathbf{y}_j \exp(-\mathrm{i}\mathbf{q}_j \cdot \mathbf{y}_j) \\ & \times \prod_{i=1}^{m} \int_{-\infty}^{-\tau} \mathrm{d}x_{i,0} \exp(-\mathrm{i}k_{i0}x_{i0}) \int \mathrm{d}^3\mathbf{x}_i \exp(\mathrm{i}\mathbf{k}_i \cdot \mathbf{x}_i) \\ & \times \sum_{\alpha\,\text{in}} \sum_{\beta\,\text{out}} \langle 0|T[\textstyle\prod_{j=1}^{n}\phi(y_j)]|\beta \text{ out}\rangle \\ & \times \langle \beta \text{ out}|A(0)|\alpha \text{ in}\rangle \langle \alpha \text{ in}|T[\textstyle\prod_{i=1}^{m}\phi(x_i)]|0\rangle \\ & + \text{remainder},\end{aligned} \tag{2.99}$$

where the remainder includes contributions from other times.

Now let us quantify the assertion above, that for times far enough in the past or future respectively, in- and out-states behave like systems of non-interacting particles. This means that for large enough τ, the matrix elements between fields and in- or out-states break up into one for each particle of definite momentum:

$$\begin{aligned}&\prod_{j=1}^{n} \int \mathrm{d}^3\mathbf{y}_j \exp(-\mathrm{i}\mathbf{q}_j \cdot \mathbf{y}_j) \langle 0|T[\textstyle\prod_{j=1}^{n}\phi(y_j)]|\{\mathbf{p}_j\}_{n'} \text{ out}\rangle \\ &\qquad = \delta_{nn'} n! \prod_{j=1}^{n} \int \mathrm{d}^3\mathbf{y}_j \exp(-\mathrm{i}\mathbf{q}_j \cdot \mathbf{y}_j) \langle 0|\phi(y_j)|\mathbf{p}_i\rangle \\ &\qquad = \delta_{nn'} n! \prod_{j=1}^{n} R^{1/2} (2\pi)^{3/2} \delta^3(\mathbf{q}_j - \mathbf{p}_j) \exp[-\mathrm{i}\omega(\mathbf{q}_j) y_{j0}].\end{aligned} \tag{2.100}$$

Here we have used translation invariance, eq. (2.26), in the single-particle matrix elements, and the definition (2.98) for R. (In a careful treatment, the single-particle states would be wave packets.) The factor of $n!$ comes from permutations of the fields. Using this result in the sum over in- and out-states in eq. (2.99), as specified by eq. (2.59), we find a result in which only the time integrals remain:

$$\begin{aligned}\bar{G}^{(A)}(q_j{}^\mu, k_i{}^\mu) &= \prod_{j=1}^{n} \frac{(2\pi)^{3/2} R^{1/2}}{2\omega(\mathbf{q}_j)} \int_\tau^\infty \mathrm{d}y_{j,0} \exp\{\mathrm{i}[q_{j0} - \omega(\mathbf{q}_j)]y_{j0}\} \\ &\times \prod_{i=1}^{m} \frac{(2\pi)^{3/2} R^{1/2}}{2\omega(\mathbf{k}_i)} \int_{-\infty}^{-\tau} \mathrm{d}x_{i,0} \exp\{-\mathrm{i}[k_{i0} - \omega(\mathbf{k}_i)]x_{i0}\} \\ &\times \langle \{\mathbf{q}_j\}_n \text{ out}|A(0)|\{\mathbf{k}_i\}_m \text{ in}\rangle + \text{remainder} \\ &= [\mathrm{i}(2\pi)^{3/2} R^{1/2}]^{m+n} \\ &\times \prod_{j=1}^{n} \frac{1}{2\omega(\mathbf{q}_j)[q_{j0} - \omega(\mathbf{q}_j)]} \prod_{i=1}^{m} \frac{1}{2\omega(\mathbf{k}_i)[k_{i0} - \omega(\mathbf{k}_i)]} \\ &\times \langle \{\mathbf{q}_j\}_n \text{ out}|A(0)|\{\mathbf{k}_i\}_m \text{ in}\rangle + \text{remainder}. \qquad (2.101)\end{aligned}$$

In the second form we have done the time integrals, and grouped all contribtions that lack a pole in at least one of the variables $k_{i0} - \omega(\mathbf{k}_i)$ and $q_{j0} - \omega(\mathbf{q}_j)$ in the remainder. Thus, we have used the expansion

$$\exp\{-\mathrm{i}[k_{i0} - \omega(\mathbf{k}_i)]\tau\} = 1 - \mathrm{i}[k_{i0} - \omega(\mathbf{k}_i)]\tau + \cdots,$$

and have kept only the first term. Using

$$[2\omega(\mathbf{q}_j)][q_{j0} - \omega(\mathbf{q}_j)] = q_j{}^2 - m^2 + \text{nonsingular terms},$$

we find the desired result, eq. (2.97), by multiplying both sides by factors of $q_j{}^2 - m^2$ and $k_i{}^2 - m^2$, and then setting all momenta on-shell.

Clearly, our argument involves some very delicate limits, especially in eq. (2.101), which we shall make no attempt to analyze here. The reduction formulas themselves were given by Lehmann, Symanzik & Zimmermann (1955) (see also Low 1955), whose formal arguments are described, for instance, in Chapter 16 of Bjorken & Drell (r1965) and Chapter 5 of Itzykson & Zuber (r1980). They allow us to formulate the S-matrix directly in terms of Heisenberg picture operators (Yang & Feldman 1950). An account of further developments is given in Chapters 13 and 14 of Bogoliubov, Logunov & Todorov (r1975). In any case, we shall accept the reduction formula (2.97), and use it to derive the S-matrix in terms of Green functions.

Exercises

2.1 Another approach to canonical quantization is define 'averaged' fields and momenta,

$$\langle\phi(x^0)\rangle_i \equiv \frac{1}{V_i^{1/2}}\int_{R_i} d^3\mathbf{x}\,\phi(\mathbf{x}, x^0),\ \langle\pi(x^0)\rangle_i \equiv \frac{1}{V_i^{1/2}}\int_{R_i} d^3\mathbf{x}\,\pi(\mathbf{x}, x^0),$$

over arbitrary spatial regions R_i. In the limit of small R_i, they become analogous to the coordinates and momenta defined in eq. (2.6a). Show that, given the continuum commutation relations (2.9), averaged fields and momenta obey the discrete commutation relations (2.2) at equal times, independently of the exact definition of the R's. In some sense, a formulation in terms of averaged operators is closer to reality, since a field can never be truly measured at a point.

2.2 Verify eqs. (2.17) from the equal-time commutation relations, at least for spatial indices. The zero components are a festival of index shuffling which the more energetic reader may wish to undertake.

2.3 Verify in the classical scalar theory, that the P_μ generate translations according to $\partial\phi(x)/\partial x_\mu = \{\phi(x), P^\mu\}_{PB}$.

2.4 Compute $\langle f|\pi(\mathbf{x}, 0)\partial^i\phi(\mathbf{x}, 0)|f\rangle$ for a single-particle momentum eigenstate $|\mathrm{f}\rangle$, and also for the wave packet (with $\gamma \gg p^2/m^4$)

$$|f(\mathbf{p}), \mathbf{y}, y_0 = 0\rangle = \left(\frac{2\gamma}{\pi}\right)^{3/4}\int \frac{d^3\mathbf{l}}{(2\omega_l)^{1/2}} \exp(\mathrm{i}\mathbf{l}\cdot\mathbf{y} - \gamma|\mathbf{l} - \mathbf{p}|^2)\, a_l^\dagger|0\rangle.$$

2.5 Let A and B be operators such that $[A, B] = B$. A useful identity is $[\exp(\alpha A)]B[\exp(-\alpha A)] = (\exp\alpha)B$. Prove this relation, and use it to construct the 'charge conjugation' operator, C, for the scalar field, which exchanges the charges of states, thus $\mathrm{C}|\{\mathbf{p}_i^{(+)}\}, \{\mathbf{q}_i^{(-)}\}\rangle = |\{\mathbf{p}_i^{(-)}\}, \{\mathbf{q}_i^{(+)}\}\rangle$.

2.6 Suppose we construct states, $|p, d\rangle$, of unit normalization, $\langle p, d'|p, d\rangle = \delta_{d'd}$, from the momentum eigenstates, eq. (2.35). Define $(T_a)_{dd'} \equiv \langle p, d'|Q_a|p, d\rangle$. Show that $[T_a, T_b] = \mathrm{i}C_{ab}{}^cT_c$ and that the states $|p, d\rangle$ transform according to the representation generated by T_a. That is, for infinitesimal transformations, show that

$$(1 + \mathrm{i}\delta\beta_a Q_a)|p, d\rangle = (\delta_{dd'} + \mathrm{i}\delta\beta_a(T_a)_{dd'})|p, d'\rangle.$$

2.7 Consider the following momentum-space Green functions for the Klein–Gordon equation:

$$G^{(\pm)}(\mathbf{p}, m) = [(p_0 \pm \mathrm{i}\epsilon)^2 - \mathbf{p}^2 - m^2]^{-1}.$$

Show that they correspond to retarded and advanced Green functions. Also, show that in the $m \to 0$ limit their Fourier transforms are $\theta(\pm x_0)\delta(x^2)$.

2.8 Show that the product of free scalar fields is singular on the 'light-cone', that is that

$$\langle 0|\phi(x)\phi(y)|0\rangle = -(2\pi^2)^{-1}(x - y)^{-2} + R(x - y, m),$$

where $(x - y)^2R(x - y, m)$ vanishes for $(x - y)^2 = 0$. Note that the leading singularity on the light cone is independent of the mass.

2.9 Equation (2.82) defines time ordering. It is often useful to define the 'time-ordered exponential' of a time-dependent operator $A(t)$ as

$$U(\tau, t) = T \exp \int_t^{\tau} A(t')\,dt'$$
$$= 1 + \int_t^{\tau} A(t')\,dt' + \int_t^{\tau} A(t_2)\,dt_2 \int_t^{t_2} A(t_1)\,dt_1$$
$$+ \cdots,$$

which clearly satisfies $dU(\tau, t)\,d\tau = A(\tau)U(\tau, t)$. (a) Show that when $[A(t_i), A(t_j)] = 0$ for all t_i and t_j, U reduces to an ordinary exponential. (b) Show that $U(\tau, t)$ obeys $U(\tau, t')U(t', t) = U(\tau, t)$. (c) In the same notation show that if $A(t)$ is hermitian, $U^{\dagger}(\tau, t) = U(t, \tau)$. From (b) and (c), the ordered exponential of an antihermitian operator is unitary.

2.10 What is the S-matrix in a free theory?

3

Path integrals, perturbation theory and Feynman rules

There is a natural solution to an interacting-field theory whenever its Lagrangian is the sum of a free (quadratic) Lagrangian and interaction terms. We associate with each interaction a multiplicative factor g_i, the *coupling constant* for that term. We then expand S-matrix elements of the interacting-field theory around those of the free-field theory, as a power series in the g_i. In this approach, the interaction is conceived as a small perturbation on the free theory; the systematic method for expanding in the coupling constant(s) is *perturbation theory*.

The development of field-theoretic perturbation theory can be carried out in a number of ways. We shall use the path integral method, initiated by Feynman (1949). The other commonly discussed approach, based on the interaction picture, is sketched in Appendix A.

The Feynman path integral is a reformulation of quantum mechanics in terms of classical quantities. Extensive discussions of the path integral in quantum mechanics may be found, for instance, in Feynman & Hibbs (r1965) and Schulman (r1981). Technical reviews include Keller & McLaughlin (r1975) and Marinov (r1980). For issues of mathematical rigor, which by and large are avoided below, see Glimm & Jaffe (r1981) and DeWitt-Morette, Maheshwari & Nelson (r1979).

3.1 The path integral

We first consider a quantum mechanical system with only one coordinate q and time-independent Hamiltonian $H(p, q)$. We recall that position operators in the Heisenberg and Schödinger pictures (Appendix A), denoted $Q_{\mathrm{H}}(t)$ and Q_{S} respectively, are related by

$$Q_{\mathrm{H}}(t) = \exp(\mathrm{i}Ht/\hbar)\, Q_{\mathrm{S}} \exp(-\mathrm{i}Ht/\hbar). \tag{3.1}$$

We shall keep track of factors of $\hbar$ in this chapter, to highlight the relation of classical to quantum quantities. In the spirit of our comments at the beginning of Section 2.3, however, we may still rescale all dimensional quantities to units of a power of action times a power of length, without any powers of velocity, by dividing them by an appropriate power of c. Then no explicit factors of c will appear in our formulas. In particular, t will be considered as identical to x_0.

Feynman kernel

The path integral is formulated to describe the time evolution of matrix elements between eigenstates of the Heisenberg position operator. Let $|q, t\rangle_{\mathrm{H}}$ denote such an eigenstate, so that

$$Q_{\mathrm{H}}(t)|q, t\rangle_{\mathrm{H}} = q|q, t\rangle_{\mathrm{H}}. \tag{3.2}$$

In the time-independent state $|q, t\rangle_{\mathrm{H}}$, t is a label. $|q, t\rangle_{\mathrm{H}}$ is not an eigenstate of $Q_{\mathrm{H}}(t')$ for $t' \neq t$. It is, however, related to the time-dependent Schrödinger eigenstates $|q(t)\rangle_{\mathrm{S}}$ of Q_{S} by

$$|q(t)\rangle_{\mathrm{S}} = \exp(-\mathrm{i}Ht/\hbar)\,|q, t\rangle_{\mathrm{H}}. \tag{3.3}$$

Now consider the inner product between two Heisenberg states separated by some finite time interval,

$$U(q'', t''; q', t') = {}_{\mathrm{H}}\langle q'', t''|q', t'\rangle_{\mathrm{H}}. \tag{3.4}$$

U is known as the *Feynman kernel*. Because the position states $|q, t\rangle_{\mathrm{H}}$ are complete in the Hilbert space, U controls the time development of all states. For instance, suppose we rewrite the kernel (3.4) in terms of Schrödinger-picture states as

$$U(q'', t''; q', t') = {}_{\mathrm{S}}\langle q''(t'')|\exp[-\mathrm{i}H(t''-t')/\hbar]|q'(t')\rangle_{\mathrm{S}}, \tag{3.5}$$

where $|q(t)\rangle_{\mathrm{S}}$ is an eigenstate of the time-independent operator Q_{S}. Then U governs the evolution of Schrödinger-picture wave functions according to

$$\psi(q'', t'') = \int \mathrm{d}q'\; U(q'', t''; q', t')\psi(q', t'), \tag{3.6}$$

where $\psi(q, t) \equiv {}_{\mathrm{S}}\langle q(t)|\psi(t)\rangle_{\mathrm{S}}$. Henceforth in this section, all states and operators will be in the Schrödinger picture, unless subscripted by 'H', and we suppress the subscript 'S'.

The path integral

In the spirit of the matrix identity (1.70), we can rewrite the exponential of the Hamiltonian in eq. (3.5) as an infinite product:

$$U(q'', t''; q', t') = \lim_{n\to\infty} \langle q''(t'')|(1 - \mathrm{i}H\,\delta t/\hbar)^n|q'(t')\rangle, \tag{3.7}$$

where $\delta t = [(t'' - t')/n] \to 0$ in the limit. This, in turn, enables us to insert the identity operator in the form of complete sets of Schrödinger position eigenstates, $I = \int \mathrm{d}q(t)\,|q(t)\rangle\langle q(t)|$, between each of the terms in the product, yielding

$$U(q'', t''; q', t') = \lim_{n\to\infty}\left[\prod_{i=1}^{n}\int_{-\infty}^{\infty}\mathrm{d}q_i\langle q_i|(1 - \mathrm{i}H\delta t/\hbar)|q_{i-1}\rangle\,\delta(q_n - q'')\right], \tag{3.8}$$

where $q_0 \equiv q'$. Here the subscript labels the time. The $n \to \infty$ limit of this expression, assuming it exists, is the path integral. The terminology may be illustrated by the picture of Fig. 3.1, in which the variables q_i, $i = 1, \ldots, n-1$ are 'frozen' and graphed against time. Each such graph specifies, in the limit, a path leading from position q' at time t' to position q'' at time t'', and, as the q_i vary, every possible path is included.

Reduction to classical quantities

The reward for expanding the infinite product in the rather complicated form of eq. (3.8) is that it enables us to eliminate the operators and matrix elements, and to write the path integral in terms of classical quantities only.

To reduce to classical quantities, we again insert unity, but this time in the form of a complete set of conjugate momentum eigenstates, $I = \int \mathrm{d}p\,|p\rangle\langle p|$, on the left of each of the matrix elements in eq. (3.8):

$$\begin{aligned} U(q'', t''; q', t') = \lim_{n\to\infty}&\left[(2\pi\hbar)^{-n/2}\prod_{i=1}^{n-1}\int \mathrm{d}q_i\mathrm{d}p_i\int \mathrm{d}p_0\,\exp(\mathrm{i}p_0q_1/\hbar)\right.\\ &\left.\times\,\langle p_i|(1 - \mathrm{i}H\delta t/\hbar)|q_i\rangle\exp(\mathrm{i}p_iq_{i+1}/\hbar)\right]\\ &\times\,\langle p_0|(1 - \mathrm{i}H\delta t/\hbar)|q'\rangle, \end{aligned} \tag{3.9}$$

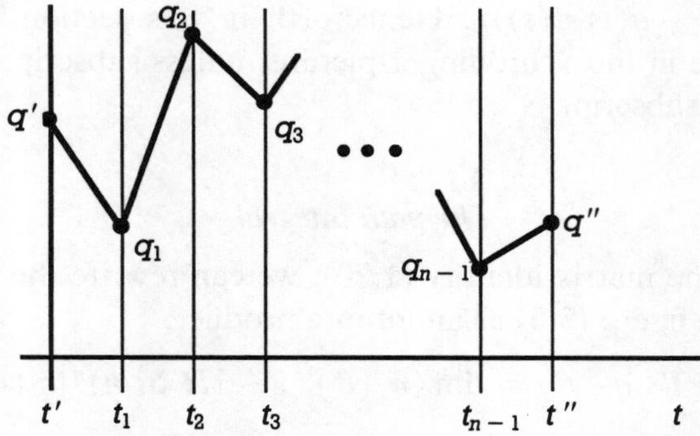

Fig. 3.1 Typical path contributing to eq. (3.8).

where we have used

$$\langle q_i | p_{i-1} \rangle = (2\pi\hbar)^{-1/2} \exp(\mathrm{i}p_{i-1}q_i/\hbar). \tag{3.10}$$

Note that, since we do this for every matrix element, there is one more p than q integral in eq. (3.9). We may now actually evaluate the matrix elements, because the operator H is a function of momentum and position operators. Since the operators Q and P obey the commutation relations of eq. (2.2), the Hamiltonian is only specified when the relative ordering of all factors of Q and P is given. Of course, this is not an issue for a Hamiltonian of the form

$$H = P^2/2m + V(Q), \tag{3.11}$$

which describes a nonrelativistic particle moving in a velocity-independent potential. On the other hand, for Hamiltonians such as $H = P^2Q^2$ and $H' = PQPQ$, which involve products of Q and P, the order matters. We may, however, put any such Hamiltonian into its *normal form*, H_{nor}, in which all position operators are commuted to the right. Normal form is related to, but not to be confused with, normal ordering (Section 2.3). For the examples just given, the normal forms are

$$H_{\mathrm{nor}} = P^2Q^2, \quad H'_{\mathrm{nor}} = P^2Q^2 + \mathrm{i}\hbar PQ. \tag{3.12}$$

For the matrix elements of eq. (3.9) we now have

$$\langle p_i | (1 - \mathrm{i}H(P, Q)\delta t/\hbar) | q_i \rangle = (2\pi\hbar)^{-1/2}[1 - \mathrm{i}h(p_i, q_i)\delta t/\hbar] \exp(-\mathrm{i}p_iq_i/\hbar), \tag{3.13}$$

where we have used the complex conjugate of eq. (3.10), and where $h(p_i, q_i)$ is now a function found by replacing the operators P and Q by the classical variables p_i and q_i in the *normal form of H*. Substituting eq. (3.13) into the kernel (3.9) now gives the purely classical expression

$$\begin{aligned} U(q'', t''; q', t') = \lim_{n\to\infty} \Bigg\{ (2\pi\hbar)^{-n} \prod_{i=1}^{n-1} \int \mathrm{d}q_i \mathrm{d}p_i \int \mathrm{d}p_0 \exp[\mathrm{i}p_0(q_1 - q')/\hbar] \\ \times [1 - \mathrm{i}h(p_i, q_i)\delta t/\hbar] \exp[\mathrm{i}p_i(q_{i+1} - q_i)/\hbar] \Bigg\} \\ \times [1 - \mathrm{i}h(p_0, q')\delta t/\hbar]. \end{aligned} \tag{3.14}$$

This may be made more manageable by going back from the infinite product to an exponential, using (see Feynman 1951 and exercise 3.1 of this volume):

$$\lim_{n\to\infty} \left\{ \prod_{i=1}^{n-1} \left[1 - \mathrm{i}h(p_i, q_i) \frac{(t'' - t')}{n\hbar} \right] \right\} = \exp\left\{ \frac{-\mathrm{i}(t'' - t')}{n\hbar} \sum_{i=1}^{n-1} h(p_i, q_i) \right\}, \tag{3.15}$$

to obtain

$$U(q'', t''; q', t') = \lim_{n\to\infty} \left\{ (2\pi\hbar)^{-n} \prod_{i=1}^{n-1} \int \mathrm{d}q_i \mathrm{d}p_i \int \mathrm{d}p_0 \right.$$
$$\left. \times \exp\left(\frac{\mathrm{i}}{\hbar} \sum_{j=0}^{n-1} \delta t \left[p_j \frac{(q_{j+1} - q_j)}{\delta t} - h(p_j, q_j) \right] \right) \right\}$$
$$\equiv \int [\mathrm{d}p][\mathrm{d}q] \exp\left(\frac{\mathrm{i}}{\hbar} \int \mathrm{d}t \, [p(t)\dot{q}(t) - H(p(t), q(t))] \right). \tag{3.16}$$

This form of the path integral is useful whenever we can perform manipulations on the exponential *for fixed paths*. We will see as we go along why this is so important. The second equality defines the symbol $\int[\mathrm{d}p][\mathrm{d}q]$, which represents the sum over paths, including normalization.

Equation (3.16) gives the path integral as a sum of paths in *phase space* (Tobocman 1956, Davies 1963, Garrod 1966), since both momenta and coordinates are integrated. For many systems, however, it is possible to revert to an integral like eq. (3.8), which is a sum of paths in coordinate space only.

Configuration space

Consider a Hamiltonian like eq. (3.11). In this case, the momentum integrals in eq. (3.16) are all Gaussian and of the form

$$\int_{-\infty}^{\infty} \mathrm{d}p \exp(-\alpha p^2 + \beta p) = (\pi/\alpha)^{1/2} \exp(\beta^2/4\alpha), \tag{3.17}$$

where $\alpha = \mathrm{i}\delta t/(2m\hbar)$ and $\beta = \mathrm{i}(q_{j+1} - q_j)/\hbar$, both pure imaginary. Using eq. (3.17) in eq. (3.16) gives

$$U(q'', t''; q', t') = \lim_{n\to\infty} \left[\left(\frac{m}{2\pi\mathrm{i}\hbar\delta t} \right)^{n/2} \prod_{i=1}^{n-1} \int \mathrm{d}q_i \right.$$
$$\left. \times \exp\left(\frac{\mathrm{i}}{\hbar} \sum_{j=1}^{n-1} \delta t \left\{ \frac{m}{2} \left[\frac{(q_{j+1} - q_j)}{\delta t} \right]^2 - V(q_i) \right\} \right) \right]$$
$$= \lim_{n\to\infty} \left[\left(\frac{m}{2\pi\mathrm{i}\hbar\delta t} \right)^{n/2} \prod_{i=1}^{n-1} \int \mathrm{d}q_i \exp\left(\frac{\mathrm{i}}{\hbar} \sum_{j=1}^{n-1} \delta t \; L(q_j, \dot{q}_j) \right) \right], \tag{3.18}$$

where $L(q, \dot{q})$ is the Lagrangian of the theory. This is the *configuration space* path integral, and it is somewhat less general than the phase-space integral, eq. (3.16). Again, the usefulness of eq. (3.18) depends on whether the integral over paths is dominated by 'smooth' paths, for which the sum in the exponential approaches an integral over the classical Lagrangian. If it does, the argument of the exponential in eq. (3.18) becomes the classical action $S(q'', t''; q', t')$ for that path,

$$U(q'', t''; q', t')$$
$$= \int_{q'}^{q''} [\mathrm{d}q] \exp\left[\frac{\mathrm{i}}{\hbar} \int_{t'}^{t''} \mathrm{d}t \; L(q(t), \dot{q}(t))\right] = \int_{q'}^{q''} [\mathrm{d}q] \exp\left[\frac{\mathrm{i}}{\hbar} S(q'', t''; q', t')\right]. \tag{3.19}$$

where $\int_{q'}^{q''}[\mathrm{d}q]$ represents the $n \to \infty$ limit of the sum over paths with the given endpoints and includes the normalization factor in eq. (3.18). It is important to realize that most paths are so irregular that the sum in the exponential does not converge to a sensible integral, even as $n \to \infty$. In fact, it is generally *not* possible to show that eq. (3.19) makes sense without further definition. One can, however, show that the limit exists when the oscillating exponential in eq. (3.19) is replaced by a decaying exponential (Glimm & Jaffe r1981, Chapter 3).

Wick rotation

Oscillations may be replaced by exponential decay in eq. (3.19) by considering $U(q'', t''; q', t')$ as an analytic function of the times t'' and t'. We make the replacements

$$t'' \to \mathrm{e}^{-\mathrm{i}\theta}\tau'', \quad t' \to \mathrm{e}^{-\mathrm{i}\theta}\tau', \quad t \to \mathrm{e}^{-\mathrm{i}\theta}\tau, \tag{3.20}$$

where τ'', τ' and τ are real, t is the integration variable and $0 < \theta < \pi/2$. As θ passes from 0 to $\pi/2$, (3.20) may be considered as an analytic continuation from 'real' to 'imaginary' time. A continuation all the way to $\theta = \pi/2$ is customarily referred to as a *Wick rotation* (Wick 1954). For a Hamiltonian of the form of eq. (3.11), $L = \frac{1}{2}m\dot{q}^2 - V(q)$, and the exponent in the path integral eq. (3.19) behaves under Wick rotation as

$$\frac{\mathrm{i}}{\hbar} \int_{t'}^{t''} \mathrm{d}t \; \mathrm{L}(q, \dot{q}) \to -\frac{1}{\hbar} \int_{\tau'}^{\tau''} \mathrm{d}\tau[m(\mathrm{d}q/\mathrm{d}\tau)^2 + V(q)], \tag{3.21}$$

which is minus the integral of the Hamiltonian, and therefore negative. In this way, we find

$$U(q'', -\mathrm{i}\tau''; q', -\mathrm{i}\tau') = \int_{q'}^{q''} [\mathrm{d}q] \exp\left\{\frac{-1}{\hbar} \int_{\tau'}^{\tau''} \mathrm{d}\tau \left[\frac{m}{2}\left(\frac{\mathrm{d}q}{\mathrm{d}\tau}\right)^2 + V(q)\right]\right\}. \tag{3.22}$$

Note that if θ were chosen negative, the exponential in eq. (3.22) would be increasing rather than decreasing. It is possible to show that eq. (3.22) exists for a wide range of potentials. When necessary, the path integral for real time is *defined* by the reverse analytic continuation from $-\mathrm{i}\tau$ to t.

Free particle

It is worthwhile to observe that there are cases where the path integral, including its normalization, can be evaluated explicitly. The free particle, with Lagrangian $L = \frac{1}{2}m\dot{q}^2$, is the simplest of these, and the result is

$$U(q + \Delta q, t + \Delta t; q, t) = (m/2\pi \mathrm{i}\Delta t)^{1/2} \exp\{-(m/2\mathrm{i}\hbar)[(\Delta q)^2/\Delta t]\}. \tag{3.23}$$

The derivation is not difficult, and is suggested as exercise 3.2. More generally, the path integral for the harmonic oscillator $L = \frac{1}{2}m\dot{q}^2 - \frac{1}{2}\omega^2 q^2 + fq$, including a 'forcing' term fq, can also be evaluated explicitly.

Semiclassical approximation

The treatment of the path integral for the harmonic oscillator is a special case of the semiclassical or Gaussian approximation, which is an expansion about the motion of the classical system. As we shall see, the semiclassical approximation is exact for the harmonic oscillator.

Consider an arbitrary path integral, in the form of eq. (3.19), and let $q_{\mathrm{cl}}(\tau)$ denote a solution to the classical equations of motion with $q_{\mathrm{cl}}(t') = q'$ and $q_{\mathrm{cl}}(t'') = q''$. Let us shift all the integration variables by the classical motion:

$$\xi(\tau) = q(\tau) - q_{\mathrm{cl}}(\tau). \tag{3.24}$$

Does such a shift make sense? From eq. (3.18), we see that the Jacobian of this change of variables is unity; it is an infinite product of factors unity, one factor for each time. Therefore, the shift does not interfere with the $n \to \infty$ limit. Thus, in eq. (3.19) $[\mathrm{d}q] = [\mathrm{d}\xi]$ and

$$U(q'', t''; q', t') = \int_0^0 [\mathrm{d}\xi] \exp\left[\frac{\mathrm{i}}{\hbar}\int_{t'}^{t''} \mathrm{d}t\, L(\xi + q_{\mathrm{cl}}, \dot{\xi} + \dot{q}_{\mathrm{cl}})\right]. \tag{3.25}$$

If the value of the integral is determined by paths that do not deviate too much from the classical path $q_{\mathrm{cl}}(\tau)$, then we may expand about $\xi(\tau) = 0$ in the exponent of eq. (3.25). To second order in $\xi(\tau)$ and its derivatives, we obtain

$$\int_{t'}^{t''} dt\, L = S_{cl}(q'', t''; q', t') + \int_{t'}^{t''} dt\, [(\partial L/\partial q_{cl})\xi + (\partial L/\partial \dot{q}_{cl})\dot{\xi}]$$
$$+ \int_{t'}^{t''} dt\, [\tfrac{1}{2}(\partial^2 L/\partial q_{cl}{}^2)\xi^2 + (\partial^2 L/\partial q_{cl}\partial \dot{q}_{cl})\xi\dot{\xi} + \tfrac{1}{2}(\partial^2 L/\partial \dot{q}_{cl}{}^2)\dot{\xi}^2]. \tag{3.26}$$

The notation $\partial L/\partial q_{cl}$ means that the derivative is evaluated at q_{cl}. For the harmonic oscillator, of course, there are no corrections to this approximation. The first term is just the classical action, and is independent of ξ. The second term vanishes by the Lagrange equations of motion and the boundary conditions $\xi(t'') = \xi(t') = 0$, after integration by parts. Here we see the usefulness of evaluating the exponent for fixed paths.

The path integral is now

$$U(q'', t''; q', t') = \exp[(i/\hbar)S_{cl}(q'', t''; q', t')] \int_0^0 [d\xi]$$
$$\times \exp\left\{(i/\hbar)\int_{t'}^{t''} dt\, [\tfrac{1}{2}(\partial^2 L/\partial q_{cl}{}^2)\xi^2 + (\partial^2 L/\partial q_{cl}\partial \dot{q}_{cl})\xi\dot{\xi} + \tfrac{1}{2}(\partial^2 L/\partial \dot{q}_{cl}{}^2)\dot{\xi}^2]\right\}. \tag{3.27}$$

The remaining integral may be carried out by returning to the discrete version of U in eq. (3.18). Then the derivatives become part of a quadratic form in the variables ξ_i:

$$U(q'', t''; q', t') = \exp[(i/\hbar)S_{cl}(q'', t''; q', t')] \lim_{n\to\infty} (m/2\pi i\hbar\delta t)^{n/2}$$
$$\times \int \prod_{i=1}^{n-1} d\xi_i \exp\left[(i/\hbar) \sum_{j,k=1}^{n-1} \xi_j A^{(n)}{}_{jk}\xi_k\right], \tag{3.28}$$

where $\xi_0 = \xi_n = 0$. The matrix $A^{(n)}{}_{jk}$ depends on the particular Lagrangian. For example, the quadratic form in eq. (3.27) can often be re-expressed as

$$L' = \tfrac{1}{2}m\dot{q}^2 - \tfrac{1}{2}f(t)q^2(\delta t)^2 \to \tfrac{1}{2}m(q_{i+1} - q_i)^2 - \tfrac{1}{2}f_i q_i{}^2,$$

with $f(t)$ some function of time. ($f(t) = $ constant corresponds to the harmonic oscillator.) $A^{(n)}{}_{ij}$ is then given by

$$A^{(n)}{}_{jk} = \tfrac{1}{2}[m/\delta t](2\delta_{jk} - \delta_{j,k+1} - \delta_{j+1,k}) - \tfrac{1}{2}f_j\delta_{jk}\delta t. \tag{3.29}$$

The integrals in eq. (3.28) can be carried out by changing variables to diagonalize $A^{(n)}{}_{ij}$. If $A^{(n)}{}_{ij}$ is real and symmetric, as above, it may be diagonalized by an orthogonal transformation:

$$\eta_i = X_{ij}\xi_j, \quad (XAX^t)_{ij} = a_i\delta_{ij}. \tag{3.30}$$

Using eq. (3.30), we reduce eq. (3.28) to an infinite product of Gaussian integrals in the η_i, with the result

$$U(q'', t''; q', t') = (m/2i\hbar\pi)^{1/2} \exp[(i/\hbar)S_{cl}(q'', t''; q', t')] \times \lim_{n\to\infty} [(m/2\delta t)^{(n-1)/2}(\delta t \det A^{(n)})^{-1/2}]. \quad (3.31)$$

The evaluation of $\det A^{(n)}$ is discussed in exercise 3.3, and in Schulman (r1981, Chapter 6) and Itzykson & Zuber (r1980, Chapter 9). Although we shall not make extensive use of the semiclassical approximation here, it is an important technique in quantum mechanics and field theory (Rajaraman r1982).

The path integral has given us a new method of evaluating inner products between states. Of course, we want the expectation values of operators as well. In the discussion that follows we shall see why the path integral method is a natural way of calculating time-ordered products in field theory, the very products related to the S-matrix via the reduction formula (Section 2.5).

Time-ordered products

Let us begin with a time-ordered product of coordinate operators between coordinate states,

$$\begin{aligned}
&{}_H\langle q'', t''|T[Q_H(t_1)Q_H(t_2)]|q', t'\rangle_H \\
&= \theta(t_2 - t_1)\int dq_1 dq_2 \langle q'', t''|Q(t_2)|q_2, t_2\rangle\langle q_2, t_2|Q(t_1)|q_1, t_1\rangle \\
&\qquad\qquad \times \langle q_1, t_1|q', t'\rangle \\
&\quad + \theta(t_1 - t_2)\int dq_1 dq_2 \langle q'', t''|Q(t_1)|q_1, t_1\rangle\langle q_1, t_1|Q(t_2)|q_2, t_2\rangle \\
&\qquad\qquad \times \langle q_2, t_2|q', t'\rangle \\
&= \theta(t_2 - t_1)\int dq_1 dq_2\, q_1 q_2 \langle q'', t''|q_2, t_2\rangle\langle q_2, t_2|q_1, t_1\rangle\langle q_1, t_1|q', t'\rangle \\
&\quad + \theta(t_1 - t_2)\int dq_1 dq_2\, q_1 q_2 \langle q'', t''|q_1, t_1\rangle\langle q_1, t_1|q_2, t_2\rangle\langle q_2, t_2|q', t'\rangle,
\end{aligned} \quad (3.32)$$

where all operators and states are in the Heisenberg picture. We have replaced the operators by integrals over the classical field values q_1 and q_2, inserting coordinate states at intermediate times. But the process of inserting complete sets of coordinate states is just the process of forming the path integral. By repeating the reasoning above, we find

$$
{}_{\mathrm{H}}\langle q'', t''|T[Q_{\mathrm{H}}(t_1)Q_{\mathrm{H}}(t_2)]|q', t'\rangle_{\mathrm{H}} = \int_{q'}^{q''}[\mathrm{d}q]\; q_1(t_1)q_2(t_2)\exp\left[\frac{\mathrm{i}}{\hbar}\int_{t'}^{t''}\mathrm{d}t\; L(t)\right], \quad (3.33)
$$

where each $q_i(t_i)$ is a number, not an operator. In the path integral, the ordering of the classical quantities q_i does not matter. Whatever the order on the right-hand side of eq. (3.33), we always get the time-ordered product on the left-hand side. This procedure is straightforward to generalize beyond two operators, so that

$$
\langle q'', t''|T[\textstyle\prod_i Q_{\mathrm{H}}(t_i)]|q', t'\rangle = \displaystyle\int_{q'}^{q''}[\mathrm{d}q]\prod_i q_i(t_i)\exp\left[\frac{\mathrm{i}}{\hbar}\int_{t'}^{t''}\mathrm{d}t\; L(t)\right]. \quad (3.34)
$$

Generating functionals

We now develop a convenient way of organizing the time-ordered products of the coordinate operator or any other operator (Schwinger 1951a). Starting with a Hamiltonian $H(P, Q)$, consider the modified Hamiltonian

$$
H_j(P, Q) = H(P, Q) + j(t)O(P, Q), \quad (3.35)
$$

with $O(P, Q)$ any function of the position and coordinate operators, and with $j(t)$ an arbitrary c-number function of time. Now consider the Feynman kernel U_j, eq. (3.9), defined by Hamiltonian H_j. Each matrix element has a term

$$
j(t' + l\delta t)\langle p_l|O(P, Q)|q_l\rangle(\mathrm{i}\delta t/\hbar) = j(t' + l\delta t)o(p_l, q_l), \quad (3.36)
$$

where $o(p_l, q_l)$ is found, just as for the usual Hamiltonian, by replacing the momentum and coordinate operators by classical quantities, in the normal form of the operator O. Taking a derivative with respect to the source at time σ, and then repeating the reasoning that leads to eq. (3.33), gives

$$
\lim_{\delta t\to 0}\{(\delta t)^{-1}\mathrm{i}\hbar[\partial/\partial j(\sigma)]U(t'', t')^{(j)}\} = {}_{\mathrm{H}}\langle q'', t''|O(P_{\mathrm{H}}(\sigma), Q_{\mathrm{H}}(\sigma))|q', t'\rangle_{\mathrm{H}}{}^{(j)}. \quad (3.37)
$$

The superscript (j) indicates that the matrix element is computed in the presence of the source. If we set the source to zero after the derivative, we find the matrix element with the original Hamiltonian. Multiple derivatives in this form clearly give time-ordered products. We next recognize that the expression $(\delta t)^{-1}(\partial/\partial j)$ is equivalent, in the continuum limit, to a variational derivative (compare eq. (2.7b)), so that we have

$$
\begin{aligned}
&{}_{\mathrm{H}}\langle q'', t''|T\textstyle\prod_i O_{\mathrm{H}}(P(t_i), Q(t_i))|q', t'\rangle_{\mathrm{H}} \\
&\quad = \int [\mathrm{d}p][\mathrm{d}q]\textstyle\prod_i o(p(t_i), q(t_i)) \exp\left[\frac{\mathrm{i}}{\hbar}\int_{t'}^{t''}\mathrm{d}t(p\dot{q} - H - jO)\right]\Bigg|_{j=0} \\
&\quad = \prod_i\left[\mathrm{i}\hbar\frac{\delta}{\delta j(t_i)}\right]\int[\mathrm{d}p][\mathrm{d}q]\exp\left[\frac{\mathrm{i}}{\hbar}\int_{t'}^{t''}\mathrm{d}t(p\dot{q} - H - jO)\right]\Bigg|_{j=0}.
\end{aligned}
\tag{3.38a}
$$

The path integral with source term $jO(P, Q)$ thus serves as a generating functional for all time-ordered products involving the operator $O(P, Q)$. Similarly, the configuration-space version of the path integral eq. (3.19) serves as a generating functional for time-ordered products of the coordinate operator $Q(t)$:

$$
\begin{aligned}
&{}_{\mathrm{H}}\langle q'', t''|T[\textstyle\prod_i Q_{\mathrm{H}}(t_i)]|q', t'\rangle_{\mathrm{H}} \\
&\quad = \prod_i\left(\mathrm{i}\hbar\frac{\delta}{\delta j(t_i)}\right)\int_{q'}^{q''}[\mathrm{d}q]\exp\left\{\frac{\mathrm{i}}{\hbar}\int_{t'}^{t''}\mathrm{d}t[L(q(t), \dot{q}(t)) - j(t)q(t)]\right\}\Bigg|_{j=0}.
\end{aligned}
\tag{3.38b}
$$

For a field theory, of course, this is just what we want, since a scheme for computing time-ordered products of fields will give us the Green functions, eq. (2.95), and ultimately the S-matrix.

Path integral for scalar fields

Needless to say, it is a big jump from a system with one degree of freedom to a field theory. All the same, it is essentially correct to use the direct field theory generalizations of eqs. (3.16) and (3.19) for phase- and configuration-space path integrals.

The coordinates of a field theory are the field values at each point of space, considered as a function of time, so that $\mathrm{d}p\mathrm{d}q$ at each time in eq. (3.16) is now replaced by an infinite product over all $\mathbf{x}$, $\prod_{\mathbf{x}}\mathrm{d}\pi(\mathbf{x}, t)\mathrm{d}\phi(\mathbf{x}, t)$, where $\pi(\mathbf{x}, t)$, $\phi(\mathbf{x}, t)$ are the conjugate momentum and field at point $\mathbf{x}$. The full set of differentials is then, schematically, $\prod_t\prod_{\mathbf{x}}\mathrm{d}\pi(\mathbf{x}, t)\mathrm{d}\phi(\mathbf{x}, t)$. The continuous limit of $\prod_t\prod_{\mathbf{x}}\int\mathrm{d}\pi(\mathbf{x}, t)\mathrm{d}\phi(\mathbf{x}, t)$ (assuming it exists) will be denoted $\int[\mathfrak{D}\pi][\mathfrak{D}\phi]$. We then propose as the generating functional for Green functions, eq. (2.95), of the free scalar field,

$$Z_{\mathscr{L}_0}[J] = \int [\mathscr{D}\pi][\mathscr{D}\phi] \exp\left\{\frac{\mathrm{i}}{\hbar} S[\phi, \pi, J]\right\},$$
$$S[\phi, \pi, J] = \int \mathrm{d}^4x(\pi\dot{\phi} - \tfrac{1}{2}\pi^2 - \tfrac{1}{2}(\nabla\phi)^2 - \tfrac{1}{2}m^2\phi^2 - J\phi) \tag{3.39a}$$

where we recall the scalar-field Hamiltonian, eq. (1.52). Note, by the way, that for these integrals, $\dot{\phi}(x)$ is not in general equal to $\pi(x)$. The $\pi(x)$ integrals, however, are Gaussian, and when they are performed we derive a 'configuration-space' expression,

$$Z_{\mathscr{L}_0}[J] = \int [\mathscr{D}\phi] \exp\left\{\frac{\mathrm{i}}{\hbar}\int_{-\infty}^{\infty} \mathrm{d}^4y[\tfrac{1}{2}(\partial_\mu\phi)^2 - \tfrac{1}{2}m^2\phi^2 - J\phi]\right\}, \tag{3.39b}$$

where $(\partial_\mu\phi)^2 \equiv (\partial_\mu\phi)(\partial^\mu\phi)$. This is perhaps the most common form in which the path integral is exhibited. We shall assume it is uniquely defined by analytic continuation from imaginary time, the Wick rotation given in (3.20). An infinitesimal Wick rotation is equivalent to adding a small negative imaginary term to the mass in the free Lagrange density, which becomes $\partial_\mu\phi\partial^\mu\phi - m^2\phi^2 + \mathrm{i}\epsilon\phi^2$. In this section, we shall ignore the boundary conditions for the fields as $t \to \pm\infty$ in vacuum-to-vacuum amplitudes; we shall simply assume that we can integrate by parts in the exponential without boundary terms. This is certainly not obvious, since the vacuum state is defined by zero energy, *not* zero fields. We shall give a more satisfactory treatment of these and related questions in the following two sections.

Free-field Green functions

The path integral of eq. (3.39b) serves as a generating functional, denoted $Z_{\mathscr{L}_0}[J]$, for the Green functions of the field,

$$G(x_1, \ldots, x_n)_{\text{free}} = \langle 0|T[\textstyle\prod_i \phi(x_i)]|0\rangle_{\text{free}} = \prod_i \left[\mathrm{i}\hbar \frac{\delta}{\delta J(x_i)}\right] Z_{\mathscr{L}_0}[J]\bigg|_{J=0}, \tag{3.40}$$

where now the variation is four-dimensional, so that

$$\delta f[\phi(y)]/\delta\phi(x) \equiv \partial f/\partial\phi|_{\phi=\phi(x)}\delta^4(y - x).$$

Evidently, $Z_{\mathscr{L}_0}[J]$ may also be represented as

$$Z_{\mathscr{L}_0}[J] \equiv \sum_{n=1}^{\infty} \frac{\mathrm{i}\hbar}{n!} \int \prod_{i=1}^{n} \mathrm{d}^4x_i\, J(x_i) G(x_1, \ldots, x_n)_{\text{free}}. \tag{3.41}$$

Equation (3.39b), though infinite-dimensional, is a Gaussian integral, and may hence be evaluated by a change of variables. We first integrate by parts in the exponential, and then change variables to

$$\phi'(y) = \phi(y) - \int d^4z\, \Delta_F(y-z)J(z). \tag{3.42}$$

This is exactly analogous to eq. (3.24), since the shift is a solution to the classical free-field equations of motion in the presence of source J. Using the propagator equation eq. (2.85), $[(\partial_\mu)^2 + m^2 - i\epsilon]\Delta_F(z) = -\delta^4(z)$, eliminates the term linear in the source and field, and gives (Fradkin 1954, 1955a, Edwards & Peierls 1954)

$$Z_{\mathcal{L}_0}[J] = \exp\left[\frac{-i}{2\hbar}\int d^4x d^4y\, J(x)\Delta_F(x-y)J(y)\right] Z_{\mathcal{L}_0}[0], \tag{3.43}$$

in which the source dependence is explicit. The '$i\epsilon$' term, required by Wick rotation, has specified the Feynman propagator, eq. (2.83), and hence the behavior of the propagator at large times (Section 2.4). The remaining factor $Z_{\mathcal{L}_0}[0]$ in eq. (3.43) is the vacuum-to-vacuum amplitude in the absence of a source. It is an infinite phase, which has no observable consequences.

Path integral for the interacting field

In an obvious generalization of the free-field form eqs. (3.39), the generator of Green functions in an interacting-field theory is (in configuration space)

$$\begin{gathered} Z_{\mathcal{L}_0}[J] = W[J]/W[0], \\ W[J] = \int[\mathcal{D}\phi]\exp\left\{\frac{i}{\hbar}\int d^4y[\tfrac{1}{2}(\partial_\mu\phi)^2 - \tfrac{1}{2}m^2\phi^2 - V(\phi) - J\phi]\right\}. \end{gathered} \tag{3.44}$$

The denominator $W[0]$ is a normalization factor, inserted to set the phase in the vacuum-to-vacuum amplitude at zero source to unity.

Perturbative expansion

The essence of perturbation theory is, as we have remarked above, the expansion of the interacting-field theory in terms of the corresponding free-field theory (Slavnov 1975, Zinn-Justin 1974). The inclusion of an external source provides a particularly convenient way of carrying this out. For, we may expand the interacting-field path integral as a power series in variations of $Z_{\mathscr{L}_0}$, eq. (3.39b), with respect to the source:

$$Z_{\mathscr{L}}[J] = W[0]^{-1}\int[\mathscr{D}\phi]\exp\left\{\frac{\mathrm{i}}{\hbar}\int \mathrm{d}^4y[\tfrac{1}{2}(\partial_\mu\phi)^2 - \tfrac{1}{2}m^2\phi^2 - V(\phi) - J\phi]\right\}$$

$$= W[0]^{-1}\exp\left[\frac{-\mathrm{i}}{\hbar}\int \mathrm{d}^4z\, V\left(\mathrm{i}\hbar\frac{\delta}{\delta J(z)}\right)\right] Z_{\mathscr{L}_0}[J]. \tag{3.45}$$

Equation (3.45) will be taken to define $Z_{\mathscr{L}}[J]$. $V(\mathrm{i}\hbar[\delta/\delta J(z)])$ is the potential with variational derivatives substituted for each factor of the field. The source dependence of $Z_{\mathscr{L}_0}[J]$, given in eq. (3.43), is thus enough to construct a perturbation series for the path integral of the interacting-field theory.

The vacuum expectation values of time-ordered products of fields may now be evaluated in precisely the same way as time-ordered products of position operators in quantum mechanics, eq. (3.38b),

$$\langle 0|T[\textstyle\prod_i\phi(x_i)]|0\rangle_{\text{interacting}} = \frac{1}{W[0]}\left[\prod_i \mathrm{i}\hbar\,\delta_J(x_i)\right] Z_{\mathscr{L}}[J]\bigg|_{J=0}$$

$$= \frac{1}{W[0]}\left[\prod_i \mathrm{i}\hbar\,\delta_J(x_i)\right]\exp\left[\frac{-\mathrm{i}}{\hbar}\int \mathrm{d}^4z\, V(\mathrm{i}\hbar\delta_J(z))\right] Z_{\mathscr{L}_0}[J]\bigg|_{J=0}, \tag{3.46}$$

where we have used eq. (3.45), and we introduce the notation

$$\delta_J(z) \equiv \delta/\delta J(z). \tag{3.47}$$

Equation (3.46) will be our starting point for the perturbative expansion of field theory. The systematics of its evaluation are the subject of Section 3.4 below. For the reader with time and interest, the following two sections discuss a more systematic approach to the field-theory path integral.

3.2 The path integral and coherent states

In Section 2.3, we saw that the free scalar field may be thought of as a collection of noninteracting harmonic oscillators, one for each wave vector **k**. Therefore, we return once again to the harmonic oscillator, and discuss its path integral in terms of the creation and annihilation operators that appear in the free-field Hamiltonian, eq. (2.38). In this we follow, with some simplifications, the approach described by Faddeev (r1981) and Faddeev & Slavnov (r1980), based on a mathematical formulation of second quantization due to Berezin (r1966). See also Sakita (1985).

In field theory, we are interested in constructing generating functionals for time-ordered products of fields. We therefore consider the Hamiltonian for a free scalar field interacting with a source. For clarity, it is useful to let t denote the zero component of x^μ, so that we write

$$H(t) = H_0(t) + \int \mathrm{d}^3\mathbf{x}\, j(\mathbf{x}, t)\phi(\mathbf{x}, t)$$
$$= H_0(t) + \int \frac{\mathrm{d}^3\mathbf{k}\, j(-\mathbf{k}, t)}{(2\pi)^{3/2} 2\omega_k} [a(\mathbf{k}, t) + a^\dagger(-\mathbf{k}, t)], \tag{3.48}$$

where H_0 is given by eq. (2.38) and $\phi(\mathbf{x}, t)$ by eqs. (2.13), and where $j(\mathbf{k}, t) = \int \mathrm{d}^3\mathbf{x} \exp(-\mathrm{i}\mathbf{k}\cdot\mathbf{x})\, j(\mathbf{x}, t)$ is the Fourier transform of $j(\mathbf{x}, t)$. The operators are all time-dependent because $j(\mathbf{k}, t)$ is time-dependent. To each wave vector $\mathbf{k}$ there corresponds an independent forced harmonic oscillator Hamiltonian

$$H_j(\mathbf{k}, t) = \tfrac{1}{2}[a^\dagger(\mathbf{k}, t)a(\mathbf{k}, t) + \hbar\omega_k]$$
$$+ [(2\pi)^{3/2} 2\omega_k]^{-1}[j(-\mathbf{k}, t)\, a(\mathbf{k}, t) + j(\mathbf{k}, t)\, a^\dagger(\mathbf{k}, t)]. \tag{3.49}$$

(Note that H_j has been put into normal form.) We shall derive a path integral in phase space, analogous to eq. (3.16), using this Hamiltonian.

Coherent states

In the momentum-space picture, $a(\mathbf{k}, t)$ plays the role of the field and $a^\dagger(\mathbf{k}, t)$ the role of its conjugate momentum. (The lack of a factor i in their commutation relations, eqs. (2.12), is a matter of definition, but will affect some of our formulas that relate to the discussion above.)

The eigenstates of $a(\mathbf{k}, t)$ are called *coherent states*. They will be labelled by an eigenvalue $\alpha(\mathbf{k}, t)$, thus

$$a(\mathbf{k}, t)|\alpha(\mathbf{k}, t)\rangle = \alpha(\mathbf{k}, t)|\alpha(\mathbf{k}, t)\rangle. \tag{3.50}$$

The solution to this eigenvalue equation is simply

$$|\alpha(\mathbf{k}, t)\rangle = \exp[\alpha(\mathbf{k}, t)a^\dagger(\mathbf{k}, t)/2\hbar\omega_k]|0\rangle, \tag{3.51}$$

where $|0\rangle$ is the usual field theory vacuum. The conjugate relations give eigenstates for $a^\dagger(\mathbf{k}, t)$ when it acts from the right:

$$\langle\alpha^*(\mathbf{k}, t)|a^\dagger(\mathbf{k}, t) = \langle\alpha^*(\mathbf{k}, t)|\alpha^*(\mathbf{k}, t), \tag{3.52a}$$

$$\langle\alpha^*(\mathbf{k}, t)| = \langle 0|\exp[\alpha^*(\mathbf{k}, t)a(\mathbf{k}, t)/2\hbar\omega_k]. \tag{3.52b}$$

Coherent states serve as generating functionals for energy eigenstates (see eq. (2.47)), as follows:

$$|m\rangle = (m!)^{-1/2}[2\hbar\omega_k\partial/\partial\alpha(\mathbf{k}, t)]^m\, |\alpha(\mathbf{k}, t)\rangle|_{\alpha(\mathbf{k},t)=0}. \tag{3.53}$$

This relation will be important when we study the S-matrix.

We shall consider the following evolution kernel between coherent states,

$$U^{(\omega_k)}(\alpha^{*\prime\prime}(\mathbf{k}), t''; \alpha'(\mathbf{k}), t') = {}_{\mathrm{H}}\langle \alpha^{*\prime\prime}(\mathbf{k}), t''|\alpha'(\mathbf{k}), t'\rangle_{\mathrm{H}}, \qquad (3.54)$$

where $\omega_k = (\mathbf{k}^2 + m^2)^{1/2}$. For the remainder of this section we suppress the label $\mathbf{k}$.

Path integral for coherent states

Starting out as for the coordinate matrix element, eq. (3.5), we go from the Heisenberg to the Schrödinger representation, expand the exponential of the Hamiltonian as an infinite product and insert unity between the factors of this product. It is not difficult to verify directly (exercise 3.5(c)) that the unit operator in the harmonic oscillator Hilbert space may be written as

$$I = \int \frac{\mathrm{d}\alpha^* \mathrm{d}\alpha}{2\pi \mathrm{i}\hbar\omega} |\alpha\rangle\langle\alpha^*| \exp(-\alpha^*\alpha/2\hbar\omega), \qquad (3.55)$$

where, with $\alpha = x + \mathrm{i}y$, we define the integrals by

$$\int \frac{\mathrm{d}\alpha^* \mathrm{d}\alpha}{2\pi\mathrm{i}} \equiv \int_{-\infty}^{\infty} \frac{\mathrm{d}x\mathrm{d}y}{\pi}. \qquad (3.56)$$

This notation has the folowing motivation. The integrands that we shall encounter will be analytic functions of α, so that it will always be possible to deform the x and y contours away from the real axis. Using this freedom, α and α^* may be considered as independent integration variables in the complex plane. Equation (3.55) when substituted into eq. (3.54) gives the path integral

$$\begin{aligned} &U^{(\omega)}(\alpha^{*\prime\prime}, t''; \alpha', t') \\ &\quad = \lim_{N\to\infty} \prod_{i=1}^{N-1} \int (\mathrm{d}\alpha^*{}_i \mathrm{d}\alpha_i/2\pi\mathrm{i}\hbar\omega)\langle\alpha^{*\prime\prime}|[1 - \mathrm{i}\delta t\, H_j(a^\dagger, a)/\hbar]|\alpha_{N-1}\rangle \\ &\qquad \times \exp(-\alpha_i{}^*\alpha_i/2\hbar\omega)\langle\alpha_i{}^*|[1 - \mathrm{i}\delta t\, H_j(a^\dagger, a)/\hbar]|\alpha_{i-1}\rangle, \end{aligned} \qquad (3.57)$$

where $\alpha_0 \equiv \alpha'$. To proceed, we denote $j_i{}^* = j(-\mathbf{k}, t_i)$ (we assume $j(\mathbf{x}, t)$ is real), so that

$$\begin{aligned} &\langle\alpha_i{}^*|H_j(a^\dagger, a)|\alpha_{i-1}\rangle \\ &= \{(\alpha_i{}^*\alpha_{i-1} + \hbar\omega)/2 + [(2\pi)^{3/2}2\omega]^{-1}(j_i{}^*\alpha_{i-1} + j_i\alpha_i{}^*)\} \exp(\alpha_i{}^*\alpha_{i-1}/2\hbar\omega), \end{aligned} \qquad (3.58)$$

which follows from the coherent state identities eqs. (3.50) and (3.52a), the form of H_j, eq. (3.49), and

$$\langle \beta^* | \alpha \rangle = \exp(\beta^* \alpha / 2\hbar\omega). \tag{3.59}$$

We can now show by the same reasoning that led to the phase-space path integral (3.16) that

$$\begin{aligned} U^{(\omega)}(\alpha^{*\prime\prime}, t''; \alpha', t') = \lim_{N\to\infty} \prod_{i=1}^{N-1} \int \mathrm{d}\alpha_i^* \mathrm{d}\alpha_i \, (2\pi \mathrm{i}\hbar\omega)^{-1} \exp(\alpha^{*\prime\prime}\alpha_{N-1}/2\hbar\omega) \\ \times \exp\Bigg((\mathrm{i}\delta t/\hbar) \sum_{i=1}^{N-1} \{ \mathrm{i}\alpha_i^* \dot{\alpha}_i / 2\omega - (\alpha_i^* \alpha_{i-1} + \hbar\omega)/2 \\ - [(2\pi)^{3/2} 2\omega]^{-1} (j_i^* \alpha_{i-1} + j_i \alpha_i^*) \} \Bigg), \end{aligned} \tag{3.60}$$

where $\dot{\alpha}_i \equiv (\alpha_i - \alpha_{i-1})/\delta t$. Once again, the factor in the exponent is the classical action, now expressed in terms of the conjugate variables α and α^*. We represent the continuum limit of eq. (3.60) as

$$\begin{aligned} U^{(\omega)}&(\alpha^{*\prime\prime}, t''; \alpha', t') \\ &= \int [\mathrm{d}\alpha^*][\mathrm{d}\alpha] \exp(\alpha^{*\prime\prime}\alpha''/2\hbar\omega) \\ &\times \exp\Bigg(\frac{\mathrm{i}}{\hbar} \int_{t'}^{t''} \mathrm{d}\tau \Bigg\{ \frac{\mathrm{i}}{2\omega} \alpha^*(\tau)\dot{\alpha}(\tau) - \frac{1}{2} [\alpha^*(\tau)\alpha(\tau) + \hbar\omega] \\ &\qquad - [(2\pi)^{3/2} 2\omega]^{-1} [j^*(\tau)\alpha(\tau) + j(\tau)\alpha^*(\tau)] \Bigg\} \Bigg), \end{aligned} \tag{3.61}$$

where α'' is integrated, while $\alpha^{*\prime\prime}$ is fixed. Equation (3.61) is the coherent state representation of the harmonic oscillator path integral.

Source dependence for the harmonic oscillator

Since eq. (3.61) corresponds to an harmonic oscillator, the semiclassical method is exact, and can be evaluated by changing variables:

$$\beta(t) \equiv \alpha(t) - \alpha_{\mathrm{cl}}(t), \quad \beta^*(t) \equiv \alpha^*(t) - \alpha^*_{\mathrm{cl}}(t). \tag{3.62}$$

Here $\alpha_{\mathrm{cl}}(t)$ and $\alpha^*_{\mathrm{cl}}(t)$ are solutions to the classical equations of motion of the forced harmonic oscillator,

$$\mathrm{i}(\mathrm{d}/\mathrm{d}t)\alpha^*_{\mathrm{cl}} = -\,\omega\alpha^*_{\mathrm{cl}} - j^*/(2\pi)^{3/2}, \quad \mathrm{i}(\mathrm{d}/\mathrm{d}t)\alpha_{\mathrm{cl}} = \omega\alpha_{\mathrm{cl}} + j/(2\pi)^{3/2}, \tag{3.63}$$

and are chosen to obey the same boundary conditions as in eq. (3.57).

These solutions are

$$\begin{aligned}\alpha^*{}_{\text{cl}}(t) &= \alpha^{*\prime\prime} \exp[\mathrm{i}\omega(t - t'')] - \mathrm{i}\int_t^{t''} \mathrm{d}\tau \exp[\mathrm{i}\omega(t - \tau)] \frac{j^*(\tau)}{(2\pi)^{3/2}},\\ \alpha_{\text{cl}}(t) &= \alpha' \exp[-\mathrm{i}\omega(t - t')] - \mathrm{i}\int_{t'}^{t} \mathrm{d}\tau \exp[-\mathrm{i}\omega(t - \tau)] \frac{j(\tau)}{(2\pi)^{3/2}}.\end{aligned} \tag{3.64}$$

Note they are not complex conjugate to each other.

It is a simple matter to substitute eq. (3.64) into the path integral, eq. (3.61), and re-express the action in terms of the new variables:

$$\begin{aligned}&U^{(\omega)}(\alpha^{*\prime\prime}, t''; \alpha', t')\\ &\qquad = \exp\Bigg(\frac{\alpha^{*\prime\prime}\alpha'}{2\hbar\omega} \exp[-\mathrm{i}\omega(t'' - t')]\\ &\qquad\quad - \frac{\mathrm{i}}{(2\pi)^{3/2} 2\hbar\omega} \int_{t'}^{t''} \mathrm{d}t \{ j(t)\alpha^{*\prime\prime} \exp[\mathrm{i}\omega(t - t'')]\\ &\qquad\quad + j^*(t)\alpha' \exp[-\mathrm{i}\omega(t - t')]\}\Bigg)\\ &\qquad\quad \times \exp\left\{-\frac{1}{(2\pi)^3 2\hbar\omega} \int_{t'}^{t''} \mathrm{d}t\, j^*(t) \int_{t'}^{t} \mathrm{d}\tau \exp[-\mathrm{i}\omega(t - \tau)]\, j(\tau)\right\}\\ &\qquad\quad \times U_0{}^{(\omega)}(0, t''; 0, t'),\end{aligned} \tag{3.65}$$

where

$$\begin{aligned}&U_0{}^{(\omega)}(0, t''; 0, t')\\ &\quad = \int [\mathrm{d}\beta^*][\mathrm{d}\beta] \exp\left(\frac{\mathrm{i}}{\hbar}\int_{t'}^{t''} \mathrm{d}\tau \left\{\frac{\mathrm{i}}{2\omega}\, \beta^*(\tau)\dot\beta(\tau)/2\omega - \frac{1}{2}[\beta^*(\tau)\beta(\tau) + \hbar\omega]\right\}\right).\end{aligned} \tag{3.66}$$

Because of the explicit forms of the classical solutions in eq. (3.62), $\beta(t)$ and $\beta^*(t)$ are not complex conjugates of each other at the outset, and eq. (3.66) it is not obviously the vacumm-to-vacuum amplitude. The integrand in eq. (3.66), however, is an analytic function of $\beta(\tau)$ and $\beta^*(\tau)$ separately or, equivalently, of $x = \frac{1}{2}(\beta + \beta^*)$ and $y = (-\frac{1}{2}\mathrm{i})(\beta - \beta^*)$. Changing to these variables, we may deform their contours back to the real axis, ensuring that β and β^* become complex conjugates. This shows that $U_0{}^{(\omega)}$ is the vacuum-to-vacuum path integral with zero source. All of the source and boundary dependence has then been factored into the exponential prefactor in eq. (3.65).

3.3 Coherent state construction of the path integral in field theory

Path integral for the free field

As we have (repeatedly) observed, the Hamiltonian of the free scalar field is a sum of harmonic oscillator Hamiltonians, one for each wave vector. An arbitrary Heisenberg-picture coherent state for the free field is thus a direct product of coherent states for each wave vector, and is denoted as $|\{\alpha(\mathbf{k})\}, t>$:

$$|\{\alpha(\mathbf{k})\}, t\rangle = \otimes_{\mathbf{k}}|\alpha(\mathbf{k}, t)\rangle. \tag{3.67}$$

We saw in eq. (3.53) that coherent states serve as generating functionals for energy eigenstates of the harmonic oscillator. Similarly, in the continuum limit (eq. (2.55)), eigenstates of the momentum and number operators for the sourceless free field can be generated by variations:

$$\begin{aligned}|\{\mathbf{p}_i\}, t\rangle &= \prod_i a^\dagger(\mathbf{p}_i)|0\rangle \\ &= \prod_i \exp[-\mathrm{i}\omega(p_i)t]\, a^\dagger(\mathbf{p}_i, t)|0\rangle \\ &= \prod_i [2\omega(\mathbf{p}_i)\hbar]\, \{\exp[-\mathrm{i}\omega(p_i)t]\, \delta/\delta\alpha(\mathbf{p}_i, t)\}|\{\alpha(\mathbf{k})\}, t\rangle|_{\{\alpha(\mathbf{k},t)=0\}}, \\ &= \prod_i [2\omega(\mathbf{p}_i)\hbar]\, [\delta/\delta\alpha(\mathbf{p}_i)]\, |\{\alpha(\mathbf{k})\}, t\rangle|_{\{\alpha(\mathbf{k},t)=0\}}.\end{aligned} \tag{3.68}$$

To be consistent with the conventions of Chapter 2, the states do not include the exponential time dependence of the free creation operator, eq. (2.40). So, as in Section 2.3, we define $a^\dagger(\mathbf{p}_i) \equiv \exp[-\mathrm{i}\omega(\mathbf{p}_i)t]\, a^\dagger(\mathbf{p}_i, t)$ as the time-independent creation operator for a free field, and similarly $\alpha(\mathbf{p}_i) \equiv \exp[\mathrm{i}\omega(\mathbf{p}_i)t]\, \alpha(\mathbf{p}_i, t)$. For ease of generalization to the interacting field, however, we keep time as a label of the Heisenberg states.

Let us assume for simplicity that the occupation number for any wave vector is one or zero. The path-integral form of matrix elements between such states is the infinite product over all $\mathbf{k}$ of eq. (3.54):

$$\begin{aligned}U_{\mathcal{L}_0}(\{\alpha^{*\prime\prime}(\mathbf{k})\}, t''; \{\alpha'(\mathbf{k})\}, t') &= \prod_{\mathbf{k}} {}_{\mathrm{H}}\langle\{\alpha^{*\prime\prime}(\mathbf{k})\}, t''|\{\alpha'(\mathbf{k})\}, t'\rangle_{\mathrm{H}} \\ &= \prod_{\mathbf{k}} U^{[(\mathbf{k}^2+m^2)^{1/2}]}(\alpha^{*\prime\prime}(\mathbf{k}), t''; \alpha'(\mathbf{k}), t').\end{aligned} \tag{3.69}$$

To derive a more explicit form than this, we combine the exponents for different values of $\mathbf{k}$ from eq. (3.65):

$$U_{\mathscr{L}_0}(\{\alpha^{*\prime\prime}(\mathbf{k})\}, t''; \{\alpha'(\mathbf{k})\}, t'; j)$$

$$= \exp\left[\int \frac{\mathrm{d}^3\mathbf{k}}{2\hbar\omega_k}\, \alpha^{*\prime\prime}(\mathbf{k})\alpha'(\mathbf{k})\right]$$

$$\times \exp\left\{-\mathrm{i}\int \frac{\mathrm{d}^3\mathbf{k}}{(2\pi)^{3/2}2\hbar\omega_k} \int_{t'}^{t''} \mathrm{d}t[j(\mathbf{k}, t)\alpha^{*\prime\prime}(\mathbf{k})\exp(\mathrm{i}\omega t) + j^*(\mathbf{k}, t)\alpha'(\mathbf{k})\exp(-\mathrm{i}\omega t)]\right\}$$

$$\times \exp\left\{\frac{-1}{(2\pi)^3\hbar}\int \frac{\mathrm{d}^3\mathbf{k}}{2\omega_k}\int_{t'}^{t''}\mathrm{d}t\; j^*(\mathbf{k}, t)\int_{t'}^{t}\mathrm{d}\tau \exp[-\mathrm{i}\omega(t-\tau)]\; j(\mathbf{k}, \tau)\right\}$$

$$\times \prod_{\mathbf{k}} U_0{}^{(\omega_k)}(0, t''; 0, t'). \tag{3.70}$$

The final factor in eq. (3.70) is isolated by setting all boundary values of $\alpha^{*\prime\prime}(\mathbf{k})$ and $\alpha'(\mathbf{k})$, as well as of the source $j(\mathbf{k}, t)$, to zero. It is thus the vacuum-to-vacuum amplitude for the sourceless free field. We may take this quantity to be unity.

We can simplify the source dependence of eq. (3.70) by re-expressing it in terms of the source in position space:

$$j(\mathbf{k}, t) = \int \mathrm{d}^3\mathbf{x} \exp(-\mathrm{i}\mathbf{k}\cdot\mathbf{x})\; J(\mathbf{x}, t). \tag{3.71}$$

$J(\mathbf{x}, t)$ is real, as is appropriate for a real field, so that $j^*(\mathbf{k}, t) = j(-\mathbf{k}, t)$. Using this, we rewrite the second and third exponents in eq. (3.70). For the second we find

$$\int \frac{\mathrm{d}^3\mathbf{k}}{(2\pi)^{3/2}2\omega_k}\int \mathrm{d}\tau\left\{j(\mathbf{k}, t)\alpha^{*\prime\prime}(\mathbf{k}, t'')\exp[\mathrm{i}\omega(t-t'')] + j^*(\mathbf{k}, t)\alpha'(\mathbf{k}, t')\exp[-\mathrm{i}\omega(t-t')]\right\}$$

$$= \int \mathrm{d}^4x\; \phi_{\mathrm{as}}(\mathbf{x}, t)\; J(\mathbf{x}, t), \tag{3.72}$$

where (the subscript 'as' means 'asymptotic')

$$\phi_{\mathrm{as}}(\mathbf{x}, t) \equiv \int \frac{\mathrm{d}^3\mathbf{k}}{(2\pi)^{3/2}2\omega_k}[\alpha'(\mathbf{k})\exp(-\mathrm{i}\omega_k t + \mathrm{i}\mathbf{k}\cdot\mathbf{x}) + \alpha^{*\prime\prime}(\mathbf{k})\exp(\mathrm{i}\omega_k t - \mathrm{i}\mathbf{k}\cdot\mathbf{x})] \tag{3.73}$$

is the classical field found from the asymptotic boundary conditions. Here

$\alpha'(\mathbf{k})$ and $\alpha^{*\prime\prime}(\mathbf{k})$ are classical analogues of the time-independent creation and annihilation operators $a(\mathbf{k})$ and $a^\dagger(\mathbf{k})$. The third exponent in eq. (3.70) may be written in terms of the Feynman propagator, eq. (2.81),

$$\int \frac{\mathrm{d}^3\mathbf{k}}{(2\pi)^3 2\omega_k} \int_{t'}^{t''} \mathrm{d}t\, j^*(\mathbf{k}, t) \int_{t'}^{t} \mathrm{d}\tau \exp[-\mathrm{i}\omega(t-\tau)]\, j(\mathbf{k}, \tau)$$
$$= \frac{\mathrm{i}}{2} \int_{t'}^{t''} \mathrm{d}^4x \mathrm{d}^4y\, J(x)\, \Delta_\mathrm{F}(x-y)\, J(y). \quad (3.74)$$

The appearance of the Feynman propagator is a direct consequence of the coherent state boundary conditions, for, by eq. (3.64), setting $\alpha^* = 0$ at $t = t''$ is the same as enforcing that the only nonvanishing waves at t'' have positive frequency. Similarly, at $t = t'$ all nonvanishing waves have negative frequency. The result, of course, is also what we obtained from the 'naive' definitions of the path integral eqs. (3.39a, b), supplemented by Wick rotation.

In summary, we can now rewrite eq. (3.70) for the free-field path integral as

$$U_{\mathcal{L}_0}(\{\alpha^{*\prime\prime}(\mathbf{k})\}, t''; \{\alpha'(\mathbf{k})\}, t')$$
$$= \exp\left[\int \mathrm{d}^3\mathbf{k}\, \frac{\alpha^{*\prime\prime}(\mathbf{k})\alpha'(\mathbf{k})}{2\hbar\omega_k}\right]$$
$$\times \exp\left[\frac{-\mathrm{i}}{\hbar}\int_{t'}^{t''} \mathrm{d}^4x\, \phi_\mathrm{as}(x)\, J(x)\right]$$
$$\times \exp\left[\frac{-\mathrm{i}}{2\hbar}\int_{t'}^{t''} \mathrm{d}^4x \mathrm{d}^4y\, J(x)\, \Delta_\mathrm{F}(x-y)\, J(y)\right]. \quad (3.75)$$

This expression will be useful when we derive the S-matrix. First, however, we will show how to relate $U_{\mathcal{L}_0}$ to an integral over fields, in the spirit of eqs. (3.39a, b).

Path integral in terms of fields

When the boundary values are zero, it is natural to re-express $U_{\mathcal{L}_0}(\{0\}, t''; \{0\}, t')$ in terms of integrals over field variables, and to re-derive eqs. (3.39), the generating functional for the free field. Going back to the coherent-state path integral for the harmonic oscillator, eq. (3.61), and taking the product over modes, eq. (3.69), we find

$$U_{\mathcal{L}_0}(\{0\}, t''; \{0\}, t') = \int [\mathrm{d}\alpha^*(\mathbf{k})]\, [\mathrm{d}\alpha(\mathbf{k})] \exp\left[\frac{\mathrm{i}}{\hbar} S(\alpha^*, \alpha, j)\right]$$

where

$$S(\alpha^*, \alpha, j) = \int_{t'}^{t''} \mathrm{d}\tau \mathrm{d}^3\mathbf{k}\, \{\mathrm{i}\alpha^*(\mathbf{k}, \tau)\dot{\alpha}(\mathbf{k}, \tau)(2\omega_k)^{-1}$$

$$- \tfrac{1}{2}[\alpha^*(\mathbf{k}, \tau)\alpha(\mathbf{k}, \tau) + \hbar\omega_k]$$

$$- [(2\pi)^{3/2}2\omega_k]^{-1} j(-\mathbf{k}, t)[\alpha(\mathbf{k}, \tau) + \alpha^*(-\mathbf{k}, \tau)]\}. \quad (3.76)$$

Here we have shifted $\mathbf{k}$ to $-\mathbf{k}$ in the very last term. Equations (2.11) and (2.13) relate ϕ and π to the creation and annihilation operators, and we can use these relations to change variables from $\alpha^*(\mathbf{k}, t)$ and $\alpha(\mathbf{k}, t)$ to classical fields and momenta. It is straightforward to write the resulting integral in the more conventional phase-space form of eq. (3.39a) (exercise 3.8).

The S-matrix

We now derive a generating functional representation for the S-matrix. Let us consider the coherent-state path integral with arbitrary boundry conditions and interacting fields in the limit that $t' \to -\infty$, $t'' \to \infty$. We let the source vanish for large $|t|$. In line with our comments in Section 2.5, we assume that there exist wave packet states that describe isolated particles of definite momentum at large times. It is thus natural to identify the states in eq. (3.68) with in- and out-states, at least up to a factor:

$$R^{-1/2}|\mathbf{p}, \pm\infty\rangle = |\mathbf{p}\ \mathrm{in}\rangle = |\mathbf{p}\ \mathrm{out}\rangle,$$

$$(3.77)$$

$$R^{-n/2}|\{\mathbf{p}_i\}_n, -\infty\rangle = |\{\mathbf{p}_i\}_n\ \mathrm{in}\rangle, \quad R^{-n/2}\langle\{\mathbf{q}_j\}_n, +\infty| = \langle\{\mathbf{q}_j\}_n\ \mathrm{out}|.$$

The constant $R^{1/2}$ is independent of $\mathbf{p}$ by Lorentz invariance, while the effective freedom of asymptotic states requires a factor $R^{n/2}$ for an n-particle state. We shall see shortly that this is the same R as was introduced in eq. (2.98). Substituting eq. (3.77), along with the relation of particle to coherent states given in eq. (3.68), into the definition of the S-matrix, eq. (2.90), we find

$$S(\{\mathbf{q}_j\}; \{\mathbf{p}_i\})$$

$$= \lim_{t'', -t' \to \infty} \prod_j [2\hbar\omega(\mathbf{q}_j)R^{-1/2}\, \delta/\delta\alpha^{*\prime\prime}(\mathbf{q}_j)] \prod_i [2\hbar\omega(\mathbf{p}_i)R^{-1/2}\delta/\delta\alpha'(\mathbf{p}_i)]$$

$$\times\, U_{\mathscr{L}}(\{\alpha^{*\prime\prime}(\mathbf{k})\}, t''; \{\alpha'(\mathbf{k})\}, t')|_{\alpha^{*\prime\prime}=\alpha'=0}. \quad (3.78)$$

That is, the coherent-state path integral in the presence of a source becomes a generating functional for the S-matrix in the limit of infinite times. Now $U_{\mathscr{L}}(\{\alpha^{*\prime\prime}(\mathbf{k})\}, t''; \{\alpha'(\mathbf{k})\}, t')$ can be treated in just the same way as

$U_{\mathscr{L}_0}$, eq. (3.69), once we use eq. (3.45) to separate the interaction in terms of variations with respect to sources. Then the generating functional for the S-matrix may be represented as

$$\begin{aligned}\mathscr{S}_{\mathscr{L}}(\{\alpha^{*\prime\prime}(\mathbf{k})\};\{\alpha'(\mathbf{k})\}) &\equiv \lim_{t'',-t'\to\infty} U_{\mathscr{L}}(\{\alpha^{*\prime\prime}(\mathbf{k})\},t'';\{\alpha'(\mathbf{k})\},t')\\ &= \exp\left[\int \mathrm{d}^3\mathbf{k}\,\frac{\alpha^{*\prime\prime}(\mathbf{k})\alpha'(\mathbf{k})}{2\hbar\omega_k}\right]\\ &\quad\times\exp\left[\frac{-\mathrm{i}}{\hbar}\int \mathrm{d}^4z\,V(\mathrm{i}\hbar\delta_J(z))\right]\\ &\quad\times\exp\left[\frac{-\mathrm{i}}{\hbar}\int \mathrm{d}^4x\,\phi_{\mathrm{as}}(x)\,J(x)\right] Z_{\mathscr{L}_0}[J], \end{aligned} \tag{3.79}$$

where $Z_{\mathscr{L}_0}[J]$ is given by (3.43). Here we have expanded perturbatively in terms of variations, as in eq. (3.45), and have used eq. (3.75) for the remaining, free-field, path integral. Next, using the propagator equation $(-\partial^2 - m^2)\Delta_{\mathrm{F}}(z) = \delta^4(z)$, eq. (2.85), $\mathscr{S}_{\mathscr{L}}$ may be written as

$$\begin{aligned}\mathscr{S}_{\mathscr{L}}(\{\alpha^{*\prime\prime}(\mathbf{k})\};\{\alpha'(\mathbf{k})\}) &= W[0]^{-1}\exp\left[\int \mathrm{d}^3\mathbf{k}\,\frac{\alpha^{*\prime\prime}(\mathbf{k})\alpha'(\mathbf{k})}{2\hbar\omega_k}\right]\\ &\quad\times\exp\left[-\int \mathrm{d}^4x\,\phi_{\mathrm{as}}(x)\,(\partial^\mu\partial_\mu + m^2)\delta_J(x)\right]\\ &\quad\times\exp\left[\frac{-\mathrm{i}}{\hbar}\int \mathrm{d}^4z\,V(\mathrm{i}\hbar\delta_J(z))\right] Z_{\mathscr{L}_0}[J]\bigg|_{J=0}. \end{aligned} \tag{3.80}$$

We shall use this form of the generating functional to rederive the reduction formulas of Section 2.5.

S-matrix for the free field

For the special case of a free field with no source, $\mathscr{S}_{\mathscr{L}}$ reduces to the first factor in eq. (3.80). Assuming that $R = 1$ for the free field, the free field S-matrix for m particles to n particles is

$$S(\{\mathbf{q}_j\};\{\mathbf{p}_i\})_{\mathrm{free}} = \delta_{mn}\sum_{\mathrm{perm}}\prod_i 2\hbar\omega_{p_i}\delta^3(\mathbf{p}_i - \mathbf{q}_j), \tag{3.81}$$

where the sum is over permutations of the momenta of the in-state. The first term in eq. (3.80) thus reproduces the unit-operator content of $S = I + \mathrm{i}T$, eq. (2.94), while the remainder of the expression generates the T-matrix.

Reduction

The expressions for Green functions in momentum space allow us to relate them directly to S-matrix elements. As we have just seen, forward scattering is generated by the first exponential in eq. (3.80). T-matrix elements (see eq. (2.94)) come from the remaining factors. A typical T-matrix element for m particles $\to$ n particles may then be written as

$$\mathrm{i}T(\{\mathbf{q}_b\};\{\mathbf{p}_a\}) = \prod_{b=1}^{n}[2\hbar\omega_{q_b}R^{-1/2}\delta/\delta\alpha^*(\mathbf{q}_b)]\prod_{a=1}^{m}[2\hbar\omega_{p_a}R^{-1/2}\delta/\delta\alpha(\mathbf{p}_a)]$$
$$\times \exp\left[-\int \mathrm{d}^4x\, \phi_{\mathrm{as}}(x)(\partial^\mu\partial_\mu + m^2)\delta_J(x)\right] Z_{\mathscr{L}}[J]\bigg|_{\alpha^{*\prime\prime}=\alpha'=J=0}, \tag{3.82}$$

where ϕ_{as} is given by eq. (3.73) and $Z_{\mathscr{L}}[J]$ by eq. (3.45). Equation (3.82), along with eq. (3.46), which relates Green functions to $Z_{\mathscr{L}}[J]$, gives

$$\mathrm{i}T(\{\mathbf{q}_b\};\{\mathbf{p}_a\}) = \prod_b \frac{\mathrm{i}}{(2\pi)^{3/2}R^{1/2}} \int \mathrm{d}^4y_b \exp(\mathrm{i}\bar{q}_b\cdot y_b)\,(\partial^\mu\partial_\mu + m^2)_b$$
$$\times \prod_a \frac{\mathrm{i}}{(2\pi)^{3/2}R^{1/2}} \int \mathrm{d}^4y_a \exp(-\mathrm{i}\bar{p}_a\cdot x_a)\,(\partial^\mu\partial_\mu + m^2)_a$$
$$\times \langle 0|T[\textstyle\prod_{b,a}\phi(y_b)\phi(x_a)]|\,0\rangle. \tag{3.83}$$

As usual $\bar{p}_a{}^\mu = (\omega(\mathbf{p}_a), \mathbf{p}_a)$ with $\omega_a = (\mathbf{p}_a{}^2 + m^2)^{1/2}$, while $\partial_a{}^\mu \equiv \partial/\partial y_{a\mu}$, and so on. Integrating by parts in the $y_b{}^\mu$ and $x_a{}^\mu$ produces explicit factors $(-\omega_b{}^2 + \mathbf{q}_b{}^2 + m^2)$ and $(-\omega_a{}^2 + \mathbf{p}_a{}^2 + m^2)$ for each value of a and b, times the Fourier transform of the Green function. The result is exactly the reduction formula, eq. (2.97), specialized to the case $A(0) = I$. We only have to check that the normalization factors R in the two cases are the same.

Normalization

The normalization factors R in eq. (3.77) relate the oscillator states $|\mathbf{p}, \pm\infty\rangle$ to in- and out-states. To compute R, we consider the normalization of the states constructed at fixed time in eq. (3.68);

$$\langle \mathbf{q}, t|\,\mathbf{p}, t\rangle = \exp(\mathrm{i}\omega_q t)\langle 0|\, a(\mathbf{q}, t)|\,\mathbf{p}, t\rangle$$
$$= (2\pi)^{-3/2}2\omega_q \exp(\mathrm{i}\omega_q t)\int \mathrm{d}^3\mathbf{x}\exp(-\mathrm{i}\mathbf{q}\cdot\mathbf{x})\langle 0|\phi^{(+)}(x)|\,\mathbf{p}, t\rangle. \tag{3.84}$$

$\phi^{(+)}(x)$ is the 'positive-frequency' part of the field (see eq. (2.13)). Applying translation invariance, as in eq. (2.26), we obtain a relation between field matrix elements and particle states,

$$\langle \mathbf{q}, t|\mathbf{p}, t\rangle = (2\pi)^{3/2} 2\omega_p \delta^3(\mathbf{q}-\mathbf{p})\langle 0|\phi(0)|\mathbf{p}, 0\rangle, \quad (3.85)$$

where we assume the negative-frequency part of the field (its creation operators) annihilates the vacuum from the right. The right-hand side of eq. (3.85) is a time-independent expression for the normalization of oscillator states. When single-particle in- and out-states are normalized to $2\hbar\omega_p\delta^3(\mathbf{q}-\mathbf{p})$, eq. (3.77) shows that R is given by eq. (2.98), $R = (2\pi)^3\hbar^{-2}|\langle 0|\phi(0)|\mathbf{p}\text{ in}\rangle|^2$.

3.4 Feynman diagrams and Feynman rules

In this section, we discuss the perturbative construction of Green functions, based on the free- and interacting-field generating functionals.

Free-field and Feynman diagrams

From the free-field generating functional in eq. (3.43), the Green functions of the free theory are

$$G_n(\{x_\alpha\})_{\text{free}} = \prod_{\alpha=1}^{n} [i\hbar\delta_J(y_\alpha)] \exp\left[\frac{-i}{2\hbar}\int d^4x d^4y\, J(x)\Delta_F(x-y)J(y)\right]\Bigg|_{J=0}$$

$$= \left(\frac{i}{2}\right)^{n/2} \sum_{P(i,j)} \hbar\Delta_F(x_i - x_j) \quad (n \text{ even}), \quad (3.86)$$

$$G_n(\{x_\alpha\})_{\text{free}} = 0 \quad (n \text{ odd}). \quad (3.87)$$

Equation (3.86) is a version of *Wick's theorem* (Wick 1950, Bjorken & Drell r1965, Chapter 17, and Itzykson & Zuber r1980, Chapter 4). $P(i, j)$ denotes the set of all ordered pairs of points x_i and x_j. Note that each term occurs $2^{n/2}$ times in the sum, corresponding to the total number of reorderings of points within pairs. Thus, we may also write

$$G_n(\{x_\alpha\}) = (i)^{n/2} \sum_{U(i,j)} \hbar\Delta_F(x_i - x_j) \quad (n \text{ even}), \quad (3.88)$$

where $U(i, j)$ denotes the set of all pairs of points in $\{x_\alpha\}$ without regard to ordering. For $n = 4$, the free-particle Green functions may be represented graphically as in Fig. 3.2, where each line corresponds to a Feynman propagator. Such a representation of the Green function is known as a *Feynman diagram* or *Feynman graph*.

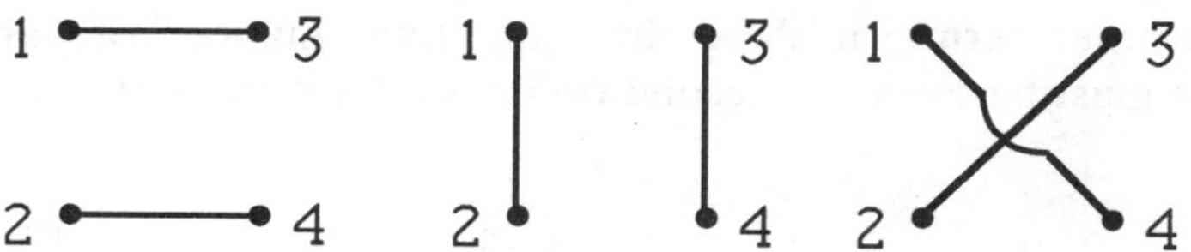

Fig. 3.2 Free-particle four-point Green function with $n = 4$ in eq. (3.88).

Interacting field

We now return to the interacting-field scalar theory, with potential $V(\phi)$, which we assume is a polynomial in the fields. By expanding the exponentials in eq. (3.45) for $Z_{\mathscr{L}}[J]$, we derive a perturbative expansion of the Green function as a power series in the coupling constant(s) of $V(\phi)$, or more generally in any interaction Lagrangian $\mathscr{L}_I(\phi, \partial_\mu\phi) \equiv \mathscr{L}_I(y)$. (We shall discuss more general interaction Lagrangians in Chapter 4.) The result is neatly summarized by comparing eq. (3.45) to the expression, eq. (3.86), for free-field Green functions:

$$G(x_1, \ldots, x_k)_{\text{interacting}} = \frac{\langle 0|T\{\prod_{i=1}^{k}\phi(x_i)\exp[(i/\hbar)\int d^4y\,\mathscr{L}_I(y)]\}|0\rangle_{\text{free}}}{\langle 0|T\{\exp[(i/\hbar)\int d^4y\,\mathscr{L}_I(y)]\}|0\rangle_{\text{free}}}. \tag{3.89}$$

This is sometimes called the Gell-Mann–Low formula (Gell-Mann & Low 1951) for Green functions. The perturbative expansion reduces the interacting-field theory Green functions to an infinite sum of free-field Green functions. According to eqs. (3.86) and (3.89), Green functions are evaluated by pairing external fields with each other, and with fields from the interaction, in all possible ways. Each such *contraction* produces a Feynman propagator, whose argument is determined by the positions of the fields.

Example

To get an explicit expression for $G(x)$ up to order n, we must evaluate each of the free-field Green functions in the nth-order expansion of eq. (3.89) in $\mathscr{L}_I(y) = -V(\phi(y))$. We are to identify all possible pairings (contractions) of fields, according to Wick's theorem, eq. (3.86). To be specific, let us compute the contributions to $G(x_1, x_2)$ in ϕ^3 theory, with $V = (g/3!)\phi^3$, up to second order in eq. (3.89).

At zeroth order, we have

$$\begin{aligned} G^{(0)}(x_1, x_2) &= (i\hbar)^2\delta_J(x_1)\delta_J(x_2)\,\frac{-i}{2\hbar}\int d^4w d^4z\; J(w)\Delta_F(w - z)J(z) \\ &= \hbar[i\Delta_F(x_1 - x_2)]. \end{aligned} \tag{3.90}$$

The first-order term vanishes, since the total number of variations in eq. (3.45) must be even. At second order, we have the rather complicated form

$$
\begin{aligned}
G^{(2)}(x_1, x_2) = {} & (\mathrm{i}\hbar)^2 \delta_J(x_1)\, \delta_J(x_2) \frac{1}{2}\left\{\frac{-\mathrm{i}}{\hbar}\int \mathrm{d}^4 y\, \frac{g}{3!}\,[\mathrm{i}\hbar\delta_J(y)]^3\right\}^2 \\
& \times \frac{1}{4!}\left[\frac{-\mathrm{i}}{2\hbar}\int \mathrm{d}^4 w \mathrm{d}^4 z\, J(w)\Delta_\mathrm{F}(w-z)J(z)\right]^4 \\
& - \left[(\mathrm{i}\hbar)^2\delta_J(x_1)\delta_J(x_2)\frac{-\mathrm{i}}{2\hbar}\int \mathrm{d}^4 w \mathrm{d}^4 z\, J(w)\Delta_\mathrm{F}(w-z)J(z)\right] \\
& \times \frac{1}{2}\left\{\frac{g}{3!}\int \mathrm{d}^4 y\, \frac{-\mathrm{i}}{\hbar}\,[\mathrm{i}\hbar\delta_J(y)]^3\right\}^2 \\
& \times \frac{1}{3!}\left[\frac{-\mathrm{i}}{2\hbar}\int \mathrm{d}^4 w \mathrm{d}^4 z\, J(w)\Delta_\mathrm{F}(w-z)J(z)\right]^3 . \qquad (3.91)
\end{aligned}
$$

The first term comes from the expansion of the numerator in eq. (3.89), and the second from the expansion of the denominator.

The contraction of fields implicit in eq. (3.91) is equivalent to the combinatorial problem of matching eight variations with eight sources. It may conveniently be illustrated as in Fig. 3.3. We represent the external fields by two arrows originating at solid points, which represent positions x_1 and x_2. Two circles with three arrows each represent the interaction Lagrangians $-(g/3!)\phi^3$. Finally, four lines with inverted arrows represent pairs of sources linked by propagators. The evaluation of eq. (3.91) is the same as the process of pairing arrows with inverted arrows in all possible ways, as shown in Figs. 3.3(a)–(c), with the final result shown by the Feynman diagrams of Fig. 3.4. Keeping all factors, we find

$$
\begin{aligned}
G^{(2)}(x_1, x_2) = -\hbar^2 g^2 \bigg\{ & \frac{1}{2}\int \mathrm{d}^4 y \mathrm{d}^4 z\, \Delta_\mathrm{F}(x_1 - y)\Delta_\mathrm{F}(x_2 - z) \\
& \quad \times \Delta_\mathrm{F}(y - z)\Delta_\mathrm{F}(z - y) \\
& + \frac{1}{2}\int \mathrm{d}^4 y \mathrm{d}^4 z \Delta_\mathrm{F}(x_1 - y)\Delta_\mathrm{F}(y - x_2)\Delta_\mathrm{F}(y - z)\Delta_\mathrm{F}(0) \\
& + \frac{1}{4}\int \mathrm{d}^4 y \Delta_\mathrm{F}(x_1 - y)\Delta_\mathrm{F}(0) \int \mathrm{d}^4 z \Delta_\mathrm{F}(x_2 - z)\Delta_\mathrm{F}(0)\bigg\}.
\end{aligned}
\qquad (3.92)
$$

These three terms correspond to Figs. 3.4(a)–(c). It is customary to refer

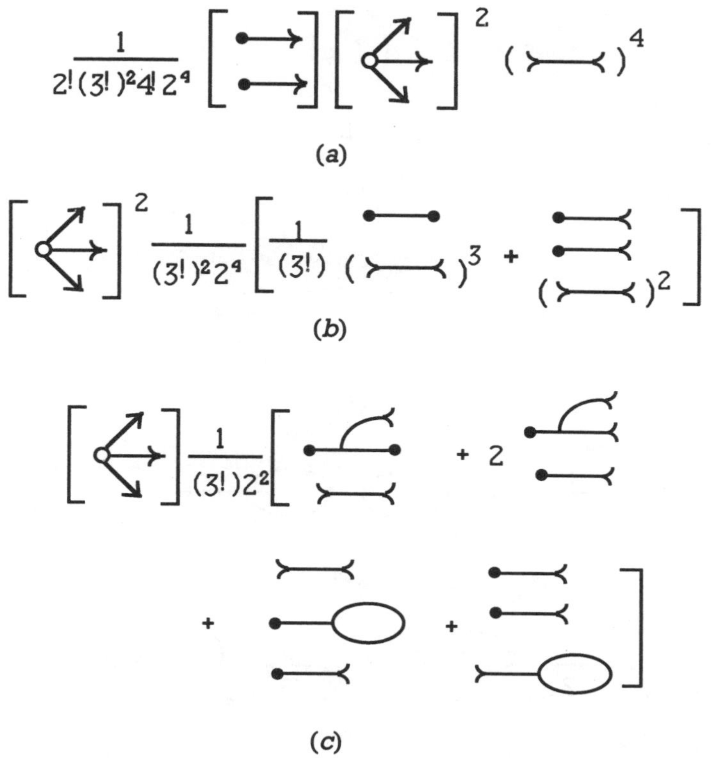

$$\frac{1}{2!(3!)^2 4! 2^4}$$

(a)

$$\frac{1}{(3!)^2 2^4}\left[\frac{1}{(3!)}\right]$$

(b)

$$\frac{1}{(3!)2^2}$$

(c)

Fig. 3.3 Three steps in the matching of variations with sources for eq. (3.91).

to the vertices that represent the interaction as *internal vertices*, and lines that connect two internal vertices as *internal lines* of the diagrams. The points x_1, x_2, arguments of fields in the matrix element, are sometimes referred to as *external vertices*, and lines connected to them as *external lines*. We note that other graphs, shown in Fig. 3.4(d), in which the external points x_1 and x_2 are not connected with the interaction, cancel between the numerator and denominator. We shall come back to this point in a moment, after we have interpreted the graphs represented in eq. (3.92) and Figs. 3.4(a)–(c).

Time ordering

Consider first Fig. 3.4(b). We recall from the discussion of Section 2.4 that the Feynman propagator $\Delta_F(y - x_1)$ describes the propagation of free particles between points x_1 and y, in which positive-energy particles always travel forward in time. This last observation is an important one, because y_0 and z_0 are integrated in eq. (3.92). The physical interpretation of $G^{(2)}(x_1, x_2)$ thus depends on the relative order of the times $x_{1,0}$, y_0, z_0 and

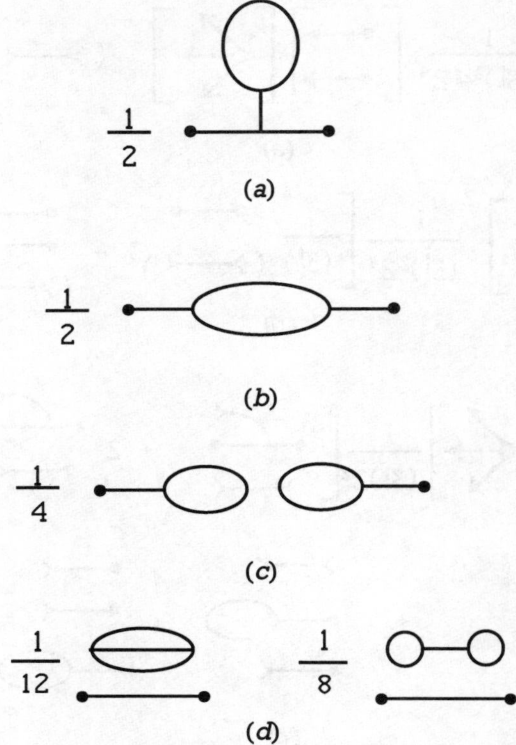

Fig. 3.4 Final results of variations in eq. (3.91).

$x_{2,0}$. For fixed $x_{i,0}$ there are 12 such orderings. In the simplest of these, shown in Fig. 3.5(a), a positive-energy particle is emitted at time $x_{1,0}$ and propagates forward in time to y_0, where the interaction acts to create a pair of particles, both of which propagate forward in time to z_0, where they are absorbed by the interaction and a single particle is re-emitted. The single particle then propagates uneventfully to $x_{2,0}$, where it is absorbed by the field. Evidently, under the influence of the interaction term, a single particle spends part of its time as a pair of particles as it propagates forward in time. This is not the only possibility, however, as shown in Fig. 3.5(b), with a time ordering where $x_{1,0} < z_0 < y_0 < x_{2,0}$. This process begins in the

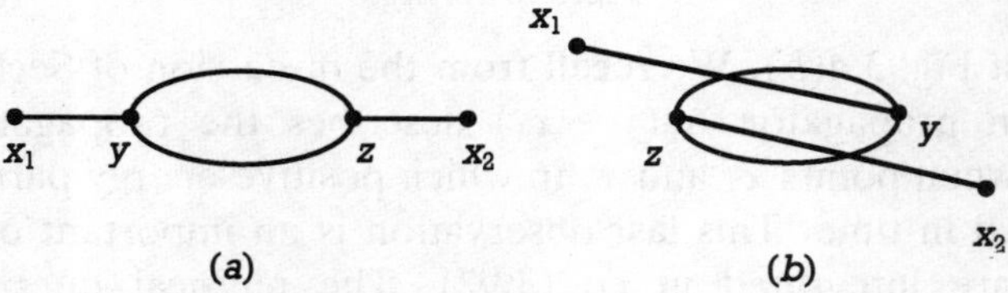

Fig. 3.5 Examples of time-ordered diagrams.

same way, but at z_0 a trio of particles is emitted by the interaction, so that for a while there are four particles moving forward in time, three of which are absorbed by the interaction at the later time y_0. These examples show that the Feynman diagrams of Fig. 3.4 summarize a large number of processes, which may be found by ordering the vertices of the diagrams according to time.

Tadpole diagrams

We can also interpret Figs. 3.4(*a*), (*c*), in which single lines circle back to the same vertex. In terms of a physical picture, external lines disappear into nothing or emerge out of nothing, depending on the ordering of the lines. Such graphs are given the less than picturesque name of *tadpole diagrams*. They will occur whenever the conserved quantum numbers of any particle are all equal to zero (otherwise conservation laws would prevent the creation or destruction of a single particle). Evidently, the vacuum state in the interacting theory is a mixture of the vacuum and one-particle states of the free theory, another effect of expanding the interacting theory in terms of the free theory.

Vacuum bubbles

Now let us return to the diagrams in Fig. 3.4(*d*), which cancelled in the example. A diagram in which a part of the graph is not attached to an external line is known as a *vacuum bubble*. They show that the vacuum state of the interacting-field theory mixes the free-field vacuum, not only with one-particle states (when possible), but also with multiparticle states. We have seen, however, that vacuum bubbles do not contribute to $G(x_1, x_2)$ up to second order. In fact, it is easy to prove that this is the case for any Green function to all orders. Thus, we may modify eq. (3.89) to read

$$G(x_1, \ldots, x_k) = \langle 0|T\left\{\prod_{j=1}^{k}\phi(x_i)\exp\left[(\mathrm{i}/\hbar)\int \mathrm{d}^4y\, \mathcal{L}_\mathrm{I}(y)\right]\right\}|0\rangle_{\text{free, sc}}. \tag{3.93}$$

Green functions are given as the sum of scattering (sc) diagrams only, diagrams in which every interaction vertex is connected to at least one external line by some sequence of lines and vertices. Let us prove this result for a general potential. At nth order, the numerator of the interacting-field generating functional, eq. (3.45), gets 'bubble' contributions whenever m interaction Lagrangians act as a unit on some set of sources. Since all the interaction Lagrangians are identical, there are

$n!/[m!(n-m)!]$ identical ways this can happen. Thus, the numerator of eq. (3.45) may be rewritten as

$$\sum_{n=0}^{\infty}\frac{1}{n!}\sum_{m=0}^{n}\frac{n!}{m!(n-m)!}\sum_{n'=0}^{\infty}\frac{1}{n'!}\sum_{m'=0}^{n'}\frac{n'!}{m'!(n'-m')!}$$
$$\times\left(\left\{\frac{\mathrm{i}}{\hbar}\int \mathrm{d}^4x\,\mathcal{L}_{\mathrm{I}}[\mathrm{i}\hbar\delta_J(x)]\right\}^m\left\{\frac{\mathrm{i}}{2\hbar}\int \mathrm{d}^4w\mathrm{d}^4z\,J(w)\Delta_{\mathrm{F}}(w-z)J(z)\right\}^{m'}\right)\Bigg|_{J=0}$$
$$\times\left(\prod_i[\mathrm{i}\hbar\delta_J(x_i)]\left\{\frac{\mathrm{i}}{\hbar}\int \mathrm{d}^4x\,\mathcal{L}_{\mathrm{I}}[\mathrm{i}\hbar\delta_J(x)]\right\}^{n-m}\right.$$
$$\left.\times\left[\frac{-\mathrm{i}}{2\hbar}\int \mathrm{d}^4w\mathrm{d}^4z\,J(w)\Delta_{\mathrm{F}}(w-z)J(z)\right]^{n'-m'}\right)\Bigg|_{\mathrm{sc}\,J=0}$$
$$= W[0]\langle 0|T\left(\prod_{j=1}^{k}\phi(x_i)\exp\left\{\frac{\mathrm{i}}{\hbar}\int \mathrm{d}^4y\,\mathcal{L}_{\mathrm{I}}[\phi(y)]\right\}\right)|0\rangle_{\mathrm{free,sc}}. \quad (3.94)$$

The normalization factor $W[0]$ thus cancels in (3.89), leaving (3.93).

Feynman rules; symmetry factors

We are now in a position to state a set of rules, called *Feynman rules*, for the perturbative expansion of any Green function $\langle 0|T[\phi(x_1)\cdots\phi(x_a)]|0\rangle$, for a Lagrangian with monomial potential $V(\phi(x))=g(\phi(x))^p/p!$. Rules for polynomial potentials are obvious generalizations.

(i) Identify all distinguishable scattering diagrams, without vacuum bubbles, in which each vertex is either an n-point internal vertex, or represents an external point x_i. Associate a factor $\hbar$ with every external point.

(ii) Associate a factor $\hbar^{p-1}(-\mathrm{i}g/p!)$ and an integral $\int \mathrm{d}^4y_j$ with the jth interaction (internal) vertex.

(iii) Associate a factor $\mathrm{i}\hbar^{-1}\Delta_{\mathrm{F}}(w-z)$ with a line connecting a vertex at point w with a field at point z, where w and z may represent one of the x_i or y_j.

(iv) Sum over all the expressions, with the combinatoric weights determined by eq. (3.84).

(3.95a)

From these four rules we can construct the entire perturbation series for the Green functions of ϕ^p-theory.

The only problematic rule above is the fourth, in which we sum over all terms. Even at low order, there are a large number of terms, and most of them are the same. For the purposes of careful calculation, it is sometimes best to work explicitly through the counting of identical terms, as in the examples above. It is often advantageous, however, to use general rules for combinatoric factors. Indeed, the factor of $1/p!$ in the definition of the ϕ^p coupling constant shows we have already assumed a pattern of this type. The general combinatoric rule may be stated as follows.

(iv′) With each diagram associate an overall factor

$$S^{-1} = s_1{}^{-1}\, s_2{}^{-1}, \tag{3.95b}$$

which depends only on the topological properties of the diagram. s_1 is given by

$$s_1 = \prod_{i \neq j} m_{ij}! \prod_k 2^{m_k}. \tag{3.95c}$$

The first product is over all distinct pairs of vertices i and j, and m_{ij} is the number of identical lines that attach vertices i and j. The second product is over all vertices k, and m_k is the number of lines that begin and end at the same vertex. There is no explicit formula for s_2. Rather, it may be defined as the number of exchanges of vertices in the graph under which any time-ordered version of the graph is invariant up to changes of labels.

Rule (iv′) may seem a little abstract. To give it substance, we note that Figs. 3.4(a) and 3.4(b) are both examples of graphs for which $s_1 = 2$, $s_2 = 1$. In the latter case, the two internal vertices are connected by two identical lines, while in the former, the single vertex of the tadpole is connected to the same line twice. Similarly, Fig. 3.6 is a graph for which $s_1 = 1$ and $s_2 = 2$, because the exchange of the two vertices labelled a and b leaves any time ordering of the graph unchanged up to an exchange of the labels a and b. The derivation of rule (iv′) is not complicated, and a fuller discussion is given in Appendix B.

Connected, truncated and one-particle irreducible diagrams

The Feynman rules eqs. (3.95) produce diagrams that contribute to the Green functions of the interacting theory. It is useful, however, to generalize the concept of diagrams to include any collection of vertices and lines,

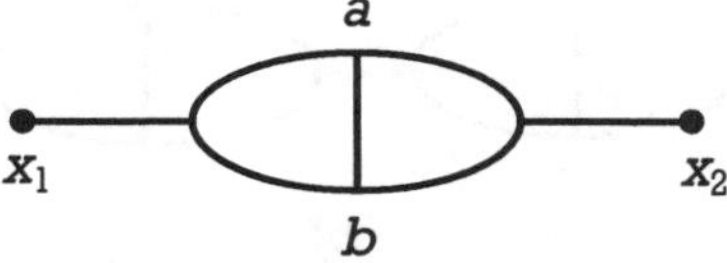

Fig. 3.6 Diagram for which $s_1 = 1$, $s_2 = 2$ in eq. (3.95b).

whether or not it corresponds directly to a Green function. This will enable us, in particular, to speak of a *subdiagram* of diagram G, which is any subset of the lines and vertices of G, and which corresponds to an expression built from the same rules, eqs. (3.95). We now introduce some useful terminology and distinctions pertaining to generalized Feynman diagrams.

A diagram is said to be *connected* if it consists of a single component connected in a topological sense. So, Figs 3.4(a) and 3.4(b) are connected, while Fig. 3.4(c) is not.

Truncated diagrams are Green function diagrams from which the propagators associated with external lines have been removed. There is no universal graphical notation for truncated diagrams, but in this book we shall represent them as in Fig. 3.7(a), which is the truncated version of Fig. 3.6. The truncated external lines are identified by bars at their ends.

Among truncated diagrams, it will be useful to distinguish those that are *one-particle irreducible* (1PI). A 1PI diagram – sometimes called a *proper* diagram – is one that cannot be separated into two disjoint subdiagrams by cutting a single line. Thus, Fig. 3.7(a) is one-particle irreducible, and Fig. 3.7(b) is 'reducible'. Clearly, a diagram that contributes to a Green function must first be truncated if it is to be 1PI.

Generating functionals

Remarkably, both connected and one-particle irreducible diagrams are summarized by generating functionals, derived from the generating functional of Green functions $Z_{\mathcal{L}}[J]$, eq. (3.45). Let $G_{c,n}(x_1, \ldots, x_n)$ be an n-point-connected Green function. Then, as we show in Appendix B, the generating functional for the $G_{c,n}$ is given by

$$G_c[J] \equiv \sum_{n=1}^{\infty} \frac{(-i/\hbar)^n}{n!} \int \prod_{i=1}^{n} d^4x_i \, J(x_i) \, G_{c,n}(x_1, \ldots, x_n) = \ln Z_{\mathcal{L}}[J]. \quad (3.96)$$

The case of 1PI functions is somewhat more subtle. Suppose we start

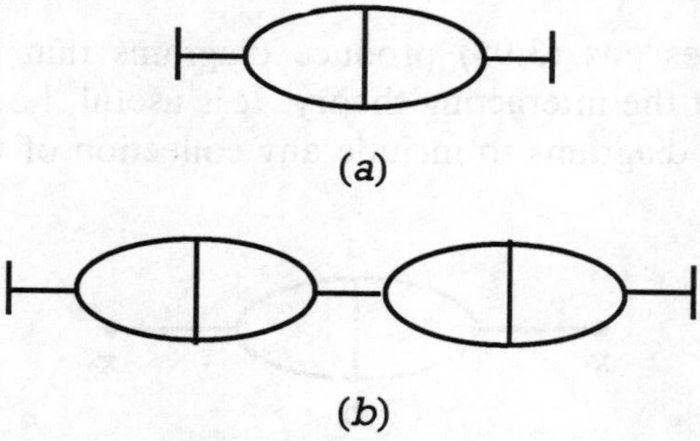

Fig. 3.7 One-particle (a) irreducible and (b) reducible diagrams.

with the diagramatic expansion of $Z_{\mathscr{L}}[J]$, in which the external lines of each diagram still end in $\int d^4x_i\, J(x_i)$. Then, as shown in Fig. 3.8, a general Feynman diagram contains (possible many) one-particle irreducible subdiagrams. To understand the nature of these subdiagrams, consider the one-point function in the presence of the source J (in natural units),

$$\phi_0(x) = \langle 0|\phi_H(x)|0\rangle^{(J)} = i\delta_J(x)\, G_c[J]. \tag{3.97}$$

By a simple application of the Feynman rules, each graphical contribution to $\phi_0(x)$ is of the form of the boxed subdiagram in Fig. 3.8. Note that $\phi_0(x)$ includes a propagator that begins at point x^μ. In fact, by making the $\phi_0(x)$ dependence of $Z_{\mathscr{L}}[J]$ explicit, we can absorb all one-particle reducible contributions into factors of $\phi_0(x)$, leaving over only one-particle irreducible diagrams. The way to do this is to produce a functional that depends only on $\phi_0(x)$. Let $\gamma_n(x_1, \ldots, x_n)$ denote the sum of one-particle irreducible diagrams with n truncated external lines, and define

$$\Gamma_2 \equiv p^2 - m^2 - i\gamma_2, \quad \Gamma_n \equiv -i\gamma_n, \quad n \geqslant 3. \tag{3.98}$$

Then (see Appendix B for proof) the functional

$$\Gamma[\phi_0] = \int \phi_0(x) J(x) d^4x - iG_c[J] \tag{3.99}$$

is the generating functional for the Γ_n,

$$\Gamma[\phi_0] = \sum_{n=1}^{\infty} \frac{1}{n!} \prod_{i=1}^{n} \int d^4x_i\, \phi_0(x_i) \Gamma_n(x_1, \ldots, x_n). \tag{3.100}$$

Equation (3.99), relating $G_c[J]$ to $\Gamma[\phi_0]$, is a Laplace transform, trading J-dependence for ϕ_0-dependence, just as the relation between the Lagrangian and the Hamiltonian in classical mechanics trades $\dot{q}$-dependence for p-dependence. Indeed, using eq. (3.97) in (3.99), we see that $\Gamma[\phi_0]$ is independent of $J(x)$:

$$\delta\Gamma[\phi_0]/\delta J = 0. \tag{3.101}$$

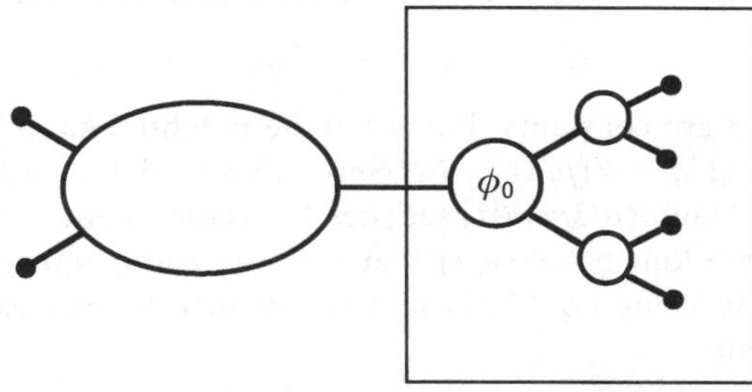

Fig. 3.8 A general one-particle reducible diagram. The box encloses a general contribution to $\phi_0(x)$.

The functional $\Gamma[\phi_0]$ defined in this way is called the *effective action*, first because at lowest order in the coupling, the terms in eq. (3.100) are given directly from the Lagrangian, and second because it satisfies

$$\delta\Gamma[\phi_0]/\delta\phi_0\Big|_J = J, \tag{3.102}$$

which is reminiscent of the classical equation of motion.

Many of these concepts and relations will be of significance later, when we discuss renormalization in field theories. In the next chapter, however, we return to the relation of the S-matrix to Green functions, to see how physical predictions can be derived from the Feynman rules.

Exercises

3.1 Verify eq. (3.15) in the case when the function $h(p_i, q_i)$ is a bounded function, by expanding the two sides of the equation and verifying that they are the same in the limit $n \to \infty$. That is, show that any term in the expansion of the exponential that is not present in the expansion of the product is suppressed by at least one power of n.

3.2 (a) Derive the free-particle equation (3.23). Hint: re-express eq. (3.23) as a function of n in eq. (3.18) by using $\Delta t = n\delta t$, and use an iterative procedure to show that the result depends on n only through Δt. (b) Evaluate the path integral when the particle moves with a velocity-dependent potential, $L = \frac{1}{2}m\dot{q}^2 - k\dot{q}$.

3.3 Suppose we define $D^{(m)} \equiv \delta t(2\delta t/m)^m|\det \mathrm{A}^{(m)}|$, where $A^{(m)}$ is an $m \times m$ matrix of the form eq. (3.29). (a) Show that the $D^{(m)}$ satisfy

$$D^{(n+1)} = [2 - (\delta t^2/m)f_n]\, D^{(n)} - D^{(n-1)}.$$

(b) Now define a continuous variable $t \equiv n\delta t$, and functions $D(t) \equiv D^{(t/\delta t)}$, $f(t) \equiv f_{t/\delta t}$. Show that $D(t)$ satisfies

$$m\mathrm{d}^2 D(t)/\mathrm{d}t^2 + f(t)D(t) = 0,$$

with boundary conditions $D(0) = 0$, $\mathrm{d}D(t)/\mathrm{d}t|_{t=0} = 1$. (c) Verify that for the harmonic oscillator $D(t) = \omega^{-1}\sin\omega(t'' - t')$.

3.4 The Lagrangian for an anharmonic oscillator may be written as

$$L = \tfrac{1}{2}m\dot{q}^2 - \tfrac{1}{4}(\lambda^{1/2}q^2 - \mu/\lambda^{1/2})^2,$$

where μ and λ are constants. For $\mu > 0$ the potential has a 'double-well' form with minima at $q = \pm(\mu/\lambda)^{1/2}$. (a) Show that in this case the 'kink solution' $q(t) = (\mu/\lambda)^{1/2}\tanh[(\mu/2m)^{1/2}t]$ satisfies the classical equations of motion with boundary conditions $q = -(\mu/\lambda)^{1/2}$ at $t = -\infty$, and $q = 0$ at $t = 0$ (Rajaraman r1982). (b) By using eq. (3.31) and the results of exercise 3.3, expand the matrix element

$${}_{\mathrm{H}}\langle q = (\mu/\lambda)^{1/2}\tanh[(\mu/2m)^{1/2}t],\, t|q = 0,\, t = 0\rangle_{\mathrm{H}}$$

about $t = 0$ to the first nonleading power in t.

3.5 (a) Compute the expectation values of energy and momentum in a coherent state. (b) Show that coherent states remain coherent states under time evolution. (c) Verify eq. (3.55).

3.6 Consider the one-dimensional quantum mechanical system with Hamiltonian

$$H = p^2/2m + b(t)p + V(q),$$

with $b(t)$ a function of time and $V(q)$ a potential. Starting from the phase-space path integral for this system, derive the configuration-space form.

3.7 The vacuum-to-vacuum amplitude $\langle 0, t''|0, t'\rangle$ for a harmonic oscillator may be expressed in terms of path integrals between coordinate eigenstates by the relation

$$\langle 0, t''|\, 0, t'\rangle = \int \mathrm{d}q''\, \psi_0{}^*(q'')_{\mathrm{H}}\langle q'', t''|q', t'\rangle\,{}_{\mathrm{H}}\psi(q'),$$

where $\psi_0(x) = \langle q', t'|0, t'\rangle = \exp(-m\omega q^2/2\hbar)$ is the vacuum wave function. Introduce source terms in the (p, q) phase-space representation of the harmonic oscillator as follows:

$$H = p^2/2m + m\omega^2 q^2/2 + Rq + Ip/m\omega,$$

where R and I are time-dependent real functions that vanish at the end-points. Show that if we define $j = R + \mathrm{i}I$, the phase-space path integral eq. (3.16) gives the same result for the vacuum expectation value as eq. (3.65), computed by coherent-state methods.

3.8 Verify that when we express eq. (3.76) in terms of field and momentum variables we rederive eq. (3.39a).

3.9 A quartic scalar Lagrange density has $V(\phi) = (\lambda/4!)\phi^4$. Using this potential, compute the two-point function to order λ in terms of Feynman propagators.

4
Scattering and cross sections for scalar fields

The elements of the S-matrix are characterized by particle momenta. Therefore, for our purposes, momentum space is the natural habitat of perturbation theory. This chapter begins with the systematics of Fourier transforms of Feynman diagrams, and their relation to the S-matrix. We then introduce the concept of a cross section. The complex scalar field, discussed at the close of the chapter, serves as a link between scalar fields and the spinor and vector fields treated in subsequent chapters. With this chapter, we bid farewell to $\hbar$, and revert to a system of units in terms of length only (Section 2.3).

4.1 Diagrams in momentum space

Fourier transform

Feynman diagrams are products of Feynman propagators, integrated over the positions of internal vertices. Here we shall discuss the Fourier transforms of expressions of this kind.

We will say that a propagator $\Delta_F(x-y)$ 'starts' at point y and 'ends' at point x. We then introduce an *incidence matrix* $\eta_{\sigma i}$, whose elements are defined by

$$\eta_{\sigma i} = +1\ (-1) \text{ if line } i \text{ ends (begins) at vertex } \sigma. \tag{4.1}$$

Following the Feynman rules of eq. (3.95), an arbitrary diagram $g(\{x_\sigma\})$, with n external lines, may be represented schematically as

$$g(x_1, \ldots, x_n) = \prod_{\substack{\text{internal}\\ \text{vertices } \tau}} \int(-\mathrm{i}g\mathrm{d}^4 y_\tau) \prod_{\text{lines } i} \mathrm{i}\Delta_F(\Sigma_\alpha \eta_{\alpha i} z_\alpha) \tag{4.2}$$

where the sum over α includes internal vertices, at positions y_τ, as well as external points x_σ. The x_σ-dependence is thus through the arguments of Feynman propagators. As stated above, factors of $\hbar$ have all been absorbed into redefinitions of dimensional quantities, in this case the couplings and $g(\{x_\sigma\})$ itself. We have suppressed possible symmetry factors.

The Fourier transform $\bar{g}(\{p_\sigma\})$ of $g(\{x_\sigma\})$ is evaluated by using eq. (2.83) for each Δ_F in eq. (4.2):

$$\bar{g}(p_1, \ldots, p_n) = \prod_{\substack{\text{external} \\ \text{points } \sigma}} \int d^4x_\sigma \exp(-ip_\sigma \cdot x_\sigma) \prod_{\substack{\text{internal} \\ \text{vertices } \tau}} \left(-ig \int d^4 y_\tau\right)$$

$$\times \prod_{\text{lines } i} \frac{1}{(2\pi)^4} \int d^4 k_i \frac{i \exp[-i\sum_\gamma k_i(\eta_{\gamma i} x_\gamma)]}{k_i^2 - m^2 + i\epsilon}. \tag{4.3}$$

k_i^μ is said to be the 'momentum of line i'. The momenta of lines that end at a vertex are said to 'flow into' that vertex, while the momenta of lines that begin at a vertex are said to 'flow out'. Exchanging position and momentum integrals, we find

$$\bar{g}(p_1, \ldots, p_n) = \prod_{\substack{\text{external} \\ \text{lines } \sigma}} \frac{i}{p_\sigma^2 - m^2} \prod_{\substack{\text{internal} \\ \text{lines } i}} \frac{1}{(2\pi)^4} \int d^4 k_i \frac{i}{k_i^2 - m^2 + i\epsilon}$$

$$\times \prod_{\substack{\text{internal} \\ \text{vertices } \gamma}} (-ig)(2\pi)^4 \delta^4(\textstyle\sum_j l_j \eta_{\gamma j}), \tag{4.4}$$

where the sum over j now includes both internal and external lines, whose momenta are denoted collectively by l. Thus, with each vertex, there is associated a momentum-conserving delta function $\delta^4(\sum_j l_j \eta_{\gamma j})$, which ensures that the total momentum flowing into the vertex equals the total momentum flowing out. Suppose graph g consists of one or more connected components h. The sum of the arguments of all the delta functions in each component h is just $\sum_{\{\sigma \text{ in } h\}} p_\sigma$, where the sum goes over the external momenta of that component. In this way, the delta functions also enforce overall momentum conservation within each connected component of the diagram. The remaining conditions reduce the number of internal momentum integrations.

As an example, consider the four-point Green function at second order in ϕ^3 theory:

$$\begin{aligned} G^{(2)}(p_1, \ldots, p_4) = (-ig)^2 \prod_{i=1}^{4} i(p_i^2 - m^2)^{-1}(2\pi)^{-4} \int d^4 k \, i(k^2 - m^2 + i\epsilon)^{-1} \\ \times (2\pi)^8 [\delta^4(p_1 + p_2 - k)\delta^4(k - p_3 - p_4) \\ + \delta^4(p_1 + p_3 - k)\delta^4(k - p_2 - p_4) \\ + \delta^4(p_1 + p_4 - k)\delta^4(k - p_2 - p_3)] \end{aligned}$$

Simplifying, we obtain

$$G^{(2)}(p_1, \ldots, p_4) = (2\pi)^4\delta(p_1+p_2+p_3+p_4)\,(-)\,\mathrm{i}\prod_{i=1}^{4}(p_i^{\,2}-m^2)^{-1}$$
$$\times\, g^2\{[(p_1+p_2)^2-m^2]^{-1}+[(p_1+p_3)^2-\mathrm{m}^2]^{-1}$$
$$+\,[(p_1+p_4)^2-m^2]^{-1}\}. \qquad (4.5)$$

This is illustrated in Fig. 4.1.

Feynman rules in momentum space

From eq. (4.4) we find the Feynman rules for momentum-space Green functions.

(i) Identify the relevant diagrams as for position-space Green functions.
(ii) Associate $(2\pi)^{-4}\mathrm{i}\int \mathrm{d}^4k_j(k_j^{\,2}-m^2+\mathrm{i}\epsilon)^{-1}$ with every line.
(iii) Associate a factor $-\mathrm{i}g(2\pi)^4\delta^4(\sum_j k_j\eta_{\tau j})$ with the τth internal vertex of the graph, and a factor $(2\pi)^4\delta^4(\mathrm{p}_\sigma-\sum_j k_j\eta_{\sigma j})$ with the σth external vertex of the graph.
(iv) Multiply by the appropriate symmetry factor (3.95b, c).

(4.6)

Using the rules, we can easily write an integral expression for any graph.

Tree and loop diagrams

In an arbitrary diagram, the number of integrals left after we have used the delta functions has an interpretation that plays an important role in our analysis of perturbation theory. Diagrams like those of Fig. 4.1, whose internal momenta are completely determined by external momenta, are called *tree diagrams*. Consider, however, a general connected diagram g with V vertices (including external vertices). As we know, the V momentum-conservation delta functions give one (four-dimensional) condition on external momenta, leaving $V-1$ independent conditions on the internal momenta. Thus, if there are N lines in the graph, and the graph has only one connected component, we have

$$L = N - V + 1 \qquad (4.7)$$

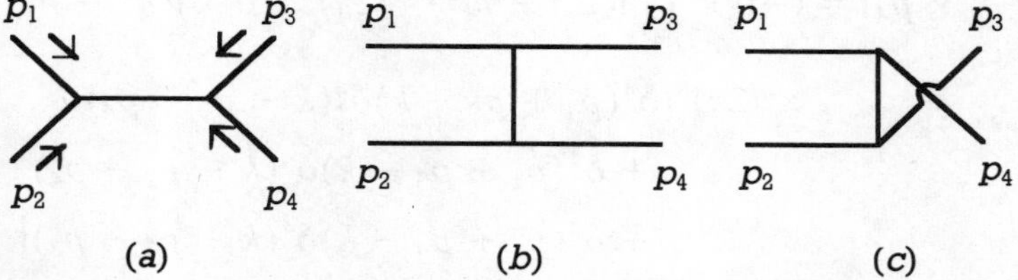

Fig. 4.1 Four-point Green function at second order in ϕ^3 theory (eq. (4.5)).

sets of four-dimensional integrals left over after the delta functions are used. This is also the number of loops in the graph, a topological result known as Euler's theorem. Figures 3.4(*a*), (*b*), for instance, are one-loop graphs, while Fig. 4.2 is a two-loop graph. There is one four-dimensional momentum integral left over for each loop in the graph.

After using all the momentum conservation delta functions, we shall have L momentum integrals left to do. Let us call these integrals $l_i{}^\mu$, $i = 1, 2, \ldots, L$. Every line momentum $k_a{}^\mu$, $a = 1, 2, \ldots, N$, will be a linear combination of the left-over internal integrals and the external momenta,

$$k_a{}^\mu = \sum_j \alpha_{aj} l_j{}^\mu + \sum_i \beta_{ai} p_i{}^\mu, \tag{4.8}$$

where α_{aj} and β_{ai} are again incidence matrices, with entries $+1$, -1 or 0.

Since momentum is conserved at each vertex, when $l_j{}^\nu$ flows into a vertex, it must also flow out. On the other hand, the $l_j{}^\nu$ can never appear as part of an external momentum, for which

$$\alpha_{aj} = 0, \quad \beta_{aj} = \delta_{aj}. \tag{4.9}$$

If we follow a given $l_j{}^\nu$ as it flows around the graph, starting at a given vertex, it must eventually return to the same vertex, because there are only a finite number of other vertices to flow through. Thus, the $l_j{}^\nu$ flow around loops in the graph, and are called *loop momenta*. The particular route that the $l_j{}^\nu$ take through the graph depends on how we use the delta functions of rule (iii) to do the integrals of rule (ii). Different orders of integration may give distinct, but equivalent, routings of the $l_j{}^\nu$.

In practice, we can determine the loop momenta directly, without going through the evaluation of delta functions. We simply identify a set of independent loops, and choose any flow consistent with momentum conservation. The external momentum can also be chosen to flow via any path consistent with momentum conservation. Going from one choice of external momentum routing to another, or one choice of loop momentum routing to another, corresponds to a change of variables. For instance, with the routing shown for the one-loop example, Fig. 4.3(*a*), we have the integral

$$\int d^4 l (l^2 - m^2 + i\epsilon)^{-1} [(p - l)^2 - m^2 + i\epsilon]^{-1}, \tag{4.10}$$

while for the routing of momenta in Fig. 4.3(*b*) we have

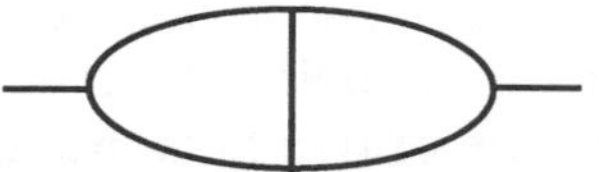

Fig. 4.2 Two-loop graph.

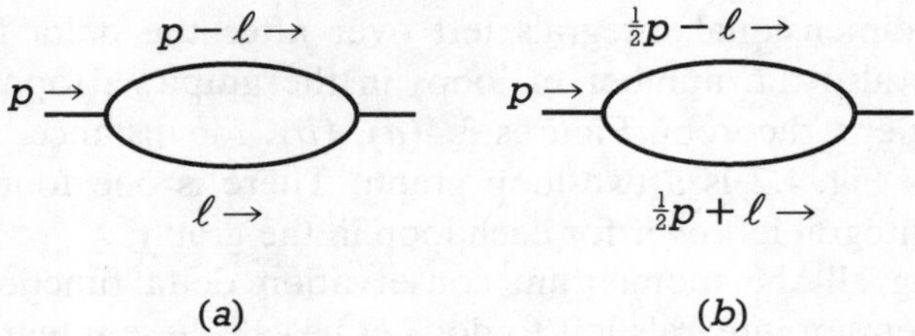

Fig. 4.3 Alternative choices of momentum routing.

$$\int d^4 l[(\tfrac{1}{2}p + l)^2 - m^2 + i\epsilon]^{-1}[(\tfrac{1}{2}p - l)^2 - m^2 + i\epsilon]^{-1}. \tag{4.11}$$

Actually, this does not express our full freedom in choosing loop momenta, since we may even route different components of the four-dimensional delta functions differently. For example, the $l_i{}^\mu$ might flow one way for $\mu = 1, 2$ and another way for $\mu = 3, 0$.

The evaluation of Green functions in momentum space, and therefore of S-matrix elements, has now been reduced to calculating multidimensional integrals of products of Feynman propagators. This task turns out to be fraught with problems, which will be the topics of Chapters 9–11. For now, however, we turn to the formal relation of Green functions to the S-matrix.

4.2 The S-matrix

The expressions for Green functions in momentum space allow us to relate them directly to T-matrix elements, via the reduction formula, eq. (2.97). In the perturbative expansion, the latter may be conveniently rewritten as

$$iT(\{\mathbf{p}'_b\}; \{\mathbf{p}_a\}) = \prod_b [R^{1/2}(2\pi)^{3/2}]^{-m-n}\, \overline{G}^{(\mathrm{T})}(\{-p'_b{}^\mu\}, \{p_a{}^\nu\}), \tag{4.12}$$

where $\overline{G}^{(\mathrm{T})}$ denotes the 'truncated' Green function, whose external propagators have been removed, and R is given by eq. (2.98). Thus, the S-matrix for $n \rightarrow m$ particles may be found by computing the Green function with $n + m$ external particles, taking its Fourier transform with opposite signs for incoming and outgoing particles and finally isolating the residue of the pole in *every one* of these particles.

In perturbation theory, multiple poles are manifest in external propagators. Referring to the position-space representation of Feynman diagrams, eq. (4.2), we can see how the product of single-particle poles arises. For any fixed space–time positions of internal vertices y^μ, the limit of infinite times $x_i{}^0$ gives rise exactly to the poles observed in the reduction formula, eq. (2.97), and justifies the use of free-field matrix elements, eq. (2.100), in the limit. Since we already have Feynman rules for Green functions, we lack only a convenient way of evaluating the matrix element $R^{1/2}$ introduced in eq. (2.98).

Normalization

Equation (2.98) for R is most easily evaluated by considering the Fourier transform of the two-point Green function,

$$\begin{aligned}\bar{G}(q) &= \int d^4x \exp(iq\cdot x)\langle 0|T[\phi(x)\phi(0)]|0\rangle, \\ &= \int d^4x \sum_n \{\theta(x_0)\exp[-i(p_n - q)\cdot x] + \theta(-x_0)\exp[i(p_n + q)\cdot x]\} \\ &\quad \times |\langle 0|\phi(0)|n\rangle|^2, \end{aligned} \tag{4.13}$$

where we have inserted a complete set of states between the two fields, and have used the definition, eq. (2.82), of time ordering, and the translation invariance expressed in eq. (2.26). The x^μ integrals can now be carried out explicitly by introducing an infinitesimal imaginary part to q_0:

$$\begin{aligned}\bar{G}(q) = \sum_n i[(q_0 - \omega_n + i\epsilon)^{-1} - (q_0 + \omega_n - i\epsilon)^{-1}](2\pi)^3\delta^3(\mathbf{q} - \mathbf{p}_n) \\ \times |\langle 0|\phi(0)|n\rangle|^2. \end{aligned} \tag{4.14}$$

This expression has singularities whenever $q_0 = \pm\omega_n$, for any state n. In the second term, we assume that $|\langle 0|\phi(0)|n\rangle|^2$ is the same for states with spatial momenta $\mathbf{q}$ and $-\mathbf{q}$. For a single-particle state, the energy is $p_n{}^0 = \omega_p = (\mathbf{p}_n{}^2 + m^2)^{1/2}$, and the sum over states is $\sum_n = \int d^3\mathbf{p}(2\omega_p)^{-1}$ (eq. (2.59)). By the 'spectral' assumptions of Section 2.5, multiparticle states have $E_n(\mathbf{p}) > \omega_p$, while the vacuum state does not contribute in eq. (4.13) because $\langle 0|\phi|0\rangle = 0$. Then $\bar{G}(q)$ develops a pole at $q_0 = \omega_q$ from single-particle states:

$$\begin{aligned}\bar{G}(q) &\approx i(2\omega_{\mathrm{q}})^{-1}[q_0 - (\mathbf{q}^2 + m^2)^{1/2} + i\epsilon]^{-1}(2\pi)^3|\langle 0|\phi(0)|\mathbf{q}\rangle|^2 \\ &= i(q_0 - \mathbf{q}^2 - m^2 + i\epsilon)^{-1}R + \text{finite remainder}, \end{aligned} \tag{4.15}$$

where we have used the definition, eq. (2.98) of R. Thus, R is the residue of the single-particle pole in the two-point Green function. By the Feynman rules, the lowest-order approximation to $\bar{G}(q)$ in perturbation theory is just i times the free propagator, so that R has the expansion

$$R = [1 + O(g^2)]. \tag{4.16}$$

As a result, at lowest order in the reduction formula, eq. (4.12), we are able to replace R by unity. To higher orders, R is found from diagrams such as Fig. 4.3, which contribute to G_2. Its explicit computation is connected with renormalization (Chapter 10).

S-matrix in the tree approximation

It is easy to derive the T-matrix from the Green functions in the approximation where diagrams with loops are neglected. This is called the *tree approximation*. The T-matrix is then found by simply truncating the propagators of the external lines of the corresponding Green function, and multiplying by $[\mathrm{i}(2\pi)^{3/2}]^{-n}$, with n the total number of external particles. For the four-point function of ϕ^3 theory in the tree approximation, eq. (4.5), we then have

$$\begin{aligned} \mathrm{i}T(\mathbf{l}_1, \mathbf{l}_2; \mathbf{p}_1, \mathbf{p}_2) &= \langle \mathbf{l}_1, \mathbf{l}_2 \text{ out} | \mathbf{p}_1, \mathbf{p}_2 \text{ in} \rangle \\ &= -\mathrm{i}(2\pi)^{-2} \delta^4(p_1 + p_2 - l_1 - l_2) \\ &\quad \times g^2[(s - m^2)^{-1} + (t - m^2)^{-1} + (u - m^2)^{-1}]. \end{aligned} \tag{4.17}$$

Here, the standard *Mandelstam variables* (Mandelstam 1958) for two-particle elastic scattering, s, t and u, are defined by

$$\begin{aligned} s &= (p_1 + p_2)^2 = E^2, \\ t &= (l_1 - p_1)^2 = -\tfrac{1}{2}(s - 4m^2)(1 - \cos\theta^*), \\ u &= (l_2 - p_1)^2 = -\tfrac{1}{2}(s - 4m^2)(1 + \cos\theta^*), \end{aligned} \tag{4.18}$$

where E is the total center-of-mass energy, and θ^*, ϕ^* are the polar and azimuthal angles between $\mathbf{p}_3$ and $\mathbf{p}_1$ in the center-of-mass system.

4.3 Cross sections

We observed in Section 2.5 that the physical content of the S-matrix is contained in its absolute value squared,

$$W_{\beta\alpha} = |S_{\beta\alpha}|^2 = |\langle \beta \text{ out} | \alpha \text{ in} \rangle|^2, \tag{4.19}$$

which is the probability for in-state α to be observed as out-state β. As it stands, eq. (4.19) is not quite adequate for our needs, because we generally take the in- and out-states to be non-normalizable plane waves, so that $S_{\beta\alpha}$ is proportional to a momentum conservation delta function, as in the example of eq. (4.17). To deal with this problem, we seek a finite quantity that does not depend on the details of the prepared initial state, but for which the concept of momentum is sensible. We would also like it to be Lorentz invariant.

The solution to our problem is facilitated in realistic scattering experiments, where there are (almost) always two particles in the initial state.

Supposing that the in-state is prepared as two wave packets near momenta $\mathbf{p}_1$ and $\mathbf{p}_2$, we define a free two-particle state by

$$|f_1, f_2 \text{ in}\rangle = \int \frac{d^3\mathbf{k}_1\, d^3\mathbf{k}_2}{2\omega(\mathbf{k}_1)2\omega(\mathbf{k}_2)} f_1(\mathbf{k}_1) f_2(\mathbf{k}_2) |\mathbf{k}_1\mathbf{k}_2 \text{ in}\rangle, \tag{4.20}$$

where $[2\omega(\mathbf{k}_i)]^{-1/2} f_i(\mathbf{k}_i)$ is a normalized function centered about $\mathbf{p}_i$ (see eqs. (2.74)–(2.78)). States such as $|f_1, f_2 \text{ in}\rangle$ define two-particle Schrödinger-picture wave functions in the usual first-quantization sense:

$$\psi(\mathbf{x}_1, \mathbf{x}_2, t) = \prod_{i=1}^{2} \int \frac{d^3\mathbf{p}_i}{[(2\pi)^3 2\omega(\mathbf{p}_i)]^{1/2}} \exp(-i\bar{p}_i \cdot x_i) f_i(\mathbf{p}_i). \tag{4.21}$$

The absolute square of this quantity is a probability density (compare eq. (2.77)).

To make the properties of ψ explicit, we introduce a large length-scale L, and write

$$\begin{gathered} \int \frac{d^3\mathbf{k}_i}{2\omega(\mathbf{k}_i)} |f_i(\mathbf{k}_i)|^2 = 1, \\ f_i(\mathbf{k}_i) \approx 0, \quad (\mathbf{p}_i - \mathbf{k}_i)^2 \gg (1/\mathrm{L})^2. \end{gathered} \tag{4.22}$$

An example of a function that satisfies eq. (4.22) is a Gaussian wave packet with width $1/L$,

$$[2\omega(\mathbf{k}_i)]^{-1} |f_i(\mathbf{k}_i)|^2 = (L/\pi)^{3/2} \exp[-(\mathbf{p}_i - \mathbf{k}_i)^2 L^2].$$

L is chosen to be very large compared with $1/|\mathbf{p}_i|$, $i = 1, 2$. The wave packet is then large enough to contain many Compton wavelengths. Equivalently, the momentum of each particle in the in-state is defined up to an uncertainty $1/L$, which is negligible compared with $|\mathbf{p}_i|$. Since our momentum measurements are of limited accuracy anyway, the wave packet is effectively a pure momentum eigenstate, even though it is normalized,

$$\langle f_1, f_2 \text{ in}|f_1, f_2 \text{ in}\rangle = 1. \tag{4.23}$$

Now consider the S-matrix element $S_{\beta\alpha}$ where, as above, α is a two-particle state, while β is an n-particle state $|\{\mathbf{l}_i\}_n \text{ out}\rangle$. Then we have, using the definition of the in-state (4.20),

$$\langle \{\mathbf{l}_i\}_n \text{ out}|f_1, f_2 \text{ in}\rangle = \int \frac{d^3\mathbf{k}_1 d^3\mathbf{k}_2}{2\omega(\mathbf{k}_1)2\omega(\mathbf{k}_2)} f_1(\mathbf{k}_1) f_2(\mathbf{k}_2) \langle \{\mathbf{l}_i\}_n \text{ out}|\mathbf{k}_1, \mathbf{k}_2 \text{ in}\rangle. \tag{4.24}$$

It is now convenient to extract the overall momentum conservation delta function, and define

$$\langle \{\mathbf{l}_i\}_n \text{ out}\,|k_1, k_2 \text{ in}\rangle \equiv i(2\pi)^4 \delta^4(k_1 + k_2 - \textstyle\sum l_i)(2\pi)^{-3(n+2)/2} M(k_1, k_2, l_i). \tag{4.25}$$

We have also factored out $(2\pi)^{-3/2}$ for each external particle in the S-matrix. M is called the *reduced* S-matrix element. It is given by the relevant truncated Feynman diagrams (times $-\mathrm{i}$). The probability of a transition from state α to state β is then given by

$$\begin{aligned} W_{\beta\alpha} &= |S_{\beta\alpha}|^2 \\ &= \prod_{i=1}^{2}\int \frac{\mathrm{d}^3\mathbf{k}_i\,\mathrm{d}^3\mathbf{k}'_i}{(2\pi)^{3/2}2\omega(\mathbf{k}_i)(2\pi)^{3/2}2\omega(\mathbf{k}'_i)} f_i(\mathbf{k}_i)f_i^*(\mathbf{k}'_i) \\ &\quad\times (2\pi)^{-3n}(2\pi)^4\delta^4(k_1+k_2-k'_1-k'_2)(2\pi)^4\delta^4(k_1+k_2-\textstyle\sum l_i) \\ &\quad\times M(k_1,k_2,l_i)M^*(k'_1,k'_2,l_i). \end{aligned} \tag{4.26}$$

This expression can be simplified by using an integral representation for the momentum-conservation delta function:

$$(2\pi)^4\delta^4(k_1+k_2-k'_1-k'_2) = \int \mathrm{d}^4x \exp[ix_\mu(k_1+k_2-k'_1-k'_2)^\mu]. \tag{4.27}$$

If it were not for the k-dependence in the amplitudes MM^* and in the factors $\omega(\mathbf{k}_i)$ in eq. (4.26) the resulting x-integrals would be the position amplitudes corresponding to the $f_i(\mathbf{k}_i)$, as in eq. (2.77),

$$\bar{f}_i(\mathbf{x}_i, t_i) = \int \frac{\mathrm{d}^3\mathbf{k}_i}{[(2\pi)^3 2\omega(\mathbf{k}_i)]^{1/2}} f_i(\mathbf{k}_i)\exp(-\mathrm{i}\bar{k}_i\cdot x_i). \tag{4.28}$$

Now the additional factors $\omega(\mathbf{k}_i)$ and M^*M depend on the $\mathbf{k}_i$ only through invariants such as $k_i\cdot l_j$ or $k_i\cdot k_j$, which we expect to change slowly, as functions of the $\mathbf{k}_i$, compared with the $f_i(\mathbf{k}_i)$ which by eq. (4.22) are sharply peaked about $\mathbf{k}_i = \mathbf{p}_i$. As a result, we may treat $\omega(\mathbf{k}_i)$ and MM^* as constants in the $\mathbf{k}_i$ integrals, up to corrections that vanish as a power of L, and replace them with their values at $\mathbf{k}_i = \mathbf{p}_i$. The result is

$$\begin{aligned} W_{\beta\alpha} &= \int \mathrm{d}^4x |\bar{f}_1(x)|^2|\bar{f}_2(x)|^2[2\omega(\mathbf{p}_1)]^{-1}[2\omega(\mathbf{p}_2)]^{-1}(2\pi)^{-3n} \\ &\quad\times (2\pi)^4\delta^4(p_1+p_2-\textstyle\sum l_i)|M(p_1,p_2,l_i)|^2[1+\mathrm{O}(1/|\mathbf{p}_i|L)]. \end{aligned} \tag{4.29}$$

By eq. (2.78), the $\bar{f}_i(x_i)$ are single-particle wave functions in the Schrödinger representation, normalized at any fixed time to unity. With this in mind, we can rewrite eq. (4.29) in differential form as

$$\begin{aligned} \mathrm{d}W_{\beta\alpha}/(\mathrm{d}^3x\mathrm{d}t) &= |\bar{f}_1(x)|^2|\bar{f}_2(x)|^2[2\omega(\mathbf{p}_1)]^{-1}[2\omega(\mathbf{p}_2)]^{-1}(2\pi)^{-3n} \\ &\quad\times (2\pi)^4\delta^4(p_1+p_2-\textstyle\sum l_i)|M(p_1,p_2,l_i)|^2, \end{aligned} \tag{4.30}$$

which is accurate up to a fraction $1/|\mathbf{p}_i|L$. $\mathrm{d}W_{\beta\alpha}/(\mathrm{d}^3\mathbf{x}\mathrm{d}t)$ is the transition probability *per unit volume and per unit time*. In general it is a function of position and time, but on distance and time scales *less than* L, it is effectively a constant. That is why the integral on the right-hand side of eq. (4.29) is sometimes said to be proportional to the 'volume of space–time'. To the extent that the wave functions $|f_i(\mathbf{x}, t)|$ are constant, it is actually proportional to L^4 in our notation, where L is the typical linear size of the wave packets in space. (This assumes, of course, that the wave packets collide.) The differential form, eq. (4.30), enables us to construct an invariant measure of the strength of the interaction called the *cross section*. The cross section has units of area, so it is something like the 'size' of the particles that collide.

Suppose we are in a frame where the momenta of the incoming particles are along a single axis. Examples are the *laboratory frame*, where one particle is stationary, and the *center-of-mass* (more accurately 'center-of-momentum') frame, where the momenta are equal and opposite. Let their relative velocity in this frame be $|\mathbf{v}_1 - \mathbf{v}_2|$.

Consider first the classical collision of point particles having velocity $\mathbf{v}_1$ and density ρ_1 with particles of finite cross sectional area σ perpendicular to $\mathbf{v}_1$ having velocity $\mathbf{v}_2$ and density ρ_2. The probability of collision per unit time and volume, $\mathrm{d}P/(\mathrm{d}t\mathrm{d}V)$, is given by

$$\mathrm{d}P/(\mathrm{d}t\mathrm{d}V) = |\mathbf{v}_1 - \mathbf{v}_2|\rho_1\rho_2\sigma. \tag{4.31}$$

Note that $1/(\rho_2\sigma)$ is the mean free path between collisions. By *analogy*, we define, in the same notation, the quantum mechanical cross section for $\alpha \rightarrow \beta$,

$$\sigma_{\beta\alpha} = (\text{probability of transition } \alpha \rightarrow \beta)/(|\mathbf{v}_1 - \mathbf{v}_2|\rho_1\rho_2), \tag{4.32}$$

or

$$\begin{aligned}\sigma_{\beta\alpha} &= \frac{\mathrm{d}W_{\alpha\beta}(\mathbf{x}, t)/(\mathrm{d}^3\mathbf{x}\mathrm{d}t)}{|\mathbf{v}_1 - \mathbf{v}_2||f_1(\mathbf{x}, t)|^2|f_2(\mathbf{x}, t)|^2} \\ &= \frac{(2\pi)^{-3n}(2\pi)^4\delta^4(p_1 + p_2 - \sum l_i)|M(p_1, p_2, l_i)|^2}{|\mathbf{v}_1 - \mathbf{v}_2|2\omega(\mathbf{p}_1)2\omega(\mathbf{p}_2)}.\end{aligned} \tag{4.33}$$

As we see, the cross section is independent of the position and time at which we form the ratio.

The advantages of the cross section include its independence of particle flux, and its appealing classical analogy. Finally, we can write it in a Lorentz invariant form by observing that when velocities are collinear,

$$|\mathbf{v}_1 - \mathbf{v}_2|2\omega(\mathbf{p}_1)2\omega(\mathbf{p}_2) = 4[(p_1\cdot p_2)^2 - m_1{}^2m_2{}^2]^{1/2}. \tag{4.34}$$

So, we *define* the fully Lorentz invariant cross section as

$$\sigma_{\beta\alpha} = \tfrac{1}{4}[(p_1 \cdot p_2)^2 - m_1{}^2 m_2{}^2]^{-1/2}(2\pi)^{-3n} \times |M(p_1, p_2, l_i)|^2 (2\pi)^4 \delta^4(p_1 + p_2 - \textstyle\sum l_i). \tag{4.35}$$

Phase space

To sum over final states, we can use the completeness relation (2.59) for free identical bosons, since out-states are effectively free. For such an n-particle final state,

$$\sigma_n = \tfrac{1}{4}[(p_1 \cdot p_2)^2 - m_1{}^2 m_2{}^2]^{-1/2} \times \frac{1}{n!}\prod_{i=1}^{n} \int \frac{d^3\mathbf{l}_i}{(2\pi)^3 2\omega(\mathbf{l}_i)} |M(p_1, p_2, l_i)|^2 (2\pi)^4 \delta^4(p_1 + p_2 - \textstyle\sum l_i). \tag{4.36}$$

The quantity

$$P^{(n)} = \frac{1}{n!}\prod_{i=1}^{n} \int \frac{d^3\mathbf{l}_i}{(2\pi)^3 2\omega(\mathbf{l}_i)} (2\pi)^4 \delta^4(p_1 + p_2 - \textstyle\sum l_i) \tag{4.37}$$

is referred to as *n-particle phase space* for identical bosons. It can be written in an invariant fashion as

$$P^{(n)} = \frac{1}{n!}\prod_{i=1}^{n} \int \frac{d^4 l_i}{(2\pi)^3} \delta_+(l_i{}^2 - m^2)(2\pi)^4 \delta^4(p_1 + p_2 - \textstyle\sum l_i). \tag{4.38}$$

Here $\delta_+(q_i{}^2 - m^2) \equiv \delta(q_i{}^2 - m^2)\theta(q_0)$ picks out the positive-energy solution in the argument of the delta function.

Total and differential cross sections

As it stands, σ_n in eq. (4.36) is the *total* cross section for two particles of fixed momenta $p_1{}^\mu$ and $p_2{}^\mu$ to produce n particles (integration has taken place over all possible momenta of the n particles). We may want more detailed information, however, such as the cross sections for producing particles with fixed momenta or at fixed directions with respect to the incoming momenta. These are *differential cross sections*. They are found by inserting delta functions that restrict the phase-space integrals in (4.36) to the desired region. Usually, the region is defined by a set of algebraic relations

$$\zeta_\gamma(l_i, p_j) = 0, \quad \gamma = 1, \ldots, a. \tag{4.39}$$

Then the corresponding differential cross section is

$$\frac{\mathrm{d}^a\sigma}{\prod_{\gamma=1}^{a}\mathrm{d}\zeta_\gamma(l_i, p_j)} = \frac{1}{4}[(p_1\cdot p_2)^2 - m_1{}^2 m_2{}^2]^{-1/2} \times \frac{1}{n!}\prod_{i=1}^{n}\int\frac{\mathrm{d}^3\mathbf{l}_i}{(2\pi)^3 2\omega(\mathbf{l}_i)}|M(p_1, p_2, l_i)|^2 \times (2\pi)^4\delta^4(p_1 + p_2 - \sum l_i)\prod_{\gamma=1}^{a}\delta[\zeta_\gamma(l_i, p_j)]. \quad (4.40)$$

Suppose we set

$$\zeta_{(i,j)} = (\mathbf{l}_i)_j - (\tilde{\mathbf{l}}_i)_j, \quad i = 1, \ldots, n; \quad j = 1, 2, 3, \qquad (4.41)$$

where the $\{\tilde{\mathbf{l}}_i\}$ are some fixed vectors; this extreme restriction defines a 'fully' differential cross section:

$$\frac{\mathrm{d}^{3n}\sigma}{\prod_{i=1}^{n}\{\mathrm{d}^3\mathbf{l}_i/[(2\pi)^3 2\omega(\mathbf{l}_i)]\}} = \frac{1}{4}[(p_1\cdot p_2)^2 - m_1{}^2 m_2{}^2]^{-1/2} \times (1/n!)|M(p_1, p_2, l_i)|^2 \times (2\pi)^4\delta^4(p_1 + p_2 - \sum l_i). \qquad (4.42)$$

Note that it is customary to associate the normalization factors $[(2\pi)^3 2\omega(\mathbf{l}_i)]^{-1}$ with the differentials in the differential cross section.

Of special interest are the *elastic* scattering cross sections in the case where $n = 2$; then $p_1 + p_2 \rightarrow l_1 + l_2$, at fixed momentum transfer $q^2 = (l_i - p_1)^2$, for i = 1 or 2. When the final-state particles are indistinguishable we must include both cases. This means inserting $\delta(q^2 - t) + \delta(q^2 - u)$ in eq. (4.42) with $n = 2$, and integrating over phase space. Equivalently, since $|M|^2$ is symmetric in l_1 and l_2, we may simply set $q^2 = (l_1 - p_1)^2 = t$ and drop the factor of $\frac{1}{2}$ in the phase-space integral. (We must, however, be careful to note that, in this case, $\int \mathrm{d}t(\mathrm{d}\sigma/\mathrm{d}t) = 2\sigma_{\mathrm{tot}}$.) This gives

$$\frac{\mathrm{d}\sigma}{\mathrm{d}t} = \frac{1}{64\pi|\mathbf{p}_{\mathrm{cm}}|^2 s}|M(s, t)|^2. \qquad (4.43)$$

The initial center-of-mass momentum $\mathbf{p}_{\mathrm{cm}}$ is related to the kinematic factor in the general cross section, eq. (4.35), by

$$|\mathbf{p}_{\mathrm{cm}}|^2 s = (p_1\cdot p_2)^2 - m_1{}^2 m_2{}^2, \qquad (4.44)$$

where we have generalized to the unequal-mass case. The general cross section at fixed scattering angle θ^* between l_1 and p_1 in the center of mass system is found by $|\mathrm{d}t/\mathrm{d}\cos\theta^*| = 2|\mathbf{k}_{\mathrm{cm}}|\,|\mathbf{p}_{\mathrm{cm}}|$, where, by direct analogy with eq. (4.44),

$$|\mathbf{k}_{\mathrm{cm}}|^2 s = [l_1\cdot l_2]^2 - \mu_1{}^2\mu_2{}^2, \qquad (4.45)$$

where μ_1, μ_2 are the outgoing masses. Then we find (again, for identical particles a factor of $\frac{1}{2}$ has to be inserted)

$$\frac{d\sigma}{d\cos\theta^*} = \frac{|\mathbf{k}_{cm}|}{32\pi s|\mathbf{p}_{cm}|}|M(s, t)|^2, \tag{4.46}$$

another useful form.

Cross section in tree approximation

The three order-g^2 contributions to G_4 were shown in Fig. 4.1. The T-matrix element is given by eq. (4.17), and M by eq. (4.25). From eq. (4.36), we find their contribution to the total cross section:

$$\begin{aligned}\sigma = &\{16[(p_1\cdot p_2)^2 - m^4]\}^{-1/2}\\ &\times \frac{1}{2}\prod_{i=1}^{2}\int \frac{d^3\mathbf{l}_i}{(2\pi)^3 2\omega(\mathbf{l}_i)}(2\pi)^4\delta^4(p_1 + p_2 - l_1 - l_2)\\ &\times g^4[(s - m^2)^{-1} + (t - m^2)^{-1} + (u - m^2)^{-1}]^2.\end{aligned} \tag{4.47}$$

Eq. (4.46) gives the differential cross section for identical particles at fixed center-of-mass angles,

$$\begin{aligned}\frac{d\sigma^{(2)}}{d\Omega^*} &= \frac{1}{64\pi^2 s}|M(\theta^*, \phi^*)|^2\\ &= (g^4/64\pi^2 s)\{(s - m^2)^{-1}\\ &\quad + [-\tfrac{1}{2}s(1 - \cos\theta^*) - 2m^2(1 - \cos\theta^*) - m^2]^{-1}\\ &\quad + [-\tfrac{1}{2}s(1 + \cos\theta^*) - 2m^2(1 + \cos\theta^*) - m^2]^{-1}\}^2,\end{aligned} \tag{4.48}$$

where the first form is for a general amplitude with all masses equal, and where $d\Omega \equiv d(\cos\theta)d\phi$. In this case, the result is independent of the azimuthal angle. The formula simplifies in the limit $m^2/s \to 0$, with angles fixed, to

$$\begin{aligned}d\sigma/d\Omega^* = &(g^4/64\pi^2 s^3)[1 - 2(1 - \cos\theta^*)^{-1} - 2(1 + \cos\theta^*)^{-1}]^2\\ &\times[1 + O(m^2/s)],\end{aligned} \tag{4.49}$$

The three terms in brackets correspond to the three graphs (a), (b) and (c) of Fig. 4.1. If we let θ^* become small (but not so small that $\theta^{*2}s \approx m^2$), we find that

$$d\sigma^{(2)}/d\Omega^* \to g^4/(64\pi^2 s^3 \sin^4 \tfrac{1}{2}\theta^*), \tag{4.50}$$

which comes entirely from the graph of Fig. 4.1(b), the 't-channel exchange graph'. This angular dependence is familiar from the classical Rutherford cross section for the scattering of charged particles.

In closing this introduction to cross sections, we recall that in our 'natural' units, all momenta and masses have units of inverse length. More typically, these quantities are specified in terms of energy. The conversion factor that allows us to calculate cross sections is just $\hbar c = 197.3\ \mathrm{MeV} \times 10^{-13}$ cm (Particle Data Group 1990).

4.4 The charged scalar field

Interactions

In the following chapters, we shall treat fields with spin, as is necessary for realistic Lagrangians. First, however, we discuss the interacting complex scalar field; this discussion will serve both as a transition to more sophisticated theories, and as a recapitulation of the foregoing developments.

The general charged scalar Lagrange density has been specified in eq. (1.17). Typical interactions involve self-interactions via four-point couplings:

$$\mathscr{L} = \partial^\mu \phi^* \partial_\mu \phi - M^2 \phi^* \phi - (\tfrac{1}{4}\lambda)(\phi^*\phi)^2. \tag{4.51}$$

Also important are Lagrangians with more than one field. We may have, for instance, a neutral scalar field ψ as well as a charged scalar field ϕ. A typical Lagrangian of this type involves a three-point coupling:

$$\mathscr{L} = \partial^\mu \phi^* \partial_\mu \phi - M^2 \phi^* \phi + \tfrac{1}{2}(\partial^\mu \psi \partial_\mu \psi - m^2 \psi^2) - g\psi(\phi^*\phi) - (\tfrac{1}{4}\lambda)(\phi^*\phi)^2, \tag{4.52}$$

where M and m are the masses of the charged and neutral fields respectively. Yet another Lagrangian couples the charged field to an *external* (classical) vector field $A^\mu(x)$:

$$\begin{aligned}\mathscr{L} &= (\partial^\mu \phi^* - \mathrm{i}eA^\mu \phi^*)(\partial_\mu \phi + \mathrm{i}eA_\mu \phi) - M^2 \phi^* \phi \\ &= \partial^\mu \phi^* \partial_\mu \phi - M^2 \phi^* \phi - \mathrm{i}eA^\mu(\phi^* \partial_\mu \phi - \phi \partial_\mu \phi^*) + e^2 A^2 \phi^* \phi.\end{aligned} \tag{4.53}$$

Here e is a new coupling constant. As yet, there is no free Lagrangian for the vector field $A^\mu(x)$, which is treated here as simply a function. In the second form, in which we separate the free Lagrange densities from the interaction Lagrange densities, the field A^μ couples linearly to the conserved current found in Section 1.3. The terms linear in $A^\mu(x)$ are *derivative couplings*, since they include gradients of ϕ.

Generating functionals

In each of these cases the construction of Green functions and the *S*-matrix proceeds in much the same way as for a neutral scalar field.

The analogy with the real scalar field may be made more explicit by expanding the charged field in terms of two real fields as in eq. (1.19),

$$\phi = 2^{-1/2}(\phi_1 + \mathrm{i}\phi_2), \quad \pi = 2^{-1/2}(\pi_1 - \mathrm{i}\pi_2),$$
$$a_\pm = 2^{-1/2}(a_1 \pm \mathrm{i}a_2). \tag{4.54}$$

This enables us to derive a phase-space form of the path integral for the generating functional of Green functions in any of our theories that is analogous to eq. (3.39a), but now includes interactions.

In the following, we combine all the interactions in eqs. (4.51)–(4.53) into a single Lagrangian. The Hamiltonian is constructed, as usual, from the Lagrange density and the conjugate momenta:

$$\pi_1 = \partial\mathcal{L}/\partial\dot{\phi}_1 = \dot{\phi}_1 - eA_0\phi_2, \quad \pi_2 = \partial\mathcal{L}/\partial\dot{\phi}_2 = \dot{\phi}_2 + eA_0\phi_1. \tag{4.55}$$

The phase-space form for the generating functional of Green functions is then

$$\begin{aligned} Z_{\mathcal{L}}[K, J_1, J_2] = \frac{1}{W[0]} \int [\mathcal{D}\pi_\psi][\mathcal{D}\psi] \exp\left\{\int \mathrm{d}^4x[\pi_\psi\dot{\psi} - \mathcal{H}_0(\pi_\psi, \psi) - K\psi]\right\} \\ \times \int [\mathcal{D}\pi_1][\mathcal{D}\phi_1][\mathcal{D}\pi_2][\mathcal{D}\phi_2] \\ \times \exp[\pi_1\dot{\phi}_1 - \mathcal{H}_0(\pi_1, \phi_1) \\ + \pi_2\dot{\phi}_2 - \mathcal{H}_0(\pi_2, \phi_2) \\ - \mathcal{H}_\mathrm{I}(\phi_1, \pi_1, \phi_2, \pi_2, \psi, \pi_\psi, A) - J^*\phi - J\phi^*], \end{aligned} \tag{4.56}$$

where $W[0]$ is chosen, as in the neutral case, to set $Z[0, 0, 0]$ to unity. In the usual way, the $\dot{\phi}_i$ are to be re-expressed in terms of the ϕ_i and π_i, by eq. (4.55). $\pi_\psi(x)$ is the momentum conjugate to the field $\psi(x)$, and $\mathcal{H}_0(\pi, \psi)$ is the free Hamiltonian for $\psi(x)$. $\mathcal{H}_\mathrm{I}(\phi_1, \pi_1, \phi_2, \pi_2, \psi, \pi_\psi, A)$ is the interaction Hamiltonian, including, in this case, all the interaction terms corresponding to eqs. (4.51)–(4.53). If we deal with (4.51) and (4.52) alone, the classical relation is $\pi_i = \dot{\phi}_i$, and $\mathcal{H}_\mathrm{I}$ is identical to the interaction Lagrangian. The contribution to $\mathcal{H}_\mathrm{I}$ from eq. (4.53), however, is

$$\begin{aligned} \mathcal{H}_\mathrm{I}{}^{(A)}(\phi_1, \pi_1, \phi_2, \pi_2, A) = \mathrm{i}e[A^0\mathrm{i}(\pi_1\phi_2 - \pi_2\phi_1) + A^k(\phi^*\partial_k\phi - \phi\partial_k\phi^*)] \\ - e^2A^kA_k\phi^*\phi, \end{aligned} \tag{4.57}$$

with $k = 1, 2, 3$. This is not the same as the interaction Lagrangian, nor is it even Lorentz invariant. Such a difference between the interaction Lagrangian and the Hamiltonian is characteristic of theories with derivative couplings. Nevertheless, when the π_1 and π_2 integrals in eq. (4.56) are carried out, we derive a manifestly invariant form, in which the interaction Lagrangian appears. Introducing a complex current $J = 2^{-1/2}(J_1 + iJ_2)$, the result may be written in terms of integrals over complex fields ϕ and ϕ^* as

$$
\begin{aligned}
Z_{\mathscr{L}}[K, J, J^*] &= \frac{1}{W[0]} \int [\mathscr{D}\psi] \exp\left\{\int d^4x [\mathscr{L}_0(\psi) - K\psi]\right\} \\
&\quad \times \int [\mathscr{D}\phi][\mathscr{D}\phi^*] \exp\left\{i \int d^4x [\mathscr{L}(\phi, \phi^*, \psi, A) - J^*(x)\phi(x) - J(x)\phi^*(x)]\right\} \\
&= \frac{1}{W[0]} \exp\left\{i \int d^4z\, \mathscr{L}_I[i\delta_J(z), i\delta_{J^*}(z), i\delta_K(z), A(z)]\right\} Z_{\mathscr{L}_0}[K, J, J^*].
\end{aligned}
\tag{4.58}
$$

$Z_{\mathscr{L}_0}$ is the free-field generating functional, which may be computed by breaking it down to real scalar fields and then using eq. (3.43):

$$
\begin{aligned}
Z_0[K, J, J^*] &= \exp\left\{-i \int d^4w\, d^4z [\tfrac{1}{2} K(w)\Delta_F(w - z)K(z) + J^*(w)\Delta_F(w - z)J(z)]\right\}.
\end{aligned}
\tag{4.59}
$$

The product of differentials $[\mathscr{D}\phi][\mathscr{D}\phi^*]$ in eq. (4.58) is just a way of writing $[\mathscr{D}\phi_1][\mathscr{D}\phi_2]$ (in much the same spirit as eq. (3.56) for coherent states). Similarly, $\delta_J(z)$ is shorthand for $2^{-1/2}(\delta_{J_1} + i\delta_{J_2})$. In the second form of (4.58), the perturbative expansion for charged scalar theories is defined in terms of the free theory generating functional, as above.

S-matrix

In this theory, an arbitrary *S*-matrix element has incoming and outgoing particles and antiparticles. Reduction formulas for the charged scalar field may be derived in just the same way as was done for the neutral field in Section 2.5. The only difference is in the breakup of the asymptotic matrix elements according to free-field theory, eq. (2.100). Here we must recall from the discussion after eqs. (2.73) that the field $\phi^\dagger(x)$ creates particles and absorbs antiparticles, while $\phi(x)$ does just the opposite. The general case is illustrated by the reduction formula for particle–antiparticle $2 \rightarrow 2$

scattering:

$$\begin{aligned}\langle \mathbf{p}_3{}^{(+)}, \mathbf{p}_4{}^{(-)} \text{ out}|\mathbf{p}_1{}^{(+)}, \mathbf{p}_2{}^{(-)} \text{ in}\rangle \\ &= [\mathrm{i}(2\pi)^{3/2} R_\phi{}^{1/2}]^{-4} \prod_{i=1}^{4} \lim_{p_i{}^2 - m^2 \to 0} (p_i{}^2 - m^2) \\ &\times \int \mathrm{d}^4 x_1\, \mathrm{d}^4 x_2\, \mathrm{d}^4 x_3\, \mathrm{d}^4 x_4 \mathrm{e}^{\mathrm{i}p_4\cdot x_4} \mathrm{e}^{\mathrm{i}p_3\cdot x_3} \mathrm{e}^{-\mathrm{i}p_2\cdot x_2} \mathrm{e}^{-\mathrm{i}p_1\cdot x_1} \\ &\times \langle 0|T[\phi^\dagger(x_4)\phi(x_3)\phi(x_2)\phi^\dagger(x_1)]|0\rangle, \qquad (4.60)\end{aligned}$$

where the factor R_ϕ is once again the residue of the two-point function at the single-particle pole:

$$R_\phi = -\mathrm{i} \lim_{q^2 - m^2 \to 0} (q^2 - m^2) \int \mathrm{d}^4 x\, \mathrm{e}^{\mathrm{i}q\cdot x} \langle 0|T[\phi^\dagger(x)\phi(0)]|0\rangle. \qquad (4.61)$$

The Green function in eq. (4.60) includes the field $\phi^\dagger$ for initial-state particles and final-state antiparticles, and the field ϕ for final-state particles and initial-state antiparticles. Otherwise, the form is the same as for the neutral scalar field.

Feynman rules for the charged field

Feynman rules for the interacting charged field summarize the expressions derived by matching variations with sources in eq. (4.58).

The fields ϕ and ϕ^*, which are associated with variations δ_{J^*} and δ_J, are always matched with J^* and J, respectively. The charged field propagator may thus be thought of as starting at ϕ^* and ending at ϕ. In enumerating diagrams, we have to take this assignment into account, and this is done graphically, as in Fig. 4.4(*a*), by associating an arrow with each line. In view of the assignment of fields to incoming and outgoing particles in eq. (4.60), particle–antiparticle scattering may be represented as in Fig. 4.4(*b*), in which arrows point from the initial state to the final state for particle lines, and from the final state to the initial state for antiparticle lines.

Since the number of factors of ϕ always equals the number of factors of ϕ^* (the action must be real), the number of arrows flowing into a vertex equals the number flowing out. This little rule is the graphical expression of charge conservation. With each scalar three- and four-point function we associate a factor $-\mathrm{i}g$ and $-\mathrm{i}\lambda$, in addition to an energy-conserving delta function. The vertex associated with the background field is somewhat more complicated. The external vector field itself contributes a factor $(2\pi)^{-4}\int \mathrm{d}^4 k\, \widetilde{A}^\mu(k)$, where

$$\widetilde{A}(k) \equiv \int \mathrm{d}^4 x\, \mathrm{e}^{-\mathrm{i}k\cdot x} A(x) \qquad (4.62)$$

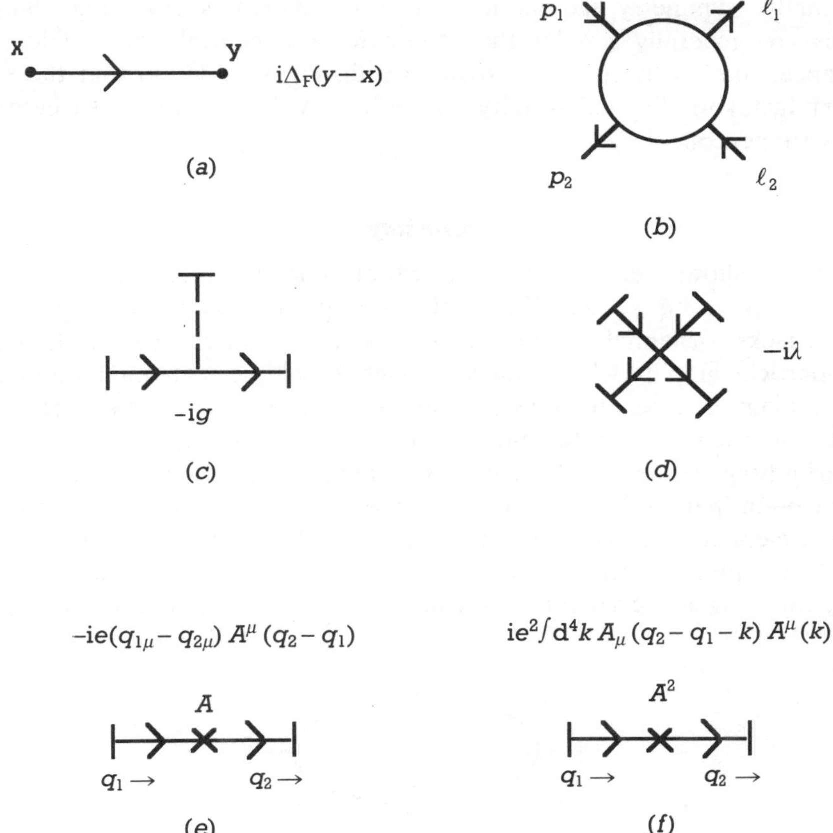

Fig. 4.4 Illustration of Feynman rules for the Lagrange densities in eqs. (4.52) and (4.53).

is the Fourier transform of the function $A^\mu(x)$. The k integral is linked with the scalar lines by the momentum-conservation delta function. At the same time, the factor $\partial^{(w)}{}_\mu\phi(w)$ becomes $\partial^{(w)}{}_\mu[\delta_{J^*}(w)]$, which will give $\partial^{(w)}{}_\mu[i\Delta_F(w-z)]$. In momentum space, using eq. (2.83), this produces an extra factor $-iq_\mu$, where q_μ is the momentum flowing *into* the vertex at w^μ along the line associated with $\phi(w)$. Analogous reasoning for ϕ^* gives the same result, and the rule of associating a derivative of the field with a factor $-i\times$ (momentum flowing in) is quite general. Taking into account all factors of i, relative signs and momentum conservation, the complete factor associated with the vertex of Fig. 4.4(e) is

$$-ie(q_{1\mu}-q_{2\mu})A^\mu(q_2-q_1). \tag{4.63}$$

The rule for the four-point interaction (sometimes called the *seagull* vertex) is given in Fig. 4.4(f). Its derivation is suggested as exercise 4.7.

Finally, symmetry factors for low-order diagrams involving charged fields are generally simpler than those for the neutral scalar field. For instance, lines with opposite arrows are distinguishable, so that the symmetry factor for Fig. 4.5 is unity, in contrast with $\frac{1}{2}$ for Fig. 4.3 where the lines are neutral.

Crossing

Figure 4.6 shows examples of $2 \to 2$ scattering for the three- and four-point interactions in eq. (4.52). The particle–antiparticle amplitude Fig. 4.6(*a*), which lacks u-channel neutral scalar exchange, is different from the particle–particle amplitude Fig. 4.6(*b*), which lacks the s-channel ('annihilation') diagram (see eq. (4.18)). Both S-matrix elements, however, arise from the same Green function, in this case the order g^2 and order λ contributions to the four-field matrix element in eq. (4.60). To change the particle–antiparticle matrix element in eq. (4.60) into a particle–particle matrix element, we need only exchange the roles of the momenta $p_2{}^\mu$ and $-p_4{}^\mu$. Graphically, this exchange switches an incoming with an outgoing line, changing at the same time the directions of their arrows. Equivalently,

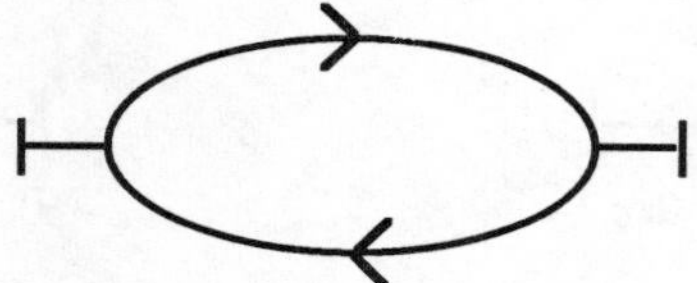

Fig. 4.5 Loop diagram involving a charged scalar field.

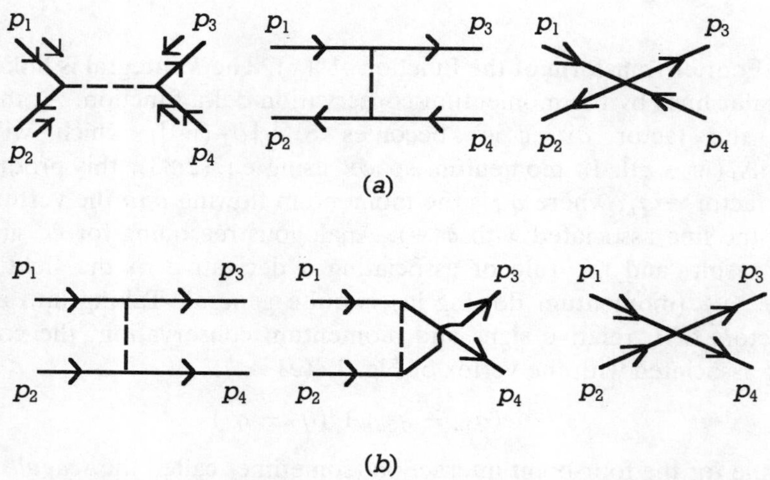

Fig. 4.6 Lowest-order $2 \to 2$ scattering diagrams for charged scalar fields.

the truncated diagrams in Fig. 4.6 are related by analytically continuing the momenta of eq. (4.60). Generalizing this reasoning, we conclude that any scalar theory S-matrix element with a particle of momentum $p_i{}^\mu$ in the initial state is related to one with an antiparticle of momentum $p_f{}^\mu$ in the final state by simply changing $p_i{}^\mu$ to $-p_f{}^\mu$ in the formula for the S-matrix. This, and similar relations are known as *crossing symmetry*.

Angle and energy behavior

For the interactions in eq. (4.52), the small-angle behavior of cross sections at high energy is dominated by neutral scalar (ψ) exchange, and is the same as eq. (4.50). The situation is different at high energy and fixed angles. Here the amplitude in the charged scalar theory is simply λ, from the four-scalar vertex, independent of energy, particle type and angle, while the cross section behaves as s^{-1}. This is in distinction to the s^{-3} behavior for a pure ϕ^3 theory, eq. (4.50).

Decay rate

Having particles of different mass, as in eq. (4.52), leads to the possibility of decay amplitudes. If, for instance, $m > 2M$, then the S-matrix element for $\psi \to \phi^{(+)}\phi^{(-)}$, $\langle \mathbf{p}_1{}^{(-)}, \mathbf{p}_2{}^{(+)} \text{ out}|\mathbf{q} \text{ in}\rangle$, is nonzero. The square of the S-matrix element can be considered once again as the total transition probability and can be used to derive a transition rate per unit volume and unit time in much the same manner as for $2 \to n$ scattering. Dividing by the density of initial-state particles gives the transition rate, or decay rate, per particle. The details of the argument are essentially the same as for the cross section in Section 4.3, and are suggested as exercise 4.8. The result for the total decay rate into two particles in the rest frame of the decaying particle (mass m) is

$$\Gamma = \frac{1}{2m} \int \prod_{i=1}^{2} \frac{\mathrm{d}^3\mathbf{p}_i}{(2\pi)^3 2\omega(\mathbf{p}_i)} (2\pi)^4 \delta^4(q - p_1 - p_2) |M(\mathbf{q}; \mathbf{p}_1{}^{(-)}, \mathbf{p}_2{}^{(+)})|^2. \tag{4.64}$$

The lowest-order reduced matrix element for $\psi \to \phi^{(+)}\phi^{(-)}$ decay, calculated from the density given in eq. (4.52), is simply the lowest-order vertex, $-ig$.

The T^* product

To complete our brief discussion of the charged scalar field and its interactions, we shall touch on an important technical point related to the computation of Green functions. The generating functional $Z[K, J, J^*]$, eq. (4.58), can be put in a 'Gell-Mann–Low' form, analogous to eq. (3.89). Thus, in the computation of Green functions, we encounter terms such as

$$\langle 0|T[\mathcal{L}_I(y_1)\mathcal{L}_I(y_2)\cdots]|0\rangle_{\text{free}} \tag{4.65}$$

In the presence of derivative couplings, we find vacuum expectation values of time-ordered time derivatives of fields. The simplest matrix element of this type is $\langle 0|T[\partial^0\phi(x)\partial^0\phi(y)]|0\rangle_{\text{free}}$. Evidently, this matrix element is represented by a factor $i\partial^0{}_x\partial^0{}_y\Delta_F(x-y)$, since in the Feynman rules the derivatives correspond to simple factors of energy in the Fourier transform, as in eq. (4.63). We note, however, that time derivatives do not commute with the time ordering defined as in eq. (2.82). In the free theory we have

$$T[\partial^0\phi(x)\partial^0\phi(y)]_{\text{free}} = \partial^0{}_x\partial^0{}_y T[\phi(x)\phi(y)]_{\text{free}} - i\delta^4(x-y), \tag{4.66}$$

where we have used $\phi = \dot{\pi}$ (exact for the free theory), and the canonical commutation relations. Why then, do we not have 'contact' terms, corresponding to this delta function in the Feynman rules? The reason is that we cannot interpret the time-ordered product in Section 2.3 literally as the time ordering that emerges from the path integral. The time-ordered product, after all, is not well defined at equal times. The product for the integrands of path integrals evidently differs from the definition (2.82) by exactly these contact terms. It is often called the 'T^*-product' and is defined by the requirement that time derivatives always appear *outside* the time ordering. So, for the T^*-product,

$$\begin{aligned}\langle 0|T^*[\partial^0\phi(x)\phi(y)]|0\rangle_{\text{free}} &\equiv i\partial^0{}_x\Delta_F(x-y),\\ \langle 0|T^*[\partial^0\phi(x)\partial^0\phi(y)]|0\rangle_{\text{free}} &\equiv i\partial^0{}_x\partial^0{}_y\Delta_F(x-y).\end{aligned} \tag{4.67}$$

This result was first derived in the interaction picture discussed in Appendix A (Mathews 1949, Nishijima 1950, Rohrlich 1950). It follows automatically in the path integral formalism above, and all perturbative matrix elements of the form of eq. (4.65) (expansions according to the Gell-Mann –Low formula, eq. (3.89)) are interpreted as T^*-products). This holds for scalar electrodynamics as well as for all the theories discussed below. So, *any* derivative acting on a field in an interaction Lagrangian corresponds to a simple factor of momentum, according to the rule above, eq. (4.63). All our T-products below will be T^*-products, but we shall not exhibit the star explicitly.

Exercises

4.1 Write down an expression for the two-loop diagram Fig. 4.2, using two different routings for the loop momenta.

4.2 Consider a real scalar quantum field that is free except for a nonvanishing source term,

$$\mathscr{L} = \tfrac{1}{2}[(\partial_\mu\phi)(\partial^\mu\phi) - m^2\phi^2] - J(x)\phi(x).$$

Compute the S-matrix element $\langle \mathbf{p}_4, \mathbf{p}_3 \text{ out} \,|\mathbf{p}_1, \mathbf{p}_2 \text{ in}\rangle$ in the case when none of the $\mathbf{p}_i$ are equal.

4.3 Repeating the reasoning that led to eq. (4.15), take into account multiparticle states in the sum in eq. (4.14) and derive the 'Källén–Lehmann' representation (Källén 1952, Lehmann 1954) for the two-point Green function in an interacting-field theory,

$$\bar{G}_{\text{int}}(q^2) = R\mathrm{i}\Delta(q^2, m^2) + \int_{4m^2}^{\infty} \mathrm{d}p^2\, \rho(p^2)\mathrm{i}\Delta(q^2, p^2),$$

where $\Delta(p^2, m^2)$ is the Feynman propagator with mass m, and $\rho(p^2)$ is a positive function (see Bjorken & Drell r1965, Chapter 16).

4.4 Let $C(x^\mu, m^2)$ represent the commutator of a free field of mass m. Derive a Källén–Lehmann representation for the commutator of the interating field of mass m, assuming, as above, that the invariant masses of multiparticle states are bounded from below by $4m^2$, i.e.,

$$[\phi(x), \phi(0)] = RC(x^\mu, m^2) + \int_{4m^2}^{\infty} \mathrm{d}p^2\, \rho(p^2)C(x^\mu, p^2),$$

where $\rho(p^2)$ is the same function as in exercise 4.3. Show that this result, combined with the canonical commutation relations, leads to the requirement $R < 1$.

4.5 Draw the lowest-order Feynman diagrams for $2 \to 3$ scattering for the neutral scalar field at energies much larger than the mass of the particles. Show that the corresponding amplitude has maxima (i) when the momentum of one final-state particle vanishes and (ii) when two final-state particles have momenta that are parallel.

4.6 Compute the total cross section for two-particle elastic scattering in ϕ^3 theory, and show that it becomes infinite in the zero-mass limit.

4.7 Derive the 'seagull' Feynman rule Fig. 4.4(f).

4.8 Derive the expression eq. (4.64) for the decay rate.

4.9 Show that, following the procedure of Section 3.3, we can define a generating functional for the charged scalar S-matrix in terms of $Z_{\mathscr{L}}[K, J, J^*]$ that is analogous to eq. (3.80):

$$\begin{aligned}
&\mathscr{S}_{\mathscr{L}}(\{\alpha^{*\prime\prime}(\mathbf{k}), \alpha^{*\prime\prime}_{\pm}(\mathbf{k})\}; \{\alpha'(\mathbf{k}), \alpha'_{\pm}(\mathbf{k})\}; K, J, J^*) \\
&\quad = \exp\left\{\mathrm{i}\int \mathrm{d}^3\mathbf{k}[\alpha^{*\prime\prime}(\mathbf{k})\alpha'(\mathbf{k}) + \alpha^{*\prime\prime}_{+}(\mathbf{k})\alpha'_{+}(\mathbf{k}) + \alpha^{*\prime\prime}_{-}(\mathbf{k})\alpha'_{-}(\mathbf{k})]\right\} \\
&\quad\quad \times \exp\left\{-\int \mathrm{d}^4x[\psi_{\text{as}}(x)(\partial^\mu\partial_\mu + m^2)\delta_K(x) + \phi_{\text{as}}(x)(\partial^\mu\partial_\mu + M^2)\delta_J(x)\right. \\
&\quad\quad\quad \left. + \phi^*_{\text{as}}(x)(\partial^\mu\partial_\mu + M^2)\delta_{J^*}(x)]\right\} \\
&\quad\quad \times \exp\left\{\mathrm{i}\int \mathrm{d}^4z\, \mathscr{L}_{\mathrm{I}}[\mathrm{i}\delta_K(z), \mathrm{i}\delta_J(z), \mathrm{i}\delta_{J^*}(z)]\right\} Z_{\mathscr{L}_0}[K, J, J^*].
\end{aligned}$$

$\mathcal{S}_{\mathcal{L}}$ is now a functional of two sets of boundary values, $\alpha'_+(\mathbf{k})$, $\alpha^{*\prime}_-(\mathbf{k})$ at $t = -\infty$ and $\alpha^{*\prime\prime}_+(\mathbf{k})$, $\alpha''_-(\mathbf{k})$ at $t = +\infty$, for the complex field, in addition to boundary values $\alpha'(\mathbf{k})$ and $\alpha^{*\prime\prime}(\mathbf{k})$ for the neutral field. The 'asymptotic' fields are defined by

$$\phi_{\text{as}}(x) = \int \frac{\mathrm{d}^3\mathbf{k}}{2\omega_k(2\pi)^{3/2}} \{\alpha'_+(\mathbf{k})\,\mathrm{e}^{-\mathrm{i}\bar{k}\cdot x} + \alpha^{*\prime\prime}_-(\mathbf{k})\,\mathrm{e}^{\mathrm{i}\bar{k}\cdot x}\},$$

$$\phi^*{}_{\text{as}}(x) = \int \frac{\mathrm{d}^3\mathbf{k}}{2\omega_k(2\pi)^{3/2}} \{\alpha'_-(\mathbf{k})\,\mathrm{e}^{-\mathrm{i}\bar{k}\cdot x} + \alpha^{*\prime\prime}_+(\mathbf{k})\,\mathrm{e}^{\mathrm{i}\bar{k}\cdot x}\},$$

and similarly for ψ_{as}.

PART II

FIELDS WITH SPIN

5

Spinors, vectors and gauge invariance

So far, we have developed the formalism almost entirely in terms of scalar fields. This has had the advantage of simplicity for techniques that are common to all theories, but it is incomplete; candidates for fundamental theories all contain fields that transform nontrivially under rotations and Lorentz boosts. In this chapter, we discuss such fields and their classical field theories. We construct equations of motion and Lagrange densities, guided by symmetry principles, especially Lorentz invariance and gauge symmetry (Section 5.4). The issue of vector particle masses turns out to be a delicate one, and we will give a brief discussion of the Higgs phenomenon, as it is thought to operate in the so-called *standard model* of fundamental interactions. Here and in the following chapters, we shall exhibit parts of the standard model Lagrangian as they are needed; a systematic description may be found in Appendix C.

5.1 Representations of the Lorentz group

The action of Lorentz transformations on classical fields is given by eq. (1.102), $\bar{\phi}_a(\Lambda x) = S_{ab}(\Lambda)\phi_b(x)$. For scalar fields, $S(\Lambda)$ is the unit matrix, but of course there are other possibilities. Nevertheless, the $S(\Lambda)$ are not arbitrary. On grounds of self-consistency, they must satisfy the group property, eq. (1.66), $S(\Lambda_1)S(\Lambda_2) = S(\Lambda_1\Lambda_2)$. The set of matrices $S_{ab}(\Lambda)$ is then a representation of the Lorentz group, and the fields $\{\phi_a\}$ are referred to as 'the tensor transforming according to the representation $S(\Lambda)$'. In this section, we outline some basic facts about representations of the Lorentz group. More details can be found in Brauer & Weyl (r1935), Lyubarskii (r1960), Bogoliubov, Logunov & Todorov (r1975), Corson (r1981) and Kim & Noz (r1986).

Single-valued representations

The most familiar possibilities are contravariant and covariant *vector fields*, $A^\alpha(x)$ and $A_\alpha(x)$, for which $S_{\alpha\beta} = \Lambda^\alpha{}_\beta$ and $\Lambda_\alpha{}^\beta$, respectively (eq.

(1.104)). Covariant and contravariant representations are said to be *equivalent*, since from eq. (1.85) they are related by the unitary transformation $A_\alpha = g_{\alpha\beta}A^\beta$.

A tensor with n contravariant and m covariant vector indices, denoted by $T^{\{\alpha_i\}}{}_{\{\beta_i\}}$, transforms by a direct generalization of eq. (1.87). Such tensor representations are obviously continuous and *single-valued*, in the sense that $S(\Lambda)$ is uniquely specified by Λ. In fact, all single-valued representations can be constructed from direct products of vector representations. They do not, however, exhaust all continuous representations of the Lorentz group. Many representations of the Lorentz group are *not* single-valued. Let us see why this already happens for the rotation group. We shall use some standard results from the nonrelativistic quantum mechanics of angular momentum.

Rotational spinors and SU(2)

A general rotation is given by eq. (1.99) with the boost parameters ω_i set to zero. It may be written as

$$R^\mu{}_\nu = [\exp(-\mathrm{i}\boldsymbol{\theta}_R \cdot \mathbf{J})]^\mu{}_\nu, \tag{5.1}$$

where the three matrices J_a are given in eq. (1.91), and the $(\boldsymbol{\theta}_R)_a$, $a = 1, 2, 3$, are the parameters that specify rotation R. The J_a obey, as in eq. (1.95), the Lie algebra of $SU(2)$. In terms of an arbitrary representation, labelled j, this Lie algebra is

$$[t_a{}^{(j)}, t_b{}^{(j)}] = \mathrm{i}\epsilon_{abc}t_c{}^{(j)}, \tag{5.2}$$

where the $t_m{}^{(j)}$ are the generators in this particular representation. In addition to eq. (5.2), we shall also need the normalization of the generators $t_m{}^{(j)}$. This is fixed by normalizing the trace of the product of generators:

$$\mathrm{tr}\,\{t_a{}^{(j)}t_b{}^{(j)}\} = T(j)\delta_{ab}. \tag{5.3}$$

We will take $T(j) = \frac{1}{2}$, for all j. The number j labels an irreducible representation of $SU(2)$, and is related to the generators $t_a{}^{(j)}$ in that representation by

$$\sum_{a=1}^{3}[t_a{}^{(j)}]^2 = j(j+1)I^{(j)}, \tag{5.4}$$

where $I^{(j)}$ is the identity matrix in the representation. Any such combination of generators, which commutes with every $t_a{}^{(j)}$, is known as a *Casimir operator*. In the language of the rotation group, j is the total angular momentum in representation j. All irreducible representations of the rotation group are found in this way. j is either half-integer or integer. For now, we are interested in the basic *spinor* representation, $j=\frac{1}{2}$, for which $\mathbf{t}_a{}^{(1/2)}=\frac{1}{2}\boldsymbol{\sigma}_a$, where $\boldsymbol{\sigma}_a$ are the Pauli matrices, eq. (1.74).

A rotational spinor $\eta^a=(\eta^1,\eta^2)$, is defined as a two-component object that transforms according to the rule

$$\bar{\eta}^b = h(R)^b{}_a\eta^a, \quad h(R)=\exp\left(-\mathrm{i}\boldsymbol{\theta}_R\cdot\tfrac{1}{2}\boldsymbol{\sigma}\right), \tag{5.5}$$

where $\boldsymbol{\theta}_R$ is the same vector as in eq. (5.1). Our use of upper and lower indices, which is analogous to the notation for Lorentz vectors, will be explained presently.

We can reconstruct the rotation matrix R (in four-dimensional notation) from $h(R)$ by the rule (exercise 5.1),

$$R^\mu{}_\nu = [\exp(-\mathrm{i}\boldsymbol{\theta}_R\cdot\mathbf{J})]^\mu{}_\nu = \tfrac{1}{2}\operatorname{tr}\{\sigma_\mu h(R)\sigma_\nu h^\dagger(R)\}, \tag{5.6}$$

where we define for future reference the vector $\sigma_\mu\equiv(\sigma_0\equiv I_{2\times2},\,\boldsymbol{\sigma})$, and where $R^0{}_\alpha=R^\alpha{}_0=\delta_{\alpha0}$. Note the placement of indices on the right-hand side of eq. (5.6). Both $h(R)$ and $-h(R)$ give the same $R^\mu{}_\nu$, so that the spinor representation is *double-valued*.

Before generalizing to relativistic transformations, we note that eq. (5.5) is not the only way to represent spinorial transformations. Equally good is $\bar{\xi}^b=h^*(R)^b{}_a\xi^a$, with $[h(R)]^*=\exp(\frac{1}{2}\mathrm{i}\boldsymbol{\theta}_R\cdot\boldsymbol{\sigma}^*)$ the complex conjugate of $h(R)$. This choice, however, is equivalent to eq. (5.5) up to a unitary transformation, since

$$h^*(R)=\sigma_2 h(R)\sigma_2, \tag{5.7}$$

and $(\sigma_2)^\dagger=\sigma_2{}^{-1}=\sigma_2$. As far as rotations are concerned, spinors transforming according to $h^*(R)$ are not independent of those transforming according to $h(R)$.

Lorentz spinors and SL(2, C)

Let us see how spinors behave under Lorentz transformations. The matrices $\frac{1}{2}\sigma_i$ in eq. (5.6) already satisfy the same $SU(2)$ Lie algebra as the J_i in eq. (1.95), which is a part of the full algebra of the Lorentz group. The remainder of the algebra in eq. (1.95) may be realized in terms of 2×2 matrices by the identification $K_i\to-\mathrm{i}\sigma_i$. We thus write for the generators in this representation,

$$K_i{}^{(1/2,0)} = -\tfrac{1}{2}\mathrm{i}\sigma_i, \quad J_i{}^{(1/2,0)}=\tfrac{1}{2}\sigma_i, \tag{5.8}$$

where the notation $(\frac{1}{2},0)$ labels the representation in a manner that will be discussed below. Equations (5.5) and (5.6) now have the following generalizations from the rotation group to the (restricted) Lorentz group:

$$h(\Lambda)=\exp\left(\tfrac{1}{2}\boldsymbol{\omega}\cdot\boldsymbol{\sigma}\right)\exp\left(-\tfrac{1}{2}\mathrm{i}\boldsymbol{\theta}\cdot\boldsymbol{\sigma}\right), \tag{5.9}$$

and

$$\Lambda^{\mu}{}_{\nu} = [\exp(\mathrm{i}\boldsymbol{\omega}\cdot\mathbf{K})\exp(-\mathrm{i}\boldsymbol{\theta}_R\cdot\mathbf{J})]^{\mu}{}_{\nu} = \tfrac{1}{2}\,\mathrm{tr}\,\{\sigma_\mu h(\Lambda)\sigma_\nu h^\dagger(\Lambda)\}, \tag{5.10}$$

where we have used the parameterization of eq. (1.99) for Λ. $h(\Lambda)$ in eq. (5.9) is an element of the group SL(2, C), the 2×2 complex matrices with unit determinant.

The $h(\Lambda)$ give a double-valued representation of the Lorentz group, but in general

$$h^*(\Lambda) \neq \sigma_2 h(\Lambda)\sigma_2. \tag{5.11}$$

In fact, $h^*(\Lambda)$ is not usually unitarily equivalent to $h(\Lambda)$, and we must admit two kinds of Lorentz spinors:

$$\bar{\eta}^b = h(\Lambda)^b{}_a\eta^a,\quad \bar{\xi}^{\dot b} = h^*(\Lambda)^{\dot b}{}_{\dot a}\xi^{\dot a}, \tag{5.12}$$

where the dots on the indices of the second transformation are a conventional notation for spinors that transform according to $h^*(\Lambda)$. They have no other significance. Spinors with undotted and dotted indices both have angular momentum $j=\frac{1}{2}$ from the point of view of spatial rotations, but they are distinguished by their behavior under boosts. They are known collectively as *Weyl spinors*.

The reader should keep in mind that there are almost as many types of notation for Weyl spinors as there are books that discuss them. The notation here is closest to Corson (r1981) and Bogoliubov, Logunov & Todorov (r1975). Other notations include, for instance, those of Jost (r1965) and Streater & Wightman (r1964), where the roles of upper and lower indices are reversed compared with those defined above.

Scalars and vectors

To use Weyl spinors, we must relate them to Lorentz scalars and vectors.

We work by analogy with our treatment of covariant and contravariant vector indices in Section 1.5 (this motivated our placement of indices in eq. (5.12)). In this spirit, we introduce matrices ϵ_{ab} and $\epsilon_{\dot a\dot b}$, which act on 'contravariant' spinors η^a and $\xi^{\dot b}$ as the metric $g_{\mu\nu}$ did on contravariant vectors. That is, we lower the indices of Weyl spinors to get the (equivalent) representations that transform according to the inverse of the transformations in eq. (5.12). Similarly, we introduce inverse matrices, denoted ϵ^{ab} and $\epsilon^{\dot a\dot b}$ (compare eq. (1.82)), to raise indices. The matrices that accomplish this are

$$\epsilon_{ab} = \begin{pmatrix} 0 & 1 \\ -1 & 0 \end{pmatrix} = \epsilon_{\dot a\dot b},\quad \epsilon^{ab} = \epsilon^{\dot a\dot b} = -\epsilon_{ab} = (\epsilon^{\mathrm{t}})_{ab}. \tag{5.13}$$

The covariant spinors

$$\eta_a = \epsilon_{ab}\eta^b, \quad \xi_{\dot{a}} = \epsilon_{\dot{a}\dot{b}}\xi^{\dot{b}}, \tag{5.14}$$

then transform according to

$$\begin{aligned} \bar{\eta}_a &= \{[h(\Lambda)^{-1}]^{\mathrm{t}}\}_a{}^b\eta_b = \eta_b[h^{-1}(\Lambda)]^b{}_a, \\ \bar{\xi}_{\dot{a}} &= \{[h^*(\Lambda)^{-1}]^{\mathrm{t}}\}_{\dot{a}}{}^{\dot{b}}\xi_{\dot{b}} = \xi_{\dot{b}}[h^*(\Lambda)^{-1}]^{\dot{b}}{}_{\dot{a}}, \end{aligned} \tag{5.15}$$

where we use the same conventional rearrangement of indices to denote the transpose of $h^{-1}(\Lambda)$ as we used for $\Lambda^\mu{}_\nu$ in eq. (1.83) (we keep the explicit '-1', however). These results follow trivially from the matrix identity

$$\epsilon\sigma_i\epsilon^{\mathrm{t}} = -\sigma_i{}^{\mathrm{t}}. \tag{5.16}$$

From eqs. (5.12) and (5.15) we find that sums over one lower and one upper index, both dotted or both undotted, are invariant under Lorentz transformations. Such quantities include (exercise 5.2)

$$(\eta')^a\eta_a, \quad (\xi^{\dot{a}})^*\eta_a, \tag{5.17}$$

and so on. To show this, we note that the complex conjugate of the spinor $\xi^{\dot{a}}$ transforms according to $\{[h^*(\Lambda)^{-1}]^{\mathrm{t}}\}^* = [h(\Lambda)^{-1}]^{\mathrm{t}}$.

Another useful fact, due to the hermiticity of the σ's, is that the matrices $[h^*(\Lambda)^{-1}]^{\mathrm{t}}$, which multiply the $\xi_{\dot{b}}$ from the left in eq. (5.15), are very close in form to $h(\Lambda)$, eq. (5.9),

$$[h^*(\Lambda)^{-1}]^{\mathrm{t}} = \exp(-\tfrac{1}{2}\boldsymbol{\omega}\cdot\boldsymbol{\sigma})\exp(-\tfrac{1}{2}\mathrm{i}\boldsymbol{\theta}\cdot\boldsymbol{\sigma}). \tag{5.18}$$

The only difference is the sign in the boost exponential. Evidently, for covariant dotted spinors, we have, by analogy with eq. (5.8), the representations

$$K_i{}^{(0,1/2)} = \tfrac{1}{2}\mathrm{i}\sigma_i, \quad \mathrm{J}_i{}^{(0,1/2)} = \tfrac{1}{2}\sigma_i, \tag{5.19}$$

where again the superscript labels the representation.

Vectors

We now discuss vectors in the language of spinors. The way to proceed is suggested by eq. (5.10), the relation between $SL(2, C)$ and the Lorentz group. For a contravariant vector v^μ, consider the matrix

$$(\underline{v})^{a\dot{b}} \equiv (v^\mu\sigma_\mu)^{a\dot{b}} = v^0(\sigma_0)^{a\dot{b}} + \mathbf{v}\cdot(\boldsymbol{\sigma})^{a\dot{b}}. \tag{5.20}$$

We define the behavior of $\underline{v}$ under Lorentz transformation Λ to be consistent with our choice of indices,

$$\bar{\underline{v}}^{a\dot{b}} = h(\Lambda)^a{}_c h^*(\Lambda)^{\dot{b}}{}_{\dot{d}}\underline{v}^{c\dot{d}} = [h(\Lambda)\underline{v}h^\dagger(\Lambda)]^{a\dot{b}}, \tag{5.21}$$

where, in the second step, we have used the definition of the hermitian conjugate. Then, using eq. (5.10), $\underline{v}^{a\dot{b}}$ transformations as

$$\bar{\underline{v}}^{a\dot{b}} = (\Lambda \underline{v})^{a\dot{b}}, \tag{5.22}$$

which is just what we want. As above, the dot on the second index has no special meaning, except as a reminder of how we have defined the matrix to transform. Note that the matrix $\boldsymbol{\sigma}_i$ will appear as $(\boldsymbol{\sigma}_i)^{a\dot{b}}$ when it represents a vector, as in eq. (5.20), as $(\boldsymbol{\sigma}_i)^a{}_b$ when it is part of the expansion of the transformation matrix $h(\Lambda)^a{}_b$ in eq. (5.9), and so on. It is the same Pauli matrix $\boldsymbol{\sigma}_i$ of eq. (1.74) in all cases. The indices of the matrix representing a vector or a transformation are raised and lowered in the usual way. Of particular interest is

$$(\underline{v})_{a\dot{b}} = \epsilon_{ac}\epsilon_{\dot{b}\dot{d}}\underline{v}^{c\dot{d}} = v^0\,(\sigma_0)_{a\dot{b}} - \mathbf{v}\cdot(\boldsymbol{\sigma}^{\mathrm{t}})_{a\dot{b}}, \tag{5.23}$$

where we have used eq. (5.16). Note the explicit transpose in the final term.

General representations

Equations (5.9) and (5.12) can be used to construct all the matrix representations of the Lorentz group, both single- and multiple-valued. Here we shall need only those we have already found, but we may note that, just as the representations of $SU(2)$ give all finite-dimensional representations of the rotation group, so the representations of $SL(2, C)$ give those of the Lorentz group.

The Lie algebra of $SL(2, C)$ is determined by the six generators in eq. (5.9) identified with the J's and K's through eq. (5.8). It is thus the same as the Lorentz group algebra, eq. (1.95). This algebra may be simplified by defining a new basis

$$M_i = \tfrac{1}{2}(J_i + \mathrm{i}K_i), \quad N_i = \tfrac{1}{2}(J_i - \mathrm{i}K_i), \tag{5.24}$$

in terms of which it becomes

$$[M_i, M_j] = \mathrm{i}\epsilon_{ijk}M_k, \quad [N_i, N_j] = \mathrm{i}\epsilon_{ijk}N_k, \quad [M_i, N_j] = 0. \tag{5.25}$$

That is, the Lie algebra of $SL(2, C)$ is $SU(2) \times SU(2)$, the direct product of two independent $SU(2)$ algebras, one for the M's and one for the N's. This result has strong consequences for the representations of $SL(2, C)$.

Since the M's and N's commute, any representation of $SL(2, C)$ is a product of two representations, one generated by the M's and the other by the N's. But the irreducible representations in this product are precisely representations of $SU(2)$, characterized by the (half) integer j of eq. (5.4). Thus, a general representation of $SL(2, C)$ is identified by a pair (j_1, j_2), corresponding to the behavior of the representation under the two independent $SU(2)$ subalgebras consisting of the M's and N's. From this point of view, we have from eqs. (5.8) and (5.19),

$$M_i^{(1/2,0)} = \tfrac{1}{2}\sigma_i, \quad N_i^{(1/2,0)} = 0; \quad M_i^{(0,1/2)} = 0, \quad N_i^{(0,1/2)} = \tfrac{1}{2}\sigma_i. \qquad (5.26)$$

Dotted and undotted spinors are labelled by $(\frac{1}{2}, 0)$ and $(0, \frac{1}{2})$ respectively (see the notation in eqs. (5.8) and (5.19)). More general representations may be found from tensors $T^{a_1 \cdots a_n, \dot{b}_1 \cdots \dot{b}_n}$, which transform according to products of the matrices $h(\Lambda)$ and $h^*(\Lambda)$, in analogy to eq. (1.87) for vector indices.

5.2 Spinor equations and Lagrangians

We are now ready to construct equations of motion and Lagrangians for the free fields that transform according to the simplest representation of $SL(2, C)$, the single-index spinors. Our considerations will lead us to Dirac's celebrated equation.

Since a spinor u_a has two components, two equations are required to describe its motion. The spinors should obey a Klein–Gordon equation, so that they may be related to particles. The very simplest possibility is

$$(\partial_\mu \partial^\mu + m^2) u_a = 0. \qquad (5.27)$$

These two equations, however, are not really restrictive enough, since they do not relate the two components of u_a. These should be related if the solutions are to describe a single particle. We shall therefore look for a matrix equation.

Weyl equation

The next equation does not suffer from this drawback, although its solutions do not admit a mass. It is known as the Weyl equation (Weyl 1929),

$$(\underline{\partial})_{a\dot{b}} u^a = (\sigma_0 \partial^0 + \boldsymbol{\sigma} \cdot \boldsymbol{\nabla})_{\dot{b}a} u^a(x) = 0, \qquad (5.28)$$

where we have used eq. (5.23) (note the reverse order of indices on the right) and $\partial^\mu = (\partial^0, -\boldsymbol{\nabla})$. If we multiply the Weyl equation by $(\underline{\partial})^{c\dot{b}} = (\sigma_0 \partial^0 - \boldsymbol{\sigma} \cdot \boldsymbol{\nabla})^{c\dot{b}}$, we obtain

$$(\partial_\mu \partial^\mu) u^c(x) = 0, \qquad (5.29)$$

so that both components of u^a obey the wave equation. This is perfectly acceptable, although not yet general enough to admit the Klein–Gordon equation. The important thing is that in eq. (5.28) the two components are linked; we have achieved this with a first-order equation. First-order equations of motion are characteristic of Lorentz spinors.

We may derive the Weyl equation from the Lagrange density

$$\mathcal{L} = (u^*)^{\dot{b}} (\underline{\partial})_{a\dot{b}} u^a, \tag{5.30}$$

(exercise 5.5) which is a Lorentz scalar. Note that the order of indices is natural, given the transpose in the definition of $(\underline{\partial})_{a\dot{b}}$, eq. (5.28).

Dirac equation and four-component spinors

The generalization of the Weyl equation to include mass is known as the Dirac equation (Dirac 1928). To construct a spinor equation that includes a mass parameter, it is necessary to include both dotted and undotted spinors, which we choose to denote η^a and $\xi_{\dot{b}}$, for a, $b = 1, 2$. Expressed in the form closest to the Weyl equation, the coupled equations are

$$\begin{aligned} \mathrm{i}(\underline{\partial})_{c\dot{b}} \eta^c(x) - m\xi_{\dot{b}}(x) &= 0, \\ \mathrm{i}(\underline{\partial})^{a\dot{d}} \xi_{\dot{d}}(x) - m\eta^a(x) &= 0. \end{aligned} \tag{5.31}$$

It is usually convenient to think of this as a single equation in which a four-by-four matrix acts on a four-component spinor. Using the matrix definitions, eqs. (5.20) and (5.23), eq. (5.31) becomes

$$\begin{pmatrix} -m\delta_{\dot{b}\dot{d}} & \mathrm{i}(\sigma_0\partial^0 + \boldsymbol{\sigma}\cdot\boldsymbol{\nabla})_{\dot{b}c} \\ \mathrm{i}(\sigma_0\partial^0 - \boldsymbol{\sigma}\cdot\boldsymbol{\nabla})^{a\dot{d}} & -m\delta_{ac} \end{pmatrix} \begin{pmatrix} \xi_{\dot{d}}(x) \\ \eta^c(x) \end{pmatrix} = 0. \tag{5.32}$$

Equation (5.32) is customarily streamlined into the elegant form in which the Dirac equation is usually presented:

$$(\mathrm{i}\partial_\mu \gamma^\mu - m)\psi(x) = 0. \tag{5.33}$$

The four-component spinor $\psi(x)$, where

$$\psi(x) \equiv \begin{pmatrix} \xi_{\dot{a}}(x) \\ \eta^c(x) \end{pmatrix}, \tag{5.34}$$

is the *Dirac field*, while the matrices γ^μ are generally referred to as *Dirac matrices*, here in the *Weyl* or *chiral representation*, which is given in terms of Pauli matrices as

$$\gamma^i = \begin{pmatrix} 0 & \sigma_i \\ -\sigma_i & 0 \end{pmatrix}, \quad \gamma^0 = \begin{pmatrix} 0 & \sigma_0 \\ \sigma_0 & 0 \end{pmatrix}. \tag{5.35}$$

In the same notation, it is also straightforward to derive the Dirac equation from the Lagrange density

$$\mathcal{L} = \bar{\psi}(\mathrm{i}\partial_\mu\gamma^\mu - m)\psi, \tag{5.36}$$

where we define

$$\bar{\psi} \equiv \psi^\dagger \gamma_0 = (\psi^*)^{\mathrm{t}} \gamma_0. \tag{5.37}$$

Note in particular that the product

$$\bar{\psi}\psi = (\xi_{\dot{a}})^* \eta^a + (\eta^a)^* \xi_{\dot{a}} \tag{5.38}$$

is a Lorentz scalar. Equation (5.36) is the standard form for the free Dirac Lagrange density, which we shall encounter again and again. In the next chapter we shall discuss solutions to the Dirac equation, and their physical significance. The following introduces the formalism that has been developed to discuss their transformation properties, which we shall need for building interacting-field theories with fermions.

Dirac matrices

Actually, the Dirac equation is much more general than the Weyl representation eq. (5.35). Any unitary transformation

$$\gamma^\mu \to U\gamma^\mu U^{-1}, \quad \psi \to U\psi, \tag{5.39}$$

gives a completely equivalent Dirac equation of the same form as eq. (5.33). Another common representation of the Dirac matrices is found by applying the transformation

$$U = 2^{-1/2}\begin{pmatrix} \sigma_0 & \sigma_0 \\ -\sigma_0 & \sigma_0 \end{pmatrix}, \tag{5.40}$$

where again, σ_0 is the 2×2 identity. This gives

$$\gamma^0 = \begin{pmatrix} \sigma_0 & 0 \\ 0 & -\sigma_0 \end{pmatrix}, \quad \gamma^i = \begin{pmatrix} 0 & \sigma_i \\ -\sigma_i & 0 \end{pmatrix}. \tag{5.41}$$

This is the *Dirac*, or *Pauli representation*, where

$$\psi = 2^{-1/2}\begin{pmatrix} \eta^a + \xi_{\dot{a}} \\ \eta^b - \xi_{\dot{b}} \end{pmatrix}, \tag{5.42}$$

in which lower dotted indices are mixed with upper undotted indices.

The defining property of Dirac matrices, valid in any representation, is that they anticommute:

$$\{\gamma^\mu, \gamma^\nu\} = \gamma^\mu\gamma^\nu + \gamma^\nu\gamma^\mu = 2g^{\mu\nu} I_{4\times 4}. \tag{5.43}$$

Equation (5.43) is easily verified in the Weyl or Dirac representations, using the multiplication properties of the Pauli matrices, $\boldsymbol{\sigma}_i\boldsymbol{\sigma}_j = i\epsilon_{ijk}\boldsymbol{\sigma}_k + \delta_{ij}I$. We will have occasion to use it a good deal. In addition, we note another representation-independent result,

$$\gamma^0\gamma^\mu\gamma^0 = (\gamma^\mu)^\dagger, \tag{5.44}$$

according to which γ^0 is hermitian, while the spatial γ's are antihermitian.

Transformations and $\sigma_{\mu\nu}$

The transformation properties of ψ are most easily studied in the Weyl representation. Given the transformation rules of eqs. (5.12) and (5.15) for the two-component spinors that make up the four-component Dirac spinor in eq. (5.34), we have

$$\bar{\psi} = S(\Lambda)\psi,$$
$$S(\Lambda) = \begin{pmatrix} [h^*(\Lambda)^{-1}]^{\mathrm{t}} & 0 \\ 0 & h(\Lambda) \end{pmatrix} \tag{5.45a}$$

Evidently, the Dirac spinor transforms according to a $(\frac{1}{2}, 0) + (0, \frac{1}{2})$ representation of $SL(2, C)$, where the sum refers to the behavior of the different spinor components of ψ. By eqs. (5.9) and (5.18), which give the two-by-two transformation matrices, an infinitesimal transformation can be written as

$$S(1 + \delta\Lambda) = \begin{pmatrix} 1 + \frac{1}{2}(-\delta\boldsymbol{\omega} - \mathrm{i}\delta\boldsymbol{\theta})\cdot\boldsymbol{\sigma} & 0 \\ 0 & 1 + \frac{1}{2}(\delta\boldsymbol{\omega} - \mathrm{i}\delta\boldsymbol{\theta})\cdot\boldsymbol{\sigma} \end{pmatrix}$$
$$= 1 - \tfrac{1}{4}\mathrm{i}\delta\lambda^{\mu\nu}\sigma_{\mu\nu}, \tag{5.45b}$$

where we define

$$\sigma_{\mu\nu} \equiv \tfrac{1}{2}\mathrm{i}[\gamma_\mu, \gamma_\nu], \tag{5.46}$$

and the parameters $\delta\lambda^{\mu\nu}$ are related to $\delta\boldsymbol{\omega}$ and $\delta\boldsymbol{\theta}$ by eq. (1.94). The $\sigma_{\mu\nu}$ form of $S(\Lambda)$ is independent of the representation chosen for the Dirac matrices. Comparing eq. (5.45b) and the parameterization for infinitesimal Lorentz transformations, eq. (1.103), we have

$$(\Sigma_{\mu\nu})_{\alpha\beta} = -\tfrac{1}{2}(\sigma_{\mu\nu})_{\alpha\beta} \tag{5.47}$$

for the Dirac field.

Form invariance

The $\sigma_{\mu\nu}$ combine with Dirac matrices in commutators according to

$$[\sigma_{\mu\nu}, \gamma_\rho] = 2\mathrm{i}(g_{\nu\rho}\gamma_\mu - g_{\mu\rho}\gamma_\nu). \tag{5.48}$$

From this result and eq. (5.44) we find the following relations between $S(\Lambda)$ and the Dirac matrices:

$$S^\dagger(\Lambda)\gamma_0 = \gamma_0 S^{-1}(\Lambda), \quad S^{-1}(\Lambda)\gamma^\mu S(\Lambda) = \Lambda^\mu{}_\nu\gamma^\nu. \tag{5.49}$$

For infinitestimal transformations, the latter result uses eqs. (1.96) and (1.97). The extension to finite transformations is straightforward, using the exponentiation formula, eq. (1.70). From eq. (5.49) we may prove the form invariance of the Dirac Lagrangian under infinitesimal Lorentz transformations. Let $\bar{\Lambda} \equiv \Lambda^{-1}$. Then we have, rewriting the Dirac density (5.36) in terms of transformed fields and coordinates,

$$\begin{aligned}
&\bar{\psi}(x)[\mathrm{i}(\partial/\partial x^\mu)\gamma^\mu - m]\psi(x) \\
&\qquad = [S(\bar{\Lambda})\psi'(x')]^\dagger \gamma_0 [\mathrm{i}\bar{\Lambda}_\mu{}^\rho(\partial/\partial x'^\rho)\gamma^\mu - m]S(\bar{\Lambda})\psi'(x') \\
&\qquad = \bar{\psi}'(x')S^{-1}(\bar{\Lambda})[\mathrm{i}\bar{\Lambda}_\mu{}^\rho(\partial/\partial x'^\rho)\gamma^\mu - m]S(\bar{\Lambda})\psi'(x') \\
&\qquad = \bar{\psi}'(x')[\mathrm{i}(\partial/\partial x'^\mu)\gamma^\mu - m]\psi'(x'). \qquad (5.50)
\end{aligned}$$

Form invariance under translations is trivial. Therefore, according to Noether's theorem, the Dirac theory possesses conserved momenta and also conserved angular momenta. We put off discussing their explicit forms until the next chapter.

Finite transformations and parity

For any finite element of the restricted Lorentz group, decomposed into a rotation and a boost as in eq. (1.99), we can use the exponentiation formula eq. (1.70) to find an explicit form for finite transformations:

$$S(\Lambda) = \exp\left(\tfrac{1}{2}\mathrm{i}\omega_i\sigma_{0i}\right)\exp\left[-\tfrac{1}{4}\mathrm{i}\theta_i\epsilon_{ijk}\sigma_{jk}\right]. \qquad (5.51)$$

The parity transformation for four-component spinors may be determined from its representation-independent action on the J_i and K_i, eq. (1.100). The natural solution is

$$\mathrm{P}^\alpha{}_\beta = (\gamma_0)^\alpha{}_\beta, \quad \psi'(\bar{x}) = \gamma_0\psi(x), \qquad (5.52)$$

where $\bar{x}^\mu = (x^0, -\mathbf{x})$ and ψ' is the transformed field. The action of parity on a Dirac spinor is best seen in the Weyl representation, eq. (5.34) and (5.35). The parity transformation interchanges undotted and dotted spinors. In fact, this result is already obvious from eq. (1.100) applied to the generators of eq. (5.24); P exchanges the M's and N's. Thus the Weyl Lagrange density (5.30) is not form invariant under parity transformations. The symmetry between the roles of dotted and undotted spinors in the Dirac Lagrange density eq. (5.36), however, ensures that the latter is invariant (exercise 5.6).

γ_5 and projection

The matrix γ_5, defined in any representation as

$$\gamma_5 \equiv \mathrm{i}\gamma^0\gamma^1\gamma^2\gamma^3 = (\mathrm{i}/4!)\epsilon_{\mu\nu\lambda\sigma}\gamma^\mu\gamma^\nu\gamma^\lambda\gamma^\sigma \qquad (5.53)$$

is of special interest (sometimes it is denoted as γ^5 to emphasize that it is the product of matrices with contravariant indices). Its explicit form is representation-dependent,

$$\gamma_5 = \begin{pmatrix} -\sigma_0 & 0 \\ 0 & \sigma_0 \end{pmatrix} \text{(Weyl)}, \quad \gamma_5 = \begin{pmatrix} 0 & \sigma_0 \\ \sigma_0 & 0 \end{pmatrix} \text{(Dirac)}. \tag{5.54}$$

From γ_5 we can construct operators that project out states with simple transformation properties. For instance, in the Weyl representation,

$$\tfrac{1}{2}(I - \gamma_5)\psi = \begin{pmatrix} \xi_{\dot{a}} \\ 0 \end{pmatrix}, \quad \tfrac{1}{2}(I + \gamma_5)\psi = \begin{pmatrix} 0 \\ \eta^a \end{pmatrix}. \tag{5.55}$$

The matrices $\frac{1}{2}(I + \gamma_5)$ and $\frac{1}{2}(I - \gamma_5)$ are called *chirality* projection operators. We shall discuss the physical significance of chirality, related to intrinsic angular momentum, in the next chapter.

The Dirac algebra

Because there are only four gamma matrices, the anticommutation relations eq. (5.43) imply that products of more than four gamma matrices can always be reduced to products of four or fewer. The complete algebra of such matrices thus contains only sixteen independent elements, one with no gamma matrices ($I_{4\times4}$), four with one gamma matrix ($\{\gamma_\mu\}$), six with two gamma matrices ($\{\sigma_{\mu\nu}\}$), four with three gamma matrices ($\{\gamma_\mu\gamma_5\}$), and one with four gamma matrices (γ_5).

Global symmetries

We close this section by noting that the free Weyl or Dirac Lagrangian, like the free Klein–Gordon Lagrangian, has an extra global $U(n)$ symmetry when it is a sum of n identical terms. In this case, we write the spinor fields with an extra index. With all indices shown, the Dirac Lagrangian, eq. (5.36), becomes

$$\mathscr{L} = \bar{\psi}_{\alpha,i}[\mathrm{i}\partial_\mu(\gamma^\mu)_{\alpha\beta} - m\delta_{\alpha\beta}]\delta_{ij}\psi_{\beta,j}(x). \tag{5.56}$$

Just as in the scalar case, it will turn out to be convenient to distinguish the invariance of eq. (5.56) under $SU(n)$ from its invariance under a change in the overall $U(1)$ phase of the Dirac fields,

$$\psi'_i(x) = [\exp(\mathrm{i}\Sigma_a\beta_a t_a)]_{ij}\psi_j(x), \quad \psi'_i(x) = [\exp(\mathrm{i}\theta)]\psi_i(x), \tag{5.57}$$

where t_a are the generators of $SU(n)$ in its defining representation. The corresponding conserved currents and charges are as follows:

$$U(1),\ j_\mu = \bar{\psi}\gamma_\mu\psi,\quad Q = \int d^3\mathbf{x}\,\bar{\psi}\gamma_0\psi;$$

$$SU(n),\ j_{\mu,a} = \bar{\psi}\gamma_\mu t_a\psi,\quad Q_a = \int d^3\mathbf{x}\,\bar{\psi}\gamma_0 t_a\psi; \tag{5.58}$$

here we have suppressed group indices on the fields. Analogous considerations apply to the Weyl Lagrangian, eq. (5.30).

5.3 Vector fields and Lagrangians

The Maxwell Lagrangian

Before studying and interpreting the solutions of the spinor equations of motion, we turn to the other classical fields that play a central role in high-energy physics, the vector fields. In this section, we shall be concerned with vector fields whose equations are linear, so that their solutions obey the principle of superposition. The vector potential A^μ of classical electrodynamics is the most familiar example. In terms of A^μ, the Maxwell Lagrangian of electrodynamics may be written as

$$\mathscr{L}(x) = -\tfrac{1}{4}F_{\mu\nu}(x)F^{\mu\nu}(x), \tag{5.59}$$

where the antisymmetric tensor field strength $F^{\mu\nu}$ is

$$F^{\mu\nu}(x) = \partial^\mu A^\nu(x) - \partial^\nu A^\mu(x). \tag{5.60}$$

(The sign convention is not universal.) The electric and magnetic fields are identified with components of $F^{\mu\nu}$ by

$$\mathbf{E}_i = -F^{0i}(x) = -\partial^0 A^i - \nabla_i A^0,\quad \mathbf{B}_k = -\tfrac{1}{2}\epsilon_{ijk}F^{ij}(x) = (\boldsymbol{\nabla}\times\mathbf{A})_k, \tag{5.61}$$

where $\nabla_i = \partial/\partial x^i$. The Lagrange equations of motion are

$$\partial_\mu(F^\mu{}_\nu) = \partial_\mu\partial^\mu A_\nu - \partial_\nu(\partial^\mu A_\mu) = 0. \tag{5.62}$$

These are equivalent to two of the source-free Maxwell equations,

$$\boldsymbol{\nabla}\times\mathbf{B} = \partial_0\mathbf{E},\quad \boldsymbol{\nabla}\cdot\mathbf{E} = 0. \tag{5.63}$$

The other two Maxwell equations,

$$\boldsymbol{\nabla}\times\mathbf{E} = -\partial_0\mathbf{B},\quad \boldsymbol{\nabla}\cdot\mathbf{B} = 0, \tag{5.64}$$

are identities that follow from eq. (5.59) for any choice of vector potential $A^\mu(x)$ (exercise 1.2). Hence the Maxwell Lagrangian generates Maxwell's equations and is true to its name. In terms of $\mathbf{E}$ and $\mathbf{B}$ in eq. (5.61), the Maxwell Lagrangian is $\tfrac{1}{2}(\mathbf{E}^2 - \mathbf{B}^2)$. Electromagnetic fields defined with this normalization are said to be measured in Heaviside–Lorentz units (Panofsky & Phillips r1962).

Gauge invariance and gauge conditions

Both the field strength and the electromagnetic Lagrange density are invariant under the *gauge transformation*

$$A^{\mu}(x) \rightarrow A^{\mu}(x) - \partial^{\mu}\alpha(x), \tag{5.65}$$

with $\alpha(x)$ an arbitrary function of dimension $(\text{length})^{-1}$. (It is trivial to check that any vector that is a total divergence identically satisfies the equation of motion.)

It is often useful, especially for the process of quantization, to abandon part or all of the gauge invariance of the Lagrange density eq. (5.59). To do this, we impose extra conditions on A^{μ}. One of the most common and useful of these is the *Lorentz condition* or *Lorentz gauge*

$$\partial^{\mu}A_{\mu}(x) = 0. \tag{5.66}$$

An arbitrary solution $A_{\mu}(x)$ to eq. (5.62) does not generally satisfy eq. (5.66). For any such field, however, we may construct a gauge-equivalent field $A'_{\mu} = A_{\mu} - \partial_{\mu}\alpha$, which satisfies the Lorentz condition – and gives the same field strengths $F^{\mu\nu}$ and action – by solving the equation

$$\partial^{\mu}\partial_{\mu}\alpha(x) = \partial^{\mu}A_{\mu}(x). \tag{5.67}$$

This is just the wave equation with a source term. It always has a solution by standard methods used in classical electrodynamics. We conclude that if we restrict ourselves to classical solutions that satisfy the Lorentz condition we shall not leave out any physically distinct solutions; we shall merely drop a multitude of solutions that are equivalent to the ones we keep.

A useful procedure is to impose the Lorentz condition from the very beginning by modifying the Maxwell density itself. Consider the new density

$$\mathcal{L} = -\tfrac{1}{4}F_{\mu\nu}F^{\mu\nu} - \tfrac{1}{2}\lambda(\partial_{\mu}A^{\mu})^2, \tag{5.68}$$

where λ is an arbitrary constant. Now, instead of eq. (5.62), the equations of motion become

$$(\partial_{\mu}\partial^{\mu})A_{\nu} - (1-\lambda)\partial_{\nu}(\partial_{\mu}A^{\mu}) = 0. \tag{5.69}$$

On taking the divergence, we find

$$\lambda(\partial^{\nu}\partial_{\nu})(\partial_{\mu}A^{\mu}) = 0. \tag{5.70}$$

If we further require that $\partial_{\mu}A^{\mu} = 0$ everywhere in space at a given time, then it will vanish everywhere at all times, again by standard analysis of the wave equation. Therefore, if we assume that the Lorentz condition is satisfied very far in the past, for instance, it will be satisfied for all times. Suppose we assume this to be the case, and take eq. (5.68) as our Lagrange density. Then, the equation of motion for each component of the A^μ field is just the wave equation

$$(\partial_\nu \partial^\nu) A^\mu = 0. \tag{5.71}$$

Other choices, leading to other conditions on the A^μ field, are possible. Equation (5.68) has the advantage of retaining form invariance under Lorentz transformations. Other choices, which do not, are permitted, however, and may be useful, depending on the situation. An example is the *axial gauge*,

$$\mathcal{L} = -\tfrac{1}{4} F_{\mu\nu} F^{\mu\nu} - \tfrac{1}{2}\lambda'(n \cdot A)^2, \tag{5.72}$$

where λ' is a constant with units of inverse length squared, and n^μ is a fixed vector with, for instance, $n^2 = \pm 1$. The analysis that connects eq. (5.72) to the condition $A_0 = 0$ is suggested as exercise 5.7.

Plane waves

With an eventual particle interpretation in mind, we are particularly interested in plane-wave solutions to the relevant equations of motion:

$$A^\mu(x) = a^\mu \exp(-\mathrm{i}k_0 x_0 + \mathrm{i}\mathbf{k} \cdot \mathbf{x}), \tag{5.73}$$

where a^μ is a constant vector, sometimes referred to as the polarization. Since the waves are lightlike,

$$k_0 = \omega(\mathbf{k}) = |\mathbf{k}|, \tag{5.74}$$

is their frequency (in units of inverse length, of course). To analyze these solutions, the Lorentz gauge is particularly appropriate. For, substituting (5.73) into the Lorentz condition eq. (5.66), we find a constraint on the form of the solution:

$$a^\mu k_\mu = 0. \tag{5.75}$$

This means that once k^μ is specified, a^μ has three, rather than four, independent components. Since $k_\mu k^\mu = 0$, one of these may be taken as proportional to k^μ, while the other two make up a two-dimensional vector, denoted ϵ^μ, where

$$a^\mu = \alpha k^\mu + \epsilon^\mu. \tag{5.76}$$

By adjusting α, the vector ϵ^μ may be taken to satisfy

$$\epsilon^0 = 0, \quad \boldsymbol{\epsilon} \cdot \mathbf{k} = 0. \tag{5.77}$$

The term αk^μ is often referred to as the *scalar polarization*, and ϵ^μ as the *transverse polarization* of the solution. Recalling the gauge invariance of the field, we at once recognize that we may remove the scalar polarization by yet another gauge transformation, this time with the choice

$$\alpha(x) = \mathrm{i}\alpha \exp(-\mathrm{i}k^\mu x_\mu). \tag{5.78}$$

We conclude from this discussion that for a plane-wave solution to (5.63), only two of the four components of the polarization are physical. We shall try to interpret this slightly puzzling (if familiar) result in the next chapter.

Massive vector field

It is natural to generalize to a vector field with mass. This can be done in a straightforward way by

$$\mathcal{L} = -\tfrac{1}{4}F_{\mu\nu}F^{\mu\nu} + m^2 A^\mu A_\mu. \tag{5.79}$$

The associated equation of motion, known as *Proca's equation* (Proca 1936), is

$$\partial_\mu\partial^\mu A_\nu - \partial_\nu(\partial^\mu A_\mu) + m^2 A_\nu = 0. \tag{5.80}$$

Taking the divergence of this equation of motion gives the Lorentz condition directly so that, for the massive vector field, eq. (5.80) is equivalent to a Klein–Gordon equation for each of the components,

$$\partial_\mu\partial^\mu A_\nu + m^2 A_\nu = 0. \tag{5.81}$$

The Lorentz condition reduces the number of independent components of the field to three from four. Unlike the massless case, however, we cannot reduce them further, because the mass term in the Lagrangian destroys the gauge invariance of eq. (5.65) present in the Maxwell Lagrangian.

Just as for scalars and spinors, vector fields may acquire an additional index indicating some internal symmetry. In realistic theories, these are generalizations of the gauge symmetry, eq. (5.65), and are associated with interaction terms in the Lagrangian. We now discuss such generalizations.

5.4 Interactions and local gauge invariance

So far, we have at our disposal quadratic Lagrangians for scalar, spinor and vector fields. Their equations of motion are linear, and all fields are free in the same sense as the free scalar field. More interesting equations of motion arise when a potential, or interaction, is added to the Lagrangian. The interaction may be of any form as long as it is invariant under the Poincaré group. We shall make no effort at completeness, but rather concentrate on those interactions relevant to our current understanding of experiment.

We have already discussed scalar interactions (Chapter 2). Consider now couplings of vector and scalar to spinor fields. We have already seen, in constructing the free spinor Lagrangians eqs. (5.30) and (5.36), how to make a scalar out of a pair of spinor fields and a vector. The most important examples of spinor–vector couplings are of this type, but now with a vector field, A^μ, taking the place of the vector derivative $(\partial/\partial x_\mu)$. So, for a Weyl spinor we have, by analogy with eq. (5.30),

$$(u^*)^{\dot{b}}(\underline{A})_{a\dot{b}}u^a, \tag{5.82a}$$

and for a Dirac spinor, by analogy with eq. (5.36),

$$\bar{\psi}A_\mu\gamma^\mu\psi. \tag{5.82b}$$

A variant of eq. (5.82b) is the *axial vector coupling*,

$$\bar{\psi}A_\mu\gamma^\mu\gamma_5\psi, \tag{5.83}$$

with γ_5 given by eq. (5.53).

Scalar particles are coupled to spinors through *Yukawa interactions* (Yukawa 1935)

$$\phi(\bar{\psi}\psi), \quad \pi(\bar{\psi}\gamma_5\psi). \tag{5.84}$$

If we demand that both these terms be invariant under the parity transformation, we must postulate the following transformation rules for the scalar fields,

$$\phi'(\bar{x}) = \phi(x), \quad \pi'(\bar{x}) = -\pi(x), \tag{5.85}$$

with, as in eq. (5.52), $\bar{x}^\mu \equiv (x_0, -\mathbf{x})$. Scalar fields that change sign under parity are referred to as *pseudoscalar*.

Other interesting interactions may involve four-vectors, or four Dirac spinors:

$$(A^\mu A_\mu)^2, \quad [\bar{\psi}\gamma^\mu(a + b\gamma_5)\psi][\bar{\psi}\gamma_\mu(a + b\gamma_5)\psi]. \tag{5.86}$$

The latter is an example of a *Fermi interaction* (Fermi 1934), which describes the low-energy weak interactions. Note that for a and b nonzero, the spinor term in eq. (5.86) is not invariant under the parity transformation in eq. (5.52).

Let us again emphasize that there are many possibilities beyond what we have written down so far. These, however, are the ones we shall need for the discussions that follow.

Local gauge invariance

We now turn to the topic of *local symmetries* in interaction terms. Local symmetries have the remarkable property that *they require the addition of interactions to preserve the symmetry* (Yang & Mills 1954). Local symmetries play a central role – perhaps *the* central role – in modern field theory.

Local symmetry makes its simplest appearance in electromagnetism. We are already familiar with the gauge symmetry of the electromagnetic field, eq. (5.65). This transformation, which leaves the Lagrangian (5.59) invariant, need not be the same at each point in space, since $\alpha(x)$ is an arbitrary function of x^μ. It is thus called a *local transformation*. Using the coupling eq. (5.82b) of a vector particle to a Dirac spinor, we can write down the full Lagrange density for a Dirac spinor particle interacting with the electromagnetic field, the Lagrangian for *quantum electrodynamics* (QED). Neglecting gauge fixing terms, it is

$$\mathscr{L} = -\tfrac{1}{4}F_{\mu\nu}F^{\mu\nu} + \bar{\psi}(\mathrm{i}\partial_\mu\gamma^\mu - m)\psi - e\bar{\psi}A_\mu\gamma^\mu\psi. \tag{5.87}$$

The first two terms are the free Lagrangians for the electromagnetic and spinor fields. The third couples the electromagnetic field to the current $\bar{\psi}\gamma_\mu\psi$ (eq. (5.58)), and has coupling constants e, which we shall identify as the electric charge of the Dirac field. Consider the effect of the gauge transformation (5.65) on (5.87). It gives

$$\mathscr{L} \to \mathscr{L} + e\bar{\psi}(\partial_\mu\alpha)\gamma^\mu\psi. \tag{5.88}$$

This change can be compensated by a change in the phase of the Dirac field,

$$\psi \to \{\exp[\mathrm{i}e\alpha(x)]\}\psi, \quad \bar{\psi} \to \bar{\psi}\{\exp[-\mathrm{i}e\alpha(x)]\}, \tag{5.89}$$

analogous to the global phase transformation eq. (1.58) for the scalar field. A more fundamental way of looking at this is that the phase of the Dirac field is not observable for a system with the Lagrangian given in eq. (5.87). Any local phase variation on the Dirac field can be compensated by an appropriate local gauge transformation on the electromagnetic field. Of course, this does not work without the interaction.

It is customary, in writing eq. (5.87), to define the combination

$$\mathrm{D}_\mu(A) = \partial_\mu + \mathrm{i}eA_\mu \tag{5.90}$$

as the *covariant derivative* with respect to the gauge transformation. The terminology, inspired by general relativity, refers to the transformation of $\mathrm{D}_\mu(A)\psi$ under local gauge transformations. Unlike $\partial_\mu\psi$, $D_\mu(A)\psi$ transform in the same way as ψ, eq. (5.89). In terms of the covariant derivative, the QED Lagrangian takes on the commonly used forms

$$\begin{aligned}\mathscr{L} &= -\tfrac{1}{4}F_{\mu\nu}F^{\mu\nu} + \bar{\psi}(\mathrm{i}\mathrm{D}_\mu(A)\gamma^\mu - m)\psi \\ &= -\tfrac{1}{4}F_{\mu\nu}F^{\mu\nu} + \bar{\psi}[\mathrm{i}(\partial_\mu + \mathrm{i}eA_\mu)\gamma^\mu - m]\psi.\end{aligned} \tag{5.91}$$

Note, by the way, that the covariant derivative was already present for scalars interacting with an external field, in eq. (4.53). A coupling through the covariant derivative is often referred to as *minimal*.

Local nonabelian gauge invariance

Now we generalize the concept of local phase invariance to Lagrange densities such as eq. (5.56), with n identical Dirac spinor fields ψ_i that posses global symmetry under the transformations of $U(n)$,

$$\psi'_i(x) = U_{ij}\psi_j(x), \quad \bar{\psi}'_i(x) = \bar{\psi}(x)_j U^\dagger{}_{ji}. \tag{5.92}$$

(Remember, the matrix U does not act on the indices of Dirac matrices.) The unitary transformation simply relabels the indices of the fields, but the fields all have identical dynamics anyway, and the relabelling has no physical effect. In the free theory, there is simply no way to distiguish between different labellings. So far, however, the fields are labelled in the same way at every point in space–time. We might ask if we have the freedom to label fields differently at different points of space and time. Such relabelling is a local $U(n)$ gauge transformation:

$$\psi'_i(x) = U_{ij}(x)\psi_j(x), \quad \bar{\psi}'_i(x) = \bar{\psi}_j(x)U^\dagger{}_{ji}(x). \tag{5.93}$$

All that has happened is that the matrix U has become a function of position. The Dirac Lagrangian (5.56), of course, is no longer invariant under such a transformation. If we start with the Dirac Lagrangian in terms of primed fields and re-express it in terms of the original fields by using eq. (5.93), we get an extra term, left over from the action of the derivative on $U(x)$:

$$\mathscr{L}_0(\bar{\psi}', \psi') \to \mathscr{L}_0(\bar{\psi}, \psi) + \bar{\psi}_i(x)\gamma_\mu[\mathrm{i}U^{-1}(x)\partial^\mu U(x)]_{ij}\psi_j(x). \tag{5.94}$$

For $n = 1$, eq. (5.94) reduces to the QED case, since a single field $\psi(x)$ transforms under a local transformation $U(1)$ according to a phase $\exp[\mathrm{i}e\alpha(x)]$, and

$$\bar{\psi}[\mathrm{i}e^{-\mathrm{i}e\alpha(x)}\partial^\mu e^{\mathrm{i}e\alpha(x)}]\gamma_\mu\psi = -e\bar{\psi}[\partial^\mu\alpha(x)]\gamma_\mu\psi. \tag{5.95}$$

For $n > 1$, the change in the Lagrangian is cancelled if we introduce a *nonabelian gauge* or *Yang–Mills field* (Yang & Mills 1954), denoted $(A^\mu)_{ij}$. $(A^\mu)_{ij}$ is a matrix in 'group' indices i, j and we define a nonabelian covariant derivative through

$$\begin{aligned}\mathscr{L}_1 &= \bar{\psi}_i\{\mathrm{i}[\mathrm{D}^\mu(A)]_{ij}\gamma_\mu - m\delta_{ij}\}\psi_j \\ &\equiv \bar{\psi}_i\{\mathrm{i}[\partial^\mu\delta_{ij} + \mathrm{i}g(A^\mu)_{ij}]\gamma_\mu - m\delta_{ij}\}\psi_j.\end{aligned} \tag{5.96}$$

$(A^\mu)_{ij}$ transforms according to the (inhomogeneous) rule

$$[A^\mu(x)]'_{ij} = [U(x)A^\mu U^{-1}(x)]_{ij} + (i/g)[\partial^\mu U(x)U^{-1}(x)]_{ij}. \tag{5.97}$$

We easily verify that this is a direct generalization of the transformation law eq. (5.65) by taking $U(x)$ to be a phase. Equation (5.96) is a generalization of minimal coupling, and $(A^\mu)_{ij}$ is referred to as the gauge field. The term 'nonabelian' refers to the underlying symmetry group, and distinguishes the cases $n > 1$ from the Maxwell (= abelian = $U(1)$) case $n = 1$.

We recall again that every group $U(n)$ can be broken down into $SU(n)$ and $U(1)$ subgroups, and we usually treat the general matrix $U(x)$ in eq. (5.97) as an element of $SU(n)$. Evidently, with a minimally coupled nonabelian field, definitions of the fermion field that differ by local $SU(n)$ transformations are not physically distinct. We should note that we need not choose $U(x)$ to be an arbitrary element of $SU(n)$. We can decree that $U(x)$ is, in any of the n-dimensional representations of any group G, large enough to have such a representation. In this case, the basic local invariance group of the Lagrangian can be arbitrary.

Infinitesimal transformations

We can find out more about the structure of the matrix gauge field $(A^\mu)_{ij}$ by looking at an infinitesimal gauge transformation. Suppose that we have chosen representation R of the group G. Then an infinitesimal transformation is of the general form of eq. (1.70) with x-dependent parameters:

$$U_{ij}(x) = \delta_{ij} + \mathrm{i}g \sum_{a=1}^{d(G)} \delta\Lambda_a(x)[t_a{}^{(R)}]_{ij} \equiv [I + \mathrm{i}g\delta\Lambda(x)]_{ij}. \tag{5.98}$$

Here, the $t_a{}^{(R)}$ are (hermitian) generators of the representation R of the group G, $d(G)$ is the dimension of G and eq. (5.98) serves as a definition of the matrix $\delta\Lambda_{ij}(x)$. We factor out an explicit g to be consistent with the convention for e in eq. (5.89) for the $U(1)$ case. The infinitesimal transformation rule for $(A^\mu)_{ij}$ is easily found by using eq. (5.98) in (5.97). It is

$$A'^\mu = A^\mu - \partial^\mu\delta\Lambda(x) + \mathrm{i}g[\delta\Lambda(x), A^\mu]. \tag{5.99}$$

For this expression to be self-consistent, the matrix A^μ should be an element of the Lie algebra, that is, a linear combination of the generators in representation R, just like $\delta\Lambda$. Thus, we have

$$[A^\mu(x)]_{ij} = \sum_{a=1}^{d(G)} A^\mu{}_a(x)[t_a{}^{(R)}]_{ij}, \tag{5.100}$$

where each component $A^\mu{}_a(x)$ is a space–time vector. The number of gauge fields is thus the dimension of the group, independent of the spinor representation R. Substituting eq. (5.100) back into (5.99), we find the transformation laws for the component fields $A^\mu{}_a$, in terms of the group structure constants defined by the Lie algebra eq. (1.71),

$$A'^\mu{}_a = A^\mu{}_a - \partial^\mu\delta\Lambda_a - gC_{abc}\delta\Lambda_b A^\mu{}_c. \tag{5.101}$$

For the special unitary groups, we may take $C_{ab}{}^c = C_{abc}$, to be antisymmetric in all three indices (exercise 5.9). In (5.101) and the following, we therefore do not distinguish between upper and lower group indices. Equations (5.99) and (5.101) give the transformation rules for the gauge field respectively treated as a matrix or expanded into components. They are equivalent, but eq. (5.99) generally gives a somewhat more compact notation.

To complete the Lagrange density $\mathcal{L}_1$ of eq. (5.96), we should supply a kinetic term for the nonabelian gauge field, a generalization of the Maxwell density, eq. (5.59). Again, gauge invariance under the new transformation, eq. (5.97), requires the addition of new interaction terms. The nonabelian gauge field Lagrangian is constructed from a new field-strength tensor that is a generalization of the abelian case, eq. (5.60). In matrix notation, we write

$$F_{\mu\nu}(x) = \partial_\mu A_\nu - \partial_\nu A_\mu + \mathrm{i}g[A_\mu, A_\nu]. \tag{5.102}$$

It can also be written in terms of group components using eq. (5.100):

$$F_{\mu\nu} = F_{\mu\nu,a}(x)t_a{}^{(R)} = \{\partial_\mu A_{\nu,a} - \partial_\nu A_{\mu,a} - gC_{abc}A_{\mu,b}A_{\nu,c}\}t_a{}^{(R)}. \tag{5.103}$$

Following common usage, we have not introduced a new symbol for the nonabelian field strength, even though it is a matrix

The gauge invariant Lagrangian for the vector field is

$$\begin{aligned} \mathcal{L} &= -\tfrac{1}{4}\sum_a F_{\mu\nu,a}F^{\mu\nu}{}_a = -[4T(R)]^{-1}\operatorname{tr}(F_{\mu\nu}F^{\mu\nu}), \\ &\operatorname{tr}[t_a{}^{(R)}t_b{}^{(R)}] = T(R)\delta_{ab}, \end{aligned} \tag{5.104}$$

where $T(R)$ is the generalization of the normalization factor $T(j)$ for $SU(2)$, eq. (5.3), to an arbitrary representation R of group G. This is possible in any semisimple Lie algebra. By convention, the factor in front of the Lagrangian in terms of components is always $\frac{1}{4}$; if it is written out in terms of the matrix field strength, the factor depends on the representation. To show that eq. (5.104) is indeed gauge invariant, it is convenient to use the behavior of the matrix $F_{\mu\nu}$ under an infinitesimal local (or global) transformation:

$$F'^{\mu\nu} = F_{\mu\nu} + \mathrm{i}g[\delta\alpha(x), F_{\mu\nu}]. \tag{5.105}$$

The invariance of $\mathcal{L}$ then follows the cyclic property of the trace:

$$\begin{aligned} \mathcal{L}(F') &= -[4T(R)]^{-1}\operatorname{tr}\{F_{\mu\nu}F^{\mu\nu} + \mathrm{i}g[\delta\alpha(x), F_{\mu\nu}]F^{\mu\nu} \\ &\quad + F_{\mu\nu}\mathrm{i}g[\delta\alpha(x), F^{\mu\nu}]\} + \mathrm{O}(\delta\alpha^2) \\ &= -[4T(R)]^{-1}\operatorname{tr}(F_{\mu\nu}F^{\mu\nu}) + \mathrm{O}(\delta\alpha^2). \end{aligned} \tag{5.106}$$

Once again, the requirement of gauge invariance has led to new interactions: in eq. (5.104) there are cubic and quartic interactions in the gauge field.

The reasoning for a scalar field with symmetry is analogous, and leads to the Lagrangian

$$\mathscr{L} = -\tfrac{1}{4}F_{\mu\nu,a}F^{\mu\nu}{}_a + [\mathrm{D}_\mu(A)\phi^*]_i[\mathrm{D}^\mu(A)\phi]_i - M^2\phi_i{}^*\phi_i, \quad (5.107)$$

where $D_\mu(A)$ is the matrix covariant derivative. This is a straightforward extension, which makes the vector field dynamical, of eq. (4.53).

Spontaneous symmetry breaking and masses

So far, each of our locally invariant Lagrangians has involved only massless vector particles, and indeed, adding $m^2A^\mu A_\mu$ as in eq. (5.79) spoils local invariance. This apparent obstacle was overcome through the realization that *spontaneous symmetry breaking* makes possible Lagrangians that have at once local gauge invariance and massive vector particles. Generically, spontaneous symmmetry breaking refers to any situation in which a system has a set of degenerate ground states, related by continuous symmetry transformations. Ideas of this sort originated in solid-state physics (Anderson 1962), in the theory of superconductivity. They were adapted to particle theory in Nambu & Jona-Lasinio (1961). The standard model for elementary interactions (Appendix C) is based on a Lagrangian with spontaneous symmetry breaking.

Actually, the underlying symmetry is not really broken, but is obscured by the nature of the ground state (Coleman r1985). In the standard model, the symmetry relates particles as different as the photon, which is massless, and the intermediate vector bosons, which are as massive as a molecule of benzene.

In the following we will show how the *Higgs mechanism* (Higgs 1964, 1966, Englert & Brout 1964, Guralnik, Hagen & Kibble 1964) for spontaneous symmetry breaking manifests itself in field theory by producing massive vector particles in a locally gauge invariant Lagrangian. Our discussion will be completely classical. The extension to quantum fields is crucial and nontrivial, but leaves the basic picture unchanged as far as masses are concerned.

Higgs mechanism in $SU(2) \times U(1)$

In the Higgs mechanism, we postulate a scalar field with internal symmetry $U(1)$, or greater, coupled to a gauge field, but with a very special kind of scalar self-interaction. This field is called a *Higgs scalar*. A typical choice is a doublet of complex scalar fields ϕ_i, $i = 1, 2$. The corresponding free scalar Lagrangian $(\partial^\mu \phi_i)^*(\partial_\mu \phi_i) - m^2 \phi_i^* \phi_i$ has a $U(2)$ invariance, that is, $SU(2)$ supplemented by a residual phase, or $U(1)$ invariance. These two subgroups are, as indicated above, treated differently. The scalar field is coupled to a triplet of $SU(2)$ gauge fields $B^\mu{}_a$, $a = 1, 2, 3$, which are invariant under $U(1)$, and a $U(1)$ gauge vector C^μ, which is a singlet under $SU(2)$. This is indicated by the notation

$$\mathrm{D}^\mu{}_{ij}(B, C) = \delta_{ij}(\partial^\mu + \tfrac{1}{2}\mathrm{i}g' C^\mu) + \mathrm{i}g B^\mu{}_a(\tfrac{1}{2}\sigma_a)_{ij}, \qquad (5.108)$$

where g and g' are the (different) couplings associated with the $SU(2)$ and $U(1)$ fields, respectively.

To incorporate local gauge invariance, we postulate the Lagrange density

$$\begin{aligned}\mathscr{L} = &-\tfrac{1}{4}F_{\mu\nu,a}(B)F^{\mu\nu}{}_a(B) - \tfrac{1}{4}F_{\mu\nu}(C)F^{\mu\nu}(C) \\ &+ [\mathrm{i}\mathrm{D}_{\mu,ij}(B, C)\phi_j]^*[\mathrm{i}\mathrm{D}^\mu{}_{ik}(B, C)\phi_k] - V(\phi_i^*\phi_i). \end{aligned} \qquad (5.109)$$

The local $SU(2) \times U(1)$ symmetry of this density is the cornerstone of the standard model for electroweak interactions (Glashow 1961, Salam 1968, Weinberg 1967). The potential term $V(\phi^*\phi)$ is taken to have both quadratic and quartic terms in the form

$$V(\phi_i^*\phi_i) = -\mu^2\phi_i^*\phi_i + \lambda(\phi_i^*\phi_i)^2, \qquad (5.110)$$

with $\mu^2 > 0$. Note the unconventional choice of sign for the mass term; this is what makes things happen. As a result of this 'negative mass squared', the classical potential no longer has a minimum at $\phi_i = 0$, but rather at any nonzero value of ϕ_i for which

$$\phi_i^*\phi_i = \mu^2/2\lambda \equiv v^2/2, \qquad (5.111)$$

where the second form is a definition of the parameter v. The unique minimum of the classical action at $\phi_i = 0$ for positive mass has been replaced by a whole subspace of minimum action defined by eq.(5.111). If we are interested in solutions close to the absolute minimum (anticipating a quantum perturbation theory), it is natural to reparameterize the field in such a way that small values of the new field variables occur near the minimum. A reparameterization that accomplishes this is of the form of a local gauge transformation. We replace the real and imaginary parts of ϕ_i, $i = 1, 2$, by four real fields: $\eta(x)$ and $\zeta_i(x)$, $i = 1, 2, 3$. This is accomplished by the change of variables

$$\boldsymbol{\phi} = \exp\{\mathrm{i}\boldsymbol{\zeta}(x)\cdot\boldsymbol{\sigma}/2v\}\begin{pmatrix} 0 \\ [v + \eta(x)]/2^{1/2} \end{pmatrix}. \qquad (5.112)$$

The fields $\boldsymbol{\zeta}(x)$ parameterize the position of $\boldsymbol{\phi}$ around the three-dimensional minimum, while $\eta(x)$ measures the distance from the minimum. The transcendental nature of the ζ part of the reparameterization might seem threatening, but by substituting eq. (5.112) into the Lagrange density eq. (5.109), we find that the components of $\zeta(x)$ act as three massless scalar fields (exercise 5.12). The occurrence of massless scalar fields is a quite general feature of spontaneously broken symmetries, a result known as *Goldstone's theorem* (Goldstone 1961, Goldstone, Salam & Weinberg 1962; see Appendix E below). In our case, it can be understood by the observation that changes in $\zeta(x)$ rotate the field around equipotential lines of V. As such, at lowest (quadratic) order in the fields, they modify the action only through the kinetic part ($|\partial_\mu \phi|^2$), which is sensitive to changes in the field, rather than to its absolute value.

The theory in terms of the new fields is drastically simplified by cancelling the nonabelian phase in eq, (5.112) by an $SU(2)$ gauge transformation

$$\phi' = U\phi, \quad B'_\mu = UB_\mu U^{-1} + (\mathrm{i}/g)(\partial_\mu U)U^{-1}, \qquad U(x) = \exp[-\mathrm{i}\zeta(x)\cdot\boldsymbol{\sigma}/2v], \tag{5.113}$$

which acts on both the scalar and vector fields. Here the gauge parameters are determined by the fields themselves. This parameterization of the density (5.109) is known as the *unitary gauge*.

After the field-dependent gauge transformation, eq. (5.113), the Lagrange density is actually independent of the fields $\zeta_i(x)$. The vector part of the density remains unchanged, but we must replace $\phi_i(x)$ everywhere by the $SU(2)$ vector $(0, [v + \eta(x)]/2v)$. The scalar part of the Lagrange density is then given by

$$\mathcal{L}_{\text{Higgs}} = \tfrac{1}{8}[g^2(B_1{}^\mu B_{1\mu} + B_2{}^\mu B_{2\mu}) + (g'C_\mu - gB_{\mu,3})^2](v+\eta)^2 + \tfrac{1}{2}(\partial^\mu\eta)(\partial_\mu\eta) - \mu^2\eta^2 - v\lambda\eta^3 - \tfrac{1}{4}\lambda\eta^4 + D, \tag{5.114}$$

where D is a constant, which we drop below. In this gauge, we have a new Lagrange density for a single, massive (mass $= \mu\surd 2$), self-coupled, real scalar field $\eta(x)$, which also couples to the vector fields. We also see terms of the form (vector field)$^2 v^2$, with v the *constant* scalar field defined in eq. (5.111). These are the mass terms we have been looking for. The structure of the transformed Lagrangian is elucidated by making another change of variables, this time linear in the vector fields:

$$\begin{aligned} W^{\mu\pm} &= (B^\mu{}_1 \mp \mathrm{i}B^\mu{}_2)/2^{1/2}, \\ Z^\mu{}_0 &= (-C^\mu \sin\theta_\mathrm{W} + B^\mu{}_3 \cos\theta_\mathrm{W}), \\ A^\mu &= (C^\mu \cos\theta_\mathrm{W} + B^\mu{}_3 \sin\theta_\mathrm{W}), \end{aligned} \tag{5.115}$$

where we define the *weak mixing angle* θ_W in terms of the couplings, by

$$\sin\theta_\mathrm{W} = g'/(g'^2 + g^2)^{1/2}. \tag{5.116}$$

In these new variables, the scalar particle density becomes

$$\begin{aligned}\mathcal{L}_{\text{Higgs}} &= \tfrac{1}{2}M_W{}^2\{|W^+|^2 + |W^-|^2\} + \tfrac{1}{2}M_Z{}^2 Z_0{}^2 \\ &\quad + \tfrac{1}{8}[g^2(|W^+|^2 + |W^-|^2) + (g'^2 + g^2)Z_0{}^2](2v\eta + \eta^2) \\ &\quad + \tfrac{1}{2}[(\partial^\mu\eta)(\partial_\mu\eta) - 2\mu^2\eta^2] - v\lambda\eta^3 - \tfrac{1}{4}\lambda\eta^4, \qquad (5.117)\end{aligned}$$

where explicit mass terms emerge for the vector fields $W^\pm$ and Z^0, while the field A remains massless. The masses are related to the scalar vacuum field v and the underlying couplings by

$$\begin{aligned}M_W &= (vg/2) \\ M_Z &= (v/2)(g'^2 + g^2)^{1/2} = M_W/\cos\theta_W.\end{aligned} \qquad (5.118)$$

In the standard model, the massive fields $W^\pm$ and Z_0 are identified with the observed massive vector bosons of electroweak theory, and the massless field A is identified with the photon. The photon, in turn, couples to the fields $W^\pm$ through electric charges $\pm e$ (exercise 5.13), given by

$$e = gg'/(g'^2 + g^2)^{1/2} = g\sin\theta_W. \qquad (5.119)$$

Let us emphasize that the Higgs phenomenon is more general than $SU(2) \times U(1)$. A general analysis of Higgs-generated spontaneous symmetry breaking in the unitary gauge has been given by Weinberg (1973a), and is summarized, for example, in Huang (r1982).

Gauge conditions for nonabelian fields

The Lagrange density eq. (5.104) retains its full gauge invariance. In the abelian case, however, we saw that abandoning all or part of the gauge invariance at the level of the Lagrangian eliminates a vast overcounting of physically equivalent classical fields. The same reasoning applies to nonabelian theories as well, and plays a crucial role in their quantization (Chapter 7). For the spontaneously broken theory, the unitary gauge is one such choice. It is natural, however, to mention some other common nonabelian gauge conditions.

For a pure, massless Yang–Mills field, eq. (5.104), it is often convenient to modify the Lagrangian by direct analogy with the abelian case. In place of eqs. (5.68) and (5.72), therefore, we take

$$\mathcal{L} = -\tfrac{1}{4}(F_{\mu\nu,a}F^{\mu\nu}{}_a) - \tfrac{1}{2}\lambda\mathcal{F}_a\mathcal{F}_a, \qquad (5.120)$$

and define nonabelian covariant and axial gauges, respectively, by

$$\mathcal{F}_a = (\partial_\mu A^\mu{}_a), \quad \mathcal{F}_a = (n_\mu A^\mu{}_a). \qquad (5.121)$$

Applied to the $W_\pm$ and Z bosons above, these same conditions give quadratic Lagrange densities that are direct generalizations of the Proca density eq. (5.79). For a Lorentz condition (Lee & Yang 1962), we have

$$\mathcal{L}_{\text{quad}} = \sum_{V=W_+,W_-,Z} [-\tfrac{1}{4}(\partial_\mu V_\nu - \partial_\nu V_\mu)^2 - M_V{}^2 V_\mu V^\mu - \tfrac{1}{2}\lambda(\partial_\mu V^\mu)^2], \quad (5.122)$$

with the masses M_V given by eq. (5.118). The limit $\lambda \to 0$ brings us back to the Proca form.

In a theory with the Higgs phenomenon, more elaborate choices are often convenient, especially for purposes of quantization. Of particular importance are the *'t Hooft* or R_ξ *gauges*, described briefly in Appendix C. Suffice it to say that in these gauges, the *quadratic* $W_\pm$ and Z boson Lagrangians are the same as in eq. (5.122).

Exercises

5.1 Verify that the Pauli matrices satisfy the completeness relation

$$\sum_{\mu=0}^{3} (\sigma_\mu)_{ab}(\sigma_\mu)_{cd} = 2\delta_{ad}\delta_{bc}$$

Use this result to prove that $\Lambda(h)^\mu{}_\nu$ as defined by eq. (5.10) satisfies the group property, $\Lambda(h_1)^\mu{}_\nu \Lambda(h_2)^\nu{}_\sigma = \Lambda(h_1 h_2)^\mu{}_\sigma$.

5.2 (a) Verify that the quantities in eq. (5.17) are indeed Lorentz scalars, and identify the remaining scalars that can be made from dotted and undotted Weyl spinors in this way. (b) Verify the following amusing feature of the spinor algebra, which follows from the antisymmetry of the 'metric' ϵ_{ab}, eq. (5.13):

$$(\eta')_a \eta^a = -(\eta')^a \eta_a,$$

where η' and η are two-component spinors. This shows that $\eta_a \eta^a = 0$.

5.3 (a) Construct a matrix representation for the infinitesimal generators of rotations for a tensor with two vector indices, $T_{\mu\nu}$. (b) Using (a), calculate the sum of the squares of the generators, as in eq. (5.4). (c) Suppose that $T_{\mu\nu}$ is symmetric and traceless, that is, $T_{\mu\nu} = T_{\nu\mu}$ and $\sum_\mu T_{\mu\mu} = 0$. Show that, on such a matrix, the sum of the squares of the generators acts as the unit (Casimir) operator with $j = 2$. In this way, we see that a traceless symmetric second-rank tensor has spin two. (d) Extend this result to arbitrary rank, and show that a kth rank, fully symmetric tensor $T_{\mu_1 \cdots \mu_k}$, traceless in all pairs of indices, i.e.

$$g^{\mu_i \mu_j} T_{\mu_1 \cdots \mu_i \cdots \mu_j \cdots \mu_k} = 0$$

has spin k.

5.4 Repeat the analysis of exercise 3 for the 'Rarita–Schwinger' field $\psi_{\mu,\sigma}$, where μ is a vector index and σ is a four-component spinor index; $\psi_{\mu,\sigma}$ transforms according to eqs. (5.45) and (1.104) and satisfies the condition $\gamma^\mu \psi_{\mu,\sigma} = 0$. Show that this field has $j = \frac{3}{2}$ (Rarita & Schwinger 1941).

5.5 Show explicitly, by considering the transformation properties of $(\underset{\smile}{\partial})_{a\dot{b}}u^a$, that the Weyl Lagrangian, eq. (5.30), is form invariant under the action of the restricted Lorentz group.

5.6 (a) Verify the transformation properties of spinors in the Weyl representation, eqs. (5.45). (b) Show that the Dirac Lagrangian is form invariant under the parity transformation eq. (5.52).

5.7 Choose $n^\mu = (1, \mathbf{0})$ in the modified vector Lagrange density eq. (5.72), and use the resulting equation of motion to relate the extra term to the condition $A_0 = 0$ in the specific Lorentz frame being considered.

5.8 Vector field propagators may be derived from a generalization of eq. (2.85) for the scalar field propagator. Let $O^{(x)}{}_{\alpha\beta}A^\beta(x) = 0$ be the equation of motion for the vector field A^β. We define the propagator by the relation

$$O^{(x)}{}_{\alpha\beta}G^{\beta\gamma}(x-y) = -g_\alpha{}^\gamma\delta^4(x-y).$$

As in the scalar case, it is easiest to determine $G^{\beta\gamma}$ in momentum space. (a) In this way, show that the momentum-space propagators for the modified massless Lagrange densities (5.68) and (5.72) are given by

$$G^{\beta\gamma}(k, \lambda) = \frac{1}{k^2}\left[-g^{\beta\gamma} + \left(1 - \frac{1}{\lambda}\right)\frac{k^\beta k^\gamma}{k^2}\right],$$

$$G^{\beta\gamma}(k, n, \lambda) = \frac{1}{k^2}\left[-g^{\beta\gamma} + \frac{n^\beta k^\gamma + k^\beta n^\gamma}{n\cdot k} - n^2\left(1 + \frac{k^2}{\lambda n^2}\right)\frac{k^\beta k^\gamma}{(n\cdot k)^2}\right],$$

respectively. (b) Show that for the unmodified Maxwell Lagrangian eq. (5.59) there is *no* solution to the propagator equation above. This is equivalent to the observation that solutions to the unmodified equations of motion are not unique. (c) For the quadratic density (5.122) for vector bosons in the standard model, show that the propagator is given by

$$G^{\beta\gamma}(k, M_V{}^2, \lambda) = \frac{1}{k^2 - M_V{}^2}\left[-g^{\beta\gamma} + \left(1 - \frac{1}{\lambda}\right)\frac{k^\beta k^\gamma}{k^2 - \dfrac{M_V{}^2}{\lambda}}\right].$$

5.9 The orthogonality property eq. (5.3) for the traces of generators may be extended from $SU(2)$ to any semisimple Lie algebra (Section 1.4):

$$\mathrm{tr}\,[t_a{}^{(R)}, t_b{}^{(R)}] = T(R)\delta_{ab},$$

where $T(R)$ is some number that depends only on the representation (see any text on group theory). Use this result and the commutation relations of the algebra to show that, independently of R,

$$C_{ab}{}^c = -C_{ac}{}^b,$$

so that, as claimed after eq. (5.101), the structure constants are completely antisymmetric.

5.10 The tensor $^*F_{\alpha\beta}$ *dual* to the field strengths $F^{\mu\nu}$, eq. (5.60) or (5.103), is defined by

$$^*F_{\alpha\beta} = \tfrac{1}{2}\epsilon_{\alpha\beta\mu\nu}F^{\mu\nu}.$$

(a) Show that the quantity $\mathcal{L}' \equiv \mathrm{tr}\,({}^*F_{\sigma\lambda}F^{\sigma\lambda})$ is invariant under the infinitesimal nonabelian gauge transformation eq. (5.99). (b) Show in addition that $\mathcal{L}' = \partial_\mu K^\mu$, with

$$K_\mu = \epsilon_{\mu\nu\rho\sigma}\,\mathrm{tr}\,\{A^\nu F^{\rho\sigma} - \tfrac{2}{3}igA^\nu[A^\rho, A^\sigma]\},$$

that is, that $\mathcal{L}'$ is a total derivative. Hence, adding an $\mathcal{L}'$ term to the nonabelian gauge theory Lagrange density does not lead to new equations of motion (see exercise 1.3).

5.11 Show that the change in $F^{\mu\nu}$ under the finite transformation of eq. (5.97) is given by $F'^{\mu\nu} = UF^{\mu\nu}U^{-1}$.

5.12 Verify by expanding the Lagrangian of eq. (5.109) that the fields $\zeta(x)$ introduced in eq. (5.112) are massless.

5.13 Verify by studying the field strengths $F_{\mu\nu,b}(B)$ that the fields $W^\pm$ couple to $A(x)$ via the charge $e = gg'/(g^2 + g'^2)^{1/2}$.

6
Spin and canonical quantization

Internal angular momentum, or spin, characterizes unitary representations of the Poincaré group, which may be constructed in terms of quantum mechanical states, or in terms of solutions to the classical equations of motion. The latter are useful in organizing the canonical quantization of spinor and vector fields.

6.1 Spin and the Poincaré group

Relativistic invariance

We discussed the requirements of relativistic invariance for a quantum field theory in Section 2.2. We need a unitary operator $U(a, \Lambda)$ that transforms states according to $|\bar{\psi}\rangle = U(a, \Lambda)|\psi\rangle$, for arbitrary elements (a, Λ) of the Poincaré group. The same operator should transform fields according to $U(a, \Lambda)\phi_a(x)U(a, \Lambda)^{-1} = S_{ab}(\Lambda)^{-1}\phi_b(\Lambda x + a)$.

$U(a, \Lambda)$ is given quite generally for infinitesimal transformations by eq. (2.18), in terms of the momentum and angular momentum operators P_μ and $J_{\lambda\sigma}$, which obey the Poincaré algebra, eq. (2.17). By constructing P_μ and $J_{\lambda\sigma}$ for the classical field theories introduced in the previous chapter, we are able to show relativistic invariance for the corresponding quantum field theories in the same manner as for the scalar field. This requires that the relevant Lagrange densities be form invariant under the Poincaré group, as verified in eq. (5.50) and exercise 5.5 for the spin one-half fields. The form invariance of the complete QED and nonabelian gauge theory Lagrange densities is also easily shown.

Conserved currents

Classical conservation for the momentum and angular momentum of fields with spin can be derived in exactly the same manner as for scalar fields. The only difference is in the value of $\delta_*\phi_a$, eq. (1.38b), which may be computed using the parameterization of infinitesimal transformations in eqs. (1.106)–(1.108). The scalar field $\delta_* \phi$ results solely from the change in the argument of the field, while for fields with spin there is also mixing of field components:

$$\delta_* \phi_a(x) = \bar{\phi}_a(x) - \phi_a(x) = -(\delta a^\mu g_\mu{}^\nu - \delta\lambda^{\mu\nu} x_\mu)(\partial\phi_a(x)/\partial x^\nu) + \tfrac{1}{2}\mathrm{i}\Sigma_{\mu\nu,ab}\delta\lambda^{\mu\nu}\phi_b(x). \tag{6.1}$$

Substituting this expression into the general conserved current (1.43b), we derive energy–momentum and angular momentum tensors. The energy–momentum tensor $T^\mu{}_\nu$ is given by the same formula, eq. (1.45), as for the scalar field, while the angular momentum tensor becomes

$$M_{\mu\tau\sigma} = x_\tau T_{\mu\sigma} - x_\sigma T_{\mu\tau} + \mathrm{i}[\partial\mathcal{L}/\partial(\partial^\mu\phi_a)](\Sigma_{\tau\sigma})_{ab}\phi_b. \tag{6.2}$$

Relative to the scalar result eq. (1.55), there is a new term proportional to $(\Sigma_{\mu\nu})_{ab}$. As discussed in Section 1.5, each matrix $(\Sigma_{\mu\nu})_{ab}$ is a generator in a representation of the Lorentz group, and satisfies the Lie algebra, eq. (1.98). For vector and Dirac spinor fields, the Σ's are given respectively by eqs. (1.105) and (5.47).

Spin

The quantities $P_\nu = \int \mathrm{d}^3\mathbf{x}\, T_{0\nu}$ and $J_{\nu\lambda} = \int \mathrm{d}^3\mathbf{x}\, M_{0\nu\lambda}$ have the usual interpretations as total momentum and angular momentum, where

$$J_{\nu\lambda} = \int \mathrm{d}^3\mathbf{x}\, M_{0\nu\lambda} = \int \mathrm{d}^3\mathbf{x}\,(x_\nu T_{0\lambda} - x_\lambda T_{0\nu}) + \mathrm{i}\int \mathrm{d}^3\mathbf{x}\,\pi_a(\Sigma_{\nu\lambda})_{ab}\phi_b, \tag{6.3}$$

with π_b the conjugate momentum of field ϕ_b, defined as in eq. (1.50). The new contribution describes the intrinsic angular momentum, or spin, carried by the field independently of its mechanical momentum. A measure of the intrinsic angular momentum of a field is given by the *Pauli–Lubański vector* (Lubański 1942),

$$W_\mu = -\tfrac{1}{2}\epsilon_{\mu\nu\lambda\sigma}J^{\nu\lambda}P^\sigma = -\tfrac{1}{2}\mathrm{i}\epsilon_{\mu\nu\lambda\sigma}\int \mathrm{d}^3\mathbf{x}\,\pi_a\Sigma^{\nu\lambda}{}_{ab}\phi_b P^\sigma. \tag{6.4}$$

Note that in the rest frame of the momentum vector, where $P^\mu = M\delta_{\mu 0}$, W_μ reduces to a three-vector, while in all frames $W_\mu P^\mu = 0$.

6.2 Unitary representations of the Poincaré group

The operator versions of P^μ and $J_{\mu\nu}$ generate unitary representations of the Poincaré group. This, of course, presupposes that their quantized versions obey the Poincaré algebra, eq. (2.17). We shall assume this to be the case, and treat the P^μ and $J_{\mu\nu}$ as hermitian operators acting on the space of states.

Casimir operators

To label irreducible representations of the Poincaré group, we use its Casimir operators. The Poincaré group has two of these, $P^2 = P^\mu P_\mu$ and $W^2 = W^\mu W_\mu$. This result follows from the Poincaré algebra eq. (1.109) alone, so it is more general than any particular representation of the algebra. Irreducible representations thus have definite invariant lengths for momentum and intrinsic angular momentum.

Method of Wigner

Wigner (1939) developed a method by which explicit representations of the Poincaré group can be constructed. Let us illustrate the method with quantum states.

To begin, consider an eigenstate $|p, \lambda\rangle$ of P^2, W^2 and P^μ, where $P^\mu|p, \lambda\rangle = p^\mu|p, \lambda\rangle$. The remaining quantum numbers are labelled by λ. The idea is to fix p^2 and then to generate basis states for all values of p^μ in terms of those for a single fixed vector. Suppose $p^2 = m^2$ (we do not require that m^2 be positive), and denote the special fixed vector by q^μ, $q^2 = m^2$. Any vector p^μ with $p^2 = m^2$ can be derived by acting on q^μ with an appropriate Lorentz transformation:

$$p^\mu = \Lambda(p, q)^\mu{}_\nu q^\nu. \tag{6.5}$$

Clearly, the transformation $\Lambda(p, q)$ is not unique, since it can be modified to

$$\hat{\Lambda}(p, q) = \Lambda(p, q)l, \tag{6.6}$$

where l is any Lorentz transformation that leaves q^ν invariant: $l^\mu{}_\nu q^\nu = q^\mu$. This ambiguity is turned into a virtue as follows.

The set of all transformations l that satisfy $lq = q$ is easily seen to be a group, which we shall denote by L. It was named by Wigner the *little group* corresponding to the vector q^μ. We can identify sets of states, with momentum q^μ, which transform according to irreducible representations of the little group:

$$U(l)|q^\mu, \lambda\rangle = \sum_\sigma D_{\lambda\sigma}(l)|q^\mu, \sigma\rangle. \tag{6.7}$$

Finding representations of L is relatively easy. For example, if $m^2 > 0$, we choose $q^\mu = (m, \mathbf{0})$, the vector at rest. Then L is the rotation group, and the states $|q^\mu, \sigma\rangle$ are most conveniently chosen with σ as the projection of the spin along a fixed axis (and, of course, total spin j^2, which turns out to be w^2/m^2). This information alone, along with the existence of $U(a, \Lambda)$, is enough to construct a unitary representation for states of any momentum, as follows.

Consider an arbitrary p^μ, $p^2 = m^2$. We have seen that the choice of Λ that takes q^μ into p^μ is ambiguous. The ambiguity can be removed, however, by making a specific choice of *Wigner boost*, $\Lambda_0(p, q)$, for each p^μ. Examples of choices for the Wigner boosts will be given shortly. We now construct basis states for $p^\mu \neq q^\mu$ by

$$|p^\mu, \lambda\rangle \equiv U(\Lambda_0(p, q))|q^\mu, \lambda\rangle, \tag{6.8}$$

where λ on the left is *defined* to be the same as λ on the right. Let us see how $|p^\mu, \lambda\rangle$ transforms under an arbitrary Lorentz transformation $\Lambda(p', p)$ that takes p^μ into some other vector p'^μ. We start by noting that

$$U(\Lambda(p', p))|p^\mu, \lambda\rangle = U(\Lambda'(p', p, q))|q^\mu, \lambda\rangle, \tag{6.9}$$

where

$$\Lambda'(p', p, q) \equiv \Lambda(p', p)\Lambda_0(p, q). \tag{6.10}$$

On the other hand, we can express $\Lambda'(p', p, q)$ in terms of the Wigner boost between q^μ and p'^μ by the relation

$$\Lambda'(p', p, q) = \Lambda_0(p', q)l(p', p, q), \tag{6.11}$$

where $l(p', p, q)$ is an element of the little group L of q^μ. To prove this, we simply note that $[\Lambda_0(p', q)]^{-1}\Lambda'(p', p, q)$ leaves q^μ invariant. Substituting eq. (6.11) into eq. (6.9) and using eq. (6.7), we find

$$\begin{aligned} U(\Lambda(p', p))|p^\mu, \lambda\rangle &= U(\Lambda_0(p', q))U(l(p', p, q))|q^\mu, \lambda\rangle \\ &= U(\Lambda_0(p', q))\sum_\sigma D_{\lambda\sigma}(l(p', p, q))|q^\mu, \sigma\rangle \\ &= \sum_\sigma D_{\lambda\sigma}(l(p', p, q))|p'^\mu, \sigma\rangle. \end{aligned} \tag{6.12}$$

Equation (6.12) shows that the states $|p^\mu, \lambda\rangle$ specified by eq. (6.8) transform according to a representation of the Poincaré group that is determined by the representation of the little group L. By construction, the $D_{\lambda\sigma}$ are unitary, so that the representation in eq. (6.12) of the Poincaré group is also unitary. This is what we set out to show. Wigner showed that all unitary representations of the Poincaré group can be generated in this way.

To discuss fields, it is convenient to work with solutions to the equations of motion. This was the point of view taken by Wigner in his original formulation (Wigner 1939), since it is closest to the interpretation of solutions to the equations of motion as wave functions (Section 1.2). See Bargmann & Wigner (1946). We will then quantize the coefficients of these solutions, as in eq. (2.12) for the scalar field.

Fields and solutions

To quantize a field, interacting or not, we expand it in terms of solutions to the *free* classical equations of motion. Such solutions, like states, may be characterized by their transformation properties under the Poincaré group. These properties are determined by the classical generators of translations and boosts, given for arbitrary classical fields in eq. (1.108). The classical generators obey the same Poincaré algebra as the quantum ones, eq. (1.109), and indeed, this is how we identified the Poincaré algebra in the first place. By analogy to eq. (6.4), then, we identify a classical Pauli–Lubański generator associated with intrinsic angular momentum:

$$\hat{w}_\mu = -\tfrac{1}{2}\epsilon_{\mu\nu\lambda\sigma}\hat{m}^{\nu\lambda}\hat{p}^\sigma. \tag{6.13}$$

$\hat{p}^2$ and $\hat{w}^2$ are Casimir operators for the representation of the Poincaré group in terms of solutions, corresponding to P^2 and W^2 for quantum states.

Basis solutions may be chosen as eigenfunctions of the generators $\hat{p}^\mu$ and $\hat{w}^2$. Therefore, a typical basis solution is given by

$$\phi_a{}^{(\pm)}(p^2, w^2, p, \lambda, x) = \xi_a{}^{(\pm)}(p^2, w^2, p, \lambda)\exp(\mp \mathrm{i}p^\mu x_\mu), \tag{6.14}$$

where $(\pm)$ refers to positive and negative energy solutions, while p^2, p^μ and w^2 are the eigenvalues of $\hat{p}^2$, $\hat{p}^\mu$ and $\hat{w}^2$, respectively (below, we suppress the arguments p^2 and w^2). We generate solutions for arbitrary p^μ in terms of those at $p^\mu = q^\mu$, with q^μ chosen as above:

$$\xi_a{}^{(\pm)}(p, \lambda) = S_{ab}(\Lambda_0(p, q))\xi_b{}^{(\pm)}(q, \lambda), \tag{6.15}$$

where $\Lambda_0(p, q)$ is the relevant Wigner boost. Equivalently, we can write

$$\phi_a{}^{(\pm)}(p, \lambda, x) = S_{ab}(\Lambda_0(p, q))\phi_b{}^{(\pm)}(q, \lambda, \Lambda_0{}^{-1}x), \tag{6.16}$$

which is consistent with the classical transformation, eq. (1.102). That $\phi_a{}^{(\pm)}(p, \lambda, x)$ is also a solution follows from the form invariance of the Lagrange density, and hence the equations of motion, under the Lorentz group. Again, we construct the $\xi_b{}^{(\pm)}(q, \lambda)$ to transform under the little group according to the unitary matrix $D_{\lambda\sigma}$:

$$S_{ab}(l)\xi_b{}^{(\pm)}(q, \lambda) = D_{\lambda\sigma}(l)\xi_a{}^{(\pm)}(q, \sigma). \tag{6.17}$$

Defined in this way, the solutions transform under the Lorentz group as

$$S_{ab}(\Lambda(p', p))\xi_b{}^{(\pm)}(p, \lambda) = D_{\lambda\sigma}(l(p', p, q))\xi_a{}^{(\pm)}(p', \sigma). \tag{6.18}$$

Note again that the transformation properties of solutions depend on the little-group element l, and hence on the choice of Wigner boost Λ_0. We will use this approach to construct basis solutions for the Dirac and Maxwell equations below.

An arbitrary field ψ is a linear combination of the ϕ's:

$$\psi(x) = \sum_{\lambda,i} \int \frac{d^3\mathbf{p}}{2\omega_p} a_\lambda{}^{(i)}(\mathbf{p}, x^0)\phi^{(i)}(p, \lambda, x), \tag{6.19}$$

where $a_\lambda{}^{(i)}(\mathbf{p}, x^0)$ is an arbitrary function and $i = \pm$. Here we assume that the basis solutions form a complete set in space. In the quantized free-field theory, the $a^{(+)}$ and $a^{(-)}$ become time-independent creation and annihilation operators.

In an interacting-field theory, the a_λ have an arbitrary time dependence, and we may absorb the time dependence of $\phi^{(i)}$ into $a_\lambda^{(i)}$. Equation (2.13) for a scalar field is a special case. To quantize, we shall specify appropriate equal-time commutation (or anticommutation – see Section 6.5) relations for the $a_\lambda{}^{(\pm)}$. From this starting point, we shall construct a path integral formalism to compute Green functions and the S-matrix, assuming a pertubation series exists about the path integral for the free theory. Therefore, we must solve the free theory first. In our case, this means constructing solutions for the free Dirac and Maxwell equations.

Physically interesting representations fall into two classes, massive ($p^2 > 0$) and massless ($p^2 = 0$), and we shall treat both of these cases below. We may also note that the method of Wigner (1939) applies as well to the unphysical cases of $p^2 < 0$ and $p^\mu \equiv 0$.

6.3 Solutions with mass

We have encountered three equations with massive plane-wave solutions: the Klein–Gordon, Dirac and Proca equations. In each of these cases we take $p^2 = m^2 > 0$, for which we choose as the special vector

$$q^\mu = (m, \mathbf{0}) \tag{6.20}$$

whose little group is the rotation group. The other invariant operator, $\hat{W}^2$ has eigenvalues $m^2 j(j+1)$, where j is the spin of the state. Let us consider the three cases in turn.

Klein–Gordon equation

The scalar field transforms trivially under the rotation group, so its solutions are simply plane waves, with no labels beyond p^μ, $p^\mu p_\mu = m^2$. The rest frame solution is, up to a constant,

$$\phi^{(\pm)}(m^2, q, x) = \exp(\mp \mathrm{i} m x_0), \tag{6.21}$$

which corresponds to $\xi = 1$ in eq. (6.14) and is invariant under any choice of Wigner boost. The transformed solutions are thus

$$\phi^{(\pm)}(m^2\, p, x) = \exp(\mp \mathrm{i} p^\mu x_\mu). \tag{6.22}$$

There is only a single irreducible representation here, since any solution can be transformed into any other by a suitable element of the Poincaré group. Even in the scalar case, however, the representation is infinite-dimensional, since there are an infinite number of vectors p^μ with $p^2 = m^2$.

Massive Dirac equation

Once again we start with solutions having $p^\mu = q^\mu = (m, \mathbf{0})$. It is most convenient to work in the Dirac representation for the gamma matrices, eq. (5.41). For a plane wave at rest,

$$\phi_a{}^{(\pm)}(q, \lambda, x) = \xi_a{}^{(\pm)}(q, \lambda) \exp(\mp \mathrm{i} q_0 x_0), \tag{6.23}$$

the Dirac equation (5.33) becomes simply

$$\begin{pmatrix} (\pm q_0 - m)\sigma_0 & 0 \\ 0 & -(\pm q_0 + m)\sigma_0 \end{pmatrix} \xi_a{}^{(\pm)}(q) = 0. \tag{6.24}$$

There are four linearly independent solutions, $u(q, \pm\frac{1}{2})\mathrm{e}^{-\mathrm{i}q\cdot x}$ and $v(q, \pm\frac{1}{2})e^{\mathrm{i}q\cdot x}$, with

$$u(q, \tfrac{1}{2}) = c \begin{pmatrix} 1 \\ 0 \\ 0 \\ 0 \end{pmatrix}, \quad u(q, -\tfrac{1}{2}) = c \begin{pmatrix} 0 \\ 1 \\ 0 \\ 0 \end{pmatrix},$$
$$v(q, -\tfrac{1}{2}) = c \begin{pmatrix} 0 \\ 0 \\ 1 \\ 0 \end{pmatrix}, \quad v(q, \tfrac{1}{2}) = c \begin{pmatrix} 0 \\ 0 \\ 0 \\ 1 \end{pmatrix}; \tag{6.25}$$

c is a normalization. The first two solutions are 'positive energy', and the remaining two 'negative energy'. The arguments $\lambda = \pm\frac{1}{2}$ are the eigenvalues of the spin generator $\hat{j}_3 \equiv -\Sigma_{12} = \frac{1}{2}\sigma_{12}$ (eq. (5.47)) for the u's and the negative of the same eigenvalues for the v's. These choices are clearly consistent with the requirements on the index λ, but the motivation for the signs will only become clear in Section 6.5, when we discuss the particle content of the free-field theory. There we will show that λ is, in fact, the third component of the spin.

Using the notation of eq. (5.37), we define $\bar{w} = w^\dagger \gamma_0$, for any Dirac spinor w. It is natural to introduce $\bar{w}$ because, as we recall from eq. (5.38),

$\bar{w}w'$ is a Lorentz scalar, for any two Dirac spinors w and w'. In these terms, the solutions obey the following elementary relations:

defining equations,

$$\begin{aligned}(\not{q} - m)u(q, \lambda) = \bar{u}(q, \lambda)(\not{q} - m) = 0,\\ (\not{q} + m)v(q, \lambda) = \bar{v}(q, \lambda)(\not{q} + m) = 0;\end{aligned} \tag{6.26a}$$

normalization,

$$\begin{aligned}\bar{u}(q, \lambda)u(q, \lambda') &= |c|^2\delta_{\lambda\lambda'} = u^\dagger(q, \lambda)u(q, \lambda),\\ \bar{v}(q, \lambda)v(q, \lambda') &= -|c|^2\delta_{\lambda\lambda'} = -v^\dagger(q, \lambda)v(q, \lambda'),\\ \bar{u}(q, \lambda)v(q, \lambda') &= \bar{v}(q, \lambda)u(q, \lambda') = 0;\end{aligned} \tag{6.26b}$$

projection operators,

$$\begin{aligned}\sum_\lambda u(q, \lambda)_a\bar{u}(q, \lambda)_b &= |c|^2\tfrac{1}{2}(\gamma_0 + I)_{ab} = (|c|^2/2m)(\not{q} + m)_{ab},\\ \sum_\lambda v(q, \lambda)_a\bar{v}(q, \lambda)_b &= |c|^2\tfrac{1}{2}(\gamma_0 - I)_{ab} = -(|c|^2/2m)(-\not{q} + m)_{ab};\end{aligned} \tag{6.26c}$$

completeness,

$$\sum_\lambda u(q, \lambda)_a\bar{u}(q, \lambda)_b - \sum_\lambda v(q, \lambda)_a\bar{v}(q, \lambda)_b = |c|^2\delta_{ab}. \tag{6.26d}$$

Here, we have introduced the common notation, for any vector a_μ,

$$\not{a} = a_\mu\gamma^\mu. \tag{6.27}$$

Note the minus signs associated with the action of γ_0 on the v's. Standard choices for the normalization of the spinors are $|c|^2 = 2m$ or unity. The conjugate versions of the momentum space spinor equations follow from the Dirac matrix identity eq. (5.44).

Having identified the solutions, we now study their transformations under the little group. The behavior of Dirac spinors under an infinitesimal rotation is given by eq. (5.51) with $\omega_i = 0$:

$$S(R) = \exp(-\tfrac{1}{4}\mathrm{i}\theta_i\epsilon_{ijk}\sigma_{jk}), \tag{6.28}$$

where σ_{jk} is given by eq. (5.46). In both Dirac and Weyl representations

$$S(R) = \begin{pmatrix}\exp(-\tfrac{1}{2}\mathrm{i}\theta_i\sigma_i) & 0\\ 0 & \exp(-\tfrac{1}{2}\mathrm{i}\theta_i\sigma_i)\end{pmatrix}. \tag{6.29}$$

As a result, under rotations the $u(q, \lambda)$ mix with each other but not with the $v(q, \lambda)$ and vice-versa. Thus the u's and v's are separate two-dimensional representations of the rotation group, and the transformation matrix

$D_{\lambda\sigma}(R)$ of eq. (6.18) is the two-by-two matrix

$$D(R) = \exp\left(-\tfrac{1}{2}\mathrm{i}\theta_i\sigma_i\right) \tag{6.30}$$

for the u's, and an analogous matrix for the v's. Having identified the basis solutions for representations of the little group, we can construct basis solutions for the corresponding representation of the full Poincaré group, by applying Wigner boosts.

Spin basis

For each choice of Wigner boost, we get a different choice of basis solutions at arbitrary p^μ. Certainly, the simplest choice is to take $\Lambda_0(p, q)$ to be a pure boost in the direction of $\hat{\mathbf{p}}$, the unit vector in the direction of $\mathbf{p}$ (see eq. (1.99)),

$$\Lambda_0(p, q) = \Lambda_s(p, q) = \exp\left(\mathrm{i}\omega\hat{\mathbf{p}}\cdot\mathbf{K}\right),\ \omega = \tanh^{-1}\left(|\mathbf{p}|/p_0\right), \tag{6.31}$$

with the generator matrices K_i given in eq. (1.91). The corresponding transformation matrix on Dirac spinors is found from eq. (5.51):

$$S(\Lambda_s(p, q)) = \exp\left(-\tfrac{1}{4}\omega\hat{\mathrm{p}}_i[\gamma_0, \gamma_i]\right) = \exp\left(\tfrac{1}{2}\omega\hat{\mathbf{p}}\cdot\boldsymbol{\alpha}\right), \tag{6.32}$$

where we define

$$\alpha_i \equiv \gamma^0\gamma^i. \tag{6.33}$$

Equation (6.32) can be evaluated using $(\hat{\mathbf{p}}\cdot\boldsymbol{\alpha})^2 = 1$, to get

$$\begin{aligned} S(\Lambda_s(p, q)) &= I\cosh\tfrac{1}{2}\omega + \hat{\mathbf{p}}\cdot\boldsymbol{\alpha}\sinh\tfrac{1}{2}\omega \\ &= \left(\frac{p_0+m}{2m}\right)^{1/2}\begin{pmatrix}\sigma_0 & (p_0+m)^{-1}\mathbf{p}\cdot\boldsymbol{\sigma} \\ (p_0+m)^{-1}\mathbf{p}\cdot\boldsymbol{\sigma} & \sigma_0\end{pmatrix}. \end{aligned} \tag{6.34}$$

Thus for the spin basis, the definition, eq. (6.15), for boosted solutions gives

$$w(p,\lambda) = S(\Lambda_s(p, q))w(q, \lambda) = [(2m)(\mathrm{p}_0 + m)]^{-1/2}(\pm\not{p} + m)w(q, \lambda) \tag{6.35}$$

$$\bar{w}(p,\lambda) = \bar{w}(q, \lambda)S^{-1}(\Lambda_s(p, q)) = [(2m)(p_0 + m)]^{-1/2}\bar{w}(q, \lambda)(\pm\not{p} + m)$$

for $w = u$ or v, where we have used eq. (5.49) to commute $S^\dagger$ with γ^0. In the explicit forms, the plus sign is for $w = u$ and the minus sign for $w = v$. From the explicit expressions for the $w(q, \lambda)$ in eq. (6.25), the basis solutions $v(p, \lambda)$ and $u(p, \lambda)$ can be read off as the columns of $S(\Lambda_s)$.

The basis solutions derived in this way are called spin solutions. The reason can be seen by studying the transformation of the solutions $u(p, \lambda)$ under rotations. Suppose $\Lambda(p', p) = R$ is a pure rotation. For the choice

$\Lambda_0 = \Lambda_s$, the boosts for p^μ and p'^μ are related by the same rotation R:

$$\Lambda_s(p', q) = R\Lambda_s(p, q)R^{-1}. \tag{6.36}$$

Then from eqs. (6.10) and (6.11), the corresponding element in the little group L is

$$l(p', p, q) = \Lambda_s(p', q)^{-1}\Lambda'(p', p, q) = R\Lambda_s(p, q)^{-1}R^{-1}R\Lambda_s(p, q) = R. \tag{6.37}$$

Now eq. (6.18) specifies the transformation properties of the boosted solutions, and we find

$$S_{ab}(R)\xi_b{}^{(\pm)}(p, \lambda) = \sum_\sigma D_{\lambda\sigma}(R)\xi_a{}^{(\pm)}(p', \sigma). \tag{6.38}$$

That is, the boosted spin solutions transform under a rotation R in exactly the same way as spin states transform in the rest frame.

It is now easy to generalize the properties of solutions, eq. (6.26), to arbitrary frames, by transforming spinors according to eq. (6.35):

defining equations,

$$(\not{p} - m)u(p, \lambda) = \bar{u}(p, \lambda)(\not{p} - m) = 0,$$
$$(\not{p} + m)v(p, \lambda) = \bar{v}(p, \lambda)(\not{p} + m) = 0;$$

normalization,

$$\begin{aligned}
&\bar{u}(p, \lambda)u(p, \lambda') = |c|^2\delta_{\lambda\lambda'},\ u^\dagger(p, \lambda)u(p, \lambda) = |c|^2(\omega_p/m),\\
&\bar{v}(p, \lambda)v(p, \lambda') = -|c|^2\delta_{\lambda\lambda'},\ v^\dagger(p, \lambda)v(p, \lambda) = |c|^2(\omega_p/m),\\
&\bar{u}(p, \lambda)v(p, \lambda') = \bar{v}(p, \lambda)u(p, \lambda') = 0;
\end{aligned} \tag{6.39a}$$

projection operators,

$$\sum_\lambda u(p, \lambda)_a\bar{u}(p, \lambda)_b = (|c|^2/2m)(\not{p} + m)_{ab},$$

$$\sum_\lambda v(p, \lambda)_a\bar{v}(p, \lambda)_b = -(|c|^2/2m)(-\not{p} + m)_{ab};$$

completeness,

$$\sum_\lambda u(p, \lambda)_a\bar{u}(p, \lambda)_b - \sum_\lambda v(q, \lambda)_a\bar{v}(q, \lambda)_b = |c|^2\delta_{ab}.$$

Other useful conditions follow from the classical parity operation, eq.

(5.52), which, acting on the explicit boosts (6.32), gives

$$u(p, \lambda) = \gamma_0 u(\bar{p}, \lambda), \quad v(p, \lambda) = -\gamma_0 v(\bar{p}, \lambda) \tag{6.39b}$$

with $\bar{p}^\mu \equiv (p_0, -\mathbf{p})$, the parity-reflected momentum. From this we find

$$u^\dagger(p, \lambda) v(\bar{p}, \lambda) = v^\dagger(p, \lambda) u(\bar{p}, \lambda) = 0. \tag{6.39c}$$

The spin basis is not the only choice. Another is the *helicity* basis (Jacob & Wick 1959) in which the Wigner boost is chosen as a boost in a fixed direction, followed by a rotation (exercise 6.1). We should also note that the relative phases of solutions in the spin basis may also be changed (see Appendix D).

Massive vector solutions

The discussion for massive vector solutions is relatively simple, now that we have developed the necessary techniques.

Recall that the Proca equation (5.80) is equivalent to a Klein–Gordon equation (5.81) for each field component, subject to the Lorentz condition $\partial^\mu A_\mu = 0$. In the rest frame this means

$$(-q_0{}^2 + m^2)\epsilon^\mu(q, \lambda) = 0, \quad q^\mu \epsilon_\mu(q, \lambda) = 0, \tag{6.40}$$

for solutions of the form $\epsilon^\mu(q, \lambda) \exp(-iq_0 x_0)$. We again pick a set of basis solutions for the representation of the rotation group by finding eigenvalues of $\hat{j}_3 \equiv -\Sigma_{12} = \hat{m}_{12}$, eq. (1.105), where

$$\hat{j}_3 = -\mathrm{i} \begin{pmatrix} 0 & 0 & 0 & 0 \\ 0 & 0 & 1 & 0 \\ 0 & -1 & 0 & 0 \\ 0 & 0 & 0 & 0 \end{pmatrix}. \tag{6.41}$$

This matrix acts directly on the vector indices of $\epsilon^\mu(q, \lambda)$. The basis solutions to (5.80) are

$$\epsilon^\mu(q, 1) = 2^{-1/2} \begin{pmatrix} 0 \\ 1 \\ \mathrm{i} \\ 0 \end{pmatrix}, \quad \epsilon^\mu(q, -1) = 2^{-1/2} \begin{pmatrix} 0 \\ 1 \\ -\mathrm{i} \\ 0 \end{pmatrix}, \quad \epsilon^\mu(q, 0) = \begin{pmatrix} 0 \\ 0 \\ 0 \\ 1 \end{pmatrix}, \tag{6.42}$$

along with their complex conjugates. The value of λ shown is the eigenvalue of $\hat{j}_3$. Analogous to eq. (6.26) are the properties of the solutions:

$$\text{normalization,} \qquad \epsilon^*(q, \lambda) \cdot \epsilon(q, \mu) = -\delta_{\mu\lambda};$$
$$\text{completeness,} \qquad \sum_\lambda \epsilon^{*\mu}(q, \lambda) \epsilon^\nu(q, \lambda) = -g^{\mu\nu} + \delta_{\mu 0}\delta_{\nu 0}. \tag{6.43}$$

Boosted solutions can be generated in the same way as for the Dirac particle. In particular, spin solutions for the vector particle are constructed using $S(\Lambda_0) = \Lambda_s$ (eq. (6.31)) in eq.(6.15). We shall not need the explicit forms of these solutions. We can boost the relations eq. (6.43), however, by simply transforming the two sides according to Λ, to find, in an arbitrary frame,

$$\text{normalization,} \qquad \epsilon^*(k, \lambda)\cdot\epsilon(k, \mu) = -\delta_{\mu\lambda};$$
$$\text{completeness,} \qquad \sum_\lambda \epsilon^{*\mu}(k, \lambda)\epsilon^\nu(k, \lambda) = -g^{\mu\nu} + k^\mu k^\nu/m^2. \tag{6.44}$$

It is worth emphasizing that the number of physical solutions is three, even though the field has four components. For the Proca equation, this condition is enforced by the condition $k^\mu\epsilon_\mu = 0$, which follows from the form of the Lagrangian. Even without assuming a particular Lagrangian, however, the additional component $\epsilon^0(k)$ does not mix with the remaining three under any Lorentz transformation. That is, by eq. (6.18), $\phi = \epsilon^0$ will transform like a Lorentz scalar, since $D = 1$ for ϵ^0 in the rest frame $k^\mu = q^\mu$, in eq. (6.17). This accounts for the name *scalar polarization* given to this component of the vector field. The separation of components into three- and one-dimensional subspaces might seem a little strange, since all components of the original field A^μ mix under Lorentz transformations, according to the matrix $\Lambda^\mu{}_\nu$. But remember, $\Lambda^\mu{}_\nu$ is not unitary in general, and if we look for unitary representations, we must treat ϵ^0 separately. If the scalar polarization were not separated by an explicit Lorentz condition built into the equations of motion, it would have to be quantized as a single-component field.

6.4 Massless solutions

Massless free Lagrangians have plane-wave solutions with $p^2 = 0$. For these vectors there is no rest frame, so we cannot pick the same special vector q^μ to generate the representation that we picked in the massive case. Rather, q^μ must itself be lightlike.

When computing with lightlike or nearly lightlike vectors, it is often useful to introduce the following notation. For any vector a^μ, we define 'plus', 'minus' and 'transverse' components by

$$a^\pm = 2^{-1/2}(a^0 \pm a^3), \quad (\mathbf{a}_\mathrm{T})_i = a^i,\ i = 1, 2. \tag{6.45}$$

When the context allows, the subscript 'T' is suppressed. In these coordinates the metric tensor $g^{\mu\nu}$ is no longer diagonal, and we have

$$g^{+-} = g^{-+} = 1, \ g^{++} = g^{--} = 0. \tag{6.46}$$

As a result, lowering or raising an index interchanges plus and minus:

$$a^+ = a_-, \quad a^- = a_+. \tag{6.47}$$

In terms of these coordinates, the square of a vector and the scalar product of two vectors are given by

$$a^2 = 2a^+a^- - \mathbf{a}^2, \quad a \cdot b = a^+b^- + a^-b^+ - \mathbf{a}\cdot\mathbf{b}. \tag{6.48}$$

Returning to the problem at hand, it is convenient to pick q^μ in the plus direction:

$$q^\mu = q^+\delta^{\mu+}. \tag{6.49}$$

The actual magnitude of the components of q^μ can be left arbitrary. Any other lightlike vector can be reached from q^μ in eq. (6.49) by the combination of a rotation and a boost.

To apply Wigner's method, we need the little group of q^μ, which is not quite as obvious as in the massive case. The three generators of the little group can be written as (exercise 6.4)

$$\hat{m}_{12}, \pi^1 \equiv \hat{m}_{13} + \hat{m}_{10} = 2^{1/2}\hat{m}_{1+}, \quad \pi^2 \equiv \hat{m}_{23} + \hat{m}_{20} = 2^{1/2}m_{2+}, \tag{6.50}$$

whose explicit forms in the vector representation may be found from eq. (1.105). Their Lie algebra is easily derived from eq. (1.98):

$$[\pi^1, \pi^2] = 0, \quad [\hat{m}_{12}, \pi^1] = \mathrm{i}\pi^2, \quad [\hat{m}_{12}, \pi^2] = -\mathrm{i}\pi^1. \tag{6.51}$$

Wigner (1939) showed that the unitary irreducible representations of this algebra fall into two classes, depending on the eigenvalue of the Casimir operator $\hat{w}^2$. For $\hat{w}^2 = 0$, the representations are one-dimensional, while for $w^2 > 0$ they are infinite-dimensional. The plane-wave solutions that we find all fall into the former category. The infinite-dimensional representations have not yet been found to occur in elementary processes.

Dirac equation

The scalar field does not really distinguish between the massless and massive cases, so we begin with spin one-half. The massless Dirac equation for a plane wave solution

$$w_a(q, \lambda) \exp(\pm \mathrm{i}q^+x^-) \tag{6.52}$$

may be written as simply

$$q^+\gamma^- w(q, \lambda) = 0. \tag{6.53}$$

In contrast to the massive case, it is most convenient to work in the Weyl representation eq. (5.35), where (6.53) becomes

$$2^{1/2}q^+\begin{pmatrix}0&0&0&0\\0&0&0&1\\1&0&0&0\\0&0&0&0\end{pmatrix}\begin{pmatrix}w_1\\w_2\\w_3\\w_4\end{pmatrix}=0. \tag{6.54}$$

Like the massive case, eq.(6.24), this equation has four solutions, two with positive energy and two with negative energy. They are given by

$$u(q,\tfrac{1}{2})=(q^+2^{1/2})^{1/2}\begin{pmatrix}0\\0\\1\\0\end{pmatrix}=v(q,-\tfrac{1}{2}),$$

$$u(q,-\tfrac{1}{2})=(q^+2^{1/2})^{1/2}\begin{pmatrix}0\\1\\0\\0\end{pmatrix}=v(q,\tfrac{1}{2}), \tag{6.55}$$

where we have chosen a convenient normalization. λ equals the eigenvalue of σ_{12} for the u's and minus this eigenvalue for the v's, in the same manner as for the massive case discussed after eq. (6.25).

The solutions eq. (6.55) have the following properties:

normalization,

$$\bar{u}(q,\lambda)u(q,\sigma)=\bar{v}(q,\lambda)v(q,\sigma)=\bar{u}(q,\lambda)v(q,\sigma)=\bar{v}(q,\lambda)u(q,\sigma)=0;$$

projection operators,

$$\begin{aligned}u_\alpha(q,\tfrac{1}{2})\bar{u}_\beta(q,\tfrac{1}{2})&=\tfrac{1}{2}[(1+\gamma_5)\not{q}]_{\alpha\beta}=v_\alpha(q,-\tfrac{1}{2})\bar{v}_\beta(q,-\tfrac{1}{2}),\\u_\alpha(q,-\tfrac{1}{2})\bar{u}_\beta(q,-\tfrac{1}{2})&=\tfrac{1}{2}[(1-\gamma_5)\not{q}]_{\alpha\beta}=v_\alpha(q,\tfrac{1}{2})\bar{v}_\beta(q,\tfrac{1}{2});\end{aligned} \tag{6.56}$$

completeness,

$$\sum_\lambda u_\alpha(q,\lambda)\bar{u}_\beta(q,\lambda)=\not{q}_{\alpha\beta}=\sum_\lambda v_\alpha(q,\lambda)\bar{v}_\beta(q,\lambda).$$

Transformation of these solutions gives results of exactly the same form for any vector p^μ, $p^2=0$, with any choice of Wigner boosts.

Helicity

We can give an interpretation to the projection operators in eq. (6.56) by looking again at the action of the generator $\hat{j}_3=\frac{1}{2}\sigma_{12}=\hat{\mathbf{q}}\cdot\hat{\mathbf{j}}$,

$$\pm\lambda w(q,\lambda)=\hat{\mathbf{q}}\cdot\hat{\mathbf{j}}\,w(q,\lambda)=\tfrac{1}{2}\mathrm{i}\gamma_1\gamma_2 w(q,\lambda), \tag{6.57}$$

where the plus refers to the positive-energy solutions $u(q, \lambda)$, and the minus to the negative-energy solutions $v(q, \lambda)$. For any one of the spinor solutions in eq. (6.55), denoted w, we have

$$\gamma^- w(q, \lambda) = 0, \tag{6.58}$$

so that also $\gamma^+\gamma^- w = 0$, from which we find

$$w(q, \lambda) = \gamma^0\gamma^3 w(q, \lambda). \tag{6.59}$$

Then, using eq. (6.59) in (6.57), and the definition, eq. (5.53), of γ_5, we get

$$\hat{\mathbf{q}} \cdot \hat{\mathbf{j}}\, w(q, \lambda) = \tfrac{1}{2}\gamma_5 w(q, \lambda) = \pm\lambda w(q, \lambda). \tag{6.60}$$

Let us again anticipate a result to be shown in the next section: λ, as defined above, is the projection of the spin along the three-axis, in the single-particle states corresponding to these solutions. (Note, this means that $\hat{\mathbf{j}}_3$ acting on $v(q, \lambda)$ gives *minus* the spin along the three-direction of the state corresponding to this solution.) The quantity $\hat{\mathbf{q}} \cdot \mathbf{s}$, where $\mathbf{s}$ is the spin vector of a particle, is called the *helicity* of that particle. As a result of eq. (6.60), $\frac{1}{2}(-1)^\delta \gamma_5$ is the helicity operator when acting on massless solutions; $\delta = 0$ for positive-energy solutions and $\delta = 1$ for negative-energy solutions. The solution with positive helicity is the state for which $\frac{1}{2}[1 + (-1)^\delta \gamma_5]$ is equivalent to unity. This solution, with the spin aligned along the direction of motion, is called the *right-handed* solution. Similarly, the solution for which $\frac{1}{2}[1 - (-1)^\delta \gamma_5]$ is equivalent to unity is called the *left-handed* solution.

Finally, we note that γ_5 anticommutes with all the γ^μ, while $S(\Lambda)$, eq. (5.51), is a power series in the $\sigma_{\mu\nu}$ and hence has only even powers of the γ's. Therefore, γ_5 commutes with all the $S(\Lambda)$. The helicity operator for massless states is thus a scalar, and massless states with different helicities do not mix under Lorentz transformation. As promised, then, the solutions to the massless Dirac equation are one-dimensional representations of the little group. The argument λ actually labels the representation. The solutions are not completely unrelated, however. The parity transformation, eq. (5.52), exchanges left- and right-handed solutions, since γ_0 anticommutes with γ_5. More generally, we often define for fermion field ψ, massless or not,

$$\psi_{\mathrm{L}} \equiv \tfrac{1}{2}(1 - \gamma_5)\psi, \quad \psi_{\mathrm{R}} \equiv \tfrac{1}{2}(1 + \gamma_5)\psi, \tag{6.61}$$

as *left-* and *right-handed fields*, respectively, corresponding to the handedness of the *positive-energy* solutions onto which the matrices project in expansions of the form of eq. (6.19) in the $m \to 0$ limit. The projection onto negative-energy solutions gives fields of handedness *opposite* to that of the solutions. In this terminology, the parity operation, eq. (5.52) exchanges left- and right-handed fields.

Massless vector

From Section 5.3, we may anticipate that only two polarizations appear in unitary representations involving the massless vector field, and indeed this turns out to be the case, independent of special considerations of gauge invariance. Without going into details, we give the following result. For massless vector fields there are two one-dimensional unitary representations, whose basis vectors may be taken as $\epsilon^\mu(q,1)$ or $\epsilon^\mu(q,-1)$, defined in eq. (6.42), that correspond to the physical transverse polarizations for momentum q^μ, eq. (6.49).

The completeness operator for massive vectors in eq. (6.43) is replaced for massless vectors by

$$P^{\mu\nu}(q) \equiv \sum_{\lambda=\pm 1} \epsilon^{*\mu}(q,\lambda)\epsilon^\nu(q,\lambda) = \delta_{\mu 1}\delta_{\nu 1} + \delta_{\mu 2}\delta_{\nu 2}. \qquad (6.62a)$$

For an arbitrary vector k^μ, this is equivalent to

$$P^{\mu\nu}(k) = -(g^{\mu\nu} + k^\mu k^\nu/|\mathbf{k}|^2)(1-\delta_{\mu 0})(1-\delta_{\nu 0}), \qquad (6.62b)$$

after Lorentz transforming (6.62a). As expected, this is not a Lorentz invariant expression.

With our basis solutions in hand, we are ready to quantize spinor and vector fields.

6.5 Quantization

Free Dirac theory

Because the Dirac Lagrangian (5.36) is linear in derivatives, the canonical momentum of field ψ is simply its conjugate field $\psi^\dagger$. The Hamiltonian of the Dirac field is then

$$H = \int \mathrm{d}^3\mathbf{x}\,(\psi^\dagger\partial_0\psi - \mathscr{L}) = \int \mathrm{d}^3\mathbf{x}\,\psi^\dagger(\mathbf{x},x_0)(-\mathrm{i}\boldsymbol{\alpha}\cdot\boldsymbol{\nabla} + \gamma^0 m)\psi(\mathbf{x},x_0), \quad (6.63)$$

with $\alpha_i \equiv \gamma^0\gamma^i$. (To get the signs right, we remember that $\gamma^\mu\partial_\mu = \gamma^0\partial_0 + \gamma^i(\partial/\partial x^i)$.) Notice that the Schrödinger equation, $H\psi = \mathrm{i}\partial_0\psi$, based on this Hamiltonian is again the Dirac equation. Thus, the Dirac equation may be thought of as a relativistic generalization of the Schrödinger equation, and this was the motivation for its original derivation (Dirac 1928). This approach has many applications, and is treated in excellent detail in many texts. Here, however, we take the second-quantization approach throughout.

We expand the field and its conjugate in terms of coefficients of the classical solutions (eq. (6.19)) at $x_0 = 0$,

$$\psi_a(\mathbf{x}, x_0) = \sum_\lambda \int \frac{d^3\mathbf{k}}{(2\pi)^{3/2} 2\omega_k} [b_\lambda(\mathbf{k}, x_0) u_a(k, \lambda) e^{i\mathbf{k}\cdot\mathbf{x}} + d_\lambda{}^\dagger(\mathbf{k}, x_0) v_a(k, \lambda) e^{-i\mathbf{k}\cdot\mathbf{x}}].$$

$$\psi^\dagger{}_a(\mathbf{x}, x_0) = \sum_\lambda \int \frac{d^3\mathbf{k}}{(2\pi)^{3/2} 2\omega_k} [b^\dagger{}_\lambda(\mathbf{k}, x_0) u^\dagger{}_a(k, \lambda) e^{-i\mathbf{k}\cdot\mathbf{x}} + d_\lambda(\mathbf{k}, x_0) v^\dagger{}_a(k, \lambda) e^{i\mathbf{k}\cdot\mathbf{x}}], \tag{6.64}$$

where we choose the normalization $|c|^2$ of the spinors (eq. (6.39a)) as $2m$.

Substituting eq. (6.64) into (6.63), and using the orthonormality of the solutions, we easily find an expression for the Dirac Hamiltonian which is analogous to eq. (2.38) for the Hamiltonian of the free scalar field:

$$H = \sum_\lambda \int d^3\mathbf{k}\, \tfrac{1}{2}[b_\lambda{}^\dagger(\mathbf{k}, x_0) b_\lambda(\mathbf{k}, x_0) - d_\lambda(\mathbf{k}, x_0) d_\lambda{}^\dagger(\mathbf{k}, x_0)]. \tag{6.65}$$

Here we have been careful to preserve the order of the coefficients, anticipating that they will be quantized shortly. Each value of $\mathbf{k}$ and λ corresponds to a pair of commuting Hamiltonians, one for the b's, one for the d's. They enter with a relative minus sign, which we must interpret.

Anticommutation relations

The $b_\lambda(\mathbf{k}, x_0)$ and $d_\lambda(\mathbf{k}, x_0)$ are coefficients of positive-energy solutions in $\psi(x)$ and $\bar{\psi}(x)$, respectively. By analogy with the charged scalar case, we expect both of them to be interpreted as annihilation operators for positive-energy states. This interpretation can be preserved if, instead of the commutation relations eq. (2.12), they obey equal-time anticommutation relations:

$$\begin{aligned}
[b_\lambda(\mathbf{k}, x_0), b_{\lambda'}{}^\dagger(\mathbf{q}, x_0)]_+ &\equiv b_\lambda(\mathbf{k}, x_0) b_{\lambda'}{}^\dagger(\mathbf{q}, x_0) + b_{\lambda'}{}^\dagger(\mathbf{q}, x_0) b_\lambda(\mathbf{k}, x_0) \\
&= \delta_{\lambda\lambda'} 2\omega_k \delta^3(\mathbf{k} - \mathbf{q}) \\
[d_\lambda(\mathbf{k}, x_0), d_{\lambda'}{}^\dagger(\mathbf{q}, x_0)]_+ &= \delta_{\lambda\lambda'} 2\omega_k \delta^3(\mathbf{k} - \mathbf{q}), \\
[b, b]_+ = [d, d]_+ = [b, d^\dagger]_+ = [d, b^\dagger]_+ &= [b^\dagger, b^\dagger]_+ = [d^\dagger, d^\dagger]_+ = 0,
\end{aligned} \tag{6.66}$$

where in the last line we have suppressed indices. These relations, as the notation indicates, are postulated for interacting fields, as well as for free fields. They were introduced by Jordan & Wigner (1928). The last line of eq. (6.66) leads, in the diagonal case ($\mathbf{k} = \mathbf{k}'$, $\lambda = \lambda'$), to Dirac-coefficient operators that are *nilpotent*, that is, whose squares vanish.

States and Fermi statistics

Even though the b's and d's satisfy anticommutation relations among themselves, their time dependence is still determined, as for any Heisenberg-picture operator, by $\dot{b} = \mathrm{i}[H, b]$ and $\dot{d} = \mathrm{i}[H, d]$. Then, given eq. (6.66), the free-field creation and annihilation operators have exponential time dependence:

$$b_\lambda(\mathbf{p}, x_0) = b_\lambda(\mathbf{p}) \exp(-\mathrm{i}\omega_p x_0), \quad b^\dagger{}_\lambda(\mathbf{p}, x_0) = b^\dagger{}_\lambda(\mathbf{p}) \exp(\mathrm{i}\omega_p x_0), \tag{6.67}$$

and similarly for the d's. The formal developments of the bosonic case can now be repeated step by step; here we shall simply give the results for the continuum eigenstates of the Hamiltonian. They are linear combinations of the basis states

$$|\{(\mathbf{k}_i, \lambda_i)^{(-)}\}, \{(\mathbf{q}_j, \mu_j)^{(+)}\}\rangle = N \prod_i b_{\lambda_i}{}^\dagger(\mathbf{k}_i) \prod_j d_{\mu_j}{}^\dagger(\mathbf{q}_j)|0\rangle, \tag{6.68}$$

with N a normalization, usually ± 1 in this case. (The absolute sign depends on the ordering of operators within the products.) The two series of quantum numbers reflect the number of b- and d-modes that have been excited in the state. The nilpotence of the raising operators forbids any pair of quantum numbers $(\mathbf{k}_i, \lambda_i)^{(-)}$ or $(\mathbf{q}_j, \mu_j)^{(+)}$ from being the same. This is the exclusion principle, and identifies a system of fermionic degrees of freedom. We can show that arbitrary Dirac theory states obey Fermi statistics in the same way that we showed scalar states obey Bose statistics (Section 2.3).

The operators b and $b^\dagger$ act according to the rules

$$\begin{aligned} &b_{\lambda'}(\mathbf{k})|\ldots, (\mathbf{k}_i, \lambda_i)^{(-)}, \ldots\rangle \\ &\quad = \pm \sum_i \delta_{\lambda'\lambda_i} 2\omega_{k_i} \delta^3(\mathbf{k} - \mathbf{k}_i)|\ldots, (\mathbf{k}_{i-1}, \lambda_{i-1})^{(-)}, (\mathbf{k}_{i+1}, \lambda_{i+1})^{(-)}, \ldots\rangle, \\ &b_{\lambda'}(\mathbf{k})|0\rangle = 0, \\ &b_{\lambda'}{}^\dagger(\mathbf{k}')|\ldots, (\mathbf{k}_i, \lambda_i)^{(-)}, \ldots\rangle = \pm|\ldots, (\mathbf{k}_i, \lambda_i)^{(-)}, \ldots (\mathbf{k}', \lambda')^{(-)}, \ldots\rangle, \end{aligned} \tag{6.69}$$

and similarly for d and $d^\dagger$. The sign is determined by the ordering of the creation operators that define the state. $b^\dagger$ and $d^\dagger$ are thus raising operators and b and d are lowering operators in the same way as are the bosonic $a_\pm{}^\dagger$ and $a_\pm$ of the charged scalar theory. The normalization of the states is easily determined from the anticommutation relations (6.66):

$$\langle(\mathbf{k}, \lambda)^{(\pm)}|(\mathbf{k}', \lambda')^{(\pm)}\rangle = \delta_{\lambda\lambda'} 2\omega_k \delta^3(\mathbf{k} - \mathbf{k}'). \tag{6.70}$$

The Hamiltonian gives an infinite – this time negatively infinite – energy for the ground state:

$$H = \sum_\lambda \int \mathrm{d}^3\mathbf{k}\, \tfrac{1}{2}\{b_\lambda{}^\dagger(\mathbf{k})b_\lambda(\mathbf{k}) + d_\lambda{}^\dagger(\mathbf{k})d_\lambda(\mathbf{k}) - [d_\lambda{}^\dagger(\mathbf{k}), d_\lambda(\mathbf{k})]_+\}. \quad (6.71)$$

As in the bosonic case, we treat this infinity by normal ordering, which eliminates the vacuum expectation value. This means moving, by anticommutation in this case, all the lowering operators to the right of all the raising operators. Since the operators anticommute, every exchange gives a minus sign. Then the normal ordered Dirac Hamiltonian is

$$\begin{aligned} :H: &= \sum_\lambda \int \mathrm{d}^3\mathbf{k}\, \tfrac{1}{2}[b_\lambda{}^\dagger(\mathbf{k})b_\lambda(\mathbf{k}) + d_\lambda{}^\dagger(\mathbf{k})d_\lambda(\mathbf{k})] \\ &\equiv \sum_\lambda \int \mathrm{d}^3\mathbf{k}\, \omega_k N(\mathbf{k}, \lambda), \end{aligned} \quad (6.72)$$

which gives zero on (annihilates) the ground or vacuum state $|0\rangle$. The Dirac *number operator* $N(\mathbf{k}, \lambda)$ counts excitations $(\mathbf{k}, \lambda)$, with b's and d's treated equally, and is analogous to eq. (2.53) for the free scalar field.

Particles, antiparticles and spin

Further insight into the nature of the excitations associated with the operators $b^\dagger$ and $d^\dagger$ may be found by deriving expressions for the remaining conserved charges of the free Dirac theory. These are the spatial momenta, the angular momenta and the charge associated with the $U(1)$ conserved current, eq. (5.58). (The latter, as expected, commutes with all the generators of the Poincaré group.) In terms of creation and annihilation operators, we find

$$P^i = \sum_\lambda \int \mathrm{d}^3\mathbf{k}(\mathbf{k}^i/2\omega_k)[b_\lambda{}^\dagger(\mathbf{k})b_\lambda(\mathbf{k}) + d_\lambda{}^\dagger(\mathbf{k})d_\lambda(\mathbf{k})], \quad (6.73a)$$

whose interpretation, in terms of the number operator, is straightforward. To derive this form we have used eq. (6.39c), $u^\dagger(p, \lambda)v(p', \lambda)\delta^3(\mathbf{p}' + \mathbf{p}) = 0$, as well as symmetric integration. It is not necessary to normal-order P^i. The angular momentum tensor is rather more complicated (exercise 6.6). Its normal-ordered expectation value in a one-particle state at rest is not difficult to calculate, however, and is interpreted as the expectation value of the angular momentum operator in that state. Specifically, we find

$$\begin{aligned} \langle(\mathbf{q}, \lambda')^{(-)}|{:}J_{ij}{:}|(\mathbf{0}, \lambda)^{(-)}\rangle &= \delta^3(\mathbf{q})u^\dagger(q, \lambda')(\tfrac{1}{2}\sigma_{ij})u(q, \lambda), \\ \langle(\mathbf{q}, \lambda')^{(+)}|{:}J_{ij}{:}|(\mathbf{0}, \lambda)^{(+)}\rangle &= \delta^3(\mathbf{q})v^\dagger(q, \lambda')(-\tfrac{1}{2}\sigma_{ij})v(q, \lambda), \end{aligned} \quad (6.73b)$$

where the difference in sign follows directly from the anticommutation relations. Up to the normalization of states, the third component of the angular momentum in each state is just the expectation value of J_{12}, and thus eq. (6.73b) justifies the identification of λ as a spin component ($\sigma_{12}u = \lambda u$, $\sigma_{12}v = -\lambda v$; see the discussion after eq. (6.25)).

Finally, for the normal-ordered $U(1)$ charge, we find

$$:Q: = \int \mathrm{d}^3\mathbf{x} :\psi^\dagger(\mathbf{x}, x_0)\psi(\mathbf{x}, x_0):$$

$$= \sum_\lambda \int \frac{\mathrm{d}^3\mathbf{k}}{2\omega_k} [b_\lambda{}^\dagger(\mathbf{k})b_\lambda(\mathbf{k}) - d_\lambda{}^\dagger(\mathbf{k})d_\lambda(\mathbf{k})]. \qquad (6.73c)$$

In this case, b- and d-modes contribute with the opposite sign, in the same manner as the oppositely charged scalars in eq. (2.73). Once again we refer to them as particle and antiparticle. The prime examples are the electron and positron. In a $U(1)$ gauge theory such as quantum electrodynamics, the current that gives rise to this charge couples to the electromagnetic field, and, in this way, becomes directly observable. In QED, the particle is the electron, with $Q = +1$ but with negative electric charge. Note that the sign superscript in states refers to $-Q$.

It is important to note that Dirac fermions need not carry electric charge. However, they still have a conserved charge Q of the type of eq. (6.73c). In this case the charge is simply the number of particles minus the number of antiparticles, the *fermion number*. Finally, just as the scalar charge, eq. (2.73), vanishes with a hermitian field, for which $a_+ = a_-$, so the fermion number vanishes identically when $b = d$. Then there is no quantum number to distinguish particles from antiparticles. Such a field is called a *Majorana fermion* (Majorana 1937).

Dirac sea

An alternative interpretation of the operators is to take the $d_\lambda{}^\dagger(\mathbf{k})$ as annihilation and operators, and the $d_\lambda(\mathbf{k})$ as creation operators, for particles of *negative* energy. This is suggested by the form of the Hamiltonian in eq. (6.65). Then, in the case of the electron field, the vacuum may be thought of as having an infinite number of electrons, with energies less than $-m$. This collection is known as the *Dirac sea*. It is not directly observable because of an *energy gap* of $2m$ between the highest (filled) state in the sea, and the lowest positive-energy state. We shall not pursue this approach here. It is, however, essentially equivalent to the interpretation using particles and antiparticles of only positive energy, because the underlying operator algebra, eq. (6.66), is the same in both cases.

Spin and statistics

The anticommutation relations, eq. (6.66) for the coefficients imply equal-time anticommutation relations for the fields:

$$[\psi_\alpha{}^\dagger(\mathbf{x}, x_0), \psi_\beta(\mathbf{x}', x_0)]_+ = \delta_{\alpha\beta}\delta^3(\mathbf{x} - \mathbf{x}'). \qquad (6.74)$$

The commutator, on the other hand, is not a c-number, and does not vanish at equal times. At first sight this result is disturbing. If the commutators of Dirac fields do not vanish at spacelike distances, what becomes of causality? We recall, however, that spinors are double-valued representations of the rotation group. As such, a spinor is not itself directly observable, since a rotation by 2π changes its sign. On the other hand, operators that are bilinear in the field – such as components of the energy–momentum tensor – do not change sign, and are observables. More generally, we may consider any operator of the form $B_i(x) = \bar{\psi}(x)O_i\psi(x)$, where O_i is some matrix, possibly combined with differential operators. We can easily show (exercise 6.7) that equal-time commutators between the B_i vanish, if the fields obey eq. (6.74).

One may ask what would happen if we chose, instead of eq. (6.66), equal-time *commutation* relations for the operators b and d, treating them as bosonic operators. Then the equal-time *commutation* relations for the fields become

$$[\psi_\alpha{}^\dagger(\mathbf{x}, x_0), \psi_\beta(\mathbf{x}', x_0)] = \int \frac{\mathrm{d}^3\mathbf{k}}{(2\pi)^3\omega_k}(\boldsymbol{\gamma}\cdot\mathbf{k} - m)_{\alpha\beta}\mathrm{e}^{\mathrm{i}\mathbf{k}\cdot\mathbf{x}}, \qquad (6.75)$$

and *neither* commutators *nor* anticommutators vanish at spacelike separations. This state of affairs is clearly unacceptable, at least if we intend to have bilinear operators observable. (But notice, other possibilities exist, under the name of *parastatistics*, in which trilinear, or other, operators are the only observables. See Greenberg & Messiah 1965.)

Equations (6.74) and (6.75) both follow from the nature of the solutions to the Dirac equation. Thus, the Dirac equation itself forces the creation and annihilation operators of the Dirac field to anticommute *and* makes the states of the Dirac theory obey Fermi statistics. This is an example of what is known as the *spin–statistics theorem* (Pauli 1940), which states that fermions must have half-integer spin, and bosons integer spin, in a relativistic theory. A more complete proof of the spin–statistics theorem from the point of view above may be found, for instance, in Bjorken & Drell (r1965). An elegant formalism, based on locality, is described in Streater & Wightman (r1964) and Bogoliubov, Logunov & Todorov (r1975).

Vector fields

Even the quantization of free vector fields is a complex subject, and we shall only scratch the surface here. The basic result is that, even in the quantized *free* theory, one must choose between manifest Lorentz covariance and a direct physical interpretation of the space of states. We have already found, in Section 5.3, that there is a mismatch between the number of components of the vector field (four), and the number of physical solutions to the linear equations of motion (three for massive, two for massless). Similarly, if we use the unmodified Lagrangians (5.59) or (5.79), the naive canonical Poisson brackets found by identifying the $A^\mu(\mathbf{x}, x_0)$ as coordinates,

$$\{A^\mu(\mathbf{x}, x_0), \pi^\nu(\mathbf{x}, x_0)\}_{\mathrm{PB}} = \delta_{\mu\nu}\delta^3(\mathbf{x} - \mathbf{x}'),$$

cannot hold. For instance, if we construct the canonical momenta for the field components A^μ we find (see eq. (5.61))

$$\pi^\mu = -F^{0\mu} = E_\mu(1 - \delta_{\mu 0}). \tag{6.76}$$

Thus there is no canonical momentum associated with A^0. Such a Lagrangian is termed *singular* (Faddeev 1969). We may still carry out our quantization as above, in terms of the coefficients of the physical solutions to equations of motion, although we must clearly give up Lorentz covariance. This is the approach first described by Fermi (1932), and described in detail in, for example, Bjorken & Drell (r1965).

We begin by defining a 'physical' field, expanded in terms of solutions associated with physical polarizations, which are constructed by boosting the solutions eq. (6.42), for either a massive or massless field:

$$A^\mu(x) = \int \frac{\mathrm{d}^3\mathbf{k}}{(2\pi)^{3/2} 2\omega_k} \sum_{\lambda=\pm 1,0} [a_\lambda(\mathbf{k}, x_0)\epsilon^\mu(k, \lambda)\mathrm{e}^{\mathrm{i}\mathbf{k}\cdot\mathbf{x}} + a_\lambda{}^\dagger(\mathbf{k}, x_0)\epsilon^{*\mu}(k, \lambda)\mathrm{e}^{-\mathrm{i}\mathbf{k}\cdot\mathbf{x}}]. \tag{6.77}$$

We leave out the longitudinal polarization $\lambda = 0$ in the massless case. In both cases, the Lorentz condition is satisfied, but in the massless case we have, in addition, $A_0 = 0$, which along with $\nabla\cdot\mathbf{A} = 0$, defines the *radiation gauge* (Bjorken & Drell r1965). As canonical commutation relations, we postulate

$$\begin{aligned} [a_\lambda(\mathbf{k}, x_0), a_{\lambda'}{}^\dagger(\mathbf{k}', x_0)] &= \delta_{\lambda\lambda'} 2\omega_k \delta^3(\mathbf{k} - \mathbf{k}') \\ [a_\lambda(\mathbf{k}, x_0), a_{\lambda'}(\mathbf{k}', x_0)] &= [a_\lambda{}^\dagger(\mathbf{k}, x_0), a_{\lambda'}{}^\dagger(\mathbf{k}', x_0)] = 0. \end{aligned} \tag{6.78}$$

The normal-ordered vector Hamiltonian is then

$$\begin{aligned} :H: &= \int \mathrm{d}^3\mathbf{x}\, \tfrac{1}{2} :[\mathbf{E}^2 + \mathbf{B}^2 + 2\mathbf{E}\cdot\nabla A^0 + m^2 A^\mu A_\mu]: \\ &= \frac{1}{2}\int \mathrm{d}^3\mathbf{k} \sum_\lambda a_\lambda{}^\dagger(\mathbf{k}, x_0) a_\lambda(\mathbf{k}, x_0), \end{aligned} \tag{6.79}$$

where again the sum is over all three components in the massive case, and only two in the massless case. This Hamiltonian is 'solvable' just as for the free Dirac and scalar cases, and the states of the free system are thus $|\{(\mathbf{k}_i, \lambda_i)\}\rangle$, with physical polarizations only.

For some purposes, especially when manifest Lorentz covariance is desirable, the 'physical' quantization just described is not particularly convenient. A covariant formulation is due to Gupta (1950) and Bleuler (1950) for QED, and reviewed in Bogoliubov & Shirkov (r1980) and in Schweber (r1961). An extensive discussion of the nonabelian case is given by Kugo & Ojima (1979).

Covariance can be built in by modifying the Lagrangian, as in eq. (5.68). Then π^0 is no longer identically zero. In particular, if we choose the parameter $\lambda = 1$ in eq. (5.68), the classical equation of motion for each field component is the wave equation (5.71). Accepting, for the time being, that we are including large sets of unphysical solutions, we expand the field in terms of a set of momentum-independent solutions that spans the full four-dimensional space of possible fields A^μ,

$$A^\mu(x) = \int \frac{\mathrm{d}^3\mathbf{k}}{(2\pi)^{3/2} 2\omega_k} \sum_{s=0}^{3} [a_s(\mathbf{k}, x_0)\epsilon^\mu(k, s)\mathrm{e}^{\mathrm{i}\mathbf{k}\cdot\mathbf{x}} + a_s{}^\dagger(\mathbf{k}, x_0)\epsilon^{*\mu}(k, s)\mathrm{e}^{-\mathrm{i}\mathbf{k}\cdot\mathbf{x}}] \tag{6.80}$$

where we may choose, for instance,

$$\epsilon^\mu(k, s) = \delta_{\mu s}. \tag{6.81}$$

We then postulate an alternate set of commutation relations, this time manifestly covariant in form:

$$[a_s(\mathbf{k}, x_0), a_{s'}{}^\dagger(\mathbf{k}', x_0)] = -g_{ss'} 2\omega_k \delta^3(\mathbf{k} - \mathbf{k}'). \tag{6.82}$$

The usual procedure can be repeated to construct states of the free system here too, which can be labelled $|\{(\mathbf{k}_i, s_i)\}\rangle$, but the unphysical nature of some of the states is manifest if we compute the norm of the state created by $a_0{}^\dagger(\mathbf{k})$:

$$\langle(\mathbf{k}, s = 0)|(\mathbf{k}', s = 0)\rangle = -2\omega_k \delta^3(\mathbf{k} - \mathbf{k}'), \tag{6.83}$$

which is negative. This is not as bad as it looks since, in the free theory, physical and unphysical states are obviously orthogonal. We have obtained manifest Lorentz invariance at the cost of enlarging the Hilbert space by an infinite set of states, which, nevertheless, do not mix with those we already had.

In an interacting-field theory with vector particles, the situation is more complicated. Here, we are interested in computing the S-matrix, which relates asymptotic states, $\langle\{(\mathbf{p}, \lambda)\}\text{ out}|\{(\mathbf{q}, \mu)\}\text{ in}\rangle$. But now, not all of the in- and out-states labelled by μ and λ are physical. For the theory to make sense, we must require that the 'physical' S-matrix, consisting of the

S-matrix restricted to the physical states only, be unitary,

$$S_{\text{phys}}(S_{\text{phys}})^{\dagger} = I. \tag{6.84}$$

This requirement is of great importance whenever gauge vector particles enter an interacting theory.

6.6 Parity and leptonic weak interactions

We close this chapter with a discussion of the quantum realization of parity, a *discrete* symmetry, which cannot therefore be generated from infinitesimal transformations. Parity, proper Lorentz transformations and time reversal make up the extended Lorentz group (Section 1.5). We refer the reader to, for example, Sakurai (r1964), Bjorken & Drell (r1965), Streater & Wightman (r1964), Bogolubov, Logunov & Todorov (r1975) and Sachs (r1987) for more complete discussions of parity and time reversal. (We shall also touch on time reversal and on charge conjugation, which relates particles and antiparticles, in Appendix D.) To allow illustration by a realistic Lagrangian, we shall introduce the standard model for the electroweak interactions of leptons. First, however, let us discuss how parity may be treated in the quantum theory.

Operator content

In the defining representation of the Lorentz group, the parity transformation is the matrix P that reverses the spatial components of vectors, but leaves their time component unchanged. With our choice of metric, $\mathrm{P}^{\mu}{}_{\nu} = g_{\mu\nu}$ (Section 1.5).

The behavior of a quantum field under parity is determined by its classical counterpart. For the classical Dirac field, for instance, we have, from eq. (5.52), $\psi'(-\mathbf{x}, x_0) = \gamma_0 \psi(\mathbf{x}, x_0)$ so for the quantum field we seek an operator U_{P} such that

$$U_{\mathrm{P}} \psi(\mathbf{x}, x_0) U_{\mathrm{P}}{}^{-1} = \gamma_0 \psi(-\mathbf{x}, x_0). \tag{6.85}$$

Now we find from (6.39b) that γ_0 exchanges $u(p, \lambda)$ and $u(\bar{p}, \lambda)$, with $\bar{\mathbf{p}} = -\mathbf{p}$. In the expansion of the field, eq. (6.64), this is the same thing as exchanging the creation operators $b_{\lambda}{}^{\dagger}(\mathbf{p})$ and $b_{\lambda}{}^{\dagger}(-\mathbf{p})$. The same happens for the v's, but with a change in sign. Thus any unitary operator U_{P} that satisfies

$$U_{\mathrm{P}} b_{\lambda}(\mathbf{k}, x_0) U_{\mathrm{P}}{}^{-1} = b_{\lambda}(-\mathbf{k}, x_0), \quad U_{\mathrm{P}} d_{\lambda}(\mathbf{k}, x_0) U_{\mathrm{P}}{}^{-1} = -d_{\lambda}(-\mathbf{k}, x_0) \tag{6.86}$$

is a good operator for parity. Here we shall simply assume that such an operator exists, without giving a construction (exercise 6.9).

At the same time, under parity we expect the spatial momentum of single-particle states to change sign, while their spin, which is an angular momentum and should behave like $\mathbf{r} \times \mathbf{p}$, remains unchanged. Thus, we look for an operator that has the action

$$U_{\mathrm{P}}|\mathbf{p}, \lambda\rangle = \mathrm{e}^{\mathrm{i}\phi}|-\mathbf{p}, \lambda\rangle, \tag{6.87}$$

where ϕ is a phase. Clearly, the operator U_{P} defined by eq. (6.86) satisfies this relation, with $\phi = 0$ for particles, and $\phi = \pi$ for antiparticles. The factor $\mathrm{e}^{\mathrm{i}\phi}$ is known as the *intrinsic parity* of the particle. Note, however, that single-particle states with fixed nonzero spatial momentum are *not* eigenstates of parity; for this we need eigenstates of angular momentum. Similar reasoning may be applied to a vector field to give

$$U_{\mathrm{P}} V^{\mu}(\mathbf{x}, x_0) U_{\mathrm{P}}^{-1} = \mathrm{P}^{\mu}{}_{\nu} V^{\nu}(-\mathbf{x}, x_0). \tag{6.88}$$

Parity conservation

The parity operator can be diagonalized if it commutes with the Hamiltonian. In particular, if U_{P} commutes with H, then it is time-independent, and it acts the same on in- as on out-states in the S-matrix, eq. (2.90),

$$\langle B \text{ out}|A \text{ in}\rangle = \langle B \text{ out}|U_{\mathrm{P}}{}^{\dagger} U_{\mathrm{P}}|A \text{ in}\rangle = \langle B_{\mathrm{P}} \text{ out}|A_{\mathrm{P}} \text{ in}\rangle, \tag{6.89}$$

where for $|n_{\mathrm{P}} \text{ in}\rangle = U_{\mathrm{P}}|n \text{ in}\rangle$ momenta are reversed, but angular momenta remain the same. Parity invariance thus relates different elements of the S-matrix. In addition, parity invariance requires that the S-matrix elements between parity eigenstates with different eigenvalues vanish. This is *conservation of parity*, which is of great use in analyzing the strong and electromagnetic decays of hadrons.

Parity invariance also places restrictions on the momentum dependence of matrix elements between momentum eigenstates. Perhaps the most useful of these are associated with the observation that if $\{p_i{}^{\mu}\}$ is a set of independent momenta, then certain contractions of these momenta with the ϵ-tensor change sign under parity. For instance, we have

$$\epsilon_{\mu\nu\lambda\sigma} p_1{}^{\mu} p_2{}^{\nu} p_3{}^{\lambda} p_4{}^{\sigma} = -\epsilon_{\mu\nu\lambda\sigma} \bar{p}_1{}^{\mu} \bar{p}_2{}^{\nu} \bar{p}_3{}^{\lambda} \bar{p}_4{}^{\sigma}, \tag{6.90}$$

where the $\bar{p}_i{}^{\mu}$ are the parity-reflected partners of the $p_i{}^{\mu}$, because three components in the product are always spatial. Parity invariance does not allow S-matrix elements to be proportional to an even function of the $p_i{}^{\mu}$ times this invariant, because such a term would change sign when the $p_i{}^{\mu}$ are replaced by $\bar{p}_i{}^{\mu}$, and hence would violate eq. (6.89). Generalizations of this result are easy to derive in specific cases.

Parity violation

Let us now discuss how parity violation can occur at the level of the Lagrangian. Consider, for instance, an interaction term that couples two different left- or right-handed fermion fields $\frac{1}{2}(1 \pm \gamma_5)\psi_l$ and $\frac{1}{2}(1 \pm \gamma_5)\psi_{l'}$ (see eq. (6.61)) to a vector field V^μ by the term $\bar{\psi}\gamma_\mu(1 \pm \gamma_5)\psi_{l'}V^\mu$. Then the action of parity on this contribution to the interaction Lagrangian is

$$\begin{aligned} U_{\mathrm{P}} \int \mathrm{d}^3\mathbf{x}\, \bar{\psi}_l(\mathbf{x}, x_0)\gamma_\mu(1 \pm \gamma_5)\psi_{l'}(\mathbf{x}, x_0)V^\mu(\mathbf{x}, x_0)U_{\mathrm{P}}^{-1} \\ = \int \mathrm{d}^3\mathbf{x}\, \psi_l^\dagger(-\mathbf{x}, x_0)\gamma_0{}^2\gamma_\mu(1 \pm \gamma_5)\gamma_0\psi_{l'}(-\mathbf{x}, x_0)g^{\mu\mu}V^\mu(-\mathbf{x}, x_0) \\ = \int \mathrm{d}^3\mathbf{x}\, \bar{\psi}_l(\mathbf{x}, x_0)\gamma_\mu(1 \mp \gamma_5)\psi_{l'}(\mathbf{x}, x_0)V^\mu(\mathbf{x}, x_0). \end{aligned} \tag{6.91}$$

As expected from classical considerations, parity reverses the role of left- and right-handed fields. This means that processes mediated by a potential that distinguishes left- from right-handed fields do not respect reflection symmetry. The simplest of these is the decay $V(q) \to l(k_1) + \bar{l}'(k_2)$. Let, for instance, $\sigma(q, s_0; \mathbf{k}_1)$ be the amplitude for a Z^0 particle of spin s_0 at rest to decay into an electron with momentum $\mathbf{k}_1$. If eq. (6.89) holds for arbitrary states, we easily prove $\sigma(q, s_0; \mathbf{k}_1) = \sigma(q, s_0; -\mathbf{k}_1)$. When it fails, we may observe a parity-violating $\mathbf{s} \cdot \mathbf{k}_1$ dependence. Thus a nonzero value for the quantity

$$A(\mathbf{k}_1) = \sigma(q, s_0; \mathbf{k}_1) - \sigma(q, s_0; -\mathbf{k}_1) \tag{6.92}$$

implies a failure of parity invariance in the underlying interaction Lagrangian, as in eq. (6.91). (Note that even if parity is violated, we need a polarized cross section to form a parity-violating term in the amplitude.) This argument generalizes to n-body decay, and is how the violation of parity was first observed (Lee & Yang 1956), in the decay of the Co^{60} nucleus and of muons (see Section 8.3).

Standard model of leptonic electroweak interactions

In the standard model, parity is violated in the manner described above, by giving right- and left-handed fermion fields different gauge transformation properties under the local $SU(2) \times U(1)$ symmetry introduced in Section 5.4. Consider, in particular, the transformation properties of leptons. The known leptons – that is, fermions that have no strong interactions – are the electron, the muon, and the tau, along with three corresponding neutrinos, ν_e, ν_μ and ν_τ. Their fields are assumed to transform under weak $SU(2)$, the gauge theory with vector boson fields $B^\mu{}_a$, $a = 1, 2, 3$ of eqs. (5.108) and (5.109), as follows. The three *left-handed* particles and their corresponding neutrinos are grouped into three $SU(2)$ doublet fields $\psi_i^{(L)}$ $(i = \mathrm{e}, \mu, \tau)$, with $I_3 = +\frac{1}{2}$ for the neutrino and $I_3 = -\frac{1}{2}$ for the particle, while the *right-handed* fields $\psi^{(R)}$ are treated as gauge singlets, which do not couple at all to the $SU(2)$ gauge field. The full set of fields is

$$\psi_i^{(L)} = \tfrac{1}{2}(1-\gamma_5)\begin{pmatrix} \nu_i \\ l_i^- \end{pmatrix}, \quad l_i^{(R)} = \tfrac{1}{2}(1+\gamma_5)l_i, \quad \nu_i^{(R)} = \tfrac{1}{2}(1+\gamma_5)\nu_i. \quad (6.93)$$

The complete electroweak couplings of these fields, and how they are related to spontaneous symmetry breaking, is described in Appendix C. Here we specify their couplings to the gauge vector fields $W^{\pm}{}_{\mu}$, Z_{μ}, and the photon field A_{μ}:

$$\begin{aligned}
\mathcal{L}_{\text{lep}} = {} & \sum_{i=\mathrm{e},\mu,\tau} [\bar{l}_i(\mathrm{i}\partial\!\!\!/ - m_i)l_i + \bar{\nu}_i \mathrm{i}\partial\!\!\!/ \nu_i] + e \sum_{i=\mathrm{e},\mu,\tau} \bar{l}_i \gamma^{\mu} A_{\mu} l_i \\
& - \frac{g}{2^{1/2}} \sum_{i=\mathrm{e},\mu,\tau} \bar{\psi}_i^{(L)} \gamma^{\mu} (\sigma_+ W^+{}_{\mu} + \sigma_- W^-{}_{\mu}) \psi_i^{(L)} \\
& - \frac{g}{2\cos\theta_{\mathrm{W}}} \sum_{i=\mathrm{e},\mu,\tau} \{\bar{\psi}_i^{(L)} \gamma^{\mu} [\sigma_3 + (1-\sigma_3)\sin^2\theta_W] \psi_i^{(L)} \\
& \qquad + 2\sin^2\theta_W \bar{l}_i^{(R)} \gamma^{\mu} l_i^{(R)}\} Z_{\mu}.
\end{aligned} \quad (6.94)$$

θ_{W} is the weak mixing angle and e the (positron) electric charge, related to the original $SU(2) \times U(1)$ couplings g and g' by eqs. (5.116) and (5.119), respectively. We also define

$$\sigma_+ = \tfrac{1}{2}(\sigma_1 + \mathrm{i}\sigma_2) = \begin{pmatrix} 0 & 1 \\ 0 & 0 \end{pmatrix}, \quad \sigma_- = \tfrac{1}{2}(\sigma_1 - \mathrm{i}\sigma_2) = \begin{pmatrix} 0 & 0 \\ 1 & 0 \end{pmatrix}, \quad (6.95)$$

which act as raising and lowering matrices for the $SU(2)$ symmetry. They ensure that charge is conserved in the couplings of the leptons and neutrinos to the $W^{\pm}$ bosons. Finally, we recall that, by construction (eq. (6.61)), any field χ is the sum of its left- and right-handed fields, $\chi = \chi^{(R)} + \chi^{(L)}$.

In summary, parity nonconservation is manifest in eq. (6.94), through the differing roles played by right- and left-handed fermion fields in their interactions with W and Z bosons. Note, however, that electromagnetic couplings, to the photon, only have vector currents and conserve parity.

Finally, because the neutrino masses are zero, the right-handed neutrino fields appear in eq. (6.94) only in a kinetic term. In this model, so long as neutrinos are exactly massless, only left-handed neutrinos (and right-handed antineutrinos) are observable. Such a theory is said to have *two-component* neutrinos, corresponding to the fact that purely left-handed (or right-handed) massless fermion fields are equivalent to Weyl spinors (see eq. (5.55)).

Exercises

6.1 An alternative to the spin basis is the helicity basis for massive particles, for which we choose a Wigner boost that consists of a pure boost fixed in the three-direction, followed by a pure rotation:

$$\Lambda_h(p, q) \equiv R(p, p^{(3)})\Lambda(p^{(3)}, q),$$

where $p^{(3)}{}_\mu \equiv (\omega_p, |\mathbf{p}|\hat{\mathbf{n}}_3)$. The rotation $R(p, p^{(3)})$ is not unique, but is set conventionally; see Jacob & Wick (1959). Let $h \equiv P_i\epsilon_{ijk}J_{jk}$ be the product of spatial momentum and angular momentum operators. Show that the states defined according to eq. (6.8), with this choice of Wigner boost, are eigenfunctions of h, i.e.,

$$h|p, \lambda\rangle_h = |\mathbf{p}|\lambda|p, \lambda\rangle_h,$$

where λ is the usual spin eigenvector for states in the rest frame. These states, therefore, possess definite helicity.

6.2 In deriving eq. (6.60) we produced the spin projection operators in eq. (6.56) for massless spin states by using the Dirac equation in its massless form. Develop an analogous trick in the massive case, to construct projection operators for the massive solutions $u(p, \pm\lambda)$ and $v(p, \pm\lambda)$ in the spin basis

$$u_b(p, \lambda)\bar{u}_a(p, \lambda) = [|c|^2(\not{p} + m)/2m][\tfrac{1}{2} + \lambda\gamma_5\not{n}(p)]_{ba},$$

$$v_b(p, \lambda)\bar{v}_a(p, \lambda) = [|c|^2(\not{p} - m)/2m][\tfrac{1}{2} + \lambda\gamma_5\not{n}(p)]_{ba},$$

respectively, where $n^\mu \equiv (\Lambda_s)^\mu{}_\nu n^{(3)\nu}$. Here the spin vector is $\lambda n^\mu \equiv \pm\frac{1}{2}n^\mu$.

6.3 Show that $u^\dagger(k, \lambda)v(k, \lambda) = (2/\omega_k)u^\dagger(q, \lambda)\mathbf{k}\cdot\boldsymbol{\gamma}v(q, \lambda)$, with $q^\mu = (m, \mathbf{0})$.

6.4 Show that the generators given in eq. (6.50) generate the little group for a massless momentum vector in the plus-direction, eq. (6.49), and verify that their algebra eq. (6.51) may be thought of as a boosted version of the algebra of the rotation group, in the limit that the velocity of the boost approaches c (Kim & Wigner 1987).

6.5 (a) Consider a Dirac field in the presence of an external electromagnetic field $A^\mu(x)$, where $A^\mu(x)$ is a classical function (Section 4.4). Show that the corresponding Dirac equation is equivalent to

$$[\mathrm{D}^\mu(A)\mathrm{D}_\mu(A) + \tfrac{1}{2}e\sigma^{\mu\nu}F_{\mu\nu}(A) + M^2]\psi(x) = 0,$$

where $\mathrm{D}_\mu(A)$ is the covariant derivative eq. (5.90), and $F_{\mu\nu}(A)$ the classical field strength. Notice that this result implies that in the absense of an electric field ($F_{0i} = 0$), the Dirac equation breaks up into two independent equations for two-component spinors, just as in the free case. (b) Consider the nonrelativistic limit of the equation for two-component spinors, with $F_{0i} = 0$, taking $\phi(x) = \exp(-\mathrm{i}[m + E_{\mathrm{nr}}]t)\ \xi(\mathbf{x})$, where E_{nr} is the nonrelativistic energy. Derive the form

$$[(-1/2m)\mathbf{D}(A)\cdot\mathbf{D}(A) - (e/2m)\boldsymbol{\sigma}\cdot\mathbf{B}]\xi(\mathbf{x}) = E_{\mathrm{nr}}\xi(\mathbf{x}),$$

which shows that the magnetic moment of the Dirac electron, defined by a contribution $\boldsymbol{\mu}\cdot\mathbf{B}$ to the energy, is $2(e/2m)\mathbf{s}$, with $\mathbf{s}$ the spin. This is conventionally described as a *gyromagnetic ratio* (coefficient of e/m times the angular momentum) of two.

6.6 Show that the normal-ordered operators $:J_{ij}:$ can be put into 'diagonal' form as

$$:J_{ij}: = \sum_{\lambda,\lambda'} \int \frac{\mathrm{d}^3\mathbf{k}\,\mathrm{d}^3\mathbf{k}'}{2\omega_k 2\omega_{k'}}$$
$$\times \Big[u^\dagger(k',\lambda')\big(\{-\mathrm{i}[k_i\nabla^{(k)}{}_j - k_j\nabla^{(k)}{}_i] + \tfrac{1}{2}\sigma_{ij}\}\delta^3(\mathbf{k}-\mathbf{k}')\big)$$
$$\times\, u(k,\lambda)b_{\lambda'}{}^\dagger(\mathbf{k}')b_\lambda(\mathbf{k}).$$
$$-\, v^\dagger(k',\lambda')\big(\{-\mathrm{i}[k_i\nabla^{(k)}{}_j - k_j\nabla^{(k)}{}_i] + \tfrac{1}{2}\sigma_{ij}\}\delta^3(\mathbf{k}-\mathbf{k}')\big)$$
$$\times\, v(k,\lambda)d_\lambda{}^\dagger(\mathbf{k})d_{\lambda'}(\mathbf{k}')\Big],$$

where $\nabla^{(k)}$ is the gradient with respect to the momentum $\mathbf{k}$, and here acts on the delta function.

6.7 Show that the equal-time commutator of two currents $j_1 = \bar{\psi}\Gamma_1\psi$ and $j_2 = \bar{\psi}\Gamma_2\psi$, where ψ is a Dirac field and the Γ_i are arbitrary Dirac matrices, is given by $[j_1(\mathbf{x},0), j_2(\mathbf{x}',0)] = \bar{\psi}(\mathbf{x},0)[\Gamma_1,\Gamma_2]\,\psi(\mathbf{x},0)\delta^3(\mathbf{x}-\mathbf{x}')$.

6.8 Consider the operator

$$U[\alpha] = \exp\left\{\int \frac{\mathrm{d}^3\mathbf{k}}{2\omega_k}\sum_\lambda [\alpha(\mathbf{k})b_\lambda{}^\dagger(\mathbf{k})d_\lambda{}^\dagger(-\mathbf{k}) + \alpha^*(\mathbf{k})b_\lambda(\mathbf{k})d_\lambda(-\mathbf{k})]\right\}.$$

Show that it transforms annihilation operators for particles into creation operators for antiparticles according to

$$U[\alpha]b_\sigma(\mathbf{k})U^{-1}[\alpha] = b_\sigma(\mathbf{k})\cos|\alpha(\mathbf{k})|$$
$$+\, [\alpha(\mathbf{k})/\alpha^*(\mathbf{k})]^{1/2} d_\sigma{}^\dagger(-\mathbf{k})\sin|\alpha(\mathbf{k})|.$$

Such operations are known as *Bogoliubov–Valatin transformations*. To prove this result, you may wish to verify the following formal identity:

$$\mathrm{e}^A B\,\mathrm{e}^{-A} = \sum_{n=0}^{\infty}\frac{1}{n!}C_n(A,B),$$

where $C_n(A,B)$ is the multiple commutator, defined by $C_0 = 1$, $C_1 = [A,B]$, $C_n = [A, C_{n-1}]$.

6.9 Construct a parity operator for the free Dirac and neutral scalar fields.

7

Path integrals for fermions and gauge fields

The path integral approach may be extended to spinor and vector fields. These applications are facilitated by the concept of integration over anticommuting classical variables. For nonabelian gauge theories, quantization leads naturally to the introduction of 'fictitious', or 'ghost', fields.

7.1 Fermionic path integrals

The realistic fermionic Lagrange densities encountered in Chapters 5 and 6 are of the form

$$\mathscr{L} = \bar{\psi} O(\partial, A) \psi, \tag{7.1}$$

where $O(\partial, A)$ represents an operator linear in derivatives, masses and Bose fields; the latter are denoted by A. As usual, we want to develop a perturbation method for expanding Green functions of the interacting-field theory about the free-field theory, and for identifying Feynman rules for Green functions and the S-matrix. All the information necessary to carry out this program is contained in the generator for fermionic Green functions. By analogy with eq. (4.58) for the charged scalar field, an obvious guess for this generator is a 'Gaussian' form with sources:

$$\begin{aligned} Z[K, \bar{K}, A] &= \frac{W[K, \bar{K}, A]}{W[0, 0, A]}, \\ W[K, \bar{K}, A] &= \int [\mathscr{D}\psi][\mathscr{D}\bar{\psi}] \exp\Big\{ \mathrm{i} \int \mathrm{d}^4 y [\bar{\psi}(y) O(\partial, A) \psi(y) \\ &\qquad - \bar{K}(y)\psi(y) - \bar{\psi}(y) K(y)] \Big\}. \end{aligned} \tag{7.2}$$

This turns out to be right, but we must pay a price, associated with the anticommutating nature of the fields. $\psi(x)$ and $\bar{\psi}(x)$ in eq. (7.2) will turn out to be rather unusual integration variables. In the following, we shall give a heuristic derivation of eq. (7.2), in the spirit of Section 3.1. In the more elaborate derivation of the path integral for scalar fields in Sections 3.2 and 3.3, we treated the free theory as an infinite ensemble of independent harmonic oscillators. The same treatment works for the free fermion field, except that now, in the classical limit, the coefficients b, $b^\dagger$, d, $d^\dagger$ still anticommuntate. If a path integral is to be constructed, it will have to be in terms of integrals over classical anticommuting numbers, whatever that means. Early discussions of this problem for path integrals were given by Mathews & Salam (1955) and Candlin (1956), but the modern treatment really begins with the work of Berezin (r1966). Building on his formalism, it is possible to retrace the reasoning of Sections 3.2 and 3.3, and to construct a path integral for anticommuting fields in precisely the form of eq. (7.2) (Faddeev r1981, Halpern, Jevicki & Senjanovic 1977, Soper 1981).

Anticommuting numbers

Anticommuting numbers, or *Grassmann variables*, define an algebra in much the same way that complex numbers do. If $\{a_1, \ldots, a_n\}$ is a set of anticommuting numbers, and $\{z\}$ labels the set of complex numbers, then the *Grassmann algebra* consists of products and sums of the a_i and z's. The product of two elements is written in the normal way as $a_i a_j$ or $a_i z$. These objects are given no special values – they are not complex numbers – but they are defined to satisfy the relations

$$a_i a_j = -a_j a_i, \; a_i z = +z a_i. \tag{7.3}$$

Similarly, multiple products are written as $a_i a_j a_k = -a_i a_k a_j$, etc. Note that eq. (7.3) requires that anticommuting c-numbers (classical numbers), just like fermionic operators, are nilpotent,

$$a_i^{\,2} = 0. \tag{7.4}$$

This means that products in the Grassmann algebra can be, at most, linear in any given variable a_i. So, if there are n Grassmann variables, the most general element of the algebra is

$$\sum_{\{\delta_i=0,1\}} z_{\delta_1 \ldots \delta_n} \prod_{i=1}^{n} a_i^{\,\delta_i}, \tag{7.5}$$

where the sum is over all possible assignments of the δ_i, each of which may be zero or unity, and $z_{\delta_1 \ldots \delta_n}$ is a complex number, the coefficient for the product $\prod_i a_i^{\,\delta_i}$ (taken in some definite order). Monomials are called even (odd) if $\sum_i \delta_i$ is even (odd). Odd elements of the algebra anticommute with each other, while even elements commute not only with each other, but also with odd elements.

The algebra of anticommuting numbers may be generalized to a differential calculus by introducing a derivative operator, $\mathrm{d}/\mathrm{d}a_i$, for each a_i. The derivative operator is defined by

$$(\mathrm{d}/\mathrm{d}a_i)a_j = \delta_{ij}, \quad (\mathrm{d}/\mathrm{d}a_i)z = 0. \tag{7.6}$$

From eqs. (7.3) and (7.6) we deduce that derivatives anticommute both with each other and with the Grassmann variables themselves (exercise 7.1). So, when there are factors to the right of a_j, we have

$$(\mathrm{d}/\mathrm{d}a_i)a_j = -a_j(\mathrm{d}/\mathrm{d}a_i) + \delta_{ij}. \tag{7.7}$$

Here and below, we always understand the derivative to act from the left, in terms of relative order.

To construct a path integral for fermions, we shall postulate that the classical limits of the fermionic coefficients b, $b^\dagger$, d, $d^\dagger$ obey the Grassmann algebra specified by letting the anticommutation relations in eq. (6.66) vanish (the $\hbar \to 0$ limit). Let us now see how to define integrals for these rather strange objects.

Grassmann integrals

Like a_i itself, the derivative operation with respect to a_i is nilpotent:

$$(\mathrm{d}/\mathrm{d}a_i)^2 = 0. \tag{7.8}$$

We are thus unable to define the integral as the inverse of the derivative. In fact, we shall take what may seem a rather desperate step, by defining the integral to be the *same* as the derivative. Only later will we see the motivation for what, at present, is only a choice of notation. So, in accordance with eqs. (7.6), and (7.7), we define

$$\int \mathrm{d}a_i\, a_j = \delta_{ij} - a_j \int \mathrm{d}a_i, \quad \int \mathrm{d}a_i\, z = 0, \tag{7.9}$$

where z is any complex number, including unity. A fundamental advantage of this definition, given by Berezin (r1966), is that it obeys a type of translation invariance: if a_i and a_j are two different Grassmann variables, and if

$$f(a_i) \equiv z_0 + z_1 a_i, \tag{7.10}$$

where z_0 and z_1 are complex numbers, then

$$\int \mathrm{d}a_i\, f(a_i) = \int \mathrm{d}a_i\, f(a_i - a_j) = z_1. \tag{7.11}$$

These results easily generalize to polynomials, eq. (7.5). In this way, we may construct multiple Grassmann integrals, noting that integrals, like derivatives, anticommute

$$\left[\int da, \int da'\right]_+ = 0, \tag{7.12}$$

both with each other and with odd elements of the algebra. The signs in eq. (7.9) are for the action of the integral from the left. For a function of m variables,

$$f(a_1, \ldots, a_m) = z^{(0)} + \sum_{i=1}^{m} z^{(1)}{}_i a_i + \sum_{i,j=1}^{m} z^{(2)}{}_{ij} a_i a_j + \ldots$$
$$+ z^{(m)} a_1 a_2 \ldots a_m, \tag{7.13}$$

the only term to survive the m-fold integral is the last one,

$$\int da_m \int da_{m-1} \ldots \int da_1 \, f(a_1, \ldots, a_m) = z^{(m)}, \tag{7.14}$$

where we have used the particular order of the integrals shown.

Integration by parts

Along with translation invariance, the integral just defined allows another fundamental manipulation, integration by parts. Because of eq. (7.7), however, the sign convention is slightly different from that for integrals of complex variables. If we decompose an arbitrary function $f(a)$ (where a denotes the set of Grassmann variables) into even and odd parts so that $f(a) = f(a)_{\mathrm{e}} + f(a)_0$, then

$$\int da_j \frac{df(a)}{da_j} g(a) = -\int da_j \left\{ [f(a)_{\mathrm{e}} - f(a)_0] \frac{dg(a)}{da_j} \right\}. \tag{7.15}$$

That is, the sign changes when the derivative of $f(a)$ acts on an odd element of the algebra.

Mixed integrals and transformations

(The author is indebted to D.E. Soper for the following argument.) As an exercise, and for future reference, we may extend the formalism to include mixed integrals, which include both anticommuting and normal variables:

$$I = \prod_{i,j} \int dy_i \int d\xi_j f(y_i, \xi_j), \tag{7.16}$$

where the y_i are commuting numbers and the ξ_j are anticommuting numbers, while f is a function of both sets. We can construct an infinitesimal transformation that mixes the commuting and anticommuting variables. For instance, if $\delta\lambda$ is an infinitesimal complex parameter, then we may define new variables x_i and η_j by

$$\xi_j = \eta_j + a_j(x, \eta)\delta\lambda,\ y_i = x_i + c_i(x, \xi)\delta\lambda, \tag{7.17}$$

where the c_i are arbitrary even functions in the Grassmann algebra and the a_j are arbitrary odd functions. Then the η_j and the x_i continue to be respectively anticommuting and commuting elements of the algebra.

We now want to re-express the integral I in terms of the new variables. To this end, we define a Jacobian,

$$J(x, \eta) = 1 + j(x, \eta)\delta\lambda, \tag{7.18}$$

such that

$$\prod_{i,j}\int dy_i \int d\xi_j\, f(y_i, \xi_j) = \prod_{i,j}\int dx_i \int d\eta_j\, J(x, \eta) f(y_i(x, \eta), \xi_j(x, \eta))$$

$$= \prod_{i,j}\int dx_i \int d\eta_j\, f(x_i, \eta_j) + O(\delta\lambda). \tag{7.19}$$

The coefficient of $\delta\lambda$ in the difference between the two forms vanishes:

$$0 = \prod_{i,j}\int dx_i \int d\eta_j [j(x, \eta) f(x, \eta) + \sum_k c_k \partial f/\partial x_k + \sum_k a_k \partial f/\partial \eta_k], \tag{7.20}$$

from which we can derive $j(x, \eta)$ to order $\delta\lambda$. Integrating by parts, using eq. (7.15), we find

$$j(x, \eta) = \sum_k (\partial c_k/\partial x_k) - \sum_{k'} (\partial a_{k'}/\partial \eta_{k'}). \tag{7.21}$$

Note the minus sign associated with the odd elements of the algebra. This is the only difference from the Jacobian for an infinitesimal transformation of commuting variables. For the generalization of this result to finite transformations, see van Nieuwenhuizen (r1981, Appendix G).

Gaussian integrals

We are now ready to construct a finite-dimensional version of the path integral, eq. (7.2). We introduce two independent sets of anticommuting variables, ψ_i and $\bar{\psi}_i$, where $i = 1, 2, \ldots n$. For definiteness, we define

$$\prod_{i=1}^{n}\int d\psi_i\, d\bar{\psi}_i \equiv \int d\psi_n\, d\bar{\psi}_n \int d\psi_{n-1}\, d\bar{\psi}_{n-1} \cdots \int d\psi_1\, d\bar{\psi}_1. \tag{7.22}$$

A general Gaussian integral in these variables is

$$I_n[M] \equiv \prod_i \int d\psi_i \, d\bar{\psi}_i \exp(-\bar{\psi}_j M_{jk} \psi_k), \tag{7.23}$$

where M_{jk} is some complex-valued matrix. Using the integral formula (7.14), we see that out of the expansion of the exponential, the *only* term that survives is $(1/n!)(\bar{\psi} M \psi)^n$, and that, even there, only those terms that are linear in every ψ_k and $\bar{\psi}_j$ contribute:

$$\begin{aligned} I_n[M] &= \prod_i \int d\psi_i \, d\bar{\psi}_i (\bar{\psi}_i \psi_i) \sum_{\{\alpha,\beta\}} \epsilon_{\alpha_1 \cdots \alpha_n} \epsilon_{\beta_1 \ldots \beta_n} \\ &\quad \times M_{\alpha_1 \beta_1} M_{\alpha_2 \beta_2} \ldots M_{\alpha_n \beta_n} \\ &= \det M \end{aligned} \tag{7.24}$$

The product of ϵ's gives the sign of the permutation necessary to put the product from eq. (7.23) into the standard form shown. This result is to be contrasted with the corresponding bosonic calculation, eqs. (3.29)–(3.31), in which we derive $[\det A]^{-1/2}$.

To complete the analogy with bosonic integrals, we introduce source terms. At this point, the translation invariance of the Grassmann integrals is indispensable, since it allows us to shift the integrals over ψ and $\bar{\psi}$ in the usual way, and to derive Gaussian integrals in the presence of anticommuting source variables $\bar{K}_j$ and K_j:

$$\prod_i \int d\psi_i \, d\bar{\psi}_i \exp[(\bar{\psi}_j M_{jk} \psi_k - \bar{K}_j \psi_j - \bar{\psi}_j K_j)] = \det M \exp[-\bar{K}_i (M^{-1})_{ij} K_j]. \tag{7.25}$$

By defining $\bar{K}_j$ and K_j to be odd elements of the algebra, we ensure that the overall argument of the exponential is even, which simplifies many of the manipulations below. The ingredients are now all in place, and we may interpret eq. (7.2) as the generator of vacuum-to-vacuum amplitudes, just as we interpreted eq. (3.39) for scalar fields. We must understand, of course, that the Grassmann integrals in (7.2) are very different from the bosonic integrals of (3.39). Nevertheless, the formula (7.25) for Gaussian integrals is all we shall need to derive a perturbation expansion for fermionic field theories whose Lagrange densities are of the form (7.1).

Fermionic field theories

A more complete discussion of fermionic path integrals would require that we construct fermionic equivalents to the scalar field coherent states of

Sections 3.2 and 3.3, defined in terms of the creation operators $b_\lambda{}^\dagger(\mathbf{k}, x_0)$, $d_\lambda{}^\dagger(\mathbf{k}, x_0)$ of Section 6.5. Let us hint at how this may be done. We introduce fermionic-oscillator coherent states, analogous to eqs. (3.51) and (3.52b). With time dependence implicit, these are

$$\begin{aligned} |\beta_\lambda(\mathbf{k})\;\rangle &= \exp[(2\omega_k)^{-1} b_\lambda{}^\dagger(\mathbf{k})\beta_\lambda(\mathbf{k})]|0\rangle \\ &= |0\rangle - (2\omega_k)^{-1}\beta_\lambda(\mathbf{k})|(\lambda, \mathbf{k})^{(-)}\rangle, \\ \langle \beta_\sigma{}^*(\mathbf{q})| &= \langle 0| \exp[(2\omega_k)^{-1}\beta_\sigma{}^*(\mathbf{q}) b_\sigma(\mathbf{q})] \\ &= \langle 0| - (2\omega_k)^{-1}\langle(\mathbf{q}, \sigma)^{(-)}|\beta_\sigma{}^*(\mathbf{q}), \end{aligned} \tag{7.26}$$

where $\beta_\lambda(\mathbf{k})$ and $\beta_\sigma{}^*(\mathbf{q})$ are anticommuting c-numbers. The nilpotence of the Grassmann variables β_λ and $\beta_\sigma{}^*$, and of the operators, ensures that the exponential has only an expansion to first order, as shown. Note the signs and orders, in eq. (7.26), which are necessary for odd elements of the Grassmann algebra. In terms of these states, we may construct a unit operator for the Hilbert space corresponding to a given particle excitation. This is the analogue of eq. (3.55) (exercise 7.2),

$$\begin{aligned} &\int \frac{\mathrm{d}\beta_\lambda{}^*(\mathbf{k})\,\mathrm{d}\beta_\lambda(\mathbf{k})}{2\omega_k} \exp\left[-\frac{\beta_\lambda{}^*(\mathbf{k})\beta_\lambda(\mathbf{k})}{2\omega_k}\right]|\beta_\lambda(\mathbf{k})\rangle\langle\beta_\lambda{}^*(\mathbf{k})| \\ &\qquad = |0\rangle\langle 0| + |(\mathbf{k}, \lambda)^{(-)}\rangle\langle(\mathbf{k}, \lambda)^{(-)}| \equiv I_{(\mathbf{k},\lambda)^{(-)}}, \end{aligned} \tag{7.27}$$

and similarly for antiparticle states. $I_{(\mathbf{k},\lambda)^{(-)}}$ is just the identity in the subspace of particles with these quantum numbers. In this form, we may see the motivation for writing the operation defined by eq. (7.9) as an integral: it generates a sum over coherent states that is formally equivalent to the sum over energy eigenstates for a fermionic system (Candlin 1956). It is hoped that the correspondence with the scalar case is clear, and we shall not pursue this approach in detail here.

7.2 Fermions in an external field

Among the simplest fermionic theories is one that combines a Dirac fermion with an external field. It will illustrate features that are characteristic of fermions in the perturbative expansion of any field theory.

A typical Lagrange density is the familiar

$$\mathcal{L} = \bar{\psi}[\mathrm{i}\gamma^\mu(\partial_\mu + \mathrm{i}eA_\mu) - m]\psi, \tag{7.28}$$

where $A^\mu(x)$ is the external field, specified once and for all as a function of space–time. In view of our discussion above, we shall accept that the generating functional for this theory is specified by eq. (7.2). The Green functions are given by

$$G_{\{a_i,b_i\}}(y_i, x_i) \equiv \langle 0|T[\textstyle\prod_{i=1}^{n}\psi_{a_i}(y_i)\bar{\psi}_{b_i}(x_i)]|0\rangle$$

$$= \prod_{i=1}^{n}\left\{\left[\mathrm{i}\frac{\delta}{\delta \bar{K}_{a_i}(y_i)}\right][-\mathrm{i}K_{b_i}(x_i)]\right\} Z[K, \bar{K}, A]\Bigg|_{K=\bar{K}=0}. \quad (7.29)$$

Here the numbers of ψ's and $\bar{\psi}$'s are the same. As we shall see, all other combinations vanish. Even though the fields and sources are anticommuting, the Lagrangian itself is a sum of commuting terms, since we take, as above, $\bar{K}(x)$ and $K(x)$ as anticommuting. Note the difference in sign between the variations with respect to $K(x)$ and $\bar{K}(y)$; to act on $K(y)$ in eq. (7.2), $\delta/\delta K$ must anticommute with $\bar{\psi}(x)$, while all other factors of Grassman variables can be paired into even elements of the algebra.

The perturbative expansion of Green functions in this, and other, fermionic theories can be carried out by direct analogy to the scalar case. Making the replacement

$$\bar{\psi}_a(x)[\not{A}(x)]_{ab}\psi_b(x) \rightarrow [-\mathrm{i}\delta_{K_a}(x)][\not{A}(x)]_{ab}[\mathrm{i}\delta_{\bar{K}_b}(x)], \quad (7.30)$$

we factor the interaction Lagrangian $-e\int \mathrm{d}^4x\, \bar{\psi}\not{A}\psi$ from the path integral, just as in eq. (3.45). (We have introduced the same shorthand notation for variational derivatives, as in eq. (3.47): $\delta_K \equiv \delta/\delta K$, etc.) The resulting expression for W is

$$W[K, \bar{K}, A] = \exp\left\{-\mathrm{i}e\int \mathrm{d}^4y\, \delta_{K_a}(y)[\not{A}(y)]_{ab}\delta_{\bar{K}_b}(y)\right\} W[K, \bar{K}, 0], \quad (7.31)$$

where $W[K, \bar{K}, 0]$ is the free-field path integral in the presence of sources,

$$W[K, \bar{K}, 0] = \int [\mathcal{D}\psi][\mathcal{D}\bar{\psi}] \exp\left\{\mathrm{i}\int \mathrm{d}^4y[\bar{\psi}(y)(\mathrm{i}\not{\partial} - m)\psi(y) - \bar{K}(y)\psi(y) - \bar{\psi}(y)K(y)]\right\}. \quad (7.32)$$

As in the scalar case, the free-field integral can be evaluated by shifting the fields:

$$\begin{aligned}\psi'(x) &= \psi(x) + \int \mathrm{d}^4z\, S_F(x-z)K(z),\\ \bar{\psi}'(x) &= \bar{\psi}(x) + \int \mathrm{d}^4w\, \bar{K}(w)S_F(w-x),\end{aligned} \quad (7.33)$$

where $S_F(w - z)$ is the functional inverse of $(i\partial\!\!\!/ - m)$:

$$(i\partial\!\!\!/ - m)S_F(w - z) = \delta^4(w - z), \tag{7.34}$$

supplemented by appropriate boundary conditions (see below). With this change of variables, we find

$$W[K, \bar{K}, 0] = \exp\left[-i\int d^4w\, d^4z\, \bar{K}(w)S_F(w - z)K(z)\right]W_0[0, 0, 0], \tag{7.35}$$

in which the source dependence is explicit.

Feynman propagator

$S_F(w - z)$ in eq. (7.35) is the Feynman propagator for the Dirac field, defined by eq. (7.34) and the boundary conditions that at large positive (negative) times, negative- (positive-) energy waves vanish (see Section 2.4). As a Fourier transform, it has an expression analogous to (2.83),

$$\begin{aligned} S_F(z) &= \frac{1}{(2\pi)^4}\int d^4k\, e^{-ik\cdot z}\,\frac{1}{k\!\!\!/ - m + i\epsilon} \\ &= \frac{1}{(2\pi)^4}\int d^4k\, e^{-ik\cdot z}\,\frac{k\!\!\!/ + m}{k^2 - m^2 + i\epsilon}. \end{aligned} \tag{7.36}$$

To show the second form we use the Dirac matrix identity,

$$k\!\!\!/ k\!\!\!/ = k^2 I_{4\times 4}, \tag{7.37}$$

which follows directly from the matrix anticommutation relations eq. (5.43). A short calculation, using the spinor completeness relations in eq. (6.39a) and the field expansion, eq. (6.64) with free-field time dependence, (6.67), shows that S_F may also be defined in terms of the time-ordered product of free fermion fields:

$$\begin{aligned} iS_F(x - y) &= \langle 0|T[\psi(x)\bar{\psi}(y)]|0\rangle_{\text{free}} \\ &\equiv \langle 0|\psi(x)\bar{\psi}(y)|0\rangle_{\text{free}}\,\theta(x_0 - y_0) \\ &\quad - \langle 0|\bar{\psi}(y)\psi(x)|0\rangle_{\text{free}}\,\theta(y_0 - x_0), \end{aligned} \tag{7.38}$$

which is to be compared to eq. (2.81) for scalar fields. Notice the minus sign in the definition for time ordering, which reflects the anticommuting nature of the fermionic fields. The time ordering of the product of an arbitrary number of fermion fields requires the specification of a 'standard' ordering, shown explicitly inside the T-operator, which is defined to have a plus sign. The sign of any other ordering is then $(-1)^\delta$, where δ is the number of exchanges required to relate the order in question to the standard order.

Gell-Mann–Low formula

In analogy to the scalar case, the computation of interacting-field Green functions is reduced by the expansion of eq. (7.31) to a perturbation series. Exactly the same reasoning as in Section 3.4 shows that the Gell-Mann–Low formula, eq. (3.89), still holds, and that it may be evaluated by computing scattering graphs only:

$$\langle 0|T[\prod_j \psi_{b_j}(y_j)\prod_i \bar{\psi}_{a_i}(x_i)]|0\rangle$$
$$= \langle 0|T(\prod_j \psi_{b_j}(y_j)\prod_i \bar{\psi}_{a_i}(x_i)\exp\{\mathrm{i}\textstyle\int \mathrm{d}^4y\,\mathcal{L}_I[\bar{\psi},\psi,A]\})|0\rangle_{\text{free,scattering}}. \tag{7.39}$$

Wick's theorem with fermions

To evaluate eq. (7.39), we develop the generalization of Wick's theorem, eqs. (3.86) and (3.87), to free fermionic fields:

$$\langle 0|T[\prod_{j=1}^n \psi_{b_j}(y_j)\prod_{i=1}^m \bar{\psi}_{a_i}(x_i)]|0\rangle_{\text{free}}$$
$$= \prod_{j=1}^n [-\,\mathrm{i}\delta/\delta\bar{K}_{b_j}(y_j)]$$
$$\times\prod_{i=1}^m[\mathrm{i}\delta/\delta K_{a_i}(x_i)]\exp[-\mathrm{i}\textstyle\int \mathrm{d}^4w\,\mathrm{d}^4z\,\bar{K}_c(w)S_{\mathrm{F},cd}(w-z)K_d(z)]$$
$$= \delta_{mn}\sum_{P(j,i)}\epsilon_{P(j,i)}\prod_{i=1}^m \mathrm{i}S_{F,b_ja_i}(y_j-x_i), \tag{7.40}$$

where $P(j,i)$ is the set of pairings $\{(y_j,x_i)\}$ and $\epsilon_{P(j,i)}$ is the sign of the permutation necessary to achieve this pairing. This is Wick's theorem for fermions, which allows us to develop Feynman rules for theories with fermionic fields. It shows explicitly that Green functions with different numbers of fields and conjugate fields vanish.

Sign of the permutation

When Wick's theorem is applied to the Gell-Mann–Low formula for an interacting-field theory, we find that fermion propagators group themselves into one or more continuous strings, often referred to as *fermion lines*. This occurs because the interaction Lagrangian is bilinear in the fermion fields, so that the indices of each internal vertex always connect exactly two propagators. Fermion lines are one of two types. They may begin at an external ψ and end at an external $\bar{\psi}$, giving an expression like

$$[\mathrm{i}\delta/\delta\bar{K}_\alpha(y)][-\mathrm{i}\delta/\delta K_\beta(x)]$$
$$\times\frac{1}{(n-1)!}\prod_{i=1}^{n-1}\int \mathrm{d}^4x_i[-\mathrm{i}\delta/\delta K_{b_i}(x_i)][-\mathrm{i}e\not{A}(x_i)]_{b_ia_i}[\mathrm{i}\delta/\delta\bar{K}_{a_i}(x_i)]$$
$$\times\frac{1}{n!}\left[-\mathrm{i}\int \mathrm{d}^4w\,\mathrm{d}^4z\,\bar{K}_c(w)S_{F,cd}(w-z)K_d(z)\right]^n$$
$$=\prod_{i=1}^{n-1}\int d^4x_i\,\mathrm{i}S_{\mathrm{F},\alpha b_{n-1}}(y-x_{n-1})[-\mathrm{i}e\not{A}(x_{n-1})]_{b_{n-1}a_{n-1}}\mathrm{i}S_{\mathrm{F},a_{n-1}b_{n-2}}(x_{n-1}-x_{n-2})$$
$$\times\cdots\times[-\mathrm{i}e\not{A}(x_1)]_{b_1a_1}\mathrm{i}S_{\mathrm{F},a_1\beta}(x_1-x)+\text{'loops'}. \tag{7.41}$$

As indicated by the term 'loops', in other contributions, some strings may close on themselves, to form *internal loops* of fermions, in the form

$$\frac{1}{m!}\prod_{i=1}^{m}\int \mathrm{d}^4x_i[-\mathrm{i}\delta/\delta K_{b_i}(x_i)][-\mathrm{i}e\not{A}(x_i)]_{b_ia_i}[\mathrm{i}\delta/\delta\bar{K}_{a_i}(x_i)]$$
$$\times\frac{1}{m!}\left[-\mathrm{i}\int \mathrm{d}^4w\,\mathrm{d}^4z\,\bar{K}_c(w)S_{\mathrm{F},cd}(w-z)K_d(z)\right]^m$$
$$=-\prod_{i=1}^{m}\int \mathrm{d}^4x_i\,\mathrm{tr}\,\{\mathrm{i}S_{\mathrm{F}}(x_m-x_{m-1})[-\mathrm{i}e\not{A}(x_{m-1})]$$
$$\times\cdots\times[-\mathrm{i}e\not{A}(x_1)]\mathrm{i}S_{\mathrm{F}}(x_1-x_m)[\mathrm{i}e\not{A}(x_m)]\}, \tag{7.42}$$

where matrix products inside the trace are understood. Notice the overall minus sign in the latter case, which follows from the anticommuting nature of the functional derivatives. Of necessity, fermion loops are all vacuum bubbles in a theory with only a background vector field (Section 3.4).

Absolute sign

To determine the sign of a given contribution to the perturbative expansion, we imagine commuting the sources and variations into a form where each fermion line is untangled from all the others, as in eqs. (7.41) or (7.42). Since the interaction Lagrangian is even, in the Grassman algebra, the only sign changes from any such reordering result from pairing the external variations corresponding to the beginning and end of each external line. The overall sign of the resulting graph is thus the product of (-1) for every fermion loop times the sign of permutation i,

$$(y_1,x_1;\,y_2,x_2;\,\ldots;\,y_n,x_n)\to(y_{i_1},x_{i_1};\,y_{i_2},x_{i_2};\,\ldots;\,y_{i_n},x_{i_n}), \tag{7.43}$$

where y_{i_1}, x_{i_1} labels the pair $\psi(y_{i_1})$ and $\bar{\psi}(x_{i_1})$ of external fields at the ends of the first external line in permutation i, and so on. The ordering of variations of the right-hand side of eq. (7.29) may be taken to define a 'standard' ordering of fields, for which the sign of the pairing $(y_1, x_1; y_2, x_2; \ldots, y_n, x_n)$ is clearly $+1$.

Feynman rules

Once again, we can conveniently summarize the above results in the form of rules for fermions in an external gauge field.

As for the charged scalar field (Section 4.4), fermion lines carry arrows, pointing from $\bar{\psi}$ field to ψ field. (According to eq. (7.36), in momentum space the corresponding propagator is the Dirac matrix $(\not{k} - m)^{-1}$ where k^μ is the momentum flowing in the direction of the arrow.)

To compute the nth-order contribution to the coordinate-space Green function $G_{\{a_i, b_j\}}(y_i, x_j)$, the Feynman rules are as follows.

(i) Identify all distinct connected nth-order graphs with the correct number of external lines, every fermion line being continuous.

(ii) Associate $-\mathrm{i}e\int \mathrm{d}^4x\, \not{A}(x)_{dc}$ with each internal vertex, where d refers to the Dirac (and possible group) index of the fermion line whose arrow points out of the vertex, and c to the index of the line whose arrow points in.

(iii) Associate a factor $\mathrm{i}S_{ba}(w - z)$ with each fermion propagator whose arrow points toward the Dirac vertex at point w with index b, and away from the Dirac vertex at point z with index a.

(iv) The absolute sign of the pairing of external fields ψ and $\bar{\psi}$ is given by the sign of the permutation in eq. (7.43), when the time-ordered product is defined as in eq. (7.29).

The connected Green functions are of the type shown in Fig. 7.1; fermion loops are always disconnected from external lines in this theory, and cancel between numerator and denominator in eq. (7.2), Also, the nontrivial symmetry factors encountered in the scalar theory (Section 3.4) cannot occur in the connected Green functions of this theory. There is no factor s_1^{-1} corresponding to multiple connections of identical lines to the same vertex; the two lines attached to any vertex differ because one arrow

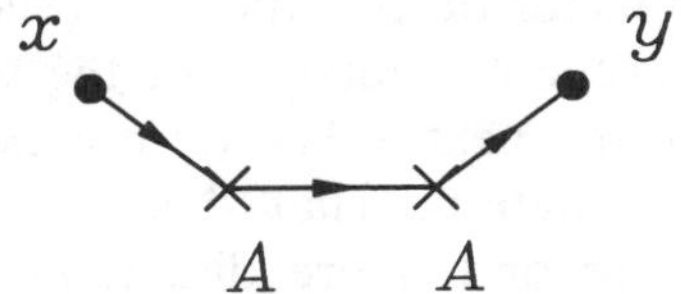

Fig. 7.1 Typical connected Green function for a fermion in an external field.

points into the vertex, and the other out. Similarly, the factor ${s_2}^{-1}$ is unity for each such graph, because the scattering graphs are not invariant under the interchange of any pair of vertices.

Fermion loops and functional determinants

The cancellation of vacuum bubbles, i.e. $W[0, 0, A]$ in eq. (7.2), follows just as in the scalar case. Nevertheless, it is still instructive to have a look at the sum of vacuum bubbles. Their computation can be carried out in two ways. First, we can use the perturbative expansion of eq. (7.31), in which the interaction term is factored from the free-field path integral. Second, we can simply observe that the integral is a Gaussian, in terms of Grassman variables, and re-express it as a determinant, as in eq. (7.25):

$$W[0, 0, A] = \det[\mathrm{i}(\mathrm{i}\not{\partial} - e\not{A} - m)]. \tag{7.44}$$

The problem in the latter case is to interpret the determinant of the Dirac operator. This we shall do by comparison with the perturbative evaluation.

Here is the explicit perturbative expansion, using eqs. (7.31) and (7.32), to second order:

$$\begin{aligned} W[K, \bar{K}, A] &= \det(-\not{\partial} - \mathrm{i}m)\exp\left\{-\mathrm{i}e\int \mathrm{d}^4y\, \delta_{K_b(y)}[\not{A}(y)]_{bc}\delta_{\bar{K}_c(y)}\right\} \\ &\quad \times \exp\left[-\mathrm{i}\int \mathrm{d}^4w\, \mathrm{d}^4z\, \bar{K}_a(w)S_{\mathrm{F},ab}(w-z)K_b(z)\right]\Big|_{K=\bar{K}=0} \\ &= \det(-\not{\partial} - \mathrm{i}m)\Bigg(1 + (-1)\mathrm{i}S_{\mathrm{F},ab}(0)\int \mathrm{d}^4x[-\mathrm{i}e\not{A}_{ba}(x)] \\ &\quad + \tfrac{1}{2}\left\{(-1)\mathrm{i}S_{\mathrm{F},ab}(0)\int \mathrm{d}^4x[-\mathrm{i}e\not{A}_{ba}(x)]\right\}^2 \\ &\quad + \tfrac{1}{2}(-1)\int \mathrm{d}^4x\, \mathrm{d}^4x'\, \mathrm{i}S_{\mathrm{F},ab}(x-x')[-\mathrm{i}e\not{A}_{bc}(x')] \\ &\quad \times \mathrm{i}S_{\mathrm{F},cd}(x'-x)[-\mathrm{i}e\not{A}_{da}(x)] + \cdots\Bigg). \end{aligned} \tag{7.45}$$

The graphical equivalent is shown in Fig. 7.2. A factor of (-1) is associated with each fermion loop, as expected. Similarly, the factor of one-half in the final term is a symmetry factor of the type ${s_2}^{-1}$ (Section 3.4 and Appendix B), associated with the cyclic permutation of the vertices $x \to x'$, $x' \to x$. The factor of one-half in the next-to-last term is perhaps less transparent, but suggestive of exponentiation. That this is indeed the case may be verified by turning to a second, more direct, method of evaluating eq. (7.44).

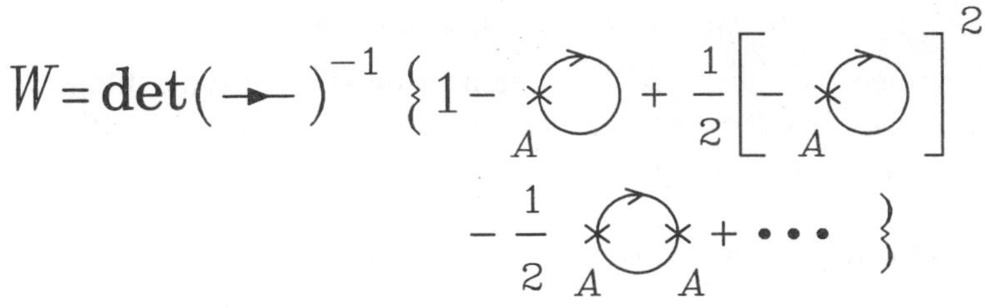

Fig. 7.2 Graphical equivalent of eq. (7.45).

In this method, we rely on two familiar matrix identities (eq. (1.76)),

$$\det M = \exp\{\mathrm{tr}\ln M\}, \quad \det\{MN\} = \det M \det N. \tag{7.46}$$

We use these relations to re-express the A-dependence of eq. (7.44) as

$$\begin{aligned}\det(-\not\partial - \mathrm{i}e\not A - \mathrm{i}m) &\\ = \det(-\not\partial - \mathrm{i}m)\exp\{\mathrm{tr}\ln[1 + (-\not\partial - \mathrm{i}m)^{-1}(-\mathrm{i}e\not A)]\} &\\ = \det(-\not\partial - \mathrm{i}m)\exp\left\{-\sum_{n=0}^{\infty}\frac{1}{n}\,\mathrm{tr}\,[\mathrm{i}(\mathrm{i}\not\partial - m)^{-1}(-\mathrm{i}e\not A)]^n\right\}, &\end{aligned} \tag{7.47}$$

where $\det(-\not\partial - \mathrm{i}m)$ contributes the same overall factor as in eq. (7.45) (exercise 7.3). This formula is to be interpreted in a functional sense, so that the inverse of the free Dirac operator is the propagator $S_{\mathrm{F}}(x-y)$. Similarly, the trace indicates integrations over positions, as well as sums over indices. The result of expanding eq. (7.47) to second order is then easily checked to be the second-order expansion of eq. (7.45). The minus signs and combinatoric factors in eq. (7.45) arise in this case from the expansion of the trace of the logarithm and the exponential.

With these results in hand, we are ready to give dynamics to the vector particles, and to discuss some features of the quantization of a full gauge theory.

7.3 Gauge vectors and ghosts

We saw in Section 6.5 that there are subtleties in the quantization of vector fields due to the singular nature of their Lagrangians. In this section, we shall discover how these problems arise in the path integral formulation, and describe a solution that will enable us to specify self-consistent Feynman rules for gauge theories.

Gauge invariance and path integrals

First, let us see why there is a problem in defining a path integral for the gauge boson density $\mathscr{L} = -\frac{1}{4}F^2$, eqs. (5.59) or (5.104). Our initial proposal for the vacuum-to-vacuum path integral for gauge bosons is

$$W[J^\mu] = \int[\mathscr{D}A(x)]\exp\left\{\mathrm{i}\int \mathrm{d}^4x[-\tfrac{1}{4}F^2(x) - J^\mu(x)A_\mu(x)]\right\}, \qquad (7.48)$$

where $[\mathscr{D}A(x)] = \prod_{\mu=0}^{3}[\mathscr{D}A^\mu]$ defines a separate functional integral over each component of the A-field. Strictly speaking, we should begin with a phase space path integral, in terms of the fields and their conjugate momenta (Faddeev 1969). We will return later to the validity of eq. (7.48) as a starting point.

Because of the gauge invariance of $F^2(x)$ under the local transformations eqs. (5.65) and/or (5.99), not all of the components of A^μ in eq. (7.48) have physical meaning, while the phase $\exp(\mathrm{i}S)$ depends only on the physical components. The action is 'flat' along the unphysical or 'gauge' directions of A^μ; there is not even a kinetic term to damp oscillations in these directions. As a result unphysical oscillations utterly dominate a path integral such as eq. (7.48). To see what this means, consider the modified Lagrange density $\mathscr{L} = -\frac{1}{4}F_{\mu\nu}F^{\mu\nu} - \frac{1}{2}\lambda(\partial\cdot A)^2$ (eq. (5.68)), which depends on unphysical components through the 'gauge-fixing' term $\frac{1}{2}\lambda(\partial\cdot A)^2$. The corresponding propagator (exercise 5.8) is

$$G_{\mathrm{F},\mu\nu}(x,\lambda) = \frac{1}{(2\pi)^4}\int \mathrm{d}^4k\,\frac{1}{k^2+\mathrm{i}\epsilon}\left[-g_{\mu\nu} + \left(1-\frac{1}{\lambda}\right)\frac{k_\mu k_\nu}{k^2+\mathrm{i}\epsilon}\right]. \qquad (7.49)$$

In the $\lambda \to 0$ limit, in which the action is independent of unphysical components, the coefficient of the gauge term $k_\mu k_\nu/k^2$ diverges. We might demand simply that λ be nonzero, but then what λ are we to choose? It is necessary to start over from the beginning.

Two-dimensional analogy

Actually, the characteristic feature of eq. (7.48) that leads to our problem is quite familiar. Consider a two-dimensional integral of the form

$$I = \int_{-\infty}^{\infty} \mathrm{d}x\,\mathrm{d}y\, f(x,y), \qquad (7.50)$$

in the case that $f(x, y)$ happens to be rotationally invariant:

$$f(x,y) = f(x\cos\theta - y\sin\theta,\, x\sin\theta + y\cos\theta) \equiv F(r). \qquad (7.51)$$

The obvious first step in evaluating eq. (7.50) is to integrate over the angle, and reduce the problem to a one-dimensional integral:

$$I = 2\pi \int_0^\infty \mathrm{d}r\, rF(r). \tag{7.52}$$

This works as long as we know an explicit form for $F(r)$. Suppose, however, we only know that $f(x, y)$ satisfies eq. (7.51), i.e., that it is invariant under the group of rotations in two dimensions. This is analogous to the case of a gauge theory, where we know that the action is invariant under the local gauge transformation, eq. (5.97), which is an invariance group at each point in space–time,

$$S[A^\mu] = S[UA_\mu U^{-1} + (\mathrm{i}/g)(\partial_\mu U)U^{-1}], \tag{7.53}$$

but we are unable to write the action explicitly in terms of a reduced number of field components. Supposing, then, that we did *not* know $F(r)$ in eq. (7.51), could we still write something like eq. (7.52), in which the number of integrals is reduced? The answer is yes, and the trick is to use the rotational invariance of the integrand.

To reduce the number of integrals, it is natural to pick out a ray in the xy-plane. In terms of the Cartesian coordinates, a ray is specified by any choice $y_\theta = 0$ where $y_\theta \equiv x \sin\theta + y\cos\theta$. What we shall do is to insert this condition without changing the value of the integral. We proceed in three steps.

(i) We define a direction-fixing condition y_θ, averaged over all angles θ between zero and 2π. The result can only be a function of $r = \surd(x^2 + y^2)$, so we may write

$$\Delta(r)\int_0^{2\pi} \mathrm{d}\theta\, \delta(y_\theta) = 1, \tag{7.54}$$

which serves as a definition of $\Delta(r)$. Explicitly, the function $\Delta(r)$ is

$$\Delta((x^2 + y^2)^{1/2}) = \left(\int \mathrm{d}y_\theta \frac{\delta(y_\theta)}{|\mathrm{d}y_\theta/\mathrm{d}\theta|}\right)^{-1} = \tfrac{1}{2}(x^2 + y^2)^{1/2}, \tag{7.55}$$

where the derivative is carried out at fixed x and y. Note the factor of one-half which results from the *two* solutions for $y_\theta = 0$, at $\theta = -\tan^{-1}(y/x)$ and at $\theta = \pi - \tan^{-1}(y/x)$.

(ii) Now we insert unity into the integral I, eq. (7.50), in the form of eq. (7.54):

$$I = \int_0^{2\pi} \mathrm{d}\theta \int_{-\infty}^{\infty} \mathrm{d}x\, \mathrm{d}y\, f(x, y)\tfrac{1}{2}(x^2 + y^2)^{1/2}\delta(x\sin\theta + y\cos\theta). \tag{7.56}$$

(iii) Finally, we change variables to factor out the θ-integral:

$$I = \int_0^{2\pi} \mathrm{d}\theta \int_{-\infty}^{\infty} \mathrm{d}x'\, \mathrm{d}y'\, f(x', y')\tfrac{1}{2}(x'^2 + y'^2)^{1/2}\delta(y'), \tag{7.57}$$

where

$$x' = x\cos\theta - y\sin\theta, \quad y' = y_\theta = x\sin\theta + y\cos\theta, \qquad (7.58)$$

and where we have used the rotational invariance of $f(x, y)$ and of $\Delta(r)$. We may then integrate over θ, and use the delta function to get

$$I = \pi\int_{-\infty}^{\infty} \mathrm{d}x'\, x' f(x', 0) = 2\pi\int_0^{\infty} \mathrm{d}x'\, x' f(x', 0) \qquad (7.59)$$

which is *exactly* the form of eq. (7.52), but in the specific direction $y' = 0$.

Note, by the way, that if we had ignored one of the solutions $y_\theta = 0$ in eq. (7.59), we would have obtained twice the correct answer. The factor 2π in front is the volume of the group $SO(2)$, the group of two-dimensional orthogonal rotations. It has been factored from the remainder of the integral by inserting eq. (7.54), and changing variables. We have indeed succeeded in reducing the integral without assuming a form for $F(r)$.

Gauge theory

Essentially, we have just described the procedure that was applied to the path integral eq. (7.48) by Faddeev & Popov (1967), with a result that has served as a cornerstone for the quantization of gauge theories.

We assume that the integrand in eq. (7.48) is invariant under the gauge transformation eq. (7.53), where U is any element of Lie group G ($U(1)$, $SU(2)$ or $SU(3)$ in the standard model), at each point in space–time. Let us go through the same steps as in the example, and factor out the volume of these groups, which produces an infinite factor that is the source of the problems noted above.

We may parameterize the group elements as

$$U(x) = \exp[\mathrm{i}\boldsymbol{\alpha}(x)\cdot\mathbf{T}], \qquad (7.60)$$

with $\{T_i\}$, the generators of G (eq. (1.70)). To anticipate, we shall need a way of integrating over the group volume. This is supplied by the *measure* of the group at each point (Hamermesh r1962, Wybourne r1974). For any compact Lie group, the product of measures is given by

$$\int[\mathcal{D}U] \equiv \int[\mathcal{D}\alpha] = \int\prod_x\left\{\prod_a \mathrm{d}\alpha_a(x)\, M(\alpha_a)\right\}. \qquad (7.61)$$

(In our example of $SO(2)$ above, the measure is simply $\mathrm{d}\theta$). The measure (7.61) has the important property of being *invariant*. That is, if all group elements are multiplied from the right, for example, by a fixed element V, then

$$\int[\mathscr{D}(UV)] = \int[\mathscr{D}U]. \tag{7.62}$$

This means we can change variables in the group integral without worrying about the measure or the limits of integration. Now we are ready to start.

First we define what it means to fix a 'ray' in the functional integral eq. (7.48), to reduce the dimensionality of the integral over $A^\mu(x)$ at each point in space–time. To do this, we demand that some arbitrary functional $g_a[A^\mu(x)]$ should take on a fixed value for each x^μ. a labels a (possible) color index. The specification of $g_a[A]$ is commonly called a *gauge choice* or *gauge-fixing condition*, since it reduces the explicit gauge freedom of the original Lagrangian (although it need not eliminate it altogether).

Typical choices of $g_a[A]$ are those introduced in Section 5.3, $g_a = \partial\cdot A_a$ or $n\cdot A_a$, but more complicated choices are possible (exercises 7.8, 7.9). Since the original action is *locally* invariant, we need not choose $g_a[A]$ to take on the same value at each point x^μ. In fact, it is convenient to let it have arbitrary values, specified by a set of functions $r_a(x)$. Thus, in analogy with $\delta(y)$ (y_θ at $\theta = 0$) for the two-dimensional integral above, we define the *functional delta function*,

$$\delta[g_a[A] - r_a] \equiv \prod_{x^\mu, a} \delta(g_a[A(x)] - r_a(x)), \tag{7.63}$$

where the product is over all space–time points and group indices. We still have the freedom to transform $g_a[A]$ according to the group at each point. So, in analogy to $\delta(y_\theta)$ we introduce

$$\delta[g_a[A_U] - r_a] = \delta[g_a[UAU^{-1} + (\mathrm{i}/g)(\partial_\mu U)U^{-1}] - r_a]. \tag{7.64}$$

The remainder of the procedure is essentially identical to the two-dimensional case.

(i) The average of $\delta[g_a[A_U] - r_a]$ over all group elements defines a gauge invariant functional, $\Delta[A]$, that is analogous to eq. (7.54). Using the definition of the invariant measure, eq. (7.61) we have

$$1 = \Delta[A]\int[\mathscr{D}\alpha_a]\delta[g_a[A_U] - r_a(x)],$$

$$\Delta[A] = \left\{\int\prod_b[\mathscr{D}g_b[A_U]]\,\frac{\delta[g_a[A_U] - r_a(x)]}{|\det(\delta g_b[A_U]/\delta\alpha_a)|}\right\}^{-1}$$

$$= \left\{\sum_{A_r}\frac{1}{|\det(\delta g_b[A_U]/\delta\alpha_a)|_{A_U = A_r}}\right\}^{-1}, \tag{7.65}$$

where $\{A_r\}$ is the set of fields that satisfy the conditions $g_b[A] = r_b(x)$, and where $U = U(\alpha)$ through eq. (7.60). Variations define the determinant, since we started with a functional delta function. These variations are evaluated by using

$$\frac{\delta A_U{}^\mu{}_b(x)}{\delta\alpha_a(y)} = \delta^4(x-y)\frac{\partial}{\partial\alpha_a(x)}\left[UA^\mu U^{-1} + \frac{i}{g}(\partial^\mu U)U^{-1}\right]_b. \tag{7.66}$$

At this point we shall make the simplifying assumption that there is only a *single* such solution A_r, so that

$$\Delta[A] = |\det(\delta g_b[A_U]/\delta\alpha_a)|_{A_U=A_r}. \tag{7.67}$$

From our example above, we know that this need not always be the case. In fact, it can be shown that, depending on the choice of $g[A]$, there may or may not be more than one solution; if there is, these are called *Gribov copies* (Gribov 1978). For the purpose of developing a perturbation theory about $A^\mu = 0$, however, the presence of Gribov copies is innocuous. This is so because if the field A^μ is sufficiently small, there is always only a single solution. Since we shall concentrate on perturbation theory, we shall simply assume eq. (7.67). As indicated $\Delta(A)$ is, in general, a function of the gauge fields. This fact will have important consequences.

(ii) We are now ready to insert eq. (7.65) into the path integral, eq. (7.48), with the result

$$W = \int[\mathscr{D}\alpha]\int[\mathscr{D}A]\exp(iS[A])|\det(\delta g_c[A_U]/\delta\alpha_d)|\delta(g_b[A_U] - r_b(x)), \tag{7.68}$$

in exact analogy with the two-dimensional integral eq. (7.56).

(iii) It is then trivial to change variables ($A' = A_U$) so that the group volume at each space–time point is factored out:

$$W = \left[\int[\mathscr{D}\alpha]\right]\int[\mathscr{D}A']\exp(iS[A'])|\det(\delta g_c[A']/\delta\alpha_d)|\delta(g_b[A'] - r_b(x)), \tag{7.69}$$

where we have used the gauge invariance of the measure $[\mathscr{D}A]$ (exercise 7.7), the action and the determinant. In eq. (7.69), we have achieved our stated aim of factoring out the product of the group volume at each point.

Effective Lagrangian

Equation (7.69) is fine as far as it goes, but unlike the two-dimensional form, the delta function does not immediately give a simple result, and we still have to interpret the determinant.

To deal with the delta function, we note that, by construction, the full expression is independent of the function $r_a(x)$. We may therefore integrate functionally over all $r_a(x)$ with an exponential weight function:

$$W = N^{-1}\int[\mathfrak{D}r_c(x)]\exp\left\{-\frac{\mathrm{i}\lambda}{2}\sum_a r_a{}^2(x)\right\}W$$

$$= N^{-1}\int[\mathfrak{D}A]\exp\left\{\mathrm{i}S[A]-\frac{\mathrm{i}\lambda}{2}\left(\sum_a g_a[A]\right)^2\right\}|\det(\delta g_b[A]/\delta\alpha_c)| \quad (7.70)$$

where $N \equiv \int[\mathfrak{D}r_c(x)]\exp\{-\frac{1}{2}\mathrm{i}\lambda\sum_c r_c{}^2(x)\}$. Up to a new normalization factor, which will cancel when we calculate Green functions, we can now absorb the delta function into the exponent.

We saw in Section 7.1 that a determinant can always be interpreted as the result of a Gaussian integral over pairs of complex Grassmann variables. Thus, we write, up to a constant factor,

$$\det(\delta g_b/\delta\alpha_a) = \int[\mathfrak{D}\eta][\mathfrak{D}\bar\eta]\exp\left\{\mathrm{i}\int \mathrm{d}^4y\,\mathrm{d}^4z\,\bar\eta_b(y)[\delta g_b(y)/\delta\alpha_a(z)]\eta_a(z)\right\}. \quad (7.71)$$

The fields $\eta_a(y)$ and $\bar\eta_b(y)$ are the Feynman–DeWitt–Faddeev–Popov *ghosts* (Feynman 1963, DeWitt 1967, Faddeev & Popov 1967). These fields have no physical counterparts. They are scalars under the Lorentz group and in the adjoint representation of the gauge group. Despite their lack of spin, they are anticommuting variables, and are thus said to have the 'wrong statistics' (Section 6.5).

Including the ghosts, the full path integral may now be written as

$$W = \frac{1}{N}\int[\mathfrak{D}A][\mathfrak{D}\eta][\mathfrak{D}\bar\eta]\exp\Bigg(\mathrm{i}\int \mathrm{d}^4y\Bigg\{-\tfrac{1}{4}(F_{\mu\nu,a})^2 - \tfrac{1}{2}\,\lambda(g_a[A(y)])^2$$

$$+\int \mathrm{d}^4z\,\bar\eta_b(y)[\delta g_b(y)/\delta\alpha_a(z)]\eta_a(z)\Bigg\}\Bigg), \quad (7.72)$$

where we denote $\sum_a g_a{}^2$ by $(g_a)^2$. The new gauge-dependent exponent defines an *effective Lagrangian* for the theory.

Covariant and physical gauges

Since the determinant is gauge invariant, we may evaluate the matrix $\delta g_b/\delta\alpha_a$ in eq. (7.72) for an infinitesimal transformation at the identity $\{\alpha_a = 0\}$. For the covariant choice $g_b = \partial_\mu A^\mu{}_b$, we find from the gauge transformation of eq. (5.101) that

$$\delta g_b(y)/\delta\alpha_a(z) = -\partial_\mu{}^{(y)}\{[\delta_{ba}\partial^{(y)\mu} + gC_{bac}A^\mu{}_c(y)]\delta^4(y-z)\}$$
$$\text{(covariant gauge).} \quad (7.73)$$

In this case, the effective Lagrange density is

$$\mathcal{L}_{\text{eff}} = -\tfrac{1}{4}(F_{\mu\nu,a})^2 - \tfrac{1}{2}\lambda(\partial\cdot A_a)^2 + (\partial_\mu\bar{\eta}_a)(\delta_{ab}\partial^\mu + gC_{abc}A^\mu{}_c)\eta_b. \quad (7.74)$$

There are a number of points to be made about this form. The gauge-fixing term ensures that the vector propagator is given by eq. (7.49). The coupling of the ghost to the vector particles is proportional to the structure constants of the group, so that in the abelian case the ghosts decouple (in this gauge; see exercise 7.9). Finally, note that the coupling of the ghosts to the vectors is independent of the gauge-fixing parameter, λ.

The role of the ghosts may be clarified by re-inserting the operator equation, (7.73), back into the determinant in eq. (7.72). Recalling the discussion of the fermionic determinant, eq. (7.47), we find that the ghosts enter the vacuum-to-vacuum amplitude as an infinite sum, of the form of Fig. 7.2, of closed loop diagrams. These diagrams may be thought of as a set of nonlocal interactions for the A-field. Indeed, at this point, the ghost fields are simply a way of organizing these interactions. They will find a more direct role when we discuss unitarity (Sections 8.5 and 11.3).

Another set of standard choices are the noncovariant axial gauges, for which

$$g_a[A] = n\cdot A_a, \quad (7.75)$$

where n^μ is some fixed vector. (Sometimes the term axial gauge is reserved for the specific choice $n^\mu = \delta_{\mu 3}$.) A convenient feature of this case is seen by computing the matrix $\delta g_b/\delta\alpha_a$:

$$\delta g_b(y)/\delta\alpha_a(z) = -n_\mu[\delta_{ba}\partial^{(y)\mu} + gC_{bac}A^\mu{}_c(y)]\delta^4(y-z) \quad \text{(axial gauge)}. \quad (7.76)$$

If we take the limit $\lambda\to\infty$, then $n\cdot A_k$ will be forced to zero in the path integral, and the ghost–vector interaction, which is proportional to $n\cdot A$, will vanish. This makes the axial gauges useful for certain purposes. To see this, consider the gauge propagator, computed, as usual, by inverting the quadratic part of the effective Lagrangian (see exercise 5.8):

$$G_{\mu\nu,ab}(k,n,\lambda) = \frac{\delta_{ab}}{k^2+i\epsilon}\left[-g_{\mu\nu} + \frac{n_\mu k_\nu + k_\mu n_\nu}{n\cdot k} - n^2\left(1+\frac{k^2}{\lambda n^2}\right)\frac{k_\mu k_\nu}{(n\cdot k)^2}\right]. \quad (7.77)$$

When $\lambda\to\infty$, we can check explicitly that

$$n^\mu G_{\mu\nu,ab}(k,n,\infty) = 0 = G_{\mu\nu,ab}(k,n,\infty)n^\nu, \quad (7.78a)$$

and that no vector particle can attach to a ghost vertex in perturbation theory (exercise 7.10). In addition, these propagators have the property that contraction with k^μ cancels the 'physical' pole at $k^2 = 0$:

$$\lim_{k^2 \to 0} k^2 k^\mu G_{\mu\nu,ab}(k, n, \lambda) = 0. \tag{7.78b}$$

As a result, axial gauges are sometimes referred to as *physical gauges*, because they have no single-particle pole when contracted with a scalar polarization. The presence of extra denominators $(n \cdot k)^{-1}$ in the propagators makes calculations in such gauges rather delicate (Leibbrandt r1987).

Before going on, we should remark on the limitations of the above approach to the quantization of gauge theories. The singular nature of the gauge theory Lagrange density makes the standard connection between the phase space and configuration space path integrals, eq. (7.48), nontrivial. This connection is gauge-dependent, because in different gauges there are different relations between fields and momenta. Faddeev (1969) showed that a full treatment leads to the same effective Lagrangian as above, eq. (7.74), in a covariant gauge. Modifications of the Faddeev–Popov results become necessary, however, in the Coulomb gauge ($g_a = \mathbf{\nabla} \cdot \mathbf{A}_a$, exercise 7.8), although not in an axial gauge (Christ & Lee 1980).

Generator

The generator of Green functions for gauge theories that couple vector particles to fermions and/or scalar particles can now be constructed in the usual fashion. For definiteness, consider an unbroken nonabelian gauge theory coupled to fermions in a covariant gauge:

$$\begin{aligned} \mathscr{L} = \bar{\psi}_i[\mathrm{i}\not{D}(A)_{ij} - m\delta_{ij}]\psi_j - \tfrac{1}{4}(F_{\mu\nu,a})^2 - \tfrac{1}{2}\lambda(\partial \cdot A_a)^2 \\ + \partial_\mu \bar{\eta}_a(\delta_{ab}\partial^\mu + gC_{abc}A^\mu{}_c)\eta_b. \end{aligned} \tag{7.79}$$

The Green functions of the theory may involve external quarks, gluons and even ghosts and/or antighosts. They are generated from the functional

$$Z[J_\mu, K, \bar{K}, \xi, \bar{\xi}] = W[J_\mu, K, \bar{K}, \xi, \bar{\xi}]/W[0, 0, 0, 0, 0], \tag{7.80}$$

with sources $J_\mu, \ldots, \bar{\xi}$, where

$$\begin{aligned} W[J_\mu, K, \bar{K}, \xi, \bar{\xi}] = \int [\mathscr{D}A_\mu][\mathscr{D}\psi][\mathscr{D}\bar{\psi}][\mathscr{D}\eta][\mathscr{D}\bar{\eta}] \\ \times \exp\left\{\mathrm{i}\left[\int \mathrm{d}^4x\, \mathscr{L}_{\text{eff}}(A, \psi, \bar{\psi}, \eta, \bar{\eta}) - (J, A)\right.\right. \\ \left.\left. - (\bar{K}, \psi) - (\bar{\psi}, K) - (\bar{\xi}, \eta) - (\bar{\eta}, \xi)\right]\right\}. \end{aligned} \tag{7.81}$$

Here we introduce the convenient and common notation,

$$(J, A) \equiv \int \mathrm{d}^4 y\, J_{\mu,a}(y) A^{\mu}{}_a(y), \tag{7.82}$$

and so forth. All source terms carry the same group indices as the fields to which they couple.

Following the familiar pattern, the rules for perturbation theory are derived by factoring interactions from eq. (7.81) in terms of functional derivatives, and performing the remaining, Gaussian, integrals by a shift of the type in eq. (7.33):

$$\begin{aligned}
W[J_\mu, K, \bar{K}, \xi, \bar{\xi}] &= \exp\left\{\mathrm{i}\int \mathrm{d}^4x\, \mathcal{L}_{\mathrm{int}}[\mathrm{i}\delta_J, -\mathrm{i}\delta_K, \mathrm{i}\delta_{\bar{K}}, -\mathrm{i}\delta_\xi, \mathrm{i}\delta_{\bar{\xi}}]\right\} W_{\mathrm{free}}[J_\mu, K, \bar{K}, \xi, \bar{\xi}], \\
W_{\mathrm{free}}[J_\mu, K, \bar{K}, \xi, \bar{\xi}] &= \int [\mathcal{D}A_\mu][\mathcal{D}\psi][\mathcal{D}\bar{\psi}][\mathcal{D}\eta][\mathcal{D}\bar{\eta}] \\
&\quad \times \exp\Bigg(\mathrm{i}\int \mathrm{d}^4x \Big\{\tfrac{1}{2}A^{\mu}{}_a[g_{\mu\nu}(\partial_\alpha)^2 - (1-\lambda)\partial_\mu\partial_\nu]A^{\nu}{}_a \\
&\quad + \bar{\psi}(\mathrm{i}\partial\!\!\!/ - m)\psi - \bar{\eta}(\partial_\alpha)^2\eta \\
&\quad - (J, A) - (\bar{K}, \psi) - (\bar{\psi}, K) - (\bar{\xi}, \eta) - (\bar{\eta}, \xi)\Big\}\Bigg) \\
&= \exp\Big\{-\mathrm{i}\int \mathrm{d}^4w\, \mathrm{d}^4z [\tfrac{1}{2}J^{\mu}{}_a(w) G_{\mathrm{F},\mu\nu,ab}(w - z, \lambda) J^{\nu}{}_b(z) \\
&\qquad + \bar{K}_a(w) S_{\mathrm{F},ab}(w - z) K_b(z) \\
&\qquad + \bar{\xi}_a(w) \Delta_{\mathrm{F},ab}(w - z)\xi_b(z)]\Big\} \\
&\quad \times W_{\mathrm{free}}[0, \ldots],
\end{aligned} \tag{7.83}$$

where, for S_{F}, a, b refer to both Dirac and group indices. The propagators $G_{\mathrm{F},\mu\nu,ab}(w - z; \lambda)$, $S_{\mathrm{F},ab}$ and $\Delta_{\mathrm{F},ab}$ are the usual ones, given by eqs. (7.49), (7.36) and (2.83), and are always proportional to the unit matrix in group space.

So far, we have only discussed the inclusion of unbroken gauge theories and massless gauge particles. Below we shall also encounter gauge bosons, which have acquired mass through the Higgs mechanism, as described in

Section 5.4. We find for each such boson ($V = W^\pm, Z_0$),

$$W_{\text{free},V}[J'_\mu] = \exp\left[-\tfrac{1}{2}\mathrm{i}J'_{\mu a}(w)G'_{\mathrm{F}}{}^{\mu\nu}{}_{ab}(w-z,\lambda,M_V)J'_{\nu b}(z)\right], \quad (7.84)$$

with G'_{F} the Fourier transform of the momentum space gauge boson propagator found in exercise 5.8:

$$G'_{\mathrm{F}}{}^{\mu\nu}{}_{ab} = \frac{\delta_{ab}}{k^2 - M_V{}^2 + \mathrm{i}\epsilon}\left[-g^{\mu\nu} + \left(1 - \frac{1}{\lambda}\right)\frac{k^\mu k^\nu}{k^2 - \dfrac{M_V{}^2}{\lambda} + \mathrm{i}\epsilon}\right], \quad (7.85)$$

with the $\mathrm{i}\epsilon$ prescriptions defined, as usual, by the requirements of Wick rotation (Section 3.1). In the full standard model of electroweak interactions, there are yet more contributions, from the Higgs and the ghost fields of $SU(2) \times U(1)$ (Appendix C). These may be incorporated similarly. We shall not, however, put them to use directly in what follows.

The perturbative expansion of eq. (7.83) and its implementation in terms of free theory Green functions, via the Gell-Mann–Low formula and Wick's theorem, is a direct generalization of our previous discussions. As usual, the constant $W_{\text{free}}[0]$ cancels between numerator and denominator in Z, eq. (7.80), as do all disconnected graphs, so that eq. (7.80) is indeed the generator of scattering Green functions, without vacuum bubbles. The resulting Feynman rules will be developed in the following chapter.

7.4 Reduction formulas and cross sections

We are now ready to specify the reduction formulas for gauge theories, as well as to introduce a few special features in the calculation of cross sections for particles with spin. As above, we shall simply quote those results that are rather straightforward generalizations of the rules for scalar fields.

Reduction for fermions and vectors

We first consider the reduction of A particles and B antiparticles from the initial state, and C particles and D antiparticles from the final state. To fix the sign, we define the relevant T-matrix element by

$$\begin{aligned}
&\mathrm{i}T_\psi{}^{(\zeta\xi)}((\mathbf{p}'_1,\lambda'_1)^{(-)},\ldots,(\mathbf{p}'_C,\lambda'_C)^{(-)};(\mathbf{q}'_1,\sigma'_1)^{(+)},\ldots,(\mathbf{q}'_D,\sigma'_D)^{(+)};\\
&\qquad(\mathbf{p}_1,\lambda_1)^{(-)},\ldots,(\mathbf{p}_A,\lambda_A)^{(-)};(\mathbf{q}_1,\sigma_1)^{(+)},\ldots,(\mathbf{q}_B,\sigma_B)^{(+)})\\
&\quad\equiv\langle\zeta\ \text{out}|b_{\lambda'_1}(\mathbf{p}'_1)^{(\text{out})}\cdots b_{\lambda'_C}(\mathbf{p}'_C)^{(\text{out})}d_{\sigma'_1}(\mathbf{q}'_1)^{(\text{out})}\cdots d_{\sigma'_D}(\mathbf{q}'_D)^{(\text{out})}\\
&\qquad\times d^\dagger{}_{\sigma_1}(\mathbf{q}_1)^{(\text{in})}\cdots d^\dagger{}_{\sigma_B}(\mathbf{q}_B)^{(\text{in})}b^\dagger{}_{\lambda_1}(\mathbf{p}_1)^{(\text{in})}\cdots b^\dagger{}_{\lambda_A}(\mathbf{p}_A)^{(\text{in})}|\xi\ \text{in}\rangle,
\end{aligned} \quad (7.86)$$

where $b^{\dagger}{}_{\lambda_1}(\mathbf{p}_1)^{(\text{in})}$ creates a particle in the in-state, and similarly for the other operators. ξ and ζ label the remaining particles in the in-and out-states. This matrix element is related to a matrix element of the time-ordered product:

$$\begin{aligned}
&G_{\psi}{}^{(\zeta\xi)}{}_{\{a'_k;b'_l;a_i;b_j\}}(p'_1, \ldots, q'_1, \ldots; p_1, \ldots, q_1, \ldots) \\
&\quad = \int \mathrm{d}^4x'_1 \cdots \mathrm{d}^4y'_1 \cdots \mathrm{d}^4x_1 \cdots \mathrm{d}^4y_1 \cdots \\
&\qquad \times \exp[\mathrm{i}(p'_1{\cdot}x'_1 + \cdots + q'_1{\cdot}y'_1 + \cdots - p_1{\cdot}x_1 - \cdots - q_1{\cdot}y_1 \cdots)] \\
&\qquad \times \langle \zeta\ \text{out}|T[\psi_{a'_1}(x'_1) \cdots \bar{\psi}_{b'_1}(y'_1) \cdots \psi_{b_1}(y_1) \cdots \bar{\psi}_{a_1}(x_1)]|\xi\ \text{in}\rangle.
\end{aligned} \tag{7.87}$$

ξ and η may or may not be vacuum states. As in the scalar case, the transition matrix element is related to the *truncated* Green function, with the propagators of external fields removed (including factors of i). In these terms, the reduction formula for fermions is given by

$$\begin{aligned}
&\mathrm{i}T_{\psi}{}^{(\zeta\xi)}((\mathbf{p}'_1, \lambda'_1)^{(-)}, \ldots, (\mathbf{p}'_C, \lambda'_C)^{(-)}; (\mathbf{q}'_1, \sigma'_1)^{(+)}, \ldots, (\mathbf{q}'_D, \sigma'_D)^{(+)}; \\
&\qquad (\mathbf{p}_1, \lambda_1)^{(-)}, \ldots, (\mathbf{p}_A, \lambda_A)^{(-)}; (\mathbf{q}_1, \sigma_1)^{(+)}, \ldots, (\mathbf{q}_B, \sigma_B)^{(+)}) \\
&\quad = [R_{\psi}{}^{-1/2}(2\pi)^{-3/2}\bar{u}_{a'_1}(p'_1, \lambda'_1)] \cdots [-(2\pi)^{-3/2}R_{\psi}{}^{-1/2}\bar{v}_{b_1}(q_1, \sigma_1)] \cdots \\
&\qquad \times G_{\psi}{}^{(\zeta\xi)}{}_{\{a'_k;b'_l;a_i;b_j\}}(\bar{p}'_1, \ldots, \bar{q}'_1, \ldots; \bar{p}_1, \ldots, \bar{q}_1, \ldots)_{\text{truncated}} \\
&\qquad \times \cdots [-(2\pi)^{-3/2}R_{\psi}{}^{-1/2}v_{b'_1}(q'_1, \sigma'_1)] \cdots [(2\pi)^{-3/2}R_{\psi}{}^{-1/2}u_{a_1}(p_1, \lambda_1)] \\
&\qquad \times \cdots,
\end{aligned} \tag{7.88}$$

with, as usual, $\bar{p}_i{}^2 = m^2$, etc. R_{ψ} is, by analogy with eq. (4.15), the residue of the full fermion two-point function at $p^2 = m^2$,

$$\int \mathrm{d}^4x\, \mathrm{e}^{\mathrm{i}p\cdot x}\langle 0|T[\psi(x)\bar{\psi}(0)]|0\rangle = \mathrm{i}R_{\psi}(\not{p} - m + \mathrm{i}\epsilon)^{-1} + \text{finite remainder.} \tag{7.89}$$

Again, as in the scalar case, the quantity R_{ψ} begins at zeroth order in the effective coupling, and we need not worry about it until we compute high-order corrections to the scattering matrix.

In summary, to derive S-matrix elements from Green functions, we truncate the external propagators (including the factor of i in $\mathrm{i}S_{\mathrm{F}}$), and then replace them by factors $R_{\psi}{}^{-1/2}u(p, \lambda)$ for incoming particles, $R_{\psi}{}^{-1/2}\bar{u}(p, \lambda)$ for outgoing particles, $-R_{\psi}{}^{-1/2}\bar{v}(q, \sigma)$ for incoming antiparticles, and $-R_{\psi}{}^{-1/2}v(q, \sigma)$ for outgoing antiparticles. The resulting sign applies when the standard ordering of operators in the time-ordered product is the same as the ordering of the corresponding creation and annihilation operators in the in- and out-states. Relative permutations of any of the fermionic operators will, in general, change the sign.

Reduction for vector particles

The procedure for vector particles is essentially identical, but simpler, since they are Bose fields, whose free states have no intrinsic ordering. Therefore, following the steps above, the reduction formula is

$$
\begin{aligned}
\mathrm{i}T_A{}^{(\zeta\xi)}(\{(\mathbf{q}'_j,\lambda'_j)\};\{(\mathbf{q}_i,\lambda_i)\}) &\equiv \langle\zeta\ \mathrm{out}|\textstyle\prod_j a_{\lambda'_j}(\mathbf{q}'_j)^{(\mathrm{out})}\prod_i a^\dagger{}_{\lambda_i}(\mathbf{q}_i)^{(\mathrm{in})}|\xi\ \mathrm{in}\rangle \\
&= \prod_j[-\,(2\pi)^{-3/2}R_A{}^{-1/2}\epsilon^{*\nu_j}(q'_j,\lambda'_j)] \\
&\quad\times G_A{}^{(\zeta\xi)}{}_{\{\nu'_j;\mu_i\}}(\{\bar{q}'_j\};\{\bar{q}_i\})_{\mathrm{truncated}} \\
&\quad\times \prod_i[-(2\pi)^{-3/2}R^{-1/2}\epsilon^{\mu_i}(q_i,\lambda_i)],
\end{aligned}
\qquad (7.90)
$$

where the full (untruncated) Green function is

$$
\begin{aligned}
&G_A{}^{(\zeta\xi)}{}_{\{\nu'_j;\mu_i\}}(\{q'_j\};\{q_i\}) \\
&\qquad = \int \mathrm{d}^4y_1\cdots\mathrm{d}^4x_1\cdots\exp[\mathrm{i}(q'_1\cdot y_1+\cdots-q_1\cdot x_1-\cdots)] \\
&\qquad\quad\times\langle\zeta\ \mathrm{out}|T[\textstyle\prod_j A_{\nu'_j}(y_j)\prod_i A_{\mu_i}(x_i)]|\xi\ \mathrm{in}\rangle.
\end{aligned}
\qquad (7.91)
$$

Here the rule for forming the T-matrix is to truncate (again including the factor of i in $G_{\mu\nu}$), and to replace the propagators by $R_A{}^{-1/2}\epsilon^\mu{}_\lambda(\mathbf{k})$ for an incoming particle, and $R_A{}^{-1/2}\epsilon^{*\nu}{}_\lambda(\mathbf{k})$ for an outgoing one, where R_A is the residue at the one-particle pole in the vector two-point function.

Example

As an example, consider the process $\mathrm{e}^-(\mathbf{k}_1,s_1)+\mathrm{e}^+(\mathbf{k}_2,s_2)\rightarrow \mathrm{e}^-(\mathbf{p}_1,\sigma_1)+\mathrm{e}^+(\mathbf{p}_2,\sigma_2)$, where $\mathrm{e}^-(\mathbf{k}_1,s_1)$ represents an electron of momentum $\mathbf{k}_1$ and spin s_1, and similarly for the positron. The corresponding S-matrix element is

$$
\begin{aligned}
&\langle(\mathbf{p}_1,\sigma_1)^{(-)},(\mathbf{p}_2,\sigma_2)^{(+)}\ \mathrm{out}|(\mathbf{k}_1,s_1)^{(-)},(\mathbf{k}_2,s_2)^{(+)}\ \mathrm{in}\rangle \\
&\quad = [R_e{}^2(2\pi)^6]^{-1}[\mathrm{i}\bar{u}_c(p_1,\sigma_1)(\not{p}_1-m)_{c\gamma}][-\mathrm{i}\bar{v}_b(k_2,s_2)(-\not{k}_2-m)_{b\beta}] \\
&\qquad\times\int \mathrm{d}^4x_1\,\mathrm{d}^4y_1\,\mathrm{d}^4x_2\,\mathrm{d}^4y_2\exp(\mathrm{i}\bar{p}_1\cdot y_1+\mathrm{i}\bar{p}_2\cdot y_2-\mathrm{i}\bar{k}_1\cdot x_1-\mathrm{i}\bar{k}_2\cdot x_2) \\
&\qquad\times\langle 0|T[\psi_\gamma(y_1)\bar{\psi}_\delta(y_2)\psi_\beta(x_2)\bar{\psi}_\alpha(x_1)]|0\rangle \\
&\qquad\times(-\not{p}_2-m)_{\delta d}[-\mathrm{i}v_d(p_2,\sigma_2)](\not{k}_1-m)_{\alpha a}[\mathrm{i}u_a(k_1,s_1)]\big|_{p_i^2=k_i^2=m^2}.
\end{aligned}
\qquad (7.92)
$$

The factors $\not{p}_i - m$ serve to truncate diagrams in the expansion of the Green function.

Cross sections and averaging

Since particles with spin and internal symmetries have more quantum numbers than scalar particles do, there are correspondingly more varied cross sections for the former than the latter. We may, for example, compute the scattering of particles with different polarizations in the initial or final states. It is often convenient or necessary, however, to leave spins or internal quantum numbers unspecified in the initial and/or final state. For instance, we may be unwilling or unable to produce initial-space particles in a definite state of polarization. We then measure an *unpolarized* cross section. Similarly, we may not measure the polarization of the particles in the final states. It is customary to summarize the treatment of these cases by the following rule: *sum over final states and average over initial states*. The sum over different final states is simply the momentum-space integral in eq. (4.38), supplemented by sums over spins and any other quantum numbers as appropriate. The average over initial states requires more justification. The cross section for a specific incoming state $|n \text{ in}\rangle$ is proportional to the expectation values:

$$\sigma(n, m) = \sum_m \langle n \text{ in}|m \text{ out}\rangle \langle m \text{ out}|n \text{ in}\rangle. \tag{7.93}$$

Now suppose any initial state can be expanded in terms of basis states of definite spin or internal symmetry, denoted by κ,

$$|n \text{ in}\rangle = \sum_{\kappa=1}^{N} c_{n\kappa}|\kappa \text{ in}\rangle, \tag{7.94}$$

where we have suppressed momentum labels. N may represent the number of independent spin states or the number of independent states in the representation of an internal symmetry group. By 'unpolarized', we mean that all $c_{n\kappa} = c^{(1)}{}_{n\kappa} + \mathrm{i}c^{(2)}{}_{n\kappa}$ are equally likely. Then the sum over all possible $|n \text{ in}\rangle$ is equivalent to a normalized integral over all possible $c_{n\kappa}$, which can be written as

$$\sum_n \sigma(n, m) = \prod_\kappa \int \mathrm{d}c^{(1)}{}_{n\kappa}\, \mathrm{d}c^{(2)}{}_{n\kappa}\, \delta(1 - \textstyle\sum_\kappa |c_{n\kappa}|^2) \displaystyle\sum_{\kappa_1=1}^{N} \sum_{\kappa_2=1}^{N} c_{n\kappa_1}(c_{n\kappa_2})^*$$
$$\times \langle \kappa_2 \text{ in}|m \text{ out}\rangle \langle m \text{ out}|\kappa_1 \text{ in}\rangle. \tag{7.95}$$

If $c_{n\kappa}$ is permissible, so is $-c_{n\kappa}$. We can therefore apply symmetric integration to eq. (7.95), and we easily find

$$\sum_n \sigma(n, m) = \sum_{\kappa_1=1}^{N} \sum_{\kappa_2=1}^{N} \frac{1}{N} \delta_{\kappa_1\kappa_2} \sum_m \langle \kappa_2 \text{ in}|m \text{ out}\rangle\langle m \text{ out}|\kappa_1 \text{ in}\rangle,$$

$$= (1/N)\textstyle\sum_\kappa \sum_m |\langle m \text{ out}|\kappa \text{ in}\rangle|^2. \tag{7.96}$$

This form exhibits the average over initial basis states explicitly.

Exercises

7.1 Verify the action of derivatives $\mathrm{d}/\mathrm{d}a_i$ from the left, eq. (7.7) and verify that derivatives with respect to anticommuting numbers anticommute with each other.

7.2 Verify the fermionic completeness relation, eq. (7.27). To treat constant factors, recall that the integral is equivalent to a derivative.

7.3 Show that the order-e^2 expansion of the determinant in eq. (7.47) gives the same result as the direct perturbative expansion of the fermionic generator with a background field, eq. (7.45).

7.4 Write down the momentum-space expression for the diagrams in Fig. 7.2, in terms of momentum integrals and Dirac traces.

7.5 Starting from the free Dirac Lagrangian with sources K and $\bar{K}$, as in the eq. (7.2), develop the analogue of the scalar particle Hamiltonian of eq. (3.48).

7.6 Following the analogue of the scalar theory in the previous problem, show that eq. (7.35) is indeed the vacuum-to-vacuum amplitude for the free Dirac theory with sources.

7.7 Show that the measure $[\mathcal{D}A]$ is invariant under the change of variables $A' = A_U$ in eqs. (7.68) and (7.69).

7.8 The *Coulomb gauge* is constructed by choosing $g_j[A] = \nabla \cdot \mathbf{A}_j$. Derive the vector propagator that follows from the Faddeev–Popov method in this gauge, and show that it is the sum of a *transverse propagator*

$$G^{\mu\nu}{}_{\text{trans}} = -(k^2)^{-1}(g^{\mu\nu} + k^\mu k^\nu/|\mathbf{k}|^2)(1 - \delta_{\mu 0})(1 - \delta_{\nu 0}),$$

and an *instantaneous Coulomb interaction*,

$$G^{\mu\nu}{}_{\text{Coul}} = \delta_{\mu 0}\delta_{\nu 0}(1/|\mathbf{k}|^2).$$

7.9 Derive the effective Lagrangian in an abelian theory with the gauge choice $g[\mathrm{A}] = \partial_\mu A^\mu + \frac{1}{2}\lambda' A^2$ (Dirac 1951), and show that it has nontrivial ghost–vector interactions.

7.10 Justify the claim after eq. (7.78a) that ghosts act as free fields in the $\lambda \rightarrow \infty$ limit of the axial gauges.

7.11 Calculate the S-matrix elements $\langle(\mathbf{p}_1, \lambda_1)^{(-)}, (\mathbf{p}_2, \lambda_2)^{(+)} \text{ out}|0 \text{ in}\rangle$, and $\langle(\mathbf{p}_2, \lambda_2)^{(-)} \text{ out}|(\mathbf{p}_1, \lambda_1)^{(-)} \text{ in}\rangle$, to order e, for the Dirac field with an external electromagnetic background, eq. (7.28).

8

Gauge theories at lowest order

The developments above bring us to a multitude of realistic cross sections based on the standard model, a selection of which we shall use to introduce elements of the perturbative expansion for gauge theories at lowest order. We shall discuss examples from the electromagnetic, weak and strong interactions.

8.1 Quantum electrodynamics and elastic fermion–fermion scattering

Lagrangian and sources

A gauge-fixed Lagrangian for quantum electrodynamics (QED) is given by eqs. (5.68) and (5.91):

$$\mathcal{L}_{\text{QED}} = \sum_f \bar{\psi}_f(x)[\mathrm{i}\gamma_\mu(\partial^\mu + \mathrm{i}eQ_f A^\mu) - m]\psi_f(x)$$
$$- \tfrac{1}{4}F^{\mu\nu}(x)F_{\mu\nu}(x) - \tfrac{1}{2}\lambda[\partial_\mu A^\mu(x)]^2, \qquad (8.1)$$

where the sum is over fundamental fermion fields, f, with particles of charge eQ_f, e being the charge of the positron. In the standard model we have leptons $f = \mathrm{e}, \mu, \tau$ (electron, muon and tau), all with $Q_f = -1$, quarks $f = \mathrm{u}$, c and t (up, charmed and top) with $Q_f = \frac{2}{3}$, and quarks $f = \mathrm{d}$, s and b (down, strange and bottom) with $Q_f = -\frac{1}{3}$. Quarks participate directly in strong interactions, while leptons do not. For the purposes of QED, however, they are all on an equal footing. Often f is referred to as the *flavor* of the field.

Feynman rules

The Feynman rules for the Green functions of QED are simple generalizations of the rules for fermions in a background field. They are found (in a covariant gauge) from the Lagrange density, eq. (8.1) in the perturbative generator eq. (7.83). The essential new feature is that fermion lines may communicate via gauge field propagators.

The rules in momentum space are now as follows.

(i) Write down all distinguishable diagrams with the correct number of external lines. The arrows of fermion lines flow into external fields ψ_f, and out of fields $\bar{\psi}_f$. Fermion lines are always continuous, with one arrow flowing in and one out at each internal vertex. An integral is constructed for each diagram according to the remaining rules.

(ii) Associate with each fermion line $\mathrm{i}(2\pi)^{-4}\int \mathrm{d}^4k[(\not{k} - m)^{-1}]_{ba}$, where the diagramatic arrow of the propagator points from the vertex with index a to the vertex with index b. The momentum k is identified with the momentum flowing in the direction of the arrow. The propagator is the identity matrix in flavor, and in any other symmetry.

(iii) Associate with each internal photon line the factor

$$\frac{\mathrm{i}}{(2\pi)^4}\int \mathrm{d}^4k \frac{1}{k^2 + \mathrm{i}\epsilon}\left[-g^{\mu\nu} + \left(1 - \frac{1}{\lambda}\right)\frac{k^\mu k^\nu}{k^2 + \mathrm{i}\epsilon}\right],$$

where the line connects vertices with vector indices μ and ν.

(iv) Associate with each interaction vertex the matrix factor

$$-\mathrm{i}eQ_f\delta_{ff'}(\gamma^\mu)_{dc}(2\pi)^4\delta^4(\textstyle\sum_i p_i),$$

where the p_i^μ label the momenta of lines flowing into the vertex. The index c is summed with the fermion propagator whose arrow points into the graph, d with the propagator whose arrow points out and μ with the photon propagator.

(v) Associate a relative minus sign between those graphs that differ by the exchange of two external fermion fields of the same flavor. The absolute sign of the Green function, if needed, should be computed according to the convention introduced after eq. (7.43).

(vi) Associate a factor (-1) with every fermion loop.

Symmetry factors may occur in vacuum bubble diagrams in QED; they can be constructed explicitly from the generating functional.

Green functions in momentum space

Consider, as an example, the four-fermion Green function in momentum space, given to lowest order by Figs. 8.1(a) and (b),

$$\left\{\int \prod_{i=1}^{4} \mathrm{d}^4y_i \exp(-\mathrm{i}k_i\cdot y_i)\right\}\langle 0|T[\psi_b(y_2)\bar{\psi}_a(y_1)\bar{\psi}_d(y_4)\psi_c(y_3)]|0\rangle$$

$$\begin{aligned} &= (2\pi)^4\delta^4(\textstyle\sum_{i=1}^4 k_i)[\mathrm{i}(-\not{k}_2 - m)^{-1}(-\mathrm{i}e\gamma^\mu)\mathrm{i}(\not{k}_1 - m)^{-1}]_{ba} \\ &\quad \times [\mathrm{i}(k_1 + k_2)^{-2}][-g_{\mu\nu} + (1 - \lambda^{-1})(k_1 + k_2)_\mu(k_1 + k_2)_\nu(k_1 + k_2)^{-2}] \\ &\quad \times [\mathrm{i}(-\not{k}_3 - m)^{-1}(-\mathrm{i}e\gamma^\nu)\mathrm{i}(\not{k}_4 - m)^{-1}]_{cd} - [(k_2, b) \leftrightarrow (k_3, c)]. \qquad (8.2) \end{aligned}$$

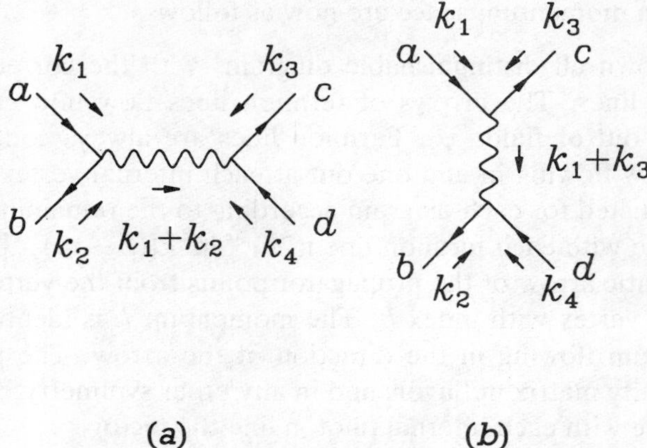

Fig. 8.1 Diagrams corresponding to eq. (8.2).

The first term is Fig. 8.1(a). The second, Fig. 8.1(b), which differs by the exchange of momenta and indices as indicated, includes a relative minus sign, according to rule (v) above.

Similarly, the two-fermion two-vector Green function in momentum space is given by

$$\left\{\int\prod_{i=1}^{4} \mathrm{d}^4 y_i \exp(-\mathrm{i}k_i\cdot y_i)\right\}\langle 0|T[\psi_b(y_3)\bar{\psi}_a(y_1)A_\beta(y_4)A_\alpha(y_2)]|0\rangle$$
$$= (2\pi)^4\delta^4(\textstyle\sum_{i=1}^4 k_i)\{\mathrm{i}(-\not{k}_3 - m)^{-1}(-\mathrm{i}e\gamma^\nu)\mathrm{i}[(\not{k}_1+\not{k}_2) - m]^{-1}$$
$$\times(-\mathrm{i}e\gamma^\mu)\mathrm{i}(\not{k}_1 - m)^{-1}\}_{ba}$$
$$\times[\mathrm{i}k_2{}^{-2}][-g_{\alpha\mu} + (1-\lambda^{-1})(k_2)_\alpha(k_2)_\mu k_2{}^{-2}]$$
$$\times[\mathrm{i}k_4{}^{-2}][-g_{\beta\nu} + (1-\lambda^{-1})(k_4)_\beta(k_4)_\nu k_4{}^{-2}] + [(k_2,\alpha)\leftrightarrow(k_4,\beta)], \tag{8.3}$$

shown in Fig. 8.2. Both these expressions are a little complicated, and,

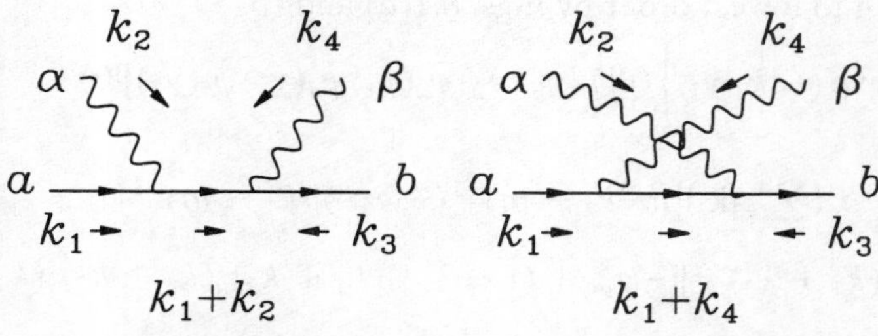

Fig. 8.2 Diagrams corresponding to eq. (8.3).

more seriously, they depend on the gauge parameter λ. Let us see what happens when we calculate S-matrix elements and cross sections.

The S-matrix for Bhabha scattering

Consider the S-matrix element for electron–positron scattering derived at $O(e^2)$ from eq. (8.2). This is known as *Bhabha scattering* (Bhabha 1936), and is the same for any fermion–antifermion pair, up to the charge of the field. It is closely related (by 'crossing', see Section 4.4) to electron–electron elastic scattering (Møller 1932), which may be evaluated from the same Green function by the same techniques (exercise 8.1).

Using the reduction formula eq. (7.88), and replacing the R's by unity (as is appropriate at lowest order), we cancel the four external propagators, replacing them by solutions to the free fermion equation of motion. This gives the simpler expression

$$\begin{aligned}
&\mathrm{i}M((\mathbf{p}_3, \sigma_3)^{(-)}, (\mathbf{p}_4, \sigma_4)^{(+)}; (\mathbf{p}_2, \sigma_2)^{(+)}, (\mathbf{p}_1, \sigma_1)^{(-)}) \\
&= \bar{v}(p_2, \sigma_2)(-\mathrm{i}e\gamma^\mu)u(p_1, \sigma_1)\bar{u}(p_3, \sigma_3)(-\mathrm{i}e\gamma^\nu)v(p_4, \sigma_4) \\
&\quad \times [\mathrm{i}(p_1 + p_2)^{-2}][-g_{\mu\nu} + (1 - \lambda^{-1})(p_1 + p_2)_\mu(p_1 + p_2)_\nu(p_1 + p_2)^{-2}] \\
&\quad - \bar{v}(p_2, \sigma_2)(-\mathrm{i}e\gamma^\mu)v(p_4, \sigma_4)\bar{u}(p_3, \sigma_3)(-\mathrm{i}e\gamma^\nu)u(p_1, \sigma_1) \\
&\quad \times [\mathrm{i}(p_3 - p_1)^{-2}][-g_{\mu\nu} + (1 - \lambda^{-1})(p_1 - p_3)_\mu(p_1 - p_3)_\nu(p_1 - p_3)^{-2}],
\end{aligned} \tag{8.4}$$

where we have relabelled the momenta in Fig. 8.1 by $p_i = k_i$, $i = 1, 2$ and $p_i = -k_i$, $i = 3, 4$. Although simpler than eq. (8.2), eq. (8.4) seems still to have gauge-parameter dependence. If we are to take this computation seriously, however, the physical S-matrix simply cannot depend on λ, since the latter is supposed to be purely a calculational convenience. And indeed, λ-dependence cancels in the S-matrix, although *not* in Green functions. Let us see how this happens here.

In the first term of eq. (8.4) (Fig. 8.1(a)), we find the factor

$$\begin{aligned}
&(1 - \lambda^{-1})(-\mathrm{i}e)\bar{v}(p_2, \sigma_2)(\not{p}_1 + \not{p}_2)u(p_1, \sigma_1) \\
&\quad = (1 - \lambda^{-1})(-\mathrm{i}e)\bar{v}(p_2, \sigma_2)[(\not{p}_2 + m) + (\not{p}_1 - m)]u(p_1, \sigma_1) = 0,
\end{aligned} \tag{8.5}$$

where we have added and subtracted m, and used the Dirac equation in momentum space, from eq. (6.39a). The same trick can be played with the factor $(p_1 - p_3)_\nu$ from Fig. 8.1(b). The terms proportional to $(1 - \lambda^{-1})$ may, therefore, be dropped, and we end up with the still more manageable expression

$$
\begin{aligned}
&\mathrm{i}M((\mathbf{p}_3,\sigma_3)^{(-)},(\mathbf{p}_4,\sigma_4)^{(+)};(\mathbf{p}_2,\sigma_2)^{(+)},(\mathbf{p}_1,\sigma_1)^{(-)})\\
&\quad = \mathrm{i}e^2[\bar{v}(p_2,\sigma_2)\gamma^\mu u(p_1,\sigma_1)][\bar{u}(p_3,\sigma_3)\gamma_\mu v(p_4,\sigma_4)][(p_1+p_2)^{-2}]\\
&\qquad - \mathrm{i}e^2[\bar{u}(p_3,\sigma_3)\gamma^\mu u(p_1,\sigma_1)][\bar{v}(p_2,\sigma_2)\gamma_\mu v(p_4,\sigma_4)][(p_1-p_3)^{-2}].
\end{aligned}
\tag{8.6}
$$

Unpolarized cross sections and traces

We can now compute cross sections for fermion–antifermion scattering. The steps leading to eq. (4.43) for scalar fields may be repeated, with some necessary changes made in the interpretation of the incoming particle wave functions, eq. (4.28). (Also, we note that identical fermions in the final state need a combinatoric factor $(n!)^{-1}$.) The result for the equal-mass unpolarized cross section at hand is then

$$\frac{\mathrm{d}\sigma}{\mathrm{d}t} = \frac{1}{64\pi s(s-4m^2)}\sum_{\text{spins}}|M(s,t)|^2, \tag{8.7}$$

where one factor of $\frac{1}{4}$ results from the average over initial-state spins, according to the discussion in Section 7.4. From eq. (8.6) we have

$$|M|^2 = e^4\big|(1/s)(\bar{v}_2\gamma^\mu u_1)(\bar{u}_3\gamma_\mu v_4) - (1/t)(\bar{u}_3\gamma^\alpha u_1)(\bar{v}_2\gamma_\alpha v_4)\big|^2, \tag{8.8}$$

where we have introduced the notation $u_i \equiv u(p_i,\sigma_i)$, and similarly for v_i, and where the invariants s, t and u are defined in eq. (4.18).

To evaluate expressions such as eq. (8.8), we need to deal with the complex conjugate of a product of gamma matrices between a spinor and a conjugate spinor. Such products can be simplified by using the identity

$$
\begin{aligned}
&(\bar{u}_i\gamma_{\alpha_1}\gamma_{\alpha_2}\cdots\gamma_{\alpha_i}\gamma_5\cdots\sigma_{\mu\nu}\cdots\gamma_{\alpha_n}u_j)^*\\
&\qquad = \bar{u}_j\gamma_{\alpha_n}\gamma_{\alpha_{n-1}}\cdots\gamma_{\alpha_i}\gamma_5\cdots\sigma_{\mu\nu}\cdots\gamma_{\alpha_1}u_i,
\end{aligned}
\tag{8.9}
$$

where we have included, for completeness, the factors $\gamma_\mu\gamma_5$ and $\sigma_{\mu\nu}$. Taking the complex conjugate of the product exchanges the roles of the two spinors, and reverses the order of the gamma matrices, although preserving the relative order within the combinations $\gamma_\mu\gamma_5$ and $\sigma_{\mu\nu}$. To prove eq. (8.9), we note that, from eq. (5.44), $(\gamma_0\gamma_\mu)^* = (\gamma_0\gamma_\mu)^{\mathrm{t}}$, so that

$$[(\bar{u}_i)_a(\gamma_\mu)_{ab}(u_j)_b]^* = (u_i)_c(\gamma_0\gamma_\mu)^*{}_{cb}(u_j{}^*)_b = (u_j{}^*)_b(\gamma_0\gamma_\mu)_{bc}(u_i)_c = \bar{u}_j\gamma_\mu u_i. \tag{8.10}$$

Products with more gamma matrices work in exactly the same fashion.

Using eq. (8.9) in (8.8) gives four terms:

$$
\begin{aligned}
|M|^2 = {} & e^4(1/s^2)(\bar{v}_2\gamma^\mu u_1)(\bar{u}_1\gamma^\nu v_2)(\bar{u}_3\gamma_\mu v_4)(\bar{v}_4\gamma_\nu u_3) \\
& + e^4(1/t^2)(\bar{u}_1\gamma^\beta u_3)(\bar{u}_3\gamma^\alpha u_1)(\bar{v}_2\gamma_\alpha v_4)(\bar{v}_4\gamma_\beta v_2) \\
& - e^4(1/st)(\bar{u}_1\gamma^\beta u_3)(\bar{u}_3\gamma_\mu v_4)(\bar{v}_4\gamma_\beta v_2)(\bar{v}_2\gamma^\mu u_1) \\
& - e^4(1/ts)(\bar{u}_1\gamma_\nu v_2)(\bar{v}_2\gamma_\alpha v_4)(\bar{v}_4\gamma^\nu u_3)(\bar{u}_3\gamma^\alpha u_1). \qquad (8.11)
\end{aligned}
$$

It is useful to think of these terms graphically, as shown in Fig. 8.3. Here we have drawn the square of the matrix element as the graphs of M linked by the final state to the graphs of M^* drawn backwards, in accordance with eq. (8.9). We term each such graphical representation a *cut graph*, where

(a)

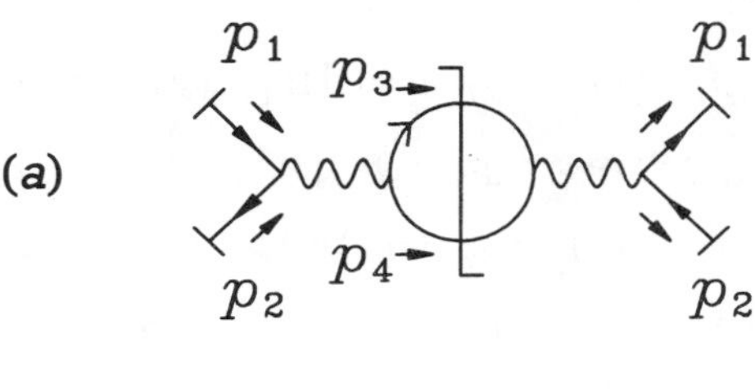

(b)

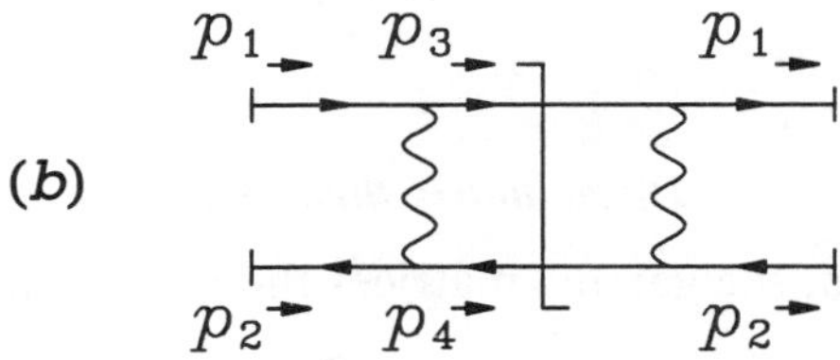

(c)

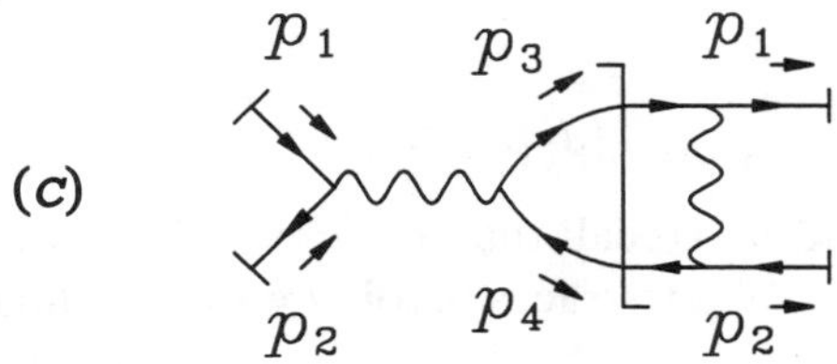

(d)

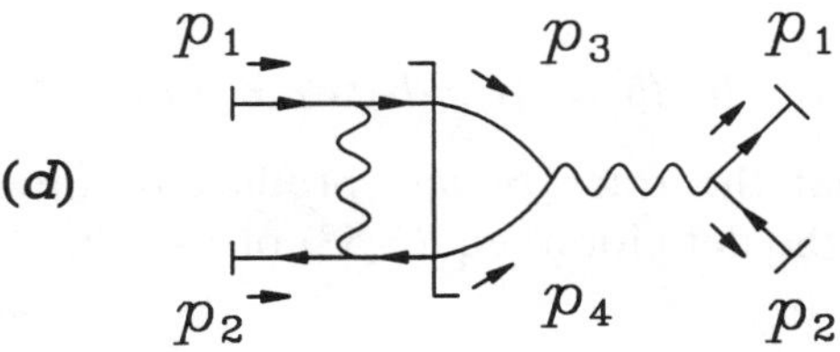

Fig. 8.3 Cut diagrams for Bhabha scattering.

'cut' refers to the final state (Kinoshita 1950, Nakanishi 1958, Veltman 1963). The final-state particles of a cut graph are intersected by a single vertical line, which divides the graph into contributions from M (on the left in the figure) and M^* (on the right). Each of the terms in eq. (8.11) corresponds to one of the cut diagrams in Fig. 8.3.

In eq. (8.11), each spinor appears with its matching conjugate in the product. We can then use the projection sum in eq. (6.39a), with $|c|^2 = 2m$, to perform the sum over spins. This leads to an expression in terms of traces of products of Dirac matrices:

$$\sum_{\text{spins}} |M|^2 = e^4[(1/s^2)T_a + (1/t^2)T_b - (1/st)(T_c + T_d)], \tag{8.12}$$

where the T_i are trace factors, given by

$$\begin{aligned} T_a &= \operatorname{tr}[\gamma^\nu(\not{p}_2 - m)\gamma^\mu(\not{p}_1 + m)]\operatorname{tr}[\gamma_\mu(\not{p}_4 - m)\gamma_\nu(\not{p}_3 + m)], \\ T_b &= \operatorname{tr}[\gamma^\beta(\not{p}_3 + m)\gamma^\alpha(\not{p}_1 + m)]\operatorname{tr}[\gamma_\alpha(\not{p}_4 - m)\gamma_\beta(\not{p}_2 - m)], \\ T_c &= \operatorname{tr}[\gamma^\beta(\not{p}_3 + m)\gamma_\mu(\not{p}_4 - m)\gamma_\beta(\not{p}_2 - m)\gamma^\mu(\not{p}_1 + m)], \\ T_d &= \operatorname{tr}[\gamma_\nu(\not{p}_2 - m)\gamma_\alpha(\not{p}_4 - m)\gamma^\nu(\not{p}_3 + m)\gamma^\alpha(\not{p}_1 + m)]. \end{aligned} \tag{8.13}$$

We must now learn how to evaluate these traces.

Dirac matrix theorems

In any representation, the gamma matrices themselves are traceless, i.e.

$$\operatorname{tr}\gamma_\mu = 0. \tag{8.14}$$

A number of identities follow the anticommutation relations, eq. (5.43), which imply

$$\{\not{a}, \not{b}\}_+ = 2a\cdot b \tag{8.15}$$

for any vectors a^μ and b^μ (recall that $\not{a} \equiv a^\mu\gamma_\mu$). We can then use the cyclic property of traces to evaluate traces involving two gamma matrices:

$$\operatorname{tr}(\gamma_\mu\gamma_\nu) = \tfrac{1}{2}[\operatorname{tr}(\gamma_\mu\gamma_\nu) + \operatorname{tr}(\gamma_\nu\gamma_\mu)] = 4g_{\mu\nu}, \tag{8.16a}$$

or equivalently,

$$\operatorname{tr}(\not{a}\not{b}) = \tfrac{1}{2}(2a\cdot b)\operatorname{tr} I = 4a\cdot b. \tag{8.16b}$$

Next, we show that the trace of any product of three gamma matrices vanishes, by using the definition, eq. (5.53) of γ_5, which anticommutes with all of the γ^μ:

$$\operatorname{tr}(\gamma^\mu\gamma^\lambda\gamma^\nu) = -(\mathrm{i}/3!)\epsilon^{\mu\lambda\nu}{}_\sigma\operatorname{tr}(\gamma_5\gamma^\sigma) = 0, \tag{8.17}$$

as follows from $\{\gamma_5, \gamma^\sigma\}_+ = 0$ and the cyclic property of the trace. Not all traces involving γ_5 vanish, however. Of special interest is

$$\operatorname{tr}(\gamma_5\gamma_\mu\gamma_\nu\gamma_\lambda\gamma_\sigma) = 4\mathrm{i}\epsilon_{\mu\nu\lambda\sigma}. \tag{8.18}$$

Equation (8.17) can be extended to any odd number of gamma matrices:

$$\operatorname{tr}[\gamma_{\alpha_1}\gamma_{\alpha_2}\cdots\gamma_{\alpha_n}] = 0, \quad n \text{ odd}. \tag{8.19}$$

To prove this, we observe that there are only four different gamma matrices, whose squares are $\pm I$. If n is odd, there are either one or three unpaired gamma matrices, and the trace vanishes, by eq. (8.14) or eq. (8.17).

Finally, for arbitrary even n, we can easily prove the general iterative relation

$$\operatorname{tr}(\not{a}_1\not{a}_2\cdots\not{a}_n) = \sum_{i=1}^{n-1}(a_n\cdot a_i)(-1)^{i-1}\operatorname{tr}(\not{a}_1\not{a}_2\cdots\hat{\not{a}}_i\cdots\not{a}_{n-1}), \quad n \text{ even} \tag{8.20}$$

where $\hat{\not{a}}_i$ indicates that the factor $\not{a}_i$ is omitted. To show eq. (8.20), it is only necessary to anticommute a_n to the left, using eq. (8.15), and to use the cyclic property of the trace. Repeated applications of eq. (8.20) allow us to evaluate the trace of any even product of gamma matrices. For instance, in the case $n = 4$ we find the often-used formula

$$\operatorname{tr}(\not{a}_1\not{a}_2\not{a}_3\not{a}_4) = 4[(a_1\cdot a_2)(a_3\cdot a_4) - (a_1\cdot a_3)(a_2\cdot a_4) + (a_1\cdot a_4)(a_2\cdot a_3)]. \tag{8.21}$$

Note the minus sign on the middle term.

For $n > 4$, the formulas are straightforward to evaluate, but relatively long. The following simple theorems, which we list for future reference, can often be applied to reduce the number of gamma matrices in a trace:

$$\gamma_\mu\not{a}\gamma^\mu = -2\not{a}, \tag{8.22a}$$

$$\gamma_\mu\not{a}\not{b}\gamma^\mu = 4a\cdot b, \tag{8.22b}$$

$$\gamma_\mu\not{a}\not{b}\not{c}\gamma^\mu = -2\not{c}\not{b}\not{a}. \tag{8.22c}$$

Each of these identities follows from the commutation relations in a straightforward way. As an example, we can prove eq. (8.22c) by using eqs. (8.15) and (8.22b):

$$\begin{aligned}\gamma_\mu\not{a}\not{b}\not{c}\gamma^\mu &= -\gamma_\mu\not{a}\not{b}\gamma^\mu\not{c} + 2\not{c}\not{a}\not{b}\\ &= -2\not{c}\{\not{a}, \not{b}\}_+ + 2\not{c}\not{a}\not{b}\\ &= -2\not{c}\not{b}\not{a}.\end{aligned} \tag{8.23}$$

With these results in hand, let us return to the evaluation of the Bhabha cross section.

Evaluation of the cross section

Consider first T_a in eq. (8.13). Terms with odd powers of the mass also have odd numbers of gamma matrices and hence vanish, by eq. (8.19), so that

$$\begin{aligned}
T_a &= \mathrm{tr}\,(\gamma^\nu \not{p}_2 \gamma^\mu \not{p}_1)\,\mathrm{tr}\,(\gamma_\mu \not{p}_4 \gamma_\nu \not{p}_3) \\
&\quad - m^2[\mathrm{tr}\,(\gamma^\mu\gamma^\nu)\,\mathrm{tr}\,(\gamma_\mu \not{p}_4 \gamma_\nu \not{p}_3) + \mathrm{tr}\,(\gamma^\nu \not{p}_2 \gamma^\mu \not{p}_1)\,\mathrm{tr}\,(\gamma_\mu\gamma_\nu)] \\
&\quad + m^4\,\mathrm{tr}\,(\gamma^\nu\gamma^\mu)\,\mathrm{tr}\,(\gamma_\mu\gamma_\nu) \\
&= 16(p_2{}^\nu p_1{}^\mu + p_1{}^\nu p_2{}^\mu - p_2\cdot p_1 g^{\nu\mu})(p_{4\mu}p_{3\nu} + p_{3\mu}p_{4\nu} - p_3\cdot p_4 g_{\mu\nu}) \\
&\quad + 8m^2(4p_3\cdot p_4 + 4p_1\cdot p_2) + 16m^4 g^{\nu\mu}g_{\mu\nu} \\
&= 8[(u-2m^2)^2 + (t-2m^2)^2] + 32m^2 s, \qquad (8.24)
\end{aligned}$$

where we have used the identities (8.16b), (8.21) and (8.22a), and the usual definitions $s=(p_1+p_2)^2$, $t=(p_3-p_1)^2$, $u=(p_4-p_1)^2$.

T_b may be found in a precisely similar manner, or by exchanging the roles of $p_2{}^\mu$ and $-p_3{}^\mu$, or equivalently the roles of s and t:

$$T_b = 8[(u-2m^2)^2 + (s-2m^2)^2] + 32m^2 t. \qquad (8.25)$$

In T_c and T_d, we have traces with up to eight gamma matrices. Here the identities (8.22b) and (8.22c) come in handy. For instance, T_d is given by

$$\begin{aligned}
T_d &= \mathrm{tr}\,(\gamma_\nu \not{p}_2 \gamma_\alpha \not{p}_4 \gamma^\nu \not{p}_3 \gamma^\alpha \not{p}_1) \\
&\quad - m^2[\mathrm{tr}\,(\gamma_\nu\gamma_\alpha \not{p}_4 \gamma^\nu \not{p}_3 \gamma^\alpha) + \mathrm{tr}\,(\gamma_\nu \not{p}_2 \gamma_\alpha \gamma^\nu \not{p}_3 \gamma^\alpha) \\
&\qquad - \mathrm{tr}\,(\gamma_\nu \not{p}_2 \gamma_\alpha \not{p}_4 \gamma^\nu \gamma^\alpha) - \mathrm{tr}\,(\gamma_\nu\gamma_\alpha\gamma^\nu \not{p}_3 \gamma^\alpha \not{p}_1) \\
&\qquad + \mathrm{tr}\,(\gamma_\nu\gamma_\alpha \not{p}_4 \gamma^\nu\gamma^\alpha \not{p}_1) + \mathrm{tr}\,(\gamma_\nu \not{p}_2 \gamma_\alpha\gamma^\nu\gamma^\alpha \not{p}_1)] \\
&\quad + m^4\,\mathrm{tr}\,(\gamma_\nu\gamma_\alpha\gamma^\nu\gamma^\alpha). \qquad (8.26)
\end{aligned}$$

There is no single way of evaluating eq. (8.26). As a first step, we may choose to apply eqs. (8.22) to the sum over the index ν in each term, to obtain

$$\begin{aligned}
T_d &= -2\,\mathrm{tr}\,(\not{p}_4 \gamma_\alpha \not{p}_2 \not{p}_3 \gamma^\alpha \not{p}_1) \\
&\quad - m^2[4p_{4\alpha}\,\mathrm{tr}\,(\not{p}_3\gamma^\alpha) + 4p_{2\alpha}\,\mathrm{tr}\,(\not{p}_3\gamma^\alpha) \\
&\qquad + 2\,\mathrm{tr}\,(\not{p}_2\gamma_\alpha \not{p}_4 \gamma^\alpha) + 2\,\mathrm{tr}\,(\gamma_\alpha \not{p}_3 \gamma^\alpha \not{p}_1) \\
&\qquad + 4p_{4\alpha}\,\mathrm{tr}\,(\gamma^\alpha \not{p}_1) + 4p_{2\alpha}\,\mathrm{tr}\,(\gamma^\alpha \not{p}_1)] \\
&\quad - 2m^4\,\mathrm{tr}\,(\gamma_\alpha\gamma^\alpha). \qquad (8.27)
\end{aligned}$$

Now using (8.22b) in the first term, and (8.22a) in the fourth and fifth terms, all the traces are reduced to two gamma matrices, giving finally

$$T_d = -8(u - 2m^2)^2 + 32m^2(u - 2m^2), \tag{8.28}$$

(here, we have used $s + t + u = 4m^2$). T_c can be derived from T_d by exchanging $p_2{}^\mu$ with $-p_3{}^\mu$. Since this leaves eq. (8.28) invariant, we conclude

$$T_c = T_d. \tag{8.29}$$

This result follows even more simply by noting that Fig. 8.3(d) represents the complex conjugate of Fig. 8.3(c), which is real. From either point of view, the use of symmetries relating different terms helps a great deal in simplifying calculations.

We can now write down the full expression for the unpolarized cross section, eq. (8.7), by combining the traces in eqs. (8.24), (8.25) (8.28) and (8.29) in the squared matrix element (8.12),

$$\begin{aligned} \mathrm{d}\sigma^{(\mathrm{ee})}/\mathrm{d}t = {} & \{2\pi\alpha^2/[s(s - 4m^2)]\} \\ & \times \big\{(1/s^2)[(u - 2m^2)^2 + (t - 2m^2)^2 + 4m^2 s] \\ & + (1/t^2)[(u - 2m^2)^2 + (s - 2m^2)^2 + 4m^2 t] \\ & + (2/st)[(u - 2m^2)^2 - 4m^2(u - 2m^2)]\big\}. \end{aligned} \tag{8.30}$$

Here, we have introduced the customary notation for the *fine-structure constant*:

$$\alpha = e^2/4\pi. \tag{8.31}$$

Comparing eq. (8.30) with the similar result, eq. (4.48), for scalar particles, we see the same general structure of s-channel and t-channel exchange terms, plus an interference term.

The behavior of the cross section in terms of the center of mass scattering angle is derived easily from eq. (8.30), by analogy with the expression for the cross section eq. (4.46), specialized to the spin-averaged case with nonidentical equal-mass particles. θ^*, the center of mass angle between $\mathbf{p}_1$ and $\mathbf{p}_3$, is related to the invariants by

$$s = 4E^2, \quad t = -4(E^2 - m^2)\sin^2 \tfrac{1}{2}\theta^*, \quad u = -4(E^2 - m^2)\cos^2 \tfrac{1}{2}\theta^*, \tag{8.32}$$

where E is the center of mass energy of the particles. Here, let us simply exhibit the behavior in the high-energy ($m/E \to 0$), and low-energy

$(m/E \to 1)$ limits:

$$d\sigma/d\Omega^* = (\alpha^2/2s)[(1 + \cos^4 \tfrac{1}{2}\theta^*)/\sin^4 \tfrac{1}{2}\theta^* \qquad (8.33a)$$
$$- 2\cos^4 \tfrac{1}{2}\theta^*/\sin^2 \tfrac{1}{2}\theta^* + \sin^4 \tfrac{1}{2}\theta^* + \cos^4 \tfrac{1}{2}\theta^*] \quad (m/E \to 0),$$

$$d\sigma/d\Omega^* = \alpha^2 m^2/[16|\mathbf{p}_1|^4 \sin^4 \tfrac{1}{2}\theta^*] \quad (m/E \to 1). \qquad (8.33b)$$

The low-energy limit is the Rutherford scattering formula.

The angular dependence of the high-energy fermion scattering cross section is similar to the scalar cross section eq. (4.49) at fixed energy. The energy dependence at fixed angle, however, is quite different. The ϕ^3 cross section decreases as g^4/s^4 in this region, where g is the coupling, which has dimensions of inverse length. The QED cross section, on the other hand, behaves as e^4/s^2. Thus, the high-energy power behavior of the QED cross section $d\sigma/dt$ is specified by its dimension, $(\text{length})^4$. Such a cross section is said to obey *dimensional counting*, a reflection of the dimensionless coupling constant e of the theory.

Mixed-flavor processes

In Bhabha scattering, the flavors of the incoming and outgoing fermions are the same. Another process of special interest is annihilation of an $(l\bar{l})$ lepton pair and the creation of a pair of different fermions, $(f\bar{f})$. Examples are e^+e^- annihilation into $\mu^+\mu^-$ or $q\bar{q}$. In these cases, only the 'annihilation' cut graph, Fig. 8.3(*a*), contributes, and the cross section has no $1/t$ dependence. In the high-energy limit, the center-of-mass differential cross section is given simply by the last two terms in eq. (8.33a), scaled by the charges of the flavors involved:

$$d\sigma_{l\bar{l}\to f\bar{f}}/d\Omega_{\text{cm}} = Q_f^2(\alpha^2/4s)(1 + \cos^2\theta^*) \quad (m_{f,\bar{f}}/E \to 0). \qquad (8.34)$$

Unlike Bhabha scattering, the total cross section for $l\bar{l}$ annihilation is finite:

$$\sigma^{(\text{tot})}{}_{l\bar{l}\to f\bar{f}} = Q_f^2(4\pi\alpha^2/3s) \quad (m_{f,\bar{f}}/E \to 0). \qquad (8.35)$$

The $1 + \cos^2\theta$ dependence in the differential cross section is characteristic of the production of a pair of spin one-half particles (exercise 8.2).

Helicity conservation

So far, we have studied spin-averaged cross sections. We shall give an example of polarization effects in massless fermion scattering, using the projection operators of eq. (6.56). For the massive case, see exercise 6.2.

Each massless helicity state forms an irreducible unitary representation of the Poincaré group, so it is quite natural, when working with massless particles, to consider cross sections with definite helicity. Suppose, then, that we consider the cross section for the scattering of a fermion of helicity λ_1 on an antifermion of helicity λ_2, into a fermion of helicity λ_3 and antifermion of helicity λ_4. Analogously to eq. (8.7), we now have

$$\frac{\mathrm{d}\sigma}{\mathrm{d}t}(\{\lambda\}) = \frac{1}{16\pi s^2}|M(\{\lambda\})|^2. \tag{8.36}$$

Note the lack of a factor $\frac{1}{4}$, since we are not averaging over helicities. The steps leading to the traces in eq. (8.13) are repeated, but now using the helicity projection matrices eq. (6.56). Again we find four trace terms:

$$|M(\{\lambda\})|^2 = e^4[(1/s^2)T_a{}' + (1/t^2)T_b{}' - (1/st)(T_c{}' + T_d{}')]. \tag{8.37}$$

The traces, however, are slightly different. Let us define

$$c(\alpha) = \tfrac{1}{2}(1 + 2\alpha\gamma_5), \tag{8.38}$$

with $\alpha = \pm\frac{1}{2}$. Then, by eq. (6.56) the $c(\alpha)$ are projection matrices for Weyl spinors, with α the helicity for particle spinors, and minus the helicity for antiparticle spinors. In this manner, we find

$$\begin{aligned}
T_a{}' &= \mathrm{tr}\,[\gamma^\nu c(-\lambda_2)\not{p}_2\gamma^\mu c(\lambda_1)\not{p}_1]\,\mathrm{tr}\,[\gamma_\mu c(-\lambda_4)\not{p}_4\gamma_\nu c(\lambda_3)\not{p}_3],\\
T_b{}' &= \mathrm{tr}\,[\gamma^\beta c(\lambda_3)\not{p}_3\gamma^\alpha c(\lambda_1)\not{p}_1]\,\mathrm{tr}\,[\gamma_\alpha c(-\lambda_4)\not{p}_4\gamma_\beta c(-\lambda_2)\not{p}_2],\\
T_c{}' &= \mathrm{tr}\,[\gamma^\beta c(\lambda_3)\not{p}_3\gamma_\mu c(\lambda_4)\not{p}_4\gamma_\beta c(-\lambda_2)\not{p}_2\gamma^\mu c(\lambda_1)\not{p}_1],\\
T_d{}' &= \mathrm{tr}\,[\gamma^\nu c(-\lambda_2)\not{p}_2\gamma_\alpha c(-\lambda_4)\not{p}_4\gamma_\nu c(\lambda_3)\not{p}_3\gamma^\alpha c(\lambda_1)\not{p}_1].
\end{aligned} \tag{8.39}$$

These traces may be simplified by using the projection property

$$c(\alpha)c(\beta) = \delta_{\alpha\beta}c(\alpha), \tag{8.40}$$

and observing that the $c(\alpha)$'s commute with any product of an even number of gamma matrices,

$$[c(\alpha_i), \not{p}_j\gamma^\nu] = 0. \tag{8.41}$$

Equations (8.40) and (8.41) together show that the argument of each trace in eq. (8.39) vanishes unless all the $c(\alpha_i)$'s within that trace are the same. This result is quite general. In any diagram involving strings of massless spinor propagators, connecting spinors $w(q, \lambda)$ and $\bar{w}'(p, \lambda')$, $\lambda = \lambda'$ if the spinors are both u's or both v's and $\lambda = -\lambda'$ if one spinor is a u and one spinor is a v. This implies that in the limit that the masses may be neglected, the helicity of an electron or positron is unchanged by electromagnetic scattering, while the helicities of a pair that annihilates must cancel. These requirements are known as *helicity conservation*.

We may now rewrite the squared matrix element, eq. (8.37) as

$$|M(\{\lambda\})|^2 = e^4[(1/s^2)T_a{}'\delta_{\lambda_1,-\lambda_2}\delta_{\lambda_3,-\lambda_4} + (1/t^2)T_b{}'\delta_{\lambda_1,\lambda_3}\delta_{\lambda_2,\lambda_4} - (1/st)(T_c{}' + T_d{}')\delta_{\lambda_1,-\lambda_2}\delta_{-\lambda_2,\lambda_3}\delta_{\lambda_3,-\lambda_4}], \tag{8.42}$$

where

$$\begin{aligned} T_a{}' &= \mathrm{tr}\,[\not{p}_1\gamma^\nu\not{p}_2\gamma^\mu c(\lambda_1)]\,\mathrm{tr}\,[\not{p}_3\gamma_\mu\not{p}_4\gamma_\nu c(\lambda_3)], \\ T_b{}' &= \mathrm{tr}\,[\not{p}_1\gamma^\beta\not{p}_3\gamma^\alpha c(\lambda_3)]\,\mathrm{tr}\,[\not{p}_2\gamma_\alpha\not{p}_4\gamma_\beta c(-\lambda_2)], \\ T_c{}' &= \mathrm{tr}\,[\not{p}_1\gamma^\beta\not{p}_3\gamma_\mu\not{p}_4\gamma_\beta\not{p}_2\gamma^\mu c(\lambda_1)], \\ T_d{}' &= \mathrm{tr}\,[\not{p}_1\gamma_\nu\not{p}_2\gamma_\alpha\not{p}_4\gamma^\nu\not{p}_3\gamma^\alpha c(\lambda_1)]. \end{aligned} \tag{8.43}$$

The evaluation of these traces is straightforward. For instance, $T_a{}'$ is the sum of four terms, the first of which is proportional to the corresponding term in the unpolarized cross section, eq. (8.24) with $m = 0$:

$$\begin{aligned} T_a{}' = \tfrac{1}{4}[T_a{}^{(m=0)} &+ 2\lambda_1\,\mathrm{tr}\,(\not{p}_1\gamma^\nu\not{p}_2\gamma^\mu\gamma_5)\,\mathrm{tr}\,(\not{p}_3\gamma_\mu\not{p}_4\gamma_\nu) \\ &+ 2\lambda_3\,\mathrm{tr}\,(\not{p}_1\gamma^\nu\not{p}_2\gamma^\mu)\,\mathrm{tr}\,(\not{p}_3\gamma_\mu\not{p}_4\gamma_\nu\gamma_5) \\ &+ 4(\lambda_1\lambda_3)\,\mathrm{tr}\,(\not{p}_1\gamma_\nu\not{p}_2\gamma^\mu\gamma_5)\,\mathrm{tr}\,(\not{p}_3\gamma_\mu\not{p}_4\gamma^\nu\gamma_5)]. \end{aligned} \tag{8.44}$$

The second and third terms do not contribute, by the trace identities, eqs. (8.18) and (8.21). For example, the second is

$$\mathrm{tr}\,(\not{p}_1\gamma^\nu\not{p}_2\gamma^\mu\gamma_5)\,\mathrm{tr}\,(\not{p}_3\gamma_\mu\not{p}_4\gamma_\nu) = 0, \tag{8.45}$$

which vanishes, since it is the product of symmetric times antisymmetric tensors in μ and ν. The fourth term is also easy to compute, using the identity

$$\epsilon^{\alpha\beta\mu}{}_\nu\epsilon_{\sigma\lambda\mu}{}^\nu = -2(g^\alpha{}_\sigma g^\beta{}_\lambda - g^\alpha{}_\lambda g^\beta{}_\sigma). \tag{8.46}$$

The complete result is

$$T_a{}' = 2(u^2 + t^2) + 4\lambda_1\lambda_3(u^2 - t^2). \tag{8.47}$$

$T_b{}'$ is found from $T_a{}'$, by exchanging (p_2, λ_2) with $(-p_3, -\lambda_3)$, so that

$$T_b{}' = 2(u^2 + s^2) - 4\lambda_1\lambda_2(u^2 - s^2). \tag{8.48}$$

$T_c{}'$ and $T_d{}'$ are even easier to evaluate, using the identities in eq. (8.22). They turn out to be independent of helicities:

$$T_c{}' = T_d{}' = \tfrac{1}{2}T_c{}^{(m=0)} = -4u^2. \tag{8.49}$$

Collecting $T_a{}'$ through $T_d{}'$, and substituting them into eqs. (8.36) and

(8.37), we find the cross section for definite helicity,

$$\begin{aligned}
\mathrm{d}\sigma(\lambda_1, \ldots, \lambda_4)/\mathrm{d}t \\
= (2\pi\alpha^2/s^2)\{(1/s^2)[(u^2 + t^2) + 2\lambda_1\lambda_3(u^2 - t^2)]\delta_{\lambda_1,-\lambda_2}\delta_{\lambda_3,-\lambda_4} \\
+ (1/t^2)[(u^2 + s^2) - 2\lambda_1\lambda_2(u^2 - s^2)]\delta_{\lambda_1,\lambda_3}\delta_{\lambda_2,\lambda_4} \\
+ (4u^2/st)\delta_{\lambda_1,-\lambda_2}\delta_{-\lambda_2,\lambda_3}\delta_{\lambda_3,-\lambda_4}\},
\end{aligned} \tag{8.50}$$

Note that the polarized cross section has a different angular dependence than the spin-averaged cross section eq. (8.30). It is easy to verify, however, that the sum over final and average over initial helicities does, in fact, reproduce the zero-mass limit of eq. (8.30).

8.2 Cross sections with photons

The S-matrix

We now study cross sections related to the Green function, eq. (8.3), involving external photons as well as electrons. The fundamental processes here are *Compton scattering*, $\mathrm{e} + \gamma \to \mathrm{e} + \gamma$, *pair annihilation*, $\mathrm{e} + \bar{\mathrm{e}} \to \gamma + \gamma$ and *pair creation* $\gamma + \gamma \to \mathrm{e} + \bar{\mathrm{e}}$, illustrated in Figs. 8.4(*a*)–(*c*). We shall use the first two for purposes of illustration. For Compton scattering, we have (choosing real polarizations)

$$\begin{aligned}
\mathrm{i}M((\mathbf{p}_2, s_2)^{(-)}, (\mathbf{k}_2, \sigma_2); (\mathbf{p}_1, s_1)^{(-)}, (\mathbf{k}_1, \sigma_1)) \\
= -\mathrm{i}e^2\epsilon^\mu(k_1, \sigma_1)\epsilon^\nu(k_2, \sigma_2) \\
\times \bar{u}(p_2, s_2)[\gamma_\nu(\not{p}_1 + \not{k}_1 - m)^{-1}\gamma_\mu + \gamma_\mu(\not{p}_1 - \not{k}_2 - m)^{-1}\gamma_\nu]u(p_1, s_1),
\end{aligned} \tag{8.51}$$

and for pair annihilation,

$$\begin{aligned}
\mathrm{i}M((\mathbf{k}_1, \sigma_1), (\mathbf{k}_2, \sigma_2); (\mathbf{p}_2, s_2)^{(+)}, (\mathbf{p}_1, s_1)^{(-)}) \\
= -\mathrm{i}e^2\epsilon^\mu(k_1, \sigma_1)\epsilon^\nu(k_2, \sigma_2) \\
\times \bar{v}(p_2, s_2)[\gamma_\nu(\not{p}_1 - \not{k}_1 - m)^{-1}\gamma_\mu + \gamma_\mu(\not{p}_1 - \not{k}_2 - m)^{-1}\gamma_\nu]u(p_1, s_1).
\end{aligned} \tag{8.52}$$

Let us first compute the cross section for Compton scattering with transversely polarized photons.

The Klein–Nishina formula

If we sum over the spins of both the incoming and outgoing electrons, we find from the square of eq. (8.51) a single trace with eight gamma matrices.

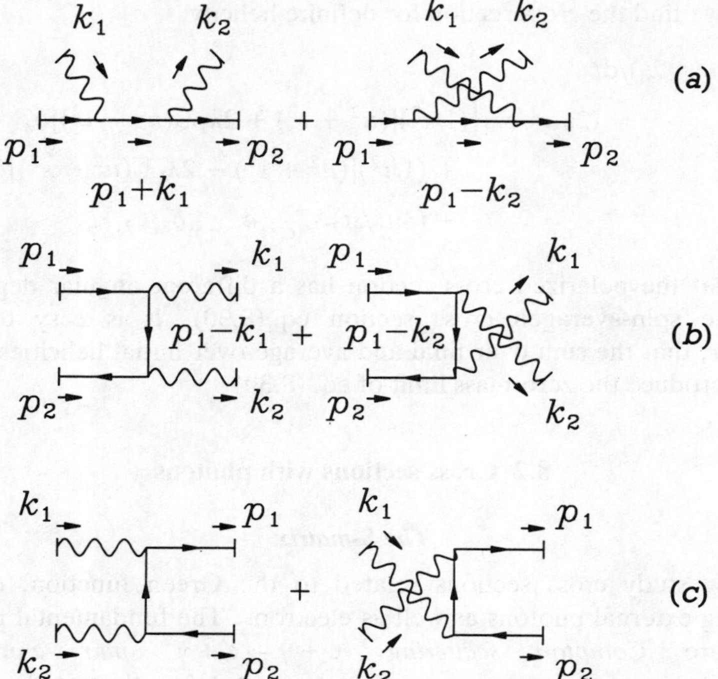

Fig. 8.4 Lowest-order photon–fermion diagrams: (*a*) Compton scattering, (*b*) pair annihilation, (*c*) pair creation.

This trace, and its evaluation, may be greatly simplified by working in the rest frame of the initial electron ($\mathbf{p}_1 = 0$, the 'lab frame'), where for transverse polarizations we have (eq. (8.15)) $\{\not\epsilon_1, \not p_1\}_+ = \{\not\epsilon_2, \not p_1\}_+ = 0$, in addition to $\{\not\epsilon_1, \not k_1\}_+ = \{\not\epsilon_2, \not k_2\}_+ = 0$, where $\epsilon_1 = \epsilon(k_1, \sigma_1)$, etc. We can easily use these relations and the Dirac equation to reduce the relevant matrix element of Fig. 8.4(*a*) to the simpler form

$$M(\epsilon_1, \epsilon_2) = \frac{ie^2 \bar{u}_2}{2}\left(\frac{\not\epsilon_2 \not\epsilon_1 \not k_1}{p_1 \cdot k_1} + \frac{\not\epsilon_1 \not\epsilon_2 \not k_2}{p_1 \cdot k_2}\right) u_1. \tag{8.53}$$

The square of this matrix element is converted to a trace using, as usual, the projection operators in eq. (6.39):

$$\sum_{\text{electron spins}} |M^2| = \frac{e^4}{4} \operatorname{tr}\left[\left(\frac{\not\epsilon_2 \not\epsilon_1 \not k_1}{p_1 \cdot k_1} + \frac{\not\epsilon_1 \not\epsilon_2 \not k_2}{p_1 \cdot k_2}\right)(\not p_1 + m) \right.$$
$$\left. \times \left(\frac{\not k_1 \not\epsilon_1 \not\epsilon_2}{p_1 \cdot k_1} + \frac{\not k_2 \not\epsilon_2 \not\epsilon_1}{p_1 \cdot k_2}\right)(\not p_2 + m)\right]$$

$$= \frac{e^4}{4}\left(2(\epsilon_1\cdot\epsilon_2)\left\{\mathrm{tr}\left[\left(\frac{\not{\epsilon}_2\not{\epsilon}_1\not{k}_1}{p_1\cdot k_1} + \frac{\not{\epsilon}_1\not{\epsilon}_2\not{k}_2}{p_1\cdot k_2}\right)\not{p}_2\right]\right.\right.$$
$$\left. + \,\mathrm{tr}\left[\left(\frac{\not{k}_1\not{\epsilon}_1\not{\epsilon}_2}{p_1\cdot k_1} + \frac{\not{k}_2\not{\epsilon}_2\not{\epsilon}_1}{p_1\cdot k_2}\right)\not{p}_2\right]\right\}$$
$$\left. + \frac{-p_1\cdot p_2 + m^2}{4(p_1\cdot k_1)(p_1\cdot k_2)}\mathrm{tr}\,(\not{\epsilon}_2\not{\epsilon}_1\not{k}_1\not{k}_2\not{\epsilon}_2\not{\epsilon}_1 + \not{\epsilon}_1\not{\epsilon}_2\not{k}_2\not{k}_1\not{\epsilon}_1\not{\epsilon}_2)\right). \tag{8.54}$$

In deriving the second form, we have eliminated $\not{p}_1$ by using the trace identity of eq. (8.20), whose application is considerably simplified in the lab frame. The last trace in eq. (8.54) still has six matrices, but it can be simplified by anticommuting factors of $\not{\epsilon}_i$ together and then using ${\not{\epsilon}_i}^2 = {\epsilon_i}^2 = -1$. Also useful are $\not{p}_2 = \not{p}_1 + \not{k}_1 - \not{k}_2$ and the lab-frame identity $-p_1\cdot p_2 + m^2 = m(|\mathbf{k}_2| - |\mathbf{k}_1|)$. Defining $|\mathbf{k}_i| \equiv k_i$, we find

$$\sum_{\text{electron spins}} |M^2| = \frac{e^4}{2}(\epsilon_1\cdot\epsilon_2)\left\{\mathrm{tr}\left[\left(\frac{\not{k}_1\not{\epsilon}_1\not{\epsilon}_2 + \not{\epsilon}_2\not{\epsilon}_1\not{k}_1}{p_1\cdot k_1} + \frac{\not{k}_2\not{\epsilon}_2\not{\epsilon}_1 + \not{\epsilon}_1\not{\epsilon}_2\not{k}_2}{p_1\cdot k_2}\right)\not{p}_1\right]\right.$$
$$\left. - \,\mathrm{tr}\left[\left(\frac{\not{k}_1\not{\epsilon}_1\not{\epsilon}_2 + \not{\epsilon}_2\not{\epsilon}_1\not{k}_1}{p_1\cdot k_1}\right)\not{k}_2 - \left(\frac{\not{k}_2\not{\epsilon}_2\not{\epsilon}_1 + \not{\epsilon}_1\not{\epsilon}_2\not{k}_2}{p_1\cdot k_2}\right)\not{k}_1\right]\right\}$$
$$- \frac{e^4(k_1 - k_2)}{4mk_1k_2}[2(\epsilon_1\cdot\epsilon_2)\,\mathrm{tr}\,(\not{\epsilon}_2\not{\epsilon}_1\not{k}_1\not{k}_2 + \not{\epsilon}_1\not{\epsilon}_2\not{k}_2\not{k}_1)$$
$$- 2\,\mathrm{tr}\,(\not{k}_1\not{k}_2)]. \tag{8.55}$$

All traces now have no more than four matrices, and it is straightforward to derive

$$\sum |M|^2 = 2e^4[k_1/k_2 + k_2/k_1 - 2 + 4(\epsilon_1\cdot\epsilon_2)^2],$$

which gives the *Klein–Nishina* formula (Klein & Nishina 1929):

$$\begin{aligned} \mathrm{d}\sigma^{(e\gamma)}/\mathrm{d}\Omega_{\mathrm{lab}} &= [\mathrm{d}\sigma^{(e\gamma)}/\mathrm{d}t](\mathrm{d}t/\mathrm{d}\cos\theta_{\mathrm{lab}})(1/2\pi) \\ &= (\alpha^2/4m^2)(k_2/k_1)^2[k_1/k_2 + k_2/k_1 - 2 + 4(\epsilon_1\cdot\epsilon_2)^2], \end{aligned} \tag{8.56}$$

where we have used eq. (4.43), and $(\mathrm{d}t/\mathrm{d}\cos\theta_{\mathrm{lab}}) = 2{k_2}^2$. Of special interest is the nonrelativistic limit, in which $k_1/m \to 0$, $k_1/k_2 \to 1$. Here the cross section approaches the classical Thomson cross section, $(\alpha^2/m^2)(\epsilon_1\cdot\epsilon_2)^2$, rather in the same way that Bhabha scattering approaches the Rutherford formula. The $m/E \to 0$ cross section is easier to calculate in the center-of-mass frame, and we shall give the result after discussing the role of unphysical photon polarizations, now in the context of pair annihilation.

Decoupling of unphysical photons

We begin by squaring the matrix element, eq. (8.52), for pair annihilation and summing over transverse photon polarizations:

$$\sum_{\sigma_1,\sigma_2} |M(\sigma_1, \sigma_2)|^2 = t_{\mu\nu} t_{\alpha\beta}{}^* P^{\mu\alpha}(k_1) P^{\nu\beta}(k_2). \tag{8.57}$$

Here $t_{\mu\nu}$ contains the Dirac matrices, and $P^{\mu\alpha}$ the sum over photon polarizations, as in eq. (6.62):

$$\begin{aligned} t_{\mu\nu} &= (-\mathrm{i}e)^2 \bar{v}_2[\gamma_\nu \mathrm{i}(\not{p}_1 - \not{k}_1 - m)^{-1}\gamma_\mu + \gamma_\mu \mathrm{i}(\not{p}_1 - \not{k}_2 - m)^{-1}\gamma_\nu]u_1, \\ P^{\mu\alpha}(k_1) &= \sum_{\sigma_1} \epsilon^\mu(k_1, \sigma_1)\epsilon^\alpha(k_1, \sigma_1), \end{aligned} \tag{8.58}$$

and similarly for $t_{\alpha\beta}$ and $P^{\nu\beta}(k_2)$. Because of the k-dependence in the P's, eq. (8.57) is not always convenient for calculation. In fact, we will now show that it is permissible to replace the physical polarization tensors $P^{\mu\nu}(k)$ in eq. (8.57) by $-g^{\mu\nu}$. This procedure gives a Lorentz invariant answer, by including the unphysical longitudinal and scalar polarizations. Nevertheless, eq. (8.57) is unaffected by their inclusion. This makes it much simpler to prove the unitarity of the physical S-matrix, eq. (6.84) (see Section 9.6 and Chapter 11). Let us see how it works for pair annihilation.

We start by re-expressing the physical polarization tensor $P^{\mu\alpha}(k)$ as

$$P^{\mu\alpha}(k) = -g^{\mu\alpha} + (\bar{k}^\mu k^\alpha + k^\mu \bar{k}^\alpha)/\bar{k}\cdot k, \tag{8.59}$$

where the vector $\bar{k}^\mu$ is the space-reflected version of k^μ,

$$\bar{k}^\mu = (k^0, -\mathbf{k}) = -k^\mu + 2g^\mu{}_0 k^0. \tag{8.60}$$

When $k^2 = 0$, $k\cdot\bar{k} = 2|\mathbf{k}|^2$. Substituting eq. (8.59) into (8.57) gives

$$\begin{aligned} \sum_{\sigma_1,\sigma_2} |M|^2 = t_{\mu\nu} t_{\alpha\beta}{}^* &[-g^{\mu\alpha}(k_1) + (\bar{k}_1{}^\mu k_1{}^\alpha + k_1{}^\mu \bar{k}_1{}^\alpha)/k_1\cdot\bar{k}_1] \\ &\times [-g^{\nu\beta}(k_2) + (\bar{k}_2{}^\nu k_2{}^\beta + k_2{}^\nu \bar{k}_2{}^\beta)/k_2\cdot\bar{k}_2]. \end{aligned} \tag{8.61}$$

On the right-hand side, the term proportional to $g^{\mu\alpha} g^{\nu\beta}$ is what we get by replacing the P's by the negative of the metric. We are going to show that all the other terms vanish.

The unwanted terms in (8.61) vanish because in each such term at least one gluon polarization $\epsilon^\mu(k_i)$ is replaced by that gluon's momentum, $k_i{}^\mu$ (scalar polarization). As we now show, every such factor vanishes identically. Consider, for example, $t_{\mu\nu} k_1{}^\mu$. Using the explicit form, eq. (8.58), for

$t_{\mu\nu}$, it may be written as

$$t_{\mu\nu}k_1{}^\mu = (-\mathrm{i}e)^2\bar{v}_2[\gamma_\nu \mathrm{i}(\not{p}_1 - \not{k}_1 - m)^{-1}\not{k}_1 + \not{k}_1 \mathrm{i}(-\not{p}_2 + \not{k}_1 - m)^{-1}\gamma_\nu]u_1. \tag{8.62}$$

The scalar $k_1{}^\mu$ produces factors of $\not{k}_1$ in the Dirac sum. These factors, in turn, may be written as the difference of two inverse propagators in two different ways, often called 'Feynman identities' (Feynman 1949):

$$\not{k}_1 = (\not{p}_1 - m) - (\not{p}_1 - \not{k}_1 - m) = (-\not{p}_2 + \not{k}_1 - m) - (-\not{p}_2 - m). \tag{8.63}$$

The first form in eq. (8.63) is substituted into the first term on the right-hand side of eq. (8.62), and the second form into the second term. This results in four terms,

$$\begin{aligned} t_{\mu\nu}k_1{}^\mu &= -\mathrm{i}e^2\bar{v}_2[\gamma_\nu(\not{p}_1 - \not{k}_1 - m)^{-1}(\not{p}_1 - m) - \gamma_\nu \\ &\quad + \gamma_\nu - (-\not{p}_2 - m)(-\not{p}_2 + \not{k}_1 - m)^{-1}\gamma_\nu]u_1 \\ &= 0. \end{aligned} \tag{8.64}$$

The first and last terms vanish by the Dirac equation, while the middle two terms cancel. Precisely the same reasoning works for all the other terms with at least one scalar polarization, and we find the hoped-for expression

$$\sum_{\sigma_1,\sigma_2} |M(\sigma_1, \sigma_2)|^2 = t_{\mu\nu}t_{\alpha\beta}{}^*(-g^{\mu\alpha})(-g^{\nu\beta}). \tag{8.65}$$

This result, that we may use $-g^{\mu\nu}$ in the polarization sum of external photons in QED, is quite general, as may be shown by repeated application of the identity of eq. (8.63) (exercise 8.3).

Zero-mass cross sections

For the spin-averaged pair annihilation cross section, represented by the cut diagrams in Fig. 8.5, we can now replace the photon polarization tensor by the metric tensor. In the zero-mass limit, we then find (compare eq. (8.12)),

$$\sum_{\sigma_1,\sigma_2} |M(\sigma_1, \sigma_2)|^2 = e^4[(1/t)^2\kappa_1 + (1/u)^2\kappa_2 + (1/ut)(\kappa_3 + \kappa_4)] \tag{8.66}$$

where

$$\begin{aligned} \kappa_1 &= \mathrm{tr}\,[\gamma^\mu(\not{p}_1 - \not{k}_1)\gamma^\nu\not{p}_2\gamma_\nu(\not{p}_1 - \not{k}_1)\gamma_\mu\not{p}_1], \\ \kappa_2 &= \mathrm{tr}\,[\gamma^\nu(\not{p}_1 - \not{k}_2)\gamma^\mu\not{p}_2\gamma_\mu(\not{p}_1 - \not{k}_2)\gamma_\nu\not{p}_1], \\ \kappa_3 &= \mathrm{tr}\,[\gamma^\nu(\not{p}_1 - \not{k}_2)\gamma^\mu\not{p}_2\gamma_\nu(\not{p}_1 - \not{k}_1)\gamma_\mu\not{p}_1], \\ \kappa_4 &= \mathrm{tr}\,[\gamma^\mu(\not{p}_1 - \not{k}_1)\gamma^\nu\not{p}_2\gamma_\mu(\not{p}_1 - \not{k}_2)\gamma_\nu\not{p}_1]. \end{aligned} \tag{8.67}$$

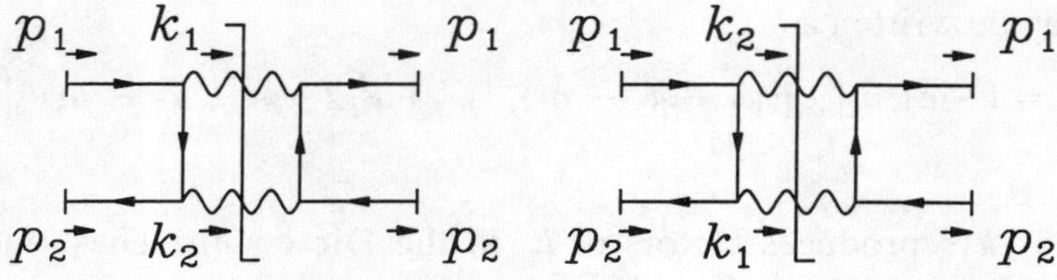

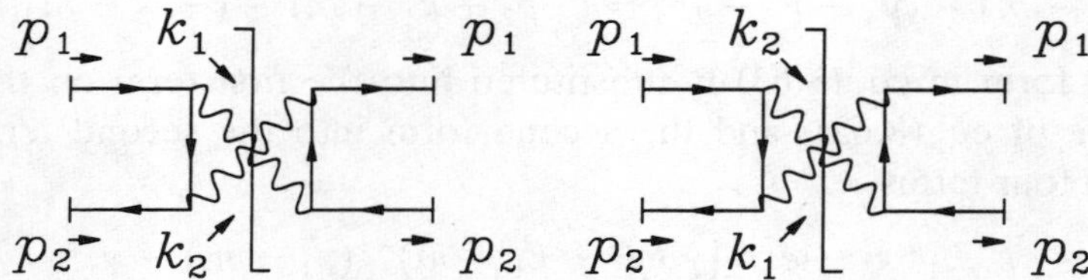

Fig. 8.5 Cut diagrams for spin-averaged annihilation.

Evaluating these traces is a simple matter, using the identities eqs. (8.21) and (8.22) and $\not{p}_i^2 = p_i^2 = \not{k}_i^2 = k_i^2 = 0$. A little algebra, and the general relation eq. (4.43), gives the spin-summed pair annihilation cross section at fixed momentum transfer (exercise 8.4),

$$\mathrm{d}\sigma/\mathrm{d}t_{e\bar{e}\to\gamma\gamma} = (2\pi\alpha^2/s^2)[(u^2+t^2)/ut] \quad (m=0). \tag{8.68}$$

In interpreting this result, we should remember that it includes contributions from both photons. The corresponding matrix element for Compton scattering, $\mathrm{e}(p_1) + \gamma(k_1) \to \mathrm{e}(p_2) + \gamma(k_2)$, can be found by simply exchanging $p_2{}^\mu$ with $-k_1{}^\mu$ (exercise 8.4), and the cross section is

$$\mathrm{d}\sigma/\mathrm{d}t_{e\gamma\to e\gamma} = (2\pi\alpha^2/s^2)[(u^2+s^2)/(-us)] \quad (m=0). \tag{8.69}$$

Magnitude of the charge

The interaction Lagrange density in eq. (8.1) is of the form $-eQ_f j(x)\cdot A(x)$, where $j_\mu(x) = \bar{\psi}\gamma_\mu\psi$ is the electromagnetic current. In natural units (Section 2.3), e is dimensionless, since A^μ has dimensions (length)$^{-1}$ and $j_\mu(x)$ has dimensions (length)$^{-3}$. Let us re-insert factors of $\hbar$ and c, and see what happens. The Lagrange density now has dimensions (action/(length)^3 $\times$ time). As mentioned in Section 5.3, the system of units implicit in our choice of electromagnetic Lagrange density is the 'Heaviside–Lorentz' system, in which the interaction is $-(1/c)A\cdot J$, where J^μ is the (classical) current density, and Gauss's law is $\nabla\cdot\mathbf{E} = J^0$ (Panofsky & Phillips r1962). In this system, the A-field has dimensions (action)$^{1/2}$ $\times$ (length $\times$ time)$^{-1/2}$ and the current density J^μ has dimensions (charge $\times$ velocity/(length)^3). If we identify e with the unit of charge, the current density in terms of fermion operators is $(1/\hbar c)\bar{\psi}\gamma_\mu\psi$, and e has dimensions (action × velocity)$^{1/2}$. The size of e is most often described in terms of the fine-structure constant (8.31), which in Heaviside–Lorentz units is $\alpha = e^2/4\pi\hbar c \approx 137.036$ (Cohen & Taylor r1987, Particle Data Group r1990).

The size of higher-order corrections

Formally, higher-order corrections to any cross section consist of extra powers of α times various factors. *If* the factors that multiply α^2 are of the same order as those that multiply α, the next-order corrections will be of the order of one per cent of the lowest order calculations. This expectation fails, however, if the coefficients are large (if, for instance, the real expansion is in e^2 rather than α), or if momentum-dependence spoils the expansion (as may happen at high energies; see Chapters 12–15). We will discuss examples of higher-order corrections in Chapter 12.

8.3 Weak interactions of leptons

Lagrangian

The leptonic weak Lagrange density is given by eq. (6.94). In terms of explicit particle fields it is

$$\begin{aligned}\mathscr{L}_{\mathrm{wk}} = &-\frac{g}{2^{1/2}}\sum_{i=\mathrm{e},\mu,\tau}[\bar{\nu}_i\gamma^\mu\tfrac{1}{2}(1-\gamma_5)W^+{}_\mu l_i + \bar{l}_i\gamma^\mu\tfrac{1}{2}(1-\gamma_5)W^-{}_\mu\nu_i]\\ &-\frac{g}{2\cos\theta_{\mathrm{W}}}\sum_{i=\mathrm{e},\mu,\tau}\{\bar{\nu}_i\gamma^\mu\tfrac{1}{2}(1-\gamma_5)Z_\mu\nu_i\\ &+\bar{l}_i\gamma^\mu\tfrac{1}{2}[\gamma_5+(4\sin^2\theta_{\mathrm{W}}-1)]Z_\mu l_i\},\end{aligned} \tag{8.70}$$

where the sum goes over the leptonic flavors, or 'generations', e, μ and τ. The Feynman rules for fermion–vector couplings that follow from this density are direct generalizations of those in QED: the only differences are in the vertices, in which γ_μ is replaced by combinations of γ_μ and $\gamma_\mu\gamma_5$, and in implicit $SU(2)$ group factors, which make some of the vertices nondiagonal in flavor. At lowest order in fermion–fermion scattering, we shall need only these vertices, plus the massive vector propagators. In unitary gauge ($\lambda \to 0$ in eq. (7.85)) the latter are $[i/(k-M^2)^2](-g_{\alpha\beta}+k_\alpha k_\beta/M^2)$ (exercise 5.8). Beyond the lowest order, the nonabelian couplings of the vector particles appear, but we shall not need these effects here (for a review, see Lynn & Wheater r1984).

The Lagrange density eq. (8.70) consists of two parts, *charged current* interactions, involving $W^{\pm}$, that couple charged leptons to neutrinos and only involve left-handed fermions, and *neutral current* interactions, coupling the Z boson to neutrinos or charged leptons. In the latter case, both left- and right-handed charged leptons contribute.

Muon decay and charged currents

In the charged current part of eq. (8.70), the W bosons couple in an identical manner to each of the lepton generations. Because the electron is lighter than the muon (and tau), the muon may decay into a muon neutrino, an electron and an electron antineutrino at second order in the weak coupling g, via the diagram shown in Fig. 8.6.

To be specific, consider the decay rate for a polarized muon, with all final-state spins unobserved (this is natural indeed for neutrinos). The rate may be calculated from the corresponding squared matrix element, in the same manner as for scalar particles, eq. (4.64). Let us suppose that the decaying muon has spin vector aligned along the n^μ-direction, $s_\mu = sn^\mu$, $n^2 = -1$, $s = \pm\frac{1}{2}$. Then the square of the matrix element is straightforward to evaluate, using the projection operators in eq. (6.39a) for the electron, eq. (6.56) for the neutrinos, and the spin basis relation $u_b(P, s)\bar{u}_a(P, s) = [(\not{P} + m_\mu)(\frac{1}{2} + s\gamma_5\not{n})]_{ba}$ for the muon (exercise 6.2). We find

$$\begin{aligned}|M(s)|^2 = {}& (g^2/8)^2 \operatorname{tr}[(\not{P} + m_\mu)(\tfrac{1}{2} + s\gamma_5\not{n})\gamma^\mu(1-\gamma_5)\not{l}\gamma^\nu(1-\gamma_5)] \\ & \times \operatorname{tr}[\gamma^{\mu'}(1-\gamma_5)(\not{p}_e + m_e)\gamma^{\nu'}(1-\gamma_5)\not{k}] \\ & \times \{(p_e + k)^2 - M_W{}^2)^{-1}[-g_{\mu\mu'} + (p_e + k)_\mu(p_e + k)_{\mu'}]M_W{}^{-2}\} \\ & \times \{(p_e + k)^2 - M_W{}^2)^{-1}[-g_{\nu'\nu} + (p_e + k)_{\nu'}(p_e + k)_\nu]M_W{}^{-2}\},\end{aligned} \tag{8.71}$$

where P, l, p_e and k are the muon, muon neutrino, electron and electron antineutrino momenta, respectively. Clearly, all these momenta are much less than M_W, and we may, to an excellent approximation, keep only the leading power in $1/M_W$. This eliminates gauge dependence automatically.

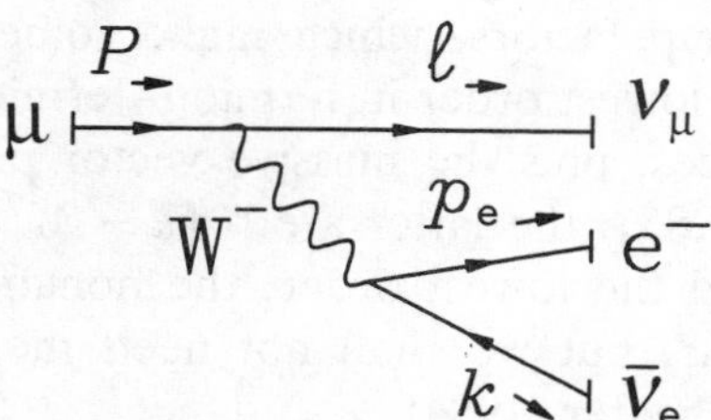

Fig. 8.6 Muon decay.

At higher orders or energies, gauge invariance is enforced through the interactions of other fields (Appendix C). Performing the resulting traces requires only the identities of eqs. (8.18) and (8.21), and the product formula for ϵ-tensors, eq. (8.46). After some calculation, we find the surprisingly simple result

$$|M(s)|^2 = 2(g^4/M_W{}^4)(l\cdot p_e)[k\cdot(P - 2sm_\mu n)], \tag{8.72}$$

which, with eq. (4.64) leads to the following differential decay rate at fixed electron momentum:

$$\begin{aligned} 2\omega_e[d\Gamma(s)/d^3\mathbf{p}_e] &= (2\pi)^{-5}(g^4/m_\mu M_W{}^4)(p_e)^\alpha(P - 2sm_\mu n)^\beta \\ &\quad \times \int d^3\mathbf{k}d^3\mathbf{l}\, l_\alpha k_\beta(4\omega_k\omega_l)^{-1}\delta^4(P - p_e - k - l). \end{aligned} \tag{8.73}$$

The phase space integral over ν_μ and $\bar{\nu}_e$ momenta is (exercise 8.6)

$$\int d^3\mathbf{k}d^3\mathbf{l}\, k_\alpha l_\beta(4\omega_k\omega_l)^{-1}\delta^4(K - k - l) = (\pi/24)(K^2 g_{\alpha\beta} + 2K_\alpha K_\beta), \tag{8.74}$$

which gives

$$\begin{aligned} 2\omega_e[d\Gamma(s)/d^3\mathbf{p}_e] &= (2\pi)^{-5}(\pi g^4/24m_\mu M_W{}^4)[(P - p_e)^2(P - 2sm_\mu n)\cdot p_e \\ &\quad + 2p_e\cdot(P - p_e)(P - 2sm_\mu n)\cdot(P - p_e)]. \end{aligned} \tag{8.75}$$

Rewriting this expression in the muon rest frame we find

$$\begin{aligned} 2\omega_e[d\Gamma(s)/d^3\mathbf{p}_e] &= (g^4/64M_W{}^4)(m_\mu/3\pi^4)\{3\omega_e(m_\mu{}^2 + m_e{}^2)/4m_\mu - \tfrac{1}{2}m_e{}^2 \\ &\quad - \omega_e{}^2 - 2\mathbf{p}_e\cdot(s\mathbf{n})[-(m_\mu{}^2 + m_e{}^2)/4m_\mu \\ &\quad - m_e{}^2/2m_\mu + \omega_e]\}, \end{aligned} \tag{8.76}$$

in which the parity violation implicit in the Lagrange density eq. (8.70) appears through the term $\mathbf{p}_e\cdot(s\mathbf{n})$, which changes sign under parity transformations (Section 6.6).

The total decay rate is now easily computed, at least in the limit that we neglect m_e/m_μ, so that the maximum momentum of the electron is $m_\mu/2$,

$$\Gamma^{(\text{tot})} = \frac{1}{2}\sum_{s=\pm 1/2}\Gamma(s) = \frac{m_\mu{}^5 G_F{}^2}{192\pi^3} \quad (m_e/m_\mu = 0). \tag{8.77}$$

Here we have introduced the *Fermi coupling* G_F, defined by

$$G_F/2^{1/2} \equiv g^2/8M_W{}^2. \tag{8.78}$$

We may note that, up to corrections with extra powers of m_μ/M_W, exactly the same differential and (hence) total decay rates follow from a four-fermion Lagrangian of the general form shown in eq. (5.86),

$$L_{4f} = (G_F/2^{1/2})[\bar{\psi}_{\nu_\mu}\tfrac{1}{2}(1-\gamma_5)\psi_\mu][\bar{\psi}_e\tfrac{1}{2}(1-\gamma_5)\psi_{\nu_e}] + \text{hermitian conjugate}, \tag{8.79}$$

in which the electron and muon and their neutrinos couple at a point. Clearly this approximation also works well for charged current scattering at low energies, whenever the W boson propagator is approximately given by $-ig_{\alpha\beta}/(-M_W{}^2)$. In any such interaction, we directly measure G_F, which involves the squared ratio of the weak coupling and the W boson mass.

Neutrino–electron elastic scattering

Neutral-current Z_0 exchange mediates neutrino–electron scattering, as illustrated in Fig. 8.7. The computation of the matrix element and the elastic scattering cross section is by now routine:

$$\begin{aligned}\Sigma|M|^2 &= [g^2/4(t-M_Z{}^2)\cos^2\theta_W]^2\,\mathrm{tr}\,[\not{k}\gamma^\beta\tfrac{1}{2}(1-\gamma_5)\not{k}'\gamma^\alpha\tfrac{1}{2}(1-\gamma_5)]\\ &\quad\times \mathrm{tr}\,[(\not{p}+m_e)\gamma_\beta(V_e-A_e\gamma_5)(\not{p}'+m_e)\gamma_\alpha(V_e-A_e\gamma_5)]\\ &= [g^2/2(t-M_Z{}^2)\cos^2\theta_W]^2[(V_e{}^2+A_e{}^2)(u^2+s^2)\\ &\quad+2A_eV_e(u^2-s^2)]\end{aligned} \tag{8.80}$$

where for simplicity, we use the notation $A_e = -\frac{1}{2}$, $V_e = -\frac{1}{2} + 2\sin^2\theta_W$, which follows from the Lagrange density eq. (8.70). Because all the neutrinos couple to the Z_0 in the same manner, this matrix element is the same for any flavor of neutrino. The traces are easily evaluated, and from eq. (4.46), adapted to nonidentical final-state particles and spin averaging, we derive the center-of-mass elastic scattering cross section:

$$\begin{aligned}\mathrm{d}\sigma^{(\nu e)}/\mathrm{d}\Omega_{\mathrm{cm}} &= (1/256\pi^2 s)[g^2/2(t-M_Z{}^2)\cos^2\theta_W]^2\\ &\quad\times[(V_e{}^2+A_e{}^2)(u^2+s^2)+2A_eV_e(u^2-s^2)].\end{aligned} \tag{8.81}$$

At very high energies ($s \gg M_Z{}^2$) this is an unexceptional cross section,

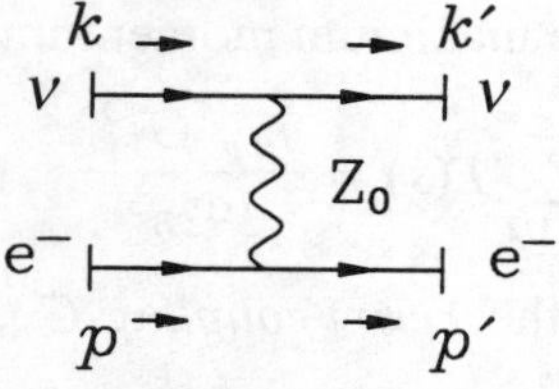

Fig. 8.7 Neutrino–electron scattering.

which behaves as $1/s$ at fixed angles, that is, according to dimensional counting. It also gives a finite total cross section, which for $s \ll M_W{}^2$ takes the form (neglecting W exchange)

$$\sigma_{tot}{}^{(\nu e)} = (G_F{}^2 s/4\pi)[1 - 4\sin^2\theta_W + \tfrac{16}{3}\sin^4\theta_W)] \quad (s \ll M_W{}^2), \quad (8.82)$$

where we have used the specific values for A_e and V_e noted above, and eq. (8.78) for G_F. Here we see that neutral current cross sections, even at low energies, measure the weak mixing angle.

Equation (8.82) also shows another interesting property of low-energy neutrino cross sections: they increase linearly with s. For four-fermion couplings, such as eq. (8.79), this increase continues to all values of s. At very high energies the four-fermion perturbation expansion fails, since eventually $G_F s$ becomes a large number. One of the great successes of the standard model is to supply a perturbative expansion for the weak interactions that is self-consistent in this sense. It does so by 'resolving' four-fermion couplings into vector boson exchanges.

Accuracy of lowest-order corrections

As in the electromagnetic case, weak-interaction cross sections are often well approximated by lowest-order calculations, essentially because the underlying couplings, g and g' of $SU(2) \times U(1)$ in eq. (5.108), are of the same order as $e = gg'/(g^2 + g'^2)^{1/2}$. The suppression that makes the weak interactions 'weak' is due to the large masses of the mediating W and Z vector bosons. This suppression is an example of the approximate *decoupling* of very massive particles at energy scales much smaller than their masses (Symanzik 1973, Appelquist & Carazzone 1975, Collins r1984). The relative suppression of higher-order corrections, however, is generally only through the couplings, and does *not* necessarily involve extra powers of the large masses. This makes higher-order corrections from the weak interactions larger – and more interesting experimentally – than might have been expected (Veltman 1977, Sirlin 1980, Marciano & Sirlin 1980, Antonelli, Consoli & Corbo 1980, Marciano & Parsa r1986).

8.4 Quantum chromodynamics and quark–quark scattering

Quantum chromodynamics (QCD) is the unbroken $SU(3)$ gauge theory of the strong interaction, whose fermionic fields, the quarks, transform according to the defining representation of the group (Fritzsch, Gell-Mann & Leutwyler 1973, Gross & Wilczek 1973, Weinberg 1973c). Its gauge vectors are called *gluons*. Taken with the weak and electromagnetic interactions, it completes the content of the standard model.

Lagrangian in SU(N)

To exhibit the group structure, it is useful to think of an unbroken $SU(N)$ gauge theory, and to specialize to $SU(3)$ only at the end. The generalization of the QED density eq. (8.1) to $SU(N)$ has been given in eq. (7.79). Recalling the definitions in eq. (5.96) for the covariant derivative and (5.100) for $A^\mu{}_a$, we have

$$\begin{aligned}\mathscr{L}_{\text{QCD}} = &\bar{\psi}_i(\mathrm{i}\gamma_\mu\{\delta_{ij}\partial^\mu + \mathrm{i}g_s A^\mu{}_a[t_a{}^{(\mathrm{F})}]_{ij}\} - m\delta_{ij})\psi_j - \tfrac{1}{4}F^{\mu\nu}{}_a F_{\mu\nu,a} \\ &- \tfrac{1}{2}\lambda[\partial_\mu A^\mu{}_a][\partial_\mu A^\mu{}_a] + \partial^\mu\bar{\eta}_a(\delta_{ab}\partial_\mu + g_s C_{abc}A_{\mu c})\eta_b .\end{aligned} \tag{8.83}$$

$\psi_i(x)$ is the quark field with color index i (a sum over quark flavors is implicit), $\eta(x)$ the ghost field, and g_s the *strong coupling*. The gluon field is $A^\mu{}_a(x)$, and the field strength (eq. (5.103)) is $F^{\mu\nu}{}_a(x) = \partial^\mu A^\nu{}_a - \partial^\nu A^\mu{}_a - g_s C_{abc} A^\mu{}_b A^\nu{}_c$, where the C_{abc} are the structure constants of $SU(N)$. The index of the gauge field always runs up to the dimension of the group (Section 5.4), which is $N^2 - 1$ for $SU(N)$ (exercise 8.7). We now summarize a *few* of the necessary facts about the Lie algebra of $SU(N)$. These, and many more, useful results may be found in Macfarlane, Sudbery & Weisz (1968) and Cutler & Sivers (1978).

The $t_a{}^{(\mathrm{F})}$ are the generators of $SU(N)$ in its defining representation. They obey a Lie algebra

$$[t_a{}^{(\mathrm{F})}, t_b{}^{(\mathrm{F})}] = \mathrm{i}C_{abc}t_c{}^{(\mathrm{F})}. \tag{8.84}$$

Recall that for $SU(N)$ C_{abc} is fully antisymmetric and we need not distinguish upper and lower indices (exercise 5.9). The structure constants C_{abc} themselves define another representation, the *adjoint* representation (exercise 8.7), by

$$[t_a{}^{(\mathrm{A})}]_{bc} \equiv -\mathrm{i}C_{abc}. \tag{8.85}$$

For most perturbative calculations, it is not necessary to know explicit forms for either the $t_c{}^{(\mathrm{F})}$ or the structure constants. More relevant are the quantities $T(R)$ and $C_2(R)$, where R labels the representation, $R = \mathrm{F}$ or A,

$$T(R)\delta_{ab} = \mathrm{tr}\,[t_a{}^{(R)}t_b{}^{(R)}], \quad C_2(R)I = \sum_{a=1}^{N^2-1} t_a{}^{(R)}t_a{}^{(R)}, \tag{8.86}$$

where I is the identity matrix in the representation R. $T(R)$ gives the normalization of the generators, and is set once and for all by a choice for the n-dimensional defining representation of $SU(N)$. The standard choice in the physics literature is

$$T(\mathrm{F}) = \tfrac{1}{2}, \quad T(\mathrm{A}) = N, \tag{8.87}$$

the second value following from the first. $C_2(R)$ is the weight of the Casimir operator formed from the sum of the squared generators once

$T(R)$ is specified. In $SU(N)$, with the normalization of eq. (8.87), we have

$$C_2(\mathrm{F}) = (N^2 - 1)/2N, \quad C_2(\mathrm{A}) = N. \tag{8.88}$$

To be specific, in quantum chromodynamics the standard choice for the generators in the defining representation is usually $t_a{}^{(\mathrm{F})} = \lambda_a/2$, where λ_a are the *Gell-Mann matrices* (Gell-Mann 1962), whose explicit forms are given, for completeness, in Appendix F.

Just as for QED, Green functions in QCD may be derived from the generating functional eq. (7.83). It is only in the specifications of the vertices, and the extra 'color' quantum numbers, that the rules for the two theories differ. As usual, the Green functions are constructed by combining propagators with a set of elementary vertices, which are in one-to-one correspondence with terms in the interaction Lagrangian. In this section, we concentrate on the quark–gluon vertex, Fig. 8.8, which we can write down immediately in momentum space by analogy with QED:

$$-\mathrm{i}g_s[t_c{}^{(\mathrm{F})}]_{ji}(\gamma_\nu)_{\beta\alpha}(2\pi)^4\delta(p + p' + q), \tag{8.89}$$

where c, j and i are color indices of the gluon line and of the quark lines whose arrows point out of and into the vertex, respectively. This is the only difference from the QED rule, and its derivation is the same. Otherwise, we only need to recall that gluon and fermion propagators are unit matrices in color space, as in eq. (7.77).

S-matrix elements

The four-point Green function is given by the same graphs as in QED, Fig. 8.1, but now with the new vertices of eq. (8.89). The same arguments as for QED show that this particular graph is gauge invariant, so we may work in the Feynman gauge. Keeping in mind that quark states are characterized by their color, we have, for the $\mathrm{q\bar{q}} \to \mathrm{q\bar{q}}$ matrix element in QCD,

$$\begin{aligned}
&\mathrm{i}M((\mathbf{p}_3,\sigma_3,\eta_3)^{(-)}, (\mathbf{p}_4,\sigma_4,\eta_4)^{(+)}; (\mathbf{p}_2,\sigma_2,\eta_2)^{(+)}, (\mathbf{p}_1,\sigma_1,\eta_1)^{(-)}) \\
&\quad = \bar{v}_2(-\mathrm{i}g_s t_c{}^{(\mathrm{F})}\gamma^\mu)u_1 \times \bar{u}_3(-\mathrm{i}g_s t_d{}^{(\mathrm{F})}\gamma^\nu)v_4 \times [-\mathrm{i}g_{\mu\nu}\delta_{cd}/(p_1 + p_2)^2] \\
&\qquad - \bar{u}_3(-\mathrm{i}g_s t_c{}^{(\mathrm{F})}\gamma^\mu)u_1 \times \bar{v}_2(-\mathrm{i}g_s t_d{}^{(\mathrm{F})}\gamma^\nu)v_4 \times [-\mathrm{i}g_{\mu\nu}\delta_{cd}/(p_1 - p_3)^2],
\end{aligned} \tag{8.90}$$

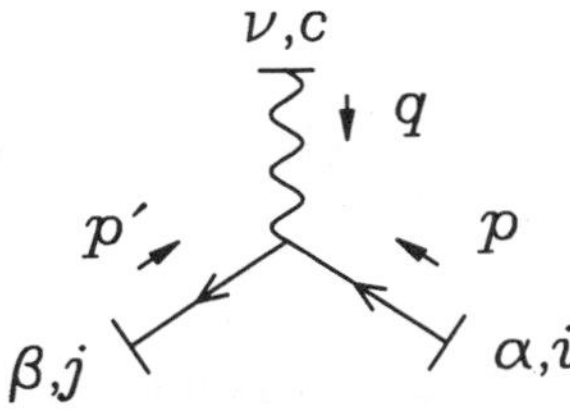

Fig. 8.8 Quark-gluon vertex.

where on the left, by analogy with QED, $(-)$ refers to a quark state, and $(+)$ to an antiquark state, while σ_i and η_i represents spin and $SU(N)$ or 'color' quantum numbers, respectively. On the right, we have used the shorthand notation $u_1 \equiv u_{j_1\alpha_1}(p_1, \sigma_1, \eta_1)$ and so on, with j_1 and α_1 color and Dirac indices, respectively. Equation (8.90) is analogous to eq. (8.6) for electromagnetic scattering.

Cross section

The unpolarized cross section for quark–antiquark scattering is found by the analogue of eq. (8.7), with $|M|^2$ specified by the square of the matrix element in eq. (8.90):

$$\frac{\mathrm{d}\sigma}{\mathrm{d}t} = \frac{1}{N^2 64\pi s(s-4m^2)} \sum_{\text{spin,color}} |M(s,t)|^2, \tag{8.91}$$

where the sum over colors refers to the states of the defining representation of $SU(N)$, in which the quarks reside, and the extra factor $1/N^2$ to the average over the colors of the incoming quarks.

As with the sum over spins, the sums over colors lead to traces, because, just like the propagator, the projection operators (eq. (6.39a)) are diagonal in $SU(N)$ color space:

$$\sum_{\sigma,\eta} u_{i\alpha}(p,\sigma,\eta)\bar{u}_{j\beta}(p,\sigma,\eta) = (\not{p} - m)_{\alpha\beta}\delta_{ij}, \tag{8.92}$$

and similarly for v's. The only new thing here is the identity matrix in color. To verify eq. (8.92), we need only assume that quark states form a complete set in the defining representation. Equation (8.92) then leads to a color trace involving the generators ${t_c}^{(\mathrm{F})}$, as well as to a trace in Dirac matrices. Since the ${t_c}^{(\mathrm{F})}$ are hermitian, the product of matrices within the complex conjugate amplitude can be replaced by the product of the same matrices in the opposite order, without the complex conjugate. The reasoning is analogous to the treatment of Dirac matrices in eq. (8.10). As a result, each Dirac trace is now multiplied by a trace of color generators, with a factor ${t_c}^{(\mathrm{F})}$ for each vertex.

The matrix element, squared and summed over spins *and* colors, now becomes

$$\sum_{\text{spins,colors}} |M|^2 = (1/s^2)T_a\Gamma_a + (1/t^2)T_b\Gamma_b - (1/st)(T_c\Gamma_c + T_d\Gamma_d), \tag{8.93}$$

where the T_i are the same trace factors given in eq. (8.13), and the Γ_i are color factors. Both the T_i and the Γ_i can be read off from the cut graphs in

Fig. 8.3. The Γ_i are given for any $SU(N)$ by (exercise 8.8)

$$\Gamma_a = \Gamma_b = \sum_{e,f} \{\mathrm{tr}\,[t_e^{(\mathrm{F})} t_f^{(\mathrm{F})}]\}^2 = (N^2 - 1)[T(\mathrm{F})]^2$$

$$\Gamma_c = \Gamma_d = \sum_{e,f} \mathrm{tr}\,[t_e^{(\mathrm{F})} t_f^{(\mathrm{F})} t_e^{(\mathrm{F})} t_f^{(\mathrm{F})}] = NC_2(\mathrm{F})[C_2(\mathrm{F}) - T(\mathrm{F})C_2(\mathrm{A})], \tag{8.94}$$

where we have used the definition (8.85) of the adjoint representation and the group matrix properties in eqs. (8.86)–(8.88). It is now a simple matter to compute the cross section by analogy with QED. For massless quarks, it is

$$\mathrm{d}\sigma/\mathrm{d}t = (2\pi\alpha_s^2/N^2s^2)\{\Gamma_a[(1/s^2)(t^2 + u^2) + (1/t^2)(s^2 + u^2)] + \Gamma_c(1/st)2u^2\}, \tag{8.95}$$

(Combridge, Kripfganz & Ranft 1977, Cutler & Sivers 1977), where Γ_a and Γ_c are given by eq. (8.94), and, where, by analogy with eq. (8.31),

$$\alpha_s \equiv g_s^2/4\pi. \tag{8.96}$$

The quark–quark cross section differs only by the exchange of s and u. Comparing eq. (8.95) with the zero-mass limit of Bhabha scattering, eq. (8.30), we see that, to this order, the effect of color is simply to supply group factors to the cross section.

8.5 Gluons and ghosts

To go further with QCD, we need to develop Feynman rules for gluon and ghost fields, and to discuss the role of the ghost field in spin-averaged cross sections.

Feynman rules

The three-gluon coupling may be found by computing the gluon three-point function at lowest order, and then removing the propagators. For simplicity, we work in the Feynman gauge, where $G_{\mathrm{F},\mu\nu,ab}(w) = -\delta_{ab}g_{\mu\nu}\Delta_{\mathrm{F}}(w)$ and $\Delta_{\mathrm{F}}(w)$ is the scalar propagator. A straightforward calculation, starting from the generating functional for QCD, gives the three-point function

$$G_3{}^{\nu_1\nu_2\nu_3}{}_{a_1a_2a_3}(x_1, x_2, x_3) = g_s C_{a_1a_2a_3} \int \mathrm{d}^4x \{ g^{\nu_1\nu_2}[\partial_{(x_1)} - \partial_{(x_2)}]^{\nu_3} + g^{\nu_2\nu_3}[\partial_{(x_2)} - \partial_{(x_3)}]^{\nu_1} + g^{\nu_3\nu_1}[\partial_{(x_3)} - \partial_{(x_1)}]^{\nu_2} \} \Delta_{\mathrm{F}}(x_1 - x)\Delta_{\mathrm{F}}(x_2 - x)\Delta_{\mathrm{F}}(x_3 - x). \tag{8.97}$$

The corresponding position-space three-gluon vertex is linear in derivatives, and when we take the Fourier transform, each $\partial_{(x)}{}^{\nu} \equiv \partial/\partial x_{\nu}$ gives a factor of $-ip^{\nu}$, where p^{ν} is the momentum flowing *into* the vertex on the corresponding line. The resulting momentum-space vertex is illustrated in Fig. 8.9(a), and may be written in the compact form

$$-g_s C_{a_1 a_2 a_3}[g^{\nu_1\nu_2}(p_1 - p_2)^{\nu_3} + g^{\nu_2\nu_3}(p_2 - p_3)^{\nu_1} + g^{\nu_3\nu_1}(p_3 - p_1)^{\nu_2}] \times (2\pi)^4\delta^4(p_1 + p_2 + p_3). \tag{8.98}$$

In the same spirit, the four-point vertex, shown in Fig. 8.9(b), is given by the following expression:

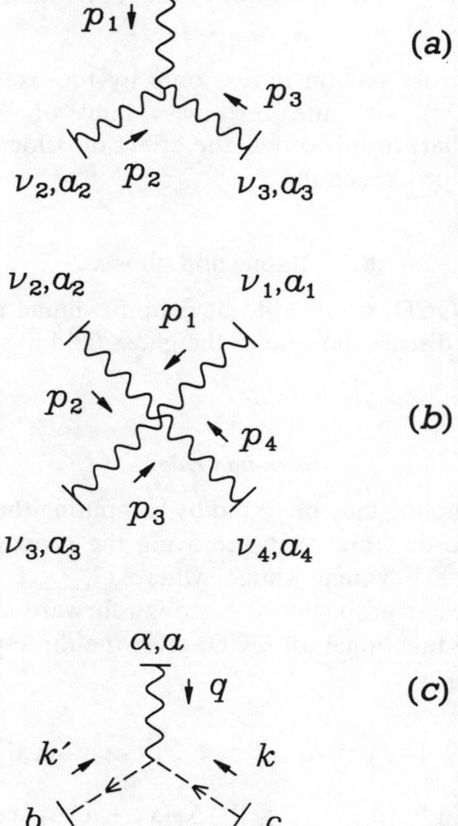

Fig. 8.9 QCD vertices: (a) three-gluon; (b) four-gluon; (c) ghost–gluon.

$$-ig_s{}^2[C_{ea_1a_2}C_{ea_3a_4}(g_{\nu_1\nu_3}g_{\nu_2\nu_4} - g_{\nu_1\nu_4}g_{\nu_2\nu_3})$$
$$+ C_{ea_1a_3}C_{ea_4a_2}(g_{\nu_1\nu_4}g_{\nu_3\nu_2} - g_{\nu_1\nu_2}g_{\nu_3\nu_4}) + C_{ea_1a_4}C_{ea_2a_3}(g_{\nu_1\nu_2}g_{\nu_4\nu_3} - g_{\nu_1\nu_3}g_{\nu_4\nu_2})]$$
$$\times (2\pi)^4\delta^4(p_1 + p_2 + p_3 + p_4). \tag{8.99}$$

In both cases, cyclic permutations give order to the welter of indices.

Finally, the gluon–ghost vertex, illustrated in Fig. 8.9(c), is

$$(2\pi)^4\delta(k + k' + q)g_sC_{abc}k'_\alpha \tag{8.100}$$

where k'_α is the momentum flowing *into* the vertex from the ghost line whose arrow points *out of* the vertex.

Along with the propagators, and eq. (8.89) for the quark–gluon vertex, the vertices above are all that is necessary to construct the full set of perturbation diagrams for QCD. To complete the Feynman rules, we note that in QCD, diagrams may have symmetry factors, which are given by eq. (3.95b, c), just as for a scalar theory. This is so because their combinatoric derivation (Appendix B) depends only on the bosonic nature of the fields. Similarly, the rules for fermionic signs are the same as in QED, and are applied to the anticommuting ghost fields as well as to the quarks. Thus, we have a factor (-1) for each ghost loop and for each quark loop.

Physical and unphysical polarizations

To illustrate these rules, we will consider quark–antiquark annihilation into two gluons. The lowest-order S-matrix for this process is found from the diagrams of Figs. 8.10(a)–(c). Figure 8.10(d) shows the nine cut graphs that result when, as after eq. (8.11), we represent contributions to $|M|^2$ by reversing the flow of momenta in the diagrams of M^* and linking them to the diagrams of M. The integer coefficients show how many times cut graphs of a given topology occur. (At this order, they are all real.) Aside from explicit complex conjugation, the only difference between the gluonic contributions to the graphs of M^* and to the graphs of M is a relative minus sign in both three-point vertices, eqs. (8.98) and (8.100). The three-gluon and ghost-gluon vertices are real, but change sign because we reverse the sense of the momentum flow in M^*.

We recall that in Section 8.2 we showed that the unpolarized cross section for pair annihilation could be simplified and made manifestly Lorentz invariant, by replacing the physical polarization tensor $P^{\mu\alpha}(k)$, eq. (8.59), with $-g^{\mu\alpha}$. What we are about to see is that the QED reasoning does not quite generalize to QCD.

Consider, then, the cross section represented by Fig. 8.10(d). In terms of

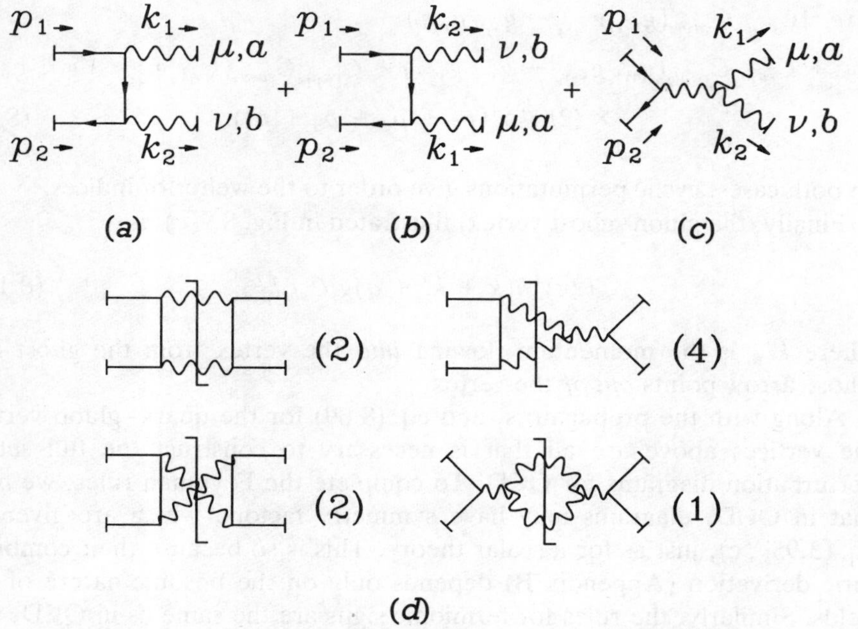

Fig. 8.10 Gluon + gluon → quark pair in QCD: (a)–(c) tree-level diagrams; (d) cut diagrams.

physical polarizations we have again, schematically,

$$\sum_{\text{spins}} |M|^2 = t_{\mu\nu,ab} t^*{}_{\alpha\beta,ab} P^{\mu\alpha}(k_1) P^{\nu\beta}(k_2), \tag{8.101}$$

where the polarization sums for gluons are diagonal in color. $t_{\mu\nu,ab}$ is made of two parts, a 'quark–gluon' contribution from Figs. 8.10(*a*) and (*b*), analogous to QED, and a 'gluon–gluon' diagram, Fig. 8.10(*c*), with a three-gluon coupling. The former is given by

$$\begin{aligned} t_{\mu\nu,ab}{}^{(\mathrm{QG})} = -\mathrm{i}g_s{}^2 \bar{v}_2[\gamma_\nu t_b{}^{(\mathrm{F})}(\not{p}_1 - \not{k}_1 - m)^{-1}\gamma_\mu t_a{}^{(\mathrm{F})} \\ + \gamma_\mu t_a{}^{(\mathrm{F})}(\not{p}_1 - \not{k}_2 - m)^{-1}\gamma_\nu t_b{}^{(\mathrm{F})}]u_1. \end{aligned} \tag{8.102}$$

As in QED, we compute the difference between eq. (8.101) and the same form with the P's replaced by $-g$'s. In the difference, at least one gluon in the graph is scalar polarized on either the right or left of the graph. Let us begin with the action of a single scalar polarized gluon on the left. Consider first the graphs with only quark–gluon couplings. These are the same graphs as in QED, but, when we repeat *exactly* the same steps as in eqs. (8.62)–(8.64), we find, instead of zero,

$$t_{\mu\nu,ab}{}^{(\mathrm{QG})} k_1{}^\mu = -\mathrm{i}g_s{}^2 \bar{v}_2 \gamma_\nu [t_a{}^{(\mathrm{F})}, t_b{}^{(\mathrm{F})}] u_1 = g_s{}^2 C_{abc} \bar{v}_2 \gamma_\nu t_c{}^{(\mathrm{F})} u_1. \tag{8.103}$$

Thus, the nonabelian nature of the coupling prevents the simple cancellation we found in QED. This is a good thing, since there is still the three-gluon diagram Fig. 8.10(c) to consider:

$$t_{\mu\nu,ab}{}^{(3G)} k_1{}^\mu = -\mathrm{i} g_s \bar{v}_2 \gamma_\sigma t_c{}^{(F)} u_1(-\mathrm{i})(k_1 + k_2)^{-2} \\ \times g_s C_{abc}[(k_1 - k_2)^\sigma g_{\mu\nu} + (2k_2 + k_1)_\mu g_\nu{}^\sigma \\ + (-2k_1 - k_2)_\nu g_\mu{}^\sigma] k_1{}^\mu, \qquad (8.104)$$

where we have used the formula for the three-gluon vertex, eq. (8.98). Carrying out the vector sums, we find

$$t_{\mu\nu,ab}{}^{(3G)} k_1{}^\mu = \mathrm{i} g_s{}^2 \bar{v}_2 \gamma_\alpha t_c{}^{(F)} u_1(-\mathrm{i})(k_1 + k_2)^{-2} \\ \times C_{abc}[-(k_1 + k_2)^2 g^\alpha{}_\nu + k_1{}^\alpha k_{2\nu}]. \qquad (8.105)$$

The first term in (8.105) cancels the quark–gluon part, (8.103), so that the complete result is

$$t_{\mu\nu,ab} k_1{}^\mu = \mathrm{i} g_s{}^2 C_{abc} \bar{v}_2 \gamma_\alpha t_c{}^{(F)} u_1(-\mathrm{i})(k_1 + k_2)^{-2} k_1{}^\alpha k_{2\nu} \\ \equiv -M_{ab}(k_1, k_2) k_{2\nu}, \qquad (8.106)$$

where

$$M_{ab}(k_1, k_2) \equiv -\mathrm{i} g_s \bar{v}_2 \gamma_\alpha t_c{}^{(F)} u_1(-\mathrm{i})(k_1 + k_2)^{-2} g_s C_{abc} k_1{}^\alpha,$$

defines M_{ab} for use below. Thus, $t_{\mu\nu,ab} k_1{}^\mu$ is not identically zero, although it *does* cancel when the second gluon has either physical *or* scalar polarization,

$$t_{\mu\nu,ab} k_1{}^\mu P^{\nu\beta}(k_2) = 0, \quad t_{\mu\nu,ab} k_1{}^\mu k_2{}^\nu = 0. \qquad (8.107)$$

We now proceed by introducing the graphical notation of Fig. 8.11(a) for $P^{\mu\alpha}$, which saves a lot of explicit calculation ('t Hooft 1971b). It is inspired by the fact that $M_{ab}(k_1, k_2)$ in eq. (8.106) is equal to the amplitude for the production of a ghost (k_1)–antighost (k_2) pair, as we see by consulting the ghost–gluon vertex, eq. (8.100). In these terms, eq. (8.106) is represented as in Fig. 8.11(b).

As shown in Fig. 8.12(a), if we expand $P^{\mu\alpha}(k_1)$, two of the three terms vanish by applying (8.107) to either the amplitude or its complex conjugate. We then expand $P^{\nu\beta}(k_2)$ to get the three terms of Fig. 8.12(b). In the first term both gluons have polarization tensors $-g^{\sigma\lambda}$; this is the complete answer in QED. Now, however, the remaining two terms do not vanish. To see this, let us calculate the second:

$$t_{\mu\nu,ab}(k_2{}^\nu \bar{k}_2{}^\beta / k \cdot \bar{k})(-g^{\mu\alpha}) t^*{}_{\alpha\beta,ab} = -M_{ab}(k_1, k_2) M^*{}_{ab}(k_2, k_1), \qquad (8.108)$$

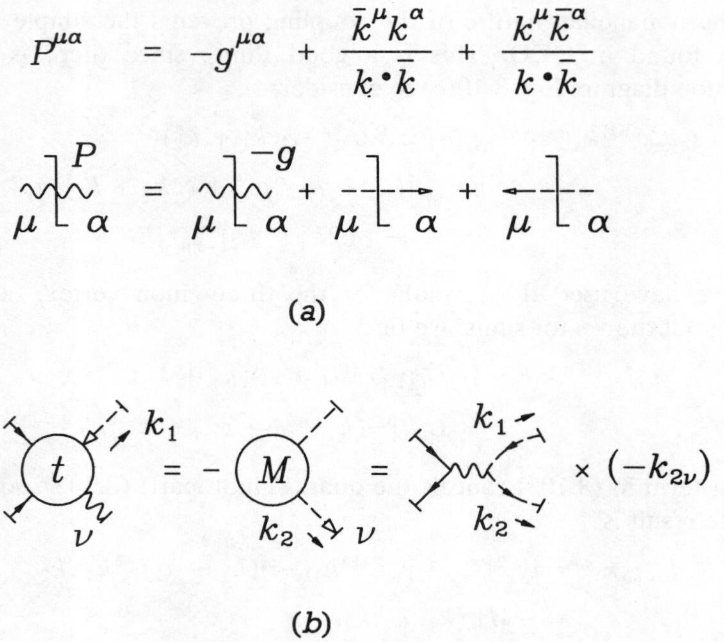

$$P^{\mu\alpha} = -g^{\mu\alpha} + \frac{\bar{k}^{\mu}k^{\alpha}}{k\cdot\bar{k}} + \frac{k^{\mu}\bar{k}^{\alpha}}{k\cdot\bar{k}}$$

(a)

(b)

Fig. 8.11 (*a*)Graphical notation relating physical polarizations to $-g^{\mu\alpha}$; (*b*) graphical representation of eq. (8.106).

which follows by applying eq. (8.106) twice. Treating the third term similarly gives

$$\begin{aligned} t_{\mu\nu,ab}t^*{}_{\alpha\beta,ab}P^{\mu\alpha}(k_1)P^{\nu\beta}(k_2) = {} & t_{\mu\nu,ab}t^*{}_{\alpha\beta,ab}(-g^{\mu\alpha})(-g^{\nu\beta}) \\ & - M_{ab}(k_1,k_2)M^*{}_{ab}(k_2,k_1) \\ & - M_{ab}(k_2,k_1)M^*{}_{ab}(k_1,k_2). \quad (8.109) \end{aligned}$$

In this form, Lorentz invariance is manifest. Since $M(k_1, k_2)$ is the tree-level quark-pair-to-ghost-pair amplitude (Fig. 8.11(*b*)), we are led to Fig. 8.12(*c*). Hence, the cross section summed over physical polarizations equals the covariant polarization sums *minus* the square of the total ghost production amplitude. Equivalently, the cross section computed with $-g^{\mu\alpha}$ gluon polarization tensors *and* with ghosts equals the cross section with physical gluons alone, when an extra (-1) is associated with the *cut* ghost loop. As with QED (Section 8.2), the lowest-order result generalizes to all orders (see 't Hooft (1971b) and Chapter 11).

These results simplify considerably the computation of the annihilation cross section. As an example, consider the contribution of the square of the three-gluon coupling graph in Fig. 8.10(*d*), now given by

$$t_{\lambda\sigma}{}^{(3G)}t^*{}_{\lambda'\sigma}{}^{(3G)}(-g^{\lambda\lambda'})(-g^{\sigma\sigma'})$$

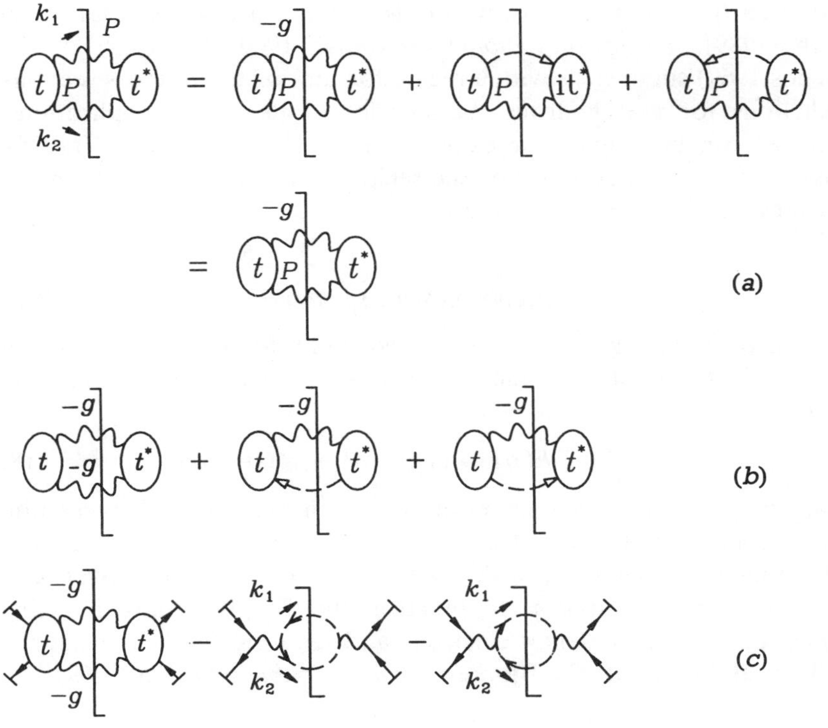

Fig. 8.12 Application of eq. (8.106) to pair annihilation in QCD.

$$
\begin{aligned}
&= \mathrm{tr}\,[(-\mathrm{i}g_s\gamma^\mu)\not{p}_1(\mathrm{i}g_s\gamma_\nu)\not{p}_2]\,\mathrm{tr}\,[t_a{}^{(\mathrm{F})}t_b{}^{(\mathrm{F})}]|-\mathrm{i}/s|^2 \\
&\quad \times (-g_sC_{adc})[(-2k_1-k_2)_\sigma g_{\mu\lambda} + (2k_2+k_1)_\lambda g_{\mu\sigma} + (-k_2+k_1)_\mu g_{\lambda\sigma}] \\
&\quad \times (g_sC_{bcd})[(-2k_1-k_2)^\sigma g^{\lambda\nu} + (2k_2+k_1)^\lambda g^{\sigma\nu} + (-k_2+k_1)^\nu g^{\lambda\sigma}] \\
&= \tfrac{1}{2}N(N^2-1)(4g^4/s^2)[-4s^2-2tu-\tfrac{7}{2}(t^2+u^2-s^2)]. \qquad (8.110)
\end{aligned}
$$

The remainder of the calculation is suggested as exercise 8.10, with the result that the complete cross section at fixed t is ($N = 3$ for QCD),

$$
\frac{\mathrm{d}\sigma}{\mathrm{d}t} = \frac{\pi\alpha_s{}^2(N^2-1)}{s^2N}\left[\frac{(N^2-1)}{2N^2}\frac{(u^2+t^2)}{ut} - \frac{u^2+t^2}{s^2}\right]. \qquad (8.111)
$$

A complete set of two-to-two tree amplitudes in QCD is given by Combridge, Kripfganz & Ranft (1977) and Cutler & Sivers (1977, 1978).

8.6 Parton-model interpretation of QCD cross sections

At this point we emerge from the perturbative formalism for a moment, to ask what use these quark and gluon cross sections can possibly be, since neither particle has been observed in isolation (Particle Data Group r1990, Smith r1989). The *parton model* (Feynman 1969a, b, r1972, Bjorken & Paschos 1969) supplies us with a recipe for turning quark cross sections into predictions for real hadrons. Summaries of its classic applications and successes can be found, for example, in Perl (r1974), Roy (r1975) and Close (r1979). We first describe the recipe of the parton model, and briefly return thereafter to its justification.

Deeply inelastic scattering

The parton model generally gives predictions for cross sections that are *inclusive* in terms of hadronic final states. An example is *deeply inelastic scattering*,

$$\mathrm{e}(k) + h(p) \to \mathrm{e}(k-q) + X(p+q), \tag{8.112}$$

in which an electron of momentum k^μ and a hadron h of momentum p^μ scatter to give an electron of momentum $(k-q)^\mu$ and a hadronic state X with momentum $(p+q)$, with $(p+q)^2 \gg {m_h}^2$. We sum over all final states at fixed q^2 and $p\cdot q$. In practice, this simply means detecting the momentum of the outgoing electron, and ignoring final-state hadrons. It will be useful to define the Bjorken *scaling variable* (Bjorken 1969),

$$x = -q^2/2p\cdot q. \tag{8.113}$$

The parton-model cross section for deeply inelastic scattering is given by

$$\frac{\mathrm{d}\sigma^{(\mathrm{e}h)}}{\mathrm{d}q^2\mathrm{d}(p\cdot q)}(s, q^2, x) = \sum_f \int_0^1 \mathrm{d}\xi\, \phi_{f/h}(\xi)\, \frac{\mathrm{d}\sigma^{(\mathrm{e}f)}}{\mathrm{d}q^2\mathrm{d}(p\cdot q)}(\xi p), \tag{8.114}$$

where $\mathrm{d}\sigma^{(\mathrm{e}f)}/\mathrm{d}q^2 d(p\cdot q)$ is the lowest-order *elastic* electron–quark cross section for quark flavor f, and *quark* momentum ξp. In the high-energy limit, we neglect all particle masses, taking $p^2 = \xi^2 p^2 = 0$ in the calculation of $\sigma^{(\mathrm{e}f)}$. The function $\phi_{f/h}(\xi)$ has the interpretation of the *probability* for the electron to encounter a 'bound' quark of flavor f and momentum ξp^μ, $1 > \xi > 0$, in hadron h; $\phi_{f/h}(\xi)$ is called the *quark distribution* for flavor f in hadron h. It is not calculable in perturbation theory. In summary, the parton-model cross section for electron–hadron scattering is given by the elastic cross section for electron–quark scattering times the probability of finding a quark with a given flavor and fractional momentum in the hadron, summed over flavors and integrated over fractional momentum. In this context, quarks are synonomous with *partons*, although the term refers to any point-like constituent of a hadron. Ignoring for the moment our ignorance of the functions $\phi_{f/h}(\xi)$, let us see what this prescription gives.

Quarks scatter from charged leptons through the QED density, eq. (8.1). The cross section for electron–quark scattering is given by the t-channel exchange of a photon (see Fig. 8.3) only. In this case, color factors occur in the combination $(1/N)\,\mathrm{tr}\, I_{N\times N} = 1$, for $SU(N)$, from the average and sum over incoming and outgoing quark colors, respectively. Then, neglecting masses, the electron–quark cross section is given by exactly the $1/t^2 = 1/(q^2)^2$ term in the $m \to 0$ limit of the Bhabha cross section, eq. (8.30), times an extra factor $Q_f{}^2$ for a quark of flavor f:

$$\mathrm{d}\sigma^{(ef)}/\mathrm{d}q^2 = Q_f{}^2(2\pi\alpha^2/\hat{s}^2)[(\hat{u}^2 + \hat{s}^2)/(q^2)^2]. \tag{8.115}$$

The circumflexes refer to the electron–quark system:

$$\hat{s} = (\xi p + k)^2, \quad \hat{u} = (\xi p + q - k)^2. \tag{8.116}$$

With massless quarks, we require $(\xi p + q)^2 = 0$, which implies

$$-q^2/2x = p\cdot q = -q^2/2\xi. \tag{8.117}$$

Therefore, the differential cross section $\mathrm{d}\sigma^{(ef)}/\mathrm{d}q^2(\xi)\mathrm{d}(p\cdot q)$ is actually overdetermined, and has a left-over delta function after the integral over two-particle phase space, which relates ξ and x,

$$\mathrm{d}\sigma^{(ef)}/\mathrm{d}q^2\mathrm{d}(p\cdot q) = [\mathrm{d}\sigma^{(ef)}/\mathrm{d}q^2]\delta(-q^2/2\xi + q^2/2x). \tag{8.118}$$

The result for the parton-model cross section is found by substituting eq. (8.118) into (8.114):

$$\frac{\mathrm{d}\sigma^{(eh)}}{\mathrm{d}q^2\mathrm{d}(p\cdot q)}(s, q^2, x) = \left(\frac{2x^2}{Q^2}\right)\frac{\mathrm{d}\sigma^{(e\mu)}}{\mathrm{d}q^2}(xs, q^2, x)\sum_f Q_f{}^2\phi_{f/h}(x), \tag{8.119}$$

with $Q^2 = -q^2 > 0$. Here, $\sigma^{(e\mu)}$ denotes the electron–muon cross section, which also proceeds entirely through t-channel exchange. When this leptonic cross section is factored out, the parton model predicts that the residual dimensionless function depends on only the variable x, independent of momentum transfer. This behavior is known as *scaling* (thus the name 'scaling variable' for x), and its (approximate) validity is interpreted as strong evidence for the existence of point-like constituent structure for the hadrons (Friedman & Kendall r1972).

In the same way, the parton model can be used for deeply inelastic scattering of a neutrino ν_i or an antineutrino $\bar{\nu}_i$, with $i = \mathrm{e}$, μ or τ. To lowest order in the weak Lagrangian (8.70), the relevant processes are, in place of eq. (8.112),

$$\begin{aligned} &\nu_i(k) + h(p) \to l_i{}^-(k - q) + X(p + q), \\ &\bar{\nu}_i(k) + h(p) \to l_i{}^+(k - q) + X(p + q), \end{aligned} \tag{8.120}$$

corresponding to the emission of a W^+ (W^-) by the neutrino (antineutrino), and its absorption by the hadronic system. Otherwise, the kinematics is just the same, and we can write

$$\frac{d\sigma^{(\nu h\to lX)}}{dq^2 d(p\cdot q)}(s, q^2, x) = \left(\frac{2x^2}{Q^2}\right)\frac{d\sigma^{(\nu l'\to l\nu')}}{dq^2}(xs, q^2, x)\sum_f \bar{Q}_f{}^2\phi_{f/h}(x), \tag{8.121}$$

where $\bar{Q}_f{}^2$ is a number that specifies the coupling of flavor f to W^+ in units of the leptonic couplings, and where $\sigma^{(\nu l'\to l\nu')}$ denotes the lepton cross section for elastic charge exchange scattering between two different leptonic flavors. Since the $\bar{Q}_f{}^2$ are generally different from the electromagnetic $Q_f{}^2$, the combination of different processes allows us to determine the parton distributions for different flavors, and to apply them in yet other reactions.

In eqs. (8.119) and (8.121), the deeply inelastic scattering cross sections are seen to *measure* the sum of quark distributions, weighted by quark charges. We can do much more, however, if we make the assumption that the quark distributions are *universal*, that is, are the same in different processes. Then the parton model acquires extra predictive power.

The Drell–Yan cross section

A natural extension of the parton model is to inclusive large-momentum transfer processes in hadron–hadron scattering. Of these, the *Drell–Yan process* (Drell & Yan 1971), is closest in spirit to deeply inelastic scattering. It is the reaction

$$h(p) + h'(p') \to l\bar{l}(q) + X(p + p' - q), \tag{8.122}$$

in which two hadrons collide to produce a lepton pair with total momentum q^μ, $q^2 > 0$. Here, the parton model prescribes a product of distributions, for quarks of flavor f in hadron h and of antiflavor $\bar{f}$ in hadron h', and vice versa, times the elastic scattering cross section for the 'partonic process', $f\bar{f}\to l\bar{l}$:

$$\frac{d\sigma^{(hh'\to l\bar{l}+X)}}{dq^2}(q^2) = \sum_f \int_0^1 d\xi d\xi'\, \phi_{f/h}(\xi)\phi_{\bar{f}/h'}(\xi')\, \frac{d\sigma^{(f\bar{f}\to l\bar{l})}}{dq^2}(\xi, \xi'), \tag{8.123}$$

where now the sum over f is defined to include antiquarks as well as quarks.

The reaction $f\bar{f}\to l\bar{l}$ proceeds by the diagonal annihilation cut graph, Fig. 8.3(a), and the cross section is closely related to the lepton-annihilation cross sections, eqs. (8.34) and (8.35). Here, however, we average over the colors of both quark and antiquark. This means an extra factor $(1/N^2)\,\mathrm{tr}\, I_{N\times N} = 1/N$ compared with the leptonic case for color $SU(N)$,

and

$$\mathrm{d}\sigma^{(f\bar{f}\to l\bar{l})}/\mathrm{d}\hat{\Omega}_{\mathrm{cm}} = Q_f^{\,2}(\alpha^2/4N\hat{s})(1+\cos^2\hat{\theta}_{\mathrm{cm}}),$$

$$\sigma_{\mathrm{tot}}{}^{(f\bar{f}\to l\bar{l})} = Q_f^{\,2}(4\pi\alpha^2/3N\hat{s}). \tag{8.124}$$

Again the circumflexes refer to the partonic system. The factor $1/N$ is the relative probability that a quark and antiquark of random colors match. As in deeply inelastic scattering, $\mathrm{d}\sigma^{(f\bar{f}\to l\bar{l})}/\mathrm{d}q^2$ is overdetermined at lowest order, and has an extra delta function:

$$\mathrm{d}\sigma^{(f\bar{f}\to l\bar{l})}/\mathrm{d}q^2 = \sigma_{\mathrm{tot}}{}^{(f\bar{f}\to l\bar{l})}\,\delta(\hat{s}-q^2). \tag{8.125}$$

Together with eq. (8.123), this gives the following for the parton-model cross section:

$$\frac{\mathrm{d}\sigma^{(hh'\to l\bar{l}+X)}}{\mathrm{d}q^2}(s,q^2) = \frac{4\pi\alpha^2}{3Nq^2s}\sum_f \int_0^1 \mathrm{d}\xi\mathrm{d}\xi'\,\phi_{f/h}(\xi)\phi_{\bar{f}/h'}(\xi')\delta\!\left(\xi\xi' - \frac{q^2}{s}\right). \tag{8.126}$$

If we take quark distributions determined from deeply inelastic scattering, this expression gives a complete prediction for the Drell–Yan cross section, including normalization.

So far, we have discussed only fully inclusive cross sections, in which we sum over all the hadronic final states. Many more parton-model predictions arise if we re-interpret outgoing quarks and gluons as *jets* of hadrons; a jet may roughly be defined as a set of hadrons moving nearly parallel to each other. Suppose, for instance, that in e^+e^- annihilation into hadrons, we define θ to be the angle between the incoming electron and the direction of maximal energy flow of hadrons in the final state. The distribution in θ simply follows the angular dependence $1+\cos^2\theta$ in the e^+e^- annihilation cross section of eq. (8.34) (Wu 1984). Extending this interpretation to lepton–quark, quark–quark, gluon–quark, and gluon–gluon elastic scattering cross sections, we can derive predictions for jet production in deeply inelastic scattering and hadron–hadron scattering (Horgan & Jacob 1981).

The manner in which parton-model predictions emerge in perturbation theory beyond the lowest order is discussed at some length in Chapter 14, and requires the full analysis of loop diagrams, which will be the subject of the intervening chapters. We shall encounter violations of pure Bjorken scaling, in which parton distributions acquire (calculable) momentum dependence. Nevertheless, scaling remains a useful concept, and it is instructive to describe briefly a heuristic argument, due to Feynman (1969a, b), for the idea of a parton density.

Consider, once again, the scattering of an electron and a proton, with the proton assumed to be made up of constituents, 'partons'. The partons interact with each other, and exist only in virtual states. Let us suppose that a typical state has lifetime τ in this frame. In the rest frame of the electron, τ is dilated to $\tau(E_p/m_p)$, with E_p the proton energy, while the proton radius r_p is Lorentz-contracted to $r_p(m_p/E_p)$. Thus, during the short time it takes the proton to pass over the electron in this frame, the partons appear 'frozen', because their self-interactions act on (dilated) time scales that are much longer than the time of collision. Since parton–parton interactions and electron–parton scattering take place on such different time scales, they cannot interfere in a quantum mechanical sense. As a result, the quantum mechanical amplitudes for the distributions of partons exhibit *incoherence*, relative to the electron–parton cross section, as if they were classical quantities. Thus it makes sense to talk about the probability for finding a parton. This probability is the parton density. We shall return in Chapter 14 to the field-theoretic realization of this heuristic, but compelling, reasoning.

Parton-model predictions, unlike lowest-order calculations in electroweak theory, are not expected to be accurate beyond, say, a factor of two. Within this range, however, they are remarkably successful in tying together a wide variety of experimental observations. (See, for instance, Grosso-Pilcher & Shochet (r1986) for a discussion of how parton-model predictions fare in the Drell–Yan process.)

Exercises

8.1 Compute $d\sigma/d\Omega_{cm}$ and $d\sigma/dt$ to order e^4 for e^-e^- elastic scattering.

8.2 Compute $d\sigma/d\Omega_{cm}$ to lowest order for $e^+e^- \to s\bar{s}$, with s a scalar field of unit electromagnetic charge. Compare the result with eq. (8.34) for the production of a spin one-half pair.

8.3 By repeated use of the identity eq. (8.63), show that the decoupling of a scalar photon, eq. (8.64), can be extended to any fermion line between on-shell spinors, with an arbitrary number of other photons attached.

8.4 Verify the $O(e^4)$ massless pair annihilation cross section, eq. (8.68), and the Compton scattering cross section, eq. (8.69).

8.5 Show that the massless $O(e^4)$ cross section $d\sigma/dt$ for $\gamma\gamma \to e\bar{e}$ is the same as the pair annihilation cross section, eq. (8.68).

8.6 Verify the phase space integral eq. (8.74). Hint: the left-hand side, $H_{\alpha\beta}$, is a second-rank tensor (why?); consider $K^\alpha K^\beta H_{\alpha\beta}$ and $H_\beta{}^\beta$.

8.7 Verify (a) that the dimension of $SU(N)$ is $N^2 - 1$, (b) that the set of 'adjoint' matrices defined by eq. (8.85) actually forms a representation of the underlying group, (c) the general relation $T(R) = C_2(R)(d_R/d_G)$, where d_R is the dimension of representation R.

8.8 Evaluate Γ_a, Γ_d in eq. (8.94).

8.9 Derive the four-gluon coupling, eq. (8.99), by calculating its contribution to the lowest-order four-gluon Green function.

8.10 Compute the lowest-order cross section $\mathrm{d}\sigma/\mathrm{d}t$, eq. (8.111), for massless quark–antiquark annihilation to two gluons. This involves the cut diagrams that result from squaring the full amplitude illustrated in Fig. 8.10. To avoid duplicating calculations, make use of the symmetries between cut diagrams and their complex conjugates and of the three-gluon coupling under interchange of two gluons.

8.11 Use the parton model to derive the differential cross section $\mathrm{d}\sigma/(\mathrm{d}q^2\mathrm{d}\cos\theta)$ for the Drell–Yan process, where θ is the angle between the incoming hadrons and the outgoing lepton, defined in the rest frame of the outgoing leptons.

PART III

RENORMALIZATION

9
Loops, regularization and unitarity

Loops are the characteristic feature of higher orders in the perturbative expansion of Green functions. In principle, loop diagrams are completely defined by the same Feynman rules that give tree diagrams; the only difference is that there are integrals left to do. In practice, the evaluation of these integrals is not completely straightforward. First, their integrands diverge at propagator poles. The integrals remain well defined, however, because of the infinitesimal term '$\mathrm{i}\epsilon$' in each propagator. We shall show below how this works, and how the '$\mathrm{i}\epsilon$-prescription' defines Green functions as analytic functions of their external momenta. Second, in most field theories, some integrands simply do not fall off fast enough at infinity for the corresponding integrals to converge. Such integrals are said to be ultraviolet divergent. The consideration of this problem will lead us, in due course, to the concepts of regularization and renormalization.

We begin our discussion with a simple example, which will illustrate both of these problems. Later in the chapter, we develop the method of time-ordered perturbation theory, and use it to verify that the perturbation expansion generates a unitary S-matrix order-by-order in perturbation theory. Unitarity is a fundamental property, necessary for a theory to make physical sense. When a new field theory is proposed, the first questions are, is it unitary, and is it renormalizable? These questions are the underlying subjects of this and the following two chapters.

9.1 One-loop example

Consider the one-loop graph Fig. 9.1, with external lines truncated. This graph contributes to the $2q$-point function in any scalar theory with a potential of the form ϕ^{q+2}. In ϕ^3 theory, for instance, it is the first

Fig. 9.1 One-loop diagram.

correction to the ϕ particle propagator, while in ϕ^4 theory it contributes to the four-point function.

Neglecting coupling constants and a factor of i (but keeping the symmetry factor), Fig. 9.1 is

$$I_n(q^2, m^2) = \tfrac{1}{2}\int d^n k(2\pi)^{-n}(k^2 - m^2 + i\epsilon)^{-1}[(q-k)^2 - m^2 + i\epsilon]^{-1}. \tag{9.1}$$

We have left the number of dimensions, n, arbitrary for now. To evaluate eq. (9.1), it is convenient to combine the two Feynman denominators by introducing an integral representation:

$$(AB)^{-1} = \int_0^1 dx[xA + (1-x)B]^{-2}. \tag{9.2}$$

Applying this to eq. (9.1), we derive

$$I_n(q^2, m^2) = \tfrac{1}{2}\int_0^1 dx \int d^n k(2\pi)^{-n}(k^2 - 2xq\cdot k + xq^2 - m^2 + i\epsilon)^{-2}. \tag{9.3}$$

This technique of combining denominators is known as *Feynman parameterization* (Feynman 1949), and x is referred to as a *Feynman parameter*. In the new form, there is only a single denominator, which is quadratic in the momenta. Linear terms in the denominator are eliminated by changing variables to $l^\mu = k^\mu - xq^\mu$, which completes the square:

$$I_n(q^2, m^2) = \tfrac{1}{2}\int_0^1 dx \int_{-\infty}^{\infty} dl_0 d^{n-1}\mathbf{l}(2\pi)^{-n}[l_0{}^2 - \mathbf{l}^2 + x(1-x)q^2 - m^2 + i\epsilon]^{-2}. \tag{9.4}$$

In this form it is particularly easy to identify points at which the integrand diverges (the first problem mentioned in the introductory comments).

Suppose that q^2 is negative, i.e., $q^2 = -Q^2$, with Q a positive number. Then the poles in the l_0-plane are at

$$l_0 = \pm[\mathbf{l}^2 + x(1-x)Q^2 + m^2 - i\epsilon]^{1/2} \equiv \pm(\lambda - i\epsilon). \tag{9.5}$$

These two poles are in the simple arrangement shown in Fig. 9.2. There is one (double) pole in the lower half-plane, near the positive real axis, and one (double) pole in the upper half-plane, near the negative real axis. We use Cauchy's theorem to replace the integral along the real l_0 axis C_R, by an integral along the imaginary axis C_I, as also shown in Fig. 9.2. This is referred to as a Wick rotation in energy (Wick 1954). It is analogous to the Wick rotation for time in the path integral, eq. (3.20), although for energy the angle of rotation is positive while for time it is negative. The integrals along the two contours C_R and C_I give the same answer because they can

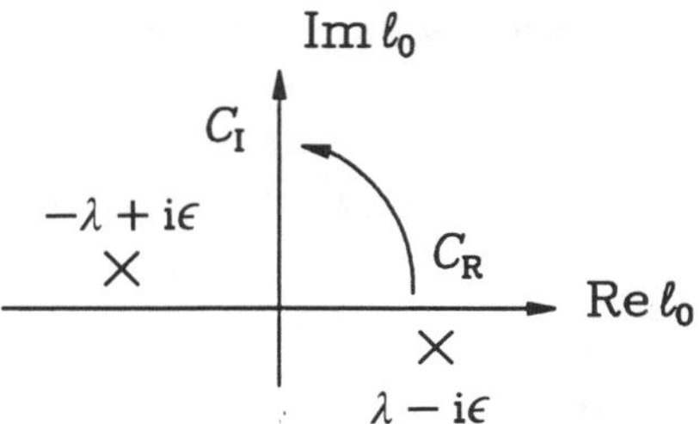

Fig. 9.2 Wick rotation.

be completed at infinity without enclosing either of the two singularities. Along C_I, l_0 is purely imaginary, and we can define a real variable l_n by

$$l_n = -il_0. \tag{9.6}$$

In terms of the new integration variable,

$$I_n(-Q^2, m^2)$$
$$= \tfrac{1}{2}i(2\pi)^{-n}\int_0^1 dx \int_{-\infty}^{\infty} dl_n d^{n-1}\mathbf{l}[-l_n{}^2 - \mathbf{l}^2 - x(1-x)Q^2 - m^2]^{-2}, \tag{9.7}$$

where we have dropped the $i\epsilon$ term, because the denominator is now negative definite. It should be emphasized that we are able to Wick-rotate only because q^μ is spacelike. For timelike q^μ, we do not always have the arrangement of poles shown in Fig. 9.2.

To get an idea of the analytic structure of $I_n(q^2, m^2)$ as a function of q^2, we can evaluate eq. (9.7) for $n = 2$. When we use polar coordinates the angular integral is trivial, and we find

$$\begin{aligned} I_2(q^2, m^2) &= \tfrac{1}{2}i(2\pi)^{-1}\int_0^1 dx \int_0^{\infty} dl\, l[l^2 + x(1-x)Q^2 + m^2]^{-2} \\ &= \frac{i}{2\pi q^2}\frac{1}{(1-4m^2/q^2)^{1/2}} \ln\left[\frac{(1-4m^2/q^2)^{1/2} - 1}{(1-4m^2/q^2)^{1/2} + 1}\right]. \end{aligned} \tag{9.8}$$

Here we have replaced $-Q^2$ by q^2. When q^2 is spacelike, this expression is well defined. It then can be analytically continued from spacelike q^2 to any value of q^2. Examining (9.8), we see that $I_2(q^2, m^2)$ is analytic in the entire q^2-plane, with the exception of a branch cut beginning at $q^2 = 4m^2$. (Its analyticity at $q^2 = 0$ may be checked by an explicit expansion.) Thus the behavior of $I_2(q^2, m^2)$ for *all* external momenta is determined by its behavior for spacelike momenta. As we shall see, this is a general feature of Feynman diagrams.

Consideration of $I_3(q^2, m^2)$ gives a slightly different formula, but exactly the same analytic structure. In four or more dimensions, however, I_n

diverges. To see this more clearly, we can re-express it in polar coordinates:

$$I_n(-Q^2, m^2) = \tfrac{1}{2}\mathrm{i}(2\pi)^{-n}\int \mathrm{d}\Omega_{n-1}\int_0^\infty \mathrm{d}\kappa\,\kappa^{n-1}\int_0^1 \mathrm{d}x[\kappa^2 + x(1-x)Q^2 + m^2]^{-2}, \tag{9.9}$$

where $\mathrm{d}\Omega_{n-1}$ represents the angular integration. The radial integral behaves as $\int^\infty \mathrm{d}\kappa\,\kappa^{n-5}$ for large κ, which is convergent only when $n-5<-1$, with a logarithmic divergence at $n=4$. This means, in particular, that Fig. 9.1 specifies a divergent integral in four dimensions, which is just where we might like to use it! For instance, in a ϕ^4 theory, Fig. 9.1 contributes to the S-matrix for two-particle elastic scattering, at second order in the coupling. Thus, a divergence in Fig. 9.1 in four dimensions shows up immediately in the S-matrix. Clearly, this is a serious problem. Let us set it aside for the moment, however.

9.2 Wick rotation in perturbation theory

In the example of Fig. 9.1, we evaluated the Feynman integral for spacelike values of its single external momentum, and then analytically continued to arbitrary complex momenta. This, in fact, is the procedure for any diagram. It is determined by the analytic form of the Feynman propagator, which we derived from the path integral (see Section 3.1). We shall show that any Feynman integral $G(p_i)$, wih external momenta p_i, may be defined as an analytic function, by evaluating it first with all $p_{i,0}$ on the *positive imaginary axis*, $\mathrm{Re}\,p_{i,0}=0$, $\mathrm{Im}\,p_{i,0}>0$. Clearly, with such a choice, all external momenta are spacelike. The values of $G(p_i)$ at all other momenta are to be found by analytic continuation. In particular, $G(p_i)$ is found for *physical* momenta – those to be used in the evaluation of S-matrix elements – by rotating any or all of the $p_{i,0}$ from the positive imaginary axis to the real axis through an angle $\pi/2$.

Consider, then, an arbitrary graph G with L loops, N lines (all of mass m) and E external momenta $\{p_i\}$, in n dimensions. At first, we take the external momenta to be real. G can be represented in the form

$$G(p_i\cdot p_j, m^2)$$
$$= \prod_{i=1}^{L}\int \mathrm{d}k_{i,0}\,\mathrm{d}^{n-1}\mathbf{k}_i \prod_{j=1}^{N}\left[\left(\sum_{n=1}^{L}\omega_{jn}k_n{}^{\mu} + \sum_{m=1}^{E}\eta_{jm}p_m{}^{\mu}\right)^2 - m^2 + \mathrm{i}\epsilon\right]^{-1}. \tag{9.10}$$

The $k_i{}^{\mu}$ are loop momenta, and we have introduced two incidence matrices, ω_{jn} and η_{jm} (by analogy with eq. (4.1)), which summarize the flow of loop and external momenta on lines j. We have split up each loop integral into a single energy, $k_{i,0}$, and $n-1$ spatial components. We have also suppressed overall constants and numerator momenta. The terms $+i\epsilon$ in each denominator define the positions of the poles of the integrand; we shall use this information to define the integrals in G. We begin by using the technique of Feynman parameterisation, generalized to N factors of arbitrary powers (exercise 9.2),

$$\prod_{i=1}^{N} A_i^{-\eta_i} = \left[\prod_{i=1}^{N} \Gamma(\eta_i)\right]^{-1} \Gamma(\textstyle\sum_{i=1}^{N}\eta_i)\int_0^1 dx_1\, x_1^{\eta_1-1} \cdots dx_N\, x_N^{\eta_N-1}$$
$$\times\, \delta(1-\textstyle\sum_{i=1}^{N}x_i)\left(\displaystyle\sum_{i=1}^{N} x_i A_i\right)^{-\Sigma_i \eta_i}, \qquad (9.11)$$

where $\Gamma(\eta)=(\eta-1)!$ when η is an integer. The extension of the functions Γ to noninteger η is given below in Section 9.3. Applied to eq. (9.10), this gives

$$G(p_i\cdot p_j, m^2) = \Gamma(N)\int_0^1 \prod_{i=1}^{N} dx_i\, \delta(1-\textstyle\sum_{i=1}^{N}x_i)\displaystyle\prod_{i=1}^{L}\int_{-\infty}^{\infty} dk_{i,0}\, d^{n-1}\mathbf{k}_i$$
$$\times\left[\sum_{j=1}^{N} x_j\left(\sum_{n=1}^{L}\omega_{jn}k_n{}^{\mu} + \sum_{m=1}^{E}\eta_{jm}p_m{}^{\mu}\right)^2 - m^2 + i\epsilon\right]^{-N}. \qquad (9.12)$$

Next, we define a new function $\bar{G}(p_i\cdot p_j, m^2, \theta)$, by

$$\bar{G}(p_i\cdot p_j, m^2, \theta) = \Gamma(N)\int_0^1 \prod_{i=1}^{N} dx_i\, \delta(1-\textstyle\sum_{i=1}^{N}x_i)\displaystyle\prod_{i=1}^{L}\int_{-\infty}^{\infty} \exp(i\theta)\, dk_{i,n}\, d^{n-1}\mathbf{k}_i$$
$$\times\left[\sum_{j=1}^{N} x_j\left(\sum_{s=1}^{L}\omega_{js}k_{s,n} + \sum_{s=1}^{E}\eta_{js}p_{s,0}\right)^2 \exp(2i\theta)\right.$$
$$\left. - \sum_{j=1}^{N} x_j\left(\sum_{s=1}^{L}\omega_{js}\mathbf{k}_s + \sum_{s=1}^{E}\eta_{js}\mathbf{p}_s\right)^2 - m^2 + i\epsilon\right]^{-N}, \qquad (9.13)$$

where the $k_{i,n}$ are real integration variables. This is just the same as G, except that we have multiplied all loop and external energies by a phase. We take $\pi/2 \geqslant \theta \geqslant 0$. The function $\bar{G}(p_i\cdot p_j, m^2, \theta)$ has the following properties:

(i) $\bar{G}(p_i\cdot p_j, m^2, 0) = G(p_i\cdot p_j, m^2)$, which follows trivially from its definition.

(ii) $\bar{G}(p_i\cdot p_j, m^2, \theta)$ is an analytic function of θ for $\pi/2 \geqslant \theta \geqslant 0$, because for all θ in this range, the imaginary part of the denominator in eq. (9.13) is positive definite for all values of loop and external momenta. Equation (9.13) is thus the integral of a bounded analytic function. It is then itself an analytic function of its parameters, at least as long as it is convergent.

(iii) By construction, $\overline{G}(p_i{\cdot}p_j, m^2, \pi/2) = G^{(\mathrm{E})}(-(p_i{\cdot}p_j)_\mathrm{E}, m^2)$, where

$$(p_i{\cdot}p_j)_\mathrm{E} \equiv \sum_{a=1}^{n} (p_i)_a (p_j)_a, \tag{9.14}$$

n being the number of dimensions and the function $G^{(\mathrm{E})}$ the same as G in eq. (9.10), but with all loop energy contours Wick-rotated up to the imaginary axis. $G^{(\mathrm{E})}$ is sometimes called the Green function in 'Euclidean space'.

These three properties show that the integral defining the Green function $G(p_i{\cdot}p_j, m^2)$ with physical momenta $\{p_i\}$ is the analytic continuation of the same integral with both contours *and* external momenta Wick-rotated. Thus, as promised, the Wick-rotated integral determines the Green function for physical momenta.

The following point is crucial. We have used the sign of the term $i\epsilon$ in eq. (9.13) in proving property (ii). Were the sense of the energy rotation to be reversed, we would have the possibility of cancellation between the $i\epsilon$ in $\overline{G}$ and the imaginary part of the rotated energies. In this case, we would *not* be able to say that $\overline{G}(p_i{\cdot}p_j, m^2, \theta)$ was a direct analytic continuation of $G(p_i{\cdot}p_j, m^2)$, since the denominator would vanish for some values of loop momenta and θ, possibly generating a singularity in the function there. Thus the direction of the Wick rotation is defined by the sign of the $i\epsilon$ in G, eq. (9.10), which is the same sign as in the Feynman denominators. This sign, in turn, was determined by the construction of the path integral, as described in Section 3.1.

In summary, we have found, by a generalization of the analysis for the one-loop example, that Feynman integrals in momentum space can always be defined as functions of their external momenta by Wick rotation. This eliminates, quite generally, a potential ambiguity in the integrals of the perturbative expansion. It is not the whole story, however. Our analysis has assumed that the loop integrals converge. As we know, however, this is not always the case.

9.3 Dimensional regularization

Let us return to the ultraviolet divergence of Fig. 9.1 in four or more dimensions. We now introduce the concept of regularization by dimensional continuation, which is one method of isolating the ultraviolet divergence in an appealing and useful fashion. Now, the reader may wonder what we can possibly mean by 'isolating' the ultraviolet divergence. Once the integral diverges, what more is there to say? Well, let us consider the ultraviolet divergence in I_n, eq. (9.9), for instance. It is much simpler than the integral itself. It comes from the region $\kappa \to \infty$, where the integrand is insensitive to the Feynman parameter and to the external momentum. It depends only on the overall power of the denominator, and on the number of dimensions in the integral. This suggests that it may be possible to study the nature of the divergence in a way that does not depend on the details of the graph.

Ultraviolet divergences

We begin by studying one-loop graphs. By using Feynman parameterization, completing the square and Wick-rotating in external as well as internal momenta, one-loop integrals can all be expressed as the form (exercise 9.4)

$$I_{n,s}(p_i\cdot p_j,\, m^2)$$
$$= S^{-1}(2\pi)^{-n}\Gamma(s)\int_0^1 \prod_{i=1}^{s} \mathrm{d}x_i\, \delta(1-\textstyle\sum_i x_i)\, i_{n,s}(M^2(p_i\cdot p_j,\, m^2,\, \{x_i\})). \quad (9.15)$$

Here s is the number of lines in the loop and $\{p_k\}$ is the set of external momenta of the loop (all 'Euclidean'). S^{-1} denotes the possible symmetry factor. The detailed structure of the diagram is contained in the function

$$i_{n,s}(M^2) \equiv \mathrm{i}\int \mathrm{d}^n l_{\mathrm{E}}(-l_{\mathrm{E}}{}^2 - M^2)^{-s} \quad (9.16)$$

where M^2 is the quadratic function of external momenta and masses that results from completing the square in the loop momentum. It is positive definite for Euclidean external momenta (exercise 9.5). The overall factor of i comes from Wick rotation. Equation (9.7) is an example of this form, with $s = 2$. Changing to polar coordinates as in eq. (9.9),

$$i_{n,s}(M^2) = \mathrm{i}\int \mathrm{d}\Omega_{n-1}\int_0^\infty \mathrm{d}\kappa\, \kappa^{n-1}(-\kappa^2 - M^2)^{-s}, \quad (9.17)$$

we recognize that $i_{n,s}$ is ultraviolet divergent for $n/2 \geqslant s$. In four dimensions the only ultraviolet divergent one-loop graphs are Figs. 9.1 and 9.3(a), the latter being convergent only when $n < 2$. In six dimensions, Fig. 9.3(b) also

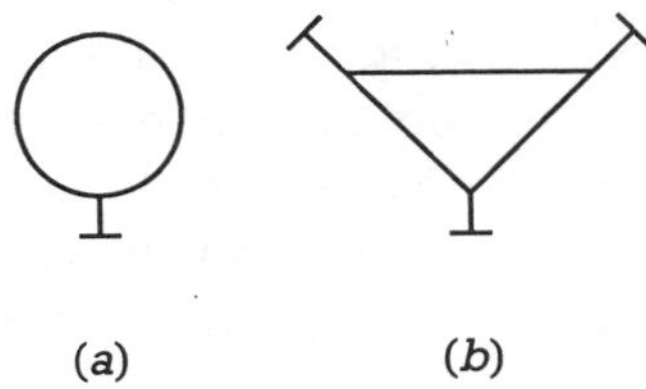

(a) (b)

Fig. 9.3 One-loop diagrams in ϕ^3 theory that are ultraviolet divergent in (a) two dimensions and (b) six dimensions.

becomes divergent. For sufficiently high dimensions, any one-loop integral is ultraviolet divergent. On the other hand, for a sufficiently small number of dimensions, every one-loop integral converges.

Regularization

A regularization of an integral is any method by which the integral is modified to make it finite. Usually, this involves the introduction of a new parameter. For instance, we might put an upper limit Λ on the radial integral in eq. (9.17) for $n/2 \geqslant s$, and define

$$i_{n,s}{}^{(\mathrm{R})}(M^2, \Lambda^2) = \mathrm{i}\int \mathrm{d}\Omega_{n-1} \int_0^{\Lambda} \mathrm{d}\kappa\, \kappa^{n-1}(-\kappa^2 - M^2)^{-s}, \qquad (9.18)$$

where $i_{n,s}{}^{(\mathrm{R})}(M^2, \Lambda^2)$ is a regulated version of $i_{n,s}(M^2)$. We recover $i_{n,s}(M^2)$ in the limit $\Lambda \to \infty$. In the case $s = 2$, $n = 4$, for instance, we easily see that

$$i_{4,2}{}^{(\mathrm{R})}(M^2, \Lambda^2) = \mathrm{i} \ln(\Lambda) \int \mathrm{d}\Omega_3 + i_{4,2}{}^{(\mathrm{F})}(M^2), \qquad (9.19)$$

where $i_{4,2}{}^{(\mathrm{F})}(M^2)$ is a finite function of M^2. In this case, the regularization gives substance to the term 'logarithmically divergent'. Cut-off regularization, appealing in its simplicity, is not generally useful beyond one loop, because it is not Lorentz invariant. In contrast, *dimensional continuation* is Lorentz invariant (Ashmore 1972, Bollini & Giambiagi 1972, 't Hooft & Veltman 1972a, Collins r1984).

Angular integrals

Dimensional continuation is suggested by the simple observation, already made above, that divergent diagrams converge when the number of dimensions, n, is reduced sufficiently. By itself, this observation does not constitute a regularization procedure, since the number of dimensions cannot (yet) be varied continuously. We shall, however, re-express eq. (9.17) in a form in which n appears only as a parameter.

As $i_{n,s}(M^2)$ stands, its angular integrals depend in a discrete fashion on the number of dimensions. When $n = 2$, for instance, $\mathrm{d}\Omega_1 = \mathrm{d}\phi$, and when $n = 3$, $\mathrm{d}\Omega_2 = \sin\theta\, \mathrm{d}\theta \mathrm{d}\phi$, with nothing in between. On the other hand, because we have completed the square in the loop momentum, the integrand does not depend on the angles. So, if we do the angular integrals first, $i_{n,s}(M^2)$ is proportional to the angular volume, Ω_{n-1} in n dimensions:

$$i_{n,s}(M^2) = \mathrm{i}(-1)^s \Omega_{n-1} \int_0^{\infty} \mathrm{d}\kappa\, \kappa^{n-1}(\kappa^2 + M^2)^{-s}. \qquad (9.20)$$

As we now show, Ω_{n-1} *can* be expressed as a function in which n appears only as a parameter. Then, once Ω_{n-1} is extended to continuous n, so is $i_{n,s}(M^2)$.

The angular integral in n Euclidean dimensions can easily be found by an iterative procedure. First, separate the nth component of each n-dimensional vector $\mathbf{k}^{(n)}$,

$$\mathbf{k}^{(n)} = (\mathbf{k}^{(n-1)}, k_n). \tag{9.21}$$

Suppose we already know how to re-express $\mathbf{k}^{(n-1)}$ in terms of $(n-1)$-dimensional polar coordinates. In addition to a radial coordinate, r_{n-1}, the vector $\mathbf{k}^{(n-1)}$ has $n-2$ angular coordinates, which we call $\theta_1, \ldots, \theta_{n-2}$. Since the coordinate k_n takes no part at all in this change of variables, we can write

$$\int_{-\infty}^{\infty} \mathrm{d}k_1 \cdots \mathrm{d}k_{n-1}\mathrm{d}k_n = \int_{-\infty}^{\infty} \mathrm{d}k_n \int_0^{\infty} \mathrm{d}\kappa_{n-1}\kappa_{n-1}{}^{n-2} \int \mathrm{d}\Omega_{n-2}, \tag{9.22}$$

where, as yet, the form of $\int \mathrm{d}\Omega_{n-2}$ is unspecified. The full set of coordinates $(k_n, \kappa_{n-1}, \theta_1, \ldots, \theta_{n-2})$ may be thought of as cylindrical coordinates in n dimensions. We can write any $\mathbf{k}^{(n)}$ in terms of cylindrical coordinates as

$$\mathbf{k}^{(n)} = (\kappa_{n-1}\mathbf{u}^{(n-1)}(\theta_1, \ldots, \theta_{n-2}), k_n), \tag{9.23}$$

where $\mathbf{u}^{(n-1)}(\theta_1, \ldots, \theta_{n-2})$ is an $(n-1)$-dimensional unit vector. Now we change variables to polar coordinates in n dimensions by introducing two new coordinates, κ_n and θ_{n-1}, $0 < \theta_{n-1} < \pi$, by the relations

$$\kappa_{n-1} = \kappa_n \sin\theta_{n-1}, \quad k_n = \kappa_n \cos\theta_{n-1}. \tag{9.24}$$

The first $n-2$ angular coordinates are kept the same. As a result, the Jacobian of the transformation is simple, and

$$\int_{-\infty}^{\infty} \mathrm{d}k_n \int_0^{\infty} \mathrm{d}\kappa_{n-1}\kappa_{n-1}{}^{n-2} \int \mathrm{d}\Omega_{n-2} = \int_0^{\infty} \mathrm{d}\kappa_n \kappa_n{}^{n-1} \int_0^{\pi} \mathrm{d}\theta_{n-1} \sin^{n-2}\theta_{n-1} \int \mathrm{d}\Omega_{n-2}, \tag{9.25}$$

which we may use to construct $\mathrm{d}\Omega_{n-1}$ iteratively. For $n = 1$, we know that

$$\int \mathrm{d}\Omega_1 = \int_0^{2\pi} \mathrm{d}\theta_1. \tag{9.26}$$

The general form that satisfies eqs. (9.25) and (9.26) is given by

$$\int \mathrm{d}\Omega_{n-1} = \int_0^{\pi} \mathrm{d}\theta_{n-1} \sin^{n-2}\theta_{n-1} \times \cdots \times \int_0^{\pi} \mathrm{d}\theta_2 \sin\theta_2 \times \int_0^{2\pi} \mathrm{d}\theta_1. \tag{9.27}$$

This is the integral we need to evaluate.

Special functions

The evaluation of eq. (9.27) is straightforward, and gives us a chance to familiarize ourselves with two special functions which will occur repeatedly in our calculations. The first is the *beta function*, which is defined by the integral representation

$$B(\mu, \nu) = \int_0^1 \mathrm{d}x\, x^{\mu-1}(1-x)^{\nu-1}. \tag{9.28}$$

The properties of the beta function are all determined by those of the Euler *gamma function* introduced in (9.11), because

$$B(\mu, \nu) = \Gamma(\mu)\Gamma(\nu)/\Gamma(\mu+\nu). \tag{9.29}$$

From eq. (9.28), we can derive, by successive changes of variable (exercise 9.6), the alternate integral representations

$$B(\mu, \nu) = \int_0^\infty \mathrm{d}y\, y^{\nu-1}(1+y)^{-\mu-\nu}, \tag{9.30a}$$

$$B(\mu, \nu) = (1-a)^\nu \int_0^1 \mathrm{d}z\, z^{\mu-1}(1-z)^{\nu-1}[1-a(1-z)]^{-\mu-\nu},\ |a|<1, \tag{9.30b}$$

$$B(\mu, \nu) = 2\int_0^{\pi/2} \mathrm{d}\theta \sin^{2\mu-1}\theta \cos^{2\nu-1}\theta. \tag{9.30c}$$

The last form, with $\nu = \frac{1}{2}$, will enable us to evaulate eq. (9.27).

The gamma function, in turn, is defined by the integral representation

$$\Gamma(z) = \int_0^\infty \mathrm{d}\beta\, \beta^{z-1} \exp(-\beta), \tag{9.31}$$

where z is any complex number with $\mathrm{Re}\, z > 0$. Important and easy-to-prove special values of the gamma function are given by

$$\Gamma(n) = (n-1)!, \quad \Gamma(\tfrac{1}{2}) = \pi^{1/2} \tag{9.32a}$$

$$\Gamma(\tfrac{1}{2}n) = (\tfrac{1}{2}n - 1)(\tfrac{1}{2}n - 2)\cdots \tfrac{1}{2}(\pi)^{1/2}, \tag{9.32b}$$

for any positive integer n. Integer and half-integer arguments are related by the *doubling formula*:

$$\Gamma(n)\Gamma(\tfrac{1}{2}) = 2^{n-1}\Gamma(\tfrac{1}{2}n)\Gamma(\tfrac{1}{2}(n+1)). \tag{9.32c}$$

We shall also need a few other general properties of the gamma function, which we now review.

$\Gamma(z)$ is an analytic function of its argument, except for zero and the negative integers. This follows very simply from the definition (9.31), which shows that $\Gamma(z)$ is convergent for $\mathrm{Re}\, z > 0$, and from the easy-to-prove recursion relation,

$$z\Gamma(z) = \Gamma(z+1). \tag{9.33}$$

Equation (9.33) may be used to define $\Gamma(z)$ for any z with a negative real part. Suppose $\mathrm{Re}\, z < 0$, but $1 > \mathrm{Re}\,(z + N_z) > 0$, where N_z is an integer. Clearly, there is a unique N_z for every z. Then we define $\Gamma(z)$ by

$$\Gamma(z) = \Gamma(1 + z + N_z)/[z(z + 1) \cdots (z + N_z)]. \tag{9.34}$$

This is a completely finite definition unless z is a negative integer. When $z \to -N_z$, $\Gamma(z)$ develops a simple pole, whose residue is $(-1)^N(1/N_z!)$.

It will be useful later to expand the gamma function into its finite terms near the negative integers. For this purpose, it is only necessary to know its expansion near unity, since eq. (9.34) determines its behavior near all the other integers from its expansion near unity. To find the expansion near unity, we return to the integral representation (9.31), and define $\epsilon = z - 1$. Then

$$\Gamma(1 + \epsilon) = 1 + \int_0^\infty \mathrm{d}x(\epsilon \ln x + \tfrac{1}{2}\epsilon^2 \ln^2 x + \cdots) \exp(-x). \tag{9.35}$$

The term proportional to ϵ is the negative of a number which is called 'Euler's constant', denoted $\gamma_\mathrm{E} = 0.544\ldots$. The expansion to order ϵ^2 turns out to be

$$\begin{aligned}\Gamma(1 + \epsilon) &= 1 - \epsilon\gamma_\mathrm{E} + \epsilon^2(\tfrac{1}{2}{\gamma_\mathrm{E}}^2 + \tfrac{1}{12}\pi^2) + \cdots \\ &= \exp(-\epsilon\gamma_\mathrm{E} + \tfrac{1}{12}\epsilon^2\pi^2) + \mathrm{O}(\epsilon^3),\end{aligned} \tag{9.36}$$

where the second equality is sometimes a useful approximation. In any case, we find, using (9.30c) and (9.32b,c), that (9.27) is given by

$$\begin{aligned}\Omega_m \equiv \int \mathrm{d}\Omega_m &= \prod_{j=1}^{m} B(\tfrac{1}{2}j, \tfrac{1}{2}) \times 2\pi \\ &= 2\pi^{(m+1)/2}/\Gamma(\tfrac{1}{2}(m + 1)) \\ &= 2^m\pi^{m/2}\Gamma(\tfrac{1}{2}m)/\Gamma(m),\end{aligned} \tag{9.37}$$

or, with the last integral left undone,

$$\int \mathrm{d}\Omega_m = 2\pi^{m/2}\frac{1}{\Gamma(\tfrac{1}{2}m)}\int_0^\pi \mathrm{d}\theta_m \sin^{m-1}\theta_m. \tag{9.38}$$

It is simple (and worthwhile) to check that these formulas give the expected results for $m = 1, 2$.

One-loop diagrams in n dimensions

Now, let us return to the evaluation of $i_{n,s}(M^2)$, eq. (9.20), which, we recall, is the general form of a momentum integral in n dimensions for a one-loop diagram. From eqs. (9.30a) and (9.37), we obtain

$$i_{n,s}(M^2) = \mathrm{i}(-1)^s \pi^{n/2} \frac{\Gamma(s - \frac{1}{2}n)}{\Gamma(s)} (M^2)^{n/2-s}. \tag{9.39}$$

This is the dimensionally continued form of the general one-loop integral. Its n-dependence is smooth except for the factor $\Gamma(s - \frac{1}{2}n)$. The integral $i_{n,s}(M^2)$ is analytic in n for $\mathrm{Re}\, n < 2s$, that is, where the radial integral in eq. (9.20) is convergent. This form provides an analytic continuation to all values of n. Its complete analytic structure is determined by the behavior of the gamma function $\Gamma(s - \frac{1}{2}n)$, which is analytic in the rest of the n-plane, except at integer values of n with $\frac{1}{2}n \geqslant s$, where it has simple poles. Thus, in the n-plane, ultraviolet divergences appear as isolated simple poles, at least for one-loop diagrams. For noninteger n, all dimensionally continued one-loop Green functions are finite!

The first ultraviolet divergence may be isolated from the rest of the formula by means of eq. (9.33):

$$i_{n,s}(M^2) = \frac{\mathrm{i}(-1)^s \pi^{n/2}}{(s - \frac{1}{2}n)} \frac{\Gamma(s - \frac{1}{2}n + 1)}{\Gamma(s)} (M^2)^{n/2-s}, \tag{9.40}$$

or, when n is near the singular dimension $2s$ (see eq.(9.36)),

$$i_{n,s}(M^2) = \frac{\mathrm{i}(-1)^s \pi^s}{(s - \frac{1}{2}n)} \frac{1}{\Gamma(s)} \times \{1 - (s - \tfrac{1}{2}n)[\gamma_{\mathrm{E}} + \ln(\pi M^2)] + \mathrm{O}((s - n/2)^2)\}. \tag{9.41}$$

In this form, we have separated the pole term from the nonsingular part of the integral. The aim of the renormalization program, to be described in the next chapter, is to eliminate these poles from the S-matrix, so that the $n \to 4$ limit of the theory is finite.

The dimensional regularization of an arbitrary one-loop graph $G(p_i \cdot p_j)$ may be summarized as follows.

(i) Express $G(p_i \cdot p_j)$ according to Feynman rules in n dimensions.
(ii) Compute the loop integral in a range of n (for example $1 < n < 4$) in which $G(p_i \cdot p_j)$ is convergent.
(iii) Analytically continue $G(p_i \cdot p_j)$ back to physical n, and isolate the ultraviolet divergence as a pole in n at the physically interesting dimension. (Note the similarity in technique to Wick rotation.)

Dimensionally continued Lagrangians

To systematize these observations, we can build dimensional continuation into the Feynman rules by modifying the Lagrangian.

Consider the Lagrange density for a real scalar field in n dimensions $\mathcal{L} = \frac{1}{2}[(\partial_\mu\phi)^2 - m^2\phi^2] - V(\phi)$, with $V(\phi) = (g/\alpha!)\phi^\alpha$. The units of the field are fixed by the requirement that the action $\int d^4x\, \mathcal{L}$ be dimensionless, in natural units (Section 2.3). From the kinetic term, we see that the field so defined has mass (or inverse length) dimension $[\phi] = \frac{1}{2}(n-2)$. Then the units of the coupling are also dependent on the number of dimensions: $[g] = n - \frac{1}{2}\alpha(n-2)$. For example, $[g] = 0$ for ϕ^4 with $n = 4$ or ϕ^3 with $n = 6$, while $[g] = 1$ for ϕ^3 with $n = 4$. Feynman integrals in n dimensions will thus contain n-dependence through factors of the coupling. It is convenient to make this dependence explicit, by introducing a new (arbitrary) mass scale μ in the definition of g. Suppose we are interested in our ϕ^α theory in n_0 dimensions, with n_0 an integer, 4 or 6, say. To study the dimensional continuation of this theory, we propose a new density,

$$\mathcal{L}(n) = \tfrac{1}{2}[(\partial_\nu\phi)(\partial^\nu\phi) - \mu^2\phi^2] - (g/\alpha!)\mu^{\delta(n,n_0)}\phi^\alpha, \tag{9.42a}$$

where

$$\delta(n, n_0) = (1 - \tfrac{1}{2}\alpha)(n - n_0), \tag{9.42b}$$

and where g now retains the standard dimensions of a ϕ^α coupling in n_0 dimensions.

In eq. (9.42a), the sum over ν formally goes from 1 to n, even though this is a little difficult to picture when n is not an integer. More specifically, however, we simply mean that in Feynman diagrams, the loop momenta have angular integrals of the form (9.37).

Dimensionally continued Lagrangians with spin also involve couplings whose units are dimension-dependent, and they are treated in the same way as for scalar interactions (exercise 9.7). For theories with spin, however, life becomes a bit more complicated when we consider the fields themselves, whose components are dimension-dependent. In principle, therefore, we ought to worry about how to define spinors and vectors in arbitrary dimensions. A more modest aim is to produce a set of Feynman rules in which – as in the scalar case – all loop momenta, in both propagators and vertices, have angular integrals of the form (9.37). This will be achieved if tensor indices are assumed to run from 1 to n, for polarization tensors ($g^{\mu\nu}$, for example), and Dirac matrices γ_μ. The self-consistency of the resulting rules for dimensional continuation is discussed at some length in Chapter 4 of Collins (r1984). Here, we shall simply assume that they are self-consistent.

With this in mind, we write the following identities in n dimensions. For the metric tensor, we have

$$g^{\mu\nu}g_{\mu\nu} = n. \tag{9.43}$$

For Dirac matrices, the identities of eq. (8.22) are replaced by

$$\begin{aligned}\gamma_\mu \not{a} \gamma^\mu &= (2-n)\not{a},\\ \gamma_\mu \not{a}\not{b}\gamma^\mu &= (n-4)\not{a}\not{b} + 4a\cdot b \qquad (9.44)\\ \gamma_\mu \not{a}\not{b}\not{c}\gamma^\mu &= -(n-4)\not{a}\not{b}\not{c} - 2\not{c}\not{b}\not{a}.\end{aligned}$$

The basic anticommutation relations $\{\gamma^\mu, \gamma^\nu\} = 2g^{\mu\nu}$ (eq. (5.43)) are left unchanged, since they define the Dirac matrices in any number of dimensions. On the other hand, true n-dimensional Dirac matrices also obey $\mathrm{tr}(\gamma_\mu\gamma_\nu) = 2^{(n-2)}g_{\mu\nu}$ in n dimensions. The factor $2^{(n-2)}$, however, will not affect the momentum integrals, and can be replaced, in accordance with our comments above, by its value at $n = 4$. So, we may continue to take $\mathrm{tr}(\gamma_\mu\gamma_\nu) = 4g_{\mu\nu}$ as n-independent. As a result, the identities of eqs. (8.14)–(8.21), which depend only on the anticommutation relations and the normalization, are all n-independent in this version of dimensional regularization. Finally, we should note that it is not possible to continue the matrix $\gamma_5 = \mathrm{i}\epsilon_{\mu\nu\lambda\sigma}\gamma^\mu\gamma^\nu\gamma^\lambda\gamma^\sigma$ to noninteger dimensions in a straightforward way, simply because the antisymmetric tensor assumes an integral number of dimensions in its definition.

For our purposes, dimensional regularization will be most useful. We shall briefly describe, however, two other regularization procedures. These procedures also preserve Lorentz invariance, and as such are more sophisticated than the simple cut-off of eq. (9.18).

Propagator subtractions

The convergence of a Feynman integral can be improved by making the replacement

$$(p^2 - m^2 + \mathrm{i}\epsilon)^{-1} \to (p^2 - m^2 + \mathrm{i}\epsilon)^{-1} - (p^2 - M^2 + \mathrm{i}\epsilon)^{-1}, \qquad (9.45)$$

for selected propagators, with M a large mass. For p^2 large compared with M^2, the new propagator behaves as p^{-4} rather than p^{-2}, but for small p^2, it is essentially the same as the standard propagator. Ultraviolet divergences will appear as logarithms or positive powers of M^2. The analytic structure of the new integral will in general be M^2-dependent, but only at momenta comparable to M. A similar procedure can be applied to fermion or vector propagators. A consistent procedure for quantum electrodynamics is based on this method, and is known as *Pauli–Villars regularization* (Pauli & Villars 1949; see also Itzykson & Zuber r1980, Chapter 8).

Functional modifications

Another approach, intermediate between propagator subtractions and dimensional regularization, is *analytic renormalization*, introduced by Speer

(1968, 1969). Here one has the same aim, of improving the high-momentum behavior of propagators, but instead of eq. (9.45), we make the substitution

$$(p^2 - m^2 + \mathrm{i}\epsilon)^{-1} \to (p^2 - m^2 + \mathrm{i}\epsilon)^{-1-\eta}, \tag{9.46}$$

for some $\eta \neq 0$. Ultraviolet divergences will appear as poles in the variable η, in a manner quite reminiscent of dimensional regularizaton.

These regularization procedures are both perfectly appropriate for scalar and Yukawa theories. They are not universally applicable, however, for reasons which will become clear in Chapter 11, when we discuss the renormalization of gauge theories.

9.4 Poles at $n = 4$

As an application, let us apply dimensional continuation to the two one-loop graphs in Figs. 9.1 and 9.3(a), which are divergent in four dimensions. As above, we suppress couplings and other overall factors. Figure 9.1 is given by eq. (9.7), and, from eqs. (9.15) and (9.39),

$$\begin{aligned} I_n(-Q^2, m^2) &= \tfrac{1}{2}(2\pi)^{-n}\int_0^1 \mathrm{d}x\, i_{n,2}(-x(1-x)q^2 + m^2) \\ &= \tfrac{1}{2}\mathrm{i}(4\pi)^{-n/2}\Gamma(2 - \tfrac{1}{2}n)\int_0^1 \mathrm{d}x[-x(1-x)q^2 + m^2]^{n/2-2}. \end{aligned} \tag{9.47}$$

Similarly, Fig. 9.3(a) is given by

$$\begin{aligned} J_n(m^2) &= \tfrac{1}{2}(2\pi)^{-n} i_{n,1}(m^2) \\ &= -\tfrac{1}{2}\mathrm{i}(4\pi)^{-n/2}\Gamma(1 - \tfrac{1}{2}n)(m^2)^{n/2-1}. \end{aligned} \tag{9.48}$$

$i_{n,1}(m^2)$ is quadratically divergent in four dimensions, and becomes convergent only when $n < 2$. When we analytically continue to $n = 4$, however, the quadratic divergence still appears as a simple pole. This is another attractive feature of dimensional regularization.

Summary of one-loop integrals

We may summarize our results so far by the fundamental integral given in eqs. (9.16) and (9.39), which may be rewritten somewhat more generally as

$$\mathrm{i}\int \mathrm{d}^n k_{\mathrm{E}}[(k^2)_{\mathrm{E}} + 2(p\cdot k)_{\mathrm{E}} + M^2]^{-s} = \mathrm{i}\pi^{n/2}\,\frac{\Gamma(s - \frac{1}{2}n)}{\Gamma(s)}\,[-(p^2)_{\mathrm{E}} + M^2]^{n/2-s}, \tag{9.49a}$$

where $\mathrm{d}^n k_{\mathrm{E}} \equiv \prod_{a=1}^4 \mathrm{d}k_a$, and, by eq. (9.14), $(p\cdot k)_{\mathrm{E}} \equiv \sum_{a=1}^4 p_a k_a$. From our discussion of eq. (9.13), we know that $M^2 > p_{\mathrm{E}}{}^2$ for any integral that results from a Feynman diagram. Thus, for our purposes, the right-hand side of eq. (9.49a) is *always* well defined. The analytic continuation back to Minkowski space ($\theta = 0$ in eq. (9.13)) is given by

$$\int \mathrm{d}^n k (k^2 + 2p\cdot k - M^2 + \mathrm{i}\epsilon)^{-s} = (-1)^s \mathrm{i}\pi^{n/2} \frac{\Gamma(s - \frac{1}{2}n)}{\Gamma(s)} (p^2 + M^2 - \mathrm{i}\epsilon)^{n/2-s}. \quad (9.49\mathrm{b})$$

Here, the sign $-\mathrm{i}\epsilon$ on the right-hand side is determined by the $+\mathrm{i}\epsilon$ on the left, whose positive sign is guaranteed in turn by eq. (9.13). Note that, in general, M^2 contains momentum dependence.

Two other forms, which can be derived from eq. (9.49) by differentiation, and which will become useful when we describe theories with spin, are

$$\mathrm{i}\int \mathrm{d}^n k_\mathrm{E}\, k_i[(k^2)_\mathrm{E} + 2(p\cdot k)_\mathrm{E} + M^2]^{-s} = -\mathrm{i}p_i\pi^{n/2} \frac{\Gamma(s - \frac{1}{2}n)}{\Gamma(s)} [-(p^2)_\mathrm{E} + M^2]^{n/2-s}, \quad (9.50\mathrm{a})$$

$$\begin{aligned}\mathrm{i}\int \mathrm{d}^n k_\mathrm{E}\, k_i k_j[(k^2)_\mathrm{E} + 2(p\cdot k)_\mathrm{E} + M^2]^{-s} &= \mathrm{i}\pi^{n/2} \frac{1}{\Gamma(s)} [-(p^2)_\mathrm{E} + M^2]^{n/2-s} \\ &\times \{p_i p_j \Gamma(s - \tfrac{1}{2}n) + \tfrac{1}{2}\delta_{ij}\Gamma(s - \tfrac{1}{2}n - 1)[-(p^2)_\mathrm{E} + M^2]\}. \quad (9.50\mathrm{b})\end{aligned}$$

Returning these formulas to Minkowski space requires that we rotate the fourth component of k^μ in the numerator. Alternately, we may differentiate eq. (9.49b) directly. In either case, we find

$$\int \mathrm{d}^n k\, k_\mu (k^2 + 2p\cdot k - M^2 + \mathrm{i}\epsilon)^{-s} = -\mathrm{i}(-1)^s p_\mu \pi^{n/2} \frac{\Gamma(s - \frac{1}{2}n)}{\Gamma(s)} (p^2 + M^2 - \mathrm{i}\epsilon)^{n/2-s}, \quad (9.51\mathrm{a})$$

$$\begin{aligned}\int \mathrm{d}^n k\, k_\mu k_\nu (k^2 + 2p\cdot k - M^2 + \mathrm{i}\epsilon)^{-s} &= \mathrm{i}(-1)^s \pi^{n/2} \frac{1}{\Gamma(s)} (p^2 + M^2 - \mathrm{i}\epsilon)^{n/2-s} \\ &\times [p_\mu p_\nu \Gamma(s - \tfrac{1}{2}n) - \tfrac{1}{2}g_{\mu\nu}\Gamma(s - \tfrac{1}{2}n - 1)(p^2 + M^2)]. \quad (9.51\mathrm{b})\end{aligned}$$

Notice, in particular, the change from δ_{ij} in (9.50b) to $-g_{\mu\nu}$ in (9.51b).

Analytic continuation in the n-plane

The outline of dimensional continuation for multiloop diagrams is a straightforward extension of the one loop method: (i) express all diagrams according to the Feynman rules in n dimensions; (ii) evaluate the diagrams for n sufficiently small that the integrals converge; (iii) analytically continue back to the physical dimension and isolate poles associated with ultraviolet divergences. A nontrivial part of the extension is to show that the analytic continuation encounters only isolated *poles* as $\mathrm{Re}\, n$ increases. Were the analytic continuation to involve branch cuts in the n plane, it would be path-dependent, and this would complicate attempts to derive a physical interpretation of the regulated theory. For single-loop diagrams, the analytic structure in n is manifest in the gamma functions of eqs. (9.49)–(9.51), so that we know there are only poles.

A general approach, which shows that at any finite number of loops there are only poles in the n plane, has been given by 't Hooft & Veltman (r1973) (see also Speer 1974). We shall not reproduce a full argument here. Instead we shall exhibit the basic technique, which is sometimes useful in practical cases. (For another method, see Itzkson & Zuber r1980, Chapter 8.)

We start with a one-loop example. Consider Fig. 9.3(a), which is divergent for $\mathrm{Re}\, n \geqslant 2$. In Euclidean space, the corresponding integral is

$$j(m^2, n) = \int \mathrm{d}^n k (k^2 + m^2)^{-1}. \tag{9.52}$$

Of course, we have already done this integral in n dimensions. Without actually performing the integral, however, we can rewrite it to make the pole structure explicit. This is done by inserting unity into the integral in the form

$$1 = \frac{1}{n} \sum_{m=1}^{n} \left(\frac{\partial}{\partial k_m} \right) k_m, \tag{9.53}$$

and then integrating by parts in k, assuming that n is small enough that this makes sense. (Note that integrating by parts with respect to each of the components of k assumes that we can exchange orders of integration, which will only be true if the integral is convergent.) This procedure gives

$$j(m^2, n) = \left(\frac{2}{n} \right) [j(m^2, n) - m^2 \int \mathrm{d}^n k (k^2 + m^2)^{-2}], \tag{9.54}$$

or, solving for $j(m^2, n)$,

$$j(m^2, n) = -m^2 \left(\frac{2}{n-2} \right) \int \mathrm{d}^n k (k^2 + m^2)^{-2}. \tag{9.55}$$

In this form, the pole at $n = 2$ in $j(m^2, n)$ is explicit, and is multiplied by an integral that is convergent for any $\mathrm{Re}\, n < 4$. Going through the same process again, we find

$$j(m^2, n) = m^4\left(\frac{8}{(n-2)(n-4)}\right)\int \mathrm{d}^n k (k^2 + m^2)^{-3}, \qquad (9.56)$$

in which the pole at $n = 4$ is also explicit, times an integral which is now convergent all the way up to $\mathrm{Re}\, n = 6$. Continuing this procedure, we can actually rederive eq. (9.48). There is no particular advantage to this approach for the one-loop diagram, of course. But, because the method does not involve actually evaluating any integrals, it can be applied to an arbitrary diagram, with any number of loops.

Multiloop diagrams and ultraviolet power counting

To treat multiloop diagrams systematically, it is necessary to have a criterion that enables us to distinguish convergent from divergent integrals. This is supplied by ultraviolet power counting. An example of the difficulties encountered in multiloop diagrams is illustrated by Fig. 9.4, from ϕ^4 theory.

Two-loop example

Figure 9.4 corresponds to the integral

$$A(p^2, m^2, n) = \frac{1}{6}\int \mathrm{d}^n k_1 \mathrm{d}^n k_2 (k_1{}^2 + m^2)^{-1}(k_2{}^2 + m^2)^{-1} \times [(p + k_1 + k_2)^2 + m^2]^{-1}, \qquad (9.57)$$

where we work in Euclidean space. Clearly, $A(p^2, m^2, n)$ is badly divergent in four dimensions. How small should we take n to make it convergent? To estimate its behavior for arbitrary n, it is useful to discuss what happens when one or more of its loop momenta squared become much larger than m^2 and p^2.

We consider representative regions from the $2n$-dimensional space of k_1 and k_2 in eq. (9.57). In region (i), $|k_1| \to \infty$, while $|k_2|$ is bounded by Lm,

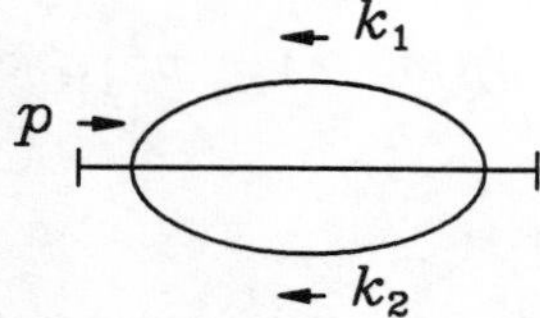

Fig. 9.4 Two-loop diagram.

with L a large but finite number. In region (ii), $|k_2| \to \infty$, while $|k_1|$ stays bounded by Lm. In region (iii), both $|k_1|$ and $|k_2|$ become large, while $|k_1 + k_2|$ stays finite. Finally, in region (iv), $|k_1|$, $|k_2|$ and $|k_1 + k_2|$ become large, while $|k_1|/|k_2|$ remains fixed. In regions (i)–(iii), the loop momentum of one of the three subdiagrams of Fig. 9.4 containing a single loop goes to infinity. In region (iv) the loop momenta diverge together.

We can derive the behavior of the integral in each region without writing down the integrals in detail, by counting powers of the large momenta for each factor in the integrand, and multiplying by the size of the volume element in that region. This is equivalent to computing the dimension of the integral involving the divergent loop momenta. In regions (i), (ii) and (iii) the volume goes as $\mathrm{d}k_i\, k_i{}^{n-1}$, where k_i is the large loop momentum. Also, since each k_i appears in two denominators, the integrand behaves as $k_i{}^{-4}$ for large k_i. Thus, in regions (i)–(iii) the k_i integral behaves as $\int \mathrm{d}k_i\, k_i{}^{n-5}$, which is logarithmically divergent at $n = 4$. In region (iv) we can change variables to $K = k_1 + k_2$. Then the volume element behaves as $\mathrm{d}K\, K^{2n-1}$, and the integrand as $(K^2)^{-3}$. This gives an overall behavior $\int \mathrm{d}K\, K^{2n-7}$, which is convergent only for $\mathrm{Re}\, n < 3$, and is quadratically divergent in four dimensions.

Superficial degree of divergence

As we see from the example, the dimension of a diagram or subdiagram determines the ultraviolet behavior of its integral. The superficial degree of divergence, $\omega(G)$, of any graph G is given by the dimension of its overall momentum integral. For a purely scalar theory, assuming that every line carries at least one loop momentum,

$$\omega(G) = nL_G - 2N_G, \tag{9.58}$$

with L_G the number of loops, and N_G the number of internal lines of the graph. Thus, for Fig. 9.4, the superficial degree of divergence is $2n - 6$, while for each of the subgraphs it is $n - 4$. When the superficial degree of divergence is negative, the integral is convergent in the region where all its loop momenta become large together. When it equals 0, +1, +2, . . . the corresponding graph has corresponding logarithmic, linear, quadratic divergence, and so on.

Weinberg's theorem

Dimensional, or equivalently, ultraviolet power counting is useful because of a theorem due originally to Weinberg (Weinberg 1960, Hahn & Zimmermann 1968 and Collins r1984, Chapter 5), which states (among other things) that if a graph and all its subgraphs have a negative superficial degree of divergence, then the Feynman integral of that graph is convergent. Thus the simple power counting given above is adequate to show that Fig. 9.4 really is convergent for Re $n < 3$.

Analytic continuation of the example

We can now apply the method of the previous section to show how the two-loop example of eq. (9.57) may be analytically continued to Re $n \geqslant 4$, and that its analytic continuation has only isolated poles. Once again, we insert unity into the Feynman integral, this time in the form

$$1 = \frac{1}{2n} \sum_{i=1}^{2} \sum_{m=1}^{n} \left(\frac{\partial}{\partial k_{i,m}} \right) k_{i,m}. \qquad (9.59)$$

Integrating by parts in eq. (9.57) then gives

$$\begin{aligned} A(p^2, m^2, n) = {} & \frac{3}{n} A(p^2, m^2, n) - \frac{1}{6n} \int \mathrm{d}^n k_1 \mathrm{d}^n k_2 \\ & \times \{ m^2 (k_1{}^2 + m^2)^{-1} + m^2 (k_2{}^2 + m^2)^{-1} \\ & + [m^2 + p\cdot(k_1 + k_2)][(p + k_1 + k_2)^2 + m^2]^{-1} \} \\ & \times (k_1{}^2 + m^2)^{-1} (k_2{}^2 + m^2)^{-1} [(p + k_1 + k_2)^2 + m^2]^{-1}, \end{aligned} \qquad (9.60)$$

or

$$A(p^2, m^2, n) = [n/(n-3)] A'(p^2, m^2, n), \qquad (9.61)$$

where the same power counting technique shows that $A'(p^2, m^2, n)$ is convergent for Re $n \leqslant 7/2$. This procedure can clearly be repeated to reach larger n, with the desired property that the only singularities in the analytic continuation are poles (exercise 9.11). Once this technique is suitably generalized, dimensional continuation provides a regularization technique for arbitrary diagrams. Of course, we have not yet solved the problem of how to interpret the regularized theory. In the next chapter, we show how to construct physical theories from dimensionally regularized ones, by the process of renormalization.

In the remainder of this chapter, we briefly leave the subject of regularization, and study convergent integrals, with the aim of deepening our understanding of the physical meaning of quantum corrections.

9.5 Time-ordered perturbation theory

As discussed in Section 2.5, the S-matrix consists of transition amplitudes between unperturbed states. In quantum-mechanical perturbation theory, such an amplitude is built up from (virtual) unperturbed states through which the system passes in making the transition (see any text on quantum mechanics). Clearly, this is also the case for Feynman diagrams, but the identification with unperturbed states is not particularly obvious, since the internal lines are usually off-shell.

In fact, Feynman integrals can be re-expressed in a form that makes the connection with on-shell, unperturbed states explicit. This form is called *time-ordered* perturbation theory, or sometimes 'old-fashioned' perturbation theory. It can be derived from the Feynman integrals by integrating over the energy components of loop momenta.

Consider an arbitrary scalar graph G with N lines, V vertices, L loops and E external momenta $\{p_i\}$. Using the notation of Section 4.1, it may be written in the form (suppressing the coupling)

$$G(p_i \cdot p_j, m^2) = (s_1 s_2)^{-1}(-\mathrm{i})^{1+V} \prod_{i=1}^{N} \int \mathrm{d}k_{i,0} \mathrm{d}^3\mathbf{k}_i \,(2\pi)^{-4} \times \mathrm{i}(k_i^2 - m^2 + \mathrm{i}\epsilon)^{-1}$$

$$\times \prod_{j=1}^{V} (2\pi)^4 \delta^4(\textstyle\sum_{n=1}^{N} \omega_{jn} k_n{}^\mu + \sum_{m=1}^{E} \eta_{jm} p_m{}^\mu). \qquad (9.62)$$

We have multiplied G by an extra $-\mathrm{i}$, so that if it is truncated according to the reduction formula eq. (2.97), it contributes to the T-matrix, eq. (2.94); s_1 and s_2 are symmetry factors (Section 3.4), $\omega(\mathbf{k}) = (\mathbf{k}^2 + m^2)^{1/2}$ and ω_{jn} and η_{jm} are incidence matrices. We now write the vertex energy delta functions in terms of Fourier transforms:

$$G(p_i \cdot p_j, m^2) = (s_1 s_2)^{-1}(-\mathrm{i})^{1+V} \prod_{i=1}^{N} \int_{-\infty}^{\infty} \mathrm{d}k_{i,0} \mathrm{d}^3\mathbf{k}_i \,(2\pi)^{-4}$$

$$\times \mathrm{i}(k_{i,0}{}^2 - \mathbf{k}_i{}^2 - m^2 + \mathrm{i}\epsilon)^{-1}$$

$$\times \prod_{j=1}^{V} \int_{-\infty}^{\infty} \mathrm{d}\tau_j \exp\left[-\mathrm{i}\tau_j \left(\sum_{n=1}^{N} \omega_{jn} k_{n,0} + \sum_{m=1}^{E} \eta_{jm} p_{m,0} \right)\right]$$

$$\times (2\pi)^3 \delta^3(\textstyle\sum_{n=1}^{N} \omega_{jn} \mathbf{k}_n + \sum_{m=1}^{E} \eta_{jm} \mathbf{p}_m). \qquad (9.63)$$

Next, we write the τ_j integrals as a sum over their relative orderings. Each ordering, labelled by s, is a permutation $\{\tau_{j(s)}\}$ of the times $\{\tau_j\}$, in which $k(s) > l(s)$ implies $\tau_{k(s)} > \tau_{l(s)}$. In these terms, we may write

$$\prod_{j=1}^{V} \int_{-\infty}^{\infty} \mathrm{d}\tau_j = \sum_s \left(\prod_{j(s)=1}^{V} \int_{-\infty}^{\tau_{j(s)+1}} \mathrm{d}\tau_{j(s)} \right), \qquad (9.64)$$

where $\tau_{V+1} \equiv \infty$. Each time order corresponds to a time-ordered diagram, in which the vertices are arranged according to their position in time as in Section 3.4. Some of these diagrams, however, may be indistinguishable. If we understand that the sum is only over time orders that give distinguishable diagrams, then we may drop the symmetry factor $s_2{}^{-1}$, which precisely compensates identical time orderings of vertices (see the discussion of eq. (3.95b)). Applied to eq. (9.63), this gives

$$
\begin{aligned}
&G(p_i\cdot p_j, m^2)\\
&\quad = (s_1)^{-1}(-\mathrm{i})^{1+V}\prod_{i=1}^{N}\int \mathrm{d}k_{i,0}\mathrm{d}^3\mathbf{k}_i\,(2\pi)^{-4}\times \mathrm{i}(k_{i,0}{}^2-\mathbf{k}_i{}^2-m^2+\mathrm{i}\epsilon)^{-1}\\
&\qquad \times\sum_s\left[\prod_{j(s)=1}^{V}\int_{-\infty}^{\tau_{j(s)+1}}\mathrm{d}\tau_{j(s)}\right]\exp\left[-\mathrm{i}\tau_{j(s)}\left(\sum_{n=1}^{N}\omega_{j(s)n}k_{n,0}+\sum_{m=1}^{E}\eta_{j(s)m}p_{m,0}\right)\right]\\
&\qquad \times(2\pi)^3\delta^3\left(\textstyle\sum_{n=1}^{N}\omega_{j(s)n}\mathbf{k}_n+\sum_{m=1}^{E}\eta_{j(s)m}\mathbf{p}_m\right).
\end{aligned}
\tag{9.65}
$$

For any time order we can now perform all the energy integrals. Suppose, for instance, that line momentum $k_{i,0}$ flows from vertex a to vertex b. Then the $k_{i,0}$ integral is the transform of the causal propagator form, eq. (2.83),

$$
\begin{aligned}
&\int_{-\infty}^{\infty}\mathrm{d}k_{i,0}\;\mathrm{i}(k_{i,0}-\mathbf{k}_i{}^2-m^2+\mathrm{i}\epsilon)^{-1}\exp[-\mathrm{i}k_{i,0}(\tau_a-\tau_b)]\\
&\qquad=[\pi/\omega(\mathbf{k}_i)]\big(\exp\{-\mathrm{i}[\omega(\mathbf{k}_i)-\mathrm{i}\epsilon](\tau_a-\tau_b)\}\,\theta(\tau_a-\tau_b)\\
&\qquad\qquad+\exp\{\mathrm{i}[-\omega(\mathbf{k}_i)+\mathrm{i}\epsilon](\tau_b-\tau_a)\}\,\theta(\tau_b-\tau_a)\big).
\end{aligned}
\tag{9.66}
$$

Of course, for a given time order in eq. (9.65) only one term of eq. (9.66) contributes for each energy integral. We then have

$$
\begin{aligned}
G(p_i\cdot p_j, m^2) &= (s_1)^{-1}(-\mathrm{i})^{1+V}\prod_{i=1}^{N}\int\frac{\mathrm{d}^3\mathbf{k}_i}{(2\pi)^3 2\omega(\mathbf{k}_i)}\sum_s\left[\prod_{j(s)=1}^{V}\int_{-\infty}^{\tau_{j(s)+1}}\mathrm{d}\tau_{j(s)}\right]\\
&\quad\times\exp\left\{-\mathrm{i}\tau_{j(s)}\left(\sum_{n=1}^{N}\alpha_{j(s)n}[\omega(\mathbf{k}_n)-\mathrm{i}\epsilon]+\sum_{m=1}^{E}\eta_{j(s)m}p_{m,0}\right)\right\}\\
&\quad\times(2\pi)^3\delta^3\left(\textstyle\sum_{n=1}^{N}\omega_{j(s)n}\mathbf{k}_n+\sum_{m=1}^{E}\eta_{j(s)m}\mathbf{p}_m\right),
\end{aligned}
\tag{9.67}
$$

where α_{jn} is defined by

$$
\begin{aligned}
\alpha_{jn} &= +1 \text{ when vertex } j \text{ is at the later end of line } n,\\
&= -1 \text{ when vertex } j \text{ is at the earlier end of line } n,\\
&= 0 \text{ otherwise.}
\end{aligned}
\tag{9.68}
$$

For a fixed time order s we can now perform the $\tau_{j(s)}$ integrals. In a time-ordered diagram, the term *intermediate state* refers to the set of lines that occur between each pair of vertices. For each time ordering of the V vertices there are $V-1$ intermediate states.

It is easiest to see the pattern in the τ integrals by considering an explicit example. Corresponding to the Feynman diagram, Fig. 9.1, there are two time-ordered diagrams, shown in Fig. 9.5(a),(b). Each has two τ integrals and one intermediate state. For Fig. 9.5(a), for instance, the integrals are of the form

$$\int_{-\infty}^{\infty} \mathrm{d}\tau_b \exp\{-\mathrm{i}\tau_b[\omega(\mathbf{p}-\mathbf{k})+\omega(\mathbf{k})-p'_0-2\mathrm{i}\epsilon]\}$$

$$\times \int_{-\infty}^{\tau_b} \mathrm{d}\tau_a \exp\{-\mathrm{i}\tau_a[p_0-\omega(\mathbf{p}-\mathbf{k})-\omega(\mathbf{k})+2\mathrm{i}\epsilon]\}$$

$$= 2\pi\delta(p_0-p'_0)\times \mathrm{i}[p_0-\omega(\mathbf{p}-\mathbf{k})-\omega(\mathbf{k})+\mathrm{i}\epsilon]^{-1}. \quad (9.69)$$

The first integral produces a denominator that is the difference between the external energy and the energy of the intermediate state consisting of the two on-shell lines with spatial momenta $\mathbf{k}$ and $\mathbf{p}-\mathbf{k}$. In the second integral all dependence on the internal momenta cancels, and we find a delta function in the external energies. More complicated examples are given by Fig. 9.6 (exercise 9.12). In general, the ith integral produces a denominator D_i, given by the 'energy deficit' of the ith state,

$$D_i = E_i - \sum_{j=1}^{N} \beta_{ij}\omega_j(\mathbf{k}_j), \quad (9.70)$$

where $\beta_{ij}=+1$ if line j is in state i, and is zero otherwise. The lines are always considered as being on-shell: the energy of a line with spatial momentum $\mathbf{k}_i$ is always $\omega(\mathbf{k}_i)$, flowing from earlier to later times. E_i is the

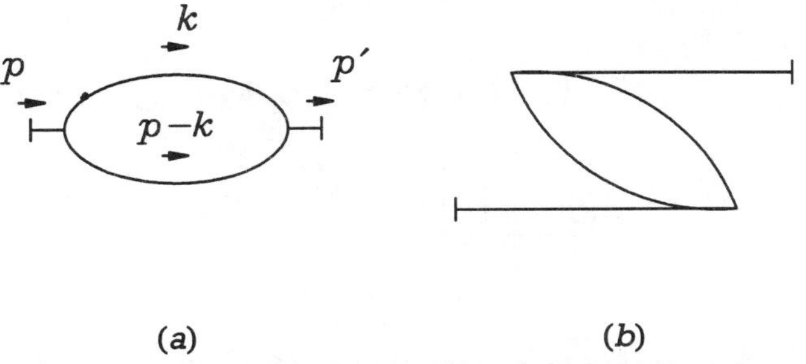

Fig. 9.5 Time-ordered diagrams corresponding to Fig. 9.1.

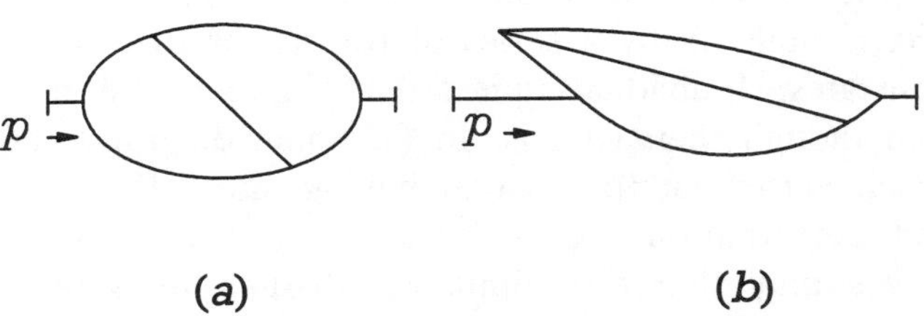

Fig. 9.6 Time-ordered diagrams at two loops.

amount of external energy that has flowed into the graph before vertex i:

$$E_i = \sum_{\substack{\text{vertices } a \\ (\tau_a \leq \tau_i)}} \left(\sum_{m=1}^{E} \eta_{am} p_{m,0} \right). \tag{9.71}$$

In the last integral of eq. (9.67), all dependence on internal momenta cancels, and we find a delta function of the external energies. In the example, we treated ϵ as a finite number, but it is infinitesimal, and we may ignore the difference between, for instance, $2\mathrm{i}\epsilon$ and $\mathrm{i}\epsilon$ for any specific integral. When dealing with multiple integrals, however, it is best to keep the same definition of ϵ throughout the calculation. It is worth checking that the imaginary part always cancels in the last integral.

Finally, we note that for theories with spin, numerator momenta do not affect the reasoning above. The only difference between a time-ordered perturbation theory numerator and the numerator of the corresponding Feynman diagram is that in the former case, energies are evaluated on-shell. This assumes that all denominators are of the Feynman form, as in eq. (9.63). Therefore, for gauge theories, we use the Feynman gauge $\lambda = 1$.

We can summarize these observations in the time-ordered perturbation theory result for the graph $G(p_i \cdot p_j, m^2)$:

$$\begin{aligned} G(p_i \cdot p_j, m^2) = {} & -2\pi\delta(E_V - E_0) \\ & \times \frac{1}{s_1} \prod_{\text{lines } i} \int \frac{\mathrm{d}^3\mathbf{k}_i}{(2\pi)^3 2\omega(\mathbf{k}_i)} \\ & \times \prod_{\text{vertices } j} (2\pi)^3 \delta^3\left(\textstyle\sum_{n=1}^{N} \omega_{jn} \mathbf{k}_n + \sum_{m=1}^{E} \eta_{jm} \mathbf{p}_m\right) \\ & \times \sum_{\text{orders } s} \prod_{\text{states } a} \left[E_a - \sum_{j=1}^{N} \beta_{aj} \omega_j(\mathbf{k}_j) + \mathrm{i}\epsilon \right]^{-1} F(p_m, k_n). \end{aligned} \tag{9.72}$$

$F(p_m, k_n)$ is a numerator factor for the graph in the case of a theory with spin. Within each term in the sum, the product is over the $V - 1$ states specified by the particular time order. In eq. (9.72) we see explicitly that the T-matrix is built up by a series of transitions through virtual unperturbed eigenstates. Individual time-ordered graphs are not Lorentz invariant, even though their sum is. In Feynman diagrams, energy is conserved at each vertex, at the cost of having lines off the mass shell. In time-ordered perturbation theory lines are on-shell, but energy is not conserved. We may therefore think of off-shell lines in the covariant formulation as summarizing the effects of virtual states.

Ultraviolet divergences

Light is shed on the nature of ultraviolet divergences by seeing how they occur in time-ordered perturbation theory. Consider the one-loop diagram Fig. 9.1, which is logarithmically divergent in four dimensions. We can use the rules of time-ordered perturbation theory to perform the energy integral of the loop. The answer is a sum of two terms, corresponding to the two time-ordered graphs of Fig. 9.5. They are contributions, for instance, to the T-matrix for two-particle elastic scattering in ϕ^4 theory. From eq. (9.72), Fig. 9.5(a) can be written as

$$-\tfrac{1}{2}(2\pi)^4\delta^4(p-p')\int\frac{\mathrm{d}^3\mathbf{k}_1}{(2\pi)^3 2\omega(\mathbf{k}_1)}\int\frac{\mathrm{d}^3\mathbf{k}_2}{(2\pi)^3 2\omega(\mathbf{k}_2)}$$
$$\times(2\pi)^3\delta^3(\mathbf{p}-\mathbf{k}_1-\mathbf{k}_2)g^2[p_0-\omega(\mathbf{p}-\mathbf{k})-\omega(\mathbf{k})+\mathrm{i}\epsilon]^{-1}. \quad (9.73)$$

The $\mathbf{k}_1$ and $\mathbf{k}_2$ integrals are sums over states, whose contributions to the diagram are weighted by the resulting energy deficit. That their integral is logarithmically divergent may be verified by counting powers of the spatial loop momenta. After using the delta function, at large $|\mathbf{k}_1|$ the integral behaves as $\int \mathrm{d}k_1/k_1$.

We now see that the ultraviolet divergence comes from having too many high-energy states. Ultraviolet divergences get worse as the dimensions increase, because the number of states increases with the dimension. From this point of view, there is nothing mysterious about ultraviolet divergences in the perturbation series. So many states are coupled by the interaction that the true states of the theory are very far from those of the free theory. The unperturbed states therefore give a very poor approximation to the true states, and a naive attempt to expand about the free states fails.

This would seem to be the end of the perturbative approach, except that it is not the whole story. In the real world, we do not observe truly unperturbed fields. Thus, we have a certain amount of freedom in defining just what we mean by the term 'free'. Also, we note that the very-high-energy states that produce divergences drastically violate energy conservation. By the uncertainty principle, they must be very short-lived. In some sense, then, they may be considered as 'local' perturbations on the fields, which, it is conceivable, could be absorbed into changes in the local interactions of the Lagrange density. Roughly speaking, these two observations are the basis of the renormalization procedure of the next chapter.

9.6 Unitarity

In this section, we use time-ordered perturbation theory to establish a fundamental result, the unitarity of the perturbative S-matrix. This proof can be given in a number of ways, some of them simpler than the one offered below. We shall prove, however, not only unitarity, but a much stronger result. For this purpose, it is useful to deal with the T-matrix, eq. (2.94), for which unitarity (eq. (2.92)) implies

$$\mathrm{i}(T - T^\dagger) = -T^\dagger T. \tag{9.74}$$

We shall consider the perturbative expansion of this expression. Our discussion deals with the 'cut diagrams' of Section 8.1, which automatically relate the S-matrix to cross sections (Veltman 1963, 't Hooft & Veltman 1973).

Cut graphs

By equating the coefficients of powers of the perturbative coupling on each side of eq. (9.74), we see that it is equivalent to an infinite number of nonlinear equations, involving the complete expressions for T at each order of the coupling. In fact, unitarity is built into the perturbative theory at an even more detailed level.

Consider the expression

$$\begin{aligned} F_\gamma(p_1, \ldots, p_m; k_1, \ldots, k_n) = \int \prod_{i=1}^{n} \mathrm{d}^4 k_i \exp(-\mathrm{i}k_i{}^\mu x_{i,\mu}) \prod_{j=1}^{m} \mathrm{d}^4 p_j \exp(\mathrm{i}p_j{}^\nu y_{j,\nu}) \\ \times \langle \gamma \,\mathrm{out}| T[\textstyle\prod_{j=1}^{m} \phi_j(y_j)]|0 \,\mathrm{in}\rangle^* \\ \times \langle \gamma \,\mathrm{out}| T[\textstyle\prod_{i=1}^{n} \phi_i(x_i)]|0 \,\mathrm{in}\rangle. \end{aligned} \tag{9.75}$$

Here $|\gamma\,\mathrm{out}\rangle$ is a momentum eigenstate, consisting of a set of particles with momenta $\{l_k{}^\mu\}$. We may isolate the poles in F_γ according to the reduction formulas of Sections 2.5 and 7.4. We then have a contribution to the right-hand side of the unitarity formula, eq. (9.74), of the form

$$\begin{aligned} &\langle p_1, \ldots, p_m \,\mathrm{in}|\gamma \,\mathrm{out}\rangle\langle \gamma \,\mathrm{out}|k_1, \ldots, k_n \,\mathrm{in}\rangle \\ &\quad = \langle p_1, \ldots, p_m \,\mathrm{in}|T^\dagger|\gamma \,\mathrm{in}\rangle\langle \gamma \,\mathrm{in}|T|k_1, \ldots, k_n \,\mathrm{in}\rangle. \end{aligned} \tag{9.76}$$

We need not restrict ourselves to T-matrix elements, however.

Contributions to eq. (9.75) can be computed using the Feynman rules. A general contribution is shown in Fig. 9.7(*a*), while a specific low-order example is shown in Fig. 9.7(*b*). A cut graph will be denoted as G_γ, where G refers to the overall graph, and γ refers to the particular final state, the out-state in (9.75). L_γ and R_γ are the subgraphs to the left and right of γ. We have retained the convention of drawing graphs with the initial state to the left, the opposite sense from the time-ordering of operators. The particles in the final state are specified by the momenta of the cut lines. The direction of the momentum flow is always from the left-hand part to the right-hand part of the graph, i.e., from T to T^*. A particular graphical

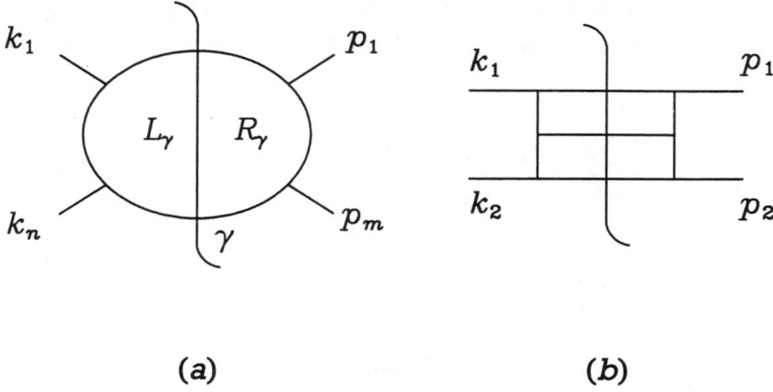

Fig. 9.7 (*a*) General cut diagram; (*b*) example.

contribution to (9.75) can be written as

$$G_\gamma(\{p_j\}, \{k_i\})$$

$$= s_\gamma{}^{-1} \prod_{\text{final-state lines } m} \int \frac{d^3\mathbf{l}_m}{(2\pi)^3 2\omega(\mathbf{l}_m)} (2\pi)^3 \delta^3(\textstyle\sum_i \mathbf{k}_i - \sum_m \mathbf{l}_m)$$
$$\times R_\gamma{}^*(\{p_j\}, \{l_m\})(2\pi)\delta(\textstyle\sum_i k_i{}^0 - \sum_m l_m{}^0) L_\gamma(\{l_m\}, \{k_i\}), \tag{9.77}$$

where $s_\gamma{}^{-1}$ is the symmetry factor associated with identical bosons in the final states. Next we use time-ordered perturbation theory to evaluate L_γ and R_γ. According to eq. (9.72), we have

$$L_\gamma(\{l_m\}, \{k_i\}) = -\sum_{\text{orders } \tau \text{ of } L} \int \prod_{\text{loops } a \text{ of } L} d^3\mathbf{q}_a (2\pi)^{-3} \prod_{\text{lines } b \text{ of } L} (2\omega_b)^{-1}$$
$$\times N_{L_\gamma}(\{k_i\}, \{l_m\}, \{q_a\})$$
$$\times \prod_{\text{states } s \text{ of } \tau} \left(E_s - \sum_{\text{lines } c \text{ of } s} \omega_c + i\epsilon\right)^{-1}, \tag{9.78}$$

$$R_\gamma{}^*(\{p_j\}, \{l_m\}) = -\sum_{\text{orders } \tau \text{ of } R} \int \prod_{\text{loops } a \text{ of } R} d^3\mathbf{q}_a (2\pi)^{-3} \prod_{\text{lines } b \text{ of } R} (2\omega_b)^{-1}$$
$$\times N_{R_\gamma}(\{p_j\}, \{l_m\}, \{q_a\})$$
$$\times \prod_{\text{states } s \text{ of } \tau} \left(E_s - \sum_{\text{lines } c \text{ of } s} \omega_c - i\epsilon\right)^{-1}.$$

As in eq. (9.71), E_s is the total external momentum that has flowed into the graph before state s. N_{L_γ} and N_{R_γ} contain the numerator factors for time-ordered perturbation theory, including possible group and symmetry factors as well as fermionic signs. Combining eqs. (9.77) and (9.78) we derive

$$G_\gamma(\{p_j\},\{k_i\}) = \sum_{\substack{\text{orders } \tau\\ \text{of} G_\gamma}} \int \prod_{\substack{\text{loops } a\\ \text{of } G}} \mathrm{d}^3\mathbf{q}_a(2\pi)^{-3} \prod_{\substack{\text{lines } b\\ \text{of } G}} (2\omega_b)^{-1} N_G(\{p_j\},\{k_i\},\{q_a\})$$
$$\times \prod_{\substack{\text{states } s\\ \text{of } R}} \left(E_s - \sum_{\substack{\text{lines } c\\ \text{of } s}} \omega_c - \mathrm{i}\epsilon\right)^{-1} (2\pi)\delta(E_\gamma - \textstyle\sum_m \omega_m)$$
$$\times \prod_{\substack{\text{states } s\\ \text{of } L}} \left(E_s - \sum_{\substack{\text{lines } c\\ \text{of } s}} \omega_c + \mathrm{i}\epsilon\right)^{-1}. \qquad (9.79)$$

To arrive at this form, we have used the three-dimensional spatial delta function of final-state momentum in eq. (9.77), and have observed that the loops of the full graph G may be identified with the union of the loops of L_γ and R_γ, along with any $n_\gamma - 1$ of the final-state momenta, where n_γ is the number of lines in the final state γ. E_γ is the total energy of the final state. Finally, as the notation suggests, the overall numerator-and-symmetry factor, N_G, is independent of the cut, γ, and is the same as the factor for the uncut graph:

$$N_G = s_\gamma{}^{-1} N_{R_\gamma} N_{L_\gamma}. \qquad (9.80)$$

Proof of equation (9.80)

The proof of eq. (9.80) requires us to treat (i) bosonic symmetry factors, (ii) Dirac algebra and (iii) fermionic signs. It is rather technical, but we describe it for completeness.

(i) Suppose we have a cut graph of the type shown in Fig. 9.8, with a final state of n identical bosons, of which $\{m_{ij}\}$ are attached to the same vertex on either side of the cut γ. The symmetry factor for the cross section is $S_\gamma = n!$, while the symmetry factor for the uncut graph is $S_G = \prod_{i,j} m_{ij}!$.

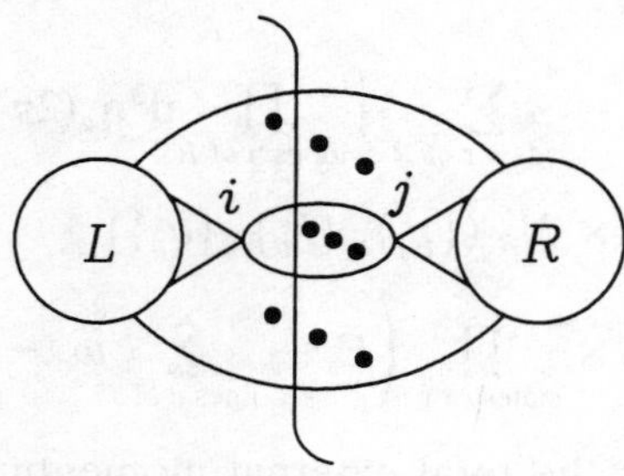

Fig. 9.8 Cut diagram involving symmetry factors.

On the other hand, all cut graphs in which two of the bosons are exchanged at fixed momenta are distinguishable, *unless* the exchanged bosons attach to the same vertices i and j on either side of the cut. Since every boson can take on any momentum, all of these distinguishable graphs contribute the same amount to the cross section at any point in phase space. Exchanging the bosons in all distinguishable ways, we thus get precisely $S_\gamma/S_G = n!/\prod_{i,j} m_{ij}!$ identical contributions, which change the symmetry factor of the cross section into that of the cut graph.

(ii) We have already shown in Chapter 8 that the order of both Dirac and matrix factors and color matrix factors is effectively reversed by taking the complex conjugate. This changes the order appropriate to graph R_γ into the order appropriate to the uncut graph G.

(iii) The last point, about fermionic signs, is the most subtle, particularly in relation to fermion loops. It is suggested as exercise 9.14 to show that the sign of a cut diagram is the same as for an uncut diagram in the absence of loops. It is important, in making these arguments, to define the Green functions on the left and right of the cut in terms of the same standard orderings of fermionic fields inside the time orders (Section 7.2).

Each *uncut*, closed fermion loop produces a minus sign in a cut diagram. When the loop is cut, this sign is absent, but we must then take into account the signs from both the pairing of Dirac fields in the final state (see eq. (7.43)), and from the polarization sum for final-state antifermions. The latter is $\sum_\lambda \overline{v}(k,\lambda)v(k,\lambda) = -(-\not{k} + m)$, the negative of the numerator in the corresponding uncut line. For cut fermion loops, the effect of these factors of (-1) is to reproduce the single minus sign of the uncut loops. This is illustrated by the example of Fig. 9.9, in which, for simplicity, the only external fermions are in a single cut loop, which crosses the final state $2n$ times ($n = 2$ in the figure). The fermionic sign of the uncut diagram is just (-1), for the single loop. The fermionic sign of the cut diagram,

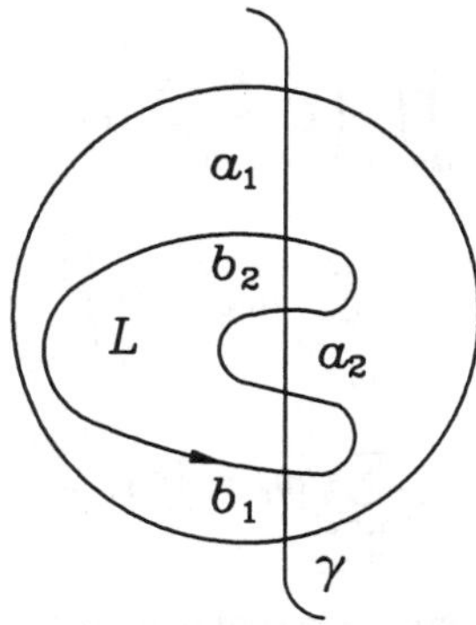

Fig. 9.9 Cut diagram with a cut fermion loop.

however, is $\delta_{p_{(L)}}\delta_{p_{(R)}}(-1)^n$, where $\delta_{p(L)}$ is the pairing sign of L_γ, $\delta_{p_{(R)}}$ the pairing sign of R_γ and $(-1)^n$ is from the n cut antifermion lines. Now supppose that in L_γ the pairing of fields is

$$\prod_{i=1}^{n} \psi_{a_i} \overline{\psi}_{b_i}, \tag{9.81}$$

with $\{\psi_{a_i}\}$ and $\{\overline{\psi}_{b_i}\}$ some permutation of fermion fields in the final state. Then in R_γ the pairing must be

$$\prod_{i=1}^{n} \psi_{b_i} \overline{\psi}_{a_{i-1}}, \tag{9.82}$$

where we define $\psi_{a_0} \equiv \psi_{a_n}$. The relative pairing sign of these two assignments is $\delta_{p_{(L)}}\delta_{p_{(R)}} = (-1)^{n+1}$, so that the overall sign associated with the loop is $(-1)^{n+1}(-1)^n = -1$, as expected from the uncut diagram. This argument may easily be generalized to more than one loop.

Sum over final states

Given eq. (9.80), we can sum eq. (9.79) over final states γ. To do so, we use the distribution identity

$$(x + \mathrm{i}\epsilon)^{-1} - (x - \mathrm{i}\epsilon)^{-1} = -2\pi\mathrm{i}\delta(x). \tag{9.83}$$

Substituting eq. (9.83) into eq. (9.79), interchanging the sum over final states and time orders, and noting that L and R are determined by γ for a fixed time order, we derive

$$\begin{aligned}
&\sum_\gamma G_\gamma(\{p_j\},\{k_i\}) \\
&\quad = \sum_{\substack{\text{orders } \tau \\ \text{of } G_\gamma}} \int \prod_{\substack{\text{loops } a \\ \text{of } G}} \mathrm{d}^3\mathbf{q}_a (2\pi)^{-3} \prod_{\substack{\text{lines } b \\ \text{of } G}} (2\omega_b)^{-1} N_G(\{p_j\},\{k_i\},\{q_a\}) \\
&\quad \times \sum_{\substack{\text{final states } \gamma \\ \text{of } \tau}} \prod_{\substack{\text{states } s \\ \text{of } R_\gamma}} \Bigg(E_s - \sum_{\substack{\text{lines } c \\ \text{of } s}} \omega_c - \mathrm{i}\epsilon \Bigg)^{-1} \\
&\quad \times (-\mathrm{i}) \Bigg[\Bigg(E_\gamma - \sum_{\substack{\text{lines } m \\ \text{of } \gamma}} \omega_m - \mathrm{i}\epsilon \Bigg)^{-1} - \Bigg(E_\gamma - \sum_{\substack{\text{lines } m \\ \text{of } \gamma}} \omega_m + \mathrm{i}\epsilon \Bigg)^{-1} \Bigg] \\
&\quad \times \prod_{\substack{\text{states } s \\ \text{of } L_\gamma}} \Bigg(E_s - \sum_{\substack{\text{lines } c \\ \text{of } s}} \omega_c + \mathrm{i}\epsilon \Bigg)^{-1} .
\end{aligned} \tag{9.84}$$

All terms in the expression cancel pairwise, in the sum over γ for fixed time order τ, except when γ is chosen to be at the extreme right or extreme left, in which the entire graph is to the left or right of the cut, as in Fig. 9.10. The result is then

$$\sum_{\gamma} G_\gamma(\{p_j\}, \{k_i\})$$

$$= \mathrm{i} \sum_{\substack{\text{orders } \tau \\ \text{of } G_\gamma}} \int \prod_{\substack{\text{loops } a \\ \text{of } G}} \mathrm{d}^3\mathbf{q}_a (2\pi)^{-3} \prod_{\substack{\text{lines } b \\ \text{of } G}} (2\omega_b)^{-1} N_G(\{p_j\}, \{k_i\}, \{q_a\})$$

$$\times \left[\prod_{\substack{\text{states } s \\ \text{of } G}} \left(E_s - \sum_{\substack{\text{lines } c \\ \text{of } s}} \omega_c + \mathrm{i}\epsilon \right)^{-1} - \prod_{\substack{\text{states } s \\ \text{of } G}} \left(E_s - \sum_{\substack{\text{lines } c \\ \text{of } s}} \omega_c - \mathrm{i}\epsilon \right)^{-1} \right], \quad (9.85)$$

or

$$\sum_{\gamma} G_\gamma(\{p_j\}, \{k_i\}) = 2 \operatorname{Im} G(\{p_j\}, \{k_i\}). \quad (9.86)$$

Recall that, on-shell, G is a contribution to the T-matrix. Thus, eq. (9.86) is the graphical analogue of the unitarity equation (9.74). As promised, however, it is much more general, since it does not require that the external lines of the cut graph be on-shell. Only cut lines need be on-shell.

Even more interestingly, the complete derivation of eq. (9.85) was carried out for *fixed* values of internal spatial loop momenta. In Chapter 13 we shall see how this stronger result may be used to prove theorems on the infrared behavior of cross sections.

It is worth noting that our proof of unitarity, although general, is not all inclusive. It assumes that all propagator poles correspond to physical states. Otherwise, eq. (9.86) is not, as it stands, an equation for the *physical*

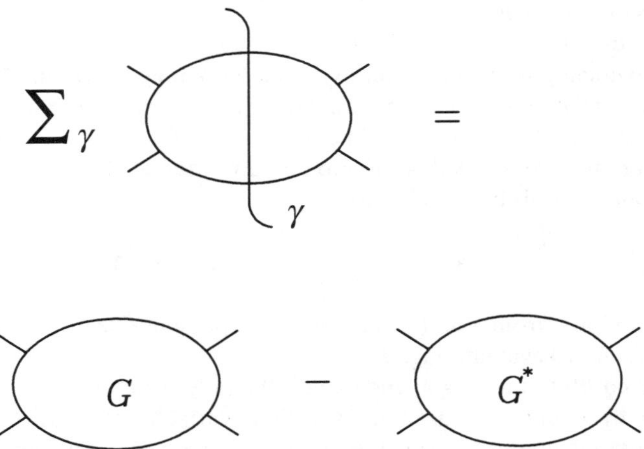

Fig. 9.10 Generalized unitarity.

S-matrix, eq. (6.84). This presents a problem for gauge theories, in which unphysical degrees of freedom propagate, and can appear in final states (Section 8.5). We shall return to this question in Chapter 11. Finally, we should also note that although (9.86) ensures unitarity on an order-by-order basis, the S-matrix calculated to any finite order in perturbation theory is never exactly unitary, simply because (9.74) is a nonlinear equation. Thus, there is nothing to prevent negative cross sections or violations of various 'unitarity bounds' on partial wave amplitudes at any finite order (see, for example Taylor r1976, Chapter 2). These results should be interpreted as requiring the inclusion of higher-order (possibly infinite-order) effects, rather than as true failures of unitarity.

Exercises

9.1 Verify that eq. (9.8) has a single branch cut at $q^2 = 4m^2$. What happens for $0 < q^2 < 4m^2$, where the arguments of the square roots are negative?

9.2 Verify eq. (9.11). Hint: use the representation (9.31) of the gamma function.

9.3 Use Wick rotation to show that an arbitrary scalar two-point Green function $\pi(q^2)$ is an analytic function in the entire q^2-plane, except for branch cuts on the real axis.

9.4 Show that any one-loop diagram can be put in the form defined by eqs. (9.15) and (9.16).

9.5 Show that M^2 in eq. (9.16) is positive definite for Euclidean external momenta and nonzero masses.

9.6 Using the definition (9.31) of the gamma function, verify eqs. (9.29), (9.30) and (9.32).

9.7 Show that the dimension of a spinor field in four dimensions is $\frac{3}{2}$. Determine the dimension of gauge couplings in n dimensions, verifying that they are dimensionless for $n = 4$.

9.8 Verify eqs. (9.44).

9.9 Use Feynman parameterization to do the loop integrals of Fig. 9.3(b), assuming Euclidean external momenta and finite masses m for internal lines. Isolate the residue of the ultraviolet pole at $n = 6$.

9.10 An alternate way of evaluating momentum integrals is by turning them into Gaussian integrals by the identity

$$A^{-\eta} = \frac{1}{\Gamma(\eta)} \int_0^\infty \mathrm{d}x\, x^{\eta-1} \exp(-xA),$$

which follows from eq. (9.31). Use this parameterization to rederive eq. (9.49a), and to evaluate Fig. 9.1.

9.11 Extend eq. (9.61) to $n = 4$, and exhibit the pole structure there.

9.12 Derive by explicit integration over time variables the expressions for the time-ordered graphs of Fig. 9.6. Draw all the remaining time-ordered diagrams corresponding to this Feynman diagram.

9.13 Verify the generalized unitarity equation (9.86) for Fig.9.1 by explicit calculation in arbitrary dimensions.

9.14 Use eq. (7.43) to show that the fermionic sign of a cut diagram is the same as for an uncut diagram in the absence of cut or uncut fermion loops.

10
Introduction to renormalization

Chapter 9 has left us in an awkward position with regard to quantum corrections. Ultraviolet divergent diagrams can be computed only by continuing to unphysical numbers of dimensions (or otherwise regularizing). In this chapter, we develop the process of renormalization, which will allow us to return many theories to physical dimensions.

It is possible to make the perturbation series of many theories finite, at the price of making certain parameters in their Lagrangians infinite. From one point of view, renormalization provides a positivistic morality play, in which, by renouncing our ability to calculate a few unobservable quantities ('bare' quantities below), we gain many predictions relating observable quantities. It may be unsettling, however, to deal with a Lagrange density that becomes infinite when the regularization is removed, even if it does give a finite perturbation series. Thus the perennial conjecture arises, that the quantum fields in nature are low-energy manifestations of an underlying finite theory. The most recent, and most promising, candidate is string theory (Green, Schwarz & Witten r1987). Here we shall simply examine renormalization as a self-consistent procedure, without asking for the ultimate origin of renormalizable theories.

We begin our discussion with the simplest case in which renormalization is necessary, ϕ^3 theory in four dimensions (${\phi^3}_4$). We then go on to use power counting to classify theories as renormalizable or nonrenormalizable. We discuss ${\phi^3}_6$ as a typical renormalizable theory, giving many details at one loop, fewer at two loops, and finally stating the basic all-order theorems. The chapter concludes with a brief introduction to the renormalization group.

10.1 ${\phi^3}_4$ and mass renormalization

From eq. (9.42a), a convenient form for the classical Lagrange density of ϕ^3 theory near $n = 4$ dimensions is given by

$$\mathcal{L}_{\text{cl}}(m, g\mu^\epsilon) = \tfrac{1}{2}[(\partial_\mu\phi)^2 - m^2\phi^2] - \tfrac{1}{6}g\mu^\epsilon\phi^3, \qquad (10.1)$$

where g is a coupling with dimensions of mass, where

$$\epsilon = 2 - \tfrac{1}{2}n, \tag{10.2}$$

and where μ is an arbitrary mass. Compared with the Feynman rules of Section 3.4, the only effect is to replace the coupling g by the combination $g\mu^\epsilon$. (We may note that, by itself, ϕ^3 is not a sensible theory either classically or quantum mechanically, since its potential is unbounded from below. The standard model, however, does include ϕ^3 interactions, and we can restrict ourselves to its ϕ^3 graphs for purposes of argument.)

The mass counterterm

In $\phi^3{}_4$ theory there are only two one-loop graphs with ultraviolet divergence, the self-energy graph, Fig. 10.1(*a*), and the 'tadpole' graph Fig. 10.1(*b*). (In fact, as we shall see, these are the only sources of ultraviolet divergence in the whole four-dimensional theory.) Let us concentrate for the moment on the self-energy graph, and come back to the tadpole diagram later. It is customary to discuss its contribution to the two-point 1PI function Γ_2. We recall (eq. (3.98)) that Γ_n is defined as $-\mathrm{i}$ times all one-particle n-point irreducible diagrams, with an extra term $p^2 - m^2$ for Γ_2. The one-loop correction to Γ_2 is then given by $-\mathrm{i}g^2\mu^{2\epsilon}$ times the one-loop integral given in eq. (9.3), evaluated via the general formula, eq. (9.49b):

$$\Gamma_2{}^{(\mathrm{U})}(p^2, m^2, \epsilon)$$
$$= p^2 - m^2 + \tfrac{1}{2}g^2\mu^{2\epsilon}(4\pi)^{-n/2}\Gamma(\epsilon)\int_0^1 \mathrm{d}x[m^2 - x(1-x)p^2 - \mathrm{i}\epsilon]^{-\epsilon}. \tag{10.3}$$

The superscript (U) denotes that Γ_2 is, as yet, an *unrenormalized* quantity, i.e., it was computed from the unmodified classical Lagrangian eq. (10.1). Near four dimensions ($\epsilon = 0$), we can expand the right-hand side of this

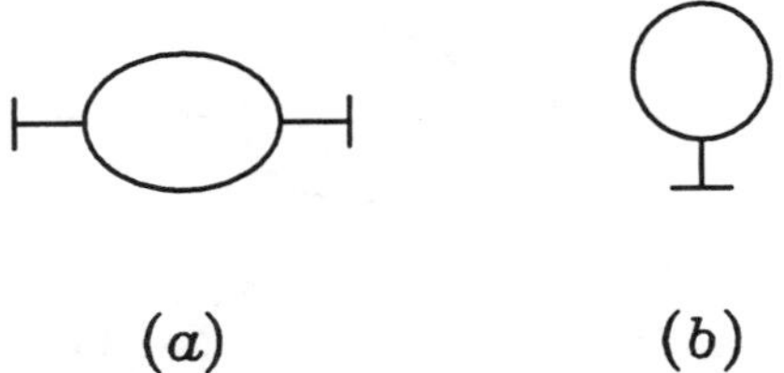

Fig. 10.1 Divergent one-loop diagrams for ϕ^3 in four dimensions.

result using eqs. (9.33) and (9.36) for $\Gamma(\epsilon)$, as well as $a^\epsilon = 1 + \epsilon \ln a + O(\epsilon^2)$, to give

$$\Gamma_2^{(U)}(p^2, m^2, \epsilon) = p^2 - m^2 + \tfrac{1}{2}g^2(4\pi)^{-2}\Bigg\{\epsilon^{-1} - \gamma_E + 2 + \ln(4\pi\mu^2/m^2) - (1 - 4m^2/p^2)^{1/2}\ln\left[\frac{(1 - 4m^2/p^2)^{1/2} + 1}{(1 - 4m^2/p^2)^{1/2} - 1}\right]\Bigg\} + O(\epsilon). \tag{10.4}$$

The ϵ^{-1} term is the ultraviolet divergence in four dimensions, and it is this divergence with which we must come to terms. To this end, we consider a *new* density for ϕ^3 theory, called the *renormalized Lagrange density*,

$$\mathcal{L}_R(m_R, g\mu^\epsilon, c_m) = \tfrac{1}{2}[(\partial_\mu\phi)^2 - m_R{}^2\phi^2] - \tfrac{1}{6}g\mu^\epsilon\phi^3 - \tfrac{1}{2}\delta m^2\,\phi^2. \tag{10.5}$$

Here m_R is a mass, to be determined later, and

$$\delta m^2 = -\tfrac{1}{2}g^2(4\pi)^{-2}(-\epsilon^{-1} + c_m), \tag{10.6}$$

where c_m is finite for $\epsilon \to 0$, but is otherwise completely arbitrary. The standard terminology for these new parameters is as follows: (i) the new term in the Lagrangian, $-\tfrac{1}{2}\delta m^2\,\phi^2$, is called a *mass counterterm*; (ii) δm^2 is called the *mass shift*; (iii) m_R is the *renormalized mass*. To make sense of the perturbation expansion generated from $\mathcal{L}_R$, we demand that m_R be a finite quantity, even at $\epsilon = 0$. This will have important consequences for the interpretation of this Lagrange density.

The Feynman rules for the new Lagrange density (10.5) are the same as for the original one, (10.1), except that the mass m is replaced by m_R, and the counterterm produces a new vertex with only two external lines, represented in Fig. 10.2(a), whose value is

$$-i(2\pi)^4\delta^4(p - p')\delta m^2 = i(2\pi)^4\delta^4(p - p')\tfrac{1}{2}g^2(4\pi)^{-2}(-\epsilon^{-1} + c_m). \tag{10.7}$$

Consider the effect of these changes in our example. Figure 10.2(b) shows the full *renormalized* 1PI function $\Gamma_2^{(R)}$ at order g^2, the two-point function

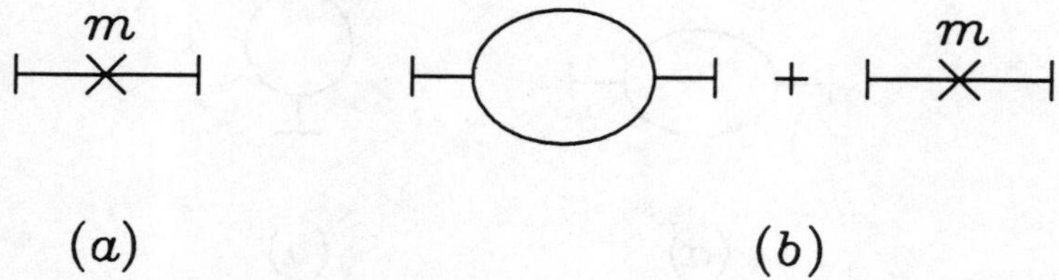

Fig. 10.2 (a) Mass counterterm; (b) renormalized 1PI two-point function.

computed with the renormalized Lagrange density $\mathscr{L}_{\mathrm{R}}$. It is given by

$$
\begin{aligned}
&\Gamma_2{}^{(\mathrm{R})}(p^2, m_{\mathrm{R}}{}^2, g, \mu, c_m, \epsilon) \\
&\quad = \Gamma_2{}^{(\mathrm{U})}(p^2, m_{\mathrm{R}}{}^2, g, \mu, \epsilon) + \Gamma_{2,\mathrm{ctr}}(g, \mu, c_m, \epsilon) \\
&\quad = p^2 - m_{\mathrm{R}}{}^2 + \tfrac{1}{2}g^2(4\pi)^{-2}\Big\{c_m - \gamma_{\mathrm{E}} + 2 + \ln(4\pi\mu^2/m_R{}^2) \\
&\qquad - (1 - 4m_{\mathrm{R}}{}^2/p^2)^{1/2}\ln\left[\frac{(1 - 4m_{\mathrm{R}}{}^2/p^2)^{1/2} + 1}{(1 - 4m_{\mathrm{R}}{}^2/p^2)^{1/2} - 1}\right]\Big\} + \mathrm{O}(\epsilon, g^4), \qquad (10.8)
\end{aligned}
$$

which is finite as $n \to 4$. Adding the counterterm has cancelled the ultraviolet divergence in the two-point Green function at order g^2.

The bare Lagrangian

To interpret the significance of this rather trivial modification of the Lagrange density, we rewrite $\mathscr{L}_{\mathrm{R}}$ of eq. (10.5) in the form

$$\mathscr{L}_{\mathrm{R}} = \tfrac{1}{2}[(\partial_\mu\phi)^2 - m_0{}^2\phi^2] - \tfrac{1}{6}g_0\phi^3, \qquad (10.9)$$

where

$$m_0{}^2 = m_{\mathrm{R}}{}^2 + \delta m^2 = m_{\mathrm{R}}{}^2 - \tfrac{1}{2}g^2(4\pi)^{-2}(-\epsilon^{-1} + c_m), \qquad (10.10a)$$

$$g_0 = g\mu^\epsilon. \qquad (10.10b)$$

Comparing eqs. (10.9) and (10.1), we see that the renormalized Lagrange density eq. (10.5) has the same form as the unrenormalized classical Lagrange density, when the latter has 'bare' mass m_0, and 'bare' coupling g_0:

$$\mathscr{L}_{\mathrm{R}}(m_{\mathrm{R}}, g\mu^\epsilon, c_m) = \mathscr{L}_{\mathrm{cl}}(m_0, g_0). \qquad (10.11)$$

In words, 'the renormalized Lagrangian equals the bare Lagrangian'. Since eq. (10.10b) for the coupling is a finite relation as $\epsilon \to 0$, we shall assume for simplicity that g_0 is a known, ϵ-independent quantity. This assumption is convenient for the following introductory discussion.

Physical parameters

By eq. (10.11), the generator of Green functions, eq. (3.44), in $\phi^3{}_4$ is the same whether the action is expressed in terms of renormalized quantities, via $\mathscr{L}_{\mathrm{R}}(m_{\mathrm{R}}, g\mu^\epsilon, c_m)$, or in terms of bare quantities, via $\mathscr{L}_{\mathrm{cl}}(m_0, g_0)$,

$$\begin{aligned} W_{\mathrm{R}}[J] &= \int [\mathscr{D}\phi] \exp\{\mathrm{i}S(m_{\mathrm{R}}, g, \mu, c_m, \epsilon) - \mathrm{i}(J, \phi)\} \\ &= \int [\mathscr{D}\phi] \exp\{\mathrm{i}S(g_0, m_0, \epsilon) - \mathrm{i}(J, \phi)\} \\ &= \mathrm{W}_0[J] \quad ({\phi^3}_4 \text{ theory}). \end{aligned} \tag{10.12}$$

Suppose we use Feynman rules to calculate a Green function in terms of the bare coupling and mass, and then re-express the bare parameters m_0, g_0 in terms of renormalized parameters m_{R}, g_{R}, via eqs. (10.10). According to eq. (10.12) we would find the same result by calculating the Green function using the renormalized Lagrangian, at least if we sum to all orders in perturbation theory.

The two forms of the generator in eq. (10.12) offer complementary information. We know that, in their renormalized forms, the Green functions are ultraviolet finite as $\epsilon \to 0$ when expressed in terms of m_{R}, g_{R}, μ and c_m (so far, to one loop). From their bare forms, however, we learn that the Green functions depend on only two independent parameters, a fact that is certainly not obvious from the renormalized Lagrangian, eq. (10.5). Evidently, we may consider μ and c_m as *arbitrary* parameters, whose values we may specify at will. Once μ and c_m are specified, the two *physical* parameters g and m_{R} distinguish physically different Lagrangians, that is, Lagrangians with different values of m_0 and g_0. Since we assume (in this theory) that we know g_0, g is already determined, by eq. (10.10b). How then can we determine m_{R}?

Suppose we know the physical mass of the ϕ-particle, that is, the position of the single-particle pole in the two-point function (see eq. (4.15)). The perturbation series for that Green function gives an equation for the physical mass m_{P} of the form $m_{\mathrm{P}} = F(m_{\mathrm{R}}, \mu, g, c_m)$ where F is some function. (In practice, of course, we shall only know the first few terms in the series for F.) We may solve this equation to find m_{R} in terms of m_{P} and the other parameters,

$$m_{\mathrm{R}} = m_{\mathrm{R}}(g, m_{\mathrm{P}}, \mu, c_m). \tag{10.13}$$

With this value in hand, all the parameters of the renormalized Lagrangian are known, and all the Green functions in the theory are uniquely determined, and may be evaluated in four dimensions ($\epsilon = 0$).

This approach generalizes to other theories. In each case, the renormalized Lagrangian includes both arbitrary and physical parameters. The arbitrary parameters are defined for convenience. Once they are defined, the physical parameters are found by determining the values of one observable quantity for each physical parameter.

The most surprising thing about this procedure is that it may be carried out by modifying the local Lagrange density. As we suggested in Section 9.5, this result finds an heuristic justification in the uncertainty principle.

The infinite-energy virtual states that produce ultraviolet divergences live only an infinitesimal time.

Renormalization schemes

An explicit procedure for defining c_m is known as a 'renormalization scheme'. It is customary to leave μ as a free parameter. Then there are two generic approaches. Either m_R is defined explicitly and c_m implicitly, or vice versa. The first approach defines 'momentum subtraction' schemes and the second 'generalized minimal subtraction' schemes.

Momentum subtraction

Momentum subtraction schemes are conventionally defined by the relation

$$\Gamma_2{}^{(R)}(-M^2, m_R{}^2, \mu^2, c_m) = -M^2 - m_R{}^2, \tag{10.14}$$

which says that when $p^2 = -M^2$, with M an arbitrary mass, Γ_2 is normalized to the value it would take for a free particle of mass m_R. Equivalently, eq. (10.14) *defines* all higher-order corrections in Γ_2 to vanish at $p^2 = -M^2$. Once we solve eq. (10.14) for c_m, the perturbation series for any Green function is uniquely determined in terms of the finite parameters g, μ and m_R. In our case, using the explicit form eq. (10.8), this gives

$$\begin{aligned} c_m = \gamma_E - 2 - \ln(4\pi\mu^2/m_R{}^2) \\ + (1 + 4m_R{}^2/M^2)^{1/2} \ln\left[\frac{(1 + 4m_R{}^2/M^2)^{1/2} + 1}{(1 + 4m_R{}^2/M^2)^{1/2} - 1}\right]. \end{aligned} \tag{10.15}$$

The corresponding form for $\Gamma_2{}^{(R)}(p^2)$ is found by substituting eq. (10.15) into eq. (10.8):

$$\begin{aligned} \Gamma_2{}^{(R)}(p^2, m_R{}^2, g, \mu, c_m, \epsilon = 0) \\ = p^2 - m_R{}^2 + \tfrac{1}{2}g^2(4\pi)^{-2} \\ \times \left\{ -(1 - 4m_R{}^2/p^2)^{1/2} \ln\left[\frac{(1 - 4m_R{}^2/p)^{1/2} + 1}{(1 - 4m_R{}^2/p^2)^{1/2} - 1}\right] \right. \\ \left. + (1 + 4m_R{}^2/M^2)^{1/2} \ln\left[\frac{(1 + 4m_R{}^2/M^2)^{1/2} + 1}{(1 + 4m_R{}^2/M^2)^{1/2} - 1}\right]\right\}. \end{aligned} \tag{10.16}$$

Generalized minimal subtraction

In generalized minimal subtraction we simply assign a convenient value to c_m, typically by the parameterization

$$c_m = \ln \kappa, \tag{10.17}$$

with κ some number. The two most common choices for κ are $\kappa = 1$ (*minimal subtraction*, conventionally denoted MS) and $\kappa = (4\pi)^{-1} \exp \gamma_E$ (*modified minimal subtraction*, denoted $\overline{\text{MS}}$). The motivation for the latter choice is that it cancels the terms $-\gamma_E + \ln 4\pi$ that occur in Γ_2, eq. (10.8) as a result of the expansion of $\Gamma(\epsilon)(4\pi)^{-n/2}$ about $\epsilon = 0$. Since similar factors are present in all the one-loop integrals, eqs. (9.49)–(9.51), $\overline{\text{MS}}$ is a popular scheme when dimensional regularization is used. To be explicit, $\Gamma_2{}^{(R)}$ is given in the $\overline{\text{MS}}$ scheme by

$$\begin{aligned} \Gamma_2{}^{(R)}(p^2, m_R{}^2, \mu, c_m, \epsilon) = p^2 - m_R{}^2 + \tfrac{1}{2} g^2 (4\pi)^{-2} \\ \times \Bigg\{ 2 + \ln(\mu^2/m_R{}^2) - (1 - 4m_R{}^2/p^2)^{1/2} \\ \times \ln\left[\frac{(1 - 4m_R{}^2/p^2)^{1/2} + 1}{(1 - 4m_R{}^2/p^2)^{1/2} - 1}\right]\Bigg\} + (\epsilon, g^4). \end{aligned} \tag{10.18}$$

The physical mass and the renormalized mass

The renormalization procedure so far has supplied us with a finite perturbation series for the Green function of $\phi^3{}_4$ in terms of g, μ and m_R. We now complete the procedure by relating m_R to an observable quantity, the physical mass, m_P. To see the relation of m_R to m_P we shall have to look at the S-matrix.

Recall (Section 2.5) that the S-matrix is derived from the residues of the single-particle poles of Green functions,

$$S(p_i) = \prod_i [R(2\pi)^3]^{-1/2} \lim_{p_i^2 \to m_P{}^2} [-i(p_i{}^2 - m_P{}^2) G(p_i)]. \tag{10.19}$$

The positions of poles in external momenta are the same, whatever the renormalization scheme and value of μ, for given bare quantities, since, by eq. (10.11), the bare quantities determine the Lagrange density. Let us look for the single-particle poles in Green functions. For the S-matrix to be well defined in eq. (10.19), these must be simple poles. The physical mass can be found directly from this observation. Now, finite-order corrections to the general Green function $G(p_i, g, m_R, \mu, c_m)$ shown in Fig. 10.3 have multiple poles at $p^2 = m_R{}^2$ (*not* $m_P{}^2$) in the p^μ-channel. Specializing to the ith external line, these graphs are of the form

$$G^{(n)}(p_i) = [i/(p_i{}^2 - m_R{}^2)][\Sigma(p^2, m_R{}^2, g)/(p_i{}^2 - m_R{}^2)]^n G^{(T)}(p_i), \tag{10.20}$$

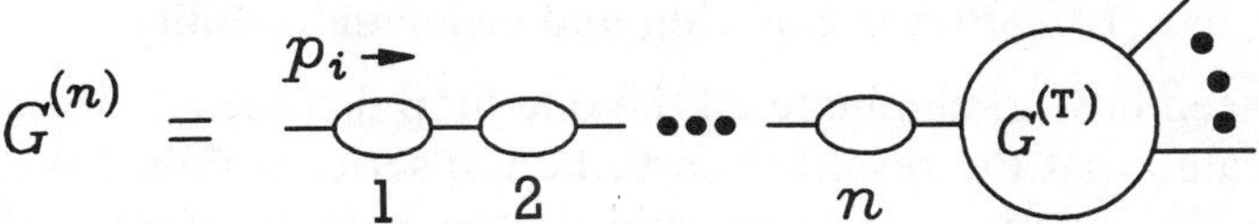

Fig. 10.3 Single-particle poles in a general Green function.

where $G^{(\mathrm{T})}$ is the truncated graph, and where Σ is defined in terms of Γ_2 by

$$\Gamma_2{}^{(\mathrm{R})} \equiv p_i{}^2 - m_\mathrm{R}{}^2 - \Sigma(p_i{}^2, m_\mathrm{R}{}^2, \mu^2, g) = \mathrm{i}[G_2{}^{(\mathrm{R})}]^{-1}. \qquad (10.21)$$

The relation to G_2 follows from a summation on n in eq. (10.20):

$$\sum_{n=0}^{\infty} G^{(n)}(p_i) = \mathrm{i}[p_i^2 - m_\mathrm{R}{}^2 - \Sigma(p_i^2, m_\mathrm{R}{}^2, \mu^2, g)]^{-1} G^{(\mathrm{T})}(p_i)$$

$$= G_2(p_i^2) G^{(\mathrm{T})}(p_i). \qquad (10.22)$$

The position of the pole in the complete Green function is then determined explicitly by the equation

$$m_\mathrm{P}{}^2 - m_\mathrm{R}{}^2 - \Sigma(m_\mathrm{P}{}^2, m_\mathrm{R}{}^2, \mu^2, g) = 0, \qquad (10.23)$$

from which we may solve for m_R, and thus uniquely determine the perturbation series.

Tadpole diagram

Now let us return to the other divergent graph in the theory, Fig. 10.1(*b*). Its value is simply

$$\mathrm{i}\tau(m_\mathrm{R}, g_\mathrm{R}, \mu, \epsilon) = -[\tfrac{1}{2}\mathrm{i}g\mu^\epsilon/(4\pi)^{2-\epsilon}]\Gamma(\epsilon - 1)(m_\mathrm{R}{}^2)^{1-\epsilon}. \qquad (10.24)$$

This term has no momentum dependence. As a result, it contributes to the vacuum expectation value of the field. In fact, a direct calculation, suggested as an exercise, shows that

$$\langle 0|\phi(0)|0\rangle = \tau/m_\mathrm{R}{}^2. \qquad (10.25)$$

Clearly this is a problem, because it means that the field includes an infinite constant vacuum expectation value in four dimensions, even in the absence of any particles. To eliminate this problem, we can add another counterterm to the Lagrangian, of the form

$$\mathcal{L}_\mathrm{tad} = -\tau\phi(x). \qquad (10.26)$$

(Note that this form has the correct dimensions.) This counterterm will have the effect of cancelling all the tadpole terms, without interfering with any of our other considerations. We shall assume that this has been done, and will not consider the tadpole diagram further.

10.2 Power counting and renormalizability

We have seen how to eliminate all the one-loop divergences in $\phi^3{}_4$, and we have indicated that the resulting perturbation series is finite. We shall now prove that this is indeed the case, and, in the process, identify those other field theories that are candidates for renormalization. For completeness and future reference, we include gauge theories. The tool we shall be using is power counting, introduced in Section 9.4. We are going to evaluate the superficial degree of divergence of an arbitrary 1PI graph. Recall that when the superficial degree of divergence of a diagram is negative, its overall integral is convergent (although subintegrals may still diverge).

Consider a 1PI graph G, with N_b (N_f) internal boson (fermion) lines, E_b (E_f) external boson (fermion) lines, L loops and V vertices. In particular, suppose it has s_η η-point pure scalar vertices, with $\eta \geqslant 3$, g_3 and g_4 three- and four-point bosonic gauge vertices, and ψ three-point fermion–vector vertices, so that

$$V = \sum_{\eta \geqslant 3} s_\eta + g_3 + g_4 + \psi. \tag{10.27}$$

The superficial degree of divergence of G in n dimensions is then given by

$$\omega(G) = nL - 2N_b - N_f + g_3, \tag{10.28}$$

where account has been taken of the momentum factor at each three-boson gauge vertex, and of the numerator factor in the fermion propagator, eq. (7.36). Of course, $\omega(G)$ is not necessarily the same as the dimension of the graph itself, to which the dimensions of scalar coupling constants may contribute. We can rewrite $\omega(G)$ in a more useful form by using eq. (10.27) and the graphical identities

$$L = N_b + N_f - V + 1,$$

$$2N_b + E_b = \sum_{s \geqslant 3} \eta s_\eta + 3g_3 + 4g_4 + \psi, \tag{10.29}$$

$$2N_f + E_f = 2\psi.$$

The first is Euler's identity, the second counts the number of ends of boson lines, and the third the ends of fermion lines. A short calculation now shows that $\omega(G)$ may also be written as

$$\begin{aligned}\omega(G) = n &+ (\tfrac{1}{2}n - 2)g_3 + (n-4)g_4 + \sum_{\eta \geqslant 3} [\tfrac{1}{2}(n-2)\eta - n]s_\eta, \\ &+ \psi(\tfrac{1}{2}n - 2) - \tfrac{1}{2}(n-2)E_b - \tfrac{1}{2}(n-1)E_f,\end{aligned} \tag{10.30}$$

or, for the particularly interesting case of $n = 4$,

$$\omega(G) = 4 - E_b - \tfrac{3}{2}E_f - \sum_{\eta \geq 3} d_\eta s_\eta, \tag{10.31}$$

where d_η is the dimension in units of mass of the η-point scalar coupling.

Super-renormalizability

Let us look first at the case of an arbitrary graph of $\phi^3{}_4$, for which $V = s_3$, $d_3 = 1$, $E_f = 0$, where

$$\omega(G) = 4 - E_b - V. \tag{10.32}$$

Here the superficial degree of divergence decreases with the order. The only nontrivial divergent graphs are $E_b = V = 1$, with $\omega(G) = 2$, and $E_b = V = 2$, with $\omega(G) = 0$. These are the lowest-order tadpole and self-energy graphs and subgraphs, as discussed above. Whenever one of these one-loop diagrams appears, so does its corresponding counterterm, and this cancels its divergent part. There are simply no other divergent graphs in the theory. Therefore, the renormalization procedure described in Section 9.1 does indeed make the theory finite. Such a theory, with only a finite number of divergent diagrams, is called *super-renormalizable*.

Renormalizability

The most interesting group of theories is *renormalizable*, when the superficial degree of divergence is non-negative for all 1PI diagrams with a limited number of external lines, for instance, $E_b + E_f \leq 4$. In a renormalizable theory, there are an infinite number of divergent graphs, but, as we shall see, they can be made finite order-by-order in perturbation theory by a generalization of the procedure used for $\phi^3{}_4$. They will require a finite number of counterterms, although each counterterm will be an infinite series in the coupling constant(s). There are only two renormalizable pure scalar theories for $n \geq 4$. By eq. (10.30), these are the theories with dimensionless couplings, $\phi^4{}_4$ and $\phi^3{}_6$. For the latter

$$\omega(G) = 6 - 2E_b \qquad (\phi^3{}_6 \text{ theory}), \tag{10.33}$$

so that G is divergent only for $E \leq 3$, while for the former

$$\omega(G) = 4 - E_b \qquad (\phi^4{}_4 \text{ theory}), \tag{10.34}$$

so that G is divergent only for $E \leq 4$.

For gauge theories in four dimensions, we have

$$\omega(G) = 4 - E_b - \tfrac{3}{2}E_f \quad (\text{gauge theory}). \tag{10.35}$$

Here, all graphs with more than four external lines are power-counting finite, and all gauge theories are power-counting renormalizable in four dimensions. We shall discuss gauge theories at length in the next chapter.

Nonrenormalizability

All other pure scalar field theories have couplings with negative dimensions, and hence, by eq. (10.31), superficial degrees of divergence that grow with the order of the graph. These field theories, such as ${\phi^4}_6$ and ${\phi^6}_4$, are called *power-counting nonrenormalizable*. Similarly, gauge theories in more than four dimensions are power-counting nonrenormalizable. Clearly, there are many more nonrenormalizable theories than renormalizable ones. The problem with nonrenormalizable theories turns out to be that they require an infinite number of different kinds of counterterms to make them finite (Section 10.4).

10.3 One-loop counterterms for ${\phi^3}_6$

In this section, we generalize the renormalization procedure developed for the super-renormalizable theory ${\phi^3}_4$ to the renormalizable theory ${\phi^3}_6$. This will lead us to a general approach, applicable to all renormalizable theories, including those with spin.

Classical Lagrangian

Equation (10.1) defines the classical Lagrangian for ${\phi^3}_n$. For $n \approx 6$, however, it is convenient to redefine

$$\epsilon = 3 - \tfrac{1}{2}n, \tag{10.36}$$

so that now the classical Lagrangian in terms of renormalized quantities is

$$\mathcal{L}_{\text{cl}}(m_{\text{R}}, g_{\text{R}}\mu^\epsilon) = \tfrac{1}{2}[(\partial_\mu \phi_{\text{R}})^2 - {m_{\text{R}}}^2{\phi_{\text{R}}}^2] - \tfrac{1}{6}g_{\text{R}}\mu^\epsilon {\phi_{\text{R}}}^3, \tag{10.37}$$

where g_{R} is dimensionless (see eq. (9.42)). The fields ϕ_{R} will also be renormalized now, for reasons that will become clear below.

Renormalization

The idea of renormalization is basically the same for ${\phi^3}_6$ as for ${\phi^3}_4$. We identify the divergent graphs generated by $\mathcal{L}_{\text{cl}}$, and add counterterms to $\mathcal{L}_{\text{cl}}$ to get a new Lagrange density, for which the divergences are cancelled in perturbation theory. The resulting density is called the renormalized Lagrange density, and we demand that it be equal to the bare density:

$$\mathcal{L}_R(m_R, g_R\mu^\epsilon, c_i) = \mathcal{L}_{cl}(m_R, g_R\mu^\epsilon) + \text{counterterms} = \mathcal{L}_{cl}(m_0, g_0), \tag{10.38}$$

where the bare density has the form of the classical density, but in terms of bare parameters and fields:

$$\mathcal{L}_{cl}(m_0, g_0) = \tfrac{1}{2}[(\partial_\mu\phi_0)^2 - m_0{}^2\phi_0{}^2] - \tfrac{1}{6}g_0\phi_0{}^3. \tag{10.39}$$

The bare and renormalized quantities will be related by the mass μ and by a set of arbitrary finite constants c_i, whose choice constitutes a renormalization scheme. As in $\phi^3{}_4$, the bare Lagrangian depends on only two parameters, and the perturbation series generated from the renormalized Lagrangian will be finite order-by-order in perturbation theory. Unlike the super-renormalizable case, however, cancelling the one-loop divergences does not make the perturbation series completely finite. Rather, renormalization will be an iterative process, extending to all loops.

We begin the discussion with the one-loop graphs of $\phi^3{}_6$.

Mass and wave function renormalization

By eq. (10.33), there are three divergent one-loop graphs for $\phi^3{}_6$, Figs. 10.4(*a*)–(*c*). Figure 10.4(*c*), the tadpole graph, is to be handled in just the same way as in four dimensions, and we ignore it below. The contribution of Fig. 10.4(*a*) to Γ_2 is given in $6 - 2\epsilon$ dimensions by

$$\Gamma_{2,a}{}^{(U)}(p^2, m_R{}^2, \epsilon) = \tfrac{1}{2}g_R{}^2\mu^{2\epsilon}(4\pi)^{\epsilon-3}\Gamma(\epsilon - 1) \times \int_0^1 dx[m_R{}^2 - x(1-x)p^2 - i\epsilon]^{1-\epsilon}. \tag{10.40}$$

Using $\Gamma(\epsilon - 1) = \Gamma(\epsilon)/(\epsilon - 1)$, we expand eq. (10.40) about $\epsilon = 0$:

$$\Gamma_{2,a}{}^{(U)}(p^2, m_R{}^2, \epsilon) = -\tfrac{1}{2}g_R{}^2(4\pi)^{-3}\Big\{[\epsilon^{-1} + 1 + \ln(4\pi\mu^2/m_R{}^2) - \gamma_E][m_R{}^2 - \tfrac{1}{6}p^2] - m_R{}^2\int_0^1 dx\, F(x)\ln F(x)\Big\}, \tag{10.41}$$

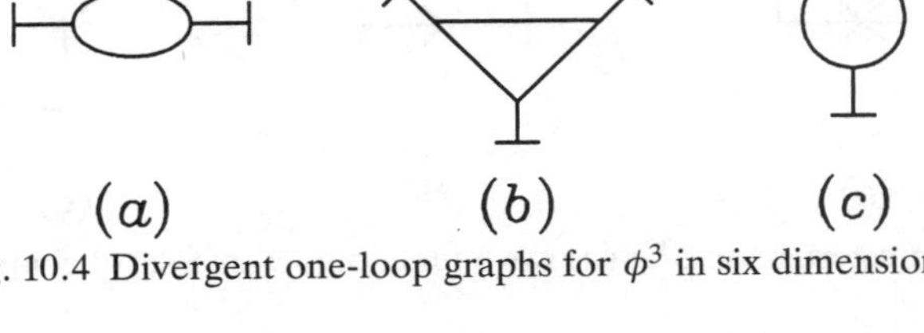

(*a*) (*b*) (*c*)

Fig. 10.4 Divergent one-loop graphs for ϕ^3 in six dimensions.

where we define

$$F(x) \equiv 1 - x(1-x)(p^2/m_R{}^2). \tag{10.42}$$

The pole term at $n = 6$ is then

$$\Gamma_{2,a}{}^{(\text{pole})} = -\tfrac{1}{2}g_R{}^2(4\pi)^{-3}(m_R{}^2\epsilon^{-1} - \tfrac{1}{6}p^2\epsilon^{-1}). \tag{10.43}$$

We must now construct counterterms to cancel these divergences. The first term in eq. (10.43) is a candidate for a mass counterterm, but the second, which is momentum-dependent, is not. It can, however, be cancelled by a counterterm proportional to the operator $(\partial_\mu\phi)^2$. We therefore add two counterterms to $\mathscr{L}_{\text{cl}}(m_R, g_R\mu^\epsilon)$ to get a new density $\mathscr{L}'(m_R, g_R\mu^\epsilon)$:

$$\begin{aligned}\mathscr{L}'(m_R, g_R\mu^\epsilon) = {}& \tfrac{1}{2}[(\partial_\mu\phi_R)^2 - m_R{}^2\phi_R{}^2] - \tfrac{1}{6}g_R\mu^\epsilon\phi_R{}^3 \\ & + \tfrac{1}{2}(\partial_\mu\phi_R)^2[Z_\phi{}^{(1)} - 1] - \tfrac{1}{2}(m_R{}^2\phi_R{}^2)(Z_\phi Z_m - 1)^{(1)}.\end{aligned} \tag{10.44}$$

The factors Z_i are known as *renormalization constants*. The superscripts (1) denote their one-loop expansions. In the notation of Section 10.1,

$$\delta m^2 = m_R{}^2(Z_\phi Z_m - 1). \tag{10.45}$$

The reason for this strange-looking notation will be described shortly. For now, the two factors $(Z_\phi - 1)$ and $(Z_\phi Z_m - 1)$ are just independent quantities, chosen to cancel the divergences of eq. (10.43). They are, by analogy with eq. (10.6), expressed as

$$Z_\phi{}^{(1)} - 1 \equiv g_R{}^2(4\pi)^{-3}(-\tfrac{1}{12})(\epsilon^{-1} + c_{\phi,1}), \tag{10.46a}$$

$$\begin{aligned}[Z_\phi Z_m - 1]^{(1)} &= [Z_\phi{}^{(1)} - 1] + [Z_m{}^{(1)} - 1] \\ &\equiv g_R{}^2(4\pi)^{-3}[(-\tfrac{1}{12})(\epsilon^{-1} + c_{\phi,1}) - \tfrac{5}{12}(\epsilon^{-1} + c_{m,1})].\end{aligned} \tag{10.46b}$$

$c_{\phi,1}$ and $c_{m,1}$ are finite, but otherwise arbitrary, and their choice defines the renormalization scheme. The counterterms give vertices denoted graphically as in Fig. 10.5(a), (b). We easily check that the contribution of the

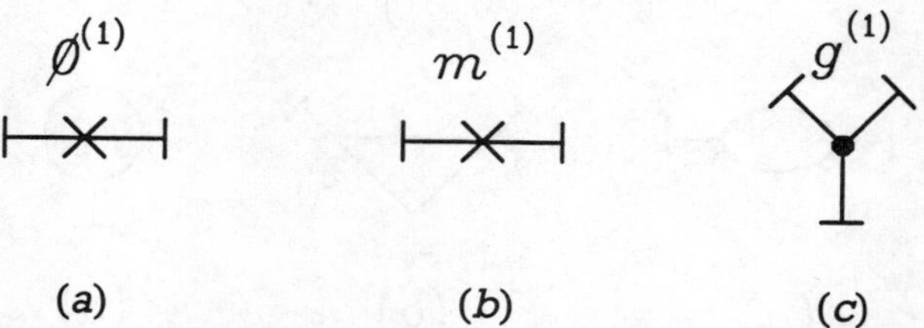

Fig. 10.5 Counterterms for $\phi^3{}_6$: (a) wave function; (b) mass; (c) coupling constant.

counterterms to Γ_2 at order $g_R{}^2$ cancels the pole terms in $\Gamma_{2,a}{}^{(U)}$:

$$\Gamma_{2,a}{}^{(R)}(p^2, m_R{}^2, \epsilon) = -\tfrac{1}{2}g_R{}^2(4\pi)^{-3}\Big\{\tfrac{1}{6}p^2 c_{\phi,1} - m_R{}^2(\tfrac{5}{6}c_{m,1} + \tfrac{1}{6}c_{\phi,1}) + [1 + \ln(4\pi\mu^2/m_R{}^2) - \gamma_E] \times (m_R{}^2 - \tfrac{1}{6}p^2) - m_R{}^2\int_0^1 dx\, F(x)\ln F(x) + O(\epsilon, g^4)\Big\}, \tag{10.47}$$

where $F(x)$ is given by eq. (10.42).

Now let us discuss the choice of notation for the counterterms. Z_ϕ and Z_m are called *field* (or *wave function*) and *mass renormalization constants*, respectively. They relate the bare and renormalized fields and mass:

$$\phi_0 = Z_\phi{}^{1/2}\phi_R, \tag{10.48a}$$

$$m_0 = Z_m{}^{1/2} m_R. \tag{10.48b}$$

Such a relation between bare and renormalized quantities is known as *multiplicative renormalization*. Similarly, the relation eq. (10.10a) for the renormalized mass is called *additive renormalization*. Substituting eqs. (10.48) into the bare Lagrange density, eq. (10.39), we find

$$\mathscr{L}_{cl}(m_0, g_0) = \tfrac{1}{2}[(\partial_\mu\phi_R)^2 - m_R{}^2\phi_R{}^2] - \tfrac{1}{6}g_0\phi_R{}^3[Z_\phi{}^{(1)}]^{3/2} + \tfrac{1}{2}(\partial_\mu\phi_R)^2[Z_\phi{}^{(1)} - 1] - \tfrac{1}{2}(m_R{}^2\phi_R{}^2)[Z_\phi{}^{(1)}Z_m{}^{(1)} - 1]. \tag{10.49}$$

This is almost equal to $\mathscr{L}'$, eq. (10.44), except for the interaction term, to whose treatment we turn.

Coupling-constant renormalization

The remaining divergent one-loop graph is the 'triangle' diagram, Fig. 10.4(*b*), whose contribution to the 1PI function Γ_3 is (as usual) $-$i times the truncated diagram:

$$\Gamma_{3,b}{}^{(U)}(p_1, p_2, p_3) = -i\frac{(-ig_R)^3\mu^{3\epsilon}}{(2\pi)^n}\int d^n k\left(\frac{i}{k^2 - m_R{}^2 + i\epsilon}\right)\left[\frac{i}{(p_1 + k)^2 - m_R{}^2 + i\epsilon}\right] \times \frac{i}{(p_2 + k)^2 - m_R{}^2 + i\epsilon}. \tag{10.50}$$

It is straightforward to perform this integral using Feynman parameterization. The corresponding contribution to Γ_3 is

$$\Gamma_{3,b}{}^{(\mathrm{U})} = -\frac{g_{\mathrm{R}}{}^3\mu^\epsilon(4\pi\mu^2)^\epsilon}{(4\pi)^3}\,\Gamma(\epsilon)\int_0^1 \frac{\mathrm{d}x\,\mathrm{d}y\,\theta(1-x-y)}{[M^2(p_i,x,y,m_{\mathrm{R}})-\mathrm{i}\epsilon]^\epsilon} \tag{10.51}$$

where

$$M^2(p_i,x,y,m_{\mathrm{R}}) = m_{\mathrm{R}}{}^2 + 2xyp_1\cdot p_2 - x(1-x)p_1{}^2 - y(1-y)p_2{}^2. \tag{10.52}$$

Next, we factor out the zeroth-order contribution to Γ_3, $-g_{\mathrm{R}}\mu^\epsilon$. Then, expanding the rest about $n=6$, we find

$$\begin{aligned}\Gamma_{3,b}{}^{(\mathrm{U})} = {} & (-g_{\mathrm{R}}\mu^\epsilon)[\tfrac{1}{2}g_{\mathrm{R}}{}^2(4\pi)^{-3}] \\ & \times\left\{\epsilon^{-1} - \gamma_{\mathrm{E}} + \ln\left(\frac{4\pi\mu^2}{m_{\mathrm{R}}{}^2}\right)\right. \\ & \left. - 2\int_0^1 \mathrm{d}x\,\mathrm{d}y\;\theta(1-x-y)\ln\left[\frac{(M^2-\mathrm{i}\epsilon)}{m_{\mathrm{R}}{}^2}\right] + \mathrm{O}(\epsilon)\right\}.\end{aligned} \tag{10.53}$$

The counterterm necessary to cancel the singular part of $\Gamma_{3,b}{}^{(\mathrm{U})}$ is simply

$$-\delta g_{\mathrm{R}}{}^{(1)}(\tfrac{1}{6}\phi_{\mathrm{R}}{}^3), \tag{10.54}$$

illustrated in Fig. 10.5(c), where

$$\delta g_{\mathrm{R}}{}^{(1)} = [-g_{\mathrm{R}}\mu^\epsilon][\tfrac{1}{2}g_{\mathrm{R}}{}^2(4\pi)^{-3}]\epsilon^{-1} + \text{finite remainder}. \tag{10.55}$$

Now we relate the renormalized to the bare coupling by a rescaling, remembering that g_{R} is dimensionless but g_0 in general is not:

$$g_0 = Z_g g_{\mathrm{R}}\mu^\epsilon. \tag{10.56}$$

We next write, by analogy with eqs. (10.48),

$$\delta g_{\mathrm{R}}{}^{(1)} = g_{\mathrm{R}}\mu^\epsilon\{[Z_\phi{}^{(1)}]^{3/2}Z_g{}^{(1)} - 1\}, \tag{10.57}$$

where we recall that g_{R} appears in $\mathscr{L}$ with three factors of the field. Then, using eqs. (10.55) and (10.46a), and a parameterization, by now standard, of the finite part of $Z_g{}^{(1)}$, we find

$$Z_g{}^{(1)} = -g_{\mathrm{R}}{}^2(4\pi)^{-3}\,[\tfrac{3}{8}(\epsilon^{-1} + c_{g,1})]. \tag{10.58}$$

Finally, using eqs. (10.48) and (10.56), we find that the bare density, eq.

(10.39) equals the (one-loop) renormalized Lagrange density:

$$\begin{aligned}\mathscr{L}_R{}^{(1)}(m_R, g_R\mu^\epsilon, c_i) = &\tfrac{1}{2}[(\partial_\mu\phi_R)^2 - m_R{}^2\phi_R{}^2] - \tfrac{1}{6}g_R\mu^\epsilon\phi_R{}^3\\ &+ \tfrac{1}{2}(\partial_\mu\phi_R)^2[Z_\phi{}^{(1)} - 1]\\ &- \tfrac{1}{2}(m_R{}^2\phi_R{}^2)[Z_\phi{}^{(1)}Z_m{}^{(1)} - 1]\\ &- \tfrac{1}{6}g_R\mu^\epsilon\phi_R{}^3\{[Z_\phi{}^{(1)}]^{3/2}Z_g{}^{(1)} - 1\}. \quad (10.59)\end{aligned}$$

When the $Z_i{}^{(1)}$ are given by eqs. (10.46) and (10.58), the perturbation theory generated by $\mathscr{L}_R{}^{(1)}$ is finite at one loop, even for $\epsilon \to 0$.

In $\phi^3{}_4$, one-loop finiteness was sufficient for the whole perturbation series. Here, however, to make n-loop graphs finite it is necessary to construct an n-loop renormalized Lagrangian $\mathscr{L}^{(n)}$ of the same form as (10.59), but with renormalization constants expanded to nth order in g^2:

$$Z_i{}^{(n)} = 1 + \sum_{k=1}^{n} z_{i,k}g^{2k}. \quad (10.60)$$

We shall see in the next section how this procedure works for $n > 1$. For now we simply assume the existence of a 'truly' renormalized Lagrangian, $\mathscr{L}_R$, which generates a finite perturbation series for all Green functions to any number of loops:

$$\begin{aligned}&\mathscr{L}_R(m_R, g_R\mu^\epsilon, c_i)\\ &\quad = \tfrac{1}{2}[(\partial_\mu\phi_R)^2 - m_R{}^2\phi_R{}^2] - \tfrac{1}{6}g_R\mu^\epsilon\phi_R{}^3 + \tfrac{1}{2}(\partial_\mu\phi_R)^2(Z_\phi - 1)\\ &\quad\quad - \tfrac{1}{2}(m_R{}^2\phi_R{}^2)(Z_\phi Z_m - 1) - g_R\mu^\epsilon\tfrac{1}{6}\phi_R{}^2(Z_\phi{}^{3/2}Z_g - 1), \quad (10.61)\end{aligned}$$

where the Z_i are given by eq. (10.60) with $n \to \infty$.

Renormalization schemes

The renormalization schemes described for $\phi^3{}_4$ may also be applied to $\phi^3{}_6$, although we no longer assume that the bare coupling is a known quantity. Each scheme defines a finite perturbation series in terms of the renormalized mass m_R and the renormalized coupling g_R, the mass μ and the finite constants c_i.

Momentum subtraction

In momentum subtraction, we use vertex functions at convenient external momenta to determine the finite terms $c_{i,j}$ in terms of the renormalized mass and coupling. Since there are three renormalization constants in $\phi^3{}_6$, corresponding to mass, coupling and field renormalization, we need three equations to fix them. These are often chosen to be

$$\Gamma_2(p^2)|_{p^2=-M^2} = -M^2 - m_R{}^2, \tag{10.62a}$$

$$[\partial\Gamma_2(p^2)/\partial p^2]|_{p^2=-M^2} = 1, \tag{10.62b}$$

$$\Gamma_3(p_1, p_2, p_3)|_{\text{sym}} = -g_R. \tag{10.62c}$$

Here the renormalization scheme is chosen to make higher-order corrections vanish at the normalization points. The notation 'sym' specifies that we choose the momenta at a *symmetric point*:

$$p_i \cdot p_j = -M^2(\delta_{ij} - \tfrac{1}{3}). \tag{10.63}$$

Note that it is not necessary, although it is customary, to choose all the scales M^2 in the three equations to be the same. Any other choice would give the same amount of information. Equations (10.62) determine the c_i as follows.

Using the Lagrange density, eq. (10.59), the complete one-loop version of $\Gamma_2(q^2)$ is given in eq. (10.47). From (10.62a), we may solve for $c_{m,1}$ in terms of $c_{\phi,1}$:

$$c_{m,1} = -\tfrac{1}{5}c_{\phi,1}(M^2/m_R{}^2 + 1) + \tfrac{6}{5}\Big((1 + \ln 4\pi - \gamma_E)(1 + M^2/6m_R{}^2) - \int_0^1 dx[1 + x(1-x)M^2/m_R{}^2]\ln\{[m_R{}^2 + x(1-x)M^2]\mu^{-2}\}\Big). \tag{10.64}$$

Next, taking the derivative with respect to p^2 and using eq. (10.62b) allows us to solve for $c_{\phi,1}$:

$$c_{\phi,1} = \ln 4\pi - \gamma_E - 6\int_0^1 dx\, x(1-x)\ln\{[m_R{}^2 + x(1-x)M^2]\mu^{-2}\}. \tag{10.65}$$

Using eqs. (10.64) and (10.65) in eq. (10.47), we can derive a not-very-simple result from which the c_i have been eliminated, and which satisfies eqs. (10.62a, b). A similar approach, applied to $\Gamma_3(p_i)$, determines c_g.

Generalized minimal subtraction

Once again, minimal subtraction is appealing in its simplicity. For instance, in MS at one loop at $\epsilon = 0$,

$$\Gamma_2(p^2, m_R, g_R, \mu) = p^2 - m_R{}^2 - \tfrac{1}{2}g_R{}^2(4\pi)^{-3}\Big((m_R{}^2 - \tfrac{1}{6}p^2) - \int_0^1 dx[m_R{}^2 - x(1-x)p^2]\ln\{[m_R{}^2 - x(1-x)p^2]\mu^{-2}\}\Big). \tag{10.66}$$

Renormalized and unrenormalized Green functions

We define renormalized and unrenormalized Green functions as the vacuum expectation values of time-ordered products of renormalized and unrenormalized fields, respectively:

$$\begin{aligned} G_R(x_1, \ldots, x_n, g_R, m_R, \mu, \epsilon) &= \langle 0| T\, [\textstyle\prod_{i=1}^n \phi_R(x_i)]|0\rangle, \\ G_0(x_1, \ldots, x_n, g_0, m_0, \epsilon) &= \langle 0| T\, [\textstyle\prod_{i=1}^n \phi_0(x_i)]|0\rangle. \end{aligned} \tag{10.67}$$

The vacuum state should have a meaning independent of renormalization. So, from the multiplicative renormalization in eq. (10.48a), we expect that renormalized and unrenormalized Green functions are related simply by a power of Z_ϕ:

$$\begin{aligned} Z_\phi{}^{n/2}(g_R, m_R, \mu, \epsilon)&G_R(x_1, \ldots, x_n, g_R, m_R, \mu, \epsilon) \\ &= G_0(x_1, \ldots, x_n, g_0, m_0, \epsilon). \end{aligned} \tag{10.68}$$

On the left-hand side of this equation, only renormalized quantities, along with μ and ϵ, appear and on the right only bare quantities. Equation (10.68) may be given a more formal proof by comparing how the generating functional $W[J]$, eq. (3.44), is expressed in terms of bare fields and renormalized fields. The bare and renormalized forms of the Lagrangian are equal, so

$$\begin{aligned} W_R[J] &= \int [\mathcal{D}\phi_R] \exp\{\mathrm{i}S[\phi_R, g_R, m_R, c_i, \epsilon] - \mathrm{i}(J, \phi_R)\} \\ &= \text{constant} \times \int [\mathcal{D}\phi_0] \exp\{\mathrm{i}S[\phi_0, g_0, m_0, \epsilon] - \mathrm{i}(JZ_\phi{}^{-1/2}, \phi_0)\} \\ &= \text{constant} \times W_0[JZ_\phi{}^{-1/2}] \qquad (\phi^3{}_6 \text{ theory}). \end{aligned} \tag{10.69a}$$

The constant cancels in the generator of Green functions, eq. (3.44), $Z_{\mathcal{L},R}[J] = W_R[J]/W_R[0] = Z_{\mathcal{L},U}[Z_\phi{}^{-1/2}J]$. Similarly, for the generator of connected Green functions, eq. (3.96), $G_{c,R}[J] = G_{c,0}[Z_\phi{}^{-1/2}J]$. Then eq. (10.68) follows by taking variations. Note the difference from the super-renormalizable case, eq. (10.12).

Pursuing this line of reasoning, we may also derive a relation between bare and renormalized 1PI functions, with generator $\Gamma[\phi]$, eq. (3.99). Defining $J' \equiv JZ_\phi{}^{-1/2}$, we find that we can express the generator of bare 1PI diagrams in terms of the generator of renormalized 1PI diagrams as follows. First, we recall that

$$\Gamma_R(\phi_c, g_R, m_R) \equiv -\mathrm{i}G_{c,R}[J, g_R, m_R] + (\phi_{c,R}, J),$$

where, by eq. (3.97), with a slight change in notation, $\phi_{c,R} \equiv \mathrm{i}\delta G_{c,R}/\delta J$. Thus, using $G_{c,R}[J] = G_{c,0}[J']$, we have

$$\Gamma_{\rm R}(\phi_{\rm c}, g_{\rm R}, m_{\rm R}) = -{\rm i}G_{\rm c,0}[J', g_0, m_0] + (Z_\phi{}^{1/2}\phi_{\rm c,R}, J'),$$

where $Z_\phi{}^{1/2}\phi_{\rm c,R} = {\rm i}\delta G_{\rm c,0}/\delta J'$. As a result, we find

$$\Gamma_{\rm R}(\phi_{\rm c}, g_{\rm R}, m_{\rm R}) = \Gamma_{\rm bare}(Z_\phi{}^{1/2}\phi_{\rm c,R}, g_0, m_0). \qquad (10.69b)$$

Here, the bare and renormalized classical fields, $\phi_{\rm c,bare}$ and $\phi_{\rm c,R}$, are related by the usual multiplicative renormalization. From eqs. (10.69b) and (3.100) we then derive the analogue of eq. (10.68) for 1PI functions:

$$Z_\phi{}^{-n/2}(g_{\rm R}, m_{\rm R}, \mu, \epsilon)\Gamma_{\rm R}(x_1, \ldots, x_n, g_{\rm R}, m_{\rm R}, \mu, \epsilon) = \Gamma_0(x_1, \ldots, x_n, g_0, m_0, \epsilon). \qquad (10.68')$$

Compared with eq. (10.68), the renormalization constants appear with inverse powers.

The S-matrix

The *S*-matrix is found by evaluating the residues of the poles of Green functions in momentum space, as in eq. (10.19). But what Green function are we going to take on the right-hand side of (10.19), renormalized or unrenormalized, and, if renormalized, with which choice of finite terms? To answer these questions, we use the momentum-independence of the multiplicative renormalization prescription (10.48b). Because $Z_\phi G_{2,{\rm R}}(p^2) = G_{2,0}(p^2)$, both the unrenormalized Green function and the renormalized Green function must have their poles at the same value of p^2. But the position of the pole in $G_2(p^2)$ is the value of the physical particle mass. This relates the residues $R_{\rm R}$ and R_0 of the renormalized and unrenormalized two-point functions (eq. (4.15)):

$$R_{\rm R} = \lim_{p^2\to m_{\rm P}{}^2} [-{\rm i}(p^2 - m_{\rm P}{}^2)G_{2,{\rm R}}(p^2)]$$

$$= Z_\phi{}^{-1} \lim_{p^2\to m_{\rm P}{}^2} [-{\rm i}(p^2 - m_{\rm P}{}^2)G_{2,0}] = Z_\phi{}^{-1}R_0. \qquad (10.70)$$

Thus the residue is renormalized by the same multiplicative process as the fields and Green functions, and the renormalization constants from the residues and the Green functions cancel in the *S*-matrix, eq. (10.19):

$$S(p_i) = \prod_{i=1}^n \left[\frac{(-{\rm i}) \lim_{p_i^2\to m_{\rm P}{}^2} (p_i{}^2 - m_{\rm P}{}^2)}{(2\pi)^{3/2} R_{\rm R}{}^{1/2}}\right] G_{n,{\rm R}}(p_i)$$

$$= \prod_{i=1}^n \left[\frac{(-{\rm i}) \lim_{p_i^2\to m_{\rm P}{}^2} (p_i{}^2 - m_{\rm P}{}^2)}{(2\pi)^{3/2} R_0{}^{1/2}}\right] G_{n,0}(p_i). \qquad (10.71)$$

This means that the S-matrix is the same for renormalized as for unrenormalized Green functions. In fact, it is the same for every consistent definition of the renormalized Green functions, that is, for every renormalization scheme and choice of μ. This statement follows in $\phi^3{}_6$ from eq. (10.71) for $n \neq 6$, where g_0 and m_0 are finite. That it remains true as $n \to 6$ ($\epsilon \to 0$) where the bare quantities diverge is not so obvious. This point is discussed in Collins (r1984, Chapter 7). We shall assume here that there is no problem in taking the limit. Then, if the theory can be renormalized, the perturbation series for the S-matrix and for other physical quantities are independent of the renormalization scheme and the scale parameter μ, since these do not appear in the bare Lagrangian.

Fixing the renormalized coupling and mass

To get unique predictions from our perturbation series, it only remains to fix g_R and m_R. This can be done by expressing any two observables, such as the physical mass of the particle and some cross section, in terms of the renormalized parameters. These relations may be inverted to determine g_R and m_R in terms of the observables, for the particular renormalization scheme and value of μ. Let us see how this process works in our case.

With $\overline{\text{MS}}$ subtraction, for example, we have from eq. (10.66) the following one-loop relation at $\epsilon = 0$ between m_R, g_R, μ and the physical mass m_P, for which $\Gamma_2(m_P{}^2) = 0$:

$$0 = m_P{}^2 - m_R{}^2 + \frac{g_R{}^2}{2(4\pi)^3}\left((m_R{}^2 - \tfrac{1}{6}m_P{}^2) + \int_0^1 \mathrm{d}x[m_R{}^2 - x(1-x)m_P{}^2] \ln\{[m_R{}^2 - x(1-x)m_P{}^2]\mu^{-2}\}\right). \tag{10.72}$$

Suppose, in addition, we assume that a given reduced S-matrix element $S(p_i, g_R, m_R, \mu)$ has value S_P. It does not matter which matrix element we pick. We can go back to the perturbation series for $S(p_i, g_R, m_R, \mu)$, compute it to one loop, and set it equal to S_P:

$$S_P = g_R{}^2 S^{(2)}(m_R{}^2, \mu^2, p_i \cdot p_j). \tag{10.73}$$

We now solve eqs. (10.72) and (10.73) for g_R and m_R as functions of S_P, m_P and μ to get relations valid to one loop:

$$g_R = g_R(S_P, m_P, \mu), \; m_R = m_R(S_P, m_P, \mu). \tag{10.74}$$

The generalization to any number of loops is straightforward. In this way, the renormalized coupling and mass may be thought of as functions of the renormalization scheme, the mass scale μ and two observables. For simplicity of notation, however, it is customary to suppress the latter dependence and write simply $g_R(\mu)$ and $m_R(\mu)$. In Chapter 12 we will give explicit examples of this procedure applied to QED and QCD.

In summary, the perturbation series for $\phi^3{}_6$ may be constructed from the classical Lagrangian by the following series of steps.

(i) Regularize graphs and construct counterterms in terms of renormalized quantities.
(ii) Select a scheme which specifies finite parts of the counterterms.
(iii) Fix renormalized parameters in terms of observables.

The result is a well-defined series, which gives explicit predictions for all remaining observables and in which the regularization may be taken away (ϵ set to zero in our case). Order-by-order, however, each series generally depends on the scheme and the mass scales μ and/or M. The equality, eq. (10.11), between the renormalized and bare forms of the Lagrange density guarantees that every scheme and set of mass scales gives the same S-matrix after the full series has been summed. Although all schemes and mass scales are formally equivalent, some may be more convenient than others. We shall exploit this observation in Section 10.5, where we discuss the renormalization group.

10.4 Renormalization at two loops and beyond

In this section, we study the construction of $\mathcal{L}_R$ to two loops, and discuss how we know that renormalization may be extended to all orders.

Two-loop renormalization of $\phi^3{}_6$

From the power-counting arguments of Section 10.2, ultraviolet divergences occur only in 1PI two- and three-point graphs in $\phi^3{}_6$. Just as the one-loop graphs of $\mathcal{L}_{cl}$ define the counterterms of $\mathcal{L}_R{}^{(1)}$, the two-loop graphs of $\mathcal{L}_R{}^{(1)}$ define second-order contributions to the counterterms of $\mathcal{L}_R{}^{(2)}$. Let us review the procedure as it applies to two-loop self-energy graphs.

The 1PI self-energy graphs generated by the Lagrangian $\mathcal{L}_R{}^{(1)}$ are shown in Fig. 10.6(a), (b). We have organized them into two sets, $\bar{\gamma}_{(a)}$ and $\bar{\gamma}_{(b)}$, corresponding to the figure, which we shall compute separately. Each has both ϵ^{-2} and ϵ^{-1} poles in dimensional regularization. In particular, we shall show that

$$-\mathrm{i}\bar{\gamma}_{(a)}(p^2, m^2)^{(\text{pole})} = p^2(a_{\phi,2}\epsilon^{-2} + a_{\phi,1}\epsilon^{-1}) + m^2(a_{\delta m,2}\epsilon^{-2} + a_{\delta m,1}\epsilon^{-1}), \tag{10.75}$$

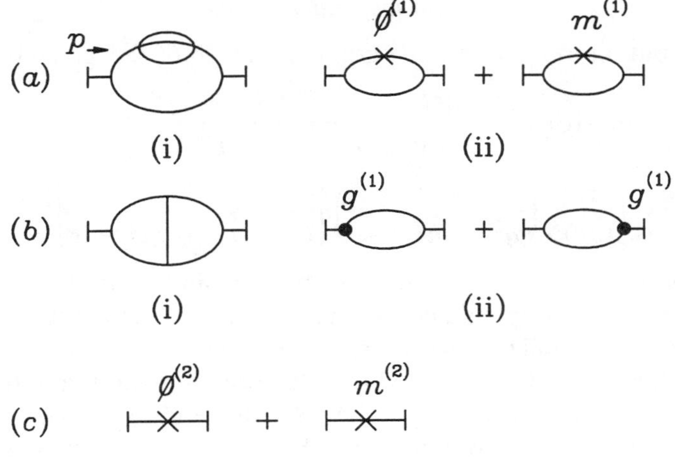

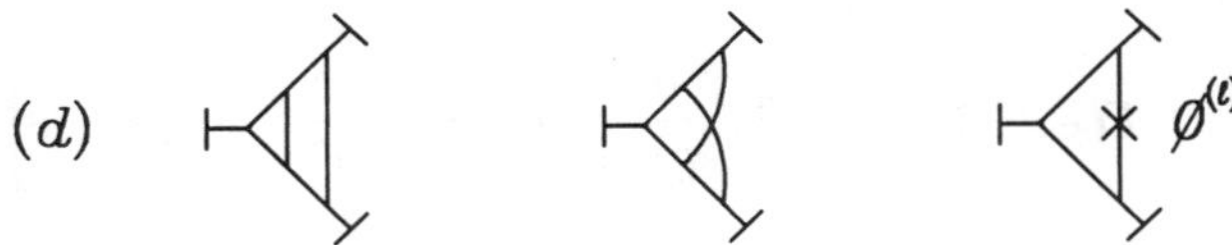

Fig. 10.6 (a), (b) Two-point diagrams generated from the one-loop renormalized Lagrange density, eq. (10.61). (c) Two-loop counterterms, (d) representative three-point diagrams from (10.61).

and similarly for $\bar{\gamma}_{(b)}$. Here $\bar{\gamma}_{(a)}$, $\bar{\gamma}_{(b)}$ stand for the truncated graphs, while $-\mathrm{i}\bar{\gamma}_{(a)}$, $-\mathrm{i}\bar{\gamma}_{(b)}$ are their contributions to Γ_2. The poles of $\bar{\gamma}_{(a)}$, for instance, may be cancelled by counterterms of the form

$$\tfrac{1}{2}(\partial_\mu\phi)^2 z_{\phi,2}[\bar{\gamma}_{(a)}] - \tfrac{1}{2}m^2\phi^2(z_m + z_\phi)_2[\bar{\gamma}_{(a)}], \tag{10.76}$$

where, in the notation of eq. (10.60),

$$\begin{aligned} z_{\phi,2}[\bar{\gamma}_{(a)}] &= -a_{\phi,2}\epsilon^{-2} - a_{\phi,1}\epsilon^{-1} + \text{finite remainder}, \\ (z_m + z_\phi)_2[\bar{\gamma}_{(a)}] &= a_{\delta m,2}\epsilon^{-2} + a_{\delta m,2}\epsilon^{-1} + \text{finite remainder}. \end{aligned} \tag{10.77}$$

As at one loop, the finite parts define the renormalization scheme. These individual counterterms are represented in Fig. 10.6(c). The complete two-loop mass and wave function counterterms are just the sums of the counterterms corresponding to $\bar{\gamma}_{(a)}$ and $\bar{\gamma}_{(b)}$, for instance,

$$z_{\phi,2} = z_{\phi,2}[\bar{\gamma}_{(a)}] + z_{\phi,2}[\bar{\gamma}_{(b)}]. \tag{10.78}$$

All this is by way of introduction. Let us turn to the explicit calculation of the graphs of Fig. 10.6.

Two-loop calculations

The contribution to Γ_2 of the first graph in Fig. 10.6(a) is given by

$$-\mathrm{i}\gamma_{(a.\mathrm{i})} = -\tfrac{1}{2}\mathrm{i}(-\mathrm{i}g)^4\mu^{4\epsilon}\int\frac{\mathrm{d}^n k}{(2\pi)^n}\left[\frac{\mathrm{i}}{(p-k)^2-m^2+\mathrm{i}\epsilon}\right]\left[\frac{\mathrm{i}^2}{(k^2-m^2+\mathrm{i}\epsilon)^2}\right]$$
$$\times\int\frac{\mathrm{d}^n q}{(2\pi)^n}\left(\frac{\mathrm{i}}{q^2-m^2+\mathrm{i}\epsilon}\right)\left[\frac{\mathrm{i}}{(k-q)^2-m^2+\mathrm{i}\epsilon}\right], \qquad (10.79)$$

where, for simplicity, we drop, for the remainder of this section, the subscript R on the coupling and the mass. They are, however, always to be regarded as renormalized. In the spirit of dimensional regularization, we should evaluate $\gamma_{(a.\mathrm{i})}$ for $n<4$, which is where the overall graph and all its subgraphs are power-counting finite. At the end, we shall analytically continue to the neighborhood of $n=6$. For $n<4$, all integrals are convergent, and we may choose to begin with whichever integral we like. We already know how to do the q integral, since it corresponds to i times Γ_2 at one loop, as given in eq. (10.40),

$$-\mathrm{i}\gamma_{(a.\mathrm{i})} = \tfrac{1}{2}\mathrm{i}g^4\mu^{4\epsilon}\Gamma(\epsilon-1)(4\pi)^{\epsilon-3}\int\frac{\mathrm{d}^n k}{(2\pi)^n}\left[\frac{1}{(p-k)^2-m^2+\mathrm{i}\epsilon}\right]$$
$$\times\frac{1}{(k^2-m^2+\mathrm{i}\epsilon)^2}\int_0^1\mathrm{d}x[x(1-x)]^{1-\epsilon}\left[\frac{m^2}{x(1-x)}-k^2-\mathrm{i}\epsilon\right]^{1-\epsilon}. \qquad (10.80)$$

The k integral can now be carried out by using Feynman parameterization, eq. (9.11). Choosing Feynman parameters w for the last factor and z for the first, we obtain

$$-\mathrm{i}\bar{\gamma}_{(a.\mathrm{i})} = \frac{g^4(4\pi\mu^2)^{2\epsilon}}{2(4\pi)^6}\Gamma(2\epsilon-1)\int_0^1\mathrm{d}x[x(1-x)]^{1-\epsilon}\int_0^1\mathrm{d}z$$
$$\times\int_0^{1-z}\mathrm{d}w\, w^{\epsilon-2}(1-w-z)$$
$$\times\left[m^2\left(1-w+\frac{w}{x(1-x)}\right)-p^2z(1-z)-\mathrm{i}\epsilon\right]^{1-2\epsilon}. \qquad (10.81)$$

Because of the factor $w^{\epsilon-2}$, this expression is, as expected, defined only for $\epsilon>1$, i.e. $n=4$. We want to put the integral into a form that we can analytically continue up to $n=6$ and we also want to evaluate the coefficients of the singular terms at $n=6$. It is these coefficients that contribute to the two-loop counterterms. We emphasize that analytic continuation is unique.

The w integral in (10.81) is already singular at $\epsilon=1$, while the x and z integrals are finite for $\epsilon>0$. Therefore, starting with $\epsilon>1$ ($n<4$), we put the w integral in a form suitable for analytic continuation by integrating by parts twice. Two integrations by parts are necessary to exhibit the poles in the w integral at both $n=4$ and $n=6$:

$$-\mathrm{i}\bar{\gamma}_{(a.\mathrm{i})} = \frac{g^4(4\pi\mu^2)^{2\epsilon}}{2(4\pi)^6}\Gamma(2\epsilon-1)\int_0^1 \mathrm{d}x[x(1-x)]^{1-\epsilon}\int_0^1 \mathrm{d}z$$
$$\times\frac{1}{\epsilon(\epsilon-1)}\left(\int_0^{1-z}\mathrm{d}w\, w^\epsilon\frac{\mathrm{d}^2}{\mathrm{d}w^2}\{(1-w-z)[E(x,z,w)]^{1-2\epsilon}\}\right.$$
$$\left.+(1-z)^\epsilon[E(x,z,1-z)]^{1-2\epsilon}\right), \qquad (10.82\mathrm{a})$$

where

$$E(x,z,w)\equiv -p^2z(1-z)+m^2\left\{1-w\left[1-\frac{1}{x(1-x)}\right]\right\}. \qquad (10.82\mathrm{b})$$

In eqs. (10.82), the poles at $n=4$ and 6 are exhibited explicitly, and the integrals are convergent all the way to $\epsilon=-1$ ($n=8$). The factor $\Gamma(2\epsilon-1)$ also diverges at $\epsilon=0$ (and at $\epsilon=\frac{1}{2}$), so there are both single and double poles at $n=6$. To find their residues, we must expand the integrals and factors in (10.82) to order ϵ. This is fairly straightforward, and we shall not give the details. A helpful hint for those wishing to verify the result is first to expand only the factor w^ϵ. In the resulting terms, the x and z integrals are all elementary, to order ϵ. (It is also useful to remember that $\int_0^1\mathrm{d}z\, z^a\ln z=-1/(a+1)^2$.) The result is

$$-\mathrm{i}\bar{\gamma}_{(a.\mathrm{i})}{}^{(\mathrm{pole})} = -\tfrac{1}{48}\epsilon^{-2}\lambda^2(\tfrac{1}{3}p^2+3m^2)$$
$$-\tfrac{1}{12}\epsilon^{-1}\lambda^2m^2\int_0^1\mathrm{d}z[F(z)-5(1-z)]\ln\left[\frac{m^2F(z)}{4\pi\mu^2}\right]$$
$$+\tfrac{1}{12}\epsilon^{-1}\lambda^2p^2(-\tfrac{23}{72}+\tfrac{1}{6}\gamma_\mathrm{E})+\tfrac{1}{12}\epsilon^{-1}\lambda^2m^2(\tfrac{9}{8}+\tfrac{3}{2}\gamma_\mathrm{E}), \qquad (10.83)$$

where $\lambda\equiv g^2/(4\pi)^3$ and, (eq. (10.42)), $F(z)=1-z(1-z)p^2/m^2$. Of special interest in eq. (10.83) are the $\epsilon^{-1}\ln F(z)$ terms. Because they are not polynomials in p^2, their Fourier transforms are not delta functions in coordinate space. Thus, they cannot be cancelled by local counterterms. Such terms *must* cancel in the sum over graphs, or our renormalization procedure fails at two loops.

Figure 10.6(a.ii) is expressed by treating the one-loop counterterm as a vertex. In minimal subtraction ($c_{m,1}=c_{\phi,1}=0$ in eq. (10.46)) we find

$$-\mathrm{i}\bar{\gamma}_{(a.\mathrm{ii})} = -\mathrm{i}(-\mathrm{i}g)^2\mu^{2\epsilon}\int\frac{\mathrm{d}^nk}{(2\pi)^n}\left[\frac{\mathrm{i}^2}{(k^2-m^2+\mathrm{i}\epsilon)^2}\right]\left[\frac{\mathrm{i}}{(p-k)^2-m^2+\mathrm{i}\epsilon}\right]$$
$$\times[\tfrac{1}{2}\mathrm{i}g^2(4\pi)^{-3}\epsilon^{-1}](m^2-\tfrac{1}{6}k^2). \qquad (10.84)$$

To evaluate this expression it is useful to rewrite the counterterm using

$$m^2 - \tfrac{1}{6}k^2 = \tfrac{5}{6}m^2 - \tfrac{1}{6}(k^2 - m^2). \quad (10.85)$$

The resulting terms in eq. (10.84) can be evaluated using Feynman parameterization. The result from the momentum integral is

$$-\mathrm{i}\bar{\gamma}_{(a.\mathrm{ii})} = \tfrac{1}{2}\epsilon^{-1}\lambda^2 m^2\Big\{\tfrac{5}{6}\Gamma(\epsilon)\int_0^1 \mathrm{d}z(1-z)\Big[\frac{m^2F(z)}{4\pi\mu^2}\Big]^{-\epsilon} + \tfrac{1}{6}\Gamma(\epsilon-1)\int_0^1 \mathrm{d}z\,F(z)\Big[\frac{m^2F(z)}{4\pi\mu^2}\Big]^{-\epsilon}\Big\}, \quad (10.86)$$

where $F(z)$ is again given by (10.42). The singular terms in (10.86) are easy to derive:

$$-\mathrm{i}\bar{\gamma}_{(a.\mathrm{ii})}{}^{(\mathrm{pole})} = \tfrac{1}{24}\epsilon^{-2}\lambda^2(\tfrac{1}{3}p^2 + 3m^2) + \tfrac{1}{12}\epsilon^{-1}\lambda^2 m^2\int_0^1 \mathrm{d}z[F(z) - 5(1-z)]\ln\Big[\frac{m^2F(z)}{4\pi\mu^2}\Big] + \tfrac{1}{72}\epsilon^{-1}\lambda^2p^2(1-\gamma_{\mathrm{E}}) - \tfrac{1}{12}\epsilon^{-1}\lambda^2m^2(1+\tfrac{3}{2}\gamma_{\mathrm{E}}). \quad (10.87)$$

Comparison with eq. (10.83) shows that the $\ln F$ terms really do cancel in the sum, and

$$-\mathrm{i}\bar{\gamma}_{(a)}{}^{(\mathrm{pole})} = -\mathrm{i}(\bar{\gamma}_{(a.\mathrm{i})} + \bar{\gamma}_{(a.\mathrm{ii})})^{(\mathrm{pole})} = \lambda^2p^2(\tfrac{1}{144}\epsilon^{-2} - \tfrac{11}{864}\epsilon^{-1}) + \lambda^2m^2(\tfrac{1}{16}\epsilon^{-2} + \tfrac{1}{96}\epsilon^{-1}). \quad (10.88\mathrm{a})$$

The evaluation of Figs. 10.6(b) is only marginally more complicated, and gives

$$-\mathrm{i}\bar{\gamma}_{(b)}{}^{(\mathrm{pole})} = \lambda^2p^2(-\tfrac{1}{24}\epsilon^{-2} + \tfrac{1}{36}\epsilon^{-1}) + \lambda^2m^2(\tfrac{1}{4}\epsilon^{-2} - \tfrac{1}{4}\epsilon^{-1}). \quad (10.88\mathrm{b})$$

Again, the answer is a polynomial in p^2, and hence consistent with renormalization. The complete two-loop pole term in Γ_2 calculated from $\mathscr{L}_{\mathrm{R}}{}^{(1)}$ is thus

$$\Gamma_2{}^{(\mathrm{pole})} = -\mathrm{i}[\bar{\gamma}_{(a)} + \bar{\gamma}_{(b)}]^{(\mathrm{pole})} = \lambda^2p^2(-\tfrac{5}{144}\epsilon^{-2} + \tfrac{13}{864}\epsilon^{-1}) + \lambda^2m^2(\tfrac{5}{16}\epsilon^{-2} - \tfrac{23}{96}\epsilon^{-1}). \quad (10.89)$$

Two-loop renormalization constants

From eqs. (10.89) and (10.77) we can read off the two-loop renormalization constants

$$z_{\phi,2} = \lambda^2(\tfrac{5}{144}\epsilon^{-2} - \tfrac{13}{864}\epsilon^{-1}),$$
$$(z_\phi + z_m)_2 = \lambda^2(\tfrac{5}{16}\epsilon^{-2} - \tfrac{23}{96}\epsilon^{-1}). \quad (10.90)$$

For completeness, we shall also quote the result (Macfarlane & Woo 1974) for the two-loop coupling constant renormalization, which may be found from diagrams like those in Fig. 10.6(d):

$$\delta g_2 = \lambda^2 g(\tfrac{5}{16}\epsilon^{-2} - \tfrac{23}{96}\epsilon^{-1}). \tag{10.91}$$

(The equality of coefficients in δg_2 and $(z_\phi + z_m)_2$ is not fortuitous; see the erratum of Macfarlane & Woo (1974).) The momentum-independence of these renormalization constants shows that the iterative program may be carried out consistently to two loops for $\phi^3{}_6$, and that we may construct $\mathcal{L}_R{}^{(2)}$, which generates a finite perturbation series to two loops.

All the MS renormalization constants are independent of μ and of the renormalization mass to two loops. This may be understood as follows (Collins & Macfarlane 1974). Since the renormalization constants are dimensionless and independent of momenta, they can depend only on the ratio μ/m_R. But μ can enter only logarithmically, through the expansion of μ^ϵ. The presence of a factor $\ln(\mu/m_R)$ in Z_i would mean that Z_i diverges in the zero-mass limit. In minimal subtraction, however, Z_i is defined by ultraviolet poles only, and these cannot depend on the propagator masses at all, since they come from regions where all loop momenta are arbitrarily large. Thus the renormalization constants remain independent of particle masses to all orders in minimal subtraction schemes. In general, this is not the case in momentum subtraction schemes.

The nature of renormalization proofs

Space does not allow a full proof here of renormalizability for $\phi^3{}_6$ and other renormalizable theories. We shall content ourselves with giving an idea of the issues involved. As we shall see, the proof of renormalizability is essentially a generalization of the two-loop procedure we have just described.

For the following discussion, there is no reason to limit ourselves to $\phi^3{}_6$, so we may consider a general $(n-1)$-loop renormalized Lagrange density of the form

$$\mathcal{L}_R{}^{(n-1)}(\phi_R, \partial_\mu\phi_R) = \mathcal{L}_{cl}(\phi_R, \partial_\mu\phi_R) + \sum_i O_i(\phi_R, \partial_\mu\phi_R)[\zeta_i{}^{(n-1)} - 1]. \tag{10.92}$$

$\mathcal{L}_{cl}$ is a classical density involving some collection of (not necessarily scalar) fields, denoted collectively by ϕ_R. The $O_i(\phi_R, \partial_\mu\phi_R)$ are local operators, and the $\zeta_i{}^{(n-1)}$ are monomials in the renormalization constants, as in eq. (10.61), calculated to order $n-1$. In perturbation theory, each

$O_i(\phi_R, \partial_\mu\phi_R)$ corresponds to a vertex:

$$O_i(\phi_R, \partial_\mu\phi_R) \to i(2\pi)^4\delta^4(\textstyle\sum_j p_j{}^\mu)V_i(\{p_i\}), \tag{10.93}$$

where $\{p_i\}$ is the set of momenta of the lines attached to the vertex. We assume that $\mathscr{L}_R{}^{(n-1)}$ generates a perturbation theory that is finite up to $n-1$ loops, and see what happens at n loops.

To begin, we compare the set of graphs generated by the full density, eq. (10.92), with the perturbation series generated by the classical part of the renormalized density alone. They differ, of course, by the set of all graphs with at least one counterterm.

Consider then, the n-loop 1PI graph $G^{(n)}$ generated from $\mathscr{L}_{cl}(\phi_R, \partial_\mu\phi_R)$. Let us define some terminology and notation.

(i) A 1PI subgraph γ of $G^{(n)}$ with superficial degree of divergence $\omega(\gamma) \geqslant 0$ is called a *renormalization part* of $G^{(n)}$.

(ii) Two subgraphs γ_i and γ_j are said to be *disjoint* if they have no lines or vertices in common.

(iii) Consider a set $u = \{\gamma_i\}$ of disjoint 1PI subgraphs γ_i of $G^{(n)}$. Then

$$\frac{G^{(n)}}{u}\prod_i v_i \tag{10.94}$$

will denote the expression that is found by replacing each subgraph γ_i in u by the *local* function v_i. What we have in mind is the replacement of renormalization parts by local counterterms.

Now, for each graph $G^{(n)}$ without counterterms, we define a new quantity $\bar{G}^{(n)}$, in which all counterterms corresponding to the renormalization parts of $G^{(n)}$ have been taken into account. $\bar{G}^{(n)}$ will be referred to as the 'internally subtracted' version of the graph $G^{(n)}$. It is defined iteratively:

$$\bar{G}^{(1)} = G^{(1)}, \tag{10.95a}$$

$$\bar{G}^{(m)} = G^{(m)} + \sum_u \frac{G^{(n)}}{u}\prod_i t(\bar{\gamma}_i), \tag{10.95b}$$

where the sum is over all sets u of *disjoint proper* renormalization parts of $G^{(m)}$. In eq. (10.95b), each proper renormalization part has been replaced by a local vertex $t(\bar{\gamma}_i)$, which is defined as the 'divergent part' of $\bar{\gamma}_i$. In dimensional regularization, this includes all the pole terms in $t(\bar{\gamma}_i)$. Its finite part depends on the renormalization scheme. Since each γ_i is a 1PI subgraph of $G^{(n)}$, it must be a graph of fewer than n loops. As a result, we may assume that the $t(\bar{\gamma}_i)$ are already known. The quantities $\bar{\gamma}_a$, $\bar{\gamma}_b$ of Fig. 10.6 are precisely the two-loop forms of eq. (10.95), applied to the self-energy graphs.

Renormalization theorems

Using the foregoing terminology, the renormalization program can be summarized in two theorems, originated by Bogoliubov, Parasiuk and Hepp (Bogoliubov & Parasiuk 1957, Bogoliubov & Shirkov r1980, Hepp 1966). Referring to these authors, the approach commonly goes by the designation BPH.

(i) The first theorem relates the internally subtracted graphs $\bar{G}^{(n)}$ to the $(n-1)$-loop renormalized Lagrange density (10.92).

Let $\{\Gamma_k{}^{(n)}\}$ be the set of all n-loop 1PI graphs with k external lines, derived from $\mathscr{L}_\mathrm{R}{}^{(n-1)}$. Then we have

$$\sum \bar{G}_k{}^{(n)} = \sum \Gamma_k{}^{(n)}. \tag{10.96}$$

That is, the sum of all graphs generated from $\mathscr{L}_\mathrm{R}{}^{(n-1)}$ including counterterms is the same as the sum of all internally subtracted n-loop graphs defined by eqs. (10.95). This was rather trivially the case at two loops in $\phi^3{}_6$, but more generally requires combinatoric arguments, found in Bogoliubov & Shirkov (r1980) and Collins (r1984). The main consequence of this theorem is that n-loop divergences may be discussed in terms of the internally subtracted graphs $\bar{G}^{(n)}$.

(ii) The second theorem is concerned with the nature of the remaining divergences in the $\bar{G}^{(n)}$ themselves. It states that the divergent part of $\bar{G}^{(n)}$ in eq. (10.95) is *also* a polynomial in the external momenta of $\bar{G}^{(n)}$, of order no greater than $\omega(\bar{G}^{(n)})$, the superficial degree of divergence of $G^{(n)}$. We may thus define a set of *local* counterterms to cancel the remaining divergences at n loops, and construct a Lagrange density $\mathscr{L}_\mathrm{R}{}^{(n)}$ that generates a finite perturbation series to n loops. Proofs of this theorem may be found in Hepp (1966), Zimmermann (r1970) and Collins (r1984).

Taken together, the BPH theorems show that, by adding a finite number of counterterms, it is possible to make any theory with a local classical Lagrangian ultraviolet finite to an arbitrary number of loops. In renormalizable theories, the superficial degree of divergence is bounded independently of the order, which means (by the second theorem) that the dimensions of counterterm operators are also bounded independently of the order. Since there are only a finite number of local operators of dimension less than any fixed number, the number of counterterms necessary to make a renormalizable theory finite to all orders is limited. By contrast, the lack of a bound on $\omega(G)$ in nonrenormalizable theories means we expect the latter to require an infinite number of counterterms. It is worth emphasizing, however, that even nonrenormalizable theories can be made finite, order-by-order, in perturbation theory. It is just that they will require an infinite number of physical parameters, one for each counterterm, to define their perturbation series.

Finally, we may note that the definition of internally subtracted graphs, eq. (10.95b), is related to a subtle point in the second theorem. This is known as the problem of *overlapping divergences*, and is illustrated at two loops by Fig. 10.7. Because the internal momenta of the overlapping subgraphs λ and μ may both be large, but on different scales, it is not obvious that a single, disjoint subtraction for each will result in an internally subtracted graph with only polynomial divergences. In fact, it does, as eq. (10.88a, b) shows. The generalization of this result to all orders is an important part of the theorem.

BPHZ renormalization

Before going on, we should add a few more words about the BPH theorems and their extension. First, the proof of theorem (ii) given by Hepp (1966) is not in terms of dimensional regularization. It is a general belief that this gap in the literature is not serious, and we shall take that point of view here. Another point concerns the definition of the operator t. The definition introduced by Bogoliubov & Parasiuk (1957) relies on the polynomial form of the divergent part of an internally subtracted graph. Then t may be defined in terms of a Taylor expansion. In a theory where counterterms involve at most two derivatives, for instance, we may define

$$\begin{aligned} \mathrm{t}\bar{G}(p_1, \ldots, p_n) &= \bar{G}(0, \ldots, 0) + \sum_i p_{i,\mu}(\partial/\partial p_{i,\mu})\bar{G}(p_1, \ldots, p_n)|_{p^\mu=0} \\ &\quad + \tfrac{1}{2}\sum_{i,j} p_{i,\mu}p_{j,\nu}(\partial^2/\partial p_{i,\mu}\partial p_{j,\nu})\bar{G}(p_1, \ldots, p_n)|_{p^\mu=0}. \end{aligned} \tag{10.97}$$

In an elegant extension of the BPH approach, Zimmermann (r1970) shows that it is possible to let the derivatives in eq. (10.97) act on the *integrands*

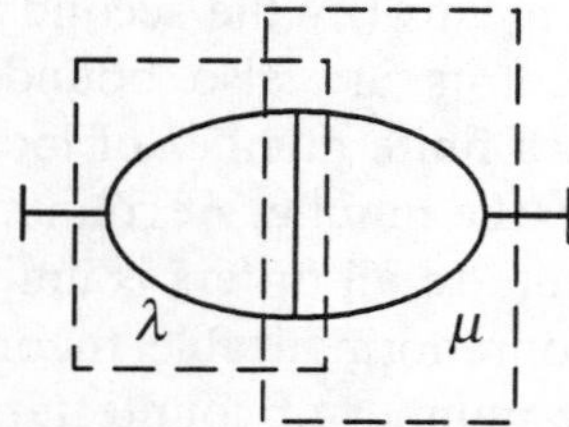

Fig. 10.7 Example of a diagram with overlapping divergences.

of the Feynman integrals of $\bar{G}$. If the different terms that make up the quantity $(1 - \mathrm{t})\bar{G}^{(n)}$ are combined before integration, the resulting integrals are actually finite *without* regularization. This regularization-free formulation is usually referred to as BPHZ renormalization, and is in contrast with procedures based on dimensional and other regularizations. Although BPHZ has many advantages, we shall find that schemes based on dimensional regularization are better adapted to gauge theories, whose renormalization we discuss in the next chapter.

10.5 Introduction to the renormalization group

We have seen that Green functions in the renormalized perturbation series are functions of the scale μ, both explicitly and through the renormalized coupling and mass, even though physical quantities are the same when computed via the renormalized or bare Lagrangian. The S-matrix, on the other hand, is independent of μ, and can, in principle, be computed in terms of only bare quantities. Then, for any physical observable $S(p_i, g_0, m_0)$ that is a function of physical particle momenta $p_i{}^\mu$, we have

$$(\mu\,\mathrm{d}/\mathrm{d}\mu)S(p_i, g_0, m_0)|_{g_0,m_0} = 0. \qquad (10.98)$$

Here S may be an S-matrix element, a cross section, or some direct functional of these quantities. Re-expressing $S(p_i, g_0, m_0)$ in terms of renormalized quantities, this means

$$(\mu\,\mathrm{d}/\mathrm{d}\mu)S(p_i, g_\mathrm{R}, m_\mathrm{R}, \mu)|_{g_0,m_0} = 0. \qquad (10.99)$$

The chain rule allows us to rewrite eq. (10.99) as

$$[(\mu\,\partial/\partial\mu)|_{g_\mathrm{R},m_\mathrm{R}} + \beta(g_\mathrm{R}, m_\mathrm{R})(\partial/\partial g_\mathrm{R})|_{\mu,m_\mathrm{R}} - \gamma_m(g_\mathrm{R}, m_\mathrm{R})(m_\mathrm{R}\,\partial/\partial m_\mathrm{R})|_{\mu,g_\mathrm{R}}]S(p_i, g_\mathrm{R}, m_\mathrm{R}, \mu) = 0, \qquad (10.100)$$

where renormalized quantities are now held fixed for the derivatives. The new functions β and γ_m are defined by

$$\beta(g_\mathrm{R}, m_\mathrm{R}) = (\mu\,\partial/\partial\mu)g_\mathrm{R}(\mu)|_{g_0,m_0} \qquad (10.101a)$$

$$\gamma_m(g_\mathrm{R}, m_\mathrm{R}) = -(1/2m_\mathrm{R}{}^2)(\mu\,\partial/\partial\mu)m_\mathrm{R}{}^2(\mu)|_{g_0,m_0}. \qquad (10.101b)$$

The advantage of eq. (10.100) over eq. (10.99) is that all derivatives with bare quantities held fixed have been absorbed into the 'universal' functions β and γ_m, which are the same for every physical quantity. Since (10.100) is an equation for a quantity that is analytic at $\epsilon = 0$, we expect the functions β and γ_m to be analytic as well (Weinberg 1973b). More detailed arguments in related situations may be found in Itzykson & Zuber (r1980, Section 13.2.2) and Callan (r1981). To make eq. (10.100) useful we still have to compute these functions.

Equation (10.100) is known as the *renormalization group equation*. The group may be thought of as all possible shifts in the value of μ.

The functions β and γ_m are computed from the renormalization constants by using the multiplicative renormalization of the mass and coupling, eqs. (10.48b) and (10.56). Then we have

$$\beta(g_R, m_R) = g_0(\mu\partial/\partial\mu)(\mu^\epsilon Z_g)^{-1}|_{g_0,m_0}, \tag{10.102a}$$

$$\gamma_m(g_R, m_R) = (m_0{}^2/2m_R{}^2)(\mu\partial/\partial\mu)Z_m{}^{-1}|_{g_0,m_0}. \tag{10.102b}$$

In general, evaluating the derivatives requires us to invert eqs. (10.48b) and (10.56) for g_R and m_R in terms of g_0, m_0 and μ and then take the indicated derivatives. The process is much simpler, however, when we use a minimal or other subtraction scheme in which the renormalization constants are independent of m_R and μ ('t Hooft 1973, Weinberg 1973b, Collins & Macfarlane 1974).

Beta function with minimal subtraction

To find the beta function with minimal subtraction, we apply the operation $\mu^{1-\epsilon}(\partial/\partial\mu)|_{g_0}$ to eq. (10.56), $g_0 = Z_g g_R \mu^\epsilon$. This gives

$$0 = \epsilon(g_R Z_g) + (\mu\partial/\partial\mu)(g_R Z_g)|_{g_0}. \tag{10.103}$$

Now we expand eq. (10.103) in powers of ϵ^{-1}, using the general form of the renormalization constant in minimal subtraction:

$$g_R Z_g = g_R\left[1 + \sum_{n=1}^{\infty} Z_{g,n}(g_R)\epsilon^{-n}\right] \equiv g_R + \sum_{n=1}^{\infty} a_n(g_R)\epsilon^{-n}. \tag{10.104}$$

where each $a_n(g_R) = g_R Z_{g,n}$ is an infinite series in the renormalized coupling. The result is

$$\epsilon\left(g_R + \sum_{n=1}^{\infty} a_n\epsilon^{-n}\right) = -(\mu\partial g_R/\partial\mu)|_{g_0}\left\{1 + \sum_{n=1}^{\infty}[\mathrm{d}a_n(g_R)/\mathrm{d}g_R]\epsilon^{-n}\right\}. \tag{10.105}$$

Every power of ϵ^{-1} must cancel, and we may solve this equation for $\beta = (\mu\partial/\partial\mu)g_R$. Since we are interested in solving eq. (10.100) at $\epsilon = 0$, we restrict ourselves to solutions in which $\beta(g, \epsilon)$ is a finite series in ϵ, $\beta(g, \epsilon) = \sum_{i=0}^{M}\beta_k\epsilon^k$ (Gross r1981). Then we find that all β_k, $k > 1$, vanish identically and the solution is simply

$$\beta(g_R, \epsilon) = -\epsilon g_R - a_1(g_R) + g_R(\mathrm{d}a_1/\mathrm{d}g_R) = -\epsilon g_R + g_R{}^2(\mathrm{d}Z_{g,1}/\mathrm{d}g_R). \tag{10.106}$$

The rest of the coefficients a_n may be computed from the recursion relation

$$g_R^2(d/dg_R)(a_{n+1}/g_R) = [-a_1(g_R) + g_R(da_1/dg_R)]da_n/dg_R. \quad (10.107)$$

From this form, we see that not only the beta function, but also the rest of the renormalization constant Z_g, is completely determined by the coefficient of the single pole in Z_g. (Of course, beyond one loop, the single pole is the hardest to calculate.) The function γ_m can be treated in just the same way.

We can now easily compute the beta functions for the scalar theories that we have considered so far. For example, applying eq. (10.106) to $\phi^3{}_6$ we find, by inspection of Z_g at one loop, eq. (10.58),

$$\beta(g, n) = -(3 - n/2)g - 3g^3/4(4\pi)^3. \quad (10.108)$$

A short calculation, suggested as an exercise, gives for $\phi^4{}_4$ with coupling g,

$$\beta(g, n) = -(2 - n/2)g + 3g^2/(4\pi)^2. \quad (10.109)$$

The qualitative difference between eqs. (10.108) and (10.109) is in the sign of the second term, which is negative for $\phi^3{}_6$ and positive for $\phi^4{}_4$. This will result in an important difference in the nature of solutions to the renormalization group equation.

Solution of the renormalization group equation

Let us assume for definiteness that S in eq. (10.100) has mass dimension $-\omega$, so that we may write

$$S(p_i{\cdot}p_j/\mu^2, m_R{}^2(\mu)/\mu^2, g_R(\mu), \mu) = \mu^{-\omega}s(p_i{\cdot}p_j/\mu^2, m_R{}^2(\mu)/\mu^2, g_R(\mu)), \quad (10.110)$$

for some dimensionless function, s. The renormalization group equation (10.100) is a single first-order homogeneous partial differential equation acting in the space of arguments of S. It may be thought of as specifying surfaces of constant values of the function. In this way, it can be used to relate changes in momenta to changes in other parameters.

To see how this comes about, we introduce a parameter t to vary the momentum scale:

$$p_i{}^\mu(t) = e^{-t}p_i{}^\mu, \quad (10.111)$$

for all i and μ. (Note that $p_i{}^2(t) = e^{-2t}p_i{}^2$, so that the masses of external particles change under this scaling.) Then since S has dimension $-\omega$,

$$[(\mu\partial/\partial\mu)|_{g_R,m_R,t} - \partial/\partial t|_{\mu,m_R,g_R} + (m_R\partial/\partial m_R)|_{\mu,g_R,t} + \omega]S(e^{-2t}p_i{\cdot}p_j/\mu^2, m_R{}^2(\mu)/\mu^2, g_R(\mu), \mu) = 0. \quad (10.112)$$

Together with eq. (10.100), this gives

$$[\partial/\partial t|_{\mu,m_R,g_R} + \beta(\partial/\partial g_R)|_{\mu,m_R,t} - (\gamma_m + 1)(m_R\,\partial/\partial m_R)|_{\mu,g_R,t} - \omega] \times S(e^{-2t}p_i\cdot p_j/\mu^2,\, m_R{}^2(\mu)/\mu^2,\, g_R(\mu),\, \mu) = 0. \quad (10.113)$$

To solve eq. (10.113) is to find surfaces of constant S. Solutions may be constructed by introducing two new quantities, called the *effective* or *running* coupling and mass, denoted respectively by $\bar{g}(g_R, t)$ and $\bar{m}(g_R, t)$. The effective coupling and mass obey the following simple equations and boundary conditions:

$$\partial\bar{g}(g_R, t)/\partial t = \beta(\bar{g}(g_R, t)), \quad \bar{g}(g_R, 0) = g_R, \quad (10.114a)$$

$$\partial\bar{m}(g_R, t)/\partial t = -[1 + \gamma_m(\bar{g}(g_R, t))]\bar{m}(g_R, t), \quad \bar{m}(g_R, 0) = m_R. \quad (10.114b)$$

The lack of direct dependence of either of these functions on the renormalized mass is a special property of minimal and other mass-independent subtraction schemes, where, as we have seen, both β and γ_m depend only on g_R. Since eq. (10.113) holds for all values of m_R and g_R, we can write (denoting $\bar{m}(t)$ for $\bar{m}(g_R, t)$, etc.)

$$\{\partial/\partial t|_{\mu,m,g} + \beta[\partial/\partial\bar{g}(t)]|_{\mu,m,t} - (\gamma_m + 1)[\bar{m}(t)\partial/\partial\bar{m}(t)]|_{\mu,g,t} - \omega\} \times S(e^{-2t}p_i\cdot p_j/\mu^2,\, \bar{m}^2(t)/\mu^2,\, \bar{g}(t),\, \mu) = 0, \quad (10.115)$$

or simply

$$(d/dt - \omega)S(e^{-2t}p_i\cdot p_j/\mu^2,\, \bar{m}^2(t)/\mu^2,\, \bar{g}(t),\, \mu) = 0. \quad (10.116)$$

This equation, with a total derivative in the scale parameter t, specifies the surfaces of constant S by

$$S(p_i\cdot p_j/(e^{2t}\mu^2),\, \bar{m}^2(t)/\mu^2,\, \bar{g}(t),\, e^t\mu) = S(p_i\cdot p_j/\mu^2,\, m_R{}^2/\mu^2,\, g_R,\, \mu). \quad (10.117a)$$

Defining $\mu' \equiv e^t\mu$, we can rewrite this result as

$$S(p_i\cdot p_j/\mu'^2,\, m_R{}^2(\mu')/\mu'^2,\, g_R(\mu'),\, \mu') = S(p_i\cdot p_j/\mu^2,\, m_R{}^2(\mu)/\mu^2,\, g_R(\mu),\, \mu), \quad (10.117b)$$

where we have revived the explicit dependence of g_R and m_R on the renormalization scale μ. Invoking the fundamental result that physical quantities are independent of μ, we can identify the running coupling and mass with the renormalized coupling and mass for the new renormalization scale $\mu' = e^t\mu$:

$$\begin{aligned} g_R(\mu') &= \bar{g}(g_R, \ln(\mu'/\mu)), \\ m_R(\mu') &= (\mu'/\mu)\bar{m}(g_R, \ln(\mu'/\mu)). \end{aligned} \quad (10.117c)$$

Note that in the equalities (10.117), the external masses, p_i^2, are once again all physical.

What is the use of eqs. (10.117)? All other things being equal, they are useful when we can pick the scale parameter t to make the effective coupling $\bar{g}(t)$ small, because then the perturbation series may converge rapidly. Let us look a little more carefully at this. The perturbative expansion of $S(p_i \cdot p_j/\mu'^2, m_R^2(\mu')/\mu'^2, g_R(\mu'), \mu')$ has the general form

$$S(p_i \cdot p_j/\mu'^2, m_R^2(\mu')/\mu'^2, g_R(\mu'), \mu')$$
$$= \sum_n S^{(n)}(p_i \cdot p_j/\mu'^2, m_R^2(\mu')/\mu'^2, \mu') g_R^{2n}(\mu'). \quad (10.118)$$

The convergence of this series depends on more than just the value of $g_R(\mu')$. The coefficients $S^{(n)}$ must also cooperate, and not become large for those values of μ' for which $g(\mu')$ is small.

We shall be interested in the high-energy behavior of field theories. In this limit, the coefficients in eq. (10.118) often tend to be large, because of the presence of large arguments $p_i \cdot p_j/\mu^2$. The occurrence of logarithms of external momenta in Feynman integrals as in eq. (10.66), is the rule rather than the exception. So, suppose that all the ratios $p_i \cdot p_j/\mu^2$ on the right-hand side of eq. (10.117b) happen to be of order Q^2/μ^2, where Q^2 is a large momentum. Then we might pick the scale parameter

$$t = \tfrac{1}{2}\ln(Q^2/\mu^2) \quad (10.119)$$

which sets $\mu' = Q$ so that all the ratios on the left-hand side of (10.117b) are of order unity. That is, to avoid large ratios involving momenta, it is natural to pick t to be large, of the order of the logarithm of these ratios. We must then ask what happens to the effective mass and coupling for large μ'. Of particular interest are theories whose couplings vanish at large t,

$$\lim_{t\to\infty} \bar{g}(g_R, t) = \lim_{\mu'\to\infty} g_R(\mu') = 0. \quad (10.120)$$

Such theories are said to be *asymptotically free*. As we shall see below, when a theory is asymptotically free,

$$\lim_{t\to\infty} \bar{m}(g_R, t) = 0. \quad (10.121)$$

For (10.117) to be useful, we must therefore *also* require that the zero-mass limit of all the $S^{(n)}(p_i \cdot p_j/\mu'^2, m_R^2/\mu'^2)$ be finite for m_R *and* for any fixed external invariant (such as $p_i^2 = m_i^2$ for an on-shell particle). This property is sometimes called *infrared safety* (see Chapter 12).

We shall now show that $\phi^3{}_6$ theory is asymptotically free while $\phi^4{}_4$ is not. Therefore, the perturbative high-energy behavior of $\phi^3{}_6$ is more accessible to the renormalization group than that of $\phi^4{}_4$.

Asymptotic behavior in $\phi^3{}_6$ and $\phi^4{}_4$

The asymptotic behavior of the coupling is determined by the solution to eq. (10.114a) with beta function given by eq. (10.108) for $\phi^3{}_6$ and by (10.109) for $\phi^4{}_4$. The solution to the equation is similar for the two. In the one-loop approximation in $\phi^3{}_6$,

$$\partial\bar{g}/\partial t = -\tfrac{3}{4}(4\pi)^{-3}\bar{g}^3, \tag{10.122}$$

which implies, using the boundary condition in (10.114a), that

$$\int_{g_R}^{\bar{g}(t)} d\bar{g}'/\bar{g}'^3 = -\tfrac{3}{4}(4\pi)^{-3}\int_0^t dt', \tag{10.123}$$

or

$$-\tfrac{1}{2}[1/\bar{g}^2(t) - 1/g_R{}^2] = -\tfrac{3}{4}(4\pi)^{-3}t, \tag{10.124}$$

which may be solved to give

$$g_R{}^2(e^t\mu) = \bar{g}^2(t) = g_R{}^2/(1 + \tfrac{3}{2}g_R{}^2(4\pi)^{-3}t). \tag{10.125}$$

Evidently, as t becomes large, the running coupling $\bar{g}^2(t)$ vanishes as t^{-1}, so that $\phi^3{}_6$ is indeed asymptotically free.

Exactly similar reasoning for $\phi^4{}_4$ gives the result

$$g_R{}^2(e^t\mu) = \bar{g}^2(t) = g_R{}^2/(1 - \tfrac{3}{2}g_R{}^2(2\pi)^{-2}t). \tag{10.126}$$

This behavior is qualitatively different, because between $t = 0$ and $t = \infty$ there is a singularity in the running coupling. On the other hand $\bar{g}^2(t)$ vanishes in $\phi^4{}_4$ as $t \to -\infty$. This is sometimes called infrared freedom. The behavior of these couplings and of their beta functions is shown in Figs. 10.8(*a*), (*b*) for the asymptotically-free and the infrared-free cases respectively.

Of course, all this has only been done to one loop. The result, however, will be unaffected by the addition of higher-loop corrections to the asymptotically-free beta function eq. (10.108), *if* we start from a renormalized coupling g_R that is sufficiently small, since the higher-order corrections are suppressed by powers of the coupling.

Fixed points

In general, values of the coupling at which the β-function vanish are known as 'fixed points' of the theory. The name is suggested by the differential equation (10.114a), since $\beta = 0$ means that the coupling is stationary with respect to the scaling variable. More formally, the solution to (10.114a) may be written as

$$\int_{g_R}^{\bar{g}(t)} \frac{dg'}{\beta(g')} = t. \tag{10.127}$$

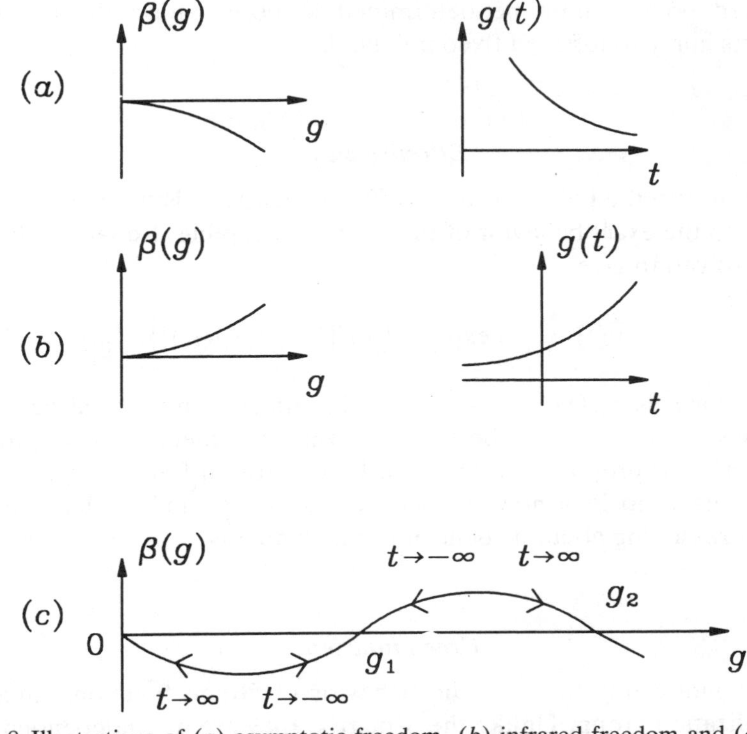

Fig. 10.8 Illustrations of (*a*) asymptotic freedom, (*b*) infrared freedom and (*c*) general beta function with ultraviolet and infrared fixed points.

So, if $\beta(g)$ vanishes at some value $g = g_0$, the corresponding t will be infinite. If we integrate eq. (10.114a), starting at some initial value g_R for increasing $|t|$, the behavior of the solution $\bar{g}(t)$ will depend on the sign of $\beta(g_R)$. For instance, for $\beta(g) < 0$, $\bar{g}(t)$ is less than g_R for positive t, and $\bar{g}(t)$ will decrease monotonically with increasing t all the way to $t = \infty$, where it attains the value of the next smallest fixed point, which is consequently known as an *ultraviolet fixed point*. This is the case for ${\phi^3}_6$, whose coupling approaches $g = 0$ as $t \to +\infty$. Alternately, if $\beta(g) > 0$, $\bar{g}(t)$ approaches the next smallest fixed point when $t \to -\infty$, as in ${\phi^4}_4$, whose fixed point at $g = 0$ is thus known as an *infrared fixed point*.

A more general example is illustrated by the model β-function and coupling of Fig. 10.7(*c*), which has several fixed points, including an ultraviolet fixed point at $g_0 = 0$. Note that our discussion of the asymptotic behavior of an asymptotically-free coupling applies only to the range where the initial renormalized coupling g_R lies between g_0 and g_1. For $g_R > g_1$, the coupling will approach another ultraviolet fixed point, g_2, as $t \to \infty$. The region $g_1 > g_R > g_0$ is referred to as the *region of attraction* of the ultraviolet fixed point g_0. In general, the extent of the region of attraction of a fixed point cannot be determined in perturbation theory. Similar comments apply to infrared fixed points.

Effective mass

We now turn to the behavior of the effective mass. This turns out to be less sensitive to the exact behavior of the effective coupling. Equation (10.114b) is easily solved to give

$$\bar{m}(t) = m_{\mathrm{R}} \exp\left\{- \int_0^t \mathrm{d}t'[1 + \gamma_m(\bar{g}(t'))]\right\}. \tag{10.128}$$

Thus, as long as $\gamma_m(t) < 1$ as $t \to \infty$, the effective mass vanishes in this limit. This should certainly be the case when the theory is asymptotically free, since γ_m is proportional to $\bar{g}^2(t)$. In an infrared-free theory, however, we really have no idea how γ_m behaves as $t \to \infty$, and we learn nothing from this reasoning about its behavior at high energy.

Green functions

It is also interesting to study the behavior of Green functions under the renormalization group. Unlike the S-matrix, they are not independent of μ, and they obey slightly different equations. For instance, 1PI functions are multiplicatively renormalized (eq. (10.68′)), while the *un*renormalized 1PI functions are μ-independent,

$$(\mu\, \mathrm{d}/\mathrm{d}\mu)\Gamma_{0,n} = 0. \tag{10.129}$$

We can derive a renormalization group equation that is analogous to eq. (10.100) (exercise 10.10):

$$[(\mu\, \partial/\partial\mu)|_{g_{\mathrm{R}},m_{\mathrm{R}}} + \beta(g_{\mathrm{R}})(\partial/\partial g_{\mathrm{R}})|_{\mu,m_{\mathrm{R}}} - \gamma_m(g_{\mathrm{R}})(m_{\mathrm{R}}\, \partial/\partial m_{\mathrm{R}})|_{\mu,g_{\mathrm{R}}} - n\gamma_\phi(g_{\mathrm{R}})]\Gamma_{\mathrm{R},n}(p_i, g_{\mathrm{R}}, m_{\mathrm{R}}, \mu) = 0. \tag{10.130}$$

Here γ_ϕ is called the *anomalous dimension* associated with the scalar field $\phi(x)$. It is given by

$$\gamma_\phi(g_{\mathrm{R}}) = \tfrac{1}{2}(\mu\, \mathrm{d}/\mathrm{d}\mu) \ln Z_\phi = \tfrac{1}{2}\beta(g_{\mathrm{R}})(\partial/\partial g_{\mathrm{R}}) \ln Z_\phi, \tag{10.131a}$$

where the second form applies to mass-independent renormalization schemes. The name 'anomalous dimension' is suggested by its role in the equation. As for $\beta(g)$ and $\gamma_m(g)$, the finiteness of the Γ_n implies that $\gamma_\phi(g)$ is also finite at $\epsilon = 0$. But the only finite term in the expansion of the right-hand side of eq. (1.131a) comes from the $-\epsilon g$ term in $\beta(g)$ (eq. (10.108) times the ϵ^{-1} term in Z_ϕ. Other terms must cancel. As a result, we have

$$\gamma_\phi(g) = -\tfrac{1}{2}(g\,\partial/\partial g)(\text{single pole term in } Z_\phi). \qquad (10.131\text{b})$$

Then, from the explicit values for the pole terms of Z_ϕ given in eqs. (10.46) and (10.90), we have for $\phi^3{}_6$,

$$\gamma_\phi(g) = \tfrac{1}{12}g^2(4\pi)^{-3} + \tfrac{13}{432}[g^2(4\pi)^{-3}]^2 + \mathrm{O}(g^6). \qquad (10.131\text{c})$$

Equation (10.130) may be solved in the same fashion as (10.100) to obtain

$$\Gamma_{\mathrm{R},n}(\mathrm{e}^{-2t}p_i\cdot p_j/\mu^2,\, \bar{m}^2(t)/\mu^2,\, \bar{g}(t))\exp\left[(4-n)t - n\int_0^t \mathrm{d}t'\,\gamma_\phi(t')\right]$$
$$= \Gamma_{\mathrm{R},n}(p_i\cdot p_j/\mu^2,\, m_\mathrm{R}{}^2/\mu^2,\, g_\mathrm{R}), \quad (10.132)$$

where $4-n$ is the mass dimension Γ_n and we have introduced the notation

$$\gamma_\phi(t') \equiv \gamma_\phi(\bar{g}(g_\mathrm{R}, t')). \qquad (10.133)$$

We have assumed, as above, a minimal subtraction scheme, in which the renormalization constants, and therefore the anomalous dimensions, are mass independent. Essentially identical reasoning may be applied to connected Green functions, to derive an equation similar to (10.130). This is suggested as exercise 10.10.

So far, we have only hinted at the practical uses of the renormalization group analysis. For now, we simply emphasize the light it sheds on the mass-scale dependence of the renormalized perturbation series. It may also be used to show *why* the basic theorems of Section 10.4 hold (Polchinski 1984, Warr 1988). Its practical power will become evident later, when we discuss high-energy behavior. First, however, we must discuss the renormalization of gauge theories.

Exercises

10.1 Verify eq. (10.8) explicitly for the two-point function in $\phi^3{}_4$ at one loop.

10.2 Calculate the tadpole diagram, Fig. 10.1(b), to get eq. (10.24), and evaluate the vacuum expectation value of the field to one loop, eq. (10.25).

10.3 Show that a theory with a four-fermion interaction is power-counting nonrenormalizable in four dimensions, but renormalizable in two. What pure scalar polynomial interactions are renormalizable in two and three dimensions?

10.4 (a) Identify and calculate, using dimensional regularization, the divergent one-loop graphs of $\phi^4{}_4$. (b) Construct mass and coupling constant counterterms that cancel the one-loop divergences of this theory. (c) Give expressions for the relevant renormalized two-point functions in a momentum subtraction and a minimal subtraction scheme.

10.5 Calculate the two-loop mass and wave function renormalization constants in $\phi^4{}_4$ using minimal subtraction.

10.6 Calculate the graphs of Fig. 10.6(b) in dimensional regularization near $n = 6$ with $m = 0$, and minimal subtraction, to order ϵ^{-2} and $\epsilon^{-1} \ln p^2$. Show that the nonlocal terms cancel (Macfarlane & Woo 1974).

10.7 Construct all local operators in $\phi^4{}_4$ of dimension equal to any integer M, thereby verifying that their number is finite.

10.8 (a) From eqs. (10.90) and (10.91) calculate the two-loop β-function in $\phi^3{}_6$. (b) Verify that the recursive identity (10.107) is satisfied to order $g_R{}^4$, that is, that the double pole in Z_ϕ at two loops may be derived from the one-loop β-function and Z_ϕ, without doing any two-loop calculations.

10.9 Derive eq. (10.109), the one-loop beta function for $\phi^4{}_4$.

10.10 Derive the analogue of (10.130) for connected Green functions.

11

Renormalization and unitarity of gauge theories

The renormalizability and unitarity of gauge theories are among the fundamental results of perturbative field theory. The renormalization procedure described in Chapter 10 may be applied to gauge theories in the same manner as to scalar theories. What is different about gauge theories is their disturbing mixture of physical and unphysical degrees of freedom.

The one-loop renormalization of gauge theories will suggest renormalized Lagrangians that incorporate classical symmetries through relations between renormalization constants. These relations may be extended to all orders in perturbation theory and imply the gauge invariance of the physical S-matrix, as well as its unitarity (Section 6.5),

$$\delta_{\alpha\beta} = \sum_{\gamma\,\mathrm{phys}} (S_{\mathrm{phys}})_{\alpha\gamma}(S_{\mathrm{phys}}{}^{\dagger})_{\gamma\beta}. \tag{11.1}$$

This program for proving unitarity works for QED, QCD and the rest of the standard model, but its applicability is not universal. The axial anomaly illustrates how a gauge symmetry may fail, through the breakdown of the classical symmetry in a quantum theory.

11.1 Gauge theories at one loop

Even at one loop, gauge theories show a rich structure. By now, the strategy is clear. We study the divergences of 1PI graphs, and construct the counterterms that define the renormalized perturbation series. We use multiplicative renormalization throughout.

From Section 8.4, the effective tree-level Lagrangian for an unbroken $SU(N)$ gauge theory, eq. (8.83), is given by

$$\begin{aligned}
&\mathscr{L}_{\mathrm{eff}}(g_{\mathrm{R}}\mu^{\epsilon}, m_{\mathrm{R}}, \lambda_{\mathrm{R}}) \\
&\quad = \bar{\psi}_{\mathrm{R}}(\mathrm{i}\not{\partial} - g_{\mathrm{R}}\mu^{\epsilon}\not{A}_{\mathrm{R},a}t_a{}^{(\mathrm{F})} - m_{\mathrm{R}})\psi_{\mathrm{R}} - \tfrac{1}{4}F_{\mu\nu,a}(A_{\mathrm{R}}, g_{\mathrm{R}}\mu^{\epsilon})F^{\mu\nu}{}_a(A_{\mathrm{R}}, g_{\mathrm{R}}\mu^{\epsilon}) \\
&\quad\quad - \tfrac{1}{2}\lambda_{\mathrm{R}}(\partial\cdot A_{\mathrm{R}})^2 - \bar{c}_{\mathrm{R},a}[(\partial_\mu)^2\delta_{ac} - g_{\mathrm{R}}\mu^{\epsilon}C_{abc}\partial_\mu A_{\mathrm{R}}{}^{\mu}{}_c]c_{\mathrm{R},b} \\
&\quad F^{\mu\nu}{}_a(A_{\mathrm{R}}, g_{\mathrm{R}}\mu^{\epsilon}) = \partial^\mu A_{\mathrm{R}}{}^{\nu}{}_a - \partial^\nu A_{\mathrm{R}}{}^{\mu}{}_a - g_{\mathrm{R}}\mu^{\epsilon}C_{abc}A_{\mathrm{R}}{}^{\mu}{}_b A_{\mathrm{R}}{}^{\nu}{}_c.
\end{aligned} \tag{11.2}$$

Here $\epsilon = 2 - \frac{1}{2}n$, where n is the dimension; the subscript R denotes renormalized quantities. Equation (11.2) is analogous to the classical Lagrangian of scalar theories.

The fermions are in the defining representation F of $SU(N)$, whose structure constants are denoted C_{abc}, normalized by $T(F) = \frac{1}{2}$, $T(\mathrm{A}) = N$ (eq. (8.87)). We denote the ghost and antighost fields by $c(x)$ and $\bar{c}(x)$, respectively, with covariant gauge-fixing; note that we allow for renormalization of the gauge-fixing parameter. For simplicity and to exhibit group structure, we will abuse the terminology a little and refer to any such $SU(N)$ theory as QCD. When the gauge group is $U(1)$ (QED), $C_{abc} = 0$, and we define $T(\mathrm{F}) = C_2(\mathrm{F}) = 1$. In this case, the ghost fields decouple even in covariant gauges (Section 7.3), and will be ignored.

Let us begin with those graphs that occur in both QED and QCD. At any number of loops, the superficial degree of divergence $\omega(G)$, eq. (10.35), is non-negative only for a limited number of 1PI functions. These are the fermion, vector and ghost self-energies, the fermion–vector and ghost–vector three-point functions, and the vector three- and four-point functions. They are all represented in Fig. 11.1, along with the four-ghost and two-ghost–two-gluon functions, which have $\omega(G) = 0$ but which are

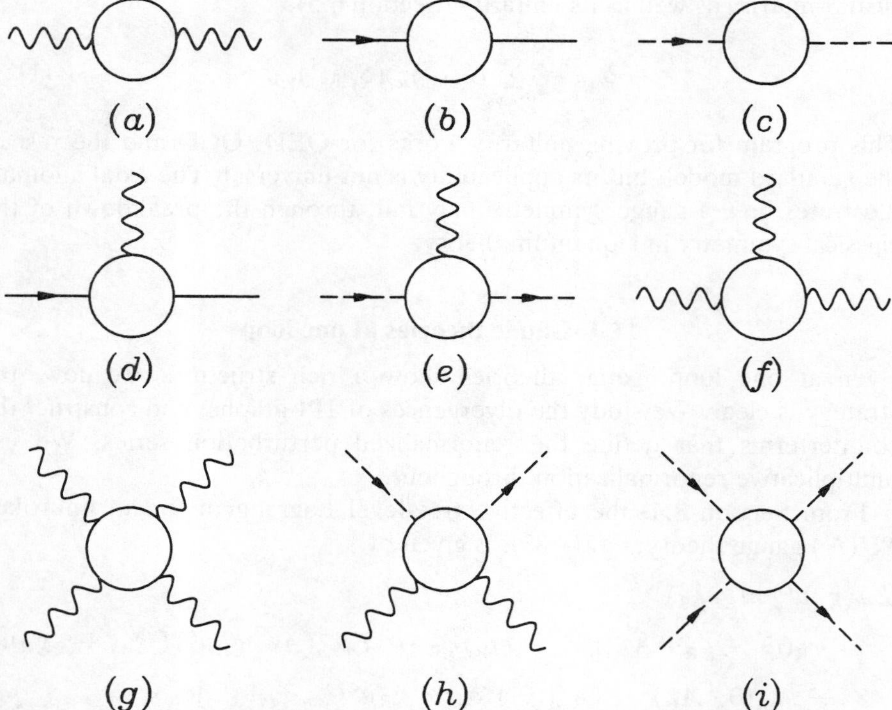

Fig. 11.1 1PI functions with non-negative superficial degree of divergence.

Table 11.1 *Coefficients of* $(\alpha/4\pi)\epsilon^{-1}$ *in the unbroken gauge theory renormalization constants*

Z_A	$C_2(\mathrm{A})[\frac{5}{3} + \frac{1}{2}(1 - \lambda^{-1})] - \frac{4}{3}T(\mathrm{F})n_f$
Z_c	$\frac{1}{2}C_2(\mathrm{A})[1 + \frac{1}{2}(1 - \lambda^{-1})]$
Z_ψ	$-C_2(\mathrm{F})\lambda^{-1}$
Z_1	$-C_2(\mathrm{A})[1 - \frac{1}{4}(1 - \lambda^{-1})] - C_2(\mathrm{F})\lambda^{-1}$
$Z_1{}'$	$C_2(\mathrm{A})[\frac{2}{3} + \frac{3}{4}(1 - \lambda^{-1})] - \frac{4}{3}T(\mathrm{F})n_f$
$Z_1{}''$	$-\frac{1}{2}C_2(\mathrm{A})\lambda^{-1}$
Z_4	$-C_2(\mathrm{A})[\frac{1}{3} - (1 - \lambda^{-1})] - \frac{4}{3}T(\mathrm{F})n_f$
$Z_m Z_\psi$	$-C_2(\mathrm{F})[4 - (1 - \lambda^{-1})]$

actually convergent, because one power of momentum is associated with the outgoing ghost line (eq. (8.100)). Let us deal with the simplest divergent cases in turn. Because we are going to give a lot of details, it will be convenient to work in Feynman gauge, $\lambda = 1$. For completeness, we list in Table 11.1 the one-loop pole parts of the renormalization constants for arbitrary λ.

Fermion self-energy

The fermion self-energy, for which $\omega(G) = 1$, is shown in Fig. 11.2(a). From the Feynman rules of Sections 8.1 and 8.4, its contribution to Γ_2 in n dimensions is (in the notation $\Gamma_2 = \not{p} - m - \Sigma$),

$$\Sigma^{2(a)} = \mathrm{i}(2\pi)^{-n}\int \mathrm{d}^n k[-\mathrm{i}g\mu^\epsilon\gamma_\mu t_b{}^{(\mathrm{F})}]\mathrm{i}(\not{p} - \not{k} + m)[(p-k)^2 - m^2 + \mathrm{i}\epsilon]^{-1}$$
$$\times [-\mathrm{i}g\mu^\epsilon\gamma_\nu t_b{}^{(\mathrm{F})}](-\mathrm{i}g^{\mu\nu})(k^2 + \mathrm{i}\epsilon)^{-1}, \tag{11.3}$$

where we have rationalized the fermion propagator. As usual, both couplings and masses are understood to be the renormalized quantities that appear in the Lagrange density. Equation (11.3) may be simplified by using the group identity eq. (8.86), the n-dimensional Dirac identity (9.44), and Feynman parameterization, eq. (9.2). The result is

$$\Sigma^{2(a)} = -\mathrm{i}C_2(F)g^2\mu^{2\epsilon}\int_0^1 \mathrm{d}x(2\pi)^{-n}$$
$$\times \int \mathrm{d}^n k[(2-n)(\not{p} - \not{k}) + nm][k^2 - 2xk\cdot p + x(p^2 - m^2) + \mathrm{i}\epsilon]^{-2} \tag{11.4a}$$

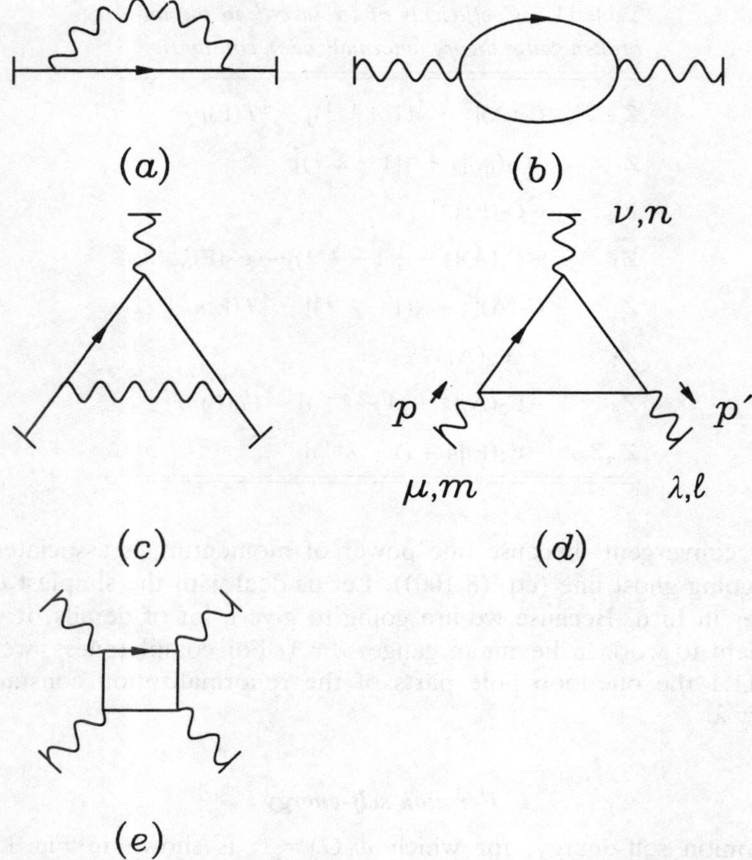

Fig. 11.2 One-loop diagrams that are superficially divergent in both QED and QCD.

The momentum integral is carried out by using (9.49b) and (9.51a),

$$\Sigma^{2(a)} = (\alpha/4\pi)C_2(F)(4\pi\mu^2/p^2)^{\epsilon}\Gamma(\epsilon) \times \int_0^1 \mathrm{d}x[(2-n)(1-x)\not{p} + nm][-x(1-x) + xm^2/p^2 - \mathrm{i}\epsilon]^{-\epsilon}, \tag{11.4b}$$

where $\alpha = g^2/4\pi$ is the (renormalized) fine-structure constant of the theory. The usual expansion about $\epsilon = 0$ then gives

$$\Sigma^{2(a)} = -(\alpha/2\pi)C_2(F)\{[\epsilon^{-1} + \ln(4\pi\mu^2/p^2) - \gamma_{\mathrm{E}}](\tfrac{1}{2}\not{p} - 2m) - \tfrac{1}{2}\not{p} + m + \int_0^1 \mathrm{d}x[\not{p}(1-x) - 2m]\ln[-x(1-x) + xm^2/p^2 - \mathrm{i}\epsilon]\}. \tag{11.5}$$

Aside from its matrix structure, eq. (11.5) is very similar to the $\phi^3{}_6$ self-energy, eq. (10.41). In particular, its divergences at $n = 4$ require both wave function and mass renormalization.

Fermion wave function and mass renormalization

To remove the divergences in the fermion self-energy at one loop, we must add two counterterms to the tree-level Lagrangian (11.2):

$$\bar{\psi}_R(\mathrm{i}\not{\partial})\psi_R[Z_\psi{}^{(1)} - 1], \quad Z_\psi{}^{(1)} = 1 - (\alpha/4\pi)C_2(F)(\epsilon^{-1} + c_{\psi,1}), \quad (11.6a)$$

and

$$-m_R\bar{\psi}_R\psi_R[Z_\psi{}^{(1)}Z_m{}^{(1)} - 1], \quad Z_m{}^{(1)} = 1 - (3\alpha/4\pi)C_2(F)(\epsilon^{-1} + c_{m,1}). \quad (11.6b)$$

As usual, c_m and c_ψ are defined by the renormalization scheme.

Schemes

The choice of an appropriate renormalization scheme depends on the theory, and often on the calculation to be carried out within that theory. For QCD, many calculations will be independent of particle masses. In this case, it is natural to choose minimal subtraction, since we shall not need to solve for m_R. We shall have occasion to follow this procedure in the final chapters. For many applications in QED, however, it is most natural to use our knowledge of the electron mass, and to choose an on-shell momentum subtraction, $m_R = m_P$ (Section 12.1). Here we leave the choice open.

Vacuum polarization

The vacuum polarization diagram, with $\omega(G) = 2$, is shown in Fig. 11.2(*b*). Its contribution to the photon 1PI two-point function is (not forgetting the minus sign for the fermion loop)

$$\begin{aligned} &-\mathrm{i}\pi_{\alpha\beta,ab}(q)^{2(b)} \\ &\quad = \mathrm{i}(2\pi)^{-n}\int \mathrm{d}^n k \,\mathrm{tr}\left\{[-\mathrm{i}g\mu^\epsilon t_a{}^{(F)}\gamma_\alpha](\not{k} + m)[-\mathrm{i}g\mu^\epsilon t_b{}^{(F)}\gamma_\beta](-\not{q} + \not{k} + m)\right\} \\ &\qquad \times \mathrm{i}(k^2 - m^2 + \mathrm{i}\epsilon)^{-1}\mathrm{i}[(q-k)^2 - m^2 + \mathrm{i}\epsilon]^{-1}. \end{aligned} \quad (11.7)$$

$\pi^{2(b)}$ is the truncated Green function, and $-\mathrm{i}\pi^{2(b)}$ the corresponding contribution to Γ_2. The only real difference from eq. (11.3) is in the group and

Dirac traces. The group identity eq. (8.86) along with Feynman parameterization now gives

$$-\mathrm{i}\pi_{\alpha\beta,ab}(q)^{2(b)} = \mathrm{i}\delta_{ab}T(\mathrm{F})g^2\mu^{2\epsilon}\int_0^1 \mathrm{d}x(2\pi)^{-n}\int \mathrm{d}^n k\, N_{\alpha\beta}(q, k) \times (k^2 - 2xk\cdot q + xq^2 - m^2 + \mathrm{i}\epsilon)^{-2}, \quad (11.8)$$

where the Dirac trace is given by

$$N_{\alpha\beta}(k, q) = 8k_\alpha k_\beta - 4(k_\alpha q_\beta + q_\alpha k_\beta) + 4g_{\alpha\beta}(k\cdot q - k^2 + m^2). \quad (11.9)$$

The evaluation of these integrals is routine, using eqs. (9.51a, b):

$$\begin{aligned}-\mathrm{i}\pi_{\alpha\beta,ab}(q)^{2(b)} &= -(q^2 g_{\alpha\beta} - q_\alpha q_\beta)\delta_{ab}T(\mathrm{F})(2\alpha/\pi)(4\pi\mu^2)^\epsilon\Gamma(\epsilon) \\ &\quad\times \int_0^1 \mathrm{d}x\, x(1-x)[m^2 - x(1-x)q^2 - \mathrm{i}\epsilon]^{-\epsilon} \\ &= -(q^2 g_{\alpha\beta} - q_\alpha q_\beta)\delta_{ab}T(\mathrm{F})(\alpha/3\pi)\epsilon^{-1} + \mathrm{O}(\epsilon^0). \end{aligned} \quad (11.10)$$

Vector wave function renormalization

The divergence in eq. (11.10) is cancelled by the counterterm

$$-\tfrac{1}{4}(\partial_\alpha A_{\mathrm{R}\beta,a} - \partial_\beta A_{\mathrm{R}\alpha,a})(\partial^\alpha A_{\mathrm{R}}{}^\beta{}_a - \partial^\beta A_{\mathrm{R}}{}^\alpha{}_a)[\zeta_A{}^{(1)} - 1], \quad (11.11)$$

with

$$\zeta_A{}^{(1)} = 1 - (\alpha/3\pi)T(\mathrm{F})(\epsilon^{-1} + \text{finite remainder}). \quad (11.12\mathrm{a})$$

In QED, this is the only photon self-energy graph, and

$$Z_A{}^{(1)} = \zeta_A{}^{(1)} = 1 - (\alpha/3\pi)(\epsilon^{-1} + c_{A,1}) \qquad (\text{QED}). \quad (11.12\mathrm{b})$$

The counterterm, eq. (11.11) is 'transverse':

$$q^\alpha(q^2 g_{\alpha\beta} - q_\alpha q_\beta) = (q^2 g_{\alpha\beta} - q_\alpha q_\beta)q^\beta = 0. \quad (11.13)$$

No counterterm need be associated with the gauge fixing or 'longitudinal' part of the quadratic Lagrangian. Nevertheless, the gauge fixing parameter, λ_{R}, must be renormalized, that is, we must put

$$Z_{\mathrm{R}}\lambda_{\mathrm{R}} = \lambda_0, \quad (11.14)$$

which follows from the requirement that the bare Lagrangian equal the renormalized Lagrangian. In the absence of a counterterm for gauge-fixing, this means that

$$\begin{aligned}\lambda_0(\partial\cdot A_0)^2 &= \lambda_{\mathrm{R}}(\partial\cdot A_{\mathrm{R}})^2 + \lambda_{\mathrm{R}}(\partial\cdot A_{\mathrm{R}})^2(Z_A Z_\lambda - 1) \\ &= \lambda_{\mathrm{R}}(\partial\cdot A_{\mathrm{R}})^2, \end{aligned} \quad (11.15)$$

or

$$Z_\lambda = Z_A{}^{-1}. \tag{11.16}$$

Thus far, of course, we have verified this identity only to one loop.

Fermion–vector vertex

The vertex correction, Fig. 11.2(c), has $\omega(G) = 0$. Its overall structure is a little more complicated than that of the self-energy graphs:

$$\begin{aligned} \Gamma_{\mu,a}{}^{2(c)} = &-\mathrm{i}(2\pi)^{-n}\int \mathrm{d}^n k[(p+k)^2 - m^2 + \mathrm{i}\epsilon]^{-1}[(p'+k)^2 - m^2 + \mathrm{i}\epsilon]^{-1} \\ &\times (k^2 + \mathrm{i}\epsilon)^{-1}(-\mathrm{i}g^{\alpha'\alpha})[-\mathrm{i}g\mu^\epsilon t_b{}^{(\mathrm{F})}\gamma_{\alpha'}]\mathrm{i}(\not{p}' + \not{k} + m) \\ &\times [-\mathrm{i}g\mu^\epsilon t_a{}^{(\mathrm{F})}\gamma_\mu]\mathrm{i}(\not{p} + \not{k} + m)[-\mathrm{i}g\mu^\epsilon t_b{}^{(\mathrm{F})}\gamma_\alpha], \end{aligned} \tag{11.17}$$

but its divergent part is almost as easy to identify. Only the term quadratic in k is ultraviolet divergent at $n = 4$. We may use the third Dirac identity in eqs. (9.44) with $n = 4$ to simplify the Dirac matrices, since corrections will be finite at $\epsilon = 0$. Combining denominators as usual, we find

$$\begin{aligned} \Gamma_{\mu,a}{}^{2(c)(\mathrm{pole})} = &-4\mathrm{i}g^3[C_2(\mathrm{F}) - \tfrac{1}{2}N]t_a{}^{(\mathrm{F})}\int_0^1 \mathrm{d}x \int_0^{1-x} \mathrm{d}x'(2\pi)^{-n}\int \mathrm{d}^n k\,(\not{k}\gamma_\mu\not{k}) \\ &\times [k^2 + 2k\cdot(xp + x'p') + xp^2 + x'p'^2 \\ &- (x+x')m^2 + \mathrm{i}\epsilon]^{-3}, \end{aligned} \tag{11.18}$$

where we have also employed the color identities of Section 8.4, commuting $t_b{}^{(\mathrm{F})}$ past $t_a{}^{(\mathrm{F})}$. The momentum integral is now easy to carry out, using (9.51b), and the pole term is

$$\Gamma_{\mu,a}{}^{2(c)(\mathrm{pole})} = -g\gamma_\mu t_a{}^{(\mathrm{F})}(\alpha/4\pi)[C_2(\mathrm{F}) - \tfrac{1}{2}N]\epsilon^{-1}. \tag{11.19}$$

This is cancelled by a counterterm

$$-g_\mathrm{R}\mu^\epsilon \bar{\psi}_\mathrm{R} A_{\mathrm{R}\mu,a}\gamma^\mu t_a{}^{(\mathrm{F})}\psi_\mathrm{R}[\zeta_1{}^{(1)} - 1], \tag{11.20}$$

where

$$\zeta_1{}^{(1)} = 1 - (\alpha/4\pi)[C_2(\mathrm{F}) - \tfrac{1}{2}N](\epsilon^{-1} + \text{finite remainder}). \tag{11.21a}$$

Charge renormalization in QED

In QED, Fig. 11.2(c) is the only one-loop 1PI correction to the photon–fermion coupling, so the corresponding renormalization constant is

$$Z_1{}^{(1)} = 1 - (\alpha/4\pi)(\epsilon^{-1} + c_{1,1}) \quad (\mathrm{QED}). \tag{11.21b}$$

For the coupling e_R, we write $Z_e e_R = e_0$. The relation between Z_e and Z_1 is

$$Z_1 = Z_A{}^{1/2} Z_\psi Z_e \quad \text{(QED)}. \tag{11.22}$$

Comparing eq. (11.21b) with eq. (11.6a) for Z_ψ in QED, we find (at least to one loop) that, in any scheme in which $c_{\psi,1} = c_{1,1}$,

$$Z_1 = Z_\psi \quad \text{(QED)}. \tag{11.23a}$$

Equivalently, from eq. (11.12b) for Z_A in QED, we find

$$Z_e = Z_A{}^{-1/2} = 1 + (\alpha/6\pi)(\epsilon^{-1} + c_{A,1}) + \mathrm{O}(\alpha^2) \quad \text{(QED)}. \tag{11.23b}$$

Equations (11.23) express the renormalization constant *Ward identity* of QED (Ward 1950). From Table 11.1, we see that it is satisfied in all covariant gauges at one loop.

Fermion loop diagrams

There are two more superficially divergent diagrams that appear in both QED and QCD. These are the fermion loop contributions to the three- and four-vector vertex functions, shown in Fig. 11.2(*d*), (*e*). Consider first the three-vector diagrams, with $\omega(G) = 1$. There are two distinguishable graphs of this kind, which differ by the exchange of the two external vectors, labelled p and p', or equivalently, by reversing the sense of the fermion arrow. Their total contribution to the 1PI vertex is

$$\begin{aligned}\Gamma_{\mu\nu\lambda,mnl}(p, p')^{2(d)} = \frac{\mathrm{i}g^3\mu^{3\epsilon}}{(2\pi)^n} \int \mathrm{d}^n k[(p + k)^2 - m^2 + \mathrm{i}\epsilon]^{-1} \\ \times [(p' + k)^2 - m^2 + \mathrm{i}\epsilon]^{-1} \\ \times (k^2 - m^2 + \mathrm{i}\epsilon)^{-1} N_{\mu\nu\lambda,mnl},\end{aligned} \tag{11.24}$$

where $N_{\mu\nu\lambda,mnl}$ contains the group and Dirac traces:

$$\begin{aligned}N_{\mu\nu\lambda,mnl} &= \mathrm{tr}\,[t_n{}^{(\mathrm{F})} t_m{}^{(\mathrm{F})} t_l{}^{(\mathrm{F})}]\,\mathrm{tr}\,[\gamma_\nu(\not{p} + \not{k} + m)\gamma_\mu(\not{k} + m)\gamma_\lambda(\not{p}' + \not{k} + m)] \\ &\quad + \mathrm{tr}\,[t_n{}^{(\mathrm{F})} t_l{}^{(\mathrm{F})} t_m{}^{(\mathrm{F})}]\,\mathrm{tr}\,[\gamma_\nu(-\not{p}' - \not{k} + m)\gamma_\lambda(-\not{k} + m)\gamma_\mu(-\not{p} - \not{k} + m)].\end{aligned} \tag{11.25}$$

Furry's theorem

The two Dirac traces in eq. (11.25) can be related by the so-called charge conjugation matrix C, defined to satisfy

$$C\gamma_\mu C^{-1} = -\gamma_\mu{}^{\mathrm{t}}. \tag{11.26}$$

(In the Dirac or Weyl representations, we may take $C = \mathrm{i}\gamma^2\gamma^0$; see Appendix D.) Inserting $C^{-1}C$ everywhere in the second Dirac trace, we find

$$\begin{aligned}
N_{\mu\nu\lambda,mnl} \\
&= \mathrm{tr}\,[t_n{}^{(\mathrm{F})}t_m{}^{(\mathrm{F})}t_l{}^{(\mathrm{F})}]\,\mathrm{tr}\,[\gamma_\nu(\not p + \not k + m)\gamma_\mu(\not k + m)\gamma_\lambda(\not p' + \not k + m)] \\
&\quad - \mathrm{tr}\,[t_n{}^{(\mathrm{F})}t_l{}^{(\mathrm{F})}t_m{}^{(\mathrm{F})}]\,\mathrm{tr}\,[(\not p + \not k + m)\gamma_\mu(\not k + m)\gamma_\lambda(\not p' + \not k + m)\gamma_\nu] \\
&= \mathrm{tr}\,\{t_n{}^{(\mathrm{F})}[t_m{}^{(\mathrm{F})}, t_l{}^{(\mathrm{F})}]\}\,\mathrm{tr}\,[\gamma_\nu(\not p + \not k + m)\gamma_\mu(\not k + m)\gamma_\lambda(\not p' + \not k + m)].
\end{aligned} \tag{11.27}$$

Being proportional to a commutator, eq. (11.27) vanishes in QED. It is easy to generalize this procedure to show that for any odd number of external photons the graphs with fermion arrows in opposite directions cancel (Fig. 11.3). The two graphs shown there are distinguishable for more than two external photons, so all photon $(2n + 1)$-point functions vanish identically. This is known as Furry's theorem (Furry 1937). Hence, no counterterm involving three photon fields is necessary. This is gratifying, since no three-photon terms appear in the bare Lagrangian of QED.

Three-gluon vertex

In QCD, the ultraviolet pole in eq. (11.24) is found by isolating numerator terms with two and three powers of k^μ. Evaluating the group trace, and using Feynman parameterization and eq. (11.27) for the numerator factor, we obtain

$$\begin{aligned}
\Gamma_{\mu\nu\lambda,mnl}(p, p')^{2(d)(\mathrm{pole})} \\
&= -C_{mln}T(\mathrm{F})g^3 2\int_0^1 \mathrm{d}x \int_0^{1-x} \mathrm{d}x' \\
&\quad \times (2\pi)^{-n}\int \mathrm{d}^n k\,\mathrm{tr}\,(\gamma_\nu \not k \gamma_\mu \not k \gamma_\lambda \not k + \gamma_\nu \not p \gamma_\mu \not k \gamma_\lambda \not k + \gamma_\nu \not k \gamma_\mu \not k \gamma_\lambda \not p') \\
&\quad \times [k^2 + 2k\cdot(xp + x'p') + xp^2 + x'p'^2 - m^2 + \mathrm{i}\epsilon]^{-3},
\end{aligned} \tag{11.28}$$

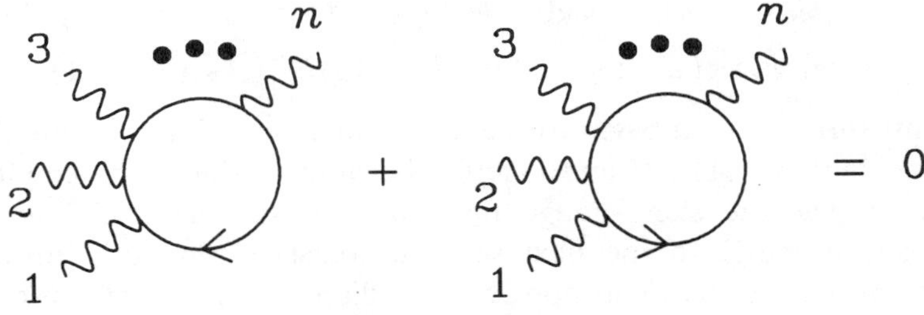

Fig. 11.3 Furry's theorem.

which yields

$$\Gamma_{\mu\nu\lambda,mnl}(p, p')^{2(d)(\text{pole})}$$
$$= \mathrm{i}gC_{mln}[(p + p')_\nu g_{\mu\lambda} + (-2p' + p)_\mu g_{\nu\lambda} + (p' - 2p)_\lambda g_{\mu\nu}](\alpha/3\pi)T(\mathrm{F})\epsilon^{-1}. \tag{11.29}$$

By comparing eq. (11.29) with the lowest-order three-gluon vertex, eq. (8.98), we see that it can be cancelled by a three-gluon counterterm of precisely the expected form.

Four-vector vertex

We shall not evaluate the four-vector vertex Fig. 11.2(*e*) explicitly. We will see in the next section, however, that in QED it, and every other fermion loop integral, is transverse. That is, if $M_{\alpha,\mu,\ldots}(q)$ represents a fermion loop and photon q^α attaches at index α, then $q^\alpha M_{\alpha,\mu,\ldots} = 0$. As a result, any loop must couple to its external photons through the field strength $F_{\mu\nu} = \partial_\mu A_\nu - \partial_\nu A_\mu$. Since this vertex has a derivative in every term, the four-photon tensor is proportional to four powers of external momenta. This means that, after combining diagrams, the true loop integral has dimension -4, rather than 0. So, there is also no necessity for a four-photon counterterm in QED.

Renormalized Lagrangian for QED

We have now exhausted all the one-loop graphs that appear in QED with non-negative superficial degree of divergence, and have generated the three relevant counterterms. The renormalized Lagrangian suggested by the one-loop calculations is (re-inserting subscripts to denote renormalized quantities),

$$\begin{aligned}\mathcal{L}_\mathrm{R}(e_\mathrm{R}, m_\mathrm{R}, \lambda_\mathrm{R}) \\ &= \bar{\psi}_\mathrm{R}(\mathrm{i}\not{\partial} - e_\mathrm{R}\mu^\epsilon \not{A}_\mathrm{R} - m_\mathrm{R})\psi_\mathrm{R} - \tfrac{1}{4}F_{\mu\nu}(A_\mathrm{R})F^{\mu\nu}(A_\mathrm{R}) \\ &\quad - \tfrac{1}{2}\lambda_\mathrm{R}(\partial \cdot A_\mathrm{R})^2 + \bar{\psi}_\mathrm{R}(\mathrm{i}\not{\partial})\psi_\mathrm{R}(Z_\psi - 1) - \bar{\psi}_\mathrm{R}(e_\mathrm{R}\mu^\epsilon \not{A}_\mathrm{R})\psi_\mathrm{R}(Z_1 - 1) \\ &\quad - m_\mathrm{R}\bar{\psi}_\mathrm{R}\psi_\mathrm{R}(Z_\psi Z_m - 1) - \tfrac{1}{4}F_{\mu\nu}(A_\mathrm{R})F^{\mu\nu}(A_\mathrm{R})(Z_A - 1).\end{aligned} \tag{11.30}$$

From this form, we can begin to see the significance of the renormalization constant identities (11.23) and (11.16); because of them, the counterterm Lagrangian has the same gauge invariance as the classical Lagrangian, before gauge-fixing. In the next section, we show that this ensures the unitarity of the theory. First, however, we discuss some divergent one-loop diagrams that are specific to QCD.

Gluon self-energy

The gluon self-energy graphs of pure Yang–Mills theory are shown in Fig. 11.4(a)–(c). All are absent in QED, in which the photon self-energy is given entirely by the fermion loop. Their ultraviolet poles are not difficult to evaluate. In the Feynman gauge, the contribution of Fig. 11.4(a) to Γ_2 is

$$-\mathrm{i}\pi_{\alpha\beta,ab}{}^{4(a)} = \tfrac{1}{2}\mathrm{i}g^2\mu^{2\epsilon}C_{acd}C_{dcb}(2\pi)^{-n}\int \mathrm{d}^n k(k^2 + \mathrm{i}\epsilon)^{-1}$$
$$\times [(q - k)^2 + \mathrm{i}\epsilon]^{-1} N_{\alpha\beta}(q, k) \qquad (11.31)$$

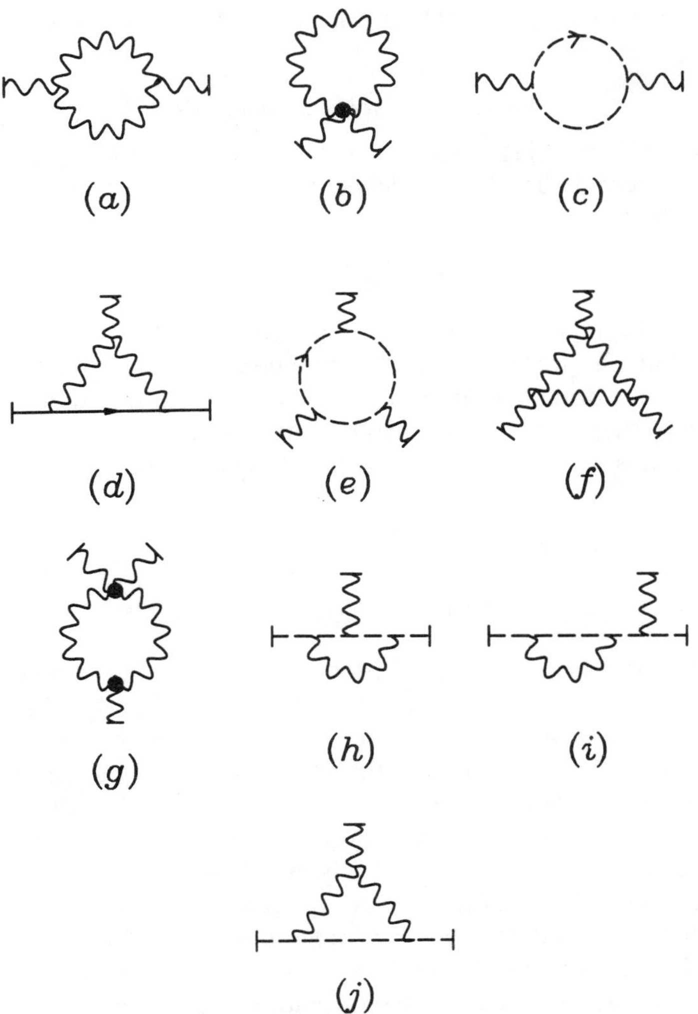

Fig. 11.4 Additional divergent one-loop diagrams in QCD.

where, as usual, we have separated the numerator momentum factor

$$N_{\alpha\beta}(q, k) = k_\alpha k_\beta(4n - 6) + q_\alpha q_\beta(n - 6) + (q_\alpha k_\beta + k_\alpha q_\beta)(-2n + 3) + g_{\alpha\beta}(2k^2 - 2q \cdot k + 5q^2). \tag{11.32}$$

The evaluation of eq. (11.31) is once again a straightforward application of eqs. (9.51a, b), while the group factor is (eqs. (8.85)–(8.87)),

$$C_{acd}C_{dcb} = -\mathrm{tr}\,(t_a{}^{(\mathrm{A})}, t_b{}^{(\mathrm{A})}) = -\,\delta_{ab}T(\mathrm{A}) = -\delta_{ab}N. \tag{11.33}$$

The pole term is then found to be

$$-\mathrm{i}\pi_{\alpha\beta,ab}{}^{4(a)(\mathrm{pole})} = (\alpha/8\pi)\delta_{ab}N(\tfrac{19}{6}q^2 g_{\alpha\beta} - \tfrac{11}{3}q_\alpha q_\beta)\epsilon^{-1}. \tag{11.34}$$

Unlike the fermion loop, eq. (11.10), this is not transverse by itself.

Vanishing of scaleless integrals

Turning to the remaining graphs, the 'tadpole' graph of Fig. 11.4(*b*) is a special case, because it is a massless integral that is also independent of external momenta:

$$-\mathrm{i}\pi_{\alpha\beta,ab}{}^{4(b)} = \delta_{ab}t_{\alpha\beta}g^2\mu^{2\epsilon}\int \mathrm{d}^n k(k^2 + \mathrm{i}\epsilon)^{-1}, \tag{11.35a}$$

where $t_{\alpha\beta}$ is the momentum-independent tensor of the four-gluon vertex, eq. (8.99). Such an integral is so far ambiguous in dimensional regularization. While it is ultraviolet finite for $n < 2$, it is also 'infrared' divergent for $n < 2$ because there are no masses to smooth out the integral as k vanishes. To deal with this problem, we may artificially split the integral as follows:

$$\int \mathrm{d}^n k(k^2 + \mathrm{i}\epsilon)^{-1} = \int \mathrm{d}^n k(k^2 - M^2 + \mathrm{i}\epsilon)^{-1} - M^2\int \mathrm{d}^n k(k^2 - M^2 + \mathrm{i}\epsilon)^{-1}(k^2 + \mathrm{i}\epsilon)^{-1}, \tag{11.35b}$$

with M an arbitrary mass. This procedure separates the infrared and ultraviolet divergences, so that the first term on the right-hand side is well defined for $n < 2$, and the second for $2 < n < 4$. Suppose we define the integral by analytically continuing the two terms from the (different) regions in which they are defined. Consistency requires that the answer be independent of M^2. Yet, by dimensional analysis, each integral must be proportional to a power of M^2. The only possibility is that the two analytic continuations cancel. A short calculation, using eq. (9.49b), shows explicitly that they do. We thus *define* the integral in eq. (11.35) to be zero, and $\pi^{4(b)} = 0$. This result may be extended to any integral that has neither internal masses nor external momenta, and any such integral vanishes in dimensional regularization ('t Hooft & Veltman r1973, Leibbrandt r1975).

Ghost loop

Finally, Fig. 11.4(c) gives

$$-\mathrm{i}\pi_{\alpha\beta,ab}{}^{4(c)} = \mathrm{i}g^2 N\delta_{ab}(2\pi)^{-n}\int \mathrm{d}^n k\,(k^2+\mathrm{i}\epsilon)^{-1}[(q-k)^2+\mathrm{i}\epsilon]^{-1}M_{\alpha\beta}(q,k),$$

$$M_{\alpha\beta}(q,k) = -k_\alpha q_\beta + k_\alpha k_\beta. \tag{11.36}$$

The minus sign associated with the ghost loop is the only subtlety associated with this graph, and we find

$$-\mathrm{i}\pi_{\alpha\beta,ab}{}^{4(c)(\mathrm{pole})} = (\alpha/8\pi)N\delta_{ab}(\tfrac{1}{3}q_\alpha q_\beta + \tfrac{1}{6}q^2 g_{\alpha\beta})\epsilon^{-1}. \tag{11.37}$$

Combining eq. (11.37) with (11.34) and (11.10), we find that the full one-loop gluon self-energy is transverse;

$$-\mathrm{i}\pi_{\alpha\beta,ab}{}^{(\mathrm{pole})} = (\alpha/3\pi)\delta_{ab}[\tfrac{5}{4}N - T(\mathrm{F})n_f](q^2 g_{\alpha\beta} - q_\alpha q_\beta)\epsilon^{-1}, \tag{11.38}$$

where n_f is the number of fermion flavors, each of which gives the same contribution in QCD.

Gluon wave function renormalization

The divergence in the gluon self-energy suggests a full one-loop counterterm analogous to eq. (11.11),

$$-\tfrac{1}{4}(\partial_\alpha A_{\mathrm{R}\beta,a} - \partial_\beta A_{\mathrm{R}\alpha,a})(\partial^\alpha A_{\mathrm{R}}{}^\beta{}_a - \partial^\beta A_{\mathrm{R}}{}^\alpha{}_a)[Z_A{}^{(1)} - 1], \tag{11.39}$$

where now

$$Z_A{}^{(1)} = 1 + (\alpha/3\pi)[\tfrac{5}{4}N - T(\mathrm{F})n_f]\epsilon^{-1} \quad (\mathrm{QCD}). \tag{11.40}$$

The transversality of the pole term means that, just as in QED, the gauge-fixing term remains unrenormalized in QCD, at least at one loop:

$$Z_\lambda = Z_A{}^{-1} \quad (\mathrm{QCD}). \tag{11.41}$$

Fermion–gluon vertex

The characteristically nonabelian contribution to the one-loop fermion–vector vertex is shown in Fig. 11.4(d). Its pole contribution is easy to calculate,

$$\Gamma_{\mu,a}{}^{4(d)(\mathrm{pole})} = -gt_a{}^{(\mathrm{F})}\gamma_\mu(3\alpha/8\pi)N\epsilon^{-1}. \tag{11.42}$$

In QCD, the full pole term consisting of eqs. (11.42) and (11.19) may then be cancelled by a counterterm of the form shown in eq. (11.20), with renormalization constant

$$Z_1 = 1 - (\alpha/4\pi)[N + C_2(\mathrm{F})](\epsilon^{-1} + c_{1,1}) + \mathrm{O}(\alpha^2) \quad (\mathrm{QCD}). \tag{11.43}$$

We are now ready to close this series of calculations with a discussion of the QCD analogue of the Ward identities eqs. (11.23). To organize our thoughts, however, it is useful to write down first the expected form for the full renormalized Lagrangian.

Renormalized Lagrangian in QCD

Let us assume that the renormalization process does not require the addition of any counterterms of a form not already in the bare Lagrangian, and that the transversality condition, eq. (11.41), generalizes to all orders. Then, expanding the field strengths in terms of fields $A_{\mu,a}$, the renormalized Lagrangian is given by the straightforward, although somewhat unwieldy, form

$$\begin{aligned}\mathscr{L}_{\mathrm{R}}(g, m, \lambda) = \mathscr{L}_{\mathrm{eff}}(g, m, \lambda) & \\ &+ \bar{\psi}(\mathrm{i}\not{\partial})\psi(Z_\psi - 1) - g\mu^\epsilon \bar{\psi}\not{A}_a t_a{}^{(\mathrm{F})}\psi(Z_1 - 1) \\ &- m\bar{\psi}\psi(Z_m Z_\psi - 1) \\ &- \tfrac{1}{4}(\partial_\alpha A_{\beta,a} - \partial_\beta A_{\alpha,a})(\partial^\alpha A^\beta{}_a - \partial^\beta A^\alpha{}_a)(Z_A - 1) \\ &+ \tfrac{1}{2} g\mu^\epsilon C_{abc} A^\alpha{}_b A^\beta{}_c(\partial_\alpha A_{\beta,a} - \partial_\beta A_{\alpha,a})(Z_1{}' - 1) \\ &- \tfrac{1}{4} g^2 \mu^{2\epsilon} C_{ebc} C_{eb'c'} A^\alpha{}_b A^\beta{}_c A_{\alpha,b'} A_{\beta,c'}(Z_4 - 1) \\ &- \bar{c}_a(\partial_\mu)^2 c_a(Z_c - 1) + g\mu^\epsilon C_{abc}\bar{c}_a \partial_\mu A^\mu{}_b c_c(Z_1{}'' - 1),\end{aligned} \tag{11.44}$$

where we have associated renormalization constants with each term in the expansion of $\mathscr{L}_{\mathrm{eff}}$, eq. (11.2). Once again, we have dropped the subscript R, but all fields and couplings are to be understood as renormalized. A priori, the constants Z_1, $Z_1{}'$, $Z_1{}''$ and Z_4 associated with the interaction terms have no connection with each other. On the other hand, the analogy with QED suggests that we might be able to find a scheme in which

$$Z_1 = Z_\psi Z_A{}^{1/2} Z_g, \quad Z_1{}' = Z_A{}^{3/2} Z_g, \quad Z_1{}'' = Z_c Z_A{}^{1/2} Z_g, \quad Z_4 = Z_A{}^2 Z_g{}^2, \tag{11.45a}$$

with the *same* coupling renormalization constant Z_g. This in turn requires

$$Z_1/Z_\psi = Z_1{}'/Z_A = Z_1{}''/Z_c = (Z_4/Z_A)^{1/2}. \tag{11.45b}$$

These are the renormalization constant identities of QCD, often called *Taylor–Slavnov identities* (Taylor 1971, Slavnov 1972). An obvious advantage of these identities is that they allow us to use the same coupling g_{R} for all the different vertices of the renormalized perturbation series, something that is not at all obvious. Even more importantly, they preserve classical gauge structure in the quantum Lagrangian, and lead eventually to unitarity (Section 11.3).

Charge renormalization at the gluon–fermion vertex

We have calculated the factors necessary to determine $Z_g^{(1)}$ from the fermion–gluon vertex in the nonabelian theory, using the relation

$$\begin{aligned} Z_g &= (Z_1/Z_\psi)Z_A^{-1/2} \\ &= 1 - (\alpha/4\pi)[\tfrac{11}{6}N - \tfrac{2}{3}T(\mathrm{F})n_f]\epsilon^{-1} + O(\alpha^2) \quad (\mathrm{QCD}), \end{aligned} \tag{11.46}$$

where we have used eqs. (11.6a), (11.40) and (11.43) for Z_ψ, Z_A and Z_1, respectively. One may check that Z_g found from this vertex at one loop is indeed the same as Z_g found from the other vertices of eq. (11.2) in minimal subtraction. The corrections involving the ghost–gluon diagrams, Figs. 11.4(h)–(j) are rather simple, and this calculation is suggested as an exercise, while the three-gluon graphs Fig. 11.4(e)–(g) are more challenging. Suffice it to say that there are no surprises.

Renormalization group for gauge theories

Renormalization group equations can be derived for gauge theories in just the same way as for scalar theories. In particular, observables, such as the S-matrix, obey exactly the same equation, (10.113), as before. A difference arises, however, with Green functions, since they depend on the gauge parameter λ_R (Section 8.1). Here it is convenient to introduce $\zeta_\mathrm{R} = \lambda_\mathrm{R}^{-1}$, and to treat ζ_R on the same footing as the renormalized coupling and mass. If we define

$$\delta(g_\mathrm{R}, \lambda_\mathrm{R}) = (\mu\partial/\partial\mu)\zeta_\mathrm{R}|_{g_0,m_0}, \tag{11.47}$$

we can easily derive the generalization to gauge theories of eq. (10.130) for 1PI functions:

$$\begin{aligned} [(\mu\partial/\partial\mu) + \beta(g_\mathrm{R})(\partial/\partial g_\mathrm{R}) - \gamma_m(g_\mathrm{R})(m_\mathrm{R}\partial/\partial m_\mathrm{R}) + \delta(g_\mathrm{R}, \lambda_\mathrm{R})(\partial/\partial\zeta_\mathrm{R}) \\ - n_A\gamma_A - n_f\gamma_\psi - n_c\gamma_c]\Gamma_{\mathrm{R},n}(g_\mathrm{R}, m_\mathrm{R}, \zeta_\mathrm{R}) = 0. \end{aligned} \tag{11.48}$$

The number of external vectors, fermions and ghosts are respectively n_A, n_f, n_c and their corresponding anomalous dimensions are γ_A, γ_f, γ_c, defined by analogy with (10.131): $\gamma_{A,\psi,c} = \frac{1}{2}(\mu\,\mathrm{d}/\mathrm{d}\mu)\ln Z_{A,\psi,c}$.

The functions $\beta(g_\mathrm{R})$ and $\gamma_m(g_\mathrm{R})$ (see eq. (10.101)) are independent of ζ_R, since they appear in the renormalization group equation, eq. (10.100) for the S-matrix. The anomalous dimensions of the fields, however, are in general gauge-dependent. Using the identities $Z_\lambda = Z_A^{-1}$, we can relate

$\delta(g_R, \zeta_R)$ to γ_A by

$$\delta(g_R, \zeta_R) \equiv (\mu\partial/\partial\mu)\zeta_R = -\zeta_R\mu(\partial \ln Z_A/\partial\mu) = -2\lambda_R{}^{-1}\gamma_A. \quad (11.49)$$

Note that in the Landau gauge, $\lambda_R \to \infty$, gauge-dependence drops out of the renormalization group equation for Green functions.

The beta function for QED and QCD

The one-loop beta functions for QED and QCD can be found from eqs. (11.23b) and (11.46) for Z_e and Z_g, and the general relation eq. (10.106). They are (Gross & Wilczek 1973, Politzer 1973, Gross r1981)

$$\begin{aligned} \beta(e, \epsilon) &= -\epsilon e + e(\alpha/3\pi) \quad \text{(QED)}, \\ \beta(g, \epsilon) &= -\epsilon g - g(\alpha/4\pi)[\tfrac{11}{3}N - \tfrac{4}{3}n_f T(\mathrm{F})] \quad \text{(QCD)}. \end{aligned} \quad (11.50)$$

Evidently, QCD is asymptotically free, while QED is not.

11.2 Renormalization and unitarity in QED

In this section, we shall prove the gauge invariance and unitarity of the physical S-matrix in quantum electrodynamics, to all orders in the renormalized perturbation theory. In this effort, the QED renormalization constant Ward identities, $Z_A = Z_\lambda{}^{-1}$ and $Z_1 = Z_\psi$, play an important role. We shall see how they imply relations between Green functions, often referred to as Ward identities as well, that reflect the underlying classical gauge invariance of the theory. We go on to prove that the renormalization-constant identities can be preserved to all orders. Finally, we shall see how Green function Ward identities are used to demonstrate gauge invariance and unitarity in the quantum theory.

Let us recall that eq. (9.86) shows that the perturbative T-matrix in any relativistic field theory satisfies the unitary relation

$$-\mathrm{i}(T - T^\dagger)_{\alpha\beta} = \textstyle\sum_{\bar\gamma}(T)_{\alpha\bar\gamma}(T^\dagger)_{\bar\gamma\beta},$$

given in eq. (9.74), where the sum is over all states found by cutting Feynman diagrams at each order. This result demonstrates unitarity for many theories, but for gauge theories it is not quite adequate. This is because a cut vector line, in Feynman gauge for example, gives $-g_{\mu\nu}\delta_+(q^2)$, rather than $P_{\mu\nu}(q)\delta_+(q^2)$, where $P_{\mu\nu}$ is the physical polarization tensor of eq. (8.59).

The decoupling of photon scalar polarizations is the key to both unitarity and gauge invariance for the S-matrix in QED. In Section 8.1, decoupling produced a gauge invariant tree-level S-matrix. In Section 8.2, it meant that tree-level inclusive cross sections could be calculated with either covariant ($-g_{\mu\nu}$) or physical ($P_{\mu\nu}$) polarization tensors. We see now that the latter result is not just a calculational convenience, but that it is necessary for unitarity.

As described in Section 8.2, the crucial decoupling of scalar photons followed at tree level from the Feynman identity, eq. (8.63), illustrated by Fig. 11.5(*a*). We now consider what becomes of this identity beyond tree level. Figure 11.5(*b*) shows two corrections, proportional to $Z_1 - 1$ and $Z_\psi - 1$, the photon–fermion charge counterterm and the fermion wave function counterterm, respectively. The graphical identity survives only if $Z_1 = Z_\psi$. This suggests that the renormalization-constant Ward identity is necessary for the decoupling of unphysical modes in the presence of loop corrections.

To implement this program, we shall assume that the Lagrange density has the form of eq. (11.30), with $Z_A = Z_\lambda^{-1}$ and $Z_1 = Z_\psi$. We shall then derive a set of Green function Ward identities, of which the Feynman identity, Fig. 11.5(*a*), is the lowest-order example. At this stage, our Green functions are regularized if necessary; we do not assume at the outset that they are finite in four dimensions. The next step is to show that QED can be renormalized in such a way that the renormalized-constant identities are satisfied order-by-order in perturbation theory, which implies that the graphical identities can be satisfied for ultraviolet finite Green functions at each order. It is from this result that we derive the gauge invariance and unitarity of the QED S-matrix in renormalized perturbation theory.

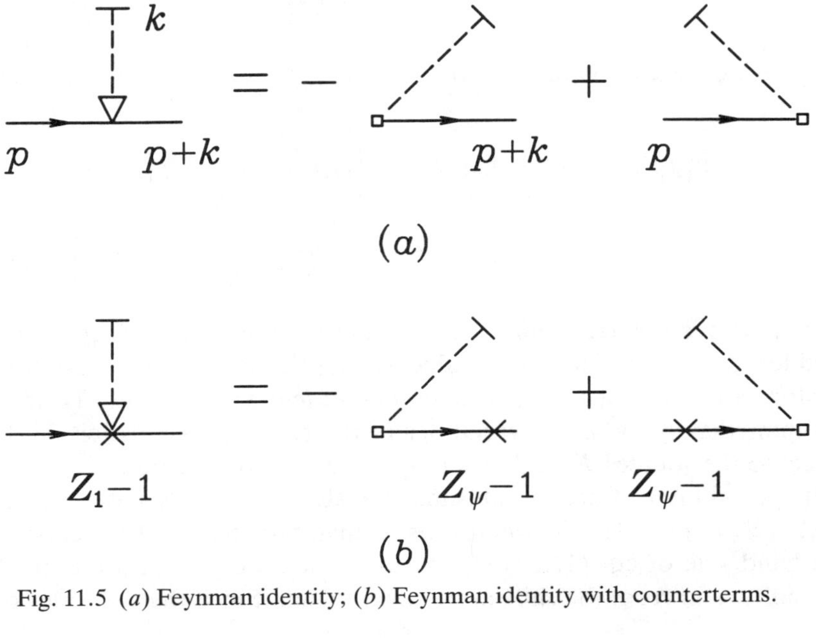

Fig. 11.5 (*a*) Feynman identity; (*b*) Feynman identity with counterterms.

Field transformations

Our first goal will be to identify the relevant Green function Ward identities. They may be derived by an elegant method based on field redefinitions. Since the method is quite general, we proceed for now without restriction to QED.

Suppose we have a theory with fields $\{\phi_i\}$ (fermionic and/or bosonic) and Lagrange density $\mathscr{L}(\phi_i)$. Now suppose we consider a transformation

$$\phi_i(x) = \phi'_i(x) + f_i(\phi'), \quad \mathscr{L}(\phi_i) = \mathscr{L}(\phi'_i(x) + f_i(\phi')), \qquad (11.51)$$

where $f_i(\phi')$ is some functional of the fields. Then the following Green function identity holds ('t Hooft & Veltman r1973, Lam 1973):

$$\langle 0|T[\phi_1(x_1)\cdots\phi_n(x_n)]|0\rangle_{\mathscr{L}(\phi_i)}$$
$$= \langle 0|T\{[\phi_1(x) + f_1(\phi)]\cdots[\phi_n(x_n) + f_n(\phi)]\}|0\rangle_{\mathscr{L}(\phi_i(x)+f_i(\phi))}, \qquad (11.52)$$

where the subscript $\mathscr{L}$ indicates that the Green function is calculated perturbatively with Lagrange density $\mathscr{L}$. In words, the vacuum expectation value of the unshifted fields, calculated with the unshifted Lagrangian, equals the vacuum expectation value of the shifted fields, calculated with the shifted Lagrangian.

For a large class of transformations, a proof of eq. (11.52) may be given very simply, by examining the path integral form of the generating functional for Green functions (eq. (3.44)),

$$Z = W[J]/W[0], \; W[J] = \int \prod_i [\mathscr{D}\phi_i] \exp\left\{ i\int d^4x\, \mathscr{L}(\phi_i) - i\sum_i (J_i, \phi_i)\right\}. \qquad (11.53)$$

Suppose we change variables to the fields ϕ'_i defined implicitly in eq. (11.51):

$$W[J] = \int \prod_i [\mathscr{D}\phi'_i]\Delta(\phi') \exp\left\{ i\int d^4x\, \mathscr{L}(\phi'_i + f_i(\phi'_j)) - \sum_i (J_i, \phi'_i + f_i(\phi'_j))\right\}, \qquad (11.54)$$

where $\Delta(\phi')$ is the Jacobian of the transformation (11.51). The change of variables does not affect the value of $Z[J]$, of course. Then for any transformation with a nonzero *field-independent* Jacobian, eq. (11.52) follows immediately by taking variations of the two expressions for $Z[J]$ with respect to the sources J_i, and dropping the primes on the right.

Of special interest are infinitesimal transformations, $\phi_i = \phi'_i + \epsilon g_i(\phi'_j)$, $\mathscr{L}(\phi_i) = \mathscr{L}(\phi'_i) + \delta\mathscr{L}(\phi'_i)$, where ϵ is a small parameter. If we expand the right-hand side of eq. (11.52) to first order in ϵ we get contributions from field shifts, and from the shift in the Lagrangian. Since the two sides of eq.

(11.52) are identically equal at $\epsilon = 0$, the $O(\epsilon)$ term on the right must vanish independently. This gives the following identity between Green functions:

$$\sum_{i=1}^{n} \langle 0|T[\phi_1(x_1)\cdots \epsilon g_i(\phi_j(x_i))\cdots \phi_n(x_n)]|0\rangle_{\mathscr{L}(\phi_i)}$$

$$+ \mathrm{i}\int d^4y\ \langle 0|T[\phi_1(x_1)\cdots \phi_n(x_n)\delta\mathscr{L}(\phi_j(y))]|0\rangle_{\mathscr{L}(\phi_i)} = 0. \quad (11.55)$$

Gauge transformations and Ward identities in QED

We now use eq. (11.55) to derive the Green function Ward identities of QED. Consider a change of field variables of the form of an infinitesimal local gauge transformation (eqs. (5.65) and (5.89)),

$$\delta_g\psi(x) = \mathrm{i}e\delta\Lambda(x)\psi(x),\ \delta_g\bar{\psi}(x) = \bar{\psi}(x)[-\mathrm{i}e\delta\Lambda(x)],\ \delta_g A^\mu(x) = -\partial^\mu\delta\Lambda(x),$$
$$\delta_g\mathscr{L}[\psi, \bar{\psi}, A] = \lambda(\partial\cdot A)(\partial_\mu\partial^\mu\delta\Lambda), \quad (11.56)$$

where $\delta\Lambda(x)$ is an arbitrary (small) function of space–time. The Jacobian of this transformation is unity, since the phases cancel between $\bar{\psi}(x)$ and $\psi(x)$ for every x, while the shift in each $A^\mu(x)$ is purely additive. Substituting eq. (11.56) into (11.55) gives the Green function identities

$$\sum_{i=1}^{n} \langle 0|T[\phi_1(x_1)\cdots \delta_g\phi_i(x_i)\cdots \phi_n(x_n)]|0\rangle_{\mathscr{L}(\phi_i)}$$

$$+ \mathrm{i}\int d^4y\langle 0|T\{\phi_1(x_1)\cdots \phi_n(x_n)\lambda[\partial\cdot A(y)][\partial_\mu\partial^\mu\delta\Lambda(y)]\}|0\rangle_{\mathscr{L}(\phi_i)} = 0, \quad (11.57)$$

where $\phi_i(x_i)$ represents any field. As long as $Z_1 = Z_2$, the *only* change in the Lagrange density is due to the gauge-fixing term. In accordance with the discussion of Section 4.4, all derivatives are understood to act outside the ('T^*') time-ordered product. After taking the variation $\delta/\delta\Lambda(z)$, eq. (11.57) becomes

$$-\lambda\partial_\mu\partial^\mu\langle 0|T\{[\partial\cdot A(z)]\textstyle\prod_a A_{\mu_a}(x_a)\prod_b\psi(y_b)\prod_c\bar{\psi}(w_c)\}|0\rangle$$

$$= \sum_d \langle 0|T\{\textstyle\prod_{a\neq d}A_{\mu_a}(x_a)\prod_b\psi(y_b)\prod_c\bar{\psi}(w_c)\}|0\rangle[\mathrm{i}\partial_{\mu_d}\delta(x_d - z)]$$

$$+ \sum_e \langle 0|T\{\textstyle\prod_a A_{\mu_a}(x_a)\prod_b\psi(y_b)\prod_c\bar{\psi}(w_c)\}|0\rangle[e\delta(y_e - z)]$$

$$+ \sum_f \langle 0|T\{\textstyle\prod_a A_{\mu_a}(x_a)\prod_b\psi(y_b)\prod_c\bar{\psi}(w_c)\}|0\rangle[-e\delta(w_f - z)]. \quad (11.58)$$

Let us assume (and we shall show in a moment) that the photon self-energy remains transverse to all orders, that is, $\pi_{\alpha\beta} = (q_\alpha q_\beta - g_{\alpha\beta}q^2)\mathrm{i}\pi(q^2)$, where $\pi(q^2)$ is a scalar function. Then, summing the geometric series in photon self-energies, as in eq. (10.22) for the scalar field, leads to the all-orders propagator

$$(G_2)_{\alpha\beta} = \frac{1}{k^2}\left[\left(-g_{\alpha\beta} + \frac{k_\alpha k_\beta}{k^2}\right)\frac{1}{1+\pi(k^2)} - \lambda^{-1}\frac{k_\alpha k_\beta}{k^2}\right], \quad (11.59\text{a})$$

in which the pole remains at $k^2 = 0$, without mass renormalization. The photon mass thus remains at zero, to all orders in perturbation theory, so long as $\pi(k^2)$ is well behaved. The operator $-\lambda\partial_\mu\partial^\mu\partial\cdot A(z)$ then gives (also to all orders) the momentum-space factor

$$\frac{-\mathrm{i}\lambda(-k^2)(\mathrm{i}k^\alpha)}{k^2}\left[\left(-g_{\alpha\nu} + \frac{k_\alpha k_\nu}{k^2}\right)\frac{1}{1+\pi(k^2)} - \lambda^{-1}\frac{k_\alpha k_\nu}{k^2}\right] = k_\nu. \quad (11.59\text{b})$$

This is the same as attaching a truncated, scalar polarized photon directly to the vertex of the current $-\mathrm{i}ej^\mu = -\mathrm{i}e\bar{\psi}\gamma^\mu\psi$, in any graph in which the operator $-\lambda\partial_\mu\partial^\mu j\cdot A(z)$ appears. In other words, it is equivalent to the replacement

$$-\lambda\partial_\mu\partial^\mu\partial\cdot A(z) \to -e\partial^\mu j_\mu(z). \quad (11.60)$$

The resulting Ward identity for the Green functions of QED is shown in Fig. 11.6. On the left-hand side, the notation represents a single scalar polarized photon attached to an arbitrary diagram in all possible ways (the left-hand side of eq. (11.58)). This equals a sum of terms in which the photon couples to external fermions at a new vertex, $\pm eI$, where I is the unit matrix in Dirac indices. The plus sign applies when the fermion arrow

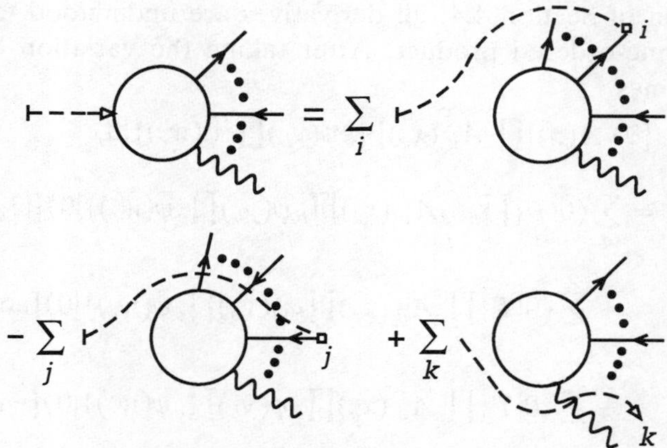

Fig. 11.6 Graphical Ward identity for QED.

ends at the vertex, the minus when it begins there. In addition, we include the possibility that the scalar polarization attaches directly to an external photon source. If there are no external fermions, these are the only terms that survive in the sum.

Before continuing our discussion of gauge invariance and unitarity, we pause to discuss two important subsidiary results: the transversality of fermion loops and the nonrenormalization of the vector current.

Graphical proof of Ward identities and transversality

It is useful to realize that the Ward identities eq. (11.58) and Fig. 11.6 may also be demonstrated directly at the level of Feynman diagrams. This graphical proof is a simple generalization of the tree-level proof given in Section 8.2.

A scalar photon attaches to fermion lines, which are either 'open', beginning and ending at external sources, or 'closed', consisting of a loop of internal propagators. For an open line with n other (virtual or real) photons attached, the application of the Feynman identity Fig. 11.5(a) leads to $2(n+1)$ terms, $2n$ of which cancel pairwise leaving two at the ends that correspond exactly to the special vertices on the right-hand side of Fig. 11.7(a).

The case of a closed fermion loop is the same, except that the two ends of the fermion line may be identified, as in Fig. 11.7(b). The two left-over

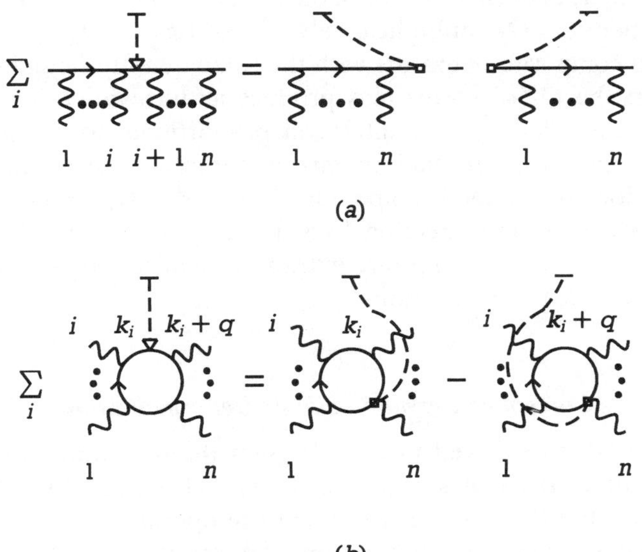

Fig. 11.7 Graphical proof of the QED Ward identity: (a) external fermion line; (b) closed fermion loop.

terms differ by a sign because in one term the external momentum q^μ of the scalar photon makes an extra circuit around the fermion loop. They cancel after the internal momentum of the loop is shifted by q^μ. If the loop integral is divergent, it must be *defined* in such a way that the shift makes sense. Dimensional regularization, for example, makes such shifts possible (Collins r1984, Chapter 4). Thus, in QED, every fermion loop decouples from a longitudinally polarized photon, and is transverse, as claimed above.

Composite operators

The operators $g_i(\phi_j)$ and $\delta\mathcal{L}(\phi_j)$ in eq. (11.55) are, in general, local products of fields, such as the current $j_\mu = \bar{\psi}\gamma_\mu\psi$. Such monomial operators, like those already in the Lagrange density, are called *composite* and introduce new local vertices, and hence new diagrams, into the theory. These new vertices may require additional renormalization for the Green functions in which they appear. They can be introduced into Green functions by adding a source term $J_O O(\phi)$ to the generating functional for each composite operator $O(\phi)$. Note that for the case in hand, the renormalization of $j_\mu(x)$, considered as an independent operator, is in principle independent of the renormalization of the operator $A^\mu(x) j_\mu(x)$ in the Lagrange density, because the latter involves another quantum field at the same point.

The BPH program (Section 10.4) applies to composite as well as elementary operators (Zimmerman r1970, Collins r1984), and we can renormalize composite operators O_{R} multiplicatively: $Z_C O_{\mathrm{R}}(\phi_{\mathrm{R}}) = O_0(\phi_0)$. If $O_{1,\mathrm{R}}$ and $O_{2,\mathrm{R}}$ are two composite operators with the same quantum numbers, then at higher orders the $O_1(\phi)$ vertex may produce an ultraviolet divergence that can only be cancelled by a counterterm proportional to $O_2(\phi)$, and vice versa. Such operators are said to mix under renormalization, and their renormalization constants become matrices: $(Z_C)_{ij} O_{j,\mathrm{R}} = O_{i,0}$. We shall encounter this situation in Section 14.5. Here, however, we shall find that the operator $j_\mu(x)$ does *not* require extra renormalization, a feature associated with its classical conservation.

Current conservation and nonrenormalization

Equation (11.60) substituted in (11.58) gives the quantum version of current conservation. It enables us to rewrite eq. (11.58) in a form from which we can deduce the behavior of the composite operator j^μ under renormalization. Since eq. (11.58) follows from the variation of the gauge-fixing term only, it holds for both renormalized and bare Green functions. If we assume multiplicative renormalization both for the fields, $Z_{\mathrm{R}}\psi_{\mathrm{R}} = \psi_0$, and the current, $Z_j j_R = j_0$, we have

$$
\begin{aligned}
-e_0\partial_\mu\langle 0|T[\psi_0(y)\bar\psi_0(w)j^\mu{}_0(z)]|0\rangle \\
&= -e_0 Z_j Z_\psi \partial_\mu\langle 0|T[\psi_R(y)\bar\psi_R(w)j^\mu{}_R(z)]|0\rangle \\
&= Z_\psi\langle 0|T[\psi_R(y)\bar\psi_R(w)]|0\rangle e_0\delta^4(y-z) \\
&\quad - Z_\psi\langle 0|T[\psi_R(y)\bar\psi_R(w)]|0\rangle e_0\delta^4(w-z), \qquad (11.61)
\end{aligned}
$$

from the bare Lagrangian. On the other hand, eq. (11.58) applied to the renormalized Lagrangian gives

$$
\begin{aligned}
-e_R\partial_\mu\langle 0|T[\psi_R(y)\bar\psi_R(w)j^\mu{}_R(z)]|0\rangle \\
&= \langle 0|T[\psi_R(y)\bar\psi_R(w)]|0\rangle e_R\delta^4(y-z)] \\
&\quad - \langle 0|T[\psi_R(y)\bar\psi_R(w)]|0\rangle e_R\delta^4(w-z). \qquad (11.62)
\end{aligned}
$$

Comparing eqs. (11.61) and (11.62), we conclude that $Z_j = 1$, and that the composite operator j^μ associated with a classically conserved current has trivial renormalization. Equation (11.62) was first proved by Fradkin (1955b) and Takahashi (1957), and the QED Green function relations eq. (11.58) are often referred to as Ward–Takahashi identities.

Renormalization-constant identities to all orders

We now return to the central issue: to prove the gauge invariance and unitarity of the S-matrix. To do so, we must first check that the identities $Z_1 = Z_\psi$ and $Z_A = Z_\lambda{}^{-1}$, from which the Green function identity eq. (11.58) follows, are consistent with renormalization. As is usual in renormalization arguments, we work inductively. Here we go.

Let us assume that we have constructed renormalization constants $Z_i{}^{(M)}$, $i = A, \psi, m, \lambda$ up to M loops, in such a way that

$$(Z_\lambda Z_A)^{(M)} = 1, \qquad (11.63a)$$

$$Z_1{}^{(M)} = Z_\psi{}^{(M)}. \qquad (11.63b)$$

Equation (11.63a) expresses the nonrenormalization of the gauge-fixing term to M loops, and we can use it to *define* $Z_\lambda{}^{(M)}$ by

$$Z_\lambda{}^{(M)} \equiv [Z_A{}^{(M)}]^{-1}. \qquad (11.64)$$

Now consider the renormalized Lagrange density $\mathscr{L}_R{}^{(M)}$, which is of the form of eq. (11.30) but with the Z_i replaced by the finite series $Z_i{}^{(M)}$ for $i = A, \psi, m$ and by eq. (11.64) for $Z_\lambda{}^{(M)}$:

$$
\begin{aligned}
\mathscr{L}_R{}^{(M)} &= \mathscr{L}_{\text{eff}}(e_R, m_R, \lambda_R) - \tfrac{1}{4}F_{\mu\nu}(A_R)F^{\mu\nu}(A_R)[Z_A{}^{(M)} - 1] \\
&\quad + \bar\psi_R i\not\partial\,\psi_R[Z_\psi{}^{(M)} - 1] - m_R\bar\psi_R\psi_R[Z_\psi{}^{(M)}Z_m{}^{(M)} - 1] \\
&\quad - e_R\mu^\epsilon\bar\psi_R\not A_R\psi_R[Z_1{}^{(M)} - 1]. \qquad (11.65)
\end{aligned}
$$

The perturbation series generated by $\mathscr{L}_R^{(M)}$ is, by construction, finite up to and including M loops. Even beyond M loops, it defines a regulated theory, one that is finite in less than four dimensions.

In particular, since $Z_1^{(M)} = Z_\psi^{(M)}$, $\mathscr{L}_R^{(M)}$, apart from the gauge-fixing term, is invariant under the 'renormalized' gauge transformation:

$$\psi_R'(x) = [1 + ie_R\delta\Lambda_R(x)]\psi_R(x), \quad \bar{\psi}_R'(x) = \bar{\psi}_R(x)[1 - ie_R\delta\Lambda_R(x)],$$
$$A^\mu{}_R'(x) = A^\mu{}_R(x) - \partial^\mu\delta\Lambda_R(x), \tag{11.66}$$

where the renormalized gauge parameter is of the form

$$\delta\Lambda_R(x) = [Z_A^{(M)}]^{-1/2}\delta\Lambda_0(x). \tag{11.67}$$

The Green function Ward identities, eq. (11.58), apply to this Lagrangian as well. They hold to *all* loops in its regulated theory, although they generally do not have a finite four-dimensional limit beyond M loops. We shall now use the identities of the regulated theory to show that $(M + 1)$-loop counterterms can be *chosen* to obey the renormalization-constant identities at $M + 1$ loops. Since we know from explicit calculation that these identities can be satisfied to one loop, this implies they can be satisfied to all loops in the renormalized theory.

Transversality of the photon self-energy

Consider first eq. (11.63a), $Z_\lambda = Z_A{}^{-1}$. If we show that up to $M + 1$ loops the photon two-point function is transverse, that is, $q^\mu\pi_{\mu\nu}{}^{(M+1)}(q) = 0$, then its tensor structure, including $(n - 4)^{-1}$ poles, has the form

$$\pi_{\mu\nu}{}^{(M+1)}(q) = (q_\mu q_\nu - g_{\mu\nu}q^2)\pi^{(M+1)}(q^2), \tag{11.68}$$

and the divergences at $(M + 1)$ loops can be cancelled by a counterterm proportional only to $(F_{\mu\nu})^2$. If so, the gauge-fixing term remains unrenormalized at $M + 1$ loops, and eq. (11.63a) is extended to this level. But eq. (11.68) follows directly from the transversality of fermion loops, or, more formally, from the Ward identity, Fig. 11.6, applied to the two-point function, as shown in Fig. 11.8. We re-emphasize that Fig. 11.8 follows simply from the form of $\mathscr{L}_R^{(M)}$, and is independent of whether or not the graphs are finite at $n = 4$, which, in general, they are not. Yet, eq. (11.68) must hold separately for the divergent and finite parts, so we may choose

= 0

Fig. 11.8 Transversality of the photon self-energy.

$Z_A{}^{(M+1)}$ to preserve the transversality structure. The BPH theorems of Section 10.4 ensure that the counterterm will be local.

The case of QED: $Z_1 = Z_\psi$

To extend eq. (11.63b) to the next order, consider the Ward identity shown in Fig. 11.9, which is just Fig. 11.6 applied to the fermion–photon Green function. When we take into account the counterterms of $\mathscr{L}_R{}^{(M)}$, divergences can only come from $(M+1)$-loop 1PI diagrams on either side of the equation. These consist of the 1PI vertex corrections and fermion self-energies. (Photon self-energies do not contribute because they are transverse.) We denote the complete pole parts as $\sum t(\mathrm{i}\bar{\Gamma}_\mu)$ and $\sum t(\mathrm{i}\bar{\Gamma}_2)$, with the sum taken respectively over all three- and two-point $(M+1)$-loop 1PI diagrams. The bars denote the internally subtracted diagrams, eq. (10.95b), each of which, by the first BPH theorem, equals the sum of *all* 1PI diagrams. From the second BPH theorem, we know that the divergent part of each internally subtracted graph is a polynomial in its external momenta and internal masses (as for the photon self-energy above). This strong restriction, along with Lorentz invariance, shows that the divergent parts of the two- and three-point functions must be of the forms

$$\sum t(\mathrm{i}\bar{\Gamma}_2) = -\mathrm{i}(\not{p} - m_\mathrm{R})b(\epsilon) + \mathrm{i}m_\mathrm{R}c(\epsilon), \tag{11.69a}$$

$$\sum t(\mathrm{i}\bar{\Gamma}_\mu) = -\mathrm{i}e_\mathrm{R}\gamma_\mu a(\epsilon), \tag{11.69b}$$

with $a(\epsilon)$, $b(\epsilon)$ and $c(\epsilon)$ polynomials in ϵ^{-1}. Writing out Fig. 11.9, we find the following relation between divergent parts:

$$\begin{aligned}
&\mathrm{i}(\not{p} + \not{q} - m_\mathrm{R})^{-1}(-\mathrm{i}e_\mathrm{R}\not{q})\mathrm{i}(\not{p} - m_\mathrm{R})^{-1}a(\epsilon)\\
&+ \mathrm{i}(\not{p} + \not{q} - m_\mathrm{R})^{-1}[-\mathrm{i}(\not{p} + \not{q} - m_\mathrm{R})b(\epsilon) + \mathrm{i}m_\mathrm{R}c(\epsilon)]\mathrm{i}(\not{p} + \not{q} - m_\mathrm{R})^{-1}\\
&\qquad\qquad\qquad\qquad \times (-\mathrm{i}e_\mathrm{R}\not{q})\mathrm{i}(\not{p} - m_\mathrm{R})^{-1}\\
&+ \mathrm{i}(\not{p} + \not{q} - m_\mathrm{R})^{-1}(-\mathrm{i}e_\mathrm{R}\not{q})\mathrm{i}(\not{p} - m_\mathrm{R})^{-1}[-\mathrm{i}(\not{p} - m_\mathrm{R})b(\epsilon)\\
&\qquad\qquad\qquad\qquad + \mathrm{i}m_\mathrm{R}c(\epsilon)]\mathrm{i}(\not{p} - m_\mathrm{R})^{-1}\\
&= -\mathrm{i}(\not{p} + \not{q} - m_\mathrm{R})^{-1}[-\mathrm{i}(\not{p} + \not{q} - m_\mathrm{R})b(\epsilon)\\
&\qquad\qquad + \mathrm{i}m_\mathrm{R}c(\epsilon)]\mathrm{i}(\not{p} + \not{q} - m_\mathrm{R})^{-1}(e_\mathrm{R})\\
&\quad + (e_\mathrm{R})\mathrm{i}(\not{p} - m_\mathrm{R})^{-1}[-\mathrm{i}(\not{p} - m_\mathrm{R})b(\epsilon) + \mathrm{i}m_\mathrm{R}c(\epsilon)]\mathrm{i}(\not{p} - m_\mathrm{R})^{-1}.
\end{aligned} \tag{11.70}$$

A short calculation shows that terms proportional to $c(\epsilon)$ cancel identically, and that eq. (11.70) reduces to

$$a(\epsilon) = -b(\epsilon). \tag{11.71}$$

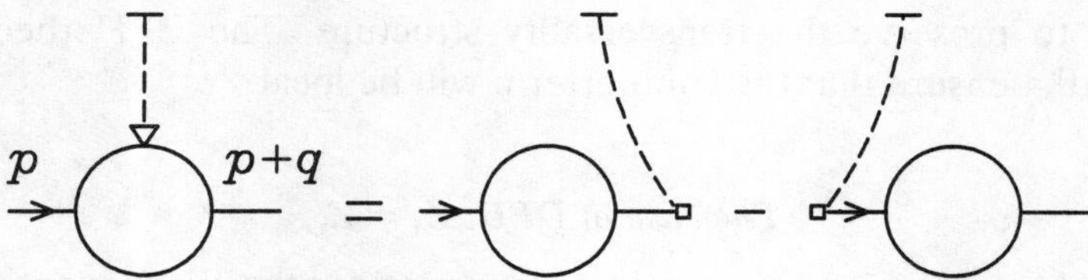

Fig. 11.9 Ward identity (Fig. 11.6) of QED applied to a general three-point vertex.

Now, following the usual inductive procedure, we want to construct $(M+1)$-loop counterterms to cancel the divergences as $\epsilon \to 0$. We easily verify that the divergences (11.69a, b) are cancelled by the following counterterms:

$$
\begin{aligned}
-e_{\mathrm{R}}\bar{\psi}_{\mathrm{R}}\not{A}_{\mathrm{R}}\psi_{\mathrm{R}}Z_1{}^{(M+1)(\mathrm{pole})}, &\quad Z_1{}^{(M+1)(\mathrm{pole})} = -a(\epsilon);\\
\bar{\psi}_{\mathrm{R}}(\mathrm{i}\not{\partial})\psi_{\mathrm{R}}Z_\psi{}^{(M+1)(\mathrm{pole})}, &\quad Z_\psi{}^{(M+1)(\mathrm{pole})} = b(\epsilon);\\
-m_{\mathrm{R}}\bar{\psi}_{\mathrm{R}}\psi_{\mathrm{R}}[Z_\psi{}^{(M+1)} + Z_m{}^{(M+1)}]^{(\mathrm{pole})}, &\quad Z_m{}^{(M+1)(\mathrm{pole})} = c(\epsilon).
\end{aligned}
\tag{11.72}
$$

From these relations we conclude that

$$Z_1{}^{(M+1)(\mathrm{pole})} = Z_\psi{}^{(M+1)(\mathrm{pole})}. \tag{11.73}$$

The pole parts of Z_1 and Z_ψ are now equal to $M+1$ loops, and we are free to choose their finite parts to be the same as well. This completes the proof by induction that the renormalization-constant identities – and hence the graphical Ward identities eq. (11.58) and Fig. 11.6 – are preserved by the renormalization of QED. Now we can verify explicitly the gauge invariance and unitarity of the physical S-matrix.

Gauge invariance of the S-matrix

For simplicity, we confine ourselves to equivalence between covariant gauges, with gauge function $g = \partial \cdot A$. Generalizations are suggested in exercise 11.6. Suppose we vary the parameter λ in the renormalized Lagrange density, eq. (11.30) by a small amount, $\lambda_{\mathrm{R}} \to \lambda_{\mathrm{R}} + 2\delta\lambda_{\mathrm{R}}$. The first-order change in the Green function $G = \langle 0|T[\prod_i \phi_i(x_i)]|0\rangle$ is given by

$$\delta_\lambda G = \mathrm{i}\int \mathrm{d}^4y\,\langle 0|T[\textstyle\prod_i \phi_i(x_i)\delta\mathscr{L}]|0\rangle, \quad \delta\mathscr{L} = \delta\lambda_{\mathrm{R}}[\partial \cdot A(y)]^2. \tag{11.74}$$

We now notice that a gauge transformation eq. (11.56), with

$$\delta\Lambda(x) = \frac{\delta\lambda_{\mathrm{R}}}{\lambda_{\mathrm{R}}}\int \mathrm{d}^4y\,\Delta(x-y)[\partial \cdot A(y)], \tag{11.75}$$

produces exactly the same $\delta\mathscr{L}$ as a change in λ_R, when $\Delta(x-y)$ is the Feynman propagator, so that $\partial_\mu\partial^\mu\Delta(x) = -\delta^4(x)$. (Here we are again using the Ward identity $Z_1 = Z_2$.) Referring to eq. (11.57), we derive from this transformation the following graphical identity:

$$\sum_i \langle 0|T[\phi_1(x_1)\cdots\delta_\lambda\phi_i(x_j)\cdots\phi_n(x_n)]|0\rangle = \delta_\lambda G, \qquad (11.76)$$

where $\delta_\lambda\phi_i$ is defined by the gauge transformation, eq. (11.56) and $\delta\Lambda$ is given by eq. (11.75). Equation (11.76) shows that the variation of the Green function under a change in λ may be expressed in terms of nonlocal gauge transformation on its external fields. This is shown graphically in Fig. 11.10(a), with one external field of each type. A dashed line represents the scalar propagator $\Delta(x-y)$ in eq. (11.75), which attaches the external field to the operator $\partial^\nu A_\nu(y)$. The derivative ∂^ν ensures that the photon $A_\nu(y)$ couples to the remainder of the diagram through its scalar polarization, just as in eq. (11.59b) and Fig. 11.6 (eq. (11.58)). We shall

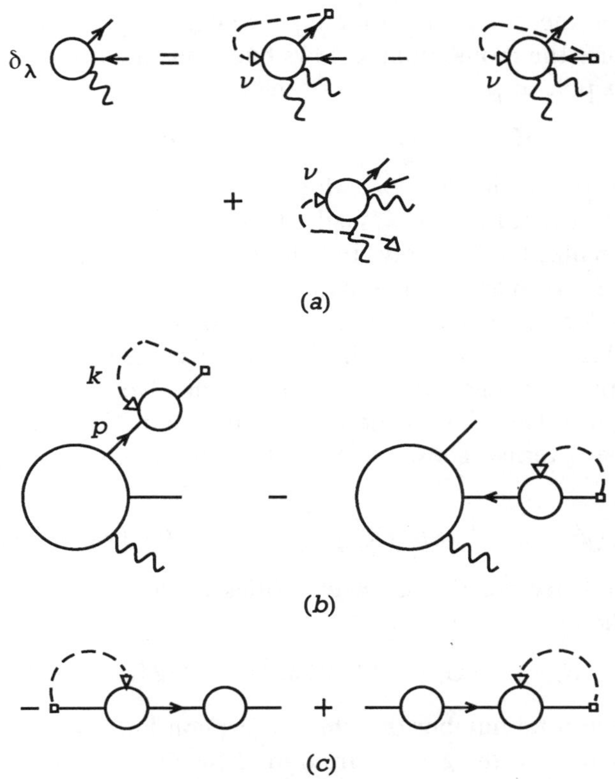

(a)

(b)

(c)

Fig. 11.10 Gauge variations in QED: (a) general Green function; (b) surviving single-pole structure from (a); (c) variation of fermion two-point function.

refer to such a scalar propagator as a 'ghost' line, by analogy to the nonabelian case, to be discussed below. As usual, an arrow represents the momentum vector of the ghost line. There is one such arrow from ∂^ν, as indicated above, and in the case of an external photon, another from the derivative in the gauge transformation.

The S-matrix is the residue of the poles of the Green function in its external momenta, multiplied by $\prod_i \xi_i[(2\pi)^3 R_i]^{-1/2}$, where R_i is the residue at the single-particle pole of the two-point function of field i, and ξ_i is a physical polarization vector or spinor. In general, both the R_i and the Green function are gauge dependent. We now show that this dependence cancels in the S-matrix.

Consider $\delta_\lambda G$, Fig. 11.10(a). Ghost lines that attach to external photon sources vanish, since the physical polarizations are orthogonal to their corresponding momenta. In the remaining diagrams of Fig. 11.10(a) an external fermion line has been cancelled, and replaced by the special vertex $\pm eI$. In general, diagrams of this form cannot contribute to the S-matrix, because they lack a full complement of poles on external lines. The exception to this rule is shown in Fig. 11.10(b), where the photon k connects to a self-energy of the external fermion p, whose propagator has been cancelled. Because the momentum k flows back into the self-energy, however, there is still a pole at $p^2 = m^2$. It is of the form

$$\delta\Omega\, \mathrm{i}(\not{p} - m)^{-1} \quad \text{or} \quad \mathrm{i}(-\not{p} - m)^{-1}\delta\bar{\Omega}, \tag{11.77}$$

with $\delta\Omega\,(\delta\bar{\Omega})$ proportional to $\delta\lambda$, where the arrow is in the same (opposite) sense as the momentum flow. Usually, this extra pole is cancelled by a zero in the renormalized self-energy, but the gauge variation of the self-energy need not have a zero at the same point.

Now consider the gauge variation of R_i for external line i. This is also found from Fig. 11.10(a), specialized to the two-point functions. The photon self-energy is gauge invariant, since it involves only closed fermion loops. The gauge variation of the fermion two-point function, on the other hand, gives two terms, shown in Fig. 11.10(c), which at the pole take on the form

$$\delta R_i\,(\not{p}_i - m)^{-1} = \delta\Omega\, R_i(\not{p}_i - m)^{-1} - R_i(\not{p}_i - m)^{-1}\,\delta\bar{\Omega}, \tag{11.78}$$

where $\delta\Omega$ and $\delta\bar{\Omega}$ are the same quantities as above. The change in the S-matrix is then proportional to

$$\delta(G/\textstyle\prod_i R_i{}^{1/2}) = (G/\prod_i R_i{}^{1/2})(\delta\Omega\,\frac{1}{2}n_f - \delta\bar{\Omega}\,\frac{1}{2}n_f - \frac{1}{2}\delta R_i\, n_f), \tag{11.79}$$

where n_f is the total number of external fermion lines. But by eq. (11.78), (11.79) vanishes, and the gauge variation of the Green function is cancelled by the variation of the residues of its external lines. The S-matrix is thus invariant under transformations between covariant gauges.

Unitarity of the S-matrix

As we have emphasized above, the graphical identity (9.86) is equivalent to unitarity only when cut lines are automatically physical. This is not the case, however, in QED, because the gauge-fixed vector propagator contains unphysical modes, and the polarization tensor in the propagator is not the physical polarization tensor of eq. (8.59),

$$P^{\mu\alpha}(k) = -g^{\mu\alpha} + (\bar{k}^\mu k^\alpha + k^\mu \bar{k}^\alpha)(\bar{k} \cdot k)^{-1},$$

where $\bar{k}^\mu = (k_0, -\mathbf{k})$. Following 't Hooft (1971a), the difference between a cut graph and its corresponding *physical* contribution will be represented as in Fig. 11.11, where the double cut stands for physical polarization tensors, and the single cut for gauge polarizations. Incoming lines (not shown) are always assumed to be physical in this discussion, because we only need to show unitarity for physical states. We are about to see that the difference shown in Fig. 11.11 vanishes, and hence that the sum over physical polarizations equals the sum over the gauge polarizations found by cutting diagrams, as in eq. (9.86). Under these conditions, the cutting equation (9.86) implies unitarity for QED.

We expand the tensor $P^{\mu\alpha}(k)$ for each cut line k^μ in the first diagram on the left-hand side. The resulting sum consists of a single term that includes all $-g^{\mu\alpha}$ polarization tensors, which cancels the second diagram, plus a number of terms in which at least one photon has a scalar polarization k^β contracted into the amplitude to the left or right of the cut. A typical term is shown on the right-hand side of Fig. 11.11, in the notation of Fig. 8.11. To these we apply the Ward identities, Fig. 11.6.

Consider the effect of going to the mass shell in all external particles, and cancelling the corresponding poles on the left-hand side of Fig. 11.6. This eliminates all terms on the right-hand side in which a photon attaches to an external fermion line, since in each of these terms there is no longer

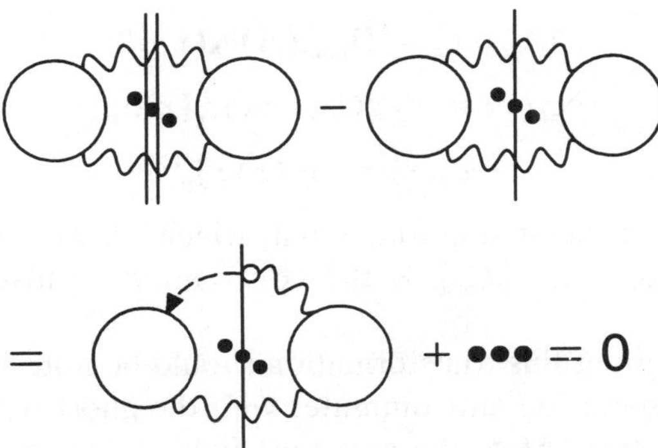

Fig. 11.11 Expression of unitarity in QED.

an isolated pole in that fermion's momentum. The only surviving terms in Fig. 11.11 are disconnected photon lines, ending in scalar polarizations at an incoming photon. (The factor $\theta(p_0)$ in $\delta_+(p^2)$ for final-state photons prevents the connection of two final-state lines.) But since the incoming photons always have physical polarization, these terms also vanish identically. Thus the right-hand side of Fig. 11.11 vanishes, and the *S*-matrix is unitary in perturbation theory. Only at this point have we really earned our confidence that quantum electrodynamics is a viable theory.

11.3 Ward identities and the *S*-matrix in QCD

In nonabelian theories, Ward identities play much the same role as in QED in proving the gauge invariance and unitarity of the *S*-matrix ('t Hooft 1971a, 't Hooft & Veltman 1972b). To simplify our discussion, we shall assume that the renormalization-constant (Taylor–Slavnov) identities, eqs. (11.45) are preserved by renormalization (Collins r1984, Frampton r1986).

Becchi–Rouet–Stora (BRS) invariance

The effective Lagrange density for Yang–Mills theory is given by eq. (11.2). The gauge invariance of its Lagrangian is broken by both the gauge-fixing term and the ghost Lagrangian. It is possible, although not particularly convenient, to generalize the discussion given for the abelian theory by studying the behavior of (11.2) under gauge transformations. A more powerful approach was originated by Becchi, Rouet & Stora (1975), who identified a transformation – involving both the ghosts and physical fields – under which the full effective Lagrangian is invariant. This 'BRS' transformation requires that we introduce an infinitesimal *anticommuting* parameter, denoted $\delta\xi$ (see Section 7.1).

Consider the Yang–Mills density eq. (11.2) without the fermions. The BRS transformations are

$$\delta A_{\mu,a}(x) = D_{\mu,ab}(A)c_b(x)\delta\xi, \tag{11.80a}$$

$$\delta c_a(x) = -\tfrac{1}{2}gC_{abc}c_b(x)c_c(x)\delta\xi, \tag{11.80b}$$

$$\delta\bar{c}_a(x) = \lambda g_a(x)\delta\xi, \tag{11.80c}$$

where $g_a(x)$ is the gauge-fixing functional, which we will take as $\partial\cdot A_a(x)$. Here $\mathrm{D}_{\mu,ab} = \delta_{ab}\partial_\mu + gA_{\mu,c}C_{cab}$ is the covariant derivative in the adjoint representation.

Several things about this transformation should be noted right away. The infinitesimal parameter $\delta\xi$ anticommutes with the ghost fields, so that their relative orders matter. Also, the antighost field does not transform as the complex conjugate of the ghost field (and need not, because they are independent). Finally, the change in the gauge field is a special gauge transformation, with infinitesimal parameter $\delta\Lambda_b(x) = -c_b(x)\delta\xi$. This suggests the following BRS variation for fermion fields with group index i:

$$\delta\psi_i(x) = \mathrm{i}g[t_b{}^{(\mathrm{F})}]_{ij}c_b(x)\delta\xi\,\psi_j(x), \tag{11.80d}$$

$$\delta\bar{\psi}_i(x) = -\mathrm{i}g\bar{\psi}_j(x)[t_b{}^{(\mathrm{F})}]_{ji}c_b(x)\delta\xi. \tag{11.80e}$$

Then the full gauge invariant part of the Lagrange density is invariant under eq. (11.80). This leaves the gauge-fixing and ghost parts.

Invariance

Recall (eq. (7.72)) that the gauge-fixing and ghost contributions to the effective Lagrange density are

$$-\tfrac{1}{2}\lambda g_a[A]g_a[A] + \bar{c}_a(\delta g_a/\delta\alpha_b)c_b, \tag{11.81}$$

where $\delta g_a/\delta\alpha_b$ is the functional variation of $g_a[A]$ under a gauge transformation. In this form, we can use eqs. (11.80a, c) and the anticommuting nature of $\delta\xi$ and the antighost field to show that the variation of the gauge-fixing term cancels that of the antighost. Meanwhile, the variation of the gauge field in the ghost Lagrangian cancels the variation due to the ghost field. With covariant gauge fixing this works as follows:

$$\begin{aligned}\delta_{\mathrm{BRS}}[\mathrm{D}_{\mu,ac}(A)c_c] &= -gC_{abc}[\mathrm{D}_{\mu,bd}(A)c_d\delta\xi]c_c \\ &\quad + \mathrm{D}_{\mu,al}(A)(-\tfrac{1}{2}gC_{lmn}c_mc_n\delta\xi) \\ &= g[C_{abc}(\partial_\mu c_b)c_c - \tfrac{1}{2}C_{amn}\partial_\mu(c_mc_n)]\delta\xi \\ &\quad - g^2[C_{abc}C_{bed}c_dc_c - \tfrac{1}{2}C_{ael}C_{lmn}c_mc_n]A_{\mu,e}\delta\xi.\end{aligned} \tag{11.82}$$

The terms linear and quadratic in the structure constants cancel independently. For the linear terms, we use the anticommuting nature of the ghosts and the antisymmetry of C_{abc}:

$$\tfrac{1}{2}C_{amn}\partial_\mu(c_mc_n) = C_{abc}(\partial_\mu c_b)c_c. \tag{11.83}$$

For the quadratic terms, it is necessary to use the Jacobi identity (exercise 1.5) as well, from which we find

$$\begin{aligned}\tfrac{1}{2}C_{ael}C_{lmn}c_mc_n &= \tfrac{1}{2}(C_{mal}C_{eln} + C_{eml}C_{aln})c_mc_n \\ &= C_{mal}C_{eln}c_mc_n \\ &= C_{abc}C_{bed}c_dc_c.\end{aligned} \tag{11.84}$$

With this invariance in hand, it is relatively easy to derive a set of Green function Ward identities for QCD from eq. (11.55), with the first-order changes in the fields given by the BRS variations, eqs. (11.80). Of course, $\delta_{\rm BRS}\mathscr{L} = 0$. To apply eq. (11.55), however, we must still check that the functional Jacobean of a BRS transformation is unity (or more precisely, field independent). Since the BRS transformation is local, it suffices to prove this at a single point.

BRS Jacobian

The BRS transformation mixes commuting and anticommuting variables in the path integral. Therefore, we refer to the discussion following eq. (7.16). According to eq. (7.21), the Jacobian of the infinitesimal transformations (11.80) is of the form $J = 1 + S$, where S is given by (van Nieuwenhuizen r1981)

$$\begin{aligned} S &= \delta A_{\mu,a}/\delta A_{\mu,a} - \delta c_a/\delta c_a - \delta \bar{c}_a/\delta \bar{c}_a - \delta\psi_i/\delta\psi_i - \delta\bar{\psi}_i/\delta\bar{\psi}_i \\ &= \{-gC_{abc}c_c\delta_{ab} + \tfrac{1}{2}gC_{abc}[-\delta_{ba}c_c + c_b\delta_{ca}] - 0 \\ &\qquad - igc_b[t_b{}^{(\rm F)}]_{ii} + igc_b[t_b{}^{(\rm F)}]_{ii}\}\delta\xi. \end{aligned} \tag{11.85}$$

But this expression vanishes, and the Jacobian is indeed unity. Now we are ready to apply eq. (11.55).

Ward identities

The resulting Ward identities are

$$\int d^4z\{\delta/\delta[\delta\xi(z)]\}\sum_{i=1}^{n}\langle 0|T[\phi_1(x_1)\cdots\delta_{\rm BRS}\phi_i(x_i)\cdots\phi_n(x_n)]|0\rangle = 0, \tag{11.86}$$

where we choose to take the variation with respect to $\delta\xi(z)$ from the left. These Ward identities are sometimes referred to as Taylor–Slavnov identities, a term which we reserve for the relation between renormalization constants. For a Green function without fermions, eq. (11.86) becomes

$$\begin{aligned} &\sum_p(-1)^{n(w_p)}\langle 0|T[\lambda g_{c_p}(w_p)\textstyle\prod_i A_{a_i}(x_i)\prod_j c_{b_j}(y_j)\prod_{k\neq p}\bar{c}_{c_k}(w_k)]|0\rangle \\ &\qquad = -\sum_m\langle 0|T\{D_{a_m}[A(x_m)]c(x_m)\textstyle\prod_{i\neq m}A_{a_i}(x_i)\prod_j c_{b_j}(y_j)\prod_k\bar{c}_{c_k}(w_k)\}|0\rangle \\ &\qquad\quad - \sum_n(-1)^{n(y_n)}\langle 0|T\{\textstyle\prod_i A_{a_i}(x_i)[-\tfrac{1}{2}gC_{b_n bc}c_b(y_n)c_c(y_n)] \\ &\qquad\qquad \times\textstyle\prod_{j\neq n}c_{b_j}(y_j)\prod_k\bar{c}_{c_k}(w_k)\}|0\rangle. \end{aligned} \tag{11.87}$$

Here $n(t_i)$ is the number of anticommuting fields in eq. (11.87) to the left of the variation at point t_i. These identities are illustrated graphically in Fig. 11.12(a). The lowest-order example, analogous to the Feynman identity Fig. 11.5(a) of QED, is shown in Fig. 11.12(b). The big difference from the QED identity, eq. (11.58), is that the ghost lines now interact with the rest of the graph, according to the Feynman rules of QCD. On the left-hand side of eq. (11.87) one antighost is replaced by a gauge function, which is equivalent to a scalar polarized gluon, as in eq. (11.59b) for the photon. On the right-hand side, external gluons and ghosts are replaced by their BRS variations. 't Hooft & Veltman (1972b) develop the analogues of eq. (11.87) in a large class of theories.

Factors of (-1) have been suppressed in Fig. 11.12(a). We can, however, observe one general rule that will be important in the proof of unitarity. Consider a graph that contributes to the left-hand sides of Fig. 11.12(a) and eq. (11.87). Its absolute sign is found from anticommuting the external fields into pairs $c(y_j)\bar{c}(w_k)$ connected by the unbroken ghost lines of the graph, and from the factor $(-1)^{n(w_p)}$. Now consider a graph that contributes to the right-hand sides of Fig. 11.12(a) and eq. (11.87), and that has the *same* pairing of external fields by ghost lines, along with a single extra pairing $D_\mu c(x_m)\bar{c}(w_p)$. This graph has the same overall sign as the first graph, because anticommuting the antighost field $\bar{c}(w_p)$ to be next to $D_\mu c(x_m)$ produces precisely the same factor $(-1)^{n(w_p)}$. The remaining product of anticommuting fields is the same in the two graphs. Thus we derive the rule that the signs of graphs that contribute to

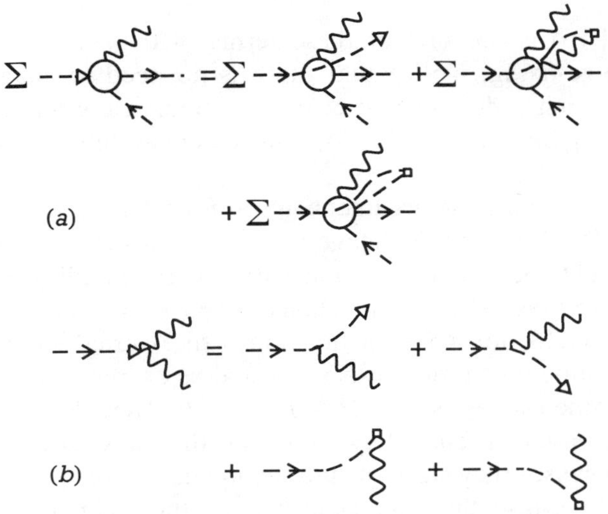

Fig. 11.12 Ward identities of QCD: (a) general Green function; (b) identity for the lowest-order three-point vertex, analogous to the QED case shown in Fig. 11.5(a).

the first term on the right-hand side of Fig. 11.12(a) are determined by the ghost lines of the left-hand side.

Referring to Section 8.5, we recognize that the calculation leading from (8.102) to (8.106), and represented in Fig. 8.11, is an illustration of eq. (11.87) at tree level.

Gauge invariance of the physical S-matrix

The proof of gauge invariance for the S-matrix follows the same pattern as in QED.

To be specific, consider linear gauge-fixing functionals, $g_a = \eta \cdot A_a$. Examples are $\eta^\mu = \partial^\mu$ for normal covariant gauge-fixing, or $\eta^\mu = n^\mu$ for the axial gauge. A change in η^μ may be generated by the following nonlocal transformation:

$$\delta A_a{}^\mu = \tfrac{1}{2}\mathrm{D}_{ab}{}^\mu(A)[\eta \cdot \mathrm{D}(A)]^{-1}{}_{bc}\delta\eta \cdot A_c \rightarrow \delta(g_a g_a) = \eta \cdot A\, \delta\eta \cdot A. \tag{11.88a}$$

The nonlocal operator $[\eta \cdot \mathrm{D}(A)]^{-1}{}_{bc} = (\delta g_b/\delta\alpha_c)^{-1}$ is the inverse of the quadratic ghost operator (see eq. (7.72)). It can be interpreted as the exact ghost propagator in a background field $A^\mu{}_a(z)$ and is schematically of the form

$$[\eta \cdot D(A)]^{-1}{}_{bc} = (\eta \cdot \partial)^{-1}\delta_{bc} - (\eta \cdot \partial)^{-1} g C_{bdc}(\eta \cdot A_b)(\eta \cdot \partial)^{-1} + \cdots. \tag{11.88b}$$

Equation (11.75) is the first of these terms, with $\eta \cdot \partial = \partial^2$ in QED for covariant gauge-fixing. Notice that there is no need to change the ghost Lagrangian at first order, since it changes automatically through the determinant $\det(\delta g_b/\delta\alpha_c)$ in eq. (7.70), which *defines* the $SU(N)$ functional integral.

We may now repeat the arguments given for QED to arrive at the same identity, eq. (11.76), for the change in any Green function, with $\delta_\lambda\phi$ now given by eq. (11.88a) for gluons and with a corresponding expression for fermions. If we make the specific choice $g_a = (\partial \cdot A_a)$, eq. (11.76) is illustrated in Fig. 11.13, for a Green function with external gluons only. (The analogous treatment of fermions is suggested as exercise 11.10.) The terms fall into the same classes as for QED, Fig. 11.10. They do not contribute to the S-matrix, either because they annihilate the physical polarization of a gluon, or because they lack a full complement of poles in external momenta, or because they are cancelled by the gauge-dependence of a gluon residue R_g, just as for fermions in QED. The gauge invariance of the S-matrix in QCD thus follows from its graphical Ward identities.

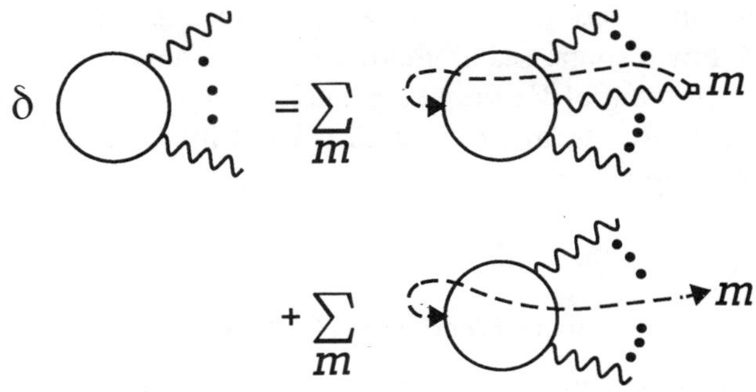

Fig. 11.13 Gauge variation in QCD.

Uncut and cut graphs

To discuss unitarity in nonabelian theories, it will be useful to derive graphical Ward identities for cut diagrams. To see how, we first give yet another instructive proof of the Green function Ward identity eq. (11.55), this time based on graphical combinatorics ('t Hooft & Veltman r1973, Lam 1973). We first expand the shifted Lagrangian, distinguishing the behavior of the quadratic, potential and source parts of the Lagrangian:

$$\mathscr{L}(\phi_n + \epsilon f_n(\phi)) - J_m(\phi_m + \epsilon f_m(\phi))$$
$$= \sum_{i,j} \left\{ \tfrac{1}{2}\phi_i M_{ij}\phi_j - V(\phi) - J_i\phi_i + \epsilon[\phi_i M_{ij} f_j(\phi) - (\delta V/\delta\phi_i) f_i(\phi) - J_i f_i(\phi)] \right\}. \tag{11.89}$$

(Here we illustrate the method with an arbitrary set of real fields.) The change in any Green function is now given by all the terms generated from eq. (11.55) that are linear in ϵ, the source term generating the shift in the external fields. A typical term linear in $\delta V/\delta\phi$ is shown schematically in Fig. 11.14(*a*). The vertex $f_i(\phi)$ may represent a number of contributions, each with different numbers of fields. For every such contribution there is another graph of the form shown in Fig. 11.14(*b*), in which the original

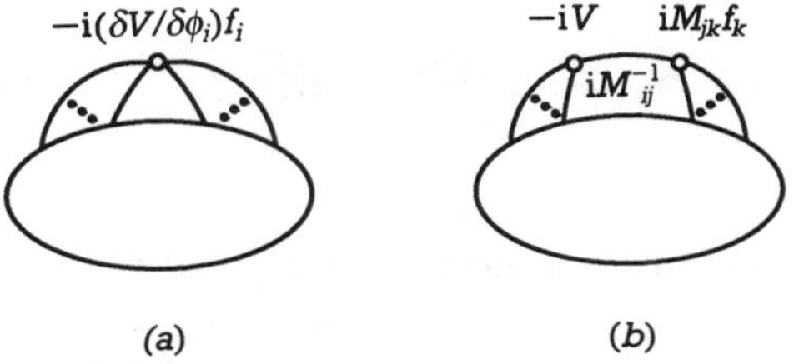

Fig. 11.14 Illustration of a combinatoric proof of Ward identities. (*a*) Variation of the potential; (*b*) variation of quadratic terms.

vertex $-\mathrm{i}V$ is connected to a ϕ_i-line, which in turn is coupled by $(\mathrm{i}M^{-1})_{ij}$ to $\sum_k \mathrm{i}M_{jk}f_k(\phi)$. For a single scalar field, we would have simply $M = k^2$. For any polynomial potential V, the two cancel. All $\mathrm{O}(\epsilon)$ contributions, including those from source terms, can be organized into cancelling pairs of this type, and the theorem is proved.

Ward identity for cut diagrams

This graphical reasoning is an elegant way of deriving Ward identities satisfied by *cut* Green functions, in which we sum over intermediate states $|\alpha\rangle$, with α lines. Using the general relation $\langle A|O|B\rangle^* = \langle B|O^\dagger|A\rangle$ for the complex conjugate T-matrix element, we can represent a cut diagram with final state $|\alpha\rangle$ as

$$\langle 0|\bar{T}[\phi^\dagger{}_{m+1}(x_{m+1}) \cdots \phi^\dagger{}_n(x_n)]|\alpha\rangle\langle\alpha|T[\phi_1(x_1) \cdots \phi_m(x_m)]|0\rangle, \tag{11.90}$$

where $\bar{T}$ represents an *anti*time ordering due to the hermitian conjugate. Exactly the same reasoning applies to the cut as to the uncut Green function, since replacing a propagator $1/k^2$ by $\delta(k^2)$ does not interfere with the cancellation necessary to prove eq. (11.55). Indeed, the cancellation shown in Fig. 11.14 is local and occurs vertex by vertex, independently on both sides of any cut, for fixed choices of the number and momenta of cut lines. It is generally necessary, however, to sum over the species of cut lines that are mixed by the transformation, gluons and ghosts in the case of BRS. This leads to the following generalization of eq. (11.55):

$$\begin{aligned}
&\sum_{i=1}^{m} \langle 0|\bar{T}[\phi^\dagger{}_{m+1}(x_{m+1}) \cdots]|\alpha\rangle\langle\alpha|T[\phi_1(x_1) \cdots \epsilon f_i(\phi(x_i)) \cdots]|0\rangle \\
&+ \sum_{j=m+1}^{n} \langle 0|\bar{T}[\phi^\dagger{}_{m+1}(x_{m+1}) \cdots \epsilon f_j{}^\dagger(\phi(x_j)) \cdots]|\alpha\rangle\langle\alpha|T[\phi_1(x_1) \cdots]|0\rangle \\
&+ \mathrm{i}\int \mathrm{d}^4y\, \langle 0|\bar{T}[\phi^\dagger{}_{m+1}(x_{m+1}) \cdots]|\alpha\rangle\langle\alpha|T[\phi_1(x_1) \cdots \delta\mathcal{L}(\phi(y))]|0\rangle \\
&- \mathrm{i}\int \mathrm{d}^4y\, \langle 0|\bar{T}[\phi^\dagger{}_{m+1}(x_{m+1}) \cdots \delta\mathcal{L}(\phi(y))]|\alpha\rangle\langle\alpha|T[\phi_1(x_1) \cdots]|0\rangle = 0.
\end{aligned} \tag{11.91}$$

A sum over all final states α (ghosts and gluons) with the same number of lines is implicit. Using eq. (11.91), we can derive graphical Ward identities for cut diagrams, by using the infinitesimal BRS transformation, eqs. (11.80), for the fields, under which $\delta_{\mathrm{BRS}}\mathcal{L} = 0$. The particular case we are interested in involves the cut diagrams

$$\langle 0|\bar{T}\Big[\textstyle\prod_j A_{\text{phys}}(x_j)A(z)\Big]|\alpha\rangle\langle\alpha|T\Big[\textstyle\prod_i A_{\text{phys}}(x_i)\bar{c}(y)\Big]|0\rangle = 0, \quad (11.92)$$

which vanish identically, because they do not preserve the number of ghost lines. The $A_{\text{phys}}(x_i)$ are physical fields, which contribute to the S-matrix on-shell. As such, the BRS variation of A_{phys} can be defined as zero. In this case, we derive from (11.80) and from the vanishing of $\delta_{\text{BRS}}\mathcal{L}$ the simple result shown in Fig. 11.15:

$$\sum_\alpha \langle 0|\bar{T}\Big[\textstyle\prod_j A_{\text{phys}}(x_j)A(z)\Big]|\alpha\rangle\langle\alpha|T\Big[\textstyle\prod_i A_{\text{phys}}(x_i)\lambda g(y)\Big]|0\rangle$$
$$+ \sum_\alpha \langle 0|\bar{T}\Big[\textstyle\prod_j A_{\text{phys}}(x_j)\mathrm{D}_\mu c(z)\Big]|\alpha\rangle\langle\alpha|T\Big[\textstyle\prod_i A_{\text{phys}}(x_i)\bar{c}(y)\Big]|0\rangle = 0. \quad (11.93)$$

With these cut-diagram identities, we can now discuss unitarity in QCD.

Unitarity

The issue in demonstrating unitarity for a nonabelian theory is the same as for QED: to show the equality of the sum over physical gluon polarizations to the sum over the gauge polarizations generated by cutting diagrams. Once we show that the two are equal, eq. (9.86), which relates the cuts of any diagram to its imaginary part, is sufficient to prove unitarity. Our aim is thus to prove the identity of Fig. 11.16, analogous to Fig. 11.11 in QED, that the spin-averaged cross section for n physically polarized gluons of fixed momenta equals the cross section for the corresponding combination of spin-averaged Feynman gauge gluons and ghosts. Cut ghost loops seem to complicate the problem, but, of course, they are really crucial to its solution, as was hinted in Section 8.5. The proof is inductive, starting with Fig. 11.16 for $n-1$ gluons. The argument follows Fig. 11.17. Step by step it retraces the example given in Section 8.5 almost exactly.

(a) The first step follows from the definition of the physical intermediate state.

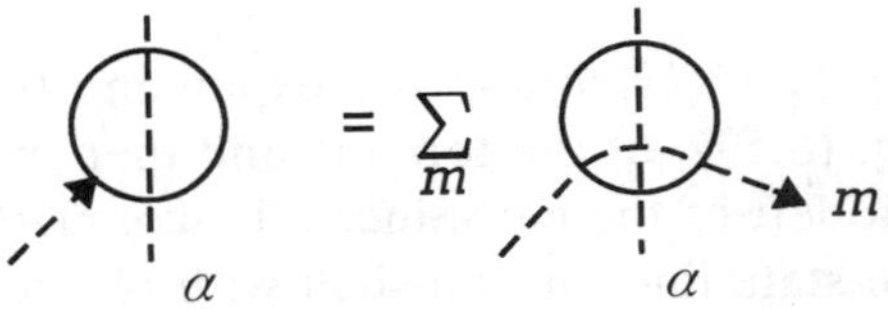

Fig. 11.15 Ward identity for cut diagrams, eq. (11.93).

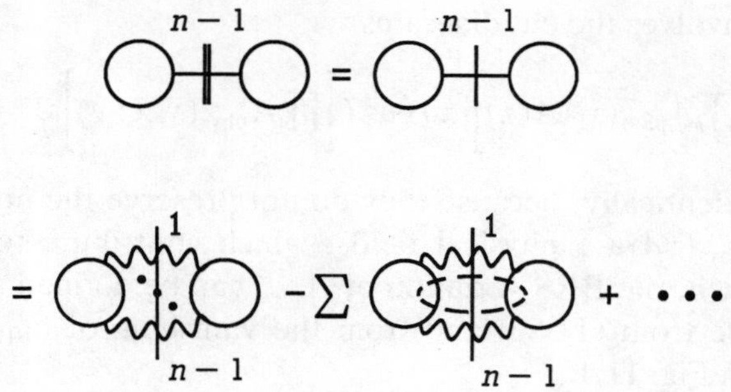

Fig. 11.16 Expression of unitarity in QCD.

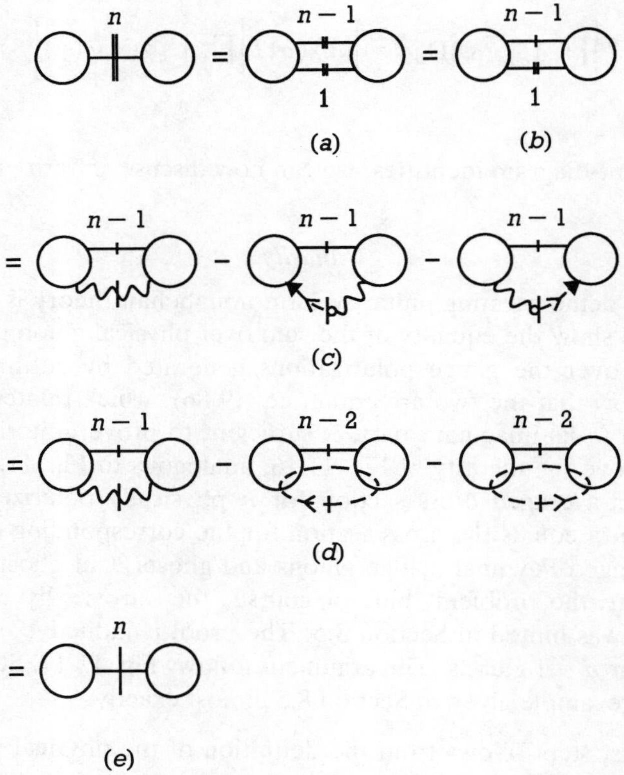

Fig. 11.17 Iterative proof of unitarity in QCD.

(b) For $n = 2$, Fig. 11.17(b) follows by expanding the physical polarization tensor, eq. (8.59), of the top line and then applying Fig. 11.12(a) to the right and left of the cut. Since all incoming lines, as well as the remaining final-state line, are on-shell with physical polarizations, the terms in the physical polarization tensor eq. (8.59), with k^μ or k^ν, vanish, and only the $g_{\mu\nu}$ term remains. For $n > 2$ this is the inductive assumption. The remainder of the argument is the same for $n = 2$ as for $n > 2$.

(c) In Fig. 11.17(*c*) we expand the physical polarization tensor of the remaining physical line k_n in the final state, according to eq. (8.59), using the notation of Fig. 8.11.

(d) Figure 11.17(*d*) is an application of the cut diagram Ward identity Fig. 11.15 to the cut amplitudes in Fig. 11.17(c), interpreting the longitudinal polarization of k_n as the BRS variation of an antighost. The new ghost line can cross the cut more than once. According to our comments following eq. (11.87), the signs of these graphs are defined by the signs in (c).

(e) Figure 11.17(*e*) is the desired result, according to Fig. 11.16, extended to n cut lines. The equality of gauge and physical polarization sums is thus proved, and unitarily established.

Renormalization and broken symmetry

Our discussion has only scratched the surface for nonabelian theories. Even for the unbroken theory, one must show that renormalization is consistent with the graphical Ward identities, derived naively above. The entire discussion, on unitarity, gauge invariance and renormalizability must also be extended to broken theories. Space allows no more than a few comments on these crucial subjects.

The renormalization of nonabelian gauge theories may be treated in much the same manner as we did above for QED, by studying the divergences of connected Green functions order by order, and noting their consistency with various Ward and Taylor–Slavnov identities. This approach (Brandt 1976), in the spirit of the original work of 't Hooft (1971a) and 't Hooft & Veltman (1972a, b), is discussed in detail by Collins (r1984) and by Frampton (r1986). Another line of proof, relating to the early work of Lee & Zinn-Justin (1972, 1973), is based on the consequences of BRS invariance for the generator of 1PI diagrams. The latter is presented, for example, in Lee (r1981) and Itzykson & Zuber (r1981).

The extension of these results to spontaneously broken gauge theories depends on showing that renormalization can be carried out when the parameters in the Lagrangian are chosen so that they do *not* produce spontaneous symmetry breaking, and that nothing unexpected happens when the parameters are varied to turn on the symmetry breaking. The unitarity of broken theories is most elegantly shown by observing that the masses of unphysical particles are gauge-dependent and can, with a specific choice of gauge (the so-called R_ξ-gauge, see Appendix C), be sent to infinity ('t Hooft 1971b, Fujikawa, Lee & Sanda 1972).

11.4 The axial anomaly

As we have just seen, it is possible for renormalized gauge theories to be unitary. Not every gauge Lagrangian, however, gives a unitary quantum theory. A crucial requirement is the existence of a regulated theory that respects the symmetries of the classical Lagrangian, and hence the Ward identities. In a complicated theory, there may be more than one set of symmetries in the classical Lagrangian and, hence, more than one set of Ward identities necessary for unitarity. As we are about to see, it can happen that the full set of classical symmetries is not consistent in the quantum theory. Such a situation is called an anomaly.

In general, if an anomaly destroys a gauge symmetry, it also destroys unitarity and, as far as is known at the time of writing, the classical theory does not correspond to a viable quantum theory (Gross & Jackiw 1972). We should mention, however, progress in quantizing some anomalous gauge theories in two dimensions (Jackiw & Rajaraman 1985, Faddeev & Shatashvili 1986).

Breaking an ungauged symmetry, on the other hand, gives no special problems. Indeed, it can give important physical consequences. An example of this is found in the simplest of the anomalies – the axial anomaly in quantum electrodynamics (Schwinger 1951b, Adler 1969, Bell & Jackiw 1969), which we describe below.

To begin, however, let us study how an ungauged, global symmetry is expected to manifest itself in a quantum theory.

Identities from global transformations

Our aim is to develop the quantum analogue of the Noether current in a classical theory (Section 1.3). Suppose that we have a Lagrange density $\mathscr{L}(\phi_i)$, where i labels space–time and internal indices of the fields. Consider a global transformation of the fields:

$$\delta\phi_i = (\partial\phi_i/\partial\omega_\alpha)\delta\omega_\alpha, \tag{11.94}$$

where the $\delta\omega_\alpha$ are a set of infinitesimal parameters. In the classical theory, we define a Noether current, $j^\mu{}_\alpha = [\partial\mathscr{L}/\partial(\partial_\mu\phi_i)][\partial\phi_i/\partial(\delta\omega_\alpha)]$, which obeys (exercise 1.7)

$$\partial_\mu j^\mu{}_\alpha = \partial\mathscr{L}/\partial\omega_\alpha. \tag{11.95}$$

If the Lagrange density is invariant under the transformation, the current is conserved.

To derive the quantum version of eq. (11.95) it is convenient to introduce position dependence into the parameters: $\omega_\alpha \to \omega_\alpha(x)$. The Lagrangian behaves under an x-dependent transformation as

$$\begin{aligned}\delta\mathscr{L} &= (\partial\mathscr{L}/\partial\omega_\alpha)\delta\omega_\alpha + [\partial\mathscr{L}/\partial(\partial_\mu\phi)](\partial\phi/\partial\omega_\alpha)\partial_\mu\delta\omega_\alpha \\ &= (\partial\mathscr{L}/\partial\omega_\alpha)\delta\omega_\alpha + j^\mu{}_\alpha\partial_\mu(\delta\omega_\alpha). \end{aligned} \tag{11.96}$$

The first term on the right is the change in $\mathscr{L}$ due to a global transformation; the second is the change due to the x-dependence of the ω_α, and comes only from the derivatives of the fields. If we assume that the Jacobian of the transformation is unity, we can apply the quantum Green function identity, eq. (11.55), to the x-dependent transformation, eq. (11.94). Taking a variational derivative with respect to $\delta\omega_\alpha(z)$, we obtain a generalization of eq. (11.58):

$$\begin{aligned}&\partial_\mu{}^{(z)}\langle 0|T[\phi_1\cdots\phi_n j^\mu{}_\alpha(z)]|0\rangle \\ &\qquad = \langle 0|T\{\phi_1\cdots\phi_n[\partial\mathscr{L}(z)/\partial\omega_\alpha]\}|0\rangle \\ &\qquad\quad -\mathrm{i}\sum_{i=1}^{n}\delta(x_i - z)\langle 0|T\{\phi_1\cdots[\delta\phi_i(z)/\delta\omega_\alpha]\cdots\phi_n\}|0\rangle. \end{aligned} \tag{11.97}$$

As usual the derivative acts outside the matrix element. Thus, the quantum Noether current obeys the classical equation, except for a set of terms resulting from the transformation of external fields.

Chiral rotation and naive identity

We are now ready to discuss our example, the behavior of the QED Lagrange density eq. (11.30) under the 'chiral' rotation

$$\psi(x) \to \mathrm{e}^{\mathrm{i}\alpha\gamma_5}\psi(x), \quad \bar\psi(x) \to \bar\psi(x)\mathrm{e}^{\mathrm{i}\alpha\gamma_5}. \tag{11.98}$$

Note the same sign in the exponentials. The rotation is called chiral because it mixes eigenfunctions of the matrices $\frac{1}{2}(1 \pm \gamma_5)$, which project out different 'handedness' or 'chirality', as discussed in Section 6.4.

The QED Lagrangian, eq. (11.30), is not invariant under (11.98). Indeed, for an infinitesimal, x-dependent transformation we find a change

$$\begin{aligned}\delta_5\mathscr{L} &= -[\partial_\mu\delta\alpha(x)]\bar\psi\gamma^\mu\gamma_5\psi - 2\mathrm{i}m\delta\alpha(x)\bar\psi\gamma_5\psi \\ &\equiv -[\partial_\mu\delta\alpha(x)]j^\mu{}_5(x) - 2\mathrm{i}m\delta\alpha(x)j_5, \end{aligned} \tag{11.99}$$

where we have introduced some standard notation in the second line. $j^\mu{}_5 \equiv \bar\psi\gamma^\mu\gamma_5\psi$, is the Noether current for chiral transformations, eq. (11.98), and j_5 is a scalar local operator. In any case, the QED Lagrangian is invariant only under global chiral rotations, and then only if the fermions are massless.

If we assume (incorrectly, as it turns out) that the Jacobian is unity for the chiral transformation, we derive a set of identities, the simplest of which is

$$\begin{aligned}\partial_\mu \langle 0|T[\bar{\psi}(y)\psi(x)j^\mu{}_5(z)]|0\rangle &= 2im\langle 0|T[\bar{\psi}(y)\psi(x)j_5(z)]|0\rangle \\ &\quad - \langle 0|T[\bar{\psi}(y)\gamma_5\psi(z)]|0\rangle\delta^4(x-z) \\ &\quad - \langle 0|T[\bar{\psi}(z)\gamma_5\psi(x)]|0\rangle\delta^4(y-z). \qquad (11.100)\end{aligned}$$

Note that the last two terms on the right-hand side do not contribute on-shell, since they lack a pole in one of the external fermion fields. In this way, for arbitrary matrix elements between physical states (or for Green functions without external fermion fields), we have the operator statement

$$\partial^\mu j_{\mu,5}(z) = 2imj_5(z). \qquad (11.101)$$

When $m = 0$, eq. (11.100) is essentially the same as eq. (11.62) for QED, and were there a gauge field associated with the chiral current, eq. (11.101) would serve much the same purpose in proving unitarity. We shall now show, however, that eq. (11.100) fails. One way to do this is by a direct evaluation of the Jacobian of the chiral transformation, as in the influential papers of Fujikawa (1979, 1980). In the following, we shall take an older route, and present a diagrammatic approach to the anomaly, following Adler (r1970). We shall find that the vector Ward identity, eq. (11.62), constrains the renormalization of the triangle diagram so tightly that it cannot satisfy the axial Ward identity, eq. (11.101) at the same time.

Coupling the axial current to fermion loops

The operator $\partial_\mu j^\mu{}_5$, the divergence of the axial vector current, is equivalent to a scalar polarized vector attached to a vertex $\gamma_\mu\gamma_5$. A perturbative argument for the naive identity, eq. (11.100), can be constructed along the same lines as for the vector Ward identity, as in Fig. 11.7. To develop the details is suggested as exercise 11.13. Here we note only that, just as for the vector Ward identity, it is an unambiguous algebraic process *except* for a shift of loop momenta around closed fermion loops. Graphs in which the axial current attaches to a fermion loop are thus the only possible source of trouble. One way of looking at the anomaly derived below is that the shifts of loop momenta necessary to prove the vector Ward identity are not consistent with those necessary to prove the axial vector identity.

Triangle diagrams

The lowest-order graph involving $j^\mu{}_5$ is a loop attaching a single vector and axial vector vertex, but this is identically zero. The next-lowest-order graph is the triangle diagram of Fig. 11.2(d), and it is in this diagram that we shall find the anomaly.

By Furry's theorem (see eq. (11.27)), when all of its vertices are vector couplings, Fig. 11.2(d) cancels the graph found by exchanging k_1 and k_2. With the extra factor of γ_5 (at the top vertex, for example) the two diagrams are equal.

Symmetry and current conservation

The calculation of Fig. 11.2(d) can be greatly simplified by first determining the tensor structure that the answer must have. We can impose two conditions on the diagram. First, it must possess Bose symmetry between the two photons:

$$T_{\sigma\rho\mu}(k_1, k_2) = T_{\rho\sigma\mu}(k_2, k_1). \tag{11.102}$$

Second, we *choose* to make it obey the vector-current Ward identity in both external photons:

$$k_1{}^\sigma T_{\sigma\rho\mu}(k_1, k_2) = k_2{}^\rho T_{\sigma\rho\mu}(k_1, k_2) = 0. \tag{11.103}$$

If eq. (11.101) were true we would conclude that, in addition,

$$(-k_1 - k_2)^\mu T_{\sigma\rho\mu}(k_1, k_2) = 2mT_{5\rho\sigma}, \tag{11.104}$$

where in $T_{5\rho\sigma}$, the vertex $(-\not{k}_1 - \not{k}_2)\gamma_5$ is replaced by γ_5. We are about to show that eqs. (11.103) and (11.104) are not consistent, and that, in fact, $(-k_1 - k_2)^\mu T_{\sigma\rho\mu}(k_1, k_2)$ will have to satisfy a different equation.

Tensor structure

$T_{\sigma\rho\mu}(k_1, k_2)$ is a third-rank tensor, odd under parity, and must be constructed from $k_{1,\alpha}$, $k_{2,\beta}$ and $\epsilon_{\alpha\beta\gamma\delta}$, times invariant functions. It is easy to see that the only other tensor available, $g_{\alpha\beta}$, cannot appear. The third-rank tensors are thus of the form $k_i{}^\alpha \epsilon_{\alpha\sigma\rho\mu}$, $i = 1, 2$, or $k_{i,\mu} k_1{}^\xi k_2{}^\tau \epsilon_{\xi\tau\rho\sigma}$ and (five) permutations of μ, ρ and σ. This is a total of eight tensors. Two of the terms with three factors of momenta may be eliminated in favor of the other four, by using the identity

$$v_\mu \epsilon_{\xi\tau\sigma\rho} = v_\xi \epsilon_{\mu\tau\sigma\rho} + v_\tau \epsilon_{\xi\mu\sigma\rho} + v_\sigma \epsilon_{\xi\tau\mu\rho} + v_\rho \epsilon_{\xi\tau\sigma\mu}, \tag{11.105}$$

which holds in four dimensions, for any vector v_μ. The most general form for $T_{\sigma\rho\mu}$ is then

$$\begin{aligned} T_{\sigma\rho\mu} = {} & A_1 k_1{}^\tau \epsilon_{\tau\sigma\rho\mu} + A_2 k_2{}^\tau \epsilon_{\tau\sigma\rho\mu} + (A_3 k_{1\rho} + A_4 k_{2\rho}) k_1{}^\xi k_2{}^\tau \epsilon_{\xi\tau\sigma\mu} \\ & + (A_5 k_{1\sigma} + A_6 k_{2\sigma}) k_1{}^\xi k_2{}^\tau \epsilon_{\xi\tau\rho\mu}. \end{aligned} \tag{11.106}$$

Power counting suggests that A_1 and A_2 are logarithmically divergent functions, while $A_3 \ldots A_6$ are finite. The values of A_1 and A_2, however, are completely determined once we impose Bose symmetry and vector current conservation. From Bose symmetry, eq. (11.102), we conclude

$$\begin{gathered} A_1(k_1, k_2) = -A_2(k_2, k_1), \\ A_4(k_1, k_2) = -A_5(k_2, k_1), \quad A_3(k_1, k_2) = -A_6(k_2, k_1), \end{gathered} \tag{11.107}$$

while from the vector Ward identity (11.103) we have

$$\begin{aligned} A_1 &= k_1 \cdot k_2 A_3 + k_2{}^2 A_4, \\ A_2 &= k_1{}^2 A_5 + k_1 \cdot k_2{}^2 A_6. \end{aligned} \tag{11.108}$$

These are the crucial results. They show that the vector Ward identity uniquely determines the ultraviolet divergent coefficients A_1, A_2 in terms of the finite coefficients $A_3 \ldots A_6$.

To check eq. (11.104) now requires only a computation of the finite quantities A_3 and A_4. In particular, from eqs. (11.103)–(11.107) we find

$$\begin{aligned} &(-k_1 - k_2)^\mu T_{\sigma\rho\mu}(k_1, k_2) \\ &\quad = -(A_1 - A_2) k_1{}^\xi k_2{}^\lambda \epsilon_{\xi\lambda\sigma\rho} \\ &\quad = -[k_2{}^2 A_4 - k_1{}^2 A_5 + k_1 \cdot k_2 (A_3 - A_6)] k_1{}^\xi k_2{}^\lambda \epsilon_{\xi\lambda\sigma\rho}. \end{aligned} \tag{11.109}$$

Finite integrals

It is quite straightforward to evaluate the finite quantities A_3 and A_4 from the integral for $T_{\sigma\rho\mu}$:

$$\begin{aligned} T_{\sigma\rho\mu} = {} & -2\mathrm{i}e^2(2\pi)^{-4}\int \mathrm{d}^4 q (q^2 - m^2 + \mathrm{i}\epsilon)^{-1}[(q + k_1)^2 - m^2 + \mathrm{i}\epsilon\,]^{-1} \\ & \times [(q - k_2)^2 - m^2 + \mathrm{i}\epsilon]^{-1} \\ & \times \operatorname{tr}[(\not{q} + \not{k}_1 + m)\gamma_\sigma(\not{q} + m)\gamma_\rho(\not{q} - \not{k}_2 + m)\gamma_\mu\gamma_5]. \end{aligned} \tag{11.110}$$

Contributions to $A_3 \ldots A_6$ occur when there are exactly three factors of the external momenta in the numerator, after Feynman parameterization and completion of the square of the loop momentum. The actual calculation is a matter of a few pages; here we quote the results:

$$\begin{aligned} A_3 &= \frac{\mathrm{i}e^2}{\pi^2}\int_0^1 \mathrm{d}x \int_0^{1-x} \mathrm{d}y\,(xy) D^{-1}, \\ A_4 &= \frac{\mathrm{i}e^2}{\pi^2}\int_0^1 \mathrm{d}x \int_0^{1-x} \mathrm{d}y\, y(1-y) D^{-1}, \\ D &= x(1-x)k_1{}^2 + y(1-y)k_2{}^2 + 2xyk_1 \cdot k_2 - m^2. \end{aligned} \tag{11.111}$$

The anomaly

Substituting eq. (11.111) into eq. (11.109), and using the symmetry between the Feynman parameters, we derive

$$(-k_1 - k_2)^\mu T_{\sigma\rho\mu}(k_1, k_2) = -\mathrm{i}(e^2/2\pi^2)k_1{}^\xi k_2{}^\lambda \epsilon_{\xi\lambda\sigma\rho} + 2mT_{5\sigma\rho}, \tag{11.112}$$

where

$$2mT_{5\sigma\rho} = -\mathrm{i}(e^2 m^2/\pi^2)k_1{}^\xi k_2{}^\lambda \epsilon_{\xi\lambda\sigma\rho}\int_0^1 \mathrm{d}x \int_0^{1-x} \mathrm{d}y\, D^{-1} \tag{11.113}$$

is exactly the right-hand side of the naive Ward identity eq. (11.104). Since the remaining, mass-independent term on the right-hand side of eq. (11.112) is a polynomial in the external momenta, it corresponds to a local interaction involving two photons. In fact, it corresponds to a new local operator, often itself referred to as the anomaly:

$$-\mathrm{i}(e^2/2\pi^2)k_1{}^\xi k_2{}^\lambda \epsilon_{\xi\lambda\sigma\rho}(\epsilon_1{}^*)^\sigma(\epsilon_2{}^*)^\rho = (\mathrm{i}e^2/8\pi^2)\langle k_1, \epsilon_1; k_2, \epsilon_2|{}^*F_{\mu\nu}(0)F^{\mu\nu}(0)|0\rangle, \tag{11.114}$$

$${}^*F_{\mu\nu}(A) \equiv \tfrac{1}{2}\epsilon_{\mu\nu}{}^{\lambda\sigma}F_{\lambda\sigma}(A).$$

We have introduced the 'dual' field strength ${}^*F_{\mu\nu}(A)$ (exercise 5.10). Evidently, the operator equation (11.101) should be modified to read

$$\partial^\mu j_{\mu,5}(z) = 2\mathrm{i}mj_5(z) + (\mathrm{e}^2/8\pi^2){}^*F_{\mu\nu}F^{\mu\nu}(z). \tag{11.115}$$

This is the axial anomaly equation, so far proved to one loop.

Let us pause a moment and ask what has happened. Nowhere have we calculated a divergent integral, and yet we have found an anomaly whose origin can only be in renormalization. The resolution of this apparent paradox is in eq. (11.108), which defines the (renormalized) ultraviolet divergent coefficients A_1 and A_2 in terms of the finite coefficients $A_3 \ldots A_6$. Whatever our regulation procedure, A_1, A_2 must be renormalized to satisfy eq. (11.108) if the vector Ward identity is to be satisfied. Once A_1, A_2 are determined, so is the axial Ward identity, via eq. (11.109), and it is no longer possible to satisfy the naive axial identity.

Higher orders and the Adler–Bardeen theorem

What about higher-order corrections with more external photons, or with internal corrections to the triangle? In the case of the triangle diagram, the worst divergence we encountered was logarithmic, because of the requirements of vector current conservation. For three or more external photons in QED, all coefficients will be finite, and all shifts necessary to prove the axial Ward identity are allowed. The second question, concerning radiative corrections to the triangle, as in Fig. 11.18, is more interesting. In fact, none of these corrections contributes to the anomaly. This is known as the Adler–Bardeen theorem (Adler & Bardeen 1969). We shall not give the details here, but the gist of the theorem is that the relevant quantity is not the superficial divergence of the graph, but the number of fermion propagators in the loop whose momenta must be shifted. Radiative corrections all have more fermion lines than the basic triangle, and when other loop integrals are regulated, the overall integral may be shifted. In summary, higher-order corrections of any kind do not affect the lowest-order result.

Gauge anomalies and the standard model

In theories such as the standard model, with left- and right-handed gauge symmetries, the analysis of anomalies is more complicated than in the case described above. The potential anomalies are known as 'gauge' anomalies. When the theory is nonabelian, the anomaly equation analogous to eq. (11.112) is also more complicated. The basic problem, however, is the same: vector and axial vector Ward identities are not necessarily consistent, and when both symmetries are gauged, there is every reason to believe that unitarity is broken, at least according to the methods currently available. It is therefore important that gauge anomalies be shown to be absent in the standard model (Gross & Jackiw 1972). Here we simply summarize the result: gauge anomalies in the standard model cancel when the number of quark and lepton generations are equal (Georgi & Glashow 1972, Bouchiat, Iliopoulos & Meyer 1972). The axial anomaly identified above remains in the standard model, but it is not gauged, since all left-handed interactions carry non-zero weak isospin.

To avoid giving the impression that the anomaly is a technical matter with only negative consequences, in Appendix E we describe the role of the anomaly in the decay of a neutral pion into two photons.

Exercises

11.1 Compute $Z_\psi{}^{(1)}$ and $Z_m{}^{(1)}$ from Fig. 11.2(a), and $Z_1{}^{(1)}$ from Fig. 11.2(c) in a general covariant gauge, $\lambda \neq 1$. Hint: it may be helpful to use the lowest-order Ward identity, Fig. 11.5(a), before doing the integrals.

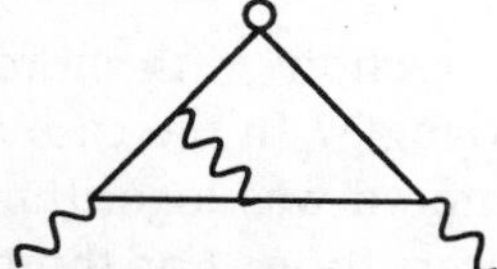

Fig. 11.18 Radiative correction that does not modify the axial anomaly.

11.2 Verify eq. (11.42).

11.3 Compute $Z_1''^{(1)}$ and $Z_c^{(1)}$ from the ghost–gluon coupling graphs, and show that the results satisfy the Taylor–Slavnov identities (11.45).

11.4 (a) Compute γ_A in minimal subtraction for QED and QCD, to one loop. (b)Determine the asymptotic behavior of an n-point function $\Gamma(\lambda p_1, \ldots, \lambda p_n)$, in the limit $\lambda \to \infty$ for both theories.

11.5 Show that the sum of all attachments of a zero-momentum photon of arbitrary polarization to a fermion loop vanishes. Hint: try to express the sum as the integral of a total derivative.

11.6 Use Ward identities to prove the gauge invariance of the S-matrix in QED for any gauge-fixing term $\lambda[g_a(A)]^2$, where $g_a(A)$ is a polynomial in the field A_μ and its derivatives.

11.7 Develop Ward identities for scalar electrodynamics.

11.8 Show from the path integral that the following Lagrange density is equivalent to eq. (11.2) (ignoring fermions):

$$\mathscr{L} = -\tfrac{1}{4}F_{\mu\nu,a}F^{\mu\nu}{}_a + \mathscr{L}_{\text{ghost}} + A_\mu\partial^\mu B + \tfrac{1}{2}\lambda B^2,$$

where $\mathscr{L}_{\text{ghost}}$ is the ghost Lagrangian, and B is an 'auxiliary' field, whose equation of motion is trivial. Show that this Lagrangian is invariant under the modified BRS transformation, specified by (11.80a, b) while replacing (11.80c) by

$$\delta\bar{c} = B\delta\xi, \quad \delta B = 0.$$

11.9 Develop the QCD Green function Ward identities for the axial gauge $g_a(A) = n \cdot A_a$, $\lambda \to \infty$. Show they are simpler than in covariant gauge.

11.10 Extend the proofs of gauge invariance and unitarity in QCD to include fermions.

11.11 By considering the transformations $\phi_i(x) \to \phi_i(x) + f_i(x)$, for arbitrary c-number functions $f_i(x)$, derive the analogue of eq. (11.97) for the equation of motion of the arbitrary field $\phi_i(x)$.

11.12 Assuming canonical anticommutation relations for the fermion fields, T^* products (Section 4.4) and the operator relation eq. (11.101), derive eq. (11.100).

11.13 Give a graphical proof of eq. (11.100). A shift in loop momenta is necessary for this proof. It is this shift, applied to the divergent triangle diagram, that is incompatable with the vector Ward identity, eq. (11.103).

11.14 Verify eqs. (11.111) and (11.113).

11.15 Suppose we postulate that the triangle diagram satisfies the axial Ward identity. Derive the resulting anomaly equation for the vector current.

PART IV

THE NATURE OF PERTURBATIVE CROSS SECTIONS

12

Perturbative corrections and the infrared problem

Renormalization makes it possible to compute perturbative corrections to lowest-order amplitudes and cross sections. The one-loop correction to the electron–photon vertex is an instructive example. Here we encounter a new kind of infinity, associated with very-long-wavelength photons. These 'infrared' divergences are well understood in quantum electrodynamics, and cancel in suitably defined cross sections. Yet another variety of on-shell infinity, the 'collinear divergence' arises in quantum chromodynamics (QCD), and in any other theory in which massless particles couple among themselves. Some of the resulting difficulties can be avoided by working with inclusive cross sections in the high-energy limit. The total and jet cross sections for e^+e^- annihilation into hadrons afford a wide range of experimental tests of QCD.

12.1 One-loop corrections in QED

Tensor structure and form factors

The fermion–photon vertex may describe the scattering of an electron or positron, or the annihilation or creation of a pair. We pick the scattering of an on-shell electron. The corresponding matrix element is $\langle (p_2, \sigma_2)^{(-)} | j_\mu(0) | (\mathrm{p}_1, \sigma_1)^{(-)} \rangle$, where j_μ is the electromagnetic current. We shall use the notation u_i for spinors $u(p_i, \sigma_i)$ below.

The most general form of the matrix element, using only parity invariance (Section 6.6), is

$$\langle (p_2, \sigma_2)^{(-)} | j_\mu(0) | (p_1, \sigma_1)^{(-)} \rangle$$
$$= \bar{u}_2[g_1 p_{1,\mu} + g_2 p_{2,\mu} + g_3 \gamma_\mu + \tfrac{1}{2} im(g_4 \sigma_{\mu\nu} p_1{}^\nu + g_5 \sigma_{\mu\nu} p_2{}^\nu)]u_1, \quad (12.1)$$

where (eq. (5.46)) $\sigma_{\mu\nu} = \frac{1}{2}\mathrm{i}[\gamma_\mu, \gamma_\nu]$, and where the *form factors* $g_i = g_i(p_1 \cdot p_2, m^2)$ are scalar functions. The terms in the square brackets represents the structure of the truncated three-point Green function, from which we have reduced the incoming and outgoing electrons.

The following relations, known as *Gordon identities* (Gordon 1928), are useful in further simplifying eq. (12.1):

$$m\bar{u}_2\gamma_\mu u_1 = p_1{}^\mu \bar{u}_2 u_1 - \mathrm{i}\bar{u}_2\sigma_{\mu\nu}p_1{}^\nu u_1, \tag{12.2a}$$

$$m\bar{u}_2\gamma_\mu u_1 = p_2{}^\mu \bar{u}_2 u_1 + \mathrm{i}\bar{u}_2\sigma_{\mu\nu}p_2{}^\nu u_1. \tag{12.2b}$$

The Gordon identities are easily proved by using the Dirac equations $mu_1 = \not{p}_1 u_1$, $\bar{u}_2 m = \bar{u}_2\not{p}_2$, and expressing the resulting products in terms of commutators and anticommutators. We can use them to eliminate the first two form factors in favor of the other three. In addition, current conservation requires $q^\mu\langle(p_2, \sigma_2)^{(-)}|j_\mu(0)|(p_1, \sigma_1)^{(-)}\rangle = 0$, where $q^\nu \equiv p_2{}^\nu - p_1{}^\nu$ is the momentum transfer. As a result, there are only two independent form factors:

$$\begin{aligned}(2\pi)^3&\langle(p_2, \sigma_2)^{(-)}|j_\mu(0)|(p_1, \sigma_1)^{(-)}\rangle \\ &= \bar{u}_2[\gamma_\mu F_1(q^2, m^2) + \tfrac{1}{2}\mathrm{i}m\sigma_{\mu\nu}q^\nu F_2(q^2, m^2)]u_1. \end{aligned} \tag{12.1$'$}$$

The one-loop corrections to the matrix element may be interpreted as corrections of order α to the lowest-order form factors, $F_1{}^{(0)} = 1$, $F_2{}^{(0)} = 0$.

In terms of the one-loop functions introduced in Section 11.1, the matrix element is given by

$$\begin{aligned}-\mathrm{i}e(2\pi)^3&\langle(p_2, \sigma_2)^{(-)}|j_\mu(0)|(p_1, \sigma_1)^{(-)}\rangle \\ &= \bar{u}_2\{\mathrm{i}\Gamma_\mu + (-\mathrm{i}e\gamma_\mu)\mathrm{i}(\not{p}_1 - m)^{-1}(-\mathrm{i}\Sigma) + \pi_{\mu\nu}(-\mathrm{i})(q^2 + \mathrm{i}\epsilon)^{-1}(-\mathrm{i}e\gamma^\nu)\}u_1. \end{aligned} \tag{12.3}$$

There is only a single fermion self-energy diagram, because in the reduction of the two electrons we divide by $R_\mathrm{e}{}^{-1}$, the residue of the electron propagator at the one-particle pole (eq. (7.88)). Up to their counterterms, the functions Γ_μ, $-\Sigma$ and $\pi_{\mu\nu}$ are defined in eqs. (11.17), (11.3) and (11.7), respectively, and the first two are to be evaluated at $p_1{}^2 = p_2{}^2 = m^2$. We shall deal with them in turn.

Vertex correction

Let us introduce the notation

$$\Gamma_\mu = -e\gamma_\mu - \Lambda_\mu, \tag{12.4}$$

where Γ_μ is the electron–photon one-particle irreducible vertex, Fig.

12.1(a), defined as in eq. (11.17). To one loop, we have

$$\begin{aligned}
&\bar{u}_2\Lambda_\mu u_1 \\
&\quad = \mathrm{i}(2\pi)^{-n}\int \mathrm{d}^n k\, [(p_2+k)^2 - m^2 + \mathrm{i}\epsilon]^{-1}[(p_1+k)^2 - m^2 + \mathrm{i}\epsilon]^{-1} \\
&\qquad \times (k^2 + \mathrm{i}\epsilon)^{-1}(-\mathrm{i}g^{\alpha'\alpha}) \\
&\qquad \times \bar{u}_2(-\mathrm{i}e\mu^\epsilon\gamma_{\alpha'})\mathrm{i}(\not{p}_2 + \not{k} + m)(-\mathrm{i}e\mu^\epsilon\gamma_\mu)\mathrm{i}(\not{p}_1 + \not{k} + m)(-\mathrm{i}e\mu^\epsilon\gamma_\alpha)u_1 \\
&\qquad + \mathrm{i}[Z_1{}^{(1)} - 1]\bar{u}_2(-\mathrm{i}e\mu^\epsilon\gamma_\mu)u_1.
\end{aligned} \tag{12.5}$$

As usual, we work in Feynman gauge with dimensional regularization. We have included the one-loop counterterm, eq. (11.21b). In accordance with the discussion of Chapters 10 and 11, e and m are the renormalized charge and mass.

Reduction of Dirac structure

To find the finite parts of eq. (12.5), we have to evaluate its Dirac structure consistently in n dimensions, noting especially that factors of $\epsilon = 2 - \frac{1}{2}n$ can combine with ϵ^{-1} poles. Thus, we are required to use the n-dimensional Dirac identities, eq. (9.44), rather than their four-dimensional counterparts.

As in Section 11.1, we shall use Feynman parameterization to perform the momentum integral. After completing the square, we find

$$\begin{aligned}
\bar{u}_2\Lambda_\mu u_1 = {}& (-\mathrm{i}e\mu^\epsilon)e^2\mu^{2\epsilon}2\int_0^1 \mathrm{d}x\int_0^{1-x}\mathrm{d}x'(2\pi)^{-n}\int \mathrm{d}^n k'\, N(p_1, p_2, k') \\
& \times [k'^2 + xx'q^2 - (x+x')^2m^2 + \mathrm{i}\epsilon]^{-3} \\
& + [Z_1{}^{(1)} - 1]\bar{u}_2(e\mu^\epsilon\gamma_\mu)u_1.
\end{aligned} \tag{12.6}$$

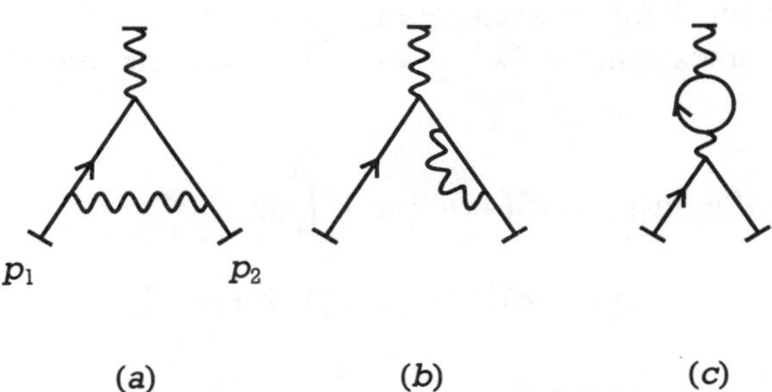

Fig. 12.1 One-loop corrections to fermion–photon interaction.

The factor $N(p_1, p_2, k')$ is given by

$$
\begin{aligned}
N&(p_1, p_2, k') \\
&= 2(\epsilon - 1)\bar{u}_2(2k'_\mu \not{k}' - k'^2\gamma_\mu)u_1 \\
&\quad + \bar{u}_2\gamma_\alpha[\not{p}_2(1 - x') - x\not{p}_1 + m]\gamma_\mu[\not{p}_1(1 - x) - x'\not{p}_2 + m]\gamma^\alpha u_1. \qquad (12.7)
\end{aligned}
$$

Here we have dropped terms which are linear in k'_ν. There is more than one way of simplifying eq. (12.7). A useful observation, however, is that the denominator in eq. (12.6) is symmetric in the Feynman parameters x and x'. As a result, we may replace factors in the numerator that are linear in x or x' by $\frac{1}{2}z$, where $z \equiv x + x'$. Then, after using the Dirac identities, eqs. (9.44), we can eliminate factors of $p_i{}^\mu$ in favor of q^μ and m by using the Dirac equation and the Gordon identities. The result of this procedure is

$$
\begin{aligned}
N(p_1, p_2, k') = {} & 2(\epsilon - 1)\bar{u}_2(2k'_\mu \not{k}' - k'^2\gamma_\mu)u_1 \\
& - \bar{u}_2\gamma_\mu u_1\{2m^2[z^2 - 2(1 - z) - \epsilon z^2] \\
& + 2q^2[(1 - x)(1 - x') - \epsilon xx']\} \\
& - 2im\bar{u}_2\sigma_{\mu\nu}q^\nu u_1[z(1 - z) - \epsilon z(2 - z)] + A(x, x'),
\end{aligned}
\tag{12.8}
$$

where $A(x, x')$ is antisymmetric in x and x', and therefore vanishes in eq. (12.6).

Equation (12.8) substituted into (12.6) gives a rather complicated expression, and, for our purposes, it is more useful to study it in a variety of limits than to give a closed form for it.

Zero momentum transfer

First, consider the limit of zero momentum transfer. The denominator in eq. (12.6) becomes simply $(k'^2 - m^2z^2)^3$, and the momentum integral gives

$$
\begin{aligned}
\bar{u}_2\Lambda_\mu u_1 = \bar{u}_2(e\mu^\epsilon\gamma_\mu)u_1\Big(& (\alpha/2\pi)(4\pi\mu^2/m^2)^\epsilon \int_0^1 \mathrm{d}z\, z^{1-2\epsilon} \\
& \times \{(1 - \epsilon)^2\Gamma(\epsilon) + \Gamma(1 + \epsilon)z^{-2}[z^2 - 2(1 - z) - \epsilon z^2]\} \\
& - (\alpha/4\pi)(\epsilon^{-1} + c_{1,1})\Big).
\end{aligned}
\tag{12.9}
$$

In (12.9) we have observed that for any function $f(z)$, where $z = (x + x')$,

$$\int_0^1 \mathrm{d}x \int_0^{1-x} \mathrm{d}x'\, f(x+x') = \int_0^1 \mathrm{d}z\, z f(z), \tag{12.10}$$

and we have used the explicit form of the counterterm eq. (11.21b). The z integrand is now simply a sum of powers, but to define the integral in the n-plane is a somewhat delicate matter. Ultraviolet poles appear, in the first term of the z integral and the counterterm, which are defined only for $\epsilon > 0$, i.e., $n < 4$. The second term in the z integral, however, is proportional to $z^{-1-2\epsilon}$, and is defined only for $\epsilon < 0$ ($n > 4$). If we combine the counterterm and the ultraviolet divergent term first, the integral is well defined in the region $n > 4$, and we find

$$\begin{aligned}\bar{u}_2 \Lambda_\mu u_1 = \bar{u}_2(e\mu^\epsilon \gamma_\mu)u_1(\alpha/4\pi)[-2(-\epsilon)^{-1} + 4 - 3\gamma_{\mathrm{E}} \\ + 3\ln(4\pi\mu^2/m^2) - c_{1,1}] + \mathrm{O}(\epsilon),\end{aligned} \tag{12.11}$$

where we have expanded in ϵ, using eq. (9.36) for the gamma function. The term $-2(-\epsilon)^{-1}$ is the infrared pole promised above. Note that $(-\epsilon)$ is positive for $n > 4$, which is where the z integral of eq. (12.9) is defined. As a result, the contribution of the infrared pole to the form factor F_1 in eq. (12.1′) is negative: it works to reduce the observed charge (by an infinite amount!).

Infrared divergence

To verify the infrared nature of the pole in eq. (12.11), we re-examine eq. (12.5) with a mass put into the photon propagator:

$$-\mathrm{i}g^{\alpha\alpha'}(k^2 + \mathrm{i}\epsilon)^{-1} \to -\mathrm{i}g^{\alpha\alpha'}(k^2 - \lambda^2 + \mathrm{i}\epsilon)^{-1}. \tag{12.12}$$

The mass λ is an ‘infrared regulator’, which softens the behavior of the photon propagator for small momenta. The right-hand side of eq. (12.12) is not the propagator of a massive photon; this includes an additional term (see eq. (7.85)). The missing term, however, is proportional to $k^\alpha k^{\alpha'}$, and therefore cancels in any gauge invariant set of graphs. In fact, a theory of ‘massive QED’ with the massive propagator in eq. (12.12) satisfies the same Ward identities as normal QED.

The effect of the replacement in eq. (12.12) on the vertex function is simply to change the momentum denominator in eq. (12.6) to

$$[k'^2 + xx'q^2 - (x+x')^2 m^2 - \lambda^2(1 - x - x') + \mathrm{i}\epsilon]. \tag{12.13}$$

This, in turn, changes a factor $(m^2 z^2)^{-1-\epsilon}$ implicit in eq. (12.9) into $[m^2 z^2 + \lambda^2(1-z)]^{-1-\epsilon}$. The $\lambda^2(1-z)$ term makes the z integral finite for all ϵ, and thus regulates the infrared divergence of the vertex. So, the infrared divergence is indeed associated with the masslessness of the photon.

On-shell renormalization

To determine the vertex correction fully, we must still choose a renormalization scheme to define the ultraviolet finite part $c_{1,1}$ of the renormalization constant $Z_1{}^{(1)}$. Because the photon is massless, it is possible to have all three external particles of the three-point vertex $\bar{u}_2\Gamma_\mu(p_2, p_1)u_1$ on-shell, by taking $p_2{}^2 = p_1{}^2 = m^2$, $q^\mu = 0$. A natural choice is thus an 'on-shell' momentum subtraction scheme (see Section 10.3), where we require that, to all orders,

$$\Gamma_\mu(p_1, p_1) = -e\gamma_\mu, \quad \Lambda_\mu = 0, \tag{12.14}$$

where e is the renormalized charge. That is, at $q^\mu = 0$ all corrections are defined to vanish. From the explicit expression, eq. (12.11), this means that the finite part is given by

$$c_{1,1} = 2\epsilon^{-1} + 4 - 3\gamma_E + 3\ln(4\pi\mu^2/m^2). \tag{12.15}$$

This is fine at $q^\mu = 0$, but we may wonder whether having an infrared divergence in $c_{1,1}$ introduces infrared divergences for nonzero momentum transfer. In fact, we shall see that this is not the case and that the infrared divergence in Z_1 will cancel against another from Z_2 (Jauch & Rohrlich 1954, r1955). For now, however, let us give eq. (12.15) our provisional acceptance, and go on with our discussion of the vertex function, renormalized on-shell.

Order-q expansion

Our next step is to expand the vertex to first order in q^μ. Since the denominator in the Feynman integral eq. (12.6) is quadratic in q^μ, the first-order contribution comes only from the numerator, and is given completely by the $\sigma_{\mu\nu}$ term in eq. (12.8). This contribution to the form factor F_2 is both infrared and ultraviolet finite, and we are free to set ϵ to zero. With on-shell renormalization, then, we can write the full three-point function as

$$\begin{aligned}\bar{u}_2\Gamma_\mu&(p_2, p_1)u_1\\ &= \bar{u}_2(-e\gamma_\mu)u_1 + 2iem(\alpha/4\pi)\bar{u}_2\sigma_{\mu\nu}q^\nu u_1\int_0^1 dz\; z(m^2z^2)^{-1}\tfrac{1}{2}z(1-z) + O(q^2)\\ &= -(e/2m)[(p+p')_\mu\bar{u}_2u_1 + (1+\alpha/2\pi)\bar{u}_2 i\sigma_{\mu\nu}q^\nu u_1] + O(q^2),\end{aligned} \tag{12.16}$$

where we have again used the Gordon identities eq. (12.2). A look at the nonrelativistic limit of eq. (12.16) shows that the second term gives a prediction for the magnetic moment of the electron. The one-loop correction (Schwinger 1948), which is proportional to $\alpha/2\pi$, is often referred to as a contribution to the 'anomalous' magnetic moment (exercise 12.2). Precise measurements and heroic calculations make it possible to compare theory and experiment to $O(e^8)$ (Kinoshita & Lindquist 1983). The spectacular success (to eleven decimal places) of this program is a classic source of confidence in quantum electrodynamics, and by implication, in quantum field theory generally.

Expanding the vertex to $O(q^2)$ is also relatively straightforward. It is the subject of exercise 12.3, in which the important physical consequences of this term are touched upon.

Infrared approximation

It is easy to isolate the infrared divergence of the vertex, eq. (12.5), explicitly for arbitrary q^ν. Factors of $\not{k}$ in the numerator may be ignored in calculating the divergence, which comes from $k^\nu = 0$. Thus we define

$$\begin{aligned}\bar{u}_2 \Lambda_\mu{}^{(\mathrm{IR})} u_1 \equiv{}& \mathrm{i}(2\pi)^{-n} \int \mathrm{d}^n k\, (2p_1\cdot k + k^2 + \mathrm{i}\epsilon)^{-1} (2p_2\cdot k + k^2 + \mathrm{i}\epsilon)^{-1} \\ &\times (k^2 + \mathrm{i}\epsilon)^{-1} (-\mathrm{i}g^{\alpha'\alpha}) \\ &\times \bar{u}_2(-\mathrm{i}e\mu^\epsilon \gamma_{\alpha'})\mathrm{i}(\not{p}_2 + m)(-\mathrm{i}e\mu^\epsilon\gamma_\mu)\mathrm{i}(\not{p}_1 + m)(-\mathrm{i}e\mu^\epsilon\gamma_\alpha)u_1 \\ &+ \mathrm{i}[Z^{(1)} - 1]^{(\mathrm{IR})}\bar{u}_2(-\mathrm{i}e\mu^\epsilon\gamma_\mu)u_1,\end{aligned} \tag{12.17a}$$

where we construct $Z_1{}^{(\mathrm{IR})}$ so that $\bar{u}_2\Lambda_\mu{}^{(\mathrm{IR})}u_1(q^2 = 0) = 0$. If we now anti-commute $(\not{p}_1 + m)$ past γ_α and $(\not{p}_2 + m)$ past $\gamma_{\alpha'}$, and use the Dirac equation, we find that $\Lambda_\mu{}^{(\mathrm{IR})}$ contributes only to the form factor $F_1(q^2)$:

$$\begin{aligned}\bar{u}(p')\Lambda_\mu{}^{(\mathrm{IR})}u(p) &= e\mu^\epsilon \bar{u}_2\gamma_\mu u_1 \alpha B(\epsilon, q^2/m^2), \\ \alpha B(\epsilon, q^2/m^2) &= -\tfrac{1}{2}(e\mu^\epsilon)^2(2\pi)^{-n}\int \mathrm{d}^n k(-\mathrm{i})(k^2 + \mathrm{i}\epsilon)^{-1} \\ &\quad\times [2p_1{}^\alpha(2p_1\cdot k + k^2 + \mathrm{i}\epsilon)^{-1} \\ &\quad - 2p_2{}^\alpha(2p_2\cdot k + k^2 + \mathrm{i}\epsilon)^{-1}]^2.\end{aligned} \tag{12.17b}$$

The cross term gives the vertex diagram, while the squares define the counterterm. (Lorentz invariance ensures that it is a function of $p_1{}^2 = p_2{}^2 = m^2$ only.) With this choice of counterterm, $\Lambda_\mu{}^{(\mathrm{IR})}$ manifestly vanishes at zero momentum transfer. In eq. (12.17b) it is particularly easy to see the origin of the infrared pole: as $k^\mu \to 0$, the integral behaves as $\int \mathrm{d}^4 k/k^4$.

Performing the loop integral by Feynman parameterization in eq. (12.17b) gives the relatively simple form

$$\alpha B(\epsilon, q^2/m^2) = -(\alpha/4\pi)(4\pi\mu^2)^\epsilon \Gamma(1+\epsilon) \times \int_0^1 dx \int_0^{1-x} dx' \{4p\cdot p[-xx'q^2 + (x+x')^2 m^2 - i\epsilon]^{-1-\epsilon} - 4m^2[(x+x')^2 m^2 - i\epsilon]^{-1-\epsilon}\}. \quad (12.18)$$

To isolate the infrared pole, the following re-expression of the parametric integral is useful:

$$\int_0^1 dx \int_0^{1-x} dx' [-xx'q^2 + (x+x')^2 m^2 - i\epsilon]^{-1-\epsilon}$$
$$= \int_0^1 dw\, w^{-1-2\epsilon} \int_0^1 du \int_0^1 du'\, \delta(1-u-u') \times [-uu'q^2 + (u+u')^2 m^2 - i\epsilon]^{-1-\epsilon}, \quad (12.19)$$

where we have defined $u = x/w$, $u' = x'/w$. A fairly straightforward calculation then gives

$$\alpha B(\epsilon, q^2/m^2) = -(\alpha/2\pi)(-\epsilon)^{-1}[4\pi\mu^2/(-q^2)]^\epsilon \times \left\{\left[\frac{(1-2m^2/q^2)}{\beta}\right]\left[\ln\frac{\beta+1}{\beta-1}\right] - 1\right\} + O(\epsilon^0),$$
$$\beta \equiv (1 - 4m^2/q^2)^{1/2} > 1. \quad (12.20)$$

For $q^2 < 0$, $\alpha B(\epsilon, q^2/m^2)$ is real, while it develops a branch cut above the two-particle threshold $q^2 = +4m^2$. It vanishes at $q^2 = 0$, as required by on-shell renormalization, and it is well defined for $\epsilon < 0$ $(n > 4)$, so it should be thought of as a negative quantity.

High-energy behavior

The dominant high-energy behavior of the vertex function can be derived from eq. (12.20) by expanding the factor $[4\pi\mu^2/(-q^2)]^\epsilon$ to first order in ϵ. Referring to eq. (12.17b) for Λ_μ and eq. (12.4) for Γ_μ (note the relative sign), we find

$$\bar{u}_2 \Gamma_\mu u_1 \xrightarrow[q^2\to\infty]{} \bar{u}_2(-e\mu^\epsilon \gamma_\mu)u_1\{1 - (\alpha/2\pi)[\ln(-q^2/m^2)\ln(-q^2/\mu^2) + (-\epsilon)^{-1}\ln(-q^2/m^2)]\}. \quad (12.21)$$

The vertex has a negative 'double' logarithm in the momentum transfer, which tends, along with the infrared pole, to decrease the lowest-order form factor. Note the singularity in the logarithm at $m = 0$. This is an example of a so-called *mass* or *collinear divergence*, which we will encounter again in QCD. With this observation, we turn to the self-energy diagrams.

Fermion self-energy

From eqs. (11.4b) and (11.6a, b), we find the complete one-loop electron self-energy, Fig. 12.1(b), which can be put into the form

$$\Gamma_2(\not p, m)$$
$$= -(\alpha/4\pi)\left\{(4\pi\mu^2)^\epsilon \Gamma(\epsilon)\int_0^1 \mathrm{d}x\, x^{-\epsilon}\frac{2(\epsilon-1)(1-x)\not p + (4-2\epsilon)m}{[(m^2-p^2)(1-x)+xm^2]^\epsilon}\right.$$
$$\left. + (\not p - m)(\epsilon^{-1} + c_{\psi,1}) - m(3\epsilon^{-1} + 3c_{m,1})\right\}. \quad (12.22)$$

To specify the self-energy completely in the on-shell renormalization scheme, we require that Γ_2 satisfy the spinor analogue of eqs. (10.62a, b) with $M = m_\mathrm{P} = m_\mathrm{R} \equiv m$,

$$(\not p - m_\mathrm{P})^{-1}\Gamma_2{}^{(\mathrm{R})}(\not p, m_\mathrm{P})u(p) = u(p). \quad (12.23)$$

That is, we choose the finite parts of Z_ψ and Z_m in order that all corrections vanish at the mass shell. This requires that we evaluate Γ_2 up to terms that vanish linearly in either $\not p - m$ or $p^2 - m^2 \approx (\not p - m)2m$. These, in turn, are found by expanding the denominator in eq. (12.22) in terms of $m^2 - p^2$, at *fixed* $\epsilon \neq 0$:

$$[(m^2-p^2)(1-x)+xm^2]^{-\epsilon} \approx (xm^2)^{-\epsilon} + \epsilon(1-x)(p^2-m^2)(xm^2)^{-1-\epsilon}. \quad (12.24)$$

The expansion in $p^2 - m^2$ produces an extra power of x in the denominator, which is regulated for $\epsilon < 0$. In fact, although eq. (12.22) is finite at $\epsilon = 0$ for $\not p = m$, its derivative with respect to $p^2 - m^2$ is not. Inserting eq. (12.24) into (12.22) gives a set of trivial integrals, and

$$\Gamma_2(\not p, m)$$
$$= (\alpha/4\pi)\{-(\not p - m)[2(-\epsilon)^{-1} - 4 + 3\gamma_\mathrm{E} - 3\ln(4\pi\mu^2/m^2) + c_{\psi,1}]$$
$$- m(4 - 3\gamma_\mathrm{E} + 3\ln(4\pi\mu^2/m^2) - 3c_{m,1})\}. \quad (12.25)$$

Demanding that the one-loop correction vanish on-shell, we find

$$c_{\psi,1} = c_{1,1} = -2(-\epsilon)^{-1} + 4 - 3\gamma_\mathrm{E} + 3\ln(4\pi\mu^2/m^2), \quad (12.26\mathrm{a})$$
$$c_{m,1} = \tfrac{4}{3} - \gamma_\mathrm{E} + \ln(4\pi\mu^2/m^2). \quad (12.26\mathrm{b})$$

Equation (12.26a) for $c_{\psi,1}$, along with eq. (12.15) for $c_{1,1}$, shows that on-shell renormalization is consistent with the renormalization-constant Ward identity $Z_1 = Z_\psi$. Then, in accordance with eq. (11.23b), we again have $e_\mathrm{R} Z_A{}^{-1/2} = e_0$.

Now let us return to the question of the infrared divergences in counterterms. The vertex and fermion wave function counterterms contribute to the matrix element, eq. (12.3), in the form

$$-\mathrm{i}e\bar{u}_2\gamma_\mu u_1[Z_1{}^{(1)} - 1] + \bar{u}_2(-\mathrm{i}e\gamma_\mu)\mathrm{i}(\not{p} - m)^{-1}[\mathrm{i}(\not{p} - m)]u_1[Z_\psi{}^{(1)} - 1]. \tag{12.27}$$

From $Z_1 = Z_\psi$, the one-loop vertex and fermion wave function counterterms cancel, including their infrared divergences. We shall show in Section 12.3 that this result generalizes to all orders.

Photon self-energy

Near $q^\mu = 0$, the one-loop photon self-energy, Fig. 12.1(*c*), plus its counterterm, is found from eqs. (11.10) and (11.12b) to be

$$\begin{aligned} &-\mathrm{i}\pi_{\alpha\beta}(q) \\ &= -(q^2 g_{\alpha\beta} - q_\alpha q_\beta)\left(\frac{2\alpha}{\pi}\right)\left\{\left(\frac{4\pi\mu^2}{m^2}\right)^\epsilon \Gamma(\epsilon)\int_0^1 \mathrm{d}x\, x(1-x) \right. \\ &\qquad\qquad \left. - \tfrac{1}{6}(\epsilon^{-1} + c_{\mathrm{A},1})\right\} + \mathrm{O}(q^2), \\ &= -(q^2 g_{\alpha\beta} - q_\alpha q_\beta)(\alpha/3\pi)[-c_{\mathrm{A},1} - \gamma_{\mathrm{E}} + \ln(4\pi\mu^2/m^2)] + \mathrm{O}(\epsilon, q^2). \end{aligned} \tag{12.28a}$$

On-shell renormalization for the photon requires, by analogy to eq. (10.62), that $\pi_{\alpha\beta}(q^2)$ vanish at $q^2 = 0$, so that $\pi_{\alpha\beta}$ makes no contribution to the residue of the pole in the photon propagator. From eq. (12.28a), this means

$$c_{A,1} = \gamma_{\mathrm{E}} + \ln(4\pi\mu^2/m^2). \tag{12.28b}$$

Z_A, and hence e, is infrared finite; $c_{A,1}$, however, has a logarithmic singularity at $m = 0$. This is another example of a mass divergence, or collinear singularity, in this limit.

To summarize, we have shown that on-shell renormalization in QED introduces no infrared divergences in the perturbation series. We have still to deal with momentum-dependent infrared divergences. In the next section, we show how this problem resolves itself in the one-loop example.

12.2 Order-α infrared bremsstrahlung

We now study photon radiation in electron scattering, or bremsstrahlung, and illustrate the resolution of the infrared problem in QED. We should note at the outset an alternative treatment in which the asymptotic states are redefined as coherent states (see Section 3.3), in such a way that infrared divergences disappear from the resulting S-matrix elements (Chung 1965, Kulish & Faddeev 1970, Contopanagos & Einhorn 1991). We shall develop a more traditional formalism, in which the S-matrix is infrared divergent in four dimensions, while appropriately constructed cross sections are nevertheless finite. This is usually referred to as the *Bloch Nordsieck mechanism* (Bloch & Nordsieck 1937).

To study 'real' (as opposed to virtual) radiative corrections, it is useful to deal directly with a cross section or decay rate. A convenient example is the decay of a neutral vector particle of mass M, which couples to electrons via the same electromagnetic current as the photon. This is given, to one loop, by the graphs of Fig. 12.2, using the 'cut graph' notation introduced in Section 8.1.

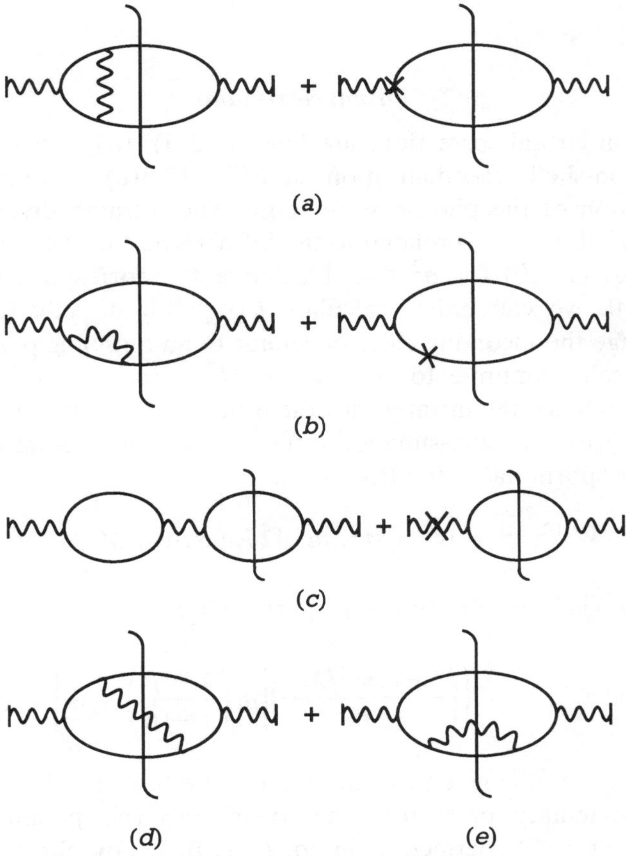

Fig. 12.2 Cut diagrams for massive vector boson decay.

Lowest order

The Born contribution to the decay width in n dimensions, averaged over the $n-1$ polarizations of the massive particle, is

$$W^{(0)}(M^2, m^2) = (n-1)^{-1} e^2 \mu^{2\epsilon} (2M)^{-1} (2\pi)^{-2n} \int \mathrm{d}^n p_1 \mathrm{d}^n p_2$$
$$\times (2\pi)^n \delta^n (q - p_1 - p_2)$$
$$\times (2\pi)\delta_+({p_1}^2 - m^2)(2\pi)\delta_+({p_2}^2 - m^2)$$
$$\times \mathrm{tr}\,[(\not{p}_2 - m)\gamma_\mu(\not{p}_1 + m)\gamma^\mu], \tag{12.29}$$

where now q^μ is the total momentum, and $q^2 = M^2$. We discard the scalar polarization of the massive vector particle, since it vanishes in the S-matrix (Section 8.1). We shall sum over final-state spins.

The remainder of our calculation will concentrate on the infrared divergent corrections to this decay rate.

Virtual corrections

The one-loop virtual corrections are Figs. 12.2(a), (b), (c). Figure 12.2(b) is zero for on-shell renormalization, and Fig. 12.3(c) is infrared finite, by our calculation of the photon self-energy. The infrared divergent part of Fig. 12.2(a), of course, is related to the infrared part of the vertex, which is given by eq. (12.20) for $q^2 < 0$. To derive the corresponding result for $q^2 = M^2 > 0$, we can either calculate $\bar{u}_2 {\Lambda_\mu}^{(\mathrm{IR})} v_1$ directly for $q^2 > 0$ or simply change the incoming electron spinor to an outgoing positron spinor, and analytically continue to ${\Lambda_\mu}^{(\mathrm{IR})}(q^2 = M^2)$. Both procedures give the same answer. Since the infrared divergent part of the vertex is now proportional to $\bar{u}_2 \gamma_\mu v_1$, the spin-summed infrared divergent virtual correction to $W(M^2)$ is proportional to the Born term:

$$W^{(\mathrm{IR,virt})} = W^{(0)}(M^2, m^2)[2\,\mathrm{Re}\,\alpha B(\epsilon, M^2/m^2)],$$
$$\alpha B(\epsilon, M^2/m^2) = -(\alpha/2\pi)(-\epsilon)^{-1}[4\pi\mu^2/M^2]^\epsilon$$
$$\times \left\{ \left(\frac{1 - 2m^2/M^2}{\beta} \right) \left[\ln\left(\frac{1+\beta}{1-\beta} \right) - \mathrm{i}\pi \right] - 1 \right\} + \mathrm{O}(\epsilon^0), \tag{12.30}$$

where $B(\epsilon, q^2/m^2)$ is given by eq. (12.20). We have exhibited an infrared divergent imaginary part from the logarithm. The parameter $0 \leqslant \beta = (1 - 4m^2/M^2)^{1/2} < 1$, defined as in eq. (12.20), is now the velocity of the fermions in the center-of-mass frame.

Photon emission

Figures 12.2(d) and (e) describe photon emission:

$$W^{(\mathrm{real})}(q^2, m^2)$$

$$= (2M)^{-1}(n-1)^{-1}(2\pi)^{-3n}\int \mathrm{d}^n p_1 \mathrm{d}^n p_2 \mathrm{d}^n k\,(2\pi)^n \delta^n(q - p_1 - p_2 - k)$$
$$\times (2\pi)\delta_+({p_1}^2 - m^2)(2\pi)\delta_+({p_2}^2 - m^2)(2\pi)\delta_+(k^2)|M|^{2(\mathrm{real})}, \quad (12.31\mathrm{a})$$

where the contribution to the squared matrix element is given by

$$\begin{aligned}\{|M|^2\}^{(\mathrm{real})} = e^4\mu^{4\epsilon}\big\{&(2p_1\cdot k)^{-2}\,\mathrm{tr}\,[\gamma^\mu(\not{p}_2 - m)\gamma_\mu(\not{p}_1 + \not{k} + m)\\ &\times \gamma_\alpha(\not{p}_1 + m)\gamma^\alpha(\not{p}_1 + \not{k} + m)]\\ &+ 2(2p_1\cdot k)^{-1}(2p_2\cdot k)^{-1}\,\mathrm{tr}\,[(\not{p}_2 - m)\gamma_\mu(\not{p}_1 + \not{k} + m)\gamma_\alpha\\ &\times (\not{p}_1 + m)\gamma^\mu(-\not{p}_2 - \not{k} + m)\gamma^\alpha]\\ &+ (2p_2\cdot k)^{-2}\,\mathrm{tr}\,[(-\not{p}_2 - \not{k} + m)\gamma_\alpha(\not{p}_2 - m)\gamma^\alpha\\ &\times (-\not{p}_2 - \not{k} + m)\gamma^\mu(\not{p}_1 + m)\gamma_\mu]\big\}. \quad (12.31\mathrm{b})\end{aligned}$$

Here p_1 is the momentum of the electron that is produced, and p_2 that of the positron. By analogy with our discussion of the vertex function, we seek an infrared approximation to eqs. (12.31). We therefore construct a partial decay rate $w(\Delta)$ that includes only states in which the photon has energy less than Δ, $\Delta \ll M$. To isolate the divergent part, we drop k^μ-dependence from the numerators in eq. (12.31b) and from the momentum-conservation delta function in eq. (12.31a). (This affects the integral only at order $|\mathbf{k}|/M$.) It is then easy to simplify the traces in eqs. (12.31), by using Dirac identities such as $(\not{p} \pm m)\gamma_\alpha(\not{p} \pm m) = 2p_\alpha$, valid for $p^2 = m^2$. This is analogous to our use of the Dirac equation to simplify the virtual correction, eq. (12.17a). Proceeding in this manner, we find a relatively simple form for $w(\Delta)$:

$$\begin{aligned} w(\Delta) &= W^{(0)}(M^2)[\alpha D(\epsilon, \Delta, M^2/m^2)],\\ \alpha D(\epsilon, \Delta, M^2/m^2) &= -e^2\mu^{2\epsilon}(2\pi)^{-n}\int \mathrm{d}^n k\,(2\pi)\delta_+(k^2)\theta(\Delta - k_0)\\ &\quad\times [{p_1}^\mu(p_1\cdot k)^{-1} - {p_2}^\mu(p_2\cdot k)^{-1}]^2. \end{aligned} \quad (12.32)$$

The function $D(\epsilon, \Delta, M^2/m^2)$ is very similar to $B(\epsilon, M^2/m^2)$, eq. (12.17b), but with the photon fixed on the mass shell, and with photon energy less than Δ. In the q^μ rest frame we have (with $k \equiv |\mathbf{k}|$)

$$\alpha D(\epsilon, \Delta, M^2/m^2) = 2e^2\mu^{2\epsilon}(2\pi)^{-n+1}\int_0^\Delta \mathrm{d}k\, k^{n-5}\int \mathrm{d}^{n-2}\Omega_k \times [(1 - 2m^2/M^2)(1 - \beta^2\cos^2\theta_k)^{-1} - (m^2/M^2)(1 - \beta\cos\theta_k)^{-2} - (m^2/M^2)(1 + \beta\cos\theta_k)^{-2}], \tag{12.33}$$

where $\cos\theta_k$ is the center-of-mass angle between the vectors $\mathbf{k}$ and $\mathbf{p}_1$. Note that in the center-of-mass frame $\mathbf{p}_1$ and $\mathbf{p}_2$ are nearly back-to-back for soft k^μ. Evaluating the integrals in eq. (12.33) shows just how similar the infrared behavior of the gluon-emission cross section is to that of the virtual correction, eq. (12.30):

$$\alpha D(\epsilon, \Delta, M^2/m^2) = (\alpha/\pi)(-\epsilon)^{-1}(4\pi\mu^2/\Delta^2)^\epsilon \times \left[\left(\frac{1 - 2m^2/M^2}{\beta}\right)\ln\left(\frac{1+\beta}{1-\beta}\right) - 1\right] + \mathrm{O}(\epsilon^0). \tag{12.34}$$

The leading large-M behavior of the gluon-emission cross section is, as a result, also similar to that of the vertex function. To one loop we have

$$W(\Delta) = W^{(0)}[1 + (\alpha/\pi)\{\ln(\Delta^2/\mu^2)\ln(M^2/m^2) + (-\epsilon)^{-1}\ln(M^2/m^2)\}]. \tag{12.35}$$

Compared with the virtual correction, eq. (12.21), the $\mathrm{O}(\alpha_s)$ pole is positive, and Δ appears in the argument of one of the logarithms.

Infrared cancellation

The sum of one-loop infrared corrections to the decay rate is given by eqs. (12.30) and (12.34):

$$W^{(\mathrm{IR},1)}(M^2, m^2, \Delta) = W^{(0)}[2\,\mathrm{Re}\,\alpha B(\epsilon, M^2/m^2) + \alpha D(\epsilon, \Delta, M^2/m^2)] = W^{(0)}\left\{(\alpha/\pi)\ln(\Delta^2/M^2)\left[\left(\frac{1 - 2m^2/M^2}{\beta}\right)\ln\left(\frac{1+\beta}{1-\beta}\right) - 1\right] + \mathrm{O}([\ln\Delta]^0)\right\}. \tag{12.36}$$

The infrared pole has cancelled in the cross section, and is replaced by a logarithm of the ratio Δ^2/M^2. Thus, although the S-matrix for exclusive processes is ill defined, the inclusive cross section, in which soft photons are unobserved, is finite, and may be compared with experiment. The correction is still negative because of the factor $\ln(\Delta^2/M^2) < 0$, which diverges as $\Delta \to 0$.

Classical analogy and energy resolution

There are a number of fundamental observations to make about infrared divergences and their cancellation. The first is that infrared divergences are associated with the classical limit.

The classical nature of the infrared divergence is seen explicitly when we examine the angular distribution of soft photons from eq. (12.33). It is exactly the same as for the soft limit of the classical bremsstrahlung radiation pattern (Jackson r1975, Chapter 15). In addition, consider the radiation energy spectrum in the frequency $I(\omega)$. Classically, $I(\omega)$ becomes independent of ω as $\omega \to 0$:

$$\lim_{\omega\to 0} \mathrm{d}I/\mathrm{d}\omega = 0. \tag{12.37}$$

By the correspondence principle, however, we can re-interpret $I(\omega)$ as

$$I(\omega) = \hbar\omega n(\omega), \tag{12.38}$$

where $n(\omega)$ is the photon number distribution. Together, (12.37) and (12.38) imply that the expected number of photons of frequency greater than or equal to any fixed ω_0 diverges logarithmically as ω_0 vanishes (Bloch & Nordsieck 1937):

$$\int_{\omega_0}^{\omega_1} \mathrm{d}\omega\, n(\omega) \xrightarrow[\omega_0\to 0]{} \frac{I(0)}{\hbar} \ln\left(\frac{\omega_1}{\omega_0}\right). \tag{12.39}$$

As a result, the probability for finding any fixed, finite number of photons is exactly zero. Equivalently, we may think of the soft classical radiation pattern as built up by many identical soft photons. The infrared divergences of the perturbation series are a reflection of this fact.

In the quantum theory, we may consider Δ as the energy resolution of a detector, so that photons with energy below Δ are not observable. In the decay process above, we lump together the exclusive electron–positron cross section with any event whose emitted photons all have energy below Δ. (Alternately, we may choose the *total* photon energy to be less than Δ.) Equation (12.36) shows that in this cross section the infrared divergent factor ϵ^{-1} is replaced by $\ln(\Delta/M)$. As long as Δ/M is large enough that $\alpha \ln(\Delta/M)$ is small compared with unity, $W^{(\mathrm{IR},1)}$, as given in eq. (12.36), is small, and higher-order corrections are expected to be even smaller. In this case, the cross section for the decay of the massive vector boson γ^*, $\gamma^* \to e^+e^-$ + soft photons, is *approximately* given by the Born approximation to $\gamma^* \to e^+e^-$, even though the true value of the latter, including higher-order corrections, is exactly zero. This seemingly paradoxical observation applies to *all* QED cross sections, and is why low-order QED calculations can be compared with experiment so successfully.

If the energy resolution is sufficiently small, correction terms proportional to $\alpha \ln(\Delta/M)$ may become large and negative, and may even overwhelm the Born cross section. This is possible, because the first-order (or any finite-order) correction to the cross section is not by itself an absolute value squared (this would require fourth-order graphs in Fig. 12.2). Hence it is not automatically positive definite. (Indeed, only the lowest-order approximation and the full sum are absolute squares.) A negative cross section simply means that higher-order corrections must be included for the perturbative cross section to be self-consistent.

12.3 Infrared divergences to all orders

An extraordinary feature of infrared divergences in QED is that they may be calculated to all orders in perturbation theory. We can explicitly implement the Bloch–Nordseick mechanism and show that, although the S-matrix is undefined, appropriately chosen cross sections are finite. For purposes of exposition, we continue with the decay process of the previous section. More detail may be found in the fundamental papers of Jauch & Rohrlich (1954), Yennie, Frautschi & Suura (1961) and Grammer & Yennie (1973). The first of these begins with an interesting review of the progress and difficulties with the 'infrared catastrophe' up to that time.

On-shell renormalization to all loops

We have seen (Section 12.1) that at one loop, Z_1 equals Z_ψ in on-shell renormalization, and that their infrared divergences cancel. To control infrared divergences at all orders, we must show that these results are general.

The renormalization-constant Ward identity, $Z_1 = Z_\psi$, can be proved to all orders for on-shell renormalization in the same way that we proved it is possible to choose $Z_1 = Z_\psi$ in *some* scheme (Section 11.2). We have already verified that $Z_1{}^{(1)} = Z_\psi{}^{(1)}$. Suppose, then, that we have successfully constructed $Z_1 = Z_\psi$ up to M loops in the on-shell scheme. Then, at $M + 1$ loops the only relevant diagrams that do not vanish at the on-shell point are the 1PI $(M + 1)$-loop vertex and self-energy corrections shown in Fig. 11.9. Equation (11.70), which applied to divergent quantities in general, now applies to the finite as well as the divergent parts of the $(M + 1)$-loop diagrams. We can thus identify $Z_1{}^{(M+1)} = -a(\epsilon)$ and $Z_\psi{}^{(M+1)} = b(\epsilon)$ in eq. (11.70), and eq. (11.71) implies that $Z_1{}^{(M+1)} = Z_\psi{}^{(M+1)}$.

We can now prove that vertex and wave function counterterms cancel each other completely to all orders in the S-matrix. Consider a graph G, in which some arbitrary vertex v has an m-loop counterterm, $Z_1{}^{(M)}$. Next consider the graph G', found by replacing $v = -ie\gamma_\mu Z_1{}^{(M)}$ by $v = -ie\gamma_\mu$, and inserting a counterterm $i(\not{p} - m)Z_\psi{}^{(M)}$ on the line immediately adjacent to v in the direction of the fermion arrow. This is illustrated in Fig. 12.3. Paired in this way, vertex counterterms cancel the wave function counterterms of all propagators, except external lines whose arrows enter the graph. The wave function renormalization counterterms of these lines are cancelled by the factors $(R_e)^{-n/2}$ in the definition of the S-matrix.

Since $Z_1 = Z_\psi$, the full vertex and fermion wave function counterterms, and not just their infrared parts, cancel. This implies that vertex and wave function *ultraviolet* divergences cancel at each order in perturbation theory, even without renormalization in QED. This has led to the 'finite QED' conjecture (Johnson & Baker 1973, Adler 1972), that quantum electrodynamics is secretly an ultraviolet finite theory.

Finally, we should check that the on-shell photon renormalization constant is also infrared finite beyond one loop. This is equivalent to the requirement that the unrenormalized photon two-point function is infrared finite at $q^2 = 0$ to all orders. The proof of this fact will be given in the next chapter. Assuming this result, we are ready to tackle the momentum-dependent infrared divergences of the vertex correction.

Infrared power counting

At one loop, infrared divergences are associated with integrals of the general form $\int_0 d^4k/k^4$. Our first goal now is to identify all such integrals, for arbitrary order.

When all the components of a photon's momentum k^μ are sufficiently small, it is natural to neglect k^2 compared with $p_1 \cdot k$ and $p_2 \cdot k$ in propagator denominators, and $\not{k}$ compared with $\not{p}$ or $\not{p}'$ in numerators. This is known as the 'eikonal approximation' (Abarbanel & Itzykson 1969, Levy & Sucher 1969). In fact, the eikonal approximation requires further justification even for soft k^μ, because we *can* have $|p \cdot k| \leq |k^2|$, if k^μ is spacelike.

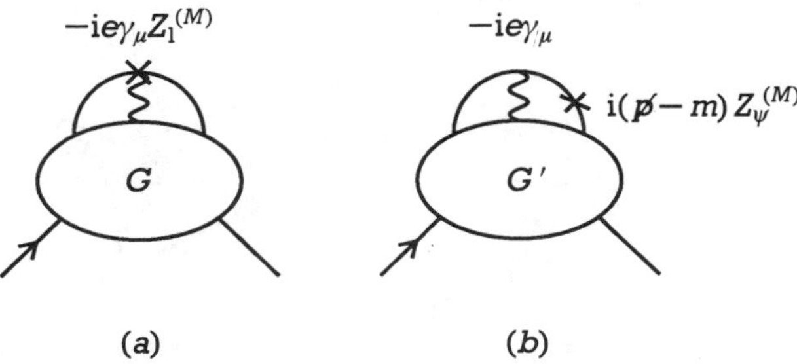

Fig. 12.3 Cancellation of vertex and fermion self-energy counterterms in QED.

We put off this justification until the next chapter, and simply assume its validity in the infrared region $k^\mu \to 0$.

We are interested in graphs of the type shown in Fig. 12.4(a). For now, we exclude internal fermion loops. Then the external fermions of momenta ${p_1}^\mu$ and ${p_2}^\mu$ are connected continuously to a unique decay vertex, labelled D, to which the massive vector is attached. The set of fermion lines connecting D with the external fermion of momentum ${p_i}^\mu$ will be referred to collectively as the 'p_i-line'.

First consider the case where all the loop momenta in graph G are small compared with all other scales, such as m and $\surd(p_1 \cdot p_2,)$ but are otherwise of comparable size. When is the integral divergent in this region? The answer is found, as for ultraviolet divergences, by combining the size of the integrand with the size of the integration volume. For each graph G, we define a 'superficial degree of infrared divergence', $\mu(G)$. Suppose γ_G is the number of photon lines, f_G the number of fermion lines and L_G the number of loops in G. In the eikonal approximation, fermion lines all contribute only a single power of soft momentum, and in n dimensions $\mu(G)$ is given in the eikonal approximation by

$$\mu(G) = nL_G - (2\gamma_G + f_G) = -2\epsilon\gamma_G - (f_G - 2\gamma_G). \tag{12.40}$$

As usual, $\epsilon = 2 - \frac{1}{2}n$, and we have used $L_G = \gamma_G$, which holds for any vertex diagram without fermion loops. With λ the scale of soft momenta, the Feynman integral then behaves near $\lambda = 0$ as

$$\int_0 \mathrm{d}\lambda\, \lambda^{\mu(G)-1} \approx [-2\epsilon\gamma_G - (f_G - 2\gamma_G)]^{-1}. \tag{12.41}$$

Equation (12.41) is divergent when $f_G \geqslant 2\gamma_G$, so zero or negative $\mu(G)$ corresponds to infrared divergence (the opposite of the ultraviolet case). In dimensional regularization, infrared divergences are regulated by going to higher, rather than lower n. Now $f_G = 2\gamma_G$ for every diagram in QED without external photons, since photons connect to fermion lines at three-point vertices. Thus, every vertex diagram is superficially divergent at $\epsilon = 0$, but (as in the one-loop case) *only* for that numerator term that is independent of photon momenta.

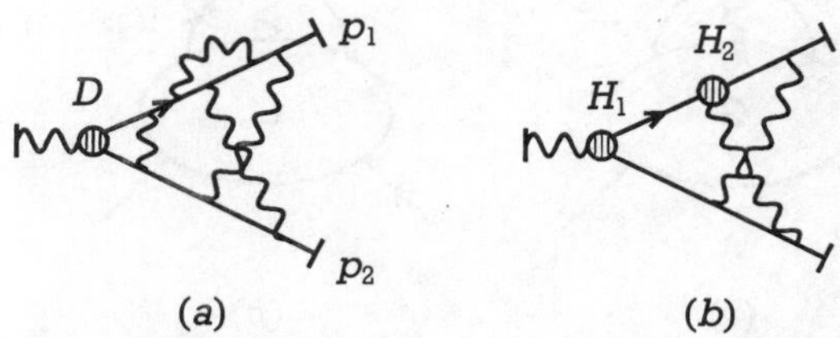

Fig. 12.4 QED diagrams with infrared divergences.

Subdivergences and reduced diagrams

To analyze infrared structure fully, we must also study what happens when some, but not all, of the photons of a graph carry zero momentum. Subdivergences that come about this way are analogous to ultraviolet subdivergences (Chapter 10).

The infrared behavior of a subset of photons, labelled S, is conveniently represented by introducing the concept of a *reduced diagram*. A reduced diagram $g(S)$ is found by contracting to a point all lines that remain off-shell when the photons in S carry zero momentum. Subdiagrams consisting of off-shell lines are referred to as 'hard' subdiagrams. An example is shown in Fig. 12.4(b), where the H_i label some possible hard subdiagrams of Fig. 12.4(a).

Hard subdiagrams with more than three external lines may be treated as constants that are independent of soft momenta for the purposes of power counting. We must, however, make exceptions for two- and three-point functions. For instance, fermion self-energies must vanish on the mass-shell, or they would introduce spurious multiple poles in propagators. Therefore, we simply keep only those hard contributions that are proportional to $\not{p} - m$, with p^μ the fermion momentum, since other dependence does not survive renormalization. Similarly, for hard three-point functions, we keep only those contributions that are proportional to γ_μ.

The evaluation of the degree of infrared divergence for a reduced diagram is now the same as for a complete diagram. A set S of vanishing photon momenta gives rise to an independent infrared divergence only if $f_{g(S)} = 2\gamma_{g(S)}$, that is, if $g(S)$ is topologically equivalent to a normal diagram in QED.

In a high-order diagram there are still many different sets of photons that give infrared divergences, and at first sight it might seem hopeless to try to compute them. In fact, just the reverse is true, and the infrared behavior of the sum of all graphs is really quite simple.

Grammer–Yennie decomposition

It is very useful that infrared divergences only arise from numerator terms without factors of photon momentum. This means that the infrared divergent parts of the fermion lines are strings of factors $(\pm\not{p}_i + m)\gamma^\alpha$. Commuting the fermion numerators $(\pm\not{p}_i + m)$ out to the external spinors, the Dirac equation may be used in the same way as after eq. (12.17a), and we find that every soft photon couples only to the external momenta ${p_1}^\alpha$ and ${p_2}^{\alpha'}$. That is, soft photons couple to each line by only a single component of their polarization. It is always possible to pick a gauge in which a particular component of the polarization vanishes, say, $p_1 \cdot A = 0$, or

$p_2 \cdot A = 0$. This suggests that infrared divergences are related to the gauge degrees of freedom of the photon.

This feature was exploited by Grammer & Yennie (1973). Consider the following replacement on the photon polarization tensor in the vertex diagram Fig.12.1(a):

$$-\mathrm{i}g^{\alpha\alpha'} \to -\mathrm{i}(p_1 \cdot p_2)k^{\alpha}k^{\alpha'}(p_1 \cdot k p_2 \cdot k)^{-1}. \tag{12.42}$$

This replacement leaves infrared behavior unchanged, because $p_{1\alpha}p_{2\alpha'}$ contracted with either side of eq. (12.42) gives the same result, $-\mathrm{i}p_1 \cdot p_2$. But the tensor on the right-hand side is made up of only scalar polarizations. In the treatment of Grammer and Yennie, Ward identities are applied to such a modified polarization tensor. We shall describe an approach based on the same observation, related to the work of Tucci (1985).

Fermion lines as Green functions

Our method will be to treat the two fermion lines as separate Green functions, embedded in a larger diagram.

First, we construct two path integrals, modified according to the reduction formula eq. (7.88), which give generating functionals for electron and positron lines that are off-shell at one end, and on-shell at the other. For an outgoing electron and positron, respectively, these are

$$\begin{aligned} Z^{(-)}[(p_2, s_2), \mathcal{L}(J_2)] &= [(2\pi)^3 R_e]^{-1/2} \int \mathrm{d}^4 y \exp(\mathrm{i}p_2 \cdot y)[-\mathrm{i}\bar{u}(p_2, s_2)](\not{p}_2 - m) \\ &\quad \times \int [\mathcal{D}\psi][\mathcal{D}\bar{\psi}]\, \psi(y)\bar{\psi}(0) \exp\{\mathrm{i}S_0[J_2{}^{\mu}]\}, \end{aligned} \tag{12.43}$$

$$\begin{aligned} Z^{(+)}[(p_1, s_1), \mathcal{L}(J_1)] &= \int \mathrm{d}^4 y \exp(\mathrm{i}p_1 \cdot y) \int [\mathcal{D}\psi][\mathcal{D}\bar{\psi}]\psi(0)\bar{\psi}(y) \exp\{\mathrm{i}S_0[J_1{}^{\mu}]\} \\ &\quad \times (-\not{p}_1 + m)[\mathrm{i}v(p_1, s_1)][(2\pi)^3 R_e]^{-1/2}. \end{aligned}$$

Here, the action S_0 is computed from the free Dirac density, plus a source term for the current $e\bar{\psi}\gamma_{\mu}\psi$:

$$S_0[J_i{}^{\mu}] = \int \mathrm{d}^4 x\ \bar{\psi}(\mathrm{i}\not{\partial} - m - e\mu^{\epsilon} J_i{}^{\mu}\gamma_{\mu})\psi. \tag{12.44}$$

Fermion lines like those in Fig. 12.4(a) are generated from $Z^{(+)}$, $Z^{(-)}$ by taking variational derivatives with respect to the source in the usual way. For instance, a positron line with one propagator is given by

$$\mathrm{i}(-\not{p}_1 + \not{k} - m)^{-1}(-\mathrm{i}e\mu^{\epsilon}\gamma_{\alpha})v(p_1) = [\delta/\delta J_1{}^{\alpha}(k)] Z^{(+)}[(p_1, s_1), \mathcal{L}(J_1)]\Big|_{J_1=0}, \tag{12.45}$$

where we define (note that $J^*(k) = J(-k)$)

$$\delta/\delta J^\alpha(k) \equiv \int d^4x \exp(-ik\cdot x)\delta/\delta J^\alpha(x). \tag{12.46}$$

Lines with more propagators simply require more variational derivatives.

Matrix element

The pair-creation matrix element without fermion loops can be found by variations on the above generating functionals:

$$\begin{aligned} -e\langle (p_1, s_1)^{(+)}, (p_2, s_2)^{(-)} | j_\mu(0) | 0\rangle \\ = Z_1 \exp\left\{\tfrac{1}{2}(2\pi)^{-n}\int d^nk\,(-i)(k^2+i\epsilon)^{-1}|\delta/\delta J_1{}^\alpha(k) + \delta/\delta J_2{}^\alpha(k)|^2\right\} \\ \times Z^{(-)}[(p_2, s_2), \mathcal{L}(J_2)](-e\mu^\epsilon\gamma_\mu)Z^{(+)}[(p_1, s_1), \mathcal{L}(J_1)]\Big|_{J_1=J_2=0}. \end{aligned} \tag{12.47}$$

The inverse factorials in the exponential compensate for the permutations of momenta within each product of variations, which give indistinguishable graphs. Z_1 renormalizes the vertex. With on-shell renormalization, $R_e = 1$, and the matrix element is given directly by the sum of the unrenormalized diagrams generated by $Z^\pm$.

Change of variables

We now change variables in the fermion generating functionals, eq. (12.43). Consider, for instance, the p_1-line. We construct a transformation that eliminates the coupling of the source to the component of $J_1{}^\alpha$ proportional to $p_1{}^\alpha$, which is responsible for the infrared divergence. This transformation is

$$\begin{aligned} \psi(x) &= \exp\{-ie\mu^\epsilon\Lambda_1[x, J_1]\}\psi'(x), \\ \Lambda_1[x, J_1] &= \int_0^\infty d\lambda\, \hat{p}_1\cdot J_1(x^\nu + \lambda\hat{p}_1{}^\nu), \end{aligned} \tag{12.48}$$

where we define

$$\hat{p}_1{}^\mu = p_1{}^\mu/m^2. \tag{12.49}$$

A precisely analogous change of variables is made in the p_2-line generating functional.

According to the theorem of eq. (11.52), the change of variables (12.48) leaves Green functions invariant when they are calculated in terms of the transformed fields, $\exp(-ie\Lambda_i)\psi'$, with the transformed Lagrangian. The transformed Lagrange densities (equal to the original ones) are

$$\mathscr{L}'_i(J_i) = \bar{\psi}'\{\mathrm{i}\partial\!\!\!/ - m - e\mu^\epsilon J\!\!\!/_i + e\mu^\epsilon \partial\!\!\!/ \Lambda_i[x, J_i]\}\psi'. \tag{12.50}$$

Because of the reduction factors in the generating functionals, eq. (12.43), nonvanishing contributions come only from the limit $y_0 \to \infty$. We shall assume that the $J_i(y)$ vanish in this limit. Then the phase changes of the products of fields in eq. (12.43) come only from the fields $\psi(0)$ and $\bar{\psi}(0)$, and the vertex function, eq. (12.47), becomes

$$\begin{aligned}
&-e\langle(p_1, s_1)^{(+)}, (p_2, s_2)^{(-)}|j_\mu(0)|0\rangle \\
&= \exp\left\{\tfrac{1}{2}(2\pi)^{-n}\int \mathrm{d}^n k\,(-\mathrm{i})(k^2 + \mathrm{i}\epsilon)^{-1}|\delta/\delta J_1{}^\alpha(k) + \delta/\delta J_2{}^\alpha(k)|^2\right\} \\
&\quad \times \exp(-\mathrm{i}e\{\Lambda_1[0, J_1] - \Lambda_2[0, J_2]\}) \\
&\quad \times Z^{(-)}[(p_2, s_2), \mathscr{L}'_2(J_2)](-e\mu^\epsilon\gamma_\mu) Z^{(+)}[(p_1, s_1), \mathscr{L}'_1(J_1)]\Big|_{J_1=J_2=0}.
\end{aligned} \tag{12.51}$$

Exponentiation of infrared divergences

We now consider the perturbation series generated by $Z^{(+)}$, $Z^{(-)}$ with the new Lagrangian. In momentum space, the interaction term gives a new momentum-dependent vertex:

$$\begin{aligned}
v_\alpha{}^{(\mathrm{i})}(k) &= -\mathrm{i}e\mu^\epsilon \delta/\delta J_i{}^\alpha(k)[J\!\!\!/_i(0) - \partial\!\!\!/ \int_0^\infty \mathrm{d}\lambda\, \hat{p}_i\cdot J_i(\lambda\hat{p}_i{}^\nu)] \\
&= -\mathrm{i}e\mu^\epsilon[\gamma_\alpha - k\!\!\!/\hat{p}_{i\alpha}(\hat{p}_i\cdot k + \mathrm{i}\epsilon)^{-1}],
\end{aligned} \tag{12.52}$$

where $\delta/\delta J_i{}^\alpha(k)$ is defined by eq. (12.46). The crucial feature of this vertex is that it anticommutes with the numerator factors $\pm p\!\!\!/_i$:

$$\{p\!\!\!/_i, v_\alpha{}^{(\mathrm{i})}(k)\}_+ \sim \{p\!\!\!/_i, \gamma_\alpha - k\!\!\!/\hat{p}_{i\alpha}(\hat{p}_i\cdot k + \mathrm{i}\epsilon)^{-1}\}_+ = 0. \tag{12.53}$$

so that

$$(\pm p\!\!\!/_i + m)v_\alpha{}^{(\mathrm{i})}(k)(\pm p\!\!\!/_i + m) = -v_\alpha{}^{(\mathrm{i})}(k)(\pm p\!\!\!/_i - m)(\pm p\!\!\!/_i + m) = 0. \tag{12.54a}$$

This is to be contrasted with the normal vertex γ_α, for which the corresponding relation is

$$(\pm p\!\!\!/_i + m)\gamma_\alpha(\pm p\!\!\!/_i + m) = \pm 2p_{i\alpha}(\pm p\!\!\!/_i + m), \tag{12.54b}$$

from which arise all infrared divergent numerator terms, as described above. The denominator in $v_\alpha{}^{(\mathrm{i})}$ does not affect power counting, because it is balanced by a numerator factor $k\!\!\!/$. Then, since a vertex $v_\alpha{}^{(\mathrm{i})}(k)$ on fermion line i anticommutes with $p\!\!\!/_i$, each $v_\alpha{}^{(\mathrm{i})}(k)$ eliminates the infrared divergence of the photon connected to it. The only photon momenta in eq.

(12.51) that are left to give infrared divergences are those both of whose variations act on the overall phase. A simple combinatoric argument shows that these variations factor out of the rest of the vertex:

$$\Gamma_\mu(p_1, p_2) = \exp\left\{-\tfrac{1}{2}(2\pi)^{-n}\int d^n k\,(-i)(k^2 + i\epsilon)^{-1} \right.$$
$$\left. \times\, e^2\mu^{2\epsilon}|\delta\Lambda_1/\delta J_1{}^\alpha(k) - \delta\Lambda_2/\delta J_2{}^\alpha(k)|^2\right\}\Gamma_\mu{}^{(H)}, \tag{12.55}$$

where every momentum integral in the expansion of the 'hard' subdiagram $\Gamma_\mu{}^{(H)}$ involves at least one vertex $v_\alpha{}^{(i)}$, and is therefore infrared finite. If we evaluate the explicit variations in eq. (12.55), we find that the infrared divergences of the vertex function *exponentiate*:

$$\Gamma_\mu(p_1, p_2) = \exp[\alpha B_{\text{eik}}(\epsilon)]\Gamma_\mu{}^{(H)}, \tag{12.56}$$

where αB_{eik} is an eikonal approximation to B, eq. (12.17b),

$$\alpha B_{\text{eik}}(\epsilon) = -\tfrac{1}{2}(e\mu^\epsilon)^2(2\pi)^{-n}\int d^n k(-i)(k^2 + i\epsilon)^{-1}$$
$$\times\,[p_1{}^\mu(p_1\cdot k + i\epsilon)^{-1} - p_2{}^\mu(p_2\cdot k + i\epsilon)^{-1}]^2. \tag{12.57}$$

Thus, as suggested above, gauge transformations like eq. (12.48), applied to the gauge-variant fermion lines, factor infrared divergences from the vertex function.

Despite its similarity to αB, eq. (12.17b), αB_{eik} includes a spurious ultraviolet divergence resulting from the lack of a k^2 term in the denominators of eq. (12.57). Since the overall matrix element is ultraviolet finite, a corresponding spurious divergence has evidently been introduced into $\Gamma_\mu{}^{(H)}$. The ultraviolet divergences of $\exp(\alpha B_{\text{eik}})$, therefore, may simply be reabsorbed into $\Gamma^{(H)}$, making the latter ultraviolet finite. A convenient method of accomplishing this is to add and subtract $\alpha B(\epsilon, q^2/m^2)$, eq. (12.20), in the exponent of eq. (12.56), giving

$$\Gamma_\mu(p_1, p_2) = \exp[\alpha B(\epsilon, M^2/m^2)]\Gamma_\mu{}^{(\text{fin})}, \tag{12.58}$$

where we define an infrared and ultraviolet finite vertex

$$\Gamma_\mu{}^{(\text{fin})} \equiv \Gamma_\mu{}^{(H)}\exp[\alpha B_{\text{eik}}(\epsilon) - \alpha B(\epsilon, M^2/m^2)]. \tag{12.59}$$

Since both Γ_μ and $B(\epsilon, q^2/m^2)$ are ultraviolet finite, all spurious ultraviolet divergences cancel in $\Gamma_\mu{}^{(\text{fin})}$.

The infrared, and leading high-energy, behavior of the vertex is now seen to be the exponential of the one-loop result, eq. (12.21),

$$\Gamma_\mu(M^2) = \exp\{-(\alpha/2\pi)[\ln(-M^2/m^2)\ln(-M^2/\mu^2) + (-\epsilon)^{-1}\ln(-M^2/m^2)]\} \times (-e\gamma_\mu + \cdots). \tag{12.60}$$

Such exponential suppression is characteristic of vertex functions in gauge theories (Sudakov 1956, Collins r1989). The elastic vertex, instead of being shifted by infrared divergences as at first order, is suppressed by them exponentially.

Fermion loops

Equation (12.58) applies only to the sum of diagrams without fermion loops. Corrections due to fermion loops are shown in Figs. 12.5(*a*) and (*b*), corresponding to photon self-energies and photon–photon scattering, respectively. As discussed in Sections 11.1 and 11.2, photons couple to fermion loops via the QED field strength, which introduces powers of k/m, where k is a typical photon momentum. Using this observation, we verify (exercise 12.11) that fermion loops cannot appear in any infrared divergent subdiagram.

Real photons

We can easily extend our results to real photons. As above, it is most convenient to work directly with cut diagrams. The relevant diagrams are of the type illustrated by Fig. 12.6. The power counting for these diagrams is carried out by treating mass-shell factors $\delta_+(k_i^2)$ in the same manner as propagators. In accordance with our comments at the end of Section 12.2, we consider the cross section for $\gamma^* \to e^+e^- + n\gamma$, when each of the n emitted photons has energy less than Δ. For Δ small enough, such a final state is indistinguishable from an exclusive one.

It is now straightforward to show (exercise 12.12) that infrared divergences from real photons exponentiate in exactly the same way as for

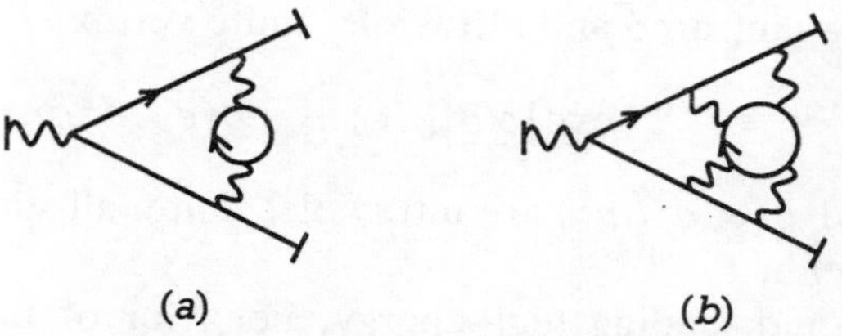

Fig. 12.5 Fermion loops in otherwise infrared divergent diagrams.

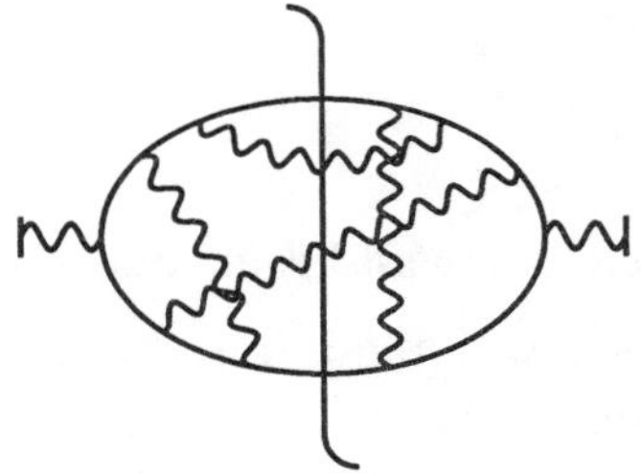

Fig. 12.6 Cut diagram with infrared divergences.

virtual photons. Neglecting overall momentum conservation, the cross section in which any photon may have an energy Δ is

$$\sigma_n = \exp[2\alpha \operatorname{Re} B(\epsilon, M^2/m^2)] \times (1/n!)[\alpha D(\epsilon, \Delta, M^2/m^2)]^n |M^{(\mathrm{fin})}(p_1, p_2)|^2, \qquad (12.61)$$

where $|M^{(\mathrm{fin})}(p_1, p_2)|^2$ is infrared (and ultraviolet) finite, and where $\alpha D(\epsilon, \Delta, M^2/m^2)$ is defined by eq. (12.32). Equation (12.61) is an (unnormalized) Poisson distribution. A Poisson distribution describes statistically independent events, and is of the form $P_n = (1/n!)\mathrm{e}^{-a}a^n$ for n such events. Apparently, the emission of each soft photon may be thought of as such an independent process. The cross section for any fixed number of real photons remains infrared divergent, and tends to zero in the $\lambda \to 0$ limit in the same way as $(\ln \lambda)^n \mathrm{e}^{-\alpha|\ln \lambda|}$ for photon mass λ. The sum over all emissions, however, is finite:

$$\sum_n \sigma_n = \exp\{\alpha[D(\epsilon, \Delta, M^2/m^2) + 2 \operatorname{Re} B^{(\mathrm{IR})}(\epsilon, M^2/m^2)]\} \times |M^{(\mathrm{fin})}(p_1, p_2)|^2. \qquad (12.62)$$

The exponent is given, to order $\ln \Delta$, by the one-loop result, eq. (12.36).

Equation (12.62) realizes our comments at the end of Section 12.2, since it shows that the total probability for photon emission is finite, even though each exclusive cross section vanishes. Notice that the argument of the exponential is negative, and diverges in the limit $\Delta \to 0$, since decreasing the allowed phase space can only decrease the cross section.

We should note that as it stands, eq. (12.62) is not really self-consistent at finite energy, because in our approximation, *all* photons can have energy Δ at the same time. For large enough n in eq. (12.61), we shall eventually run out of energy, and the approximation of neglecting soft photon momenta in the matrix element will fail. A more realistic cross section restricts the *sum* of soft photon momenta to be less than Δ. For the exact form in this case, see Jauch & Rohrlich (1954, r1955) or Grammer &

Yennie (1973). The qualitative features of eq. (12.62) remain in the more careful treatment.

12.4 Infrared safety and renormalization in QCD

In this section, we shall see why the infrared problem is more severe in an unbroken nonabelian gauge theory such as QCD than in QED. We shall discuss why on-shell renormalization is not an option in QCD. These considerations will suggest the importance of cross sections that are mass-independent at high energy. As an example, we show how the gauge coupling may, in principle, be determined from the total cross section for e^+e^- annihilation into hadrons.

On-shell behavior

In QED, on-shell renormalization is possible, because at zero momentum transfer the vertex correction cancels the electron self-energy exactly, and because the photon self-energy is infrared finite. Neither of these properties survives the generalization to a nonabelian theory. For instance, the QCD group structures of Figs. 12.1(a) and 12.1(b) are no longer the same (see eqs. (11.6a) and (11.21a)). In addition, we must include the three-gluon coupling of Fig. 11.4(d). Of course, nothing stops us from choosing $c_{1,1}$ and $c_{\psi,1}$ to cancel the infrared divergences of the different diagrams at zero momentum transfer. If we do, however, the resulting counterterms will not cancel each other ($Z_1 \neq Z_\psi$ in QCD). Their difference will then produce infrared divergences in every off-shell Green function. We can only conclude that on-shell renormalization is inappropriate for QCD.

Our inability to define a perturbative coupling from on-shell scattering is not entirely unwelcome, because we know from experiment that the asymptotic states of QCD (protons, pions, etc.) must be thought of as bound states of the elementary quanta of the theory (quarks, gluons). This is the property of 'confinement'. We shall take confinement as given, and refer the reader elsewhere for the supporting empirical evidence (Smith r1989) and theoretical evidence (Creutz r1983), both of which are beyond the range of perturbation theory. The lesson we wish to draw is that, even more than in QED, we shall have to interpret carefully the results of perturbation theory in order to derive physical predictions.

Physical parameters

Having abandoned on-shell renormalization, we still have to develop a practical method of implementing the program of Section 10.3: to define the perturbation series in terms of observable quantities. We may use either off-shell momentum subtraction or a generalized minimal subtraction. The latter is generally simpler, and is the choice we shall make.

In QCD, with a single quark flavor, there are two physical parameters, the gauge coupling g_s and the quark mass. $g_{s,\mathrm{R}}$ and m_R may be determined by solving two implicit equations of the type of eq. (10.74), in terms of some physical quantities. As a result of confinement, however, the determination of the renormalized quark mass must, of necessity, be somewhat indirect (Gasser & Leutwyler 1975, r1982, Langaker & Pagels 1979). For this reason, it is useful to determine the coupling independently of the quark mass.

High-energy and zero-mass limits: infrared safety

To fix the renormalized coupling directly, we consider cross sections, or related physical quantities, that are independent of particle massses in perturbation theory. Of course, any cross section is formally dependent on quark masses, at least through virtual corrections. There is, however, a class of cross sections for which the mass dependence disappears at high energy.

Suppose, for instance, that σ is a cross section that depends on a single large invariant q^2 in addition to the renormalized mass and coupling. The zero-mass limit of σ exists if, for $m^2 \ll q^2$,

$$\sigma(q^2/\mu^2, g_s, m^2/\mu^2) = \bar{\sigma}(q^2/\mu^2, g_s) + \mathrm{O}[(m^2/q^2)^b], \qquad (12.63)$$

where $\bar{\sigma}(q^2/\mu^2, g_s)$ is a finite function and $b > 0$. Here g_s and m are the renormalized charge and mass and, as such, are μ-dependent. $\bar{\sigma}(q^2/\mu^2, g_s)$ may be thought of as either the high-energy or zero-mass limit of σ. In an asymptotically free theory $m_\mathrm{R}(\mu)$ vanishes as $\mu \to \infty$ (Section 10.5), and the two limits are essentially equivalent. Any function that satisfies eq. (12.63) is said to be *infrared safe*.

Once we have computed $\bar{\sigma}(q^2/\mu^2, g_s)$ in perturbation theory, we can invert eq. (12.63) to find an approximate expression for $g_s(\mu)$, accurate up to $\mathrm{O}[(m^2/q^2)^b]$, in which the renormalized mass does not enter. In the following, we give a low-order example of this approach in QCD. We shall compute the one-loop correction to the cross section for $\mathrm{e}^+\mathrm{e}^-$ annihilation into hadrons, and use it to determine the coupling.

Born cross section for e^+e^- annihilation at high energy

At lowest order, the process $\mathrm{e}^+\mathrm{e}^- \to$ hadrons is described by Fig. 12.7, in which all vertices are electromagnetic. We have already computed this annihilation cross section in eqs. (8.34), (8.35). We must sum over the cross-sections for each quark pair, specified by color and flavor. In terms of

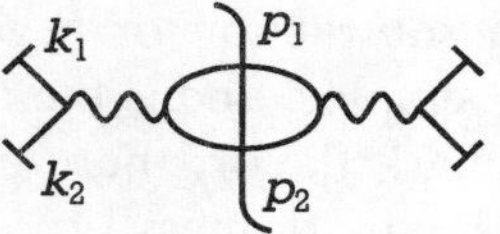

Fig. 12.7 Lowest-order cut diagram for $e^+e^- \to$ hadrons.

the center-of-mass polar scattering angle θ^*, the lowest-order cross section for a set of massless quarks with charges $Q_f e$ in color $SU(N)$ is

$$d\sigma/d\cos\theta^* = N\pi\alpha^2(1/2s)\sum_f Q_f^2(1+\cos^2\theta^*), \tag{12.64}$$

where $s=q^2$, the virtual-photon invariant mass squared. This gives the total Born cross section

$$\sigma_{\text{tot}} = N(4\pi\alpha^2/3q^2)\left(\sum_f Q_f^2\right). \tag{12.65}$$

Of course, the on-shell lines in Fig. 12.7 are quarks. This is a useful result when we interpret it as the first term in the perturbative series for the total cross section. For the elastic quark-pair production cross section, on the other hand, we expect higher orders to overwhelm completely the lowest-order result. This is very similar in spirit to the QED calculation, where the lowest-order cross section is a good approximation to the cross section for pair production plus soft photon emission, not for exclusive pair production.

Equation (12.65) is zeroth order in the QCD coupling. For large center-of-mass energy $\sqrt{q^2}$, asymptotic freedom suggests that $g_s(\mu=\sqrt{q^2})$ is small (Section 10.5). As a result, higher-order corrections may be small, even negligible, at high enough energy. This, however, assumes that the perturbation series is well defined in the high-energy or zero-mass limit. We shall check that this is indeed the case for $e^+e^- \to$ hadrons at $O(g_s^2)$ below, and prove it to all orders in the next chapter.

A standard way of presenting the lowest-order result (12.65) for $e^+e^- \to$ hadrons is in terms of its ratio to the total cross section for $e^+e^- \to \mu^+\mu^-$, which is computed from the same lowest-order diagram, Fig. 12.7. The only difference in the calculations is that there is no sum over color or flavor for the outgoing muon pair. The ratio, which is commonly labelled R, is thus given in the Born approximation by

$$R = \frac{\sigma_{\text{tot}}(e^+e^- \to \text{hadrons})}{\sigma_{\text{tot}}(e^+e^- \to \mu^+\mu^-)} = N\left(\sum_f Q_f^2\right). \tag{12.66}$$

This result can be tested experimentally (Particle Data Group r1990, p. III.74). The presence of the color quantum number results in a total cross section that is N (= 3) times greater than it otherwise would be. This simple observation constitutes important evidence for the color quantum number in strong interactions.

We now turn to the first-order correction, at order $\alpha\alpha_s$ ($\alpha \equiv e^2/4\pi$, $\alpha_s \equiv g_s{}^2/4\pi$). The relevant cut graphs are the same as in Fig. 12.2, with leptons in the initial state (except for Fig. 12.2(c), which only contributes at $O(\alpha^2)$). All these diagrams contribute to the cross section in the general form.

$$\sigma_{\text{tot}}(q^2) = [e^2\mu^{2\epsilon}/2(q^2)^3](k_1{}^\mu k_2{}^\nu + k_2{}^\mu k_1{}^\nu - k_1 \cdot k_2 g^{\mu\nu}) H_{\mu\nu}(q^2),$$
$$H_{\mu\nu}(q^2) = e^2\mu^{2\epsilon}\sum_n \langle 0|j_\mu(0)|n\rangle\langle n|j_\nu(0)|0\rangle(2\pi)^4\delta^4(p_n - q), \tag{12.67}$$

where k_1 and k_2 are the momenta of the incoming lepton pair ($q = k_1 + k_2$) and the sum is over hadronic states n. The contribution from the purely hadronic part of the diagram, in our case, the cut quark loop, is $H_{\mu\nu}(q)$. After integrating over phase space, the only vector on which $H_{\mu\nu}(q)$ can depend is q^μ. Current conservation is also satisfied by $H_{\mu\nu}(q)$: $q^\mu H_{\mu\nu} = H_{\mu\nu}q^\nu = 0$. It is therefore transverse:

$$H_{\mu\nu}(q) = (q_\mu q_\nu - q^2 g_{\mu\nu})H(q^2), \tag{12.68}$$

where $H(q^2)$ is a scalar function. Combining eqs. (12.67) and (12.68), we easily derive the general relation

$$\begin{aligned}\sigma_{\text{tot}}&(q^2)\\ &= [e^2\mu^{2\epsilon}/2(q^2)^2][(1-\epsilon)/(3-2\epsilon)][-g^{\mu\nu}H_{\mu\nu}(q^2)],\\ &= [e^2\mu^{2\epsilon}/2(q^2)^2][(\epsilon-1)/(3-2\epsilon)]e^2\mu^{2\epsilon}\sum_n\langle 0|j_\mu(0)|n\rangle\langle n|j^\mu(0)|0\rangle(2\pi)^4\\ &\qquad\times\delta^4(p_n - q).\end{aligned} \tag{12.69}$$

The 'trace' $g^{\mu\nu}H_{\mu\nu}$ is easier to evaluate than $H_{\mu\nu}$ itself. We shall now carry out this calculation for a quark with electric charge eQ_f.

In terms of invariant matrix elements, the real-gluon contributions $H_{\mu\nu}{}^{(\text{r})}$, Figs. 12.2($d$), ($e$), may be written

$$\begin{aligned}-g^{\mu\nu}H_{\mu\nu}{}^{(\text{r})} &= \int \mathrm{d}^n p_1\,\mathrm{d}^n p_2\,\mathrm{d}^n k\,(2\pi)^{-2n}\delta^n(q - p_1 - p_2 - k)\\ &\qquad\times 2\pi\delta_+(p_1{}^2)2\pi\delta_+(p_2{}^2)2\pi\delta_+(k^2)\sum_{\text{spins}}|M(p_1, p_2, k)|^2,\end{aligned} \tag{12.70}$$

where k^μ is the gluon momentum and $p_1{}^\mu$ and $p_2{}^\mu$ are the momenta of the quark pair. We can take $|M(p_1, p_2, k)|^2$ to be a function of the energies of

$p_1{}^\mu$ and k^μ, which we denote as p_1 and k, and of the cosine of the angle between their spatial components, which we denote by u. The remaining variables describe the $(n-2)$-dimensional angular integral for $p_1{}^\mu$ and the $(n-3)$-dimensional 'azimuthal' angular integral for k^μ (see eq. (9.27)):

$$\begin{aligned}-g^{\mu\nu}H_{\mu\nu}{}^{(\mathrm{r})} &= [4(2\pi)^{2n-3}]^{-1}\Omega_{n-2}\Omega_{n-3}\int_0^\infty \mathrm{d}p_1\, p_1{}^{n-3}\int_0^\infty \mathrm{d}k\, k^{n-3} \\ &\quad\times \int_{-1}^1 \mathrm{d}u\,(1-u^2)^{-\epsilon}\,\delta((q-p_1-k)^2) \\ &\quad\times \sum_{\text{spins}} |M(p_1, q-p_1-k, k)|^2. \end{aligned} \tag{12.71}$$

We have used the expression for the angular integral in m dimensions, eq. (9.38), to get the form of the u integral. We can simplify matters still further by working in the center-of-mass frame, where the mass-shell delta function for $p_2{}^\mu$ becomes (recall $s \equiv q^2$)

$$\delta((q-p_1-k)^2) = \delta(s - 2s^{1/2}(p_1+k) + 2p_1k(1-u)). \tag{12.72}$$

The matrix element in eq. (12.71) is not difficult to calculate, using the symmetry between mirror graphs:

$$\begin{aligned}&\sum_{\text{spins}} |M(p_1, p_2, k)|^2 \\ &\quad = 16NC_2(\mathrm{F})Q_f{}^2e^2g_s{}^2\mu^{4\epsilon}(1-\epsilon)[q^2\tau/t_1t_2 - \epsilon + (1-\epsilon)(t_2/t_1)],\end{aligned} \tag{12.73}$$

where we define

$$\begin{aligned}\tau &\equiv 2p_1\cdot p_2 = 2p_1[s^{1/2} - k(1-u)], \\ t_1 &\equiv 2p_1\cdot k = 2p_1k(1-u), \\ t_2 &\equiv 2p_2\cdot k = 2k[s^{1/2} - p_1(1-u)].\end{aligned} \tag{12.74}$$

Inserting eqs. (12.72) and (12.73) in eq. (12.71), we find

$$\begin{aligned}&-g^{\mu\nu}H_{\mu\nu}{}^{(\mathrm{r})} \\ &= 4NC_2(\mathrm{F})Q_f{}^2e^2g_s{}^2\mu^{4\epsilon}(2\pi)^{-5+4\epsilon}(1-\epsilon)\Omega_{n-2}\Omega_{n-3}(s^{1/2})^{1-2\epsilon} \\ &\quad\times\int_0^{s^{1/2}/2}\mathrm{d}k\,k^{1-2\epsilon}\int_{-1}^1\mathrm{d}u(1-u^2)^{-\epsilon}(s^{1/2}-2k)^{1-2\epsilon}[2s^{1/2}-2k(1-u)]^{-2+2\epsilon} \\ &\quad\times\left\{\frac{[s^{1/2}-(1-u)k]^2}{k^2(1-u^2)} - \epsilon + (1-\epsilon)\frac{s^{1/2}(1+u)}{(s^{1/2}-2k)(1-u)}\right\}.\end{aligned} \tag{12.75}$$

The second term in the braces, $-\epsilon$, does not contribute in the limit $n \to 4$, and we drop it below.

The remaining integrals in eq. (12.75) can easily be performed by changing variables to

$$y = (1-u)/2, \quad z = 2k/s^{1/2}, \tag{12.76}$$

after which we have

$$\begin{aligned} -g^{\mu\nu} H_{\mu\nu}{}^{(r)} &= \tfrac{1}{2} N C_2(\mathrm{F}) Q_f{}^2 e^2 g_s{}^2 \mu^{4\epsilon} (2\pi)^{-5+4\epsilon} s(4s^{-2})^{\epsilon} \Omega_{n-2}\Omega_{n-3}(1-\epsilon) \\ &\quad \times \int_0^1 \mathrm{d}z\, z^{1-2\epsilon}(1-z)^{1-2\epsilon} \int_0^1 \mathrm{d}y [y(1-y)]^{-\epsilon}(1-zy)^{-2+2\epsilon} \\ &\quad \times \left[\frac{(1-zy)^2}{z^2 y(1-y)} + (1-\epsilon)\frac{1-y}{(1-z)y} \right]. \end{aligned} \tag{12.77}$$

The integrals over y and z separately give factors of $1/\epsilon$. The variable z is proportional to the energy of the gluon, and the pole at $z \to 0$ is just the infrared divergence familiar from QED. The divergence in the y integral, on the other hand, is absent when the fermion has a mass. Tracing backwards, we easily find that the factors y and $1-y$ come from the angular dependence of the two denominators:

$$(p_1+k)^2 = 4p_1 k y, \; (p_2+k)^2 = 2(Q-2ky)^{-1} Q^2 k(1-y). \tag{12.78}$$

$y=0$ and $y=1$ thus correspond to $\mathbf{k}$ parallel to $\mathbf{p}_1$ and $\mathbf{p}_2$, respectively. In each case one of the two denominators in eq. (12.77) vanishes. Thus, the resulting divergences are descriptively known as *collinear*. Since they are associated with the masslessness of the quark, they are sometimes referred to as mass divergences (see eqs. (12.21) and (12.28)).

Using the integral formulas of eqs. (9.30), the result of the integrations is

$$\begin{aligned} -g^{\mu\nu} H_{\mu\nu}{}^{(r)} &= \tfrac{1}{2} N C_2(\mathrm{F}) Q_f{}^2 e^2 g_s{}^2 (2\mu^2/q^2)^{2\epsilon} (2\pi)^{-5+4\epsilon} q^2 (1-\epsilon)\Omega_{n-2}\Omega_{n-3} \\ &\quad \times [B(-\epsilon,-\epsilon)B(-2\epsilon, 2-\epsilon) \\ &\quad + (1-\epsilon)B(2-\epsilon,-\epsilon)B(2-2\epsilon, 1-\epsilon)]. \end{aligned} \tag{12.79}$$

The angular volumes may be found from eq. (9.37). It is a simple matter to use eq. (9.36) to expand the gamma functions, and find the $\epsilon \to 0$ behavior of (12.79):

$$\begin{aligned} -g^{\mu\nu} H_{\mu\nu}{}^{(r)} &= 2NC_2(\mathrm{F}) Q_f{}^2 (\alpha\alpha_s/\pi) q^2 (4\pi\mu^2/q^2)^{2\epsilon} [(1-\epsilon)/\Gamma(2-2\epsilon)] \\ &\quad \times [\epsilon^{-2} + \tfrac{3}{2}\epsilon^{-1} - \tfrac{1}{2}\pi^2 + \tfrac{19}{4} + \mathrm{O}(\epsilon)]. \end{aligned} \tag{12.80}$$

Note in particular the ϵ^{-2} terms, which are a combination of infrared ($z \to 0$) and collinear ($y \to 0, 1$) behavior. They are positive in sign, like the infrared pole of real photon emission in QED.

Virtual correction

Since the basic interaction is electromagnetic, the vertex and fermion self-energy corrections have the same group structure, and their counterterms cancel. In addition, because the fermions are massless, the fermion self-energy without counterterm is a scaleless integral, which vanishes in dimensional regularization (Section 11.1). Note, by the way, that even though the self-energy vanishes, its counterterm is required by the Ward identity to be nonzero. Of the virtual graphs, this leaves only Fig. 12.2(a), calculated as if there were no counterterm. The fully massless vertex correction can be found from eqs. (12.6) and (12.8), by changing the kinematics and spinors, setting $m = 0$ and replacing e by g_s. This gives

$$\bar{u}(p_2)\Lambda_\mu v(p_1) = (-\mathrm{i}e\mu^\epsilon){g_s}^2\mu^{2\epsilon}C_2(\mathrm{F})2\int_0^1 \mathrm{d}x \int_0^{1-x} \mathrm{d}x'(2\pi)^{-n}$$
$$\times \int \mathrm{d}^n k'\,\bar{u}(p_2)\{2(\epsilon-1)(2k'_\mu \not{k}' - k'^2\gamma_\mu)$$
$$+ 2q^2\gamma_\mu[(1-x)(1-x') - \epsilon xx']\}v(p_1)$$
$$\times (k'^2 + xx'q^2 + \mathrm{i}\epsilon)^{-3}, \tag{12.81}$$

where we have inserted the group factor $C_2(\mathrm{F})$. The evaluation of eq. (12.81) follows the normal rules, and we find

$$\bar{u}(p_2)\Lambda_\mu v(p_1) = (-\mathrm{i}e\mu^\epsilon)\bar{u}(p_2)\gamma_\mu v(p_1)\gamma(q^2),$$
$$\gamma(q^2) = -(\alpha_s/2\pi)C_2(\mathrm{F})[4\pi\mu^2/(-q^2)]^\epsilon[\Gamma^2(1-\epsilon)\Gamma(1+\epsilon)/\Gamma(1-2\epsilon)]$$
$$\times (\epsilon^{-2} + \tfrac{3}{2}\epsilon^{-1} + 4). \tag{12.82}$$

The leading double logarithm in q^2 of this form, found by expanding $[4\pi\mu^2/(-q^2)]^\epsilon$ to second order in ϵ, is $(\alpha/4\pi)C_2(\mathrm{F})\ln^2 q^2$. Aside from the factor $C_2(\mathrm{F})$, this is one-half the coefficient of $\ln^2 q^2$ in the QED vertex function eq. (12.21), calculated with a massive fermion and a massless photon. Apparently, part of the logarithmic momentum dependence in eq. (12.21) comes from scales that are less than or of the order of the fermion mass. In an asymptotically free theory, the perturbative effective coupling diverges at low momentum scales (eq. (10.125)). This is one of the reasons why in QCD we are primarily interested in quantities that are independent of particle masses altogether.

Since the sum of virtual corrections, eq. (12.82), is proportional to the lowest-order vertex, its effect on the total cross section is purely multiplicative. Let us define

$$-g^{\mu\nu}H_{\mu\nu}{}^{(\mathrm{v})} = -g^{\mu\nu}\pi_{\mu\nu}{}^{(0)}2\,\mathrm{Re}\,\gamma(q^2), \tag{12.83}$$

where $\pi_{\mu\nu}{}^{(0)}$ is the hadronic tensor in the Born approximation,

$$
\begin{aligned}
-g^{\mu\nu}\pi_{\mu\nu}{}^{(0)} &= -\ NQ_f{}^2e^2\mu^{2\epsilon}(2\pi)^{-2+2\epsilon}\int \mathrm{d}^n p_1\,\mathrm{d}^n p_2\,\delta_+(p_1{}^2)\delta_+(p_2{}^2) \\
&\qquad \times\ \delta^n(p_1+p_2-k_1-k_2)\,\mathrm{tr}\,(\gamma_\mu \not{p}_1\gamma^\mu\not{p}_2) \\
&= 2\alpha Q_f{}^2N(4\pi\mu^2/q^2)^\epsilon(1-\epsilon)[\Gamma(1-\epsilon)/\Gamma(2-2\epsilon)]q^2. \quad (12.84)
\end{aligned}
$$

Substituting eqs. (12.82) and (12.84) into eq. (12.83), and expanding in ϵ, we obtain

$$
\begin{aligned}
-g^{\mu\nu}H_{\mu\nu}{}^{(\mathrm{v})} &= -2NC_2(\mathrm{F})Q_f{}^2(\alpha\alpha_s/\pi)q^2(4\pi\mu^2/q^2)^{2\epsilon}[(1-\epsilon)/\Gamma(2-2\epsilon)] \\
&\qquad \times [\epsilon^{-2}+\tfrac{3}{2}\epsilon^{-1}-\tfrac{1}{2}\pi^2+4+\mathrm{O}(\epsilon)]. \qquad (12.85)
\end{aligned}
$$

This expression has double poles that are infrared and collinear divergences, which cancel those of the real contributions, eq. (12.80). As in QED, the divergent contribution of virtual diagrams is negative. Finally, combining eqs. (12.80) and (12.85) with (12.69) gives, for the full $\mathrm{O}(\alpha_s)$ cross section, the surprisingly simple result (summed over quark flavors)

$$
\sigma_{\mathrm{tot}}(q^2) = N(4\pi\alpha^2/3q^2)\sum_f Q_f{}^2[1+(\alpha_s/\pi)\tfrac{3}{4}C_2(\mathrm{F})] + \mathrm{O}(\alpha\alpha_s{}^2). \quad (12.86\mathrm{a})
$$

Thus, we have verified that the cross section is indeed finite in the zero-mass and high-energy limits at one loop, and is hence suitable for determining α_s (Appelquist & Georgi 1973, Zee 1973). This calculation has now been extended to $\mathrm{O}(\alpha\alpha_s{}^3)$ (Surguladze & Samuel 1991, Gorishny, Kateev & Larin 1991).

Determination of α_s

In principle, once σ_{tot} is measured at a given q^2, α_s is determined approximately by solving eq. (12.86a). The solution is a renormalized coupling, and can be used to calculate other cross sections to a comparable degree of accuracy. Typical values of α_s found in this way are about 0.1 for q^2 in the range of 10^2–10^4 GeV^2 (Komamiya, Le Diberder *et al.* 1990, OPAL Collaboration 1990). The first approximation to $\alpha_s(\mu^2)$ is independent of the renormalization scheme and the value of μ. This is a general feature, since at lowest order in α_s, any cross section is independent of the finite parts of the QCD renormalization constants. (Equation (12.82) is a one-loop correction to the *electromagnetic* vertex, and therefore does not involve QCD renormalization.) At two loops, however, the series for the cross section is of the form

$$\sigma_{\text{tot}}(q^2) = N(4\pi\alpha^2/3q^2)\sum_f Q_f{}^2[1 + (\alpha_s/\pi)\tfrac{3}{4}C_2(\text{F}) + (\alpha_s/\pi)^2 A_2(q^2/\mu^2)] + \text{O}(\alpha_s{}^3), \qquad (12.86\text{b})$$

where A_2 *is* scheme-dependent and μ-dependent. The value $\alpha_s(q^2)$ found by solving eq. (12.86b) is correspondingly renormalization-scheme-dependent. This dependence would disappear if we knew the expansion to all orders, but, of course, this is not the case. The renormalization group equation (10.100) for the high-energy cross section,

$$[\mu\partial/\partial\mu + \beta(g)\partial/\partial g]\sigma_{\text{tot}}(q^2, \mu^2, \alpha_s(\mu^2)) = 0, \qquad (12.87)$$

is a statement of the equality of the total cross section for different values of μ. To simplify higher-order corrections, it is natural to choose $\mu^2 = q^2$, so that all logarithms of q^2/μ^2 vanish. In an asymptotically free theory, σ_{tot} will then be given by a series in the small coupling $\alpha_s(q^2)$ times numbers that do not grow with q^2. Let us emphasize that the usefulness of eq. (12.87) follows from the infrared safety of the cross section, eq. (12.63). Residual leading-power mass dependence would appear as logarithms of mass over μ^2 or q^2. With such mass-dependence, no single choice of μ^2 would eliminate all logarithmic behavior in energy.

We can use eq. (12.86b) to relate the expressions for $\alpha_s(q^2)$ found in two different schemes or values of μ. The difference is $\text{O}(\alpha_s{}^2)$:

$$\alpha_s = \alpha_s{}' + [(\alpha_s{}')^2/\pi][A_2{}'(1) - A_2(1)] + \text{O}(\alpha_s{}')^3. \qquad (12.88)$$

This kind of ambiguity in the determination of α_s – one order higher than the calculation of the cross section – is unavoidable at any finite order of perturbation theory. Since our estimate of the size of uncalculated higher-order corrections depends on α_s, and since A_2 can be large in particular renormalization schemes, extra care must be used in the interpretation of low-order calculations in QCD (Duke & Roberts r1985). In QED the corresponding problem is much less severe, because the QED fine-structure constant is smaller than α_s by a factor of more than ten.

The running coupling and the QCD scale parameter

Given a determination of $\alpha_s(q^2)$ at any q^2, we can use the β-function of QCD to define the running coupling $\bar{\alpha}_s(\mu^2)$ through the equation

$$(\mu\partial/\partial\mu)(\bar{\alpha}_s/\pi) = (\bar{g}/2\pi^2)\beta(\bar{g}_s) = (\bar{\alpha}_s/\pi)^2(\beta_1/2) + (\bar{\alpha}_s/\pi)^3(\beta_2/8), \qquad (12.89)$$

which is an expansion of eq. (10.101a), in this case to two loops. β_1 and β_2 are conventional coefficients in the expansion of $\beta(g_s)$:

$$\beta(g_s) \equiv g[(\alpha_s/4\pi)\beta_1 + (\alpha_s/4\pi)^2\beta_2 + \cdots]. \qquad (12.90)$$

From eq. (11.50) for the one-loop beta function, and from the calculations of Caswell (1974) and Jones (1974), we have

$$\beta_1 = -\tfrac{11}{3}C_2(\mathrm{A}) + \tfrac{4}{3}T(\mathrm{F})n_f,$$
$$\beta_2 = -\tfrac{34}{3}C_2{}^2(\mathrm{A}) + [\tfrac{20}{3}C_2(\mathrm{A}) + 4C_2(\mathrm{F})]T(\mathrm{F})n_f. \qquad (12.91)$$

Solving eq. (12.89) gives an effective coupling $\alpha_s(\mu^2)$ that depends on both an initial condition $\alpha_s(\mu_0{}^2)$ and the ratio $\mu^2/\mu_0{}^2$. Other choices of μ_0, and hence $\alpha_s(\mu_0{}^2)$, must give the same effective coupling. It is therefore convenient and desirable to absorb the dependence on the two parameters $\alpha_s(\mu_0{}^2)$ and μ_0 into a single parameter. This parameter is known as the QCD scale parameters Λ (or Λ_{QCD}), and may be introduced as follows.

We can easily integrate eq. (12.89) by partial fractions. Denoting $\bar{\alpha}_s(\mu)/\pi$ by a, the result is

$$1/a + (\beta_2/4\beta_1)\ln[-a\beta_1/4(1 + \beta_2 a/4\beta_1)] = -(\beta_1/2)\ln(\mu/\Lambda), \qquad (12.92)$$

where Λ has been chosen to have the less than obvious form

$$\Lambda = \mu_0 \exp\{(2/\beta_1)[1/a_0 - (\beta_2/4\beta_1)\ln[(1 + \beta_2 a_0/4\beta_1)/a_0] + (\beta_2/4\beta_1)\ln(-\beta_1/4)]\}. \qquad (12.93)$$

Other choices of Λ are possible; the advantage of this particular choice is seen by solving eq. (12.92) at small a to get an expansion of $\bar{\alpha}_s(\mu^2)$ in terms of μ and Λ only:

$$\bar{\alpha}_s(\mu^2)/\pi = \frac{-2}{\beta_1 \ln(\mu/\Lambda)} - (\beta_2/\beta_1{}^3)\frac{\ln\ln(\mu^2/\Lambda^2)}{\ln^2(\mu/\Lambda)} + \mathrm{O}\left(\frac{1}{\ln^3(\mu^2/\Lambda^2)}\right). \qquad (12.94)$$

The motivation for (12.93) is simply that it eliminates terms proportional to a constant times $\ln^{-2}(\mu^2/\Lambda^2)$ in (12.94) (Bardeen, Buras, Duke & Muta 1978).

Comparison of QED and QCD

Our treatments of the infrared problem for QED and QCD have been rather different. In both cases, however, our aim was to find an appropriate definition of the coupling, and then to identify a set of cross sections that are computable in perturbation theory. In both cases, we found that finite-order cross sections must be re-interpreted as to their particle content.

In QED we developed an on-shell renormalization program, based on the finiteness of the coupling at zero momentum transfer and the measurability of the electron mass. To obtain finite cross sections, we found that is was necessary to sum over soft, unobserved photons. The resulting cross sections are well approximated by lowest-order calculations, in which these photons are ignored.

In QCD, the coupling could not be defined on-shell (and the quarks cannot be observed directly). This difficulty is addressed by considering cross sections that have finite high-energy, or zero-mass, limits. To $O(\alpha_s)$, we verified that this is indeed the case for the total e^+e^- annihilation cross section. Here again, it is necessary to re-interpret the low-order calculations, since, although they are carried out in terms of a few quarks and gluons, the result is a prediction for the sum over *all* hadronic states.

12.5 Jet cross sections at order α_s in e^+e^- annihilation

The one-loop annihilation cross section with massless quarks involves the cancellation of both collinear and infrared divergences. In QED, it is not necessary to sum over the emission of all photons to derive infrared-finite cross sections; it is only necessary to sum over soft photons. Similarly, in the limit of massless photons or gluons, it is not necessary to sum over all final states to cancel collinear divergences. The following heuristic considerations lead us to the concept of jet cross sections, mentioned first in Section 8.6 in connection with the parton model, as a direct generalization of the Bloch–Nordsieck mechanism in QED.

Indistinguishable states

The physical origin of collinear divergences is easily understood. An on-shell lightlike particle of momentum p^μ can decay into two lightlike particles, of momenta αp^μ and $(1-\alpha)p^\mu$, $0<\alpha<1$ (or vice versa). Such collinear rearrangements, however, do not affect the overall flow of momentum (Kinoshita & Sirlin 1958). The conserved quantum numbers of the two states are the same, and they would be indistinguishable to an observer whose detectors have finite *angular* resolution. As in the case of soft photon emission, collinear divergences are a sign that many particles are produced. Although the probability for any fixed number of particles may be small in perturbation theory, the *total* probability for all states with the same momentum flow may nevertheless remain finite and calculable.

Angular resolutions and jet cross sections

Consider the process $e^+e^- \to \gamma^* \to$ hadrons, as in Fig. 12.2. A *jet cross section*, $\sigma_{\text{jet}}(\Omega_i, E_i, \delta E_i, \mu, g_s, m)$, is the inclusive cross section for states in which total energies E_i flow into angular regions or 'cones' Ω_i in phase space, up to an energy resolution δE_i for each region. Figure 12.8 shows typical angular regions for a three-jet cross section. The definition is frame

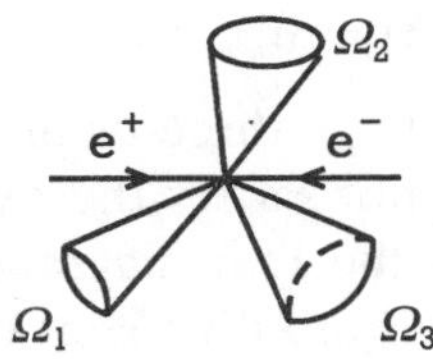

Fig. 12.8 Angular regions for a three-jet cross section.

dependent, of course; we shall work in the center-of-mass frame. As usual μ denotes the renormalization scale, g_s the strong coupling, and m the quark masses (renormalized). Each angular region serves as a 'calorimeter', and counts those events for which the total energy flowing into that region is within the specified range. For small Ω_i, the particles flowing into each region are all nearly parallel, and are referred to collectively as a *jet*. Many other definitions are possible (see Bethke, Kunszt, Soper & Stirling 1992) but the one discussed here (Sterman & Weinberg 1977) is closest to the Bloch–Nordsieck approach.

In accordance with our comments above, we expect all such jet cross sections to be finite in the zero-mass limit. In Section 13.5, we shall prove this to all orders for jet cross sections in e^+e^- annihilation. High-energy jet cross sections may then be calculated in massless perturbation theory, where they satisfy the trivial renormalization-group equation

$$[\mu\partial/\partial\mu + \beta(g)\partial/\partial g]\sigma_{\text{jet}}(\Omega_i, E_i, \delta E_i, \mu, g_s) = 0,$$

just as for the total e^+e^- annihilation cross section, eq. (12.87).

Two-jet cross section

The simplest example of a jet cross section in the process $e^+e^- \to$ hadrons has two jets, defined by two circular angular regions (cones) of half-angle δ (Sterman & Weinberg 1977). Half the energy flows into each region, up to a resolution ϵQ, where $Q = \sqrt{q^2}$ is the center-of-mass energy and $\epsilon \ll 1$. (In the above notation, $E_1 = E_2 = Q/2$, and $\delta E_1 + \delta E_2 = \epsilon Q$.) The cones are back-to-back, but we allow their common axis to be arbitrary. The Born approximation for the two-jet cross section is actually the total cross section at zeroth order, eq. (12.65),

$$\sigma_{2j}{}^{(0)}(Q, \epsilon, \delta) = N\left(\sum_f Q_f{}^2\right)\frac{4\pi\alpha^2}{3Q^2}, \tag{12.95}$$

since a two-particle final state always satisfies the two-jet criteria.

The $O(\alpha_s)$ correction to eq. (12.95), $\sigma_{2j}{}^{(1)}$, is closely related to σ_{tot} at $O(\alpha_s)$, eq. (12.86a). In fact, we have (Stevenson 1978, Weeks 1979)

$$\sigma_{2j}{}^{(1)}(Q, \epsilon, \delta) = \sigma_{\text{tot}}{}^{(1)} - \sigma_{3j}{}^{(1)}(Q, \epsilon, \delta), \tag{12.96}$$

where $\sigma_{3j}{}^{(1)}(Q, \epsilon, \delta)$ is the $O(\alpha_s)$ 'three-or-more-jet' cross section, consisting of all final states that do not satisfy the two-jet criteria. $\sigma_{3j}{}^{(1)}(Q, \epsilon, \delta)$ gets contributions only from the real diagrams, Figs. 12.2(*d*), (*e*). In the zero-mass limit, these diagrams are divergent only when the gluon is either infrared or collinear to one of the quarks, and this part of phase space contributes only to $\sigma_{2j}{}^{(1)}$. The cross section $\sigma_{3j}{}^{(1)}$ accounts for the rest of phase space, and is manifestly infrared and collinear finite to this order. It may therefore be calculated directly in four dimensions.

The actual calculation of $\sigma_{3j}{}^{(1)}$ is carried out by going through the same steps as for the total cross section above. To begin with, Figs. 12.2(*d*), (*e*) together satisfy current conservation. Therefore, by analogy with eq. (12.69),

$$\sigma_{3j}{}^{(1)} = \tfrac{1}{3}(e^2/2Q^4)[-g^{\mu\nu}h_{\mu\nu}{}^{(1)}(Q^2)], \tag{12.97}$$

where $h_{\mu\nu}{}^{(1)}$ is the purely three-jet hadronic part. It is given by eq. (12.75), but with the gluon phase space integral restricted to that region that contributes to the three-jet cross section. By construction, this means

$$k > \epsilon Q, \quad u \equiv \cos\theta_{pk} < 1 - 2\delta^2, \quad \cos\theta_{p'k} < 1 - 2\delta^2. \tag{12.98}$$

These conditions ensure that the gluon is sufficiently energetic, and sufficiently separated from the quark and antiquark in direction (minimum angle 2δ), to qualify as a third 'jet'. In principle, we should also require $\omega(p)$, $\omega(p') > \epsilon Q$, and $\cos\theta_{pp'} > 1 - 2\delta^2$, to make sure that the quark pair produces two separate jets. There are no enhancements in the matrix element in these limits, however, and corrections due to these requirements are small, i.e. of order of the energy and angular resolutions.

Of the three conditions in eq. (12.98), the first two may be implemented directly in the integral of eq. (12.75), while the third is equivalent to (exercise 12.15)

$$u > -1 + 2\delta^2(1 - 2k^2/Q^2). \tag{12.99}$$

The resulting expression for $-g^{\mu\nu}h_{\mu\nu}{}^{(1)}$ with a single quark flavor may be read directly from eq. (12.75) at $\epsilon = 0$:

$$-g^{\mu\nu}h_{\mu\nu}{}^{(1)}$$
$$= \frac{8NC_2(\mathrm{F})Q_f{}^2e^2g_s{}^2}{(2\pi)^3}\, Q\int_{\epsilon Q}^{Q} \mathrm{d}k\, k(Q-2k)\int_{u_1}^{u_2} \mathrm{d}u[2Q - 2k(1-u)]^{-2}$$
$$\times \left\{\frac{[Q-(1-u)k]^2}{k^2(1-u^2)} + \frac{Q(1+u)}{(Q-2k)(1-u)}\right\}, \tag{12.100}$$

where $u_1 = -1 + 2\delta^2(1 - 2k^2/Q^2)$, $u_2 = 1 - 2\delta^2$. The integrals are all elementary. Dropping terms proportional to the resolutions, and using eq.

(12.97), we find that the lowest-order contribution to the three-jet cross section is

$$\sigma_{3j}{}^{(1)} = N\sum_f Q_f{}^2(4\pi\alpha^2/3Q^2)C_2(\mathrm{F})(\alpha_s/\pi)(4\ln\delta\ln 2\epsilon + 3\ln\delta + \tfrac{1}{3}\pi^2 - \tfrac{7}{4}). \tag{12.101}$$

We then derive the two-jet cross section at order α_s from eq. (12.96) for $\sigma_{2j}{}^{(1)}$, using eq. (12.86a) for $\sigma_{\mathrm{tot}}{}^{(1)}$:

$$\begin{aligned}\sigma_{2j}{}^{(1)}(Q, \epsilon, \delta) = N\sum_f Q_f{}^2(4\pi\alpha^2/3Q^2) \\ \times \{1 - [C_2(\mathrm{F})\alpha_s/\pi][4\ln(\delta^{-1})\ln(\tfrac{1}{2}\epsilon^{-1}) \\ - 3\ln(\delta^{-1}) + \tfrac{1}{3}\pi^2 - \tfrac{5}{2}]\}.\end{aligned} \tag{12.102}$$

As observed above, it is natural to choose $\mu \approx Q$, to avoid factors of $\ln(Q/\mu)$ at higher orders. Then asymptotic freedom ensures that the first-order correction is relatively small, and that high-energy jet cross sections may be calculated reliably by perturbation theory, as an expansion in the small coupling $\alpha_s(Q^2)$. Note that in the limit of large Q^2, the total cross section is dominated by the two-jet cross section, as is experimentally observed (Wu r1984).

Equation (12.102) has a structure similar to the infrared finite QED decay rate, eq. (12.36), except that there are now two resolutions instead of one. Note that the high-energy ($\beta \to 1$) limit of eq. (12.36) is divergent, precisely because it does not include collinear photons with energies greater than Δ. The $\mathrm{O}(\alpha_s)$ correction in eq. (12.102) is negative, since a stricter definition of a jet can only reduce the cross section. When $\alpha_s \ln\delta\ln\epsilon$ is no longer a small number, the complete one-loop cross section can go negative, another type of behavior familiar from the infrared problem in QED. As in the QED infrared problem, it is possible to exponentiate the dependence on the resolutions (Mueller r1981, Mukhi & Sterman 1982). Various jet cross sections have also been computed to $\mathrm{O}(\alpha_s{}^2)$ (see Kramer & Lampe 1989). For the present, however, the important point is that for sufficiently loose definition of jets, the cross section for jets may be computed approximately from lowest order, 'as if' the jets were simply made up of single quarks. This is similar to the situation for the total annihilation cross section, described in Section 12.4.

Differential jet cross sections

The relation between quarks, gluons and jets is deepened by studying jet angular dependence. We may, for instance, fix the axes of the two angular regions of the two-jet cross section at an angle θ^* to the incoming e^+e^- direction in the center-of-mass system. Any e^+e^- annihilation cross section is of the form $(k_1{}^\mu k_2{}^\nu + k_2{}^\mu k_1{}^\nu - g^{\mu\nu} k_1 \cdot k_2) H_{\mu\nu}$, (eq. (12.67)), where the $k_i{}^\mu$ are the lepton momenta, and $H_{\mu\nu}$ is the hadronic tensor. The θ^*-dependence in the two-jet cross section is determined by the tensor structure of $H_{\mu\nu}$. The lowest-order, $1 + \cos^2\theta^*$, dependence in eq. (12.64) follows from the form $\mathrm{tr}\,(\not{p}_1\gamma_\mu\not{p}_2\gamma_\nu)$ in the hadronic part, where the p_i are the outgoing quark and antiquark momenta (Section 8.1). This Dirac structure is preserved at one loop for σ_{2j}, and the one-loop correction enters simply as an overall factor times the lowest-order result (exercise 12.16). The cross section with cones centered on the axis θ^* is then given by

$$\begin{aligned}\bar\sigma_{2j}{}^{(1)}(Q, \epsilon, \delta, \theta^*) = (\pi\delta^2)[\mathrm{d}\sigma^{(0)}/\mathrm{d}\cos\theta^*]\\ \times \{1 - [C_2(\mathrm{F})\alpha_s/\pi][4\ln(\delta^{-1})\ln(\tfrac{1}{2}\epsilon^{-1})\\ \times 3\ln(\delta^{-1}) + \tfrac{1}{3}\pi^2 - \tfrac{5}{2}]\},\end{aligned} \tag{12.103}$$

where

$$\mathrm{d}\sigma^{(0)}/\mathrm{d}\cos\theta = \sum_f Q_f{}^2(N\pi\alpha^2/2Q^2)(1 + \cos^2\theta^*)$$

is the differential Born cross section, eq. (12.64). The $1 + \cos^2\theta^*$ distribution is observed experimentally (Wu r1984), and is one of the persuasive arguments for the spin of the quark (exercise (8.2)).

Exercises

12.1 Verify eqs. (12.6)–(12.8).

12.2 Verify that the $\bar u_2\sigma_{\mu\nu}q^\nu u_1$ term in (12.16) describes a correction to the interaction of the electron with a magnetic field (see exercise 6.5).

12.3 If we expand the vertex $\bar u_2\Lambda_\mu u_1$ and the photon self-energy to order q^2, to (12.16) we must add the expression

$$\bar u_2\gamma_\mu u_1(\alpha/3\pi)(q^2/m^2)\{\ln(m/\lambda) + 31/120\},$$

where λ is the infrared regulator mass. Show that this term, along with a corresponding term from the vacuum polarization, leads to a new interaction term in Coulomb's law, proportional to $\delta^3(\mathbf{r})$. In an atomic system, we expect λ to be replaced by R^{-1}, where R is the atomic length beyond which fields are shielded. Such a term, which is concentrated at the origin, distinguishes between otherwise degenerate atomic states whose behavior at the origin is different, such as the $2S_{1/2}$ and $2P_{1/2}$ states of hydrogen. This level displacement is known as the Lamb shift (Lamb & Retherford 1947, Bethe 1947), and its observation was a crucial step in the development of modern quantum field theory. For a more extensive discussion of the Lamb shift, see Bjorken & Drell (r1964, Chapters 4 and 8).

12.4 Evaluate the leading $p^2 \to \infty$ behavior of $\Gamma_2(\not{p}, m)$ and $\pi_{\alpha\beta}(p^2)$. Show that both are proportional to a single logarithm of p^2.

12.5 Show that in Landau gauge ($\lambda = \infty$ in eq. (7.79)) $Z_1 = Z_2 = 1$ at one loop. This result generalizes to all orders.

12.6 Elastic potential scattering is described by the vertex eq. (12.3), with $(p_2 - p_1)^2 = q^2 < 0$. To this we may add corrections due to the emission of a photon into the final state. Show that infrared divergences cancel in the sum of $O(\alpha)$ real and virtual corrections to potential scattering.

12.7 Show that the infrared divergences of αB and αD (eqs. (12.17b) and (12.32)) are gauge invariant, although their infrared finite parts are not.

12.8 Study the one-loop infrared behavior of Yukawa theory with pseudoscalar and scalar exchange. Show that the former is infrared finite, while the latter has infrared divergences which, however, vanish at high energy.

12.9 Use power counting to extend the results of exercise (12.8) to all orders for the vertex function.

12.10 Show that the anomalous magnetic moment is infrared finite to all orders in perturbation theory in QED.

12.11 Use current conservation to show that fermion loops do not affect the form (12.58) for the exponentiation of infrared divergences. Consider in particular the case of the photon self-energy.

12.12 Verify eq. (12.61) for real photon cross sections.

12.13 Consider the scattering of a scalar boson with charge e by a potential. Suppose its elastic amplitude $\Gamma_{\text{el}}(p_1{}^2, p_2{}^2, p_1 \cdot p_2)$ depends only on the invariants shown. The corresponding one-photon bremsstrahlung amplitude, $\Gamma^\mu(p_1, p_2, k)$, shown in Fig. 12.9, may be written as

$$\begin{aligned}\Gamma^\mu(p_1, p_2, k) = {} & [e(2p_1 - k)^\mu/(-2p_1 \cdot k)]\Gamma_{\text{el}}((p_1 - k)^2, p_2{}^2, (p_1 - k)\cdot p_2) \\ & + \Gamma_{\text{el}}(p_1{}^2, (p_2 + k)^2, p_1 \cdot (p_2 + k))[e(2p_2 + k)^\mu/(2p_2 \cdot k)] \\ & + \Gamma^\mu{}_{\text{int}}(p_1, p_2, k),\end{aligned}$$

where we have isolated emission from the two external lines. Suppose that all these functions can be expanded about $k^\mu = 0$, except for the explicit $1/(2p_i \cdot k)$ factors due to external line emission. (This is appropriate when Γ_{el} is computed in a theory with no massless particles except for the photon.) Use current conservation, Taylor expansion and the chain rule to determine

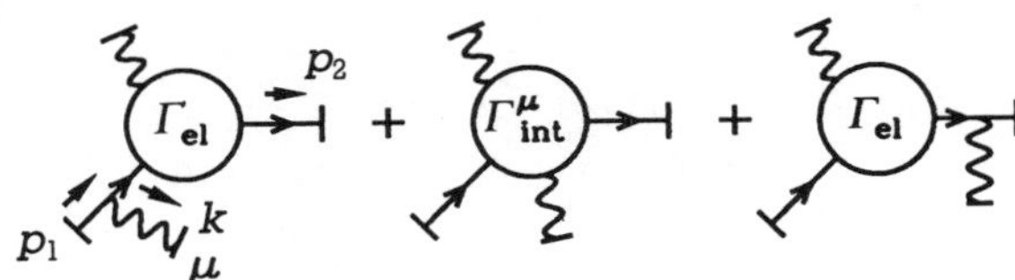

Fig. 12.9 One-photon bremsstrahlung amplitude.

$\Gamma^\mu{}_{\rm int}(p_1, p_2, k = 0)$ and find the following expression for Γ^μ:

$$\begin{aligned}\Gamma^\mu(p_1, p_2, k) = {} & e[(2p_1 - k)^\mu/(-2p_1\cdot k) + (2p_2 + k)^\mu/(2p_2\cdot k)] \\ & \times \Gamma_{\rm el}(p_1{}^2, p_2{}^2, p_1\cdot p_2) \\ & + e[p_1{}^\mu(p_2\cdot k/p_1\cdot k) - p_2{}^\mu + p_2{}^\mu(p_1\cdot k/p_2\cdot k) - p_1{}^\mu] \\ & \times [\partial/\partial(p_1\cdot p_2)]\Gamma_{\rm el}(p_1{}^2, p_2{}^2, p_1\cdot p_2) \\ & + {\rm O}(k^\mu).\end{aligned}$$

This result shows that up to $\mathrm{O}(k^0)$, the bremsstrahlung amplitude is determined by the elastic scattering amplitude. This is *Low's theorem* (Low 1958), specialized to a charged scalar particle. The basic result applies to any spin for cross sections (Burnett & Kroll 1968, Bell & van Royen 1969, Del Duca 1990), although the term 'Low's theorem' is often applied specifically to the first ($\mathrm{O}(1/k)$) term.

12.14 Compute the infrared divergence of Fig. 11.4(*d*) at $q^\mu = 0$, and verify that at zero momentum transfer, infrared divergences do not cancel in QCD. Show also that the anomalous magnetic moment is not finite at one loop in QCD.

12.15 Verify that eq. (12.99) is equivalent to the last restriction in eq. (12.98).

12.16 Show that the Dirac structure that gives the leading behavior of the one-loop jet cross section is the same as that for the Born cross section.

12.17 Compute the $\mathrm{O}(g^2)$ two-jet cross section in a Yukawa theory. Show that no infrared resolution is necessary.

13

Analytic structure and infrared finiteness

The zero-mass limit is intimately related to high-energy behavior in field theory. As we have seen in Section 12.4, cross sections develop collinear, as well as infrared, divergences as masses vanish. These divergences, however, cancel in certain perturbative quantities, jet cross sections in e^+e^- annihilation among them. Proofs of finiteness for these and other 'infrared safe' quantities require a formalism to treat infrared and collinear divergences to all orders in perturbation theory. Such a formalism can be developed on the basis of the analytic structure of Feynman diagrams, which we shall supplement by a power-counting method for estimating the strength of singularities in massless perturbation theory. We shall use these tools to prove a number of important results, including the infrared finiteness of Wick-rotated Green functions and of the e^+e^- total cross section, as well as of e^+e^- jet cross sections. The finiteness of the latter, in turn, is a variant of the famous KLN theorem, which states that suitably averaged transition probabilities are finite in the zero-mass limit for any unitary theory.

13.1 Analytic structure of Feynman diagrams

The calculations of Section 12.5 are suggestive and encouraging, but we must still determine whether jet cross sections are really mass-independent at higher orders in perturbation theory. We need to know, for example, whether the collinear and infrared regions are the only sources of divergence in the zero mass limit. One approach to these questions is to analyze amplitudes as complex functions of their momenta and masses. (For an extensive review of analytic structure in scattering amplitudes, see Eden, Landshoff, Olive & Polkinghorne r1966.)

We showed in Section 9.2 that for Euclidean external momenta and nonzero masses, any Feynman integral is an analytic function of momenta. We found, however, branch points for certain physical momenta associated with particle thresholds (see, for instance, eq. (9.8) or (12.30)). Given this relation, we can deduce a close connection between branch points in the complex space of momentum invariants and zero-mass divergences. Any on-shell momentum is *always* at threshold for the emission of a particle with zero momentum, and any on-shell massless line is always at threshold for the production of two or more collinear-moving massless lines. From this point of view, infrared and collinear divergences are special cases of the larger class of singularities of Feynman amplitudes in the complex plane, to which we now turn.

Singularities

An arbitrary Feynman diagram $G(\{p_s{}^\mu\})$ with external momenta $\{p_s{}^\mu\}$ may be put in Feynman-parameterized form (eq. (9.11)),

$$\begin{aligned} G(\{p_s{}^\mu\}) &\\ = \prod_{\text{lines } i} \int_0^1 \mathrm{d}\alpha_i\, \delta(\textstyle\sum_i \alpha_i - 1) \prod_{\text{loops } r} \int \mathrm{d}^n k_r\, D(\alpha_i, k_r, p_s)^{-N} F(\alpha_i, k_r, p_s), & \quad (13.1)\\ D(\alpha_i, k_r, p_s) = \sum_j \alpha_j [l_j{}^2(p,k) - m_j{}^2] + \mathrm{i}\epsilon. & \end{aligned}$$

$F(\alpha, k, p)$ represents constant and numerator factors that do not affect our arguments. In gauge theories, we work in the Feynman gauge. α_j is the Feynman parameter of the jth line, and $l_j{}^\mu(p, k)$ its momentum, a linear function of loop momenta $\{k_r\}$ and external momenta $\{p_s\}$.

Our goal is to find the positions of all poles and branch points of $G(\{p_s{}^\mu\})$ as a function of the $\{p_s{}^\mu\}$. These singularities must arise from zeros of the denominator $D(\alpha_i, k_r, p_s)$. In the absence of such zeros, the integrand in eq. (13.1) is bounded and analytic everywhere in the integration region, and $G(\{p_s{}^\mu\})$ is an analytic function of the $p_s{}^\mu$.

Since the line momenta $l_j{}^\mu$ are functions of the external momenta $p_s{}^\mu$, the positions of the zeros of $D(\alpha_i, k_r, p_s)$ depend on the values of the $p_s{}^\mu$. In particular, we know from Section 9.2 that if all the $p_s{}^0$ are Wick rotated to positive imaginary energies, the $k_r{}^\mu$ and α_i integrals encounter no poles along the real axis. For real energies, however, poles may migrate to the real axis, making a zero of $D(\alpha_i, k_r, p_s)$ possible in the integration range of eq. (13.1). The mere presence of a zero of D, however, is not enough to produce a singlarity in $G(\{p_s\})$. A much stronger condition is necessary, because the integrals of (13.1) are contour integrals in complex (k, α) space.

Poles, singular surfaces and contour integrals

Consider a pole, labeled $P(z)$, at real point $z = \{\bar{k}_r{}^\mu, \bar{\alpha}_i\}$ in (k, α) space. Since the $k_r{}^\mu$ and α_i integrals are contour integrals in complex space, they can be deformed from one path to another by using Cauchy's theorem.

Multidimensional integrals of complex variables are difficult to picture. In our discussion, therefore, we shall always do the integrals one at a time. For example, if $P(z)$ is an isolated pole in the complex plane of *any one* integration variable $\zeta \epsilon \{k_r{}^\mu, \alpha_i\}$, the ζ contour can be deformed away from $P(z)$, as shown schematically in Fig. 13.1. Along the new contour, D does not vanish, and the integrand is everywhere an analytic function of the external momenta. Thus, isolated poles of the integrand do not produce singularities in the integral $G(\{p_s\})$.

Of course, it is not always possible to avoid poles by contour deformation. Consider a general function of parameter w, defined as the integral over complex variable ζ of the function $F(\zeta, w)$ between fixed end-points ζ_a and ζ_b:

$$I(w) = \int_{\zeta_a}^{\zeta_b} \mathrm{d}\zeta\, F(\zeta, w). \tag{13.2}$$

Compared with eq. (13.1), ζ is one of the variables in (k, α) space, w stands for the remaining variables and $F(\zeta, w)$ stands for D^{-N}. Thus we suppose that $F(\zeta, w)$ is a rational function of ζ, which has, at worst, isolated poles in that variable whose positions $\xi_i(w)$ depend on w.

As long as the integration contour encouters no poles, $I(w)$ is an analytic function of w. Even when one of the poles, $\xi_i(w)$, moves into the path of the contour, no singularity occurs as long as the contour can be deformed away, as in Fig. 13.1. Singularities in $I(w)$ are generated only when contour deformations can no longer avoid the pole. For an integral of the form eq. (13.2), this can happen in two ways.

(i) End-point singularity. One of the poles may migrate to ζ_a or ζ_b: $\xi_i(w_0) = \zeta_a$, say, for $w = w_0$. Then the pole is at one of the end-points of the integral and $I(w)$ is undefined. This is called an *end-point singularity*, and $w = w_0$ is a branch point of $I(w)$, as illustrated in Fig. 13.2(a), which shows both the w- and ζ-planes. Suppose we want to analytically continue the function $I(w)$ from point w_1 to point w_2 along path R_j, which avoids w_0. Any such continuation corresponds to a path ρ_j of pole $\xi_i(w)$ in the ζ-plane from $\xi_i(w_1)$ to $\xi_i(w_2)$. The latter paths are of two types. Paths like ρ_1 go around the contour, while those like ρ_2 cross the contour, which then encloses the extra pole at $\xi_i(w_2)$. The two analytic continuations thus differ by $2\pi \mathrm{i} Z_i(w_2)$, where $Z_i(w_2)$ is the residue of the pole of $F(\zeta, w)$ at $\zeta = \xi_i(w_2)$. In w-space, the corresponding continuations R_1 and R_2 enclose

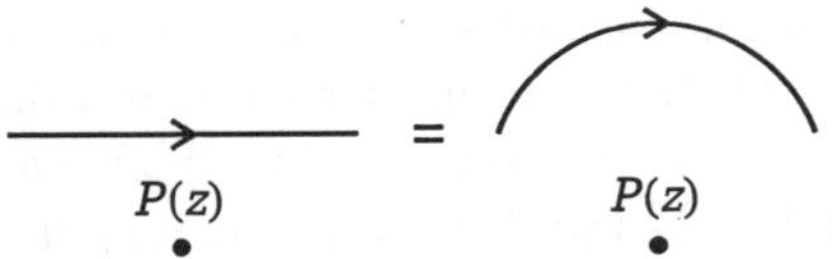

Fig. 13.1 Avoiding an isolated pole by contour deformation.

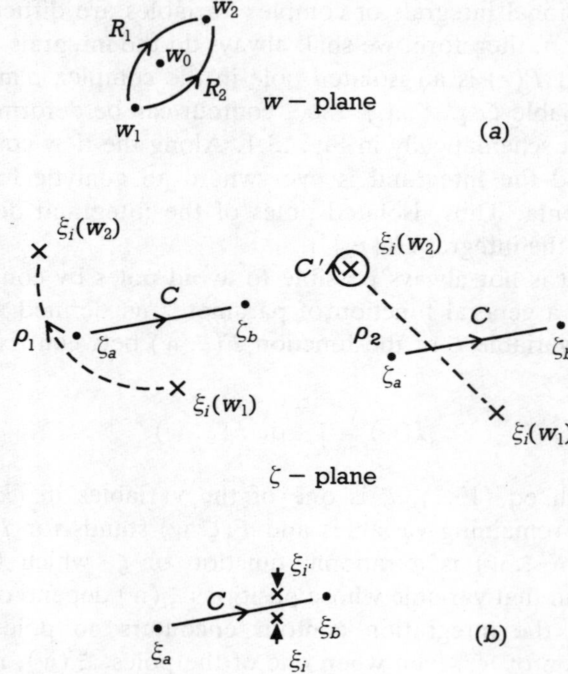

Fig. 13.2 Illustration of singularity structure of a typical integral, eq. (13.2): (*a*) migration of an isolated pole in the complex ζ-plane as a function of a variable w; (*b*) pinch in the ζ-plane.

the branch point w_0. A suggested exercise is to implement this scenario in a number of simple cases (exercises 13.1, 13.2).

(ii) Pinch singularities. Two (or more) poles $\xi_i(w)$ and $\xi_{i'}(w)$ may coalesce at point $\xi_i(w_0) = \xi_{i'}(w_0)$, one on either side of the ζ contour, as in Fig. 13.2(*b*). This gives a 'pinch' singularity at $w = w_0$, which can be shown to be a branch point, in a manner analogous to the end-point case.

Now let us return to eq. (13.1). $D(\alpha_i, k_r, p_s)$ vanishes on sets of points $S = \{\bar{k}, \bar{\alpha}\}$ that are surfaces in (k, α) space. Suppose we identify such a surface, S. We want to know whether surface S produces a singlarity; we shall refer collectively to those surfaces that do as *pinch surfaces* (in the original momentum space, there are no true end-point singularities, since all integrals are unbounded). According to our comments above, we shall look at each variable in (k, α) space in turn.

Let ζ be any one of the variables $\{k_i{}^\mu, \alpha_j\}$, with all other variables fixed at values for which $D = 0$ for some set of real values $\{\xi\}$ of ζ. It may happen that the set $\{\xi\}$ is continuous, in which case D in eq. (13.1) is independent of ζ on S (i.e. the ζ-direction is tangent to S). In this case ζ need not be trapped on S.

The more interesting case occurs when the points $\{\xi\}$ are isolated in the ζ-plane. Then, on the basis of our discussion above, a necessary condition for S to be a pinch surface is that every such ζ contour must be trapped at each $\zeta = \xi$. Landau (1959) put this condition into an explicit form, which we now review.

Landau equations

First, suppose that $\zeta = k_i{}^\nu$, a loop momentum variable. The denominator is independent of $k_i{}^\nu$ only if

$$\alpha_m = 0 \text{ for every line } m \text{ in loop } i. \tag{13.3}$$

Otherwise, the $k_i{}^\nu$ contour must be trapped. Now the denominator in eq. (13.1) is quadratic in $k_i{}^\nu$. For fixed values of the other variables, the function D^{-N} is analytic in the $k_i{}^\nu$-plane, except for two poles, found by solving the quadratic equation

$$\sum_{\text{lines } i} \alpha_i[l_i{}^2(p,k) - m_i{}^2] + \mathrm{i}\epsilon = 0,$$

$$l_i{}^\mu(p,k) = \sum_r \eta_{ir} k_r{}^\mu + \sum_s \hat{\eta}_{is} p_s{}^\mu, \tag{13.4a}$$

where η and $\hat{\eta}$ are incidence matrices for loop and external momenta, respectively. The $k_j{}^\nu$ contour is trapped only if these two solutions are equal. This will happen when the derivative of D vanishes at $D = 0$:

$$0 = (\partial/\partial k_j{}^\nu)\left\{\sum_{\text{lines } i} \alpha_i[l_i{}^2(p,k) - m_i{}^2] + \mathrm{i}\epsilon\right\} = 2\sum_{\text{lines } i} \eta_{ij}(\alpha_i l_i{}^\mu). \tag{13.4b}$$

Now suppose $\zeta = \alpha_i$, a Feynman parameter. $D(\alpha_i, k_r, p_s)$ is independent of α_i when $l_i^2 = m_i{}^2$. Alternatively, since D is linear in each α_i, the α_i contour can be trapped only at its end point $\alpha_i = 0$.

These considerations are conveniently summarized as the *Landau equations*. S is a pinch surface only if the following conditions hold for each point $\{\bar{k}_r{}^\mu, \bar{\alpha}_i\}$ on S. For lines with $l_i^2 = m_i{}^2$, we have, for every loop j that includes line i,

$$\sum_i \eta_{ij}\alpha_i l_i{}^\mu = 0. \tag{13.5a}$$

On the other hand, for lines with $l_i^2 \neq m_i{}^2$, we have

$$\alpha_i = 0. \tag{13.5b}$$

Physical pictures and reduced diagrams

Having derived the Landau equations (13.5), we can give them a surprisingly physical interpretation (Coleman & Norton 1965). Suppose a set of lines $\{i\}$ go on-shell at pinch surfaces S. We think of each on-shell line as representing a particle propagating freely between the vertices which it connects. The Feynman parameter α_i is then interpreted as the Lorentz invariant ratio of the time of progagation to the energy for particle i. Then, in any frame the vector quantity

$$\Delta x_i{}^\mu \equiv \alpha_i l_i{}^\mu \tag{13.6}$$

is the space–time separation between the point where line i begins and the point where it ends.

We now imagine that the vertices $\{a\}$ represent points in space–time, specified by position vector $x_a{}^\mu$. Suppose, then, that line 1 begins at vertex a and ends at vertex b. From eq. (13.6), we conclude that

$$x_b{}^\mu - x_a{}^\mu = \alpha_1 l_1{}^\mu. \tag{13.7}$$

For this separation to be well defined, it must be independent of the line or lines used to compute it. For instance, suppose vertex a and vertex b are connected by another set of lines, with momenta $l_2, \ldots, l_n$, as shown in Fig. 13.3. Using the analogue of eq. (13.7) for each of the momenta $l_j{}^\mu$, we find

$$x_b{}^\mu - x_a{}^\mu = \sum_{j=2}^{n} \alpha_j l_j{}^\mu. \tag{13.8}$$

For eq. (13.8) to be consistent with eq. (13.7), we must have

$$\alpha_1 l_1{}^\mu + \sum_{j=2}^{n} \alpha_j(-l_j{}^\mu) = 0. \tag{13.9}$$

But this is just a restatement of eq. (13.5a) for loop k in Fig. 13.3. We can repeat this argument for every path connecting every pair of vertices. The interpretation of vertices as points in space–time connected by freely propagating lines is then consistent if, and only if, the Landau equations (13.5) are satisfied.

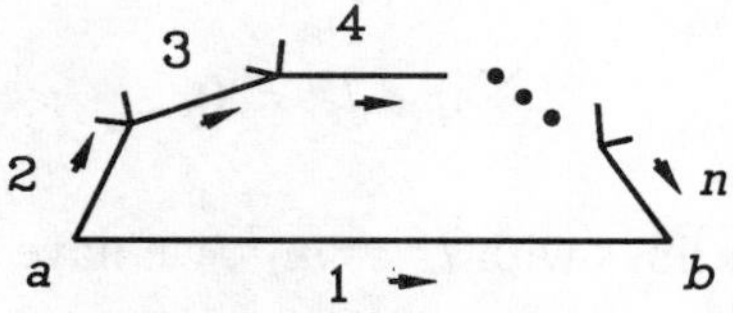

Fig. 13.3 Arbitrary loop illustrating eq. (13.8).

An attractive feature of the foregoing analysis is that each off-shell line, $l_i^2 \neq m_i^2$, automatically has $\alpha_i = 0$, corresponding to no propagation at all. We can draw a picture of the physical process corresponding to the singular surface by contracting all the off-shell lines to points. The resulting graph is a reduced diagram of the type introduced in Section 12.3. We shall see that physical pictures and reduced diagrams are very convenient tools for identifying pinch surfaces.

In the following section, we illustrate the foregoing results in the simplest of cases, the two-point function.

13.2 The two-point function

The two-point function and normal thresholds

Let us first consider an arbitrary one-particle irreducible two-point diagram $G(q)$ in a theory with only one species of line, of mass m. Our aim is to find all the pinch surfaces of $G(q)$, and hence its singularities in q^μ. We already know from our analysis of Wick rotation that $G(q)$ has no branch points for any external momentum with $q^2 = q_0^2 \exp(2i\theta) - \mathbf{q}^2$, $\pi/2 > \theta > 0$. In fact, we shall show that its only singularities are at

$$q^2 = n^2 m^2, \tag{13.10}$$

where n is an integer. These values are known as 'normal thresholds', at which there is just enough energy available to produce a set of particles that are at rest relative to one another.

Let S denote a singular surface of G, at which a set of lines $\{l_i^\mu\}$ goes on-shell. We can always choose to work in the rest frame, $q^\mu = (q_0, \mathbf{0})$. In this frame, the special role of normal thresholds, eq. (13.10), follows easily from the Landau equations. First, it is clear that there are pinch surfaces at the normal thresholds. These are of the type shown in Fig. 13.4 for $n = 3$. A set of particles is created at rest at some point in space–time, and they remain at rest, interacting in various combinations at later times but always in the same place until, finally, they annihilate. It is trivial to find solutions to the Landau equations on such singular surfaces.

What happens if we are not at a normal threshold? Consider an arbitrary reduced diagram R, at which one or more of the on-shell particles has a non-zero spatial momentum component, say in the z-direction. Since the external momentum q^μ has zero z-component, z-momentum must flow

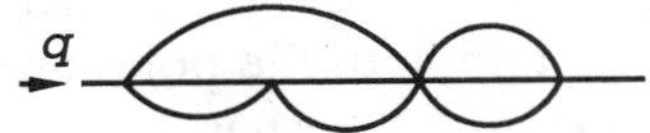

Fig. 13.4 Reduced diagram for a normal threshold at $q^2 = 9m^2$.

around a loop j of R, always in the same direction. The z-components of the momentum of loop j cannot satisfy the Landau equations with $\alpha \neq 0$, because all the terms $\eta_{ij} l_{i,z}$ in eq. (13.5a) have the same sign. Thus, there are no nontrivial reduced diagrams that satisfy the Landau equations in two-point functions away from normal thresholds. As a result, two-point functions are analytic except at normal thresholds.

We can see examples of this behavior in the one-loop calculations of eqs. (12.20) and (12.22). In each case there is a single branch point in the graph as a function of external momentum, at the point $q^2 = (m_1 + m_2)^2$, where the m_i are the masses of the internal lines.

One-loop example and light-cone variables

As an example, we consider the ϕ^3 self-energy diagram, Fig. 10.1(a), which is given in n dimensions by eq. (10.3). We shall work in light-cone coordinates, $v^{\pm} = 2^{-1/2}(v^0 \pm v^3)$, and $\boldsymbol{v}_{\mathrm{T}} = (v_1, v_2)$ (eq. (6.45)). In terms of light-cone variables, a timelike total momentum q may be written in the center-of-mass frame as

$$q^{\mu} = (q^+, q^-, \mathbf{q}_{\mathrm{T}}) = (2^{-1/2}Q, 2^{-1/2}Q, \mathbf{0}_{\mathrm{T}}), \tag{13.11}$$

where $q^2 = Q^2$. In these terms, the integral for this diagram is

$$\begin{aligned} \Gamma_2(q^2) = \frac{-ig^2}{(2\pi)^4} \int_{-\infty}^{\infty} & \mathrm{d}k^+ \mathrm{d}k^- \mathrm{d}^2\mathbf{k}_{\mathrm{T}} (2k^+k^- - \mathbf{k}_{\mathrm{T}}{}^2 - m^2 + \mathrm{i}\epsilon)^{-1} \\ & \times [2(2^{-1/2}Q - k^+)(2^{-1/2}Q - k^-) - \mathbf{k}_{\mathrm{T}}{}^2 - m^2 + \mathrm{i}\epsilon]^{-1}. \end{aligned} \tag{13.12}$$

We start by performing the k^- integral in eq. (13.12) by contour integration. There are poles in the k^- plane at

$$\begin{aligned} \kappa_1 &= (\mathbf{k}_{\mathrm{T}}{}^2 + m^2 - \mathrm{i}\epsilon)/2k^+, \\ \kappa_2 &= 2^{-1/2}Q - (\mathbf{k}_{\mathrm{T}}{}^2 + m^2 - \mathrm{i}\epsilon)/[2(2^{-1/2}Q - k^+)]. \end{aligned} \tag{13.13}$$

Since Feynman denominators are linear in light-cone coordinates, there is only one pole per denominator. This is a simplification compared with energy integrals, which have two. The poles migrate from one half-plane to the other, depending on the value of k^+, and we get very different results for the k^- integral depending on their positions. In particular, whenever both poles are in the same k^- half-plane, the k^- integral vanishes. To show this, we note that the integrand in eq. (13.12) behaves as $(k^-)^{-2}$ for large k^-. As a result, we can complete the k^- contour at infinity in the half-plane without poles, as in Fig. 13.5(a), which shows the arrangement of poles when $k^+ > 2^{-1/2}Q$, and both the poles are in the lower half-plane. The completed contour integral vanishes by Cauchy's theorem. It also vanishes when $k^+ < 0$, and both poles are in the upper half-plane. Thus,

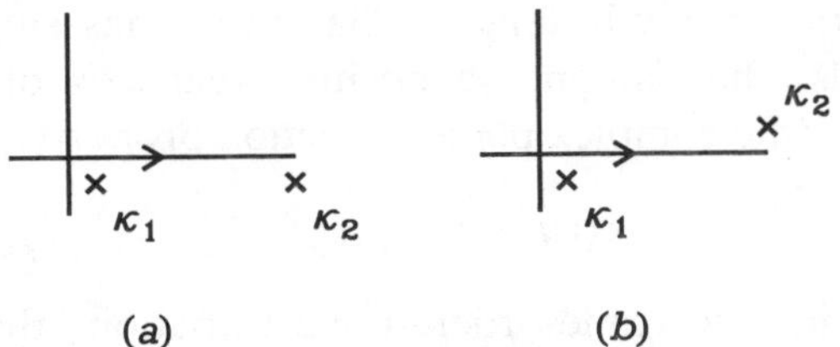

Fig. 13.5 Typical configurations of poles κ_i in eq. (13.13).

the k^- integral is only nonzero for $2^{-1/2}Q > k^+ > 0$ so that κ_2 is in the upper half-plane, and κ_1 is in the lower half-plane, as shown in Fig. 13.5(*b*). If the k^- contour is closed at infinity in the lower half-plane in this range, the contour picks up only the κ_1 pole. The result is easily put in the form

$$\Gamma_2(q^2) = \frac{-g^2}{(2\pi)^2 2^{1/2} Q} \int_0^\infty \mathrm{d}\mathbf{k}_\mathrm{T}^2 \int_0^{2^{-1/2}Q} \mathrm{d}k^+ [2k^+(2^{-1/2}Q - k^+) - \mathbf{k}_\mathrm{T}^2 - m^2 + \mathrm{i}\epsilon]^{-1}. \tag{13.14}$$

After the k^- integral, there is only a single denominator left, and a singularity in $\Gamma_2(q^2)$ requires end-point or pinch singularities in both k^+ and $\mathbf{k}_\mathrm{T}^2$. Since the denominator is linear in $\mathbf{k}_\mathrm{T}^2$, it can only have an end-point singularity, at $\mathbf{k}_\mathrm{T}^2 = 0$. The zeros of the denominator in eq. (13.14) are at

$$k^+ = 2^{-3/2} Q\{1 \pm [1 - 4(\mathbf{k}_\mathrm{T}^2 + m^2)/Q^2]^{1/2} \pm \mathrm{i}\epsilon\}, \tag{13.15}$$

For the two k^+ poles to pinch the k^+ contour at $\mathbf{k}_\mathrm{T}^2 = 0$, we must have $Q^2 = 4m^2$, which is just the condition for the two-particle threshold. Note that substituting the value of k^+ at $\mathbf{k}_\mathrm{T}^2 = 0$ back into eq. (13.13) shows that the k^- integral is also pinched at the normal threshold.

Dispersion relations

The analytic structure of the two-point function leads to a particularly interesting integral representation, in terms of its imaginary part, called a dispersion relation. To be specific, we shall derive a dispersion relation for the 1PI photon self-energy:

$$\begin{aligned} \rho_{\mu\nu}(q) &= (q_\mu q_\nu - g_{\mu\nu} q^2)\pi(q^2) \\ &= \mathrm{i}e^2 \int \mathrm{d}^4 x\, \mathrm{e}^{\mathrm{i}q\cdot x} \langle 0| T[j_\mu(x) j_\nu(0)] |0\rangle, \end{aligned} \tag{13.16}$$

where $\pi(q^2)$ is a scalar function. We have used, as before, the transversality of the photon self-energy.

Reduced diagram analysis tells us that $\pi(q^2)$ has singularities only at normal thresholds, that is, on the positive real axis of the q^2 complex plane. $\pi(q^2)$ obeys the complex plane reflection property

$$\pi((q^2)^*) = [\pi(q^2)]^*. \tag{13.17}$$

This results from the time-ordered perturbation theory expression, eq. (9.72), applied to the two-point function. In that formula, the only explicit factors of i are in the iϵ's of the energy denominators. When the external momentum carries a finite imaginary part, we can ignore these infinitesimal contributions, and eq. (13.17) follows immediately. Finally, $\pi(q^2)$ is real for any real $q^2 < q^2_{\text{min}}$, where q^2_{min} is the position of the branch point furthest to the left in the q^2-plane. Let q^2 be a fixed value, real but with $q^2 < q^2_{\text{min}}$. We can then use Cauchy's theorem to express the quantity $\pi(q^2) - \pi(0)$ in terms of a contour integral;

$$\pi(q^2) - \pi(0) = \frac{1}{2\pi \mathrm{i}}\left[\int_{C_1} \mathrm{d}\eta^2(\eta^2 - q^2)^{-1}\pi(\eta^2) - \int_{C_0} \mathrm{d}\xi^2(\xi^2)^{-1}\pi(\xi^2)\right]. \tag{13.18}$$

Here C_1 is any counterclockwise contour that encloses q^2 but none of the branch points of $\pi(q^2)$ while C_0 encloses $q^2 = 0$, as in Fig. 13.6(a). Cauchy's theorem tells us that these contours are arbitrary as long as they do not enclose the branch points. Therefore, we may deform them both into the contour C', which encloses the branch cuts running from q^2_{min} to $+\infty$ along the real axis, Fig. 13.6(b). Assuming that $\pi(q^2)$ behaves at worst as a constant as $q^2 \to \infty$, we can neglect the contour at infinity, and derive

$$\begin{aligned}\pi(q^2) - \pi(0) = \frac{1}{2\pi \mathrm{i}}\int_{q^2_{\text{min}}}^{\infty} \mathrm{d}\eta^2[(\eta^2 - q^2)^{-1} - (\eta^2)^{-1}]\pi(\eta^2 + \mathrm{i}\epsilon)\\ - \frac{1}{2\pi \mathrm{i}}\int_{q^2_{\text{min}}}^{\infty} \mathrm{d}\eta^2[(\eta^2 - q^2)^{-1} - (\eta^2)^{-1}]\pi(\eta^2 - \mathrm{i}\epsilon).\end{aligned} \tag{13.19}$$

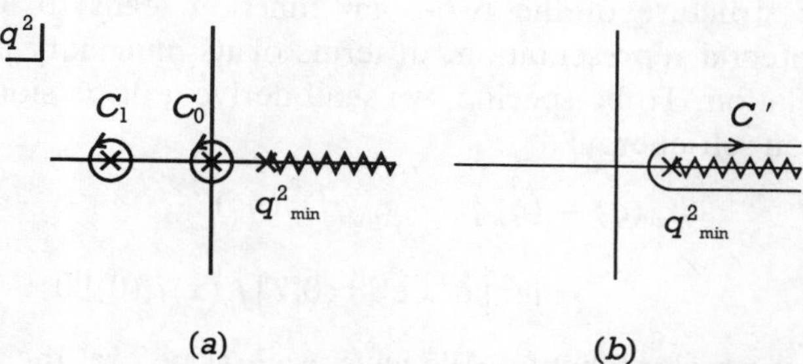

Fig. 13.6 Use of Cauchy's theorem to derive the dispersion relation, eq. (13.20).

Finally, using eq. (13.17), we find the representation we are looking for,

$$\pi(q^2) - \pi(0) = \frac{1}{\pi}\int_{q^2_{\min}}^{\infty} \mathrm{d}\eta^2[(\eta^2 - q^2)^{-1} - (\eta^2)^{-1}]\,\mathrm{Im}\,\pi(\eta^2). \tag{13.20}$$

Up to a constant, the two-point function is determined by its imaginary part. Equation (13.20) is an example of a *subtracted dispersion relation*. The η-integral is finite in eq. (13.20), even if the imaginary part goes to a constant at high energy. This is because of the *subtraction* terms, $\pi(0)$. In this way, the subtraction plays the role of renormalization in the determination of $\pi(q^2)$.

The imaginary part of $\pi(q^2)$ is directly related to the total annihilation cross section of e^+e^- into hadrons, σ_{tot}. For perturbative unitarity, eq. (9.86), implies the following relation between the sum over hadronic states and the two-current Green function:

$$2\,\mathrm{Im}\,\rho_{\mu\nu}(q) = e^2\sum_n \langle 0|j_\mu(0)|n\rangle\langle n|j_\nu(0)|0\rangle(2\pi)^4\delta^4(q - p_n). \tag{13.21a}$$

Then we find from eq. (12.69) for σ_{tot}

$$\sigma_{\text{tot}}(q^2) = (e^2/q^2)\,\mathrm{Im}\,\pi(q^2). \tag{13.21b}$$

This result deepens the physical content of the dispersion relation. It shows that the off-shell Green function is determined by the physical cross section, up to a constant. (The relation between cross sections and Green functions is actually quite general (Feynman 1963).)

Infrared finiteness of σ_{tot}

The unitarity relation, eq. (13.21b), has further consequences for the high-energy behavior of the total annihilation cross section for e^+e^- into hadrons. If we show that $\rho_{\mu\nu}(q^2)$ is finite in the zero-mass limit, then its imaginary part will be too, and the zero-mass finiteness of σ_{tot} will be demonstrated, along with its corresponding mass-independent renormalization group equation, (12.87), at high energy. It will also verify our claim (Section 12.3) that the photon self-energy in QED is infrared finite, even at $q = 0$. This proof is easy to carry out with the tools at hand.

Since it is a two-point function, $\rho_{\mu\nu}(q^2)$ has no branch points aside from normal thresholds. In the massless limit, all branch points collapse to the origin, $q^2 = 0$. Then the massless $\rho_{\mu\nu}(q^2)$ is analytic for all $q^2 > 0$, because its Landau equations, (13.5), have solutions only at normal thresholds. More accurately, away from normal thresholds they have no solutions with finite-energy lines. When all lines are massive, this is sufficient to eliminate *all* pinch surfaces, and hence all possible sources of divergence in the zero-mass limit. In the zero-mass limit, however, the Landau equations are trivially solved for any pinch surface with loops made entirely of *zero-momentum* lines, as shown in the reduced diagram of Fig. 13.7, where H represents all off-shell lines and S represents the zero-momentum lines. Massless zero-momentum lines are on-shell, and we must consider zero-momentum pinch surfaces, even though they are independent of external momenta. This means that, as for the QED vertex function of Section 12.3, we must evaluate their degree of infrared divergence. These pinch surfaces are *less* singular than those in the QED vertex function, however, because they lack fermion lines. Indeed, in eq. (12.40), the superficial degree of infrared divergence for a pinch surface with no on-shell fermions is $2(1-\epsilon)\gamma_G$, where γ_G is the number of photons and $\epsilon = 2 - n/2$. This is positive definite in more than two dimensions, and corresponds to an infrared convergent integral. It suggests, however, that we take a closer look at the special features of pinch surfaces with zero mass internal lines.

13.3 Massless particles and infrared power counting

For diagrams consisting entirely of massless lines, a number of general results are very easy to derive, and to express in the language of reduced diagrams. Nevertheless, these results give important constraints on the variety of pinch singularities that can occur at high energy. For many cross sections of physical interest, they will make it possible to catalogue *all* pinch singular surfaces in a natural way. Once pinch surfaces are identified, it is important to have criteria for determining when a Feynman integral is infrared finite. Power-counting criteria for this purpose, which are the infrared analogues of Weinberg's theorem for ultraviolet behavior (Section 9.4), are also developed below.

First, let us introduce some terminology.

Jet and soft subdiagrams

Within a reduced diagram R, a connected set of massless lines q_i^{μ}, which are on-shell with finite energy, and all of whose momenta are proportional to some lightlike momentum p^{μ},

$$q_i^{\mu} = \beta_i p^{\mu}, \quad \beta_i > 0, \quad p^2 = 0, \tag{13.22}$$

Fig. 13.7 Reduced diagram with zero-momentum lines only.

is called a *jet*. The constants of proportionality β_i are positive, in order that all energies in the jet may have the same sign. A jet may consist of a single particle or a complicated diagram.

Similarly, a set of on-shell massless lines with momenta that vanish in all four components, $q_i^\mu = 0$, will be referred to as the *soft subdiagram*. We encountered soft subdiagrams in our discussion of infrared divergences. Unlike jets, soft subdiagrams need not be connected.

Hard and soft vertices

A vertex of a reduced diagram to which lines from two or more jets attach is called a *hard* vertex. Other vertices, to which only lines from a single jet and/or the soft subdiagram attach are referred to as *soft*. A soft vertex may be *elementary*, if it is one of the vertices in the interaction Langrangian, or *composite*, if the vertex results from the contraction of a number of loops carrying finite momentum. With these definitions, every line in a reduced diagram is either a jet line or a soft line, and every vertex is either a hard vertex or a soft vertex.

Pinch surfaces with massless lines

Suppose that S_R is a pinch surface with reduced diagram R, and that J is a jet subdiagram of R. Let R' be the reduced diagram found by replacing J with any other jet subdiagram J', as in Fig. 13.8(a), where J' connects to

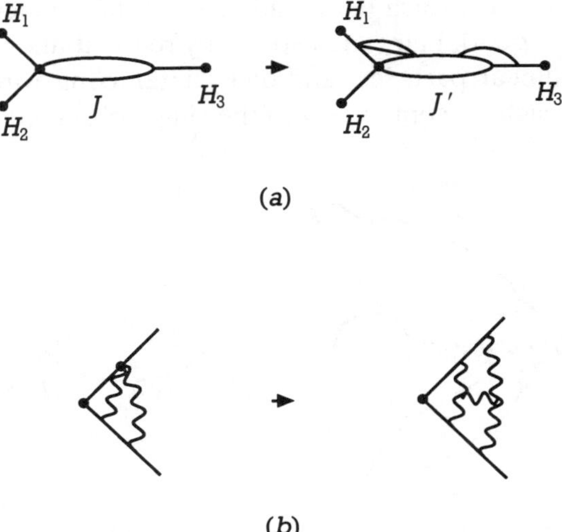

Fig. 13.8 Modifications of (a) jet subdiagrams and (b) soft subdiagrams that produce new pinch surfaces.

the same hard vertices H_i as J. Then the resulting reduced diagram R' also describes a possible physical process, and hence corresponds to a pinch surface of the resulting graph. Similarly, any modification of the soft subdiagram S (as illustrated in Fig. 13.8(b)) also gives a pinch surface.

These properties follow immediately from the Coleman–Norton analysis. Each of a set of massless collinear particles has vanishing invariant mass, even when the particles' energies are nonzero. In terms of the Coleman–Norton physical picture, the particles in a jet are all moving in the same direction at the same velocity – the speed of light – and may interact an arbitrary number of times during the lifetime of the jet. In the same way, the inclusion of lines with momenta for which $q_i{}^\mu = 0$ does not affect the Landau equations (13.5) at all. Their Feynman parameters α_i in eq. (13.5) are completely arbitrary. It is natural to interpret soft lines as excitations of infinite wavelength, which may therefore attach to vertices separated by any distance within a reduced diagram.

Application: jets in decay processes

A particularly simple and useful application of the above reasoning is given by the decay of an off-shell particle into massless particles, as in e^+e^- annihilation jet cross sections. Let us begin with virtual corrections to the simplest final state, a (massless) quark pair, and analyze its physical pictures at all orders.

Figure 13.9(a) shows the most general physical picture. At vertex D, a point-like current acts, and produces two jets of particles and possibly a set of zero-momentum lines. Each jet already has all the momentum of one of the members of the quark pair, and evolves by redistributing its momentum among sets of collinear particles, and/or by interacting with the soft subdiagram. Finally, each jet combines into the single observed particle (either

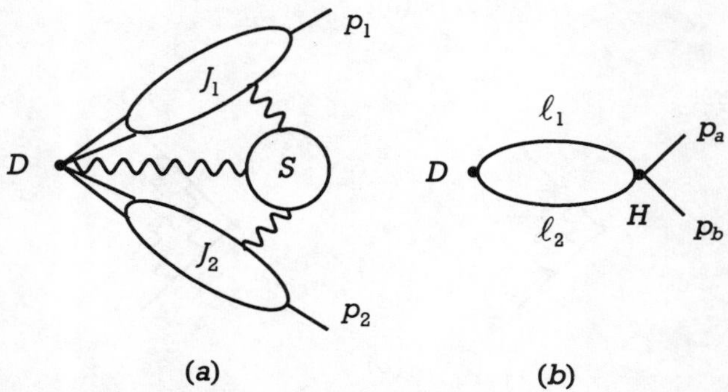

Fig. 13.9 (a) Most general pinch surface for the process $e^+e^- \to$ quark pair; (b) reduced diagram that does not correspond to a pinch surface.

a quark or antiquark for e^+e^- annihilation). This class of physical processes exhausts all possibilities. In particular, if there are two particles in the final state, there are only two internal jets of on-shell particles at any pinch surface. These results are shown as follows.

In every possible reduced diagram, a set of jets emerges from the vertex at D. There must be more than one jet because the invariant mass of a jet is always zero, while the invariant mass of the system is, by assumption, positive. To produce a new set of jets, there would have to be another hard vertex, at which two or more lines from different jets collide. Yet, since the jets are propagating freely in different directions, starting at the same point, they can *never* collide at a later time. They can only interact through the exchange of zero-momentum particles. Thus, there can be only one hard vertex, and every jet produced at the hard vertex appears in the final state. For the two-particle final state there are two and only two virtual jets, each with precisely the total momentum of the final-state particle into which it evolves.

Figure 13.9(b) shows an example of a reduced diagram that is *not* allowed at a pinch singular point. The reason is that the Landau equations for the loop formed by l_1 and l_2,

$$\alpha_1 l_1{}^\mu - \alpha_2 l_2{}^\mu = 0, \tag{13.23}$$

have no solution unless l_1 and l_2 are proportional. For these processes, then, we can answer the question posed at the start of Section 13.1. Collinear and infrared regions *are* the only sources of divergence due to vanishing masses.

Application: physical pictures in leptoproduction

Another very interesting class of processes is those of high-momentum-transfer lepton–hadron scattering, in the one-photon approximation, illustrated in Fig. 13.10(a). We consider the *deeply inelastic* region where

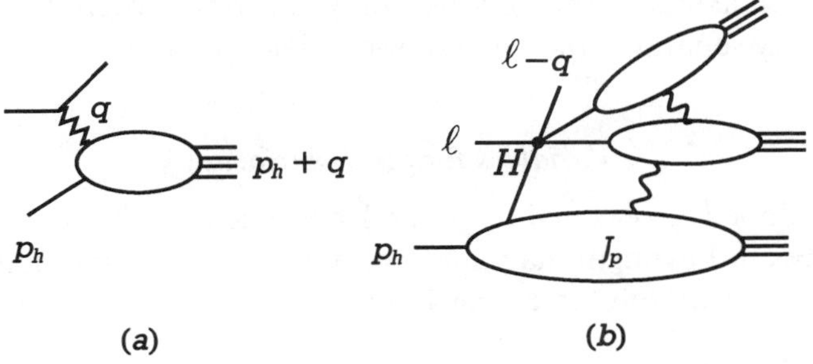

Fig. 13.10 (a) Amplitude for deeply inelastic scattering; (b) typical reduced diagram corresponding to a pinch surface.

$(p_h + q)^2 = O(q^2) \gg m^2$, with m the mass of the initial-state hadron. The pinch surfaces of the massless diagrams correspond to singular contributions to the amplitude in the limit $q^2/m^2 \to \infty$. The physical processes are as easy to identify as in the case of e^+e^- annihilation. The generic picture is illustrated by Fig. 13.10(b). Incoming at the left we have a single lepton of momentum l, and an incoming or 'forward' jet J_p consisting of 'constituents' of the hadron, with momentum p^μ. In the zero-mass limit these constituents approach the mass shell as they become parallel to the incoming momentum. (For the zero-mass limit, of course, we cannot work in the rest frame of the incoming hadron.) At the hard vertex H the lepton interacts with the initial jet by the exchange of a single off-shell vector boson. The exchanged boson is included in H. This hard scattering sends the final-state lepton, of momentum $l - q$, off in a new direction, and produces various final-state jets, while the unscattered portion of the forward jet also passes into the final state. Since the jets propagate freely away from the space–time point H, they cannot collide at any later point in space–time. Thus, as in the timelike case, all jets produced at the hard interaction emerge into the final state in their original directions, each with its original total momentum. As usual, we can attach any number of soft lines to the jets. The picture, then, is very similar to the general timelike case, Fig. 13.9(a), except for the presence of the forward jet.

Initial- and final-state interactions

Cross sections in e^+e^- annihilation and in deeply inelastic scattering are dependent on a single hard scattering. It is conventional to characterize other interactions in the process by the time relative to this hard interaction. Those that occur after the hard interaction are 'final-state interactions'; those that occur before are 'initial-state interactions'. In e^+e^- annihilation, at the lowest order in α, all (strong) interactions are final state. Deeply inelastic scattering involves both initial state interactions (incoming jet only) and final-state interactions. Feynman diagrams are a mixture of initial- and final-state interactions; only in time-ordered perturbation theory are they separated on a diagram-by-diagram basis.

Variables for pinch surfaces

Having developed a reasonably general approach to the identification of pinch surfaces in Feynman diagrams, we now describe a method of estimating, or more accurately of giving bounds for, the contributions of these surfaces.

Consider an arbitrary pinch surface γ of diagram G. It is specified by setting certain line momenta on-shell, and requiring that some lines are part of jets and others are soft. All these requirements are algebraic conditions on loop momenta. γ is thus an algebraic surface embedded within the loop momentum space of G. To discuss the behavior of the integral near γ, it is convenient to distinguish between the internal coordinates of γ and those of the space normal to γ. Small variations in internal coordinates leave all the momenta $\{k_i{}^{\mu}\}$ on-shell, while small changes in normal coordinates take them out of γ, so that one or more lines go off-shell. It is thus the normal coordinates that determine the behavior of the graph near the pinch surface. Within these constraints, the choice of normal and internal coordinates is arbitrary.

Homogeneous integral

We denote normal variables by $\{k_i\}$, internal variables by $\{l_j\}$ and external momenta by $\{p\}$. Schematically, the integral near pinch surface γ has the form

$$G_\gamma = \int \prod_j \mathrm{d}l_j \int_0 \prod_i \mathrm{d}k_i \, I_\gamma(k_i, l_j, p), \tag{13.24}$$

where the notation '$\int_0$' means that the normal variables are all very small compared with the scale of the external momenta and the l_j. To estimate the behavior of the integral near γ we compute the analogue of the superficial degree of infrared divergence introduced in Section 12.3. We assume that each factor (denominator or numerator) in $I_\gamma(k_i, l_j, p)$ is a polynomial in normal coordinates, with coefficients depending, in general, on the internal coordinates. Each such polynomial is approximated by its monomials of lowest order in the normal coordinates. The resulting expressions will be referred to as the *homogeneous integral* $\bar{G}_\gamma$ and *integrand* $\bar{I}_\gamma$ for γ:

$$\bar{G}_\gamma = \int \prod_j \mathrm{d}l_j \int_0 \prod_i \mathrm{d}k_i \, \bar{I}_\gamma(k_i, l_j, p). \tag{13.25}$$

As an example, consider the infrared pinch surfaces of the QED vertex (Fig. 12.4). The normal coordinates are the four-momenta of the soft photons. On-shell photon denominators are already homogeneous expressions of order two in the normal coordinates. Fermion denominators, however, are of the form $2p\cdot k + k^2 + \mathrm{i}\epsilon$, which is replaced by $2p\cdot k + \mathrm{i}\epsilon$. Similarly, we approximate exact numerator factors $(p + k)^\mu$ by p^μ. In this case, the homogeneous integral is the eikonal approximation.

As in the ultraviolet case, the superficial degree of infrared divergence $\mu(\gamma)$ of pinch surface γ is defined by combining the volume of integration with the size of the integrand, as all the k_i vanish according to a scaling variable λ having the dimensions of mass:

$$\bar{G}_\gamma = \int\prod_j \mathrm{d}l_j \int_0^\infty \mathrm{d}\lambda^2 \int_0 \prod_i \mathrm{d}k_i\, \delta(\lambda^2 - \textstyle\sum_i |k_i|^2)\bar{I}_\gamma(k_i, l_j, p)$$
$$= 2\int\prod_j \mathrm{d}l_j \int_0^\infty \mathrm{d}\lambda\,(\lambda)^{\mu(\gamma)-1} \int_0 \prod_i \mathrm{d}k'_i\, \delta(1 - \textstyle\sum_i |k'_i|^2)\bar{I}_\gamma(k'_i, l_j, p). \tag{13.26}$$

In the second line, $k'_i = k_i/\lambda$. The integral is superficially infrared finite for $\mu(\gamma) > 0$. Just as in the ultraviolet case, the superficial degree of divergence is just the beginning of the story, since the remaining integral over the k'_i may still contain pinch surfaces, which are the 'subdivergences' of the infrared problem.

Finiteness condition

A sufficient condition for the infrared finiteness of a Feynman diagram is that the integral over a neighborhood of every point on every pinch surface should be finite. Since the number of pinch surfaces is finite, any diagram with this property is infrared finite. Corresponding to Weinberg's theorem (Section 9.4), we may formulate a test for infrared convergence. For any Feynman diagram, it is possible to choose coordinates for the normal space of each pinch surface in such a way that the only pinch surfaces of the resulting homogeneous integral correspond to pinch surfaces of the original integral. (As we shall see, it is sometimes, but not always, necessary to use different variables in different regions of the normal space.) Then the diagram is convergent if the superficial degree of infrared divergence for every set of variables on every pinch surface is positive. At the time of writing, there is no formal proof of this theorem in the literature, but it is implicit in all work on the subject of infrared divergences.

As our first application of these results, we can show the infrared finiteness of massless Green functions in Euclidean space (Poggio & Quinn 1976, Sterman 1976).

Pinch surfaces for nonexceptional Euclidean momenta

The external momenta of a Green function $G(\{p_i\})$ are said to be *non-exceptional* if for every proper subset A of $\{p_i\}$,

$$\left(\sum_{i\in A} p_i{}^\mu\right)^2 \neq 0. \tag{13.27}$$

Suppose we work in Euclidean space, where $k^2 = 0 \rightarrow k^\mu = 0$. Then pinch surfaces in massless perturbation theory can only involve zero-momentum internal lines. When external momenta are nonexceptional, the requirement of a physical picture implies that in massless perturbation theory there is only *a single hard vertex* in the reduced diagram of any pinch surface, since by eq. (13.27), no proper subset of the external lines can conserve momenta at any vertex. The most general pinch surface of a graph with nonexceptional Euclidean external momenta is thus shown in Fig. 13.7. Here, soft lines attach directly to H and interact via the subdiagram S, all of whose internal on-shell lines have vanishing momenta. The physical picture corresponding to this reduced diagram is one in which all the external momenta meet at a single point in Euclidean space–time, where they also interact with the 'cloud', S, of zero-momentum soft particles.

Euclidean infrared power counting

Let us now show that Euclidean Green functions are infrared finite. We begin with the simplifying assumption that all the internal vertices of subgraph S are elementary, rather than composite, and we work in four dimensions. The normal variables for any pinch surface are the complete loop momenta of its reduced diagram. Then, clearly, the only pinch surfaces in the corresponding homogeneous integral are (exactly) those in the original diagram. Thus, in accordance with our comments above, it is not necessary to choose more than one set of normal variables around any pinch surface to test for finiteness.

The evaluation of the superficial degree of divergence $\mu(S)$ is straightforward and follows the same pattern as in the ultraviolet case, Section 10.2, and as in the infrared power counting of Section 12.3. There are $4L$ normal variables, where L is the number of loops in the reduced diagram. Boson and fermion denominators contribute -2 and -1 to the degree of divergence respectively, while soft three-gluon couplings contribute $+1$. Suppose that e_{b} soft bosons and e_{f} soft fermions attach to the hard subdiagram in Fig. 13.7 and that, in addition to gauge couplings, there are v_d d-scalar vertices in the soft part. Then we find, after a brief calculation,

$$\mu(S) = e_{\mathrm{b}} + \tfrac{3}{2}e_{\mathrm{f}} + \sum_{d \geqslant 3} (d-4)v_d. \tag{13.28}$$

Thus, $\mu(S) > 0$ as long as there are no super-renormalizable couplings involving massless scalars.

The last step is to remove the restriction that the soft subdiagram includes only elementary soft vertices. Suppose there is a K-point composite vertex within S. For simplicity, let all its external lines be vectors in a gauge theory. If all the loop momenta within the vertex are fixed, then its behavior is given by dimensional counting, and

$$\mu(S) = e_{\mathrm{b}} + \tfrac{3}{2}e_{\mathrm{f}} + \sum_{d \geqslant 3} (d-4)v_d + (K-4). \tag{13.29}$$

That is, composite vertices act just like higher-order scalar interactions, and as long as they have four or more external lines, they do not decrease the superficial degree of divergence. In the special cases $K = 2, 3$ we keep only those contributions to the composite vertex that survive renormalization. Then the only scale upon which the composite vertices can depend is their external momenta, and the complete two- and three-point functions behave in the same way as elementary vertices for power-counting purposes. We conclude that Euclidean Green functions are infrared finite for nonexceptional external momenta.

As a second application, we can justify the eikonal approximation in QED (Section 12.3), and discuss its range of applicability.

Eikonal approximation in QED

In proving the exponentiation of infrared divergences in QED, we gave a power-counting argument for the finiteness of the vertex function when all infrared regions are suppressed, in the case of the modified vertex, eq. (12.52). By construction, the degree of divergence is positive at any pinch surface in $\Gamma_\mu{}^{(\mathrm{fin})}$ of eq. (12.59), so we only need to verify that there are no new pinch surfaces in the homogeneous integrals of $\Gamma_\mu{}^{(\mathrm{fin})}$. This is equivalent to verifying that the eikonal approximation is valid. The only problematic part of the homogeneous integral, and the eikonal approximation, is the replacement $2p\cdot k + k^2 + \mathrm{i}\epsilon \rightarrow 2p\cdot k + \mathrm{i}\epsilon$ in the fermion denominators. It is not difficult to check that this replacement produces no new pinch surfaces.

This is most easily seen by noting that a pinch surface for the homogeneous integral still corresponds to a physical picture in which the fermions have infinite mass, in order that k^2 may always be neglected compared with $2p\cdot k$, and k^μ may be neglected compared with p^μ. The physical pictures associated with fermions of finite or infinite mass are essentially the same, however. Once a pair is created with nonzero relative velocity, it can never come together at a point at a later time, nor can either fermion emit or absorb finite-energy photons and stay on-shell. The only pinch surfaces of the homogeneous integral are therefore the usual ones, at zero photon momentum.

Threshold singularities and enhancements

At the threshold $s = 4m^2$, the above reasoning breaks down, and this shows why it is sometimes necessary to choose more than one set of coordinates in the normal space. When $s = 4m^2$ we can work in a frame where both fermions remain at rest. In the eikonal approximation, they then remain in contact, even while exchanging off-shell photons of arbitrary spatial momenta but zero energy. As a result, the homogeneous integral has a whole new set of pinch surfaces, $k_i{}^0 = 0$, $\mathbf{k}_i \neq 0$, which were absent from the exact integral. Hence the set $k_i{}^\mu$, $i = 0, 1, 2, 3$, is a good set of normal variables in only part of the normal space for pair production at threshold. The power-counting argument for the finiteness of the quantity $\Gamma_\mu{}^{(\mathrm{fin})}$ in eq. (12.59) fails at threshold, because it is based upon power counting in terms of the set of normal variables $k_i{}^\mu$, $i = 0, 1, 2, 3$. In the eikonal approximation the k_0 integral behaves as $\int_0 \mathrm{d}k_0/k_0{}^2$ at threshold, and is power divergent. In fact, near threshold it behaves as $1/\beta$, where β is the relative velocity of the fermion and antifermion. This may be seen explicitly from the imaginary factor $(\alpha/2\pi)(-\mathrm{i}\pi/\beta)$ in eq. (12.30).

13.4 The three-point function and collinear power counting

Analysis of the three-point function at high energies leads us naturally to a general discussion of collinear divergences. We shall begin this section, however, with a few observations on the pinch structure of the three-point function with nonzero masses.

Normal thresholds

The reduced diagrams of the three-point function can be classified into three cases according to how the external lines attach: those in which the external lines all attach to the same vertex, as in Fig. 13.11(a) (zero-momentum only), those in which two external lines attaches to a single vertex, while the remaining line attaches to another, as in Fig. 13.11(b) (normal thresholds), and those in which all three external lines attach to separate vertices, as in Figs. 13.11(c), (d). Of these cases, only Fig. 13.11(c) is new, compared with the two-point case. Singularities of this type are called anomalous thresholds.

Anomalous thresholds

Consider the one-loop reduced diagram shown in Fig. 13.11(e). The whole process depicted by the figure takes place in one spatial dimension and in time. Two particles, k_1 and k_2, are produced at vertex v_1, one at rest, and the other moving in the x-direction. After time t, particle k_1 hits a 'brick wall' at vertex v_2 and bounces off elastically, to give a particle with momentum k_3, which has the same energy as k_1, but precisely opposite spatial momentum $\mathbf{k}_3 = -k_{1,x}\hat{\mathbf{x}}$. It travels back in the $-x$-direction for exactly the same time t, until it hits particle k_2, at vertex v_2, where it is absorbed. The Landau equations are satisfied for

$$\alpha_1 = \alpha_3 = \alpha_2 m/(2(k_{1,x}{}^2 + m^2)^{1/2}). \tag{13.30}$$

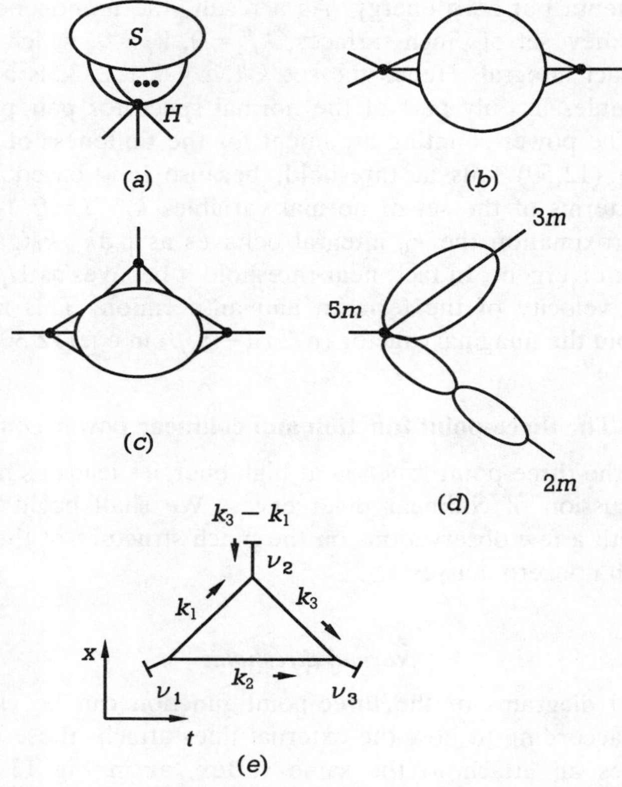

Fig. 13.11 Reduced diagrams for three-point functions.

The external momentum at ν_2 is spacelike, with invariant length $(k_3 - k_1)^2 = -4k_{1,x}{}^2$. Other than this, there is nothing particularly 'anomalous' about anomalous thresholds.

For four- and higher-point functions the situation gets vastly more complicated. We shall not try to give a complete discussion here. (For a review, see Eden, Landshoff, Olive & Polkinghorne r1966.) We shall turn instead to issues that are specific to the zero-mass limit in Minkowski space. We shall not limit power counting to quantities that have finite high-energy and zero-mass limits, but shall also discuss the origin of high-energy logarithms in Feynman diagrams. We shall be interested particularly in the collinear divergences that arise from the coupling of massless particles with finite momentum.

Power counting for one-loop diagrams

Consider the one-loop vertex correction to the timelike electromagnetic vertex in the zero-mass limit. In Fig. 13.12 its pinch surfaces with two or

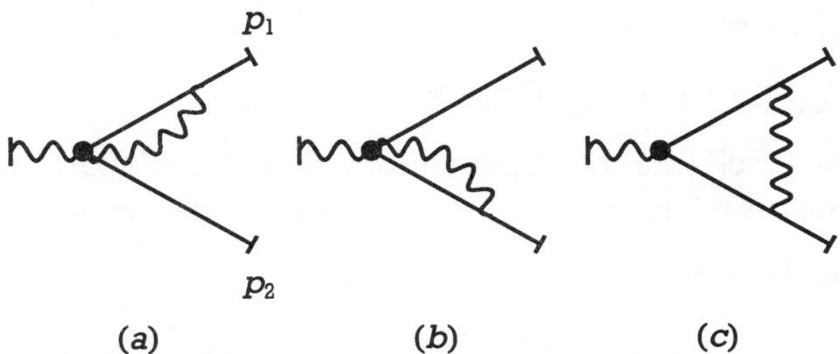

Fig. 13.12 Reduced diagrams corresponding to pinch surfaces of the fermion–vector three-point function.

three lines are exhibited. Figures 13.12(a), (b) include jets, consisting of the vector line along with one of the fermions. Figure 13.12(c) is the infrared reduced diagram, which is the same as the original diagram.

We have already calculated this diagram in Chapter 12. It is useful, however, to see how power counting gives results that are consistent with the explicit calculations. We are already familiar with the behavior of the integral near the infrared pinch surfaces, Fig. 13.12(c). Therefore, let us look at the collinear pinch surfaces.

To be specific, we work in the rest frame of the timelike photon $q^\mu = (Q, \mathbf{0})$, with the outgoing fermions in the $\pm z$-direction,

$$p_1{}^\mu = \delta_{\mu+} Q/2^{1/2}, \quad p_2{}^\mu = \delta_{\mu-} Q/2^{1/2}. \tag{13.31}$$

Now consider the integral near the pinch surface where the photon momentum, k^μ, becomes collinear with $p_1{}^\mu$. For this surface we choose

$$\begin{aligned} &\text{normal variables } k^-,\ |\mathbf{k}|^2, \\ &\text{internal variables } k^+,\ \phi, \end{aligned} \tag{13.32}$$

where ϕ is the azimuthal angle of $\mathbf{k}$ about the z-axis. With this choice, the k^2 and $(p_1 + k)^2$ denominators are both linear in the normal variables, while the $(p_2 + k)^2$ denominator and the numerator are zeroth order in normal variables. With dimensional regularization, the homogeneous integral corresponding to this choice can then be put in the form

$$\begin{aligned} &g^3 \mu^{3\epsilon} (2\pi)^{-n} \Omega_{n-3} \int \mathrm{d}k^+ (2^{1/2} Q k^+)^{-1} N_\mathrm{h} \\ &\times \int_0 \mathrm{d}k^- \mathrm{d}|\mathbf{k}|^2\, |\mathbf{k}|^{-2\epsilon} (2k^+ k^- - |\mathbf{k}|^2 + \mathrm{i}\epsilon)^{-1} \\ &\times [-2(2^{-1/2} Q - k^+) k^- - |\mathbf{k}|^2 + \mathrm{i}\epsilon]^{-1}, \end{aligned} \tag{13.33}$$

where the homogeneous numerator N_h is given by

$$N_\mathrm{h} = C_2(\mathrm{F})\bar{u}(p_2)\gamma^\alpha(p_2{}^-\gamma^+ + k^+\gamma^-)\gamma_\mu[2(-p_1 + k)^+\delta_{\beta+}]v(p_1)g_{\alpha\beta},$$
$$= -C_2(\mathrm{F})2Q^2\bar{u}(p_2)\gamma_\mu v(p_1)(1 - 2^{1/2}k^+/Q). \qquad (13.34)$$

Here we have used the Dirac equation. The 'hard' part in this simple expression consists of the $p_2 + k$ line and its adjacent vertices:

$$H_\mu{}^\alpha(p_2, k) = H_\mu{}^-(p_2, k)\delta_{\alpha-}$$
$$= -\mathrm{i}g^2\mu^{2\epsilon}\bar{u}(p_2)\gamma^-(p_2{}^-\gamma^+)\gamma_\mu(2^{1/2}Qk^+)^{-1}\delta_{\alpha-}. \qquad (13.35)$$

Equation (13.33) exhibits a logarithmic divergence at $\epsilon = 0$ associated with the normal variables at the pinch surface $k^- = |\mathbf{k}|^2 = 0$, and in dimensional regularization a pole in ϵ produces a logarithm of Q^2 in the finite part. From eq. (13.35), we see that the collinear gluon k^μ couples to $H_\mu{}^\alpha$ by the plus component of its polarization only, and is thus effectively scalar polarized at the pinch surface. This scalar polarization is characteristic of collinear gluons that attach to a hard subdiagram in Feynman gauge (see below).

It is important to observe that we do not have to worry about cancellations between terms internal to the k^2 and $(p_1 + k)^2$ denominators, since these singularities do not correspond to pinch surfaces in either the original or the homogeneous integral.

In eq. (13.33) there is an additional logarithmic divergence when the internal variable k^+ also vanishes. This overlap of the collinear and infrared pinch surfaces produces the double pole and a double logarithm of Q^2 in the explicit form, eq. (12.85).

Jet power counting beyond one loop

Now let us generalize power counting to arbitrary order (Sterman 1978, Ellis, Georgi, Machacek, Politzer & Ross 1979) for the timelike vertex correction, with external momenta given by eq. (13.31). The on-shell external lines may be either fermions, scalars or transversely polarized vectors. For gauge theories, we work in the Feynman gauge and allow self-interacting scalars.

The arbitrary pinch surface has already been illustrated in Fig. 13.9(*a*), with two jets J_1, J_2 a single hard part D and a soft subdiagram S. Normal variables are chosen by direct generalization from the one-loop case. First, there are soft momenta $k_i{}^\mu$, which vanish in all four components at the pinch surface. The corresponding normal variables are any independent set of internal and external loop momenta of S. In addition there are the loop momenta of the jets J_i, which become lightlike and parallel to $p_i{}^\mu$ at the pinch surface. For instance, let us choose $\{k_j{}^-, |\mathbf{k}_j|^2\}$ as normal variables from the loop momenta of J_1, where the k_j are an independent set of the internal loops of J_1. All remaining momenta are internal variables of the pinch surface. Soft line denominators are quadratic, and jet denominators linear, in these normal variables. Soft (but not jet) fermions contribute an extra normal variable to the numerator.

With this choice the general homogeneous integral has no extra pinch surfaces (exercise 13.7), and the superficial degree of infrared divergence for an arbitrary jet pinch surface γ is

$$\mu(\gamma) = [4L_S - 2N_\mathrm{b} - N_\mathrm{f} + v^{(3)}{}_S] + \sum_{i=1}^{2} (2L_i - N_i + t_i), \quad (13.36a)$$

where L_S and L_i are the numbers of loops in S and J_i respectively. N_b and N_f are the numbers of boson and fermion lines in the soft subdiagram, and N_i is the total number of lines in jet i. $v^{(3)}{}_S$ is the suppression factor in the soft subdiagram from three-point vertices, while t_i is the contribution of numerator momenta in J_i, which we shall determine below. The four terms in square brackets are $\mu(S)$, eq. (13.29), and we find

$$\mu(\gamma) = \sum_{i=1}^{2} (2L_i - N_i + b_i + \tfrac{3}{2}f_i + t_i), \quad (13.36b)$$

where b_i and f_i are the numbers of soft bosons and soft fermions attached to J_i. Now, let us evaluate t_i. We assume that no soft lines attach directly to the hard part, since such an arrangement clearly has fewer on-shell lines than when they are attached to jets, and is hence suppressed.

Accounting for both three-gluon couplings and fermion propagators, the number of factors of numerator momenta in J_i is $u^{(i)}{}_3$, where $u^{(i)}{}_3$ is the number of three-point vertices in J_i (we asssume no super-renormalizable couplings). The total momentum of jet i is lightlike, and may be taken in the plus direction without loss of generality. Then the plus component of each jet loop momentum is an internal variable, but when any two momentum factors contract to form an invariant, their product is linear in normal variables, because $g_{++} = g_{--} = 0$. This does not always have to happen, however, since the momenta may 'escape' the jet by contracting against the polarization tensors of the soft vectors of S or of jet vectors attached to D (see our discussion of the one-loop example, eq. (13.33) above). The suppression factor t_i is then bounded from below by

$$t_i \geqslant \max\{\tfrac{1}{2}[u^{(i)}{}_3 - w_i - v_i], 0\} \quad \text{(Feynman gauge)}, \quad (13.37a)$$

where w_i is the number of scalar polarized jet vector lines attached to H from jet i and v_i is the number of soft vectors attached to J_i. Actually, this bound is gauge-dependent. In axial gauges, the gluon propagator is given by eq. (7.77), and it is suppressed when contracted with any vector proportional to its momentum (eq. (7.78b)). In this case,

$$t_i \geqslant \max\{\tfrac{1}{2}[u^{(i)}{}_3 - v_i], 0\} \quad \text{(axial gauge)}. \tag{13.37b}$$

Returning to Feynman gauge, we find for $\mu(\gamma)$,

$$\mu(\gamma) \geqslant \sum_{i=1}^{2} \{2L_i - N_i + b_i + \tfrac{3}{2}f_i - \tfrac{1}{2}v_i - \tfrac{1}{2}w_i + \tfrac{1}{2}u^{(i)}{}_3 + \tfrac{1}{2}[v_i + w_i - u^{(i)}{}_3]\theta(v_i + w_i - u^{(i)}{}_3)\}. \tag{13.38}$$

To proceed further, we use Euler's identity and the relation between the number of lines and vertices in J_i. Let h_i be the number of fermion, scalar and/or transversely polarized vector jet lines attached to the hard part. Then the relation is

$$3u^{(i)}{}_3 + 4u^{(i)}{}_4 + (h_i + w_i) = 2N_i + b_i + f_i + 1. \tag{13.39}$$

Here we have used the fact that J_i attaches to the hard vertex by $h_i + w_i$ lines, and that it has $b_i + f_i + 1$ external lines. Jet lines attached to the hard part are considered internal to the jet. It is now easy to show that

$$\mu(\gamma) \geqslant \sum_{i=1}^{2} \{\tfrac{1}{2}(h_i - 1) + f_i + \tfrac{1}{2}(b_i - v_i) + \tfrac{1}{2}[v_i + w_i - u^{(i)}{}_3]\theta(v_i + w_i - u^{(i)}{}_3)\}. \tag{13.40}$$

$\mu(\gamma)$ is positive semidefinite, as long as $h_i \geqslant 1$. If $h_i = 0$, however, the jet attaches to the hard part by scalar-polarized vectors only at the pinch surface. Such a polarization configuration, although it may be nonzero in a given diagram, vanishes in a gauge invariant set of diagrams by the graphical Ward identities of the theory, such as Fig. 11.6 in QED. It is important to realize, however, that divergences can be worse than logarithmic in individual diagrams in the Feynman gauge.

By inspection of eq. (13.40), pinch surface γ is associated with a logarithmic divergence in Feynman gauge only if: (i) a single fermion, or a scalar-polarized *or* physically polarized vector attaches the jet to the hard part, (ii) the only additional lines attaching the jet to the hard part are scalar-polarized vectors, (iii) soft lines attach only to the jets, and not to the hard part and (iv) for each jet the number of external soft vectors plus the number of jet vectors attached to the hard part is not greater than the total number of three-point vertices. This most general logarithmically divergent pinch surface is illustrated by the fermionic vertex in Fig. 13.13(a); here the gluon lines ending in arrows are scalar polarized at the pinch surface. As mentioned above, scalar polarizations are suppressed in axial gauge, and the logarithmic pinch surfaces in that gauge are somewhat simpler (exercise 13.8), as illustrated in Fig. 13.13(b).

We also learn from eq. (13.40) that in purely massless theories without vector fields, pinch surfaces with soft lines are power-counting finite in four

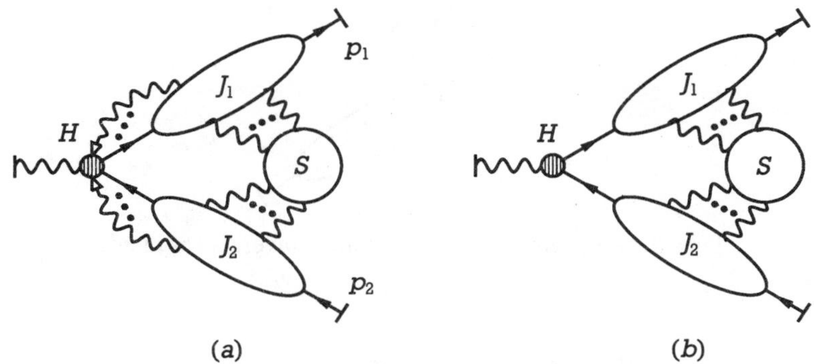

Fig. 13.13 Logarithmically divergent pinch surfaces in zero-mass gauge theories: (*a*) Feynman gauge, (*b*) axial gauge.

dimensions, and massless Yukawa or ϕ^4 theories in four dimensions (or ϕ^3 in six dimensions), for instance, lack infrared divergences. By the same token, for such theories there are no logarithms of energy over mass from soft lines in the high-energy limit. Collinear divergences, however, are associated with particles of any spin, as long as they possess three- or four-point interactions. It should be noted that massless scalar lines *may* give rise to infrared divergences when coupled by scalar interactions to massive fermions (exercise 13.10).

The electromagnetic vertex in ϕ^4

The absence of infrared divergences in ϕ^4 has strong consequences for the behavior of its electromagnetic vertex, dressed to all orders by scalar self-interactions. This amplitude depends on both the ratios q^2/μ^2 and m^2/μ^2, where m is the renormalized scalar particle mass. The electromagnetic vertex is a renormalization group invariant, obeying $(\mu \mathrm{d}/\mathrm{d}\mu) G_\mu = 0$. This relation alone, however, is not enough to determine the high energy behavior of $G_\mu(q^2/\mu^2, m^2/\mu^2)$. Nonvanishing m-dependence at high energy still occurs in the form $\ln(\mu^2/m^2)$, which diverges in the zero-mass limit. From the above, we learn that logarithms of mass may come only from collinear, and not infrared, divergences, and then only from configurations where the scalar 'jets' are connected to the hard scattering vertex by only one scalar line each. As a result, in ϕ^4 theory collinear divergences in the vertex function come from self-energies only. After reduction (which cancels one self-energy on-shell), the relevant reduced diagrams are of the form shown in Fig. 13.14, in which mass and momentum dependence are 'factorized'. The one-particle irreducible vertex Γ_μ is mass-independent at high energy (finite in the zero-mass limit), while all mass-dependence

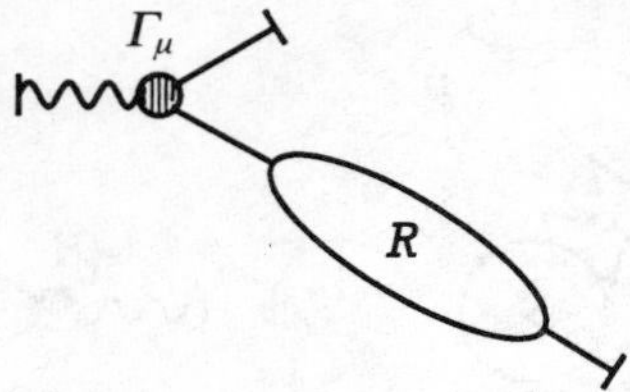

Fig. 13.14 Pinch surface with collinear divergences in ϕ^4.

resides in the on-shell residue of the Green function:

$$G_\mu(q^2/\mu^2, \mu^2/m^2, \lambda(\mu)) = \Gamma_\mu(q^2/\mu^2, \lambda(\mu))R(\mu^2/m^2, \lambda(\mu)). \quad (13.41)$$

λ is the renormalized coupling. Γ_μ obeys the renormalization group equation (10.130):

$$[\mu\partial/\partial\mu + \beta(\lambda)(\partial/\partial\lambda) - 2\gamma_\phi]\Gamma_\mu(q^2/\mu^2, \lambda(\mu)) = 0, \quad (13.42\text{a})$$

whose solution is eq. (10.132),

$$\Gamma_\mu(q^2/\mu^2, \lambda) = \Gamma_\mu(1, \bar{\lambda}(q^2, \lambda)) \exp\left[-2\int_0^{\ln(q^2/\mu^2)} \mathrm{d}t'\, \gamma_\phi(t')\right]. \quad (13.42\text{b})$$

The momentum-dependence of the electromagnetic vertex in ϕ^4 is thus determined by the anomalous dimension of the field (Creutz & Wang 1974, Marques 1974, Callan & Gross 1975, Thacker & Weisberger 1976). This is a much more moderate dependence at high energy than the exponential of double logarithms in gauge theories, as in eq. (12.60), which is due to the interplay of collinear and infrared divergences. Similar results may also be derived for ϕ^3 in six dimensions (exercise 13.11). The extension of this kind of reasoning to gauge theories generalizes the leading logarithm high-energy behavior of eq. (12.60). It has been carried out variously by Sudakov (1956), Mueller (1979), Collins (1980), Collins & Soper (1981), Sen (1981) and Korchemsky & Radyushkin (1986), as reviewed in Collins (r1989). The electromagnetic vertex, of course, remains infrared divergent and is not directly observable in four dimensions.

Power counting for cross sections and cut diagrams

It is quite straightforward to generalize power counting to an arbitrary amplitude. Of more physical interest, however, are cross sections, and it is to these that we now turn. A great simplification is achieved when we recall (Section 8.1) that a cross section is a sum of cut diagrams. For example, a typical contribution to e^+e^- annihilation is shown in Fig. 13.15. To analyze

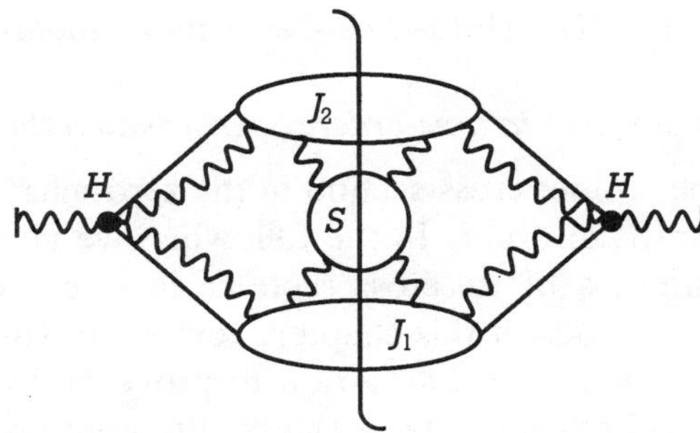

Fig. 13.15 Pinch surface of a general cut diagram that gives a logarithmically divergent contribution to a two-jet cross section.

the behavior of cut diagrams, and hence cross sections, in the zero-mass limit, it is only necessary to identify the pinch surfaces of the amplitude and its complex conjugate at each position in phase space, and then to perform the appropriate power counting.

For e^+e^- annihilation, we already know the pinch surfaces of the amplitudes to the left and right of the cut. We can construct reduced diagrams of a cut graph in the same way as for an uncut graph. In general, there is a jet, appearing on both sides of the cut, associated with each final-state particle, or set of final-state particles, moving in a given direction. In addition, there is an overall soft subdiagram, extending to both sides of the cut, that includes zero-momentum particles in the final state. Finally, there will be two (and only two) hard subdiagrams, one on each side of the cut.

Power counting for the generic pinch surface of a cut diagram, such as Fig. 13.15, proceeds as above. We split up the loop momenta of the overall cut diagram, including phase space integrals, into internal loop momenta of the jets and hard parts, and soft loop momenta. The mass-shell delta functions of cut lines act, for power-counting purposes, just like propagators, since each delta function can be used to eliminate a single normal variable. Here we shall concentrate on cross sections in which all final-state momenta are integrated over at least part of their range. Such *semi-inclusive* cross sections have the same dimensions as the total cross section. Examples are the jet cross sections of Section 12.5. A rather straightforward calculation then shows that for semi-inclusive e^+e^- annihilation cross sections, the superficial degree of infrared divergence is exactly the same as for the vertex function. Contributions to semi-inclusive cross sections are, at worst, logarithmically divergent in the zero-mass limit, or equivalently, grow logarithmically – but no worse – with the energy. It is these logarithms that must cancel in any cross section with a finite zero-mass limit.

13.5 The Kinoshita–Lee–Nauenberg theorem

Pinch surfaces in time-ordered perturbation theory

The finiteness of the two-jet cross section in the zero-mass limit was easy to derive at lowest nontrivial order. In the following, we give a general proof of the infrared finiteness of jet cross sections in e^+e^- annihilation. The ingredients are the methods of this chapter, combined with the generalized unitarity equation, (9.85). We shall go on to prove an even more general result, the Kinoshita–Lee–Nauenberg (KLN) theorem on the finiteness of averaged transition probabilities.

Our discussion of zero-mass behavior has been given thus far in terms of Feynman integrals. The transition to time-ordered perturbation theory, in terms of which generalized unitarity has been stated above, is relatively simple. A general time-ordered diagram with K states is of the form (eq. (9.72))

$$\prod_a \int_0^1 \mathrm{d}\beta_a\, \delta\left(\sum \beta_a - 1\right) \int \prod_{\text{loops } i} \frac{\mathrm{d}^3\mathbf{q}_i}{(2\pi)^3} \prod_{\text{lines } j} \frac{1}{2\omega(\mathbf{k}_j)}$$
$$\times \left\{ \sum_{\text{states } a} \beta_a \Big[E_a - \sum_{\text{lines } j \text{ in } a} \omega(\mathbf{k}_j) + \mathrm{i}\epsilon\Big]^{-1} \right\}^{-K}, \tag{13.43}$$

where we have Feynman-parameterized the state denominators, and, as usual, suppressed numerators and overall constants. The pinch surfaces of this integral may be analyzed in the same way as for the Feynman form eq. (13.1), with exactly parallel results (exercise 13.12). If we interpret β_a as the time elapsed in state a, then a pinch surface for the time-ordered integral has the same physical interpretations as for Feynman diagrams. The vertices of the time-ordered reduced diagram, made up of on-shell states, can consistently be assigned positions in space–time, which are connected by the free propagation of on-shell particles. Thus, just as for Feynman diagrams, we can identify pinch surfaces by the criterion that they correspond to physical processes.

Finiteness of jet cross sections

The proof that jet cross sections are finite is straightforward (Sterman 1978). It is essentially the same as the proof of unitarity in Section 9.6, and again the main problem is simply notation. To make our equations a bit less unwieldy, we introduce the definitions

$$S^{(a)} = \sum_{\text{lines } j \text{ in } a} \omega(\mathbf{k}_j), \tag{13.44a}$$

for the on-shell energy of state a, and

$$\mathrm{d}\tau_G = \int \prod_{\text{loops } j \text{ of } G} \frac{\mathrm{d}^3\mathbf{q}_i}{(2\pi)^3} \prod_{\text{lines } i \text{ of } G} \frac{1}{2\omega_i}, \tag{13.44b}$$

for the time-ordered perturbation theory (TOPT) integrals of graph G.

Let G_{jet} be the contribution of cut diagram G to an arbitrary jet cross section. Each G_{jet} is a sum over cuts, and for each cut we can evaluate the diagrams on either side of the cut in time-ordered perturbation theory. This gives

$$G_{\text{jet}} = \sum_C \int \mathrm{d}\tau_G \left[\sum_{t'(C)} \prod_{b \text{ in } t'(C)} (E - S_b - \mathrm{i}\epsilon)^{-1} \right] 2\pi\delta(E - S_C) \times \left[\sum_{t(C)} \prod_{a \text{ in } t(C)} (E - S_a + \mathrm{i}\epsilon)^{-1} \right], \tag{13.45}$$

where the sum over $t(C)$ ($t'(C)$) is over all time orderings of the diagram on the left (right) of the cut C. Exchanging the summations over C with those over $t(C)$ and $t'(C)$, G_{jet} can be written as

$$G_{\text{jet}} = \sum_T G_{\text{jet}}(T), \tag{13.46a}$$

where the sum is now over all orderings T of the vertices of the full diagram G such that one or more cuts of the resulting diagram contribute to the jet cross section. This is equivalent to

$$G_{\text{jet}}(T) = \sum_{c \text{ in } C_T} \int \mathrm{d}\tau_G \prod_{b>c} (E - S_b - \mathrm{i}\epsilon)^{-1} 2\pi\delta(E - S_c) \prod_{a<c} (E - S_a + \mathrm{i}\epsilon)^{-1}, \tag{13.46b}$$

where C_T is the set of all cuts of the diagram ordered according to T that contribute to the jet cross section. '$a < c$' means that state a is before the cut in the overall ordering and '$b > c$' means that b is after the cut.

Pinch surfaces in eq. (13.46b) arise when one or more of the state denominators vanish. We know from the physical interpretation that every state that is on-shell at such a pinch surface will have the same set of jets, since these are the physical pictures encountered in the cut Feynman diagrams. We label each such pinch surface S_γ, where γ is its TOPT reduced diagram. Then in region R_γ near S_γ we can write

$$G_\gamma(T) = \sum_{c \text{ in } C_T} \int_{R_\gamma} \mathrm{d}\tau_G \, H_{\mathrm{R}}{}^* \prod_{b>c} (E - S_b - \mathrm{i}\epsilon)^{-1} 2\pi\delta(E - S_c) \times \prod_{a<c} (E - S_a + \mathrm{i}\epsilon)^{-1} H_{\mathrm{L}}, \tag{13.47}$$

where H_L and $H_R{}^*$ are 'hard parts', consisting of various off-shell denominators, which are contracted into the annihilation vertices to the left and right of the cut, as in Fig. 13.15. Since S_γ is a pinch surface, there can be no hard scattering between states in the sets $\{a\}$ or $\{c\}$. As a result, at S_γ every on-shell cut in C_T has the same set of jets, and contributes to the jet cross section. Thus, in region R_γ the sum over cuts may be performed explicitly, and we find, by eq. (9.85), that

$$G_\gamma(T) = \mathrm{i}\int_{R_\gamma} \mathrm{d}\tau_G \, H_R{}^* \left\{ \prod_a (E - S_a + \mathrm{i}\epsilon)^{-1} - \prod_b (E - S_b - \mathrm{i}\epsilon)^{-1} \right\} H_L, \tag{13.48}$$

where the products now go over all states of γ. On the right-hand side of eq. (13.48) the only pinch surfaces are those of the *uncut* diagram γ, which has no nontrivial pinch surfaces at all, since it is a two-point function. The only possibilities are pinch surfaces involving only zero-momentum lines, and those associated with endpoints of the region of integration. The former give only finite contributions, by the power counting of Section 13.3. The latter occur whenever the jets of the pinch surface are at the boundary of phase space defined by the jet cross section. For instance, one or more jets may be precisely at the boundary of one of the jet cones. But, as we have seen in Section 13.4, divergences in jet cross sections are at worse logarithmic. Fixing the precise direction of one or more jets decreases the volume in phase space, and the integral is power-counting convergent at the possible end-point singularities. We therefore conclude that jet cross sections in e^+e^- annihilation are finite on a graph-by-graph basis in time-ordered perturbation theory.

An extension of this reasoning (Ore & Sterman 1980) shows that any e^+e^- annihilation cross section has a finite zero-mass limit, which does not distinguish states that differ only by the emission or absorption of zero-momentum particles, and/or the rearrangement of momenta within any jet.

Initial-state interactions and the KLN theorem

To close this chapter, we come back briefly to the question of initial-state interactions involving massless particles. Such interactions are absent in e^+e^- annihilation (to lowest order in QED, of course), but are certainly present in the zero-mass limit of deeply inelastic scattering, for instance, or, more generally in any cross section in which particles not treated at lowest order appear in the initial state.

We have just seen that divergences associated with final-state interactions cancel in the sum over final states. It is thus natural to ask whether initial-state interactions might also cancel when initial states are summed over. The affirmative answer to this question is known as the Kinoshita–

Lee–Nauenberg (KLN) theorem (Kinoshita 1962, Lee & Nauenberg 1964). For applications to high-energy scattering, its importance has thus far been more conceptual than practical, but it is a fundamental theorem of quantum mechanics and puts many specific results in perspective.

The theorem is stated in terms of transition probabilities rather than cross sections. For a given theory (i.e. a given Lagrangian), let $D(E_0)$ denote the set of states whose energies E lie in the range

$$E_0 - \epsilon < E < E_0 + \epsilon. \tag{13.49}$$

Let p_{ij} be the probability density per unit volume of space–time for a transition to state j from state i. We shall assume that renormalization does not introduce any infrared divergences. Then at any order of perturbation theory, the quantity

$$P(E_0, \epsilon) = \sum_{i,j \in D(E_0)} p_{ij} \tag{13.50}$$

is free of singularities in the zero-mass limit.

By analogy with the discussion of the jet cross section, let us introduce the following notation. Capital letters will stand for particular cuts of diagrams, and corresponding lower-case letters to the states defined by assigning momenta to the particles of that cut. For example, $G_{I,J}(i, j)$ stands for the cut J of time-ordered diagram G_I, with momenta j for lines in the final state, and momenta i for lines in the intial state. An example from deeply inelastic scattering (with leptonic parts omitted) is shown in Fig. 13.16(a). The contribution of $G_{I,J}(i, j)$ to the probability p_{ij} is

$$G_{I,J}(i, j) = \int \mathrm{d}\tau_{G_I} \prod_{B>J} (E - S_B(b) - \mathrm{i}\epsilon)^{-1} 2\pi\delta(E - S_J(j)) \prod_{A<J} (E - S_A(a) + \mathrm{i}\epsilon)^{-1}, \tag{13.51a}$$

$$p_{ij} = \sum_{G_{I,J}} G_{I,J}(i, j), \tag{13.51b}$$

where $S_A(a)$ is given by eq. (13.44a), again with A labelling the cut and a the momenta of the lines of the cut. Once again we neglect overall constants and numerator momenta. As a result of the delta function in (13.51a), if i is in $D(E_0)$, j is too. Thus, we can sum over final state cuts J with momenta j for fixed initial state i, to arrive in the usual way at the analogue of eq. (13.48):

$$\sum_{J,j} G_{I,J}(i, j) = \mathrm{i}\int \mathrm{d}\tau_{G_I} \left\{ \prod_A [S_I(i) - S_A(a) + \mathrm{i}\epsilon]^{-1} - \prod_A [S_I(i) - S_A(a) - \mathrm{i}\epsilon]^{-1} \right\}$$
$$= \mathrm{i}[G_I(i) - G_I{}^*(i)], \tag{13.52a}$$

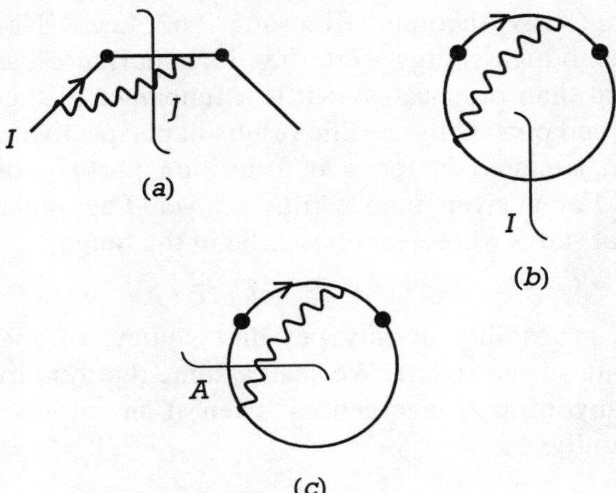

Fig. 13.16 (*a*) Deeply inelastic scattering diagram (leptonic part omitted); (*b*) circular diagram found from (*a*); (*c*) related circular diagram.

where the products are over all intermediate states of the uncut TOPT diagram G_I. Using eqs. (13.50) and (13.51b) we have, for the full probability,

$$P(E_0, \epsilon) = \mathrm{i} \sum_{i \in D(E_0)} \left\{ \sum_{G_I} [G_I(i) - G_I{}^*(i)] \right\}. \tag{13.52b}$$

Let us now sum over initial states for a given $G_I(i)$. This is done in two steps. First, we introduce phase space integrals for all the lines of the initial state I. This enables us to sum explicitly over i for fixed I:

$$G_I{}^{(D)} = \int_{i \in D} \mathrm{d}\tau_I \, G_I(i) = \int_D \mathrm{d}\tau_G \prod_A [S_I(i) - S_A(a) + \mathrm{i}\epsilon]^{-1}. \tag{13.53a}$$

Here the notation $\int_{i\in D}\mathrm{d}\tau_I$ indicates that we restrict the integral to those initial states i with energies in the range (13.49). $\int_D \mathrm{d}\tau_G \equiv \int_{i\in D}\mathrm{d}\tau_I \int \mathrm{d}\tau_{G_I}$ is the full integration measure for the 'circular' diagram G, shown for our example in Fig. 13.16(*b*), which is found from Fig. 13.16(*a*) by identifying the initial-state line(s) of the ordered diagram G_I. The circular diagram G looks like a vacuum bubble (but is not; it is defined by eq. (13.53a)). The vertical line indicates the initial-state cut I, which has no energy denominator in eq. (13.53a). In the notation of eq. (13.53a), eq. (13.52b) becomes

$$P(E_0, \epsilon) = \mathrm{i} \sum_{G_I} \{ G_I{}^{(D)} - [G_I{}^{(D)}]^* \}. \tag{13.53b}$$

The second step is to sum over the set $\{G_A{}^{(D)}\}$, which is generated from $G_I{}^{(D)}$ by letting each cut A play the role of the initial state cut in turn.

$G_I^{(D)} \in \{G_A^{(D)}\}$, and each $G_A^{(D)}$ is a contribution to the probability $P(E_0, \epsilon)$. A typical $G_A^{(D)}$ is illustrated in Fig. 13.16(c), which is related to Fig. 13.16(b) by simply rotating the cut, the relative order of all vertices and lines being preserved. In this case, two lines appear in the (new) initial state. The sum over initial states is taken by summing over all cuts B found by rotating the cut around the diagram. The result is

$$\sum_B G_B{}^{(D)} = \int \mathrm{d}\tau_G \sum_B \theta(\tau_B, E_0, \epsilon) \prod_{C \neq B} [S_B(b) - S_C(c) + \mathrm{i}\epsilon]^{-1}, \quad (13.54)$$

where the factor $\theta(\tau_B, E_0, \epsilon)$ is nonzero only when the cut B has on-shell energy within the range (13.49), in which case it is unity. In general, each of the terms in the sum over B is divergent in a massless theory as a result of the usual collinear and infrared divergences associated with points where $S_B = S_C$ for a set of the C's.

Consider eq. (13.54) in a region of momentum space where $\theta(\tau_B, E_0, \epsilon) = 1$ for every state in the sum over B. At such a singularity, all intermediate states are on-shell. In this case, we can rewrite the sum as an integral by using Cauchy's theorem:

$$\int \mathrm{d}\tau_G \sum_B \prod_{C \neq B} [S_B(b) - S_C(c) + \mathrm{i}\epsilon]^{-1}$$

$$= \frac{-1}{2\pi \mathrm{i}} \int \mathrm{d}\tau_G \int_{-\infty}^{\infty} \mathrm{d}x \prod_{\text{all } F} [x - S_F(f) + \mathrm{i}\epsilon]^{-1} = 0, \quad (13.55)$$

where F ranges over every cut of the circular diagram. The integral vanishes because we can always close the x contour in the upper half-plane, while all the poles are in the lower half-plane. Thus, the contribution from this region is exactly zero. In other regions, where some, but not all, of the intermediate states are in $D(E_0)$, nonleading singularities cancel by the same method.

Finally, we observe that the sum over diagrams G_I in eq. (13.53b) can be replaced by a sum over circular diagrams G and a sum over initial-state cuts B of each G:

$$P(E_0, \epsilon) = -\sum_G \frac{1}{K_G} \sum_B 2\,\mathrm{Im}\, G_B{}^{(D)}. \quad (13.56)$$

The only subtlety here is the weight factor $1/K_G$, which is introduced when a set of initial-state cuts $\{B_i\}$, $i = 1, \ldots, K_G$ give identical cut graphs, and hence identical quantities $\{G_{B_i}{}^{(D)}\}$. In this case, it is easy to see that G is periodic with period K_G and that *all* its cuts come in sets of order K_G. An example with $K = 2$ is given in Fig. 13.17. The weight factor $K_G{}^{-1}$ thus eliminates the double counting associated with such diagrams. From eq. (13.56) we conclude that since $\sum_B G_B{}^{(D)}$ has no singularities it is finite. Then $P(E_0, \epsilon)$ is also finite, and the theorem is proved.

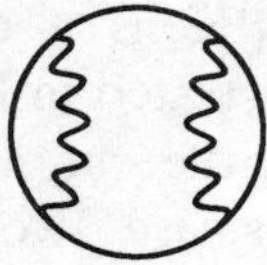

Fig. 13.17 Uncut circular diagram with $K_G = 2$ in eq. (13.56).

If we re-expand the $G_B{}^{(D)}$ of eq. (13.56) according to eqs. (13.52) and (13.53), $P(E_0, \epsilon)$ is expressed as the sum of terms in which the circular diagrams such as Fig. 13.16(b) are cut twice, once for the initial, and once for the final states. Figure 13.18 shows two of these 'double-cut' diagrams (Nakanishi 1958) that occur in our example.

Interpretation

From the point of view of the KLN theorem, it is something of an accident that, in QED, infrared divergences cancel by summing over only final-state, not initial-state, photons. In fact, in QCD, the naive generalization of the Bloch–Norseick mechanism fails when applied to soft gluons (Doria, Frenkel & Taylor 1980, Di'Lieto, Gendron, Halliday & Sachrajda 1981). (Note, however, that it is reinstated in the very-high-energy limit (Labastida 1984, Collins, Soper & Sterman r1989).) The KLN theorem, of course, still applies (Nelson 1981).

It is also important to note that the KLN theorem does not make scattering cross sections with hadrons in the initial state reliably calculable in perturbation theory. The basic problem in applying the KLN theorem to these cross sections is the lack of an average over the initial states. This is a subtle point, however, because it is (as mentioned in Section 12.3) possible to re-interpret the S-matrix in terms of 'coherent' states, which are *constructed* to include averages over initial states. Indeed, physical cross sections may be thought of this way (Contopanagos & Einhorn 1992). Cross sections with initial-state hadrons therefore remain infrared finite in a

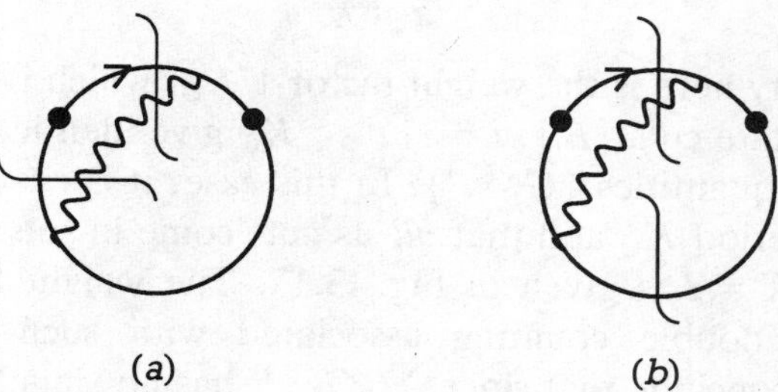

Fig. 13.18 Double-cut diagrams.

carefully formulated perturbation theory, as a result of the KLN theorem. They will still depend, however, on fixed momentum scales of the physical initial state. Such cross sections, even though finite in the zero-mass limit, are not infrared safe in the sense of eq. (12.63). In terms of the KLN theorem, this can be understood as residual dependence on the parameter ϵ in eq. (13.49), which defines the set of states to be averaged over. In QCD, ϵ-dependence cannot be computed reliably unless ϵ is large compared with the scale of the effective coupling, Λ_{QCD}, (eq. (12.93)). In summary, it is important to distinguish between the concepts of infrared finiteness and infrared safety. Many infrared finite quantities are not profitably calculable in perturbation theory, because they still depend on momentum scales at which the effective coupling is large.

Exercises

13.1 Analyze the integral $\int_0^1 d\zeta\,(\zeta - w)^{-1}$ in terms of end-point singularities, and use it to derive the cut-plane structure of the logarithm.

13.2 Analyze the cut-plane structure of $\int_0^1 d\zeta\,(\zeta^2 - w^2)^{-1}$ in terms of end-point and pinch singularities.

13.3 Study the scalar triangle diagram in terms of light-cone variables, when two internal lines have mass m and one is massless. Derive the analogue of eq. (13.14) for this case, and identify the singularities leading to a normal threshold when the appropriate external invariant reaches $4m^2$.

13.4 Classify pinch surfaces in the following processes: (a) two-particle elastic scattering at fixed angle (i.e. $|t| = O(s)$); (b) two-particle inelastic scattering with two or more jets produced at large angles. Show that in each case more than one disconnected hard subdiagram is possible, but that all hard subdiagrams are at the same position in space–time in the corresponding physical picture.

13.5 Consider the diagram of Fig. 13.19 in the massless limit. There is a pinch surface at $k^\mu = 0$ and $l_i^\mu = \alpha_i p_i^\mu$, $i = 1, 2$. Show that the four components $\{k^\mu\}$ do not make a good set of normal variables for this singular surface. That is, show that the resulting homogeneous integral has an extra pinch surface.

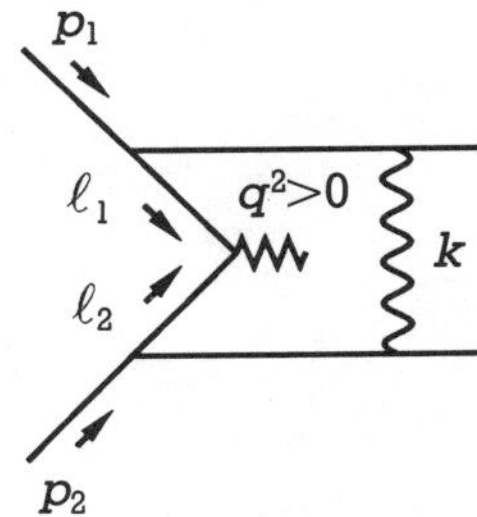

Fig. 13.19 Diagram with pinch surface at $k^\mu = 0$.

13.6 Verify by power counting that near threshold the singularity in the one-loop vertex correction in QED behaves as β^{-1}, where β is the relative velocity of the produced fermion pair.

13.7 Show that in a general pinch surface such as Fig. 13.9(a), normal variables may be chosen as in Section 13.4, without introducing extra pinch surfaces in the homogeneous integral.

13.8 Using the suppression of scalar polarizations, eq. (7.78b), carry out power counting for the timelike vertex in axial gauge, and verify that the leading behavior is associated with reduced diagrams of the form shown in Fig. 13.13(b).

13.9 Carry out power counting for the pinch surfaces illustrated in Fig. 13.10, and show that deeply-inelastic-scattering semi-inclusive cross sections are no worse than logarithmically divergent in the high-energy limit in Feynman gauge.

13.10 Show that in *scalar* Yukawa theory, with coupling $\bar{\psi}\psi\phi$, there are infrared divergences from the exchange of soft scalar lines but that this is not the case for a *pseudoscalar* interaction $\bar{\psi}\gamma_5\psi\phi$.

13.11 Carry out power counting for jets in the vertex function for $\phi^3{}_6$, and verify the absence of infrared divergences. Using the anomalous dimension, eq. (10.131c), derive the high-energy behavior of the electromagnetic form factor in this theory.

13.12 Analyze the pinch surfaces of eq. (13.43), and derive the time-ordered perturbation theory analogue of the Landau equations (13.5).

14

Factorization and evolution in high-energy scattering

Most zero-mass and high-energy cross sections are not directly calculable in perturbation theory, because of the presence of large logarithms of energy over mass (see the comments at the end of Chapter 13). Nevertheless, events with large momentum transfer are the result of violent short-distance collisions, which one can isolate quantitatively. The separation of calculable short-distance from incalculable long-distance effects is known as *factorization*. Deeply inelastic scattering cross sections illustrate this property. They may be used, in turn, to compute a wide class of other inclusive and semi-inclusive cross sections with large momentum transfers. The *evolution* in momentum transfer of deeply inelastic and related cross sections may also be determined by methods related to both the renormalization group and the parton model. The operator product expansion gives an alternate interpretation, which generalizes factorization beyond leading power behavior, at least in deeply inelastic scattering.

The examples of this chapter, in which the above results are derived and discussed, are drawn primarily from QCD, but the techniques of factorization and the operator product expansion are generally applicable in field theory. They transcend low-order calculations, by systematically organizing contributions from arbitrary orders of the perturbative expansion.

14.1 Deeply inelastic scattering

As an introduction, and to facilitate calculation, we first discuss tensor analysis and kinematics for the leptoproduction amplitude (Roy r1975, Close r1979), in which a lepton scatters from a hadron of momentum p^μ to produce an arbitrary hadronic state with momentum p_n. The reaction is deeply inelastic if ${p_n}^2 \gg p^2$.

The deeply inelastic scattering cross section is not infrared safe. It retains its mass dependence through initial-state interactions of the hadron. In the language of the KLN theorem (Section 13.5) there is an insufficient average over initial states to cancel all logarithmic enhancements in perturbation theory.

Short of calculating the deeply inelastic cross section directly, we appeal for inspiration to the parton model (Section 8.6). In the parton model, a large class of hard scattering cross sections may be understood in terms of a limited number of nonperturbative distributions of elementary constituents (partons). It will be our goal to see how the parton model arises, and may be improved upon, in a perturbative picture of deeply inelastic scattering.

Amplitudes

The general form of a deeply inelastic lepton–hadron interaction to lowest order in the electroweak interaction, but to all orders in the strong interaction, is shown in Fig. 13.10. The interaction is mediated by the exchange of a single vector boson of momentum q^μ. Since q^2 is negative, it is often convenient to define $Q^2 = -q^2 > 0$. Denoting the leptonic spins as λ and λ', and the initial hadronic spin as σ, the S-matrix element for this process may be written as

$$S_{l+h\to l'+n}(q, p, \lambda, \lambda', \sigma)$$
$$= \bar{u}_{\lambda'}(k')[\mathrm{i}\Gamma_\mu{}^{(i)}]u_\lambda(k)[-\mathrm{i}g^{\mu\mu'}/(q^2 - M^2)]\ \langle n\,\mathrm{out}|\mathrm{i}g^{(i)}J_{\mu'}{}^{(i)}(0)|h(p, \sigma)\ \mathrm{in}\rangle, \tag{14.1}$$

where $\mathrm{i}\Gamma_\mu{}^{(i)}$ is the electroweak vertex connecting leptons $l(k^\mu, \lambda)$ and $l'(k'^\mu, \lambda')$, and $\mathrm{i}g^{(i)}J_\mu{}^{(i)}$ is the hadronic current coupling to the exchanged boson, with the coupling constant separated. $|h(p, \sigma)\ \mathrm{in}\rangle$ and $|n\ \mathrm{out}\rangle$ are the initial and final hadronic states, and M is the mass of the vector boson that couples to vertex $\Gamma_\mu{}^{(i)}$. At high energy, we ignore the lepton masses, and the $g^{\mu\mu'}$ term is the only part of the vector boson propagator, eq. (7.85), that contributes.

Deeply inelastic scattering is of fundamental conceptual importance, because it provides direct information about hadron (in practice nucleon) structure. Since the vector boson carries no hadronic quantum numbers, the state $|n\ \mathrm{out}\rangle$ may be thought of as one of the virtual states of the target hadron, realized by the absorption of a vector boson.

The hadronic tensor

It is convenient to separate the leptonic part of the cross section from its much more complex hadronic part. A short calculation, starting from eq. (14.1), gives the following high-energy formula for the spin-averaged cross section:

$$\mathrm{d}\sigma = (\mathrm{d}^3\mathbf{k}'/|\mathbf{k}'|)[1/2s(Q^2 - M^2)^2]L^{\mu\nu(i)}(k, k')W_{\mu\nu}{}^{(i)}(p, q), \tag{14.2a}$$

$$L^{\mu\nu(i)}(k, k') = \{[g^{(i)}]^2/8\pi^2\}\,\mathrm{tr}\,[\not{k}\Gamma^{\mu(i)}\not{k}'\Gamma^{\nu(i)}], \tag{14.2b}$$

$$W_{\mu\nu}{}^{(i)} \equiv \frac{1}{8\pi} \sum_{\sigma,n} \langle h(p,\sigma)\ \text{in}|J_\mu{}^{(i)\dagger}(0)|n\ \text{out}\rangle \langle n\ \text{out}|J_\nu{}^{(i)}(0)|h(p,\sigma)\ \text{in}\rangle$$
$$\times (2\pi)^4 \delta^4(p_n - q - p),$$
$$= \frac{1}{8\pi} \sum_\sigma \int \mathrm{d}^4x\, \mathrm{e}^{iq\cdot x} \langle h(p,\sigma)\ \text{in}|J_\mu{}^{(i)\dagger}(x) J_\nu{}^{(i)}(0)|h(p,\sigma)\ \text{in}\rangle, \tag{14.2c}$$

where the coupling constant of the hadronic electroweak current, $g^{(i)}$, has been absorbed into the leptonic tensor, $L^{\mu\nu(i)}$. We should note that the hadronic tensor, $W_{\mu\nu}{}^{(i)}$, is defined in the literature with a variety of normalizations.

To be specific, we work with the electromagnetic current, $J_\mu{}^{(\text{em})} \equiv j_\mu$. With this choice, parity is conserved and $W_{\mu\nu}{}^{(\text{em})}$ is expressible in terms of the metric tensor $g_{\mu\nu}$ and a combination of the vectors p^μ or q^μ (Section 6.6). Quite generally, then, we can write

$$W_{\mu\nu}{}^{(\text{em})}(p,q) = p_\mu p_\nu g_1(p,q) + p_\mu q_\nu g_2(p,q) + q_\mu p_\nu g_3(p,q)$$
$$+ q_\mu q_\nu g_4(p,q) + g_{\mu\nu} q^2 g_5(p,q), \tag{14.3}$$

where the $g_i(p,q)$ are scalar functions. On eq. (14.3) we impose the requirement of current conservation:

$$q^\mu W_{\mu\nu}{}^{(\text{em})} = W_{\mu\nu}{}^{(\text{em})} q^\nu = 0. \tag{14.4}$$

This reduces the four independent functions to two, usually written as

$$W_{\mu\nu}{}^{(\text{em})}(p,q) = -(g_{\mu\nu} - q_\mu q_\nu/q^2) W_1{}^{(\text{em})}(x, q^2, m^2)$$
$$+ [p_\mu - q_\mu(p\cdot q/q^2)][p_\nu - q_\nu(p\cdot q/q^2)]$$
$$\times W_2{}^{(\text{em})}(x, q^2, m^2). \tag{14.5}$$

$W_1{}^{(\text{em})}$ and $W_2{}^{(\text{em})}$ are known as structure functions, for whose arguments we have taken the standard choices q^2, p^2 $(= m^2)$ and

$$x = -q^2/2p\cdot q = Q^2/2p\cdot q; \tag{14.6a}$$

x is the Bjorken scaling variable, eq. (8.113). In terms of x, we have

$$p_n{}^2 = (p+q)^2 = m^2 + (2p\cdot q)(1-x). \tag{14.6b}$$

At high energy, the criterion for deeply inelastic scattering, $p_n{}^2 \gg m^2$, is equivalent to $x \neq 1$. Another common notation, which we shall employ below, is

$$\nu = p\cdot q = m(\omega(\mathbf{k}) - \omega(\mathbf{k}')). \tag{14.7}$$

The second form is special to the rest frame of a massive target hadron; ν is proportional to the energy loss of the lepton.

Electroweak left-handed currents break parity invariance, and in the neutrino–nucleon tensor $W_{\mu\nu}{}^{(\mathrm{vh})}$ another invariant tensor is allowed, the totally antisymmetric tensor with four indices, $\epsilon_{\mu\nu\lambda\sigma}$. Then there is a third structure function in the tensor for unpolarized scattering:

$$\begin{aligned} W_{\mu\nu}{}^{(\mathrm{vh})}(p,q) = &-(g_{\mu\nu} - q_\mu q_\nu/q^2) W_1{}^{(\mathrm{vh})}(x,q^2,m^2) \\ &+ [p_\mu - q_\mu(p\cdot q/q^2)][p_\nu - q_\nu(p\cdot q/q^2)] W_2{}^{(\mathrm{vh})}(x,q^2,m^2) \\ &- (\mathrm{i}\epsilon_{\mu\nu}{}^{\lambda\sigma}/2m^2) q_\lambda p_\sigma W_3{}^{(\mathrm{vh})}(x,q^2,m^2). \end{aligned} \tag{14.8}$$

For the remainder of the discussion, we return to the electromagnetic current, $j_\mu(x)$, and suppress the superscript (em).

The hadronic tensor $W_{\mu\nu}$, eq. (14.2c), is reminiscent of eq. (12.67) for the total cross section of $\mathrm{e}^+\mathrm{e}^-$ annihilation into hadrons, with the difference that the latter is an expectation value in the vacuum, the former in the single-particle state $|h(p,\sigma)\rangle$. As with the annihilation cross section, deeply inelastic scattering structure functions are the imaginary part of a forward scattering amplitude, in this case forward 'Compton' scattering of the hadron with an electroweak boson:

$$\begin{aligned} W_{\mu\nu}(x,q^2) &= 2\,\mathrm{Im}\, T_{\mu\nu}(\omega,q^2), \\ T_{\mu\nu}(\omega,q^2) &= \frac{\mathrm{i}}{8\pi}\int \mathrm{d}^4x\, \mathrm{e}^{\mathrm{i}q\cdot x} \sum_\sigma \langle h(p,\sigma)|T[j_\mu(x)j_\nu(0)]|h(p,\sigma)\rangle, \\ &= -\left(g_{\mu\nu} - \frac{q_\mu q_\nu}{q^2}\right)T_1(\omega,q^2) + \left[p_\mu - q_\mu\left(\frac{p\cdot q}{q^2}\right)\right]\left[p_\nu - q_\nu\left(\frac{p\cdot q}{q^2}\right)\right]T_2(\omega,q^2), \end{aligned} \tag{14.9}$$

where we define $\omega \equiv 2p\cdot q/Q^2$, the inverse of the Bjorken scaling variable x. (From now on we suppress the 'in' and 'out' labels on states and the m^2-dependence in structure functions.) The perturbative proof of eq. (14.9) is given in Fig. 14.1. $W_{\mu\nu}$ is a sum of cut diagrams, Fig. 14.1(a). Assuming the state $|h(p,\sigma)\rangle$ is stable under strong interactions, the overall diagram in Fig. 14.1(a) has only cuts with total momentum ('channel') $p+q$ or $p-q$ (but not both) at fixed q^μ. Therefore, it is twice the imaginary part of the uncut diagram, Fig. 14.1(b), by unitarity, eq. (9.86). The explicit factor of $+\mathrm{i}$ in the definition of $T_{\mu\nu}$ is chosen so that the first line holds. (Note the lack of explicit factors of $-\mathrm{i}$ times the currents.)

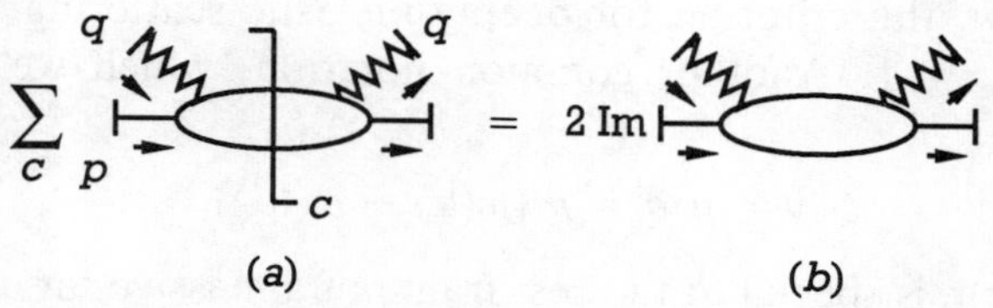

Fig. 14.1 Graphical expression of eq. (14.9).

A fixed $q^2 < 0$, $T_{\mu\nu}(p, q)$ is analytic for all ω, except for normal thresholds at $(p \pm q)^2 = \kappa^2$, where κ is the mass of any physical state (exercise 14.1). In the ω-plane this translates to branch points at

$$\omega = \pm[1 + (\kappa^2 - p^2)/Q^2], \tag{14.10}$$

where the branch cuts can always be chosen to run along the real axis as shown in Fig. 14.2(a). The $T_a(\omega, Q^2)$, $a = 1, 2$, of eq. (14.9) obey dispersion relations analogous to (13.20) for the vacuum expectation value, but now with two branch cuts.

Only on the real axis does the tensor $T_{\mu\nu}(\omega, Q^2)$ have the direct interpretation, eq. (14.9), of a matrix element. The structure functions at negative and positive ω are numerically related for any complex ω,

$$T_a(-\omega, Q^2) = T_a(\omega, Q^2), \tag{14.11}$$

as may be checked easily from eq. (14.9) (exercise 14.2).

From eq. (14.2), after a straightforward calculation, we find, for the cross section,

$$\begin{aligned}\frac{\mathrm{d}\sigma}{\mathrm{d}E_{k'}\mathrm{d}\Omega_{k'}} &= \frac{4\alpha^2 E_{k'}}{sQ^4}\left[Q^2 W_1 + \tfrac{1}{2}(s^2 - 2sp\cdot q + \tfrac{1}{2}q^2 m^2)W_2\right],\\ &= \frac{\alpha^2}{4mE_k{}^2 \sin^4 \frac{1}{2}\theta}\left(2\sin^2 \tfrac{1}{2}\theta\, W_1 + m^2 \cos^2 \tfrac{1}{2}\theta\, W_2\right),\end{aligned} \tag{14.12}$$

where the second form applies to the rest frame of the massive target. Thus, the structure functions may be determined directly by measuring the angular distribution of the scattered lepton. Finally, it is often convenient to introduce dimensionless structure functions $F_a(x, Q^2)$:

$$F_1(x, Q^2) \equiv W_1(x, Q^2), \tag{14.13a}$$

$$F_2(x, Q^2) \equiv \nu W_2(x, Q^2), \tag{14.13b}$$

where ν is given by eq. (14.7).

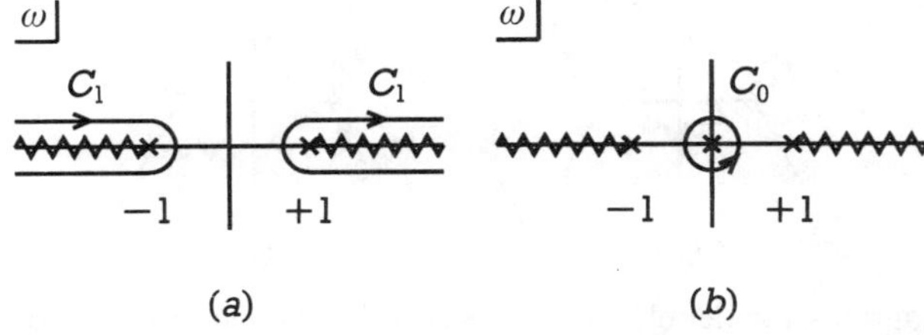

Fig. 14.2 (a) Analytic structure of $T_{\mu\nu}$; (b) illustration of eq. (14.107).

14.2 Deeply inelastic scattering for massless quarks

We now turn to an explicit evaluation of the structure functions for massless quarks, at low order in perturbation theory. This cross section is not for direct application, not least because confinement prevents us from preparing isolated quarks as targets. Despite this, we shall recognize a usable result at the end, which will lead us to the parton model.

Born cross section and scaling

For a massless quark with charge $Q_f e$, the hadronic tensor in the Born approximation is given by Fig. 14.3(a):

$$\begin{aligned} W_{\mu\nu}{}^{(f,0)} &= \frac{1}{8\pi}\int \frac{\mathrm{d}^3\mathbf{p}'}{(2\pi)^3 2E_{p'}} Q_f{}^2 \,\mathrm{tr}\,(\gamma_\mu \not{p}' \gamma_\nu \not{p})(2\pi)^4 \delta^4(p' - p - q) \\ &= -\frac{1}{2}\left(g_{\mu\nu} - \frac{q_\mu q_\nu}{q^2}\right) Q_f{}^2 \delta(1 - x) \\ &\quad + \left(p_\mu - q_\mu \frac{p\cdot q}{q^2}\right)\left(p_\nu - q_\nu \frac{p\cdot q}{q^2}\right) Q_f{}^2 \frac{\delta(1 - x)}{\nu}. \end{aligned} \tag{14.14}$$

By comparison with the general form of eq. (14.5), we then have

$$W_1{}^{(f)(0)} = \tfrac{1}{2} Q_f{}^2 \delta(1 - x), \quad W_2{}^{(f,0)} = Q_f{}^2[\delta(1 - x)]/\nu. \tag{14.15}$$

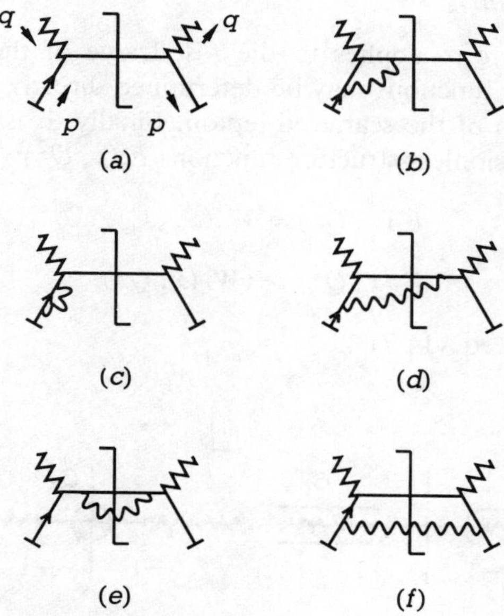

Fig. 14.3 Cut diagrams for deeply inelastic scattering of a quark: (a) Born approximation; (b)–(f), order-α_s corrections.

The F's, eqs. (14.13), are then functions of x only and are independent of Q^2:

$$2F_1^{(f,0)}(x, Q^2) = F_2^{(f,0)}(x, Q^2) = Q_f^2\delta(1-x). \tag{14.16}$$

Structure functions that depend only on the dimensionless variable x are said to *scale* (see also Section 8.6). Scaling is an approximate property of the experimentally observed structure functions of nucleons (Friedman & Kendall r1972). Q^2-dependence in the F's is called *scale breaking*; this will appear in one-loop corrections.

Quark structure functions at one loop

The graphs that contribute to the quark structure functions at one loop are shown in Figs. 14.3(b)–(f). As above, we work with massless quarks, and deal with infrared divergences by dimensional regularization.

Since we know there are only two independent functions in the tensor $W_{\mu\nu}$, it is simpler to compute two contractions of $W_{\mu\nu}$ (here, and in many other details below, we follow Altarelli, Ellis & Martinelli 1979):

$$-g^{\mu\nu}W_{\mu\nu} = (1-\epsilon)(F_2/x) - (3-2\epsilon)[(F_2/2x) - F_1], \tag{14.17a}$$

$$p^\mu p^\nu W_{\mu\nu} = (Q^2/4x^2)[(F_2/2x) - F_1]. \tag{14.17b}$$

We begin with $-g^{\mu\nu}W_{\mu\nu}$, which is the more challenging of the two. For massless quarks, the virtual corrections, Figs. 14.3(b), (c) (along with their mirror diagrams and counterterms) are proportional to the lowest-order vertex, so that we may use eq. (14.16) with (14.17a) to find

$$-g^{\mu\nu}W_{\mu\nu}{}^{(v)(1)} = 2\gamma(q^2)(1-\epsilon)\delta(1-x), \tag{14.18}$$

where $\gamma(q^2)$ is the full vertex given by eq. (12.82). The explicit form has double poles:

$$\frac{\alpha_s}{\pi}[-g^{\mu\nu}W_{\mu\nu}{}^{(v)(1)}] = -\frac{\alpha_s}{\pi}Q_f^2C_2(F)\left(\frac{4\pi\mu^2}{Q^2}\right)^\epsilon(1-\epsilon)\frac{\Gamma(1+\epsilon)\Gamma^2(1-\epsilon)}{\Gamma(1-2\epsilon)} \times (\epsilon^{-2} + \tfrac{3}{2}\epsilon^{-1} + 4)\delta(1-x), \tag{14.19}$$

where we use the notation $f(\alpha) \equiv \sum(\alpha/\pi)^n f^{(i)}$, for any expansion in α_s.

The real-gluon corrections are given in Figs. 14.3(d)–(f):

$$\frac{\alpha_s}{\pi}[-g^{\mu\nu}W_{\mu\nu}{}^{(r)(1)}] = \frac{1}{8\pi}\int d^np'd^nk\,(2\pi)^{-n}\delta^n(p+q-p'-k) \times (2\pi)\delta_+(p'^2)(2\pi)\delta_+(k^2)|M(p,q,k)|^2{}_r, \tag{14.20}$$

where $|M(p,q,k)|^2{}_r$ is the color-averaged invariant amplitude squared:

$$|M(p,q,k)|^2{}_{\mathrm{r}} = 8g^2\mu^{2\epsilon}Q_f{}^2C_2(\mathrm{F})(1-\epsilon)$$
$$\times\left[(1-\epsilon)\left(\frac{s}{2p\cdot k}+\frac{2p\cdot k}{s}\right)+\frac{(4p\cdot p')Q^2}{(2p\cdot k)s}+2\epsilon\right], \quad (14.21)$$

where $s \equiv (p+q)^2$. If we choose to evaluate eq. (14.20) in the center-of-mass frame of $(q+p)^\mu$, we find $p_0 = \frac{1}{2}Q[x(1-x)]^{-1/2}$, and the invariants are

$$s = (Q^2/x)(1-x),\ \ 2p\cdot k = (Q^2/x)(1-y),\ \ 2p\cdot p' = (Q^2/x)y, \quad (14.22)$$

where (compare eq. (12.76)), $y = \frac{1}{2}(1+\cos\theta_{pk})$. Then we may pick θ_{pk} as the polar angle in phase space, and eq. (14.20) becomes, using the dimensional regularization formula, eq. (9.38),

$$(\alpha_s/\pi)[-g^{\mu\nu}W_{\mu\nu}{}^{(\mathrm{r})(1)}]$$
$$= (\alpha_s/\pi)Q_f{}^2C_2(\mathrm{F})(1-\epsilon)(\pi\mu^2)^\epsilon[1/\Gamma(1-\epsilon)]$$
$$\times\int_0^1 \mathrm{d}y\,[y(1-y)]^{-\epsilon}\int_0^\infty \mathrm{d}k\,k^{1-2\epsilon}\,\delta(\tfrac{1}{2}s - s^{1/2}k)$$
$$\times\left[\left(\frac{1-x}{1-y}+\frac{1-y}{1-x}\right)(1-\epsilon)+\frac{2xy}{(1-x)(1-y)}+2\epsilon\right]. \quad (14.23)$$

Using the delta function for the k-integral and eqs. (9.30) for the y-integrals, and employing the identity $z\Gamma(z) = \Gamma(1+z)$ a number of times, we derive

$$(\alpha_s/\pi)[-g^{\mu\nu}W_{\mu\nu}{}^{(\mathrm{r})(1)}]$$
$$= (\alpha_s/2\pi)Q_f{}^2C_2(\mathrm{F})(1-\epsilon)(4\pi\mu^2/s)^\epsilon[\Gamma(1-\epsilon)/\Gamma(1-2\epsilon)]$$
$$\times\left\{-\frac{1-\epsilon}{\epsilon}\left[1-x+\left(\frac{2x}{1-x}\right)\left(\frac{1}{1-2\epsilon}\right)\right]+\frac{(1-\epsilon)}{2(1-2\epsilon)(1-x)}+\frac{2\epsilon}{1-2\epsilon}\right\}. \quad (14.24)$$

The complete one-loop result is the sum of eqs. (14.19) and (14.24):

$$(\alpha_s/\pi)[-g^{\mu\nu}W_{\mu\nu}{}^{(1)}]$$
$$= (\alpha_s/2\pi)Q_f{}^2C_2(\mathrm{F})(1-\epsilon)(4\pi\mu^2/Q^2)^\epsilon[\Gamma(1-\epsilon)/\Gamma(1-2\epsilon)]$$
$$\times\Bigg(-\Gamma(1+\epsilon)\Gamma(1-\epsilon)(2\epsilon^{-2}+3\epsilon^{-1}+8)\delta(1-x)$$
$$+\left[\frac{x}{(1-x)}\right]^\epsilon\left\{-\frac{1-\epsilon}{\epsilon}\left[1-x+\left(\frac{2x}{1-x}\right)\left(\frac{1}{1-2\epsilon}\right)\right]\right.$$
$$\left.+\frac{1-\epsilon}{2(1-2\epsilon)(1-x)}+\frac{2\epsilon}{1-2\epsilon}\right\}\Bigg). \quad (14.25)$$

What are we going to make of this rather untransparent formula? It is singular as $x \to 1$, as well as $n \to 4$. We have already seen, in the discussion of e^+e^- annihilation, that $1/\epsilon$ poles in the renormalized vertex are due to infrared and/or collinear divergences. From the discussion of Section 13.4, we expect the same for deeply inelastic scattering.

Simple kinematic considerations show that when $x \to 1$, the momentum k^μ of the emitted gluon is either infrared or collinear to p'^μ. In either case the invariant mass of the photon–quark system vanishes in the limit. On the other hand, the gluon is exactly collinear to p^μ for any x when $y = 1$ in the integral of eq. (14.23). Thus, the factors ϵ^{-1} in eq. (14.25) arise when k^μ is collinear to p^μ, and the factors $1/(1-x)$ arise when k^μ is either collinear to p'^μ or infrared. In eq. (14.25), in contrast with the case of annihilation, the real and virtual singularities do not cancel, as is obvious for any $x \neq 1$.

We can simplify eq. (14.25) by treating the terms proportional to $(1-x)^{-1-\epsilon}$ as *distributions*, that is, singular functions defined by their integrals with smooth functions. The Dirac delta function, $\delta(x)$, is the best-known example of a distribution. Consider the integral

$$I[f] = \int_z^1 \mathrm{d}x\,(1-x)^{-1-\epsilon} f(x), \tag{14.26}$$

where $f(x)$ is a differentiable function. Suppose further that $\epsilon < 0$, so that $I[f]$ is well defined. Then $I[f]$ may be rewritten as

$$I[f] = (-\epsilon)^{-1} f(1) + \int_z^1 \mathrm{d}x[f(x) - f(1)][1/(1-x)] + f(1)\ln(1-z) + \mathrm{O}(\epsilon), \tag{14.27}$$

assuming that $f(x) - f(1)$ vanishes as a power at $x = 1$. In this form, we have isolated the $1/\epsilon$ pole in $I[f]$. The second term, which is finite in the $\epsilon \to 0$ limit, is commonly expressed in terms of the distribution $[1/(1-x)]_+$, defined by

$$\int_z^1 \mathrm{d}x\, f(x)[1/(1-x)]_+ = \int_z^1 \mathrm{d}x[f(x) - f(1)][1/(1-x)] + f(1)\ln(1-z). \tag{14.28}$$

Since $f(x)$ is arbitrary, we may consider eq. (14.27) as a statement of the distribution identity:

$$(1-x)^{-1-\epsilon} = (-\epsilon)^{-1}\delta(1-x) + [1/(1-x)]_+ + \mathrm{O}(\epsilon). \tag{14.29}$$

In an exactly similar way, we can expand to the next order in ϵ to derive

$$(-\epsilon)^{-1}[1-x]^{-1-\epsilon} = \epsilon^{-2}\delta(1-x) + (-\epsilon)^{-1}[1/(1-x)]_+ + \{[\ln(1-x)]/(1-x)\}_+ + \mathrm{O}(\epsilon). \tag{14.30}$$

It is worth making the technical point that $\int_0^1 dx[1/(1-x)]_+ = 0$.

We now see that terms in eq. (14.25) proportional to $\epsilon(1-x)^{-1-\epsilon}$ contribute to $W_{\mu\nu}$ at $n = 4$, while those proportional to $\epsilon(1-x)^{-\epsilon}$ or $\epsilon^2(1-x)^{-1-\epsilon}$ do not. Therefore, we may simplify eq. (14.25) up to corrections that vanish as $\epsilon \to 0$:

$$
\begin{aligned}
&(\alpha_s/\pi)[-g^{\mu\nu}W_{\mu\nu}{}^{(1)}] \\
&= (\alpha_s/2\pi)Q_f{}^2 C_2(\mathrm{F})(1-\epsilon)(4\pi\mu^2/Q^2)^\epsilon[\Gamma(1-\epsilon)/\Gamma(1-2\epsilon)] \\
&\quad \times \Bigg\{ -\Gamma(1+\epsilon)\Gamma(1-\epsilon)(2\epsilon^{-2}+3\epsilon^{-1}+8)\delta(1-x) \\
&\qquad + x^\epsilon\left[-\frac{1}{\epsilon}\left(\frac{1+x^2}{(1-x)^{1+\epsilon}}\right) - \left(\frac{3+7\epsilon}{2(1-x)^{1+\epsilon}}\right) + \frac{(3-x)}{(1-x)^\epsilon}\right]\Bigg\}. \qquad (14.31)
\end{aligned}
$$

Using the distribution identities eqs. (14.29) and (14.30), and expanding the gamma functions where necessary with eq. (9.36), we now write the one-loop correction to $-g^{\mu\nu}W_{\mu\nu}$ as a distribution:

$$
\begin{aligned}
&(\alpha_s/\pi)[-g^{\mu\nu}W_{\mu\nu}{}^{(1)}] \\
&= (1-\epsilon)(\alpha_s/2\pi)Q_f{}^2(4\pi\mu^2/Q^2)^\epsilon \times \Bigg[-\epsilon^{-1}P_{qq}(x)[\Gamma(1-\epsilon)/\Gamma(1-2\epsilon)] \\
&\quad + C_2(\mathrm{F})\Bigg((1+x^2)\{[\ln(1-x)]/(1-x)\}_+ - \tfrac{3}{2}[1/(1-x)]_+ \\
&\qquad - (1+x^2)(\ln x)/(1-x) + 3 - x - [\tfrac{9}{2}+\tfrac{1}{3}\pi^2]\delta(1-x)\Bigg)\Bigg], \qquad (14.32)
\end{aligned}
$$

where the *evolution kernel* or *splitting function*, P_{qq} (Gribov & Lipatov 1972a, Lipatov 1974, Altarelli & Parisi 1977, Altarelli r1982) is defined by

$$P_{qq}(x) \equiv C_2(\mathrm{F})\{(1+x^2)[1/(1-x)]_+ + \tfrac{3}{2}\delta(1-x)\}. \qquad (14.33)$$

The double poles have all cancelled, and the remaining singular behavior is proportional to the evolution kernel, whose additional significance will become clearer below. The (collinear) pole in eq. (14.32) is a manifestation of the perturbative evolution of the initial quark state, as anticipated above. The singular behavior is interesting not so much for its own sake, since it is infinite in four dimensions, but because it gives rise to a scale-breaking logarithm, $(\alpha_s/\pi)C_2(\mathrm{F})\ln(Q^2/\mu^2)P_{qq}(x)$, found by expanding the factor $(4\pi\mu^2/Q^2)^\epsilon$ in eq. (14.32). Thus far, the rest of eq. (14.32) contains no physical information, because it may (and does) depend on the infrared regularization scheme. These finite corrections can be useful, however. To see how, let us complete our calculation of $W_{\mu\nu}{}^{(1)}$ by computing $p^\mu p^\nu W_{\mu\nu}$, which, by eq. (14.17b), is proportional to $(F_2/2x) - F_1$.

The calculation of $p^\mu p^\nu W_{\mu\nu}{}^{(1)}$ is much simpler than that of $g^{\mu\nu} W_{\mu\nu}{}^{(1)}$, because use of the Dirac equation for the initial-state massless quark eliminates Figs. 14.3(a)–(e). Only Fig. 14.3(f) is not proportional to a factor such as $\not p u(p) = 0$. The remaining squared matrix element is simply

$$-g_s^2 \mu^{2\epsilon} Q_f{}^2 C_2(\mathrm{F})(2p\cdot k)^{-2}\,\mathrm{tr}\,[\not p \gamma_\alpha(\not p - \not k)\not p \not p' \not p(\not p - \not k)\gamma^\alpha]$$
$$= 4Q_f{}^2 C_2(\mathrm{F}) g_s^2 \mu^{2\epsilon}(1-\epsilon)(Q^2 y/x), \quad (14.34)$$

where we have used eq. (14.22) for the momentum invariants. The integration of (14.34) over phase space is trivial, and gives

$$(\alpha_s/\pi)[p^\mu p^\nu W_{\mu\nu}{}^{(1)}] = Q_f{}^2 C_2(\mathrm{F})(\alpha_s/4\pi)(Q^2/2x), \quad (14.35)$$

or, using eq. (14.17b) to relate this result to the structure functions,

$$(\alpha_s/\pi)(F_2 - 2xF_1)^{(f,1)} = (\alpha_s/\pi) Q_f{}^2 C_2(\mathrm{F}) x^2. \quad (14.36)$$

Using eqs. (14.17), (14.32) and (14.36), we can summarize the one-loop results by

$$(\alpha_s/\pi)F_2{}^{(f,1)}(x,Q^2)$$
$$= (\alpha_s/2\pi)Q_f{}^2 x\Big[-(4\pi\mu^2/Q^2)^\epsilon[\Gamma(1-\epsilon)/\Gamma(1-2\epsilon)]\epsilon^{-1}P_{qq}(x)$$
$$+ C_2(\mathrm{F})\Big((1+x^2)\{[\ln(1-x)]/(1-x)\}_+ - \tfrac{3}{2}[1/(1-\mathrm{x})]_+$$
$$- (1+x^2)(\ln x)/(1-x) + 3 + 2x - [\tfrac{9}{2} + \tfrac{1}{3}\pi^2]\delta(1-\mathrm{x})\Big)\Big],$$
$$(14.37)$$
$$(\alpha_s/\pi)F_1{}^{(f,1)}(x,Q^2) = (\alpha_s/\pi)F_2{}^{(f)(1)}(x,Q^2)/2x - Q_f{}^2 C_2(\mathrm{F})(\alpha_s/2\pi)x.$$

A striking feature of this result is that although F_1 and F_2 are not individually infrared safe, their difference is, at least to one loop. This suggests that we might use an experimental determination of $F_1(x, Q^2)$ to give a theoretical prediction of $F_2(x, Q^2)$, or vice versa. In a sense, perturbation theory provides half of the tensor $W_{\mu\nu}$, if we have experimental input for the other half. In the next section, we shall see how this comes about.

14.3 Factorization and parton distributions

Parton-model structure functions

As indicated above, we shall look to the parton model for an interpretation of eq. (14.37). The parton-model formula for the deeply inelastic cross section, eq. (14.12) may, following the general form eq. (8.114), be written as

$$\frac{d\sigma^{(lh)}}{dE_{k'}d\Omega_{k'}} = \sum_f \int_0^1 d\xi \frac{d\sigma^{(lf)}}{dE_{k'}d\Omega_{k'}} \phi_{f/h}(\xi), \tag{14.38}$$

where $\phi_{f/h}(\xi)$ is the distribution of quark flavor f in hadron h. To derive parton-model relations for hadronic structure functions, we recognize that the invariant normalization $1/2s$ in the lepton–hadron cross section, eq. (14.2a), becomes $1/2\xi s$ in a lepton–quark cross section. All other factors relating the structure functions to cross sections are common on both sides of eq. (14.38), and the parton-model formula for the hadronic tensor is

$$W_{\mu\nu}{}^{(h)}(p, q) = \sum_f \int_0^1 (d\xi/\xi) W_{\mu\nu}{}^{(f,0)}(\xi p, q)\phi_{f/h}(\xi), \tag{14.39}$$

where $W_{\mu\nu}{}^{(f,0)}$ is the zeroth-order hadronic tensor for quark f, eq. (14.14). Equation (14.39) implies the following parton-model relations for the structure functions F_a defined in eqs. (14.13):

$$F_2{}^{(h)}(x) = \sum_f \int_0^1 d\xi\, F_2{}^{(f,0)}(x/\xi)\phi_{f/h}(\xi), \tag{14.40a}$$

$$F_1{}^{(h)}(x) = \sum_f \int_0^1 (d\xi/\xi) F_1{}^{(f,0)}(x/\xi)\phi_{f/h}(\xi), \tag{14.40b}$$

where $x/\xi = -q^2/2\xi p\cdot q$. Equation (14.40a) and eq. (14.16) for the Born structure functions immediately give the famous direct relation between $F_2(x)$ and the quark distributions:

$$F_2{}^{(h)}(x) = \sum_f Q_f{}^2 x \phi_{f/h}(x). \tag{14.41}$$

Factorization theorem

The following theorem for $F_a{}^{(h)}(x)$, $a = 1, 2$, where h is a nucleon *or* a parton, generalizes eqs. (14.40) to include perturbative corrections:

$$\begin{aligned} F_2{}^{(h)}(x, Q^2) &= \sum_i \int_0^1 d\xi\, C_2{}^{(i)}(x/\xi, Q^2/\mu^2, \alpha_s(\mu^2))\phi_{i/h}(\xi, \epsilon, \alpha_s(\mu^2)) \\ &\quad + O(m^2/Q^2), \\ F_1{}^{(h)}(x, Q^2) &= \sum_i \int_0^1 (d\xi/\xi) C_1{}^{(i)}(x/\xi, Q^2/\mu^2, \alpha_s(\mu^2))\phi_{i/h}(\xi, \epsilon, \alpha_s(\mu^2)) \\ &\quad + O(m^2/Q^2), \end{aligned} \tag{14.42}$$

where the $C_a^{(i)}(x/\xi,\ Q^2/\mu^2,\ \alpha_s(\mu^2))$ are *infrared-safe functions that are independent of the external hadron h*. The sum over partons i includes quarks, antiquarks and the gluon. $\phi_{i/h}(\xi, \epsilon, \alpha_s(\mu^2))$ is the distribution of parton i in hadron h, and $C_a^{(i)}(x/\xi,\ Q^2/\mu^2,\ \alpha_s(\mu^2))$ is its corresponding *coefficient function*. We assume a renormalization scale μ, and allow the distributions and coefficient functions to depend upon it, although, of course, the structure functions themselves do not.

The fundamental content of the factorization theorem is that all the short-distance (Q^2) dependence is in the coefficient functions, while all long-distance dependence (denoted ϵ) resides in the parton distributions. The coefficient functions are infrared safe, and may be calculated perturbatively. Corrections to factorization are suppressed by a power of Q^2, as indicated.

Technical derivations of eqs. (14.42), in the spirit of our discussion below, may be found in Amati, Petronzio & Veneziano (1978), Ellis, Georgi, Machacek, Politzer & Ross (1979), Libby & Sterman (1978) and Collins, Soper & Sterman (r1989). An alternate approach, based on the Feynman parameter representation, eq. (9.11), is described by Efremov & Radyushkin (1980b).

Factorization in perturbation theory

For consistency with perturbation theory we require, in the case of $h =$ parton j,

$$\phi_{i/j}(\xi, \epsilon, \alpha_s(\mu^2)) = \delta_{ij}\delta(1-\xi) + \mathrm{O}(\alpha_s). \tag{14.43a}$$

This states that the parton remains itself in the absence of interaction. Then the normalization of the Born structure functions, eq. (14.16) requires, in turn, that

$$\begin{aligned} C_2^{(i,0)}(x/\xi, Q^2/\mu^2, \alpha_s(\mu^2)) &= 2C_1^{(i,0)}(x/\xi, Q^2/\mu^2, \alpha_s(\mu^2)) \\ &= Q_i^2\delta(1-x/\xi). \end{aligned} \tag{14.43b}$$

To first order in α_s, eq. (14.43b) in eq. (14.42) gives, for parton j,

$$\begin{aligned} &F_2^{(j)}(x, Q^2) \\ &= Q_j^2\delta(1-x) + (\alpha_s/\pi)\Big[\sum_i Q_i^2 x\phi_{i/j}^{(1)}(x,\epsilon) + C_2^{(j,1)}(x, Q^2/\mu^2)\Big] + \mathrm{O}(\alpha_s^2) \end{aligned} \tag{14.44}$$

$$\begin{aligned} &F_1^{(j)}(x, Q^2) \\ &= \tfrac{1}{2}Q_j^2\delta(1-x) + (\alpha_s/\pi)\Big[\tfrac{1}{2}\sum_i Q_i^2\phi_{i/j}^{(1)}(x,\epsilon) + C_1^{(j,1)}(x, Q^2/\mu^2)\Big] + \mathrm{O}(\alpha_s^2) \end{aligned}$$

For quark scattering, i.e., $h = f$, only $i = f$ contributes at $O(\alpha_s)$ in eq. (14.44). The additive nature of the right-hand side of eq. (14.44) ensures that factorization is consistent with our one-loop calculation of $F_1^{(f)}$, $F_2^{(f)}$, eqs. (14.37). We have only to absorb the collinear pole terms $(1/\epsilon)P_{qq}$ into $\phi_{f/f}{}^{(1)}(x, \epsilon)$. The allocation of the remaining, infrared finite terms between $\phi_{f/f}{}^{(1)}$ and $C_2{}^{(f,1)}$ is a matter of definition. Before giving such a definition, let us describe the perturbative basis of factorization.

The following arguments draw heavily upon methods developed in Chapter 13. Readers who wish to avoid these technicalities may proceed directly to the discussion of eq. (14.52).

Final-state interactions and factorization

In the parton model, factorization is a statement of incoherence between long- and short-distance effects. How can perturbation theory incorporate incoherence? This question may be answered most easily in terms of the Compton scattering tensor $T_{\mu\nu}$, eq. (14.9), whose imaginary part gives the hadronic tensors of deeply inelastic scattering. We recall from Section 13.2 that infrared and collinear divergences arise from pinch surfaces, whose reduced diagrams describe physical processes. As shown in Fig. 13.10, the pinch surfaces of $W_{\mu\nu}$ are rather complicated, and involve both initial- and final-state interactions. The latter are particularly troublesome, since they have no analogue in the parton model, where all nonperturbative information is contained in parton distributions. The pinch surfaces of $T_{\mu\nu}$, however, are much simpler, as we now show.

We shall call a pinch surface *leading* if its superficial degree of infrared divergence (Section 13.4) indicates logarithmic behavior. Figure 14.4 shows the reduced diagrams of 'leading' pinch surfaces for our Compton scattering process. There is only a *single* jet $\varkappa_{i/h}$, collinear to the incoming hadron, which interacts via partons i with the off-shell photons at a *single* hard vertex. No other finite-energy lines are consistent with a physical picture at a pinch surface for this process. Most importantly, final-state jets may not

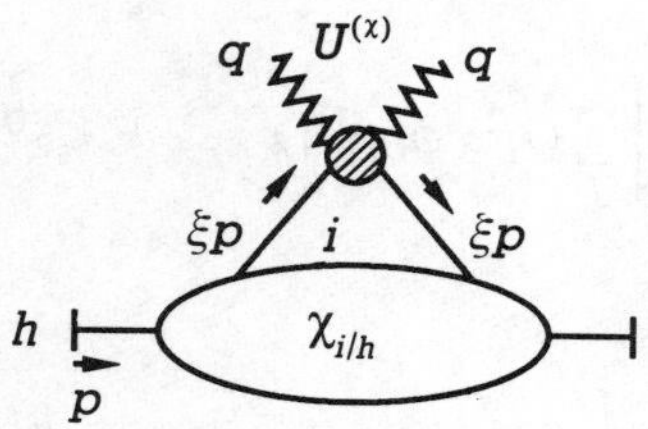

Fig. 14.4 Leading pinch surfaces for forward Compton scattering.

appear since, exactly as in e^+e^- annihilation, they could never meet again to reform the outgoing photon and hadron. Therefore, there is no analogue of the on-shell final-state interactions in Fig. 13.10 in the Compton scattering amplitude.

Power counting, as described in Section 13.4, shows that a leading pinch surface can have no soft lines attached to the hard subdiagram $U^{(\varkappa)}$ (exercise 14.3). As usual, in Feynman gauge, scalar-polarized vectors may attach the incoming jet with the hard subdiagram (not shown in Fig. 14.4), while they are absent in a physical gauge. Since the structure functions W_a ($a = 1, 2$) are given by the imaginary parts of the T_a (eq. 14.9), all final-state contributions to W_a must cancel as well. This leaves only the initial-state interactions of the target with itself, which are conceptually consistent with the parton model, and therefore with factorization.

Parton interpretation of a pinch surface

We now describe how factorization is realized in the diagrams themselves. Our aim is to gain familiarity with standard methods, as well as to motivate our eventual choice of parton distributions. We shall restrict ourselves to a single pinch surface in a physical gauge, treating it as if it were the only source of leading infrared divergence in the entire diagram.

We have shown that a leading pinch surface of T_2 is characterized by (i) a single jet $\varkappa_{i/h}$, all of whose lines are on-shell and collinear to p^μ, the momentum of the external hadron h, and (ii) a single hard part $U_2{}^{(\varkappa_i)}$, all of whose lines are off-shell by $O(Q^2)$. The jet is connected to the hard part by a pair of on-shell parton lines of type i (quark or gluon). Since the diagram describes forward scattering, one of the lines carries collinear momentum $k^\mu = \xi p^\mu$ into $U_2{}^{(\varkappa_i)}$, while the other carries exactly the same momentum back to the jet. Here we already see the parton picture emerge. We can do even better: in any forward scattering diagram, we have $|\xi| \leqslant 1$, so that at the pinch surface, the lines connected to the hard part carry a parton-model fractional momentum. This result follows from a simple contour integration argument.

Suppose the external hadron is lightlike, and moves in the plus direction:

$$p^\mu = p^+ \delta_{\mu+}. \tag{14.45}$$

Then if $|\xi| > 1$, we can *always* find at least one loop that includes the parton lines i and whose plus momentum flows in a circuit, always with the same sign in the direction of the loop (exercise 14.4). Around such a loop the *minus* loop integral has all its poles on the same side of the real axis. By completing the minus integration contour at infinity in the half-plane with no poles, we find that the minus integral vanishes. Thus, the diagram itself is zero unless $|\xi| \leqslant 1$. For an explicit example that illustrates this pattern, see the two-point function of eq. (13.12).

In summary, near pinch surface $\varkappa_i$ the integral takes on the form

$$T_2^{(\varkappa_i)} = \int_{n(\varkappa_i)} \mathrm{d}^4 k\, \theta(p^+ - |k^+|) \mathcal{U}_2^{(\varkappa_i)}(q, k) \varkappa_{i/h}(k, p, \epsilon), \quad (14.46)$$

where an implicit sum over the spin indices of parton i (with momentum $k^\mu \approx \xi p^\mu$) links the functions $\mathcal{U}_2$ and $\varkappa_{i/h}$, and where we take the integral to cover only a neighborhood $n(\varkappa_i)$ near the pinch surface in question. Since every line in $\mathcal{U}_2^{(\varkappa_i)}$ is off-shell by $O(Q^2)$ in $n(\varkappa_i)$, we can expand (14.46) about $k^\mu = \xi p^\mu$:

$$T_2^{(\varkappa_i)} = \int_{-1}^{1} \mathrm{d}\xi\, \mathcal{U}_2^{(\varkappa_i)}(q, \xi p) \int_{n(\varkappa_f)} \mathrm{d}^4 k\, \delta(\xi - k^+/p^+) \varkappa_{i/h}(k, p, \epsilon), \quad (14.47)$$

where corrections contribute only to subdivergences. For $i = f$, a quark of flavor f, negative ξ has the interpretation of $i = \bar{f}$, while for $i = g$, a gluon, the integral will be symmetric about $\xi = 0$ after a sum over diagrams. Equation (14.47) is thus a factorized form, similar to eq. (14.42), although the hard part and the jet are still linked by spin sums.

Factorization of spin matrices

We can show that spin structure is also consistent with factorization, proceeding in the same heuristic fashion. First, consider the quark case, $i = f$. The jet subdiagram $\varkappa_{i/h}$ is a matrix in Dirac space, and its leading behavior is of the form

$$\left[\int_{n(\varkappa_f)} \mathrm{d}^4 k\, \delta(\xi - k^+/p^+) \varkappa_{f/h}(k, p, \epsilon)\right]_{ab} = (p^+ \gamma^-)_{ab} \lambda_{f/h}(\xi, p, \epsilon), \quad (14.48)$$

where a and b label Dirac indices and where $\lambda_{f/h}$ is a scalar function that is logarithmically divergent in four dimensions (exercise 14.3). Any other Dirac structure would either be proportional to a vector in some other direction, or would involve γ_5. The only vector on which the jet may depend is p^μ, while γ_5 would require either parity noninvariance, or a polarized matrix element. As a result, corrections to eq. (14.48) are not leading at this pinch surface. We now use $\mathrm{tr}\,(\gamma^+ \gamma^-) = 4$ to isolate this leading behavior in eq. (14.47), and to reduce the product of spin matrices to a product of functions. The result is

$$T_2^{(\varkappa_f)} = \int_{-1}^{1} \mathrm{d}\xi\, \mathcal{T}_2^{(\varkappa_f)}(q, \xi p) \psi_{f/h}^{(\varkappa_f)}(\xi), \quad (14.49)$$

where we define

$$\mathcal{T}_2^{(\varkappa_f)}(\xi p, k) \equiv \mathrm{tr}\,[\mathcal{U}_2^{(\varkappa_f)}(q, \xi p)\tfrac{1}{2}\gamma^- p^+], \tag{14.50a}$$

$$\psi_{f/h}{}^{(\varkappa_f)}(\xi) \equiv \int_{n(\varkappa_f)} \mathrm{d}^4 k\, \delta(\xi p^+ - k^+)\,\mathrm{tr}\,[\tfrac{1}{2}\gamma^+ \varkappa_{f/h}(k, p, \epsilon)]. \tag{14.50b}$$

$\mathcal{U}_2{}^{(\varkappa_f)}(q, \xi p)$ and $\psi_{f/h}{}^{(\varkappa_f)}(\xi)$ contribute directly to the hard-scattering and jet functions in the factorized form of the structure function T_2. Similar arguments may be given for i = gluon (exercise 14.5). The factorized form for the Compton scattering structure function T_2, analogous to eqs. (14.42), may thus be written as

$$T_2(\omega, Q^2) = \sum_{i=f,g} \int_{-1}^{1} \mathrm{d}\xi\, \mathcal{T}_{2,i}(\xi\omega, Q^2/\mu^2, \alpha_s(\mu^2))\,\psi_{i/h}(\xi, \epsilon, \alpha_s(\mu^2)) + \cdots, \tag{14.51}$$

where the remainder is suppressed by a power of Q^2 for ω fixed. Factorization for T_a implies a similar result for its imaginary part, $W_a(x, Q^2)$. (In fact, $\psi_{i/h}$ in eq. (14.51) even turns out to be equal to the parton distribution $\psi_{i/h}$ encountered in the factorization of $W_a(x, Q^2)$ (Collins & Soper 1982, Jaffe 1983), but we need not insist on this point here.)

In summary, the incoherence of the parton model, which separates long- and short-distance time scales, is an intrinsic feature of perturbation theory for deeply inelastic scattering at high energy.

We emphasize that our treatment above is heuristic, and that a formal proof of factorization for the $T_a(\omega, Q^2)$ and/or the $W_a(x, Q^2)$ must deal with subdivergences systematically (Collins, Soper & Sterman r1989). The basic observations are those given here, however.

Minimal subtraction distributions

Now let us return to the question of what to choose for the quark distribution. A natural choice for the one-loop quark distribution $(\alpha_s/\pi)\phi_{f/f}{}^{(1)}$ is related to minimal subtraction (MS) (Section 10.1):

$$(\alpha_s/\pi)\phi_{f/f}{}^{(1)}(x) = (\alpha_s/2\pi)(-\epsilon)^{-1} P_{qq}(x). \tag{14.52}$$

This MS quark distribution (Curci, Furmanski & Petronzio 1980, Collins & Soper 1982) absorbs only the collinear pole in eq. (14.37), expanded according to eq. (14.44).

The MS quark distribution can be given a very convenient definition in terms of matrix elements, suggested by our expression, eq. (14.50b), for the integral of the Compton amplitude near a pinch surface. We boldly extend the k^- and k_T integrals in eq. (14.50) to infinity, to arrive at

$$\begin{aligned}
&\phi_{f/h}(\xi, m^2/\mu^2) \\
&= (2\pi)^{-1} \int_{-\infty}^{\infty} dy^- \exp(-i\xi p^+ y^-) \\
&\quad \times \tfrac{1}{2} \sum_\sigma \langle h(p,\sigma) | \bar{q}_f(0, y^-, \mathbf{0}) \tfrac{1}{2}\gamma^+ q_f(0) | h(p,\sigma) \rangle_{A^+=0}, \\
&= \tfrac{1}{2} \sum_\sigma \sum_n \langle h(p,\sigma) | \bar{q}_f(0) \tfrac{1}{2}\gamma^+ | n \rangle \langle n | q_f(0) | h(p,\sigma) \rangle_{A^+=0} \delta(p_n{}^+ - (1-\xi)p^+),
\end{aligned} \tag{14.53a}$$

where $q_f(x)$ is the field for quark flavor f. For h = a parton, an average over color is understood, and over spin is shown. In the first expression, the two quark operators are related by a lightlike vector $y^\mu = y^- \delta_{\mu-}$. As the second expression shows, this distribution includes outgoing states $|n\rangle$ with arbitrarily large transverse and minus momenta. The unbounded integrals make it necessary to renormalize $\phi_{f/h}$. Other definitions of the quark density, for which renormalization is not necessary, are possible (exercise 14.6). This form, however, has the advantage that it can be used directly in the proof of factorization.

The matrix element in eq. (14.53a) is not gauge invariant, and we have indicated the use of the *light-cone* gauge, $A^+ = 0$. One reason for this definition is that (14.53a) then has a natural extension to a gauge invariant form

$$\begin{aligned}
&\phi_{f/h}(\xi, m^2/\mu^2) \\
&= (2\pi)^{-1} \int_{-\infty}^{\infty} dy^- \exp(-i\xi p^+ y^-) \\
&\quad \times \tfrac{1}{2} \sum_\sigma \langle h(p,\sigma) | \bar{q}_f(0, y^-, \mathbf{0}) P \exp\left[ig \int_0^{y^-} dy^- A^+(y^- u^\mu)\right] \tfrac{1}{2}\gamma^+ q_f(0) | h(p,\sigma) \rangle,
\end{aligned} \tag{14.53b}$$

in which the two quark fields are connected by a 'path-ordered' exponential of the gluon field (exercise 14.8) along the direction

$$u^\alpha = \delta_{\alpha-}. \tag{14.54}$$

The gauge variation of the path-ordered exponential precisely compensates that of the quark fields.

Another reason that $A^+ = 0$ is a natural gauge for parton distributions is related to the Lorentz properties of spin. A physical polarization s^μ in the p^μ rest frame generally has a finite component s^+. The center-of-mass frame is reached by a boost in the plus direction, and the s^+ component becomes correspondingly large. Of course, the plus component of the gluon's momentum does too. The largest component of the center-of-mass spin s'^{μ} is thus proportional to its momentum, and hence unphysical. The choice $A^+ = 0$ simply eliminates this possibility.

Note that in the hadron rest frame, the quark fields in eqs. (14.53) may be separated by large distances on the light-cone $y^2 = 0$. As a result, deeply inelastic scattering is often referred to as being light-cone dominated. In the center of mass of $(q+p)^{\mu}$, however, the distance between the fields is quite small, because of the rapidly oscillating exponential, $\exp(-\mathrm{i}\xi p^+ y^-)$.

Equations (14.53) serve for antiquark distributions as well, if we define (Collins & Soper 1982)

$$\bar{q}_f(x) \equiv \gamma^+ q_f(x), \quad q_{\bar{f}}(x) \equiv \bar{q}_f(x). \tag{14.55}$$

The role of the Dirac matrix $\frac{1}{2}\gamma^+$ in eqs. (14.53) is already clear at zeroth order in the strong coupling when h is a quark of flavor f:

$$\phi_{f/f}{}^{(0)}(\xi) = (2p^+)^{-1} N^{-1} \operatorname{tr} I_{N\times N} \operatorname{tr}(p^+\gamma^- \tfrac{1}{2}\gamma^+)\delta(1-\xi) = \delta(1-\xi), \tag{14.56}$$

where we note the $SU(N)$ color average and trace. Any other choice of Dirac matrix in eq. (14.53) would give a vanishing Dirac trace.

Gluon distribution

The complete factorization formula eq. (14.42) requires that we define a distribution for gluons as well as for quarks. An MS gluon distribution, normalized at zeroth order as in eq. (14.43a), is given in $A^+ = 0$ gauge by

$$\phi_{g/h}(\xi, m^2/\mu^2) = (2\pi\xi p^+)^{-1} \int_{-\infty}^{\infty} \mathrm{d}x^- \exp(-\mathrm{i}\xi p^+ x^-)$$
$$\times \tfrac{1}{2} \sum_{\sigma} \langle h(p,\sigma)| F^+{}_{\mu}(0, x^-, \mathbf{0}) F^{\mu+}(0) |h(p,\sigma)\rangle_{A^+=0}, \tag{14.57}$$

(Curci, Furmanski & Petronzio 1980, Collins & Soper 1982).

One-loop density

As described in Section 7.3, the light-cone gauge is generated by choosing a gauge fixing term $\frac{1}{2}\lambda(u\cdot A) = \frac{1}{2}\lambda(A^+)^2$, where u^{μ} is given in eq. (14.54), and then letting $\lambda \to \infty$. In this limit ghosts decouple, and, according to eq.

(7.77), the gluon propagator is

$$G_{\mu\nu,ab}(k, u) = \delta_{ab}(k^2 + i\epsilon)^{-1}(-g_{\mu\nu} + k_\mu u_\nu/(u \cdot k) + u_\mu k_\nu/(u \cdot k)). \tag{14.58}$$

It should be noted that the gauge denominators lead to extra pinch surfaces in diagrams calculated in this gauge. The standard way of defining these denominators is through the principle-value prescription. This approach develops practical problems in explicit two-loop calculations (Liebbrandt r1989). These difficulties, however, do not appear in a one-loop calculation of $\phi_{f/f}$.

At one loop in the light-cone gauge, the nonsinglet quark distribution for external quark f is given by the two diagrams of Fig. 14.5 (plus a counter-term). The real-gluon contribution is

$$\begin{aligned}(\alpha_s/\pi)&\phi_{f/f}{}^{(r,1)}(\xi, \epsilon, \alpha_s(\mu^2)) \\ &= -\tfrac{1}{4}N^{-1}\,\mathrm{tr}\,[t_a{}^{(F)}t_a{}^{(F)}]g_s{}^2\mu^{2\epsilon} \\ &\quad\times \int \frac{d^n k}{(2\pi)^n}\left\{\frac{2\,\mathrm{tr}\,[\not{p}\gamma^+(\not{p}-\not{k})\gamma^+]}{(u\cdot k)(p-k)^2} + \frac{\mathrm{tr}\,[\not{p}\gamma_\mu(\not{p}-\not{k})\gamma^+(\not{p}-\not{k})\gamma^\mu]}{[(p-k)^2]^2}\right\} \\ &\quad\times 2\pi\delta_+(k^2)\delta(k^+ - (1-\xi)p^+) - (\alpha_s/\pi)Z_q{}^{(1)},\end{aligned} \tag{14.59}$$

where $Z_q{}^{(1)}$ is a minimal subtraction counterterm, which we are about to determine. The delta functions reduce eq. (14.59) to

$$\begin{aligned}(\alpha_s/\pi)\phi_{f/f}{}^{(r,1)}(\xi, \mu^2) &= C_2(F)\alpha_s\mu^{2\epsilon}(2\pi)^{2-n} \\ &\quad\times\left[\frac{2(1+\xi^2)}{1-\xi} - 2\epsilon(1-\xi)\right]\int\frac{d^{n-2}\mathbf{k}_T}{\mathbf{k}_T{}^2} + \left(\frac{\alpha_s}{\pi}\right)Z_q{}^{(1)}.\end{aligned} \tag{14.60}$$

The k_T integral is zero in dimensional regularization, but its ultraviolet pole may be isolated by the simple manipulation introduced in eq. (11.35b):

$$\begin{aligned}\int\frac{d^{n-2}\mathbf{k}_T}{\mathbf{k}_T{}^2} &= \int\frac{d^{n-2}\mathbf{k}_T}{(\mathbf{k}_T{}^2 + M^2)} + M^2\int\frac{d^{n-2}\mathbf{k}_T}{\mathbf{k}_T{}^2(\mathbf{k}_T{}^2 + M^2)} \\ &= \pi\epsilon^{-1} + \pi(-\epsilon)^{-1} + \ldots = 0.\end{aligned} \tag{14.61}$$

where M is an arbitrary mass, $\pi\epsilon^{-1}$ is the ultraviolet pole, and $\pi(-\epsilon)^{-1}$ the infrared pole. The MS counterterm is then determined by eqs. (14.60) and

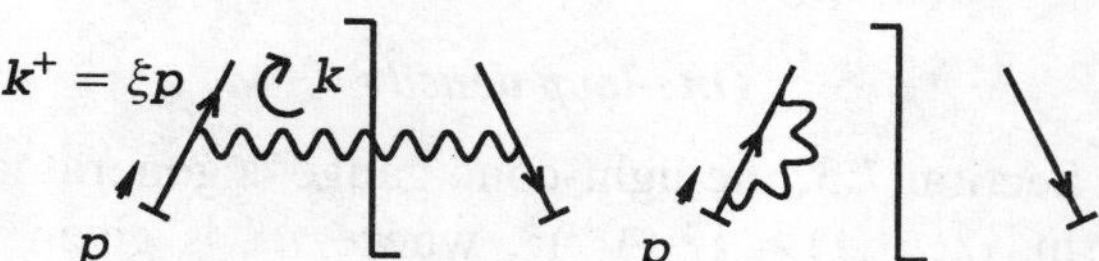

Fig. 14.5 Quark distribution at order α_s.

Table 14.1 *One-loop evolution kernels in QCD*
One-loop MS distributions are given by $(\alpha_s/\pi)\phi^{(1)}{}_{i/j}(x) = (\alpha_s/2\pi)(-\epsilon)^{-1}P_{ij}(x)$.

$$P_{qq}(x) = P_{\bar{q}\bar{q}}(x) = C_2(\mathrm{F})[(1 + x^2)[1/(1 - x)]_+ + \tfrac{3}{2}\delta(1 - x)]$$

$$P_{\bar{q}q}(x) = P_{q\bar{q}}(x) = 0$$

$$P_{qg}(x) = P_{\bar{q}g}(x) = T(\mathrm{F})[z^2 + (1 - z)^2]$$

$$P_{gq}(x) = P_{g\bar{q}}(x) = C_2(\mathrm{F})\{x^{-1}[1 + (1 - x)^2]\}$$

$$P_{gg}(x) = 2C_2(\mathrm{A})\{x[1/(1 - x)]_+ + (z^{-1} + z)(1 - z)\} + \{\tfrac{11}{6}C_2(\mathrm{A}) - \tfrac{2}{3}n_f T(\mathrm{F})\}\delta(1 - z)$$

(14.61), and matches eq. (14.52) for $\xi < 1$. A similar calculation for the virtual graph gives the rest of $\phi_{f/f}{}^{(1)}$. In the same manner, all the one-loop distributions, $\phi_{g/f}{}^{(1)}$, $\phi_{f/g}{}^{(1)}$ and $\phi_{g/g}{}^{(1)}$ can be computed (Altarelli & Parisi 1977). The complete set is summarized in Table 14.1.

Note that the minimal subtraction defined by this procedure is distinct from renormalization through local counterterms, because the coefficient of the pole in eq. (14.60) is ξ-dependent in a nonpolynomial way. This is due to the nonlocal nature of the operators that define the distributions in eqs. (14.53). The procedure is, however, self-consistent (Collins & Soper 1982). The scale associated with the parton distribution therefore need not be the same as that used for renormalization of the theory. For this reason it is often referred to as the *factorization scale*.

Coefficient functions

Given the one-loop distribution, eq. (14.52), we can find the $\mathrm{O}(\alpha_s)$ coefficient functions $C_a{}^{(\mathrm{f},1)}$ by inspection of eqs. (14.37) and (14.44):

$$\begin{aligned}(\alpha_s/\pi)&C_2{}^{(\mathrm{f},1)}(x, Q^2)\\ &= (\alpha_s/2\pi)Q_f{}^2 x\Big[[\ln(Q^2/\mu^2) + \gamma_\mathrm{E} - \ln 4\pi]P_{qq}(x)\\ &\qquad + C_2(\mathrm{F})\Big((1 + x^2)\{[\ln(1 - x)]/(1 - x)\}_+ - \tfrac{3}{2}[1/(1 - x)]_+\\ &\qquad - (1 + x^2)(\ln x)/(1 - x) + 3 + 2x\\ &\qquad - (\tfrac{9}{2} + \tfrac{1}{3}\pi^2)\delta(1 - x)\Big)\Big] + \mathrm{O}(\epsilon),\end{aligned} \tag{14.62}$$

$$(\alpha_s/\pi)C_1{}^{(\mathrm{f},1)}(x, Q^2) = (\alpha_s/\pi)C_2{}^{(\mathrm{f},1)}(x, Q^2)/2x - Q_f{}^2 C_2(\mathrm{F})(\alpha_s/2\pi)x + \mathrm{O}(\epsilon).$$

Here γ_E is the Euler constant, introduced in (9.36). As expected, the $C_a^{(f,1)}$ depend only on the parton that participates in the hard scattering, and not on the external hadron. They also depend on the operator definition of the parton distributions, but *not* on the particular infrared regularization employed to calculate those distributions in perturbation theory. The explicit logarithm of Q^2 exhibits scale breaking at lowest order. The combination $\gamma_E - \ln 4\pi$ may be absorbed into a shift in μ; defined with this μ, the distribution is termed '$\overline{\text{MS}}$' (Section 10.1).

Using factorization

Since they are infrared safe, the coefficient functions may be applied to the structure functions of nucleons, putting $h = \text{n}, \text{p}$ in the factorization formula, eq. (14.42). This allows us to use deeply inelastic scattering to *measure* the quark distributions by inverting (14.42) (Close r1979). Once determined, the physical parton distributions may be applied to compute any cross section that enjoys a factorization formula similar to eq. (14.42). The accuracy of such a calculation depends, of course, upon the order to which the coefficient functions have been computed in deeply inelastic scattering. To order $(\alpha_s)^0$, the structure functions F_1 and F_2 already afford an example of this procedure, since by eqs. (14.42) and (14.43b), the *Callan–Gross relation* $F_2 = 2xF_1$ (Callan & Gross 1969), holds for hadrons as well as for partons and, exactly as in the parton model, a measurement of F_2 gives a prediction for F_1. The perturbative expansion of coefficient functions allows us to compute systematic corrections to this, and other, parton-model predictions, including various semi-inclusive jet processes in deeply inelastic scattering (Craig & Llewellyn Smith 1978, Libby & Sterman 1978).

The procedure for using factorization is reminiscent of renormalization, in which one or more physical measurements define renormalized couplings and masses, and hence the full perturbation series. Here, the parton distributions play the role of the couplings and masses, and are fixed by experiment through the deeply-inelastic-scattering structure functions. It should be emphasized, however, that this choice follows history and convenience rather than necessity, and that any other factorized cross section could, in principle, serve the same purpose.

Factorization in the Drell–Yan process and universality

The potential of factorization is realized by following the parton model yet further, and deriving factorization forms for hard scattering processes with two hadrons in the initial state (Politzer 1977, Sachrajda 1978). This opens the door to a world of applications at high energy, involving the production of leptons, jets and new particles (Altarelli r1989). The classic extension of factorization is to the (unpolarized) Drell–Yan process introduced in Section 8.6:

$$\frac{d\sigma_{AB}}{dQ^2}(\tau, Q^2) = \sum_{a,b}\int_0^1 d\xi_a\, d\xi_b\, \phi_{a/A}(\xi_a, \epsilon) H_{ab}(\tau/\xi_a\xi_b, Q^2, \epsilon)\phi_{b/B}(\xi_b, \epsilon) + O(m^2/Q^2),$$
$$\tau \equiv Q^2/s, \tag{14.63}$$

where $\phi_{a/A}$ is the distribution of parton a in hadron A, and the sum is over all parton types, including quarks, antiquarks and the gluon (we have suppressed μ-dependence). $\tau/\xi_a\xi_b$ is the ratio of the large invariants in the hard scattering. We specialize here to lepton pair production through a virtual photon, but the same analysis describes W and Z production. Corrections, as usual, are suppressed by a power of Q^2. Again, H_{ab} is a short-distance coefficient function that is independent of the external hadrons and hence *may be calculated in perturbation theory with external partons*. To do so, we use MS, or other, perturbative definitions for the distributions of partons in partons. The parton distributions that absorb long-distance dependence in the physical Drell–Yan cross section, with external hadrons, are then taken as the same functions determined phenomenologically in deeply inelastic scattering. In this sense the distributions $\phi_{i/h}$ are said to be *universal*. Together with H_{ab}, they fully determine $d\sigma/dQ^2$.

At lowest order, H_{ab} is given by the Born cross sections for a quark–antiquark pair to annihilate into a lepton pair through a single photon. $H_{f\bar{f}}$ is most easily computed by using the transversality of the quark and lepton loops, as we did for the total cross section for e^+e^- annihilation in Section 12.4. The result in n dimensions is

$$\begin{aligned} H_{f\bar{f}}^{(0)}(\tau/\xi_a\xi_b, Q^2, \epsilon) &= (1/8N^2Q^2)(1/Q^2)^2[16\pi N Q_f{}^2\alpha(1-\epsilon)Q^2\mu^{2\epsilon}] \\ &\quad \times[-\bar{\pi}^\mu{}_\mu(\alpha, Q^2, \epsilon)/(3-2\epsilon)][(1/Q^2)\delta(1-\tau/\xi_a\xi_b)] \\ &\equiv \hat{H}_{f\bar{f}}^{(0)}(\tau, Q^2, \epsilon)\delta(\tau-\xi_a\xi_b) + O(\epsilon). \end{aligned} \tag{14.64}$$

In the first equality, the factors in square brackets represent the overall normalization, the photon propagators, the quark loop, the final-state lepton loop (with $\bar{\pi}^\mu{}_\mu$ given explicitly by the trace of eq. (12.84) with $N = Q_f = 1$) and the phase space delta function. Even $H_{f\bar{f}}^{(0)}$ must be computed beyond $O(\epsilon^0)$, because in perturbation theory the distributions are poles in dimensional regularization.

The proof of factorization for the Drell–Yan process is complicated by interactions between the two incoming hadrons. For discussions of the difficulties and their resolution, see Bodwin (1985) and Collins, Soper & Sterman (1988, r1989).

One-loop corrections in the Drell–Yan process

Higher-order corrections to H_{ab} are calculated in the same manner as for deeply inelastic scattering (Altarelli, Ellis & Martinelli 1979, Kubar-André & Paige 1979, Harada, Kaneko & Sakai 1979, Humpert & van Neerven 1979). For instance, to calculate $H_{f\bar{f}}{}^{(1)}$, we first calculate $(\mathrm{d}\sigma/\mathrm{d}Q^2)^{(1)}$ with massless quarks in dimensional regularization. We then expand the one-loop partonic cross section, using eq. (14.43) for $\phi_{f/f}{}^{(0)} = \phi_{\bar{f}/\bar{f}}{}^{(0)}$ and (14.52) for $\phi_{f/f}{}^{(1)} = \phi_{\bar{f}/\bar{f}}{}^{(1)}$ in eq. (14.63):

$$\frac{\mathrm{d}\sigma_{f\bar{f}}{}^{(1)}}{\mathrm{d}Q^2}(\tau, Q^2) = \sum_{a=\{f,\bar{f}\}} [H_{aa}{}^{(0)}(\tau, Q^2, \epsilon)(-\epsilon)^{-1} P_{qq}(\tau) + H_{a\bar{a}}{}^{(1)}(\tau, Q^2)]. \tag{14.65}$$

The zeroth-order hard part is eq. (14.64), and in $\sigma_{f\bar{f}}$ there is no mixing of flavors, to $\mathrm{O}(\alpha_s)$. Equation (14.65) enables us to determine $H_{f\bar{f}}{}^{(1)}$ by inspection, once the one-loop quark–antiquark cross section has been computed. The resulting hard part can then be used in eq. (14.63), with physical parton distributions from deeply inelastic scattering, to give the $\mathrm{O}(\alpha_s)$ quark–antiquark correction to the Drell–Yan cross section. At this order, quark–gluon corrections also contribute, but we shall not compute them here.

The cut diagrams that contribute to $\mathrm{d}\sigma_{f\bar{f}}{}^{(1)}/\mathrm{d}Q^2$ are shown in Fig. 14.6. The cross section is closely related to the $\mathrm{e}^+\mathrm{e}^-$ total annihilation cross section of Section 12.4, and we may simply take over many of the results found there. For instance, the cut diagrams of Fig. 14.6(*a*) are equal to those for Fig. 12.2(*a*), (*b*). The only differences between the virtual corrections to $\mathrm{d}\sigma_{f\bar{f}}{}^{(1)}/\mathrm{d}Q^2$ and σ_{tot} for $\mathrm{e}^+\mathrm{e}^-$ are from the delta function that fixes Q^2, and from color averaging. We have

$$\mathrm{d}\sigma_{f\bar{f}}{}^{(\mathrm{v})}/\mathrm{d}Q^2 = H_{f\bar{f}}{}^{(0)}(\tau, Q^2, \epsilon)[2\,\mathrm{Re}\,\gamma(Q^2)], \tag{14.66}$$

where $\gamma(Q^2)$ is the vertex given by eq. (12.82). The real-gluon corrections, Fig. 14.6(*b*), are only slightly more complicated to derive. The transversality of both the hadronic and leptonic Dirac traces leads to the following expression:

$$\begin{aligned}\frac{\alpha_s}{\pi}\left(\frac{\mathrm{d}\sigma_{f\bar{f}}}{\mathrm{d}Q^2}\right)^{(\mathrm{r},1)} = \frac{1}{8N^2 sQ^4}\left[-\frac{\bar{\pi}^\mu{}_\mu(q^2)}{(3-2\epsilon)}\right]&\int \mathrm{d}^n k\, \mathrm{d}^n q\,(2\pi)^{n-1} \\ &\times \delta^n(p_1 + p_2 - q - k)\delta_+(k^2)\delta(q^2 - Q^2) \\ &\times |M(p_1, p_2, k)|^2{}_{\mathrm{r}}.\end{aligned} \tag{14.67}$$

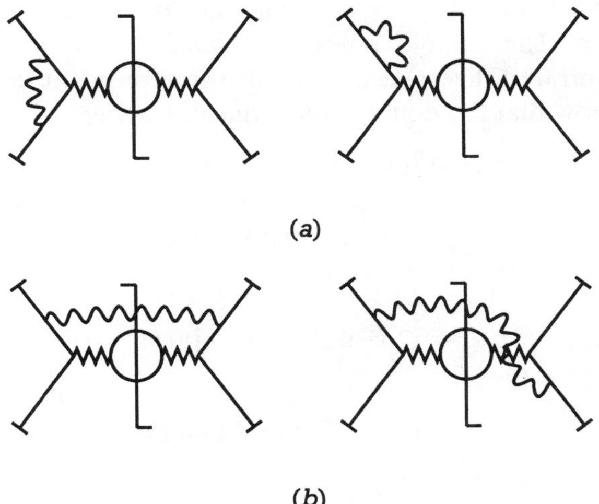

Fig. 14.6 Cut diagrams for the Drell–Yan process at order α_s.

As above, $\bar{\pi}^{\mu}{}_{\mu}(q^2)$ is the trace of the cut leptonic loop, while $|M(p_1, p_2, k)|^2{}_{\mathrm{r}}$ is given exactly by the spin-summed annihilation amplitude squared, eq. (12.73), since this quantity is invariant under $p_1 \to -p_1$, $p_2 \to -p_2$:

$$|M(p_1, p_2, k)|^2{}_{\mathrm{r}} = 16(1-\epsilon)C_2(\mathrm{F})Q_f{}^2 e^2 g_s{}^2 \mu^{4\epsilon} \times \left[\frac{q^2(2p_1\cdot p_2)}{(2p_1\cdot k)(2p_2\cdot k)} - \epsilon + (1-\epsilon)\frac{2p_2\cdot k}{2p_1\cdot k}\right]. \quad (14.68)$$

The kinematic expressions for the invariants are

$$2p_1\cdot p_2 = s,\ 2p_1\cdot k = s^{1/2}k(1-u),\ 2p_2\cdot k = s^{1/2}k(1+u), \quad (14.69)$$

where $k \equiv |\mathbf{k}|$ and $u \equiv \cos\theta_{pk}$. The calculation of the one-loop Drell–Yan cross section is a matter of evaluating the integrals in eq. (14.67). This process (exercise 14.9) is actually simpler than in deeply inelastic scattering. After using the delta functions, the only nontrivial integrals are those over u, and they may be reduced to the standard beta function form, eq. (9.30a), by the usual change of variables, $y = \frac{1}{2}(1+u)$. The result, with real and virtual contributions separated, can be put into the form

$$\begin{aligned}(\alpha_s/\pi)(\mathrm{d}\sigma_{f\bar{f}}/\mathrm{d}Q^2)^{(1)} &= (\alpha_s/\pi)H_{f\bar{f}}{}^{(0)}(\tau, Q^2, \epsilon)C_2(\mathrm{F})(4\pi\mu^2/Q^2)^{\epsilon}[\Gamma(1-\epsilon)/\Gamma(1-2\epsilon)](1-\epsilon) \\ &\quad\times\Big\{[-\epsilon^{-2} - \tfrac{3}{2}\epsilon^{-1} - 4 + \tfrac{1}{3}\pi^2]\delta(1-\tau) \\ &\quad - \epsilon^{-1}[(1-\tau)^{1-2\epsilon}\tau^{\epsilon} + 2\tau^{1+\epsilon}(1-\tau)^{-1-2\epsilon}] + \mathrm{O}(\epsilon)\Big\}. \quad (14.70)\end{aligned}$$

We have used eq. (9.36) to expand gamma functions, leaving in front a common factor. The double poles, as usual, are from the overlap of collinear and infrared divergences. The distribution identities, eqs. (14.29) and (14.30), show that the double poles cancel, leaving

$$\begin{aligned}(\alpha_s/\pi)(\mathrm{d}\sigma_{f\bar{f}}/\mathrm{d}Q^2)^{(1)} = H_{f\bar{f}}{}^{(0)}(\tau, Q^2, \epsilon)(\alpha_s/\pi) \\ \times \Big[(-\epsilon)^{-1} P_{qq}(\tau) + [\ln(Q^2/\mu^2) + \gamma_{\mathrm{E}} - \ln 4\pi] P_{qq}(\tau) \\ + C_2(\mathrm{F})\Big(2(1+\tau^2)\{[\ln(1-\tau)]/(1-\tau)\}_+ \\ - [(1+\tau^2)\ln\tau]/(1-\tau) + (\tfrac{1}{3}\pi^2 - 4)\delta(1-\tau)\Big) \\ + \mathrm{O}(\epsilon)\Big].\end{aligned} \tag{14.71}$$

Comparing this result with eq. (14.65), we can read off $H_{f\bar{f}}{}^{(1)}$. It is, as promised, infrared finite. The factorization theorem therefore holds at order α_s. This is a nontrivial result, depending as it does on the cancellation of the double poles and the matching of single-pole coefficients to the perturbative quark distribution, eq. (14.52).

Factorization has also been explicitly verified, and hard parts computed, at two loops in the Drell–Yan process in QCD (Matsuura, van der Marck & van Neerven 1989).

Other processes

Factorization theorems such as eq. (14.63) hold for a sizeable class of cross sections with a large momentum transfer such as those for the production of heavy vector bosons, heavy flavors or jets in hadron–hadron collisions, as well as those for the 'crossed' version of deeply inelastic scattering, single-particle inclusive annihilation, shown in Fig. 14.7 (Gribov & Lipatov 1972b, Mueller 1974, 1978, Collins & Sterman 1981). The basic techniques for analyzing these cross sections are along the lines discussed above. Equation (14.63) may also be usefully applied to such very-high-energy QED cross sections as $e^+e^- \rightarrow Z^0$ (Fadin & Kuraev 1985, Bovincini 1990). Here, one can actually calculate the parton densities, since the QED coupling may be renormalized on-shell (Section 12.1). The factorization formula then organizes powers of logarithms such as $\ln(M_Z{}^2/m_e{}^2)$, which is really quite a large number.

In QCD, the numerical values of both one- and two-loop corrections to hard scattering functions in the Drell–Yan process, and in many other

Fig. 14.7 Amplitude for the single-particle inclusive process in e^+e^- annihilation.

processes, are substantial at accessible energies. This has led to an ongoing effort toward further reorganization of the perturbation series by various generalizations of the factorization program (Gribov, Levin & Ryskin 1983, Caifaloni 1989, Dokshitzer, Khoze & Troyan 1989, Lipatov 1989, Catani & Trentadue 1989, Levin & Ryskin 1990, Magnea 1991).

Choice of scale

The one-loop corrections to coefficient functions for both deeply inelastic scattering and the Drell–Yan process include scale breaking terms proportional to $\alpha_s(\mu^2)\ln(Q^2/\mu^2)$. This suggests $\mu = Q$ as the choice of renormalization scale, to eliminate the logarithm and to take advantage of asymptotic freedom. This is fine as far as it goes, but it requires that we choose $\mu = Q$ in the physical parton distributions as well. We are thus led to investigate the Q^2-dependence of parton distributions and cross sections.

14.4 Evolution

The prediction of one structure function in terms of another is an instructive application of factorization, but it is not the only one. Equally remarkable is our ability to calculate Q^2-dependence for structure functions. Specifically, if we measure $W_{\mu\nu}(x, Q^2)$ over a range $x_0 \leqslant x \leqslant 1$ at one value of Q^2, we can use perturbation theory to predict $W_{\mu\nu}(x, Q'^2)$ for $Q'^2 > Q^2$ over the same range in x. The variation of the structure functions with Q^2 is known as their *evolution*. The only restriction is that the lower value, Q^2, should be large enough that an expansion in $\alpha_s(Q^2)$ makes sense. Extrapolations of this sort play a central role in predictions of very-high-energy cross sections (Eichten, Hinchcliff, Lane & Quigg 1984). Measurements of parton distributions in deeply inelastic scattering at relatively low energy are evolved to give predictions for hadron–hadron cross sections at much higher energies.

Nonsinglet structure functions

To simplify our discussion, we introduce a *flavor nonsinglet* structure function

$$F_a{}^{(\mathrm{p-n})}(x, Q^2) \equiv F_a{}^{(\mathrm{p})}(x, Q^2) - F_a{}^{(\mathrm{n})}(x, Q^2), \qquad (14.72)$$

where p and n are the proton and neutron. Any combination such as this, in which flavor-independent components cancel, is known as a nonsinglet quantity. Other combinations are referred to as *singlet*. In perturbation theory, the only diagrams that survive in a nonsinglet structure function are of the form of Fig. 14.8(a), in which the external quark line flows through the entire diagram to the electromagnetic vertex. Diagrams like Fig. 14.8(b) in which the electromagnetic vertex couples to an internal quark loop cancel in $F_a{}^{(\mathrm{p-n})}$, because the flavor-symmetric strong interactions produce gluons equally in the proton and the neutron, and do not distinguish quark and antiquark in the upper part of the diagram. Similarly, the nonsinglet cross section has no gluonic contributions. The factorization theorem for a nonsinglet distribution is almost identical to eqs. (14.42):

$$\begin{aligned} F_1{}^{(\mathrm{p-n})}(x, Q^2) = \sum_f \int_0^1 &(\mathrm{d}\xi/\xi) C_1{}^{(\mathrm{ns},f)}(x/\xi, Q^2/\mu^2, \alpha_s(\mu^2)) \\ &\times [\phi_{f/\mathrm{p}}{}^{(\mathrm{val})}(\xi, \epsilon, \alpha_s(\mu^2)) - \phi_{f/\mathrm{n}}{}^{(\mathrm{val})}(\xi, \epsilon, \alpha_s(\mu^2))] \\ &+ \mathrm{O}(m^2/Q^2). \end{aligned} \tag{14.73a}$$

The short-distance function $C_1{}^{(\mathrm{ns},f)}$ is flavor-independent except for an overall constant $Q_f{}^2$, while we define the *valence distributions* by

$$\phi_{f/h}{}^{(\mathrm{val})}(\xi, \epsilon, \alpha_s(\mu^2)) \equiv \phi_{f/h}(\xi, \epsilon, \alpha_s(\mu^2)) - \phi_{\bar{f}/h}(\xi, \epsilon, \alpha_s(\mu^2)). \tag{14.73b}$$

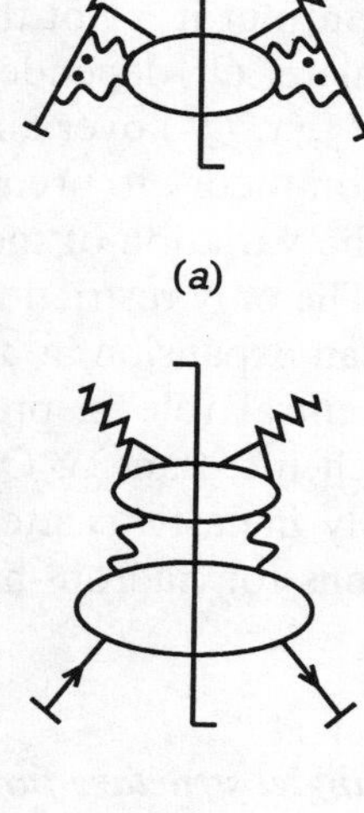

Fig. 14.8 (a) Diagrams that contribute to the nonsinglet structure function. eq. (14.72); (b) additional diagrams that contribute to singlet structure functions.

The term 'valence' refers to an assumption that the entire antiquark content of the external hadron is the result of the creation of virtual quark–antiquark pairs. The valence distributions may be experimentally determined through linear combinations of electromagnetic and weak structure functions (exercise 14.10, Close r1979). We shall refer below, for instance, to $F_1{}^{(\mathrm{ns},f)}$ as that part of $F_1{}^{(\mathrm{p-n})}$ due to the valence quark distribution of flavor f.

Moments

The deeply-inelastic-scattering factorization forms, eqs. (14.42) and (14.73), can be simplified by taking moments with respect to the variable x, defined for any function $F(x)$ by

$$\bar{F}(x) = \int_0^1 \mathrm{d}x\, x^{n-1} F(x). \tag{14.74}$$

Convolutions reduce to products under these moments. The moments of $F_1{}^{(\mathrm{ns},f)}$, for instance, are given by

$$\bar{F}_1{}^{(\mathrm{ns},f)}(n, Q^2) = \bar{C}_1{}^{(\mathrm{ns},f)}(n, \alpha_s(\mu^2), Q^2/\mu^2)\bar{\phi}_f{}^{(\mathrm{ns})}(n, \epsilon, \alpha_s(\mu^2)). \tag{14.75}$$

Here $\bar{\phi}_f{}^{(\mathrm{ns})}$ refers to the linear combination of valence distributions in eq. (14.73a).

Renormalization group

The moments $\bar{F}_a{}^{(\mathrm{ns},f)}$ are physical quantities, and must therefore be independent of our choice of the factorization scale, μ^2:

$$(\mu\, \mathrm{d}/\mathrm{d}\mu)\bar{F}_a{}^{(\mathrm{ns},f/h)}(n, Q^2) = 0. \tag{14.76}$$

Applied to eq. (14.75), this gives a renormalization group equation for the moments of the valence densities and the hard scattering functions,

$$\begin{aligned}(\mu\, \mathrm{d}/\mathrm{d}\mu) \ln [\bar{\phi}^{(\mathrm{ns})}(n, \epsilon, \alpha_s(\mu^2))] &= -(\mu\, \mathrm{d}/\mathrm{d}\mu) \ln [\bar{C}_1{}^{(\mathrm{ns})}(n, \alpha_s(\mu^2), Q^2/\mu^2)] \\ &\equiv -\gamma_n(\alpha_s(\mu^2)),\end{aligned} \tag{14.77a}$$

which is independent of f and h (we therefore suppress the flavor labels in nonsinglet quantities below). The nonsinglet anomalous dimensions γ_n are functions only of $\alpha_s(\mu^2)$, because μ is the only dimensional variable that $\bar{\phi}^{(\mathrm{val})}$ and $\bar{C}_2{}^{(\mathrm{ns})}$ have in common. Exactly similar equations can be derived for the singlet case, but now the anomalous dimensions make up a matrix

$[\gamma^{(s)}{}_n]_{ij}$, where i and j label quarks, antiquarks and the gluon:

$$(\mu \mathrm{d}/\mathrm{d}\mu)\bar{\phi}_{i/h}(n, \epsilon, \alpha_s(\mu^2)) = -[\gamma_n{}^{(s)}]_{ij}\bar{\phi}_{j/h}(n, \epsilon, \alpha_s(\mu^2)) \quad (14.77b)$$

(Gross & Wilczek 1974, Georgi & Politzer 1974).

One-loop anomalous dimensions

The γ_n are calculated explicitly at one loop by the same trick used to calculate the coefficient functions. Consider the nonsinglet structure function for a single quark of flavor f. Then γ_n may be computed by acting with $\mu \mathrm{d}/\mathrm{d}\mu$ on moments of the one-loop quark distribution, eq. (14.52), using (see eqs. (14.43a) and (14.73b)) $\bar{\phi}_{f/h}{}^{(\mathrm{val})(0)}(n) = \bar{\phi}_{f/h}{}^{(0)}(n) = 1$ for all n. Since the only μ-dependence in $\bar{\phi}$ is in the running coupling $\alpha_s(\mu^2)$, we find

$$(\alpha_s/\pi)\gamma_n{}^{(1)} = -\beta^{(0)}(g_s)(\partial/\partial g_s)(\alpha_s/\pi)\bar{\phi}^{(\mathrm{val})(1)}(n), \quad (14.78)$$

where $\beta^{(0)}(g)g = -\epsilon g$, by eq. (11.50). In terms of the evolution function P_{qq}, $\gamma_n{}^{(1)}$ is then

$$\begin{aligned}(\alpha_s/\pi)\gamma_n{}^{(1)} &= -(\alpha_s/\pi)\int_0^1 \mathrm{d}x\, x^{n-1} P_{qq}(x) \\ &= (\alpha_s/2\pi)C_2(\mathrm{F})\left[4\sum_{m=2}^{n}\frac{1}{m} - \frac{2}{n(n+1)} + 1\right]. \quad (14.79)\end{aligned}$$

The same method may be used to calculate any of the singlet anomalous dimensions from the one-loop distributions of Table 14.1.

The significance of the anomalous dimensions is found by solving eq. (14.77a):

$$\begin{aligned}\bar{\phi}^{(\mathrm{val})}(n, \epsilon, \alpha_s(Q^2)) &= \bar{\phi}^{(\mathrm{val})}(n, \epsilon, \alpha_s(m^2)) \\ &\quad \times \exp\left[-\frac{1}{2}\int_0^{\ln(Q^2/m^2)} \mathrm{d}t\gamma_n(\alpha_s(m^2\mathrm{e}^t))\right],\end{aligned} \quad (14.80)$$

for any reference mass m. This also determines the Q^2-dependence of moments of the structure functions, eq. (14.75):

$$\begin{aligned}\bar{F}_1{}^{(\mathrm{ns})}(n, Q^2) &= \bar{C}_1{}^{(\mathrm{ns})}(n, \alpha_s(Q^2))\exp\left[-\tfrac{1}{2}\int_0^{\ln(Q^2/m^2)} \mathrm{d}t\gamma_n(\alpha_s(m^2\mathrm{e}^t))\right] \\ &\quad \times \bar{\phi}^{(\mathrm{val})}(n, \epsilon, \alpha_s(m^2)).\end{aligned} \quad (14.81)$$

The actual scale breaking that results from this expression depends on the renormalization group behavior of the theory (Section 10.5). There are two cases in which we can compute its asymptotic behavior: when the theory has an ultraviolet fixed point, $\alpha_s(\mu^2) \to \alpha_0 \neq 0$, or, like QCD, is asymptotically free, $\alpha_s(\mu^2) \approx 4\pi[|\beta_1| \ln(\mu^2/\Lambda^2)]^{-1}$ (eq. (12.94)). Thus we have for an ultraviolet fixed point

$$\bar{F}_1^{(\mathrm{ns})}(n, Q^2) = \bar{C}_2^{(\mathrm{ns})(0)}(n)\bar{\phi}^{(\mathrm{val})}(n, \epsilon, \alpha_s(m^2)) \times (Q^2/m^2)^{-(\alpha_0/2\pi)\gamma_n^{(1)}} + R, \tag{14.82a}$$

and for asymptotic freedom,

$$\bar{F}_1^{(\mathrm{ns})}(n, Q^2) = \bar{C}_2^{(\mathrm{ns})(0)}(n)\bar{\phi}^{(\mathrm{val})}(n, \epsilon, \alpha_s(m^2)) \times \left[\frac{\ln(Q^2/\Lambda^2)}{\ln(m^2/\Lambda^2)}\right]^{-2\gamma_n^{(1)}/|\beta_1|} + R, \tag{14.82b}$$

where $\gamma_n \equiv (\alpha/\pi)\gamma_n^{(1)} + \dots$. The corrections R are suppressed by a power of Q^2 in the first case and by a logarithm in the second (exercise 14.11).

Equations (14.82) gives a set of predictions, for the Q^2-dependence of moments of deeply-inelastic-scattering structure functions, that can be compared directly with experiment. (It should be pointed out that higher-order corrections (Altarelli r1982, Yndurain r1983) are important in practical comparisons of theory with experiment (Mishra & Sciulli r1989).) Without going into details, we note a characteristic qualitative feature of eqs. (14.82) that is experimentally observed: since the $\gamma_n^{(1)}$ increase with n, higher moments show a faster decrease in Q^2. This corresponds to a 'softening', a shift to lower values of x in the structure function, due to increased radiation at larger momentum transfer. For singlet structure functions, scale breaking is dominated asymptotically by the eigenvalue of $[\gamma_n^{(s)}]_{ij}$ that leads to the slowest decrease with Q^2 (Gross & Wilczek 1974, Georgi & Politzer 1974).

Evolution equation

The moment analysis described above is attractive in its simplicity, but for many purposes it is convenient to deal more directly with the structure functions and parton densities themselves. For this purpose, we can invert the moment equation (14.77a) to recover a convolution:

$$(\mu\,\mathrm{d}/\mathrm{d}\mu)\phi_{f/h}^{(\mathrm{val})}(x, \epsilon, \alpha_s(\mu^2)) = \int_x^1 (\mathrm{d}\xi/\xi) P_f(x/\xi, \alpha_s(\mu^2))\phi_{f/h}^{(\mathrm{val})}(\xi, \epsilon, \alpha_s(\mu^2)), \tag{14.83}$$

where $P_f(x/\xi, \alpha_s(\mu^2))$ satisfies

$$P_f(\zeta, \alpha_s) = \int_{-i\infty}^{i\infty} (\mathrm{d}n'/2\pi \mathrm{i}) \zeta^{-n'+1} \gamma^{(\mathrm{ns})}(n', \alpha_s),$$
$$\int_0^1 \mathrm{d}\zeta\, \zeta^{n-1} P_f(\zeta, \alpha_s) = -\gamma^{(\mathrm{ns})}(n, \alpha_s). \tag{14.84}$$

Thus $P_f(\zeta, \alpha_s)$ is the inverse Mellin transform of $\gamma(n, \alpha_s)$ for $\mathrm{Re}\, n > 0$, the extension of $\gamma_n(\alpha_s)$, eq. (14.79), to continuous values of n. At lowest order, $P_f(x, \alpha_s)$ is just the evolution function,

$$(\alpha_s/\pi) P_f(x)^{(1)} = (\alpha_s/\pi) P_{qq}(x). \tag{14.85}$$

Equation (14.83) is the *evolution equation* for the valence distribution. Similar (matrix) evolution equations hold jointly for the full quark and gluon distributions (Altarelli & Parisi 1977). The formal content of eq. (14.77) for the moments and eq. (14.83) for the distribution is the same. The latter, however, is closer in spirit to the physical picture of scale breaking and parton evolution.

Physical basis of scale breaking

In the parton model, and at lowest order in perturbation theory, a structure function $F(x, Q^2)$ measures the total density of partons with fractional momentum x. In perturbation theory, however, only those partons that are off-shell by less than Q^2 contribute; beyond Q^2, they must be considered as a coherent part of the hard scattering, and are suppressed in an asymptotically free theory. Now, if the constituents of the external hadron are off-shell by no more than some fixed value ${Q_0}^2$, then once $Q^2 \gg {Q_0}^2$, the parton density counts all virtual partons, and the structure function $F(x, Q^2) = F(x, {Q_0}^2)$ scales. This, however, is *not* the case in a renormalizable field theory, where virtual particles of arbitrarily high invariant masses are present. As Q^2 increases, the structure function includes contributions from more and more partons, and consequently changes. The strength of this scale breaking, or evolution, depends on the population of states with large invariant masses, and is a direct test of the short-distance structure of the theory.

As we have seen, factorization realizes the parton-model insight that scattering at large Q^2 is independent of long-time initial-state interactions, which produce the parton density. These two processes are incoherent, and contribute to $F(x, Q^2)$ in the convolution form, eqs. (14.42). This same observation, however, makes it possible to calculate scale breaking in $F(x, Q^2)$. We simply observe that incoherence applies to *any* pair of widely separated momentum scales ${Q_1}^2 \gg {Q_2}^2$, even if ${Q_1}^2 \ll Q^2$. The arguments for factorization in Section 14.3 apply in this case as well, and these two scales contribute to the cross section in a convolution form like eq. (14.42), in which the long-distance part is now the evolution to scale $Q_2{}^2$, and the hard part the evolution at $Q_1{}^2$ and beyond. This suggests that ordering according to mass scales is a characteristic of evolution. Let us see how to organize these observations quantitatively.

Generalized ladder structure

The production of a highly virtual (and therefore short-lived) parton results from a series of perturbative transitions, starting at lower (and therefore longer-lived) mass scales (Dokshitzer, Dyakonov & Troyan 1980). Diagrams that describe these transitions, as they occur in a structure function, can be organized into the cut 'ladder' graphs shown in Fig. 14.9, in which the 'rungs' are taken to be two-particle irreducible in the vertical channel; thus we have

$$F(x, Q^2)_N = \prod_{j=1}^{N} \int d^4 k_j \, R_j(k_j, p - \textstyle\sum_{i=1}^{j-1} k_i, p - \sum_{i=1}^{j} k_i)$$
$$\times \, \delta((1 - x) - \textstyle\sum_{i=1}^{n} k_i{}^+/p^+), \qquad (14.86)$$

where we take $p^\mu = p^+ \delta_{\mu +}$. We suppress overall constants and spin indices, which separate in the same manner as in the factorization theorem above. The momentum $k_i{}^\mu$ flows through each ladder to the top, as shown, and we define $k_0{}^\mu \equiv 0$. We shall group with the ith rung the two vertical lines directly above it. The self-energy diagrams of the vertical lines are split, however, half being associated with the ladder below and half with the ladder above. For purposes of renormalization, each rung then acts like

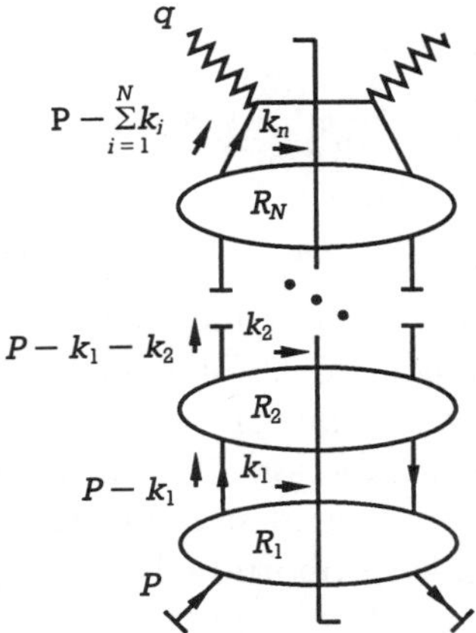

Fig. 14.9 Cut ladder diagrams related to evolution.

an S-matrix element, and

$$\mu \mathrm{d}R_j/\mathrm{d}\mu = 0. \tag{14.87}$$

We define the fractional momentum carried just above the ith rung as

$$\zeta_j = \left(p^+ - \sum_{i=1}^{j} k_i{}^+\right)\left(p^+ - \sum_{i=1}^{j-1} k_i{}^+\right)^{-1}. \tag{14.88}$$

In cut diagrams, the ζ_j satisfy

$$0 < \zeta_j < 1, \quad \prod_{j=0}^{N} \zeta_j = x, \tag{14.89}$$

since each cut rung describes the emission of real quanta.

To implement a parton picture, we consider the *strong-ordering* region in transverse momenta, defined by

$$Q^2 \gg |k_{N,\mathrm{T}}{}^2| \gg |k_{N-1,T}{}^2| \gg \ldots \gg |k_{1,\mathrm{T}}{}^2|. \tag{14.90}$$

In the strong-ordering region, the ith ladder, in which lines are typically off-shell by order $-k_{i,\mathrm{T}}{}^2$, is insensitive to momenta $k_j{}^\mu$, $j < i$. More specifically, we shall make approximation

$$R_j(k_j, p - \textstyle\sum_{i=1}^{j-1} k_i, p - \sum_{i=1}^{j} k_i) \approx R_j(k_j, \prod_{i=1}^{j-1} \zeta_i p, \prod_{i=1}^{j-1} \zeta_i p - k_j). \tag{14.91}$$

In this approximation, the integrals $k_j{}^-$ and $k_{j,\mathrm{T}}$ act only on R_j, and may be carried out independently of all the other rungs. Now, the quantity $\int \mathrm{d}k_j{}^- R_j$ behaves under boosts in the three-direction like the minus component of a momentum, or equivalently, like the inverse of a plus component. Similarly, in $\int \mathrm{d}k_j{}^- R_j$ there is no fixed minus momentum to form an invariant with the external plus momentum $\prod_{i=1}{}^{j-1} \zeta_i p^+$. As a result, the form of this integral is

$$\int \mathrm{d}k_j{}^- R_j(k_j, \textstyle\prod_{i=1}^{j-1} \zeta_i p, \prod_{i=1}^{j-1} \zeta_i p - k_j)$$

$$= \left(\prod_{i=1}^{j-1} \zeta_i p^+\right)^{-1} (k_{j,\mathrm{T}}{}^2)^{-1} P_j(\zeta_j, \alpha_s(k_{j,\mathrm{T}}{}^2)), \tag{14.92}$$

where we have used eq. (14.87) to set $\mu^2 = k_{j,\mathrm{T}}{}^2$, and thereby avoid logarithms of $k_{j,\mathrm{T}}{}^2/\mu^2$. The $k_{j,\mathrm{T}}$-dependence follows from the overall dimension of R_j. The dimensionless function $P(\zeta_j, \alpha_s(k_\mathrm{T}{}^2))$ describes the 'splitting' of a parton by the 'emission' of a cut rung of transverse momentum $k_{j,\mathrm{T}}$, which also carries a fraction $1 - \zeta_j$ of the parton's longitudinal momentum. Apart from its dependence through the running coupling, the probability for this event is proportional to $1/k_{j,\mathrm{T}}{}^2$, and the form of eq. (14.92) is completely general for any renormalizable theory. At lowest order, the rung R_j is given by the emission of a single parton, and

$P_j(\zeta_j, \alpha_s(k_{j,\mathrm{T}}^2))$ is given, up to a normalization, by one of the evolution functions introduced above.

We now change variables from k_i^+ to ζ_i in eq. (14.86), and extend the approximation of eq. (14.91) to the entire region where the $k_{i,\mathrm{T}}^2$ are ordered simply by $k_{i,\mathrm{T}}^2 \geqslant k_{i-1,T}^2$. The result is

$$F(n, Q^2/m^2) = \sum_{N=0}^{\infty} \prod_{i=0}^{N} \int_{m^2}^{k_{i+1,\mathrm{T}}^2} \frac{\mathrm{d}k_{i,\mathrm{T}}^2}{k_{i,\mathrm{T}}^2} \int_0^1 \mathrm{d}\zeta_i \, P_j(\zeta_i, \alpha_s(k_{i,\mathrm{T}}^2)) \times \delta(x - \textstyle\prod_{j=0}^{N} \zeta_j), \tag{14.93}$$

where we define $k_{N+1,\mathrm{T}} = Q^2$ to limit the ultraviolet region, while m^2 serves as an infrared cutoff with $m > \Lambda$. This procedure does not affect the overall logarithmic behavior of the integral, and our expression contains the correct leading logarithmic behavior at each order. Equation (14.93) explicitly shows the evolution of partons before the hard scattering, up to the point when the struck parton is off-shell by a scale of nearly Q^2. A close relation to the one-loop distribution, eq. (14.60), is clear. The scale μ in the distribution corresponds in the structure function to a natural cut-off in transverse momentum. In the parton distribution, μ is an adjustable parameter, the factorization scale, while in the structure function, the cut-off is automatic when $k_\mathrm{T} \geqslant Q$.

From eq. (14.93) we can rederive the quantitative scale breaking results of eqs. (14.82). To be specific, we specialize to the evolution of a fermion through the emission of vector particles. Then, to lowest order, $P_j = (\alpha_s/2\pi) P_{qq}$ as above. Under moments of x, the ζ-integrals in (14.93) factorize:

$$\int_0^1 \mathrm{d}x \, x^{n-1} F(n, Q^2/m^2) = \sum_{N=0}^{\infty} \prod_{i=0}^{N} \int_{m^2}^{k_{i+1,\mathrm{T}}^2} \frac{\mathrm{d}k_{i,\mathrm{T}}^2}{k_{i,\mathrm{T}}^2} \left(\frac{\alpha_s(k_{i,\mathrm{T}}^2)}{2\pi} \right) \int_0^1 \mathrm{d}\zeta_i \, \zeta_i^{n-1} P_j(\zeta_i). \tag{14.94}$$

Once again, the two cases of interest are the fixed-point case, $\alpha_s(k_{i,\mathrm{T}}^2) \to \alpha_0$, and the case where there is asymptotic freedom, $\alpha_s(k_{i,\mathrm{T}}^2) \approx 4\pi[|\beta_1| \ln(k_{i,\mathrm{T}}^2)/\Lambda^2)]^{-1}$. For both, the transverse integrals can be done explicitly:

$$\prod_{i=0}^{N} \int_{m^2}^{k_{i+1,\mathrm{T}}^2} \frac{\mathrm{d}k_{i,\mathrm{T}}^2}{k_{i,\mathrm{T}}^2} \left(\frac{\alpha_0}{2\pi} \right) = \left(\frac{\alpha_0}{2\pi} \right)^N \frac{1}{N!} \ln^N \left(\frac{Q^2}{m^2} \right),$$

$$\prod_{i=0}^{N} \int_{m^2}^{k_{i+1,\mathrm{T}}^2} \frac{\mathrm{d}k_{i,\mathrm{T}}^2}{k_{i,\mathrm{T}}^2} \left(\frac{\alpha_s(k_{i,\mathrm{T}}^2)}{2\pi} \right) = \left(\frac{2}{|\beta_1|} \right)^N \frac{1}{N!} \left\{ \ln \left[\frac{\ln(Q^2/\Lambda^2)}{\ln(m^2/\Lambda^2)} \right] \right\}^N. \tag{14.95}$$

By substituting these results in eq. (14.94), and summing over N, we derive the Q^2-dependence of eqs. (14.82). The weaker scale breaking in an asymptotically free theory is due to the effective coupling $\alpha_s(k_T{}^2)$. When $\alpha_s(k_T{}^2)$ decreases with $k_T{}^2$, it weakens the basic splitting that drives evolution. Without this suppression, highly virtual states are more strongly populated, and there is stronger scale breaking.

14.5 The operator product expansion

Many of the results of the factorization program outlined above, especially in deeply inelastic scattering, were originally derived by a different technique, the operator product expansion (Wilson 1969). This concept has found wide-ranging applications, and it is more than worth the effort to derive (a part of) the operator product expansion, by assuming our factorization theorem.

Operator product expansion for T_2

In Section 14.1, we observed that the forward Compton structure functions $T_a(\omega, Q^2)$, eq. (14.9), have the analytic structure shown in Fig. 14.2(*a*) in the $\omega = 1/x$ plane. Since $T_2(\omega)$, for example, is analytic at $\omega = 0$, it can be Taylor-expanded there:

$$T_2(\omega, Q^2) = \sum_{n=0}^{\infty} (1/n!)\omega^n(\mathrm{d}^n T_2(\omega, Q^2)/\mathrm{d}\omega^n)|_{\omega=0}. \qquad (14.96)$$

In addition, $T_2(\omega, Q^2)$ obeys the factorization eq. (14.51). Then the expansion of $T_2(\omega, Q^2)$ in ω about $\omega = 0$ is given by the expansion of the hard scattering functions, $\mathcal{T}_{2,i}(\xi\omega, Q^2)$ in eq. (14.51). So, we can rewrite eq. (14.96) as

$$T_2(\omega, Q^2) = \sum_{n=0}^{\infty} \sum_{i=f,g} E_{n,i}(q, \mu, \alpha_s(\mu^2))(p^+)^n v_{n,i}(\epsilon, \alpha_s(\mu^2)) + \bar{R}(\omega, Q), \qquad (14.97\text{a})$$

where we define

$$E_{n,i}(q, \mu, \alpha_s(\mu^2)) \equiv (2q^-/Q^2)^n(1/n!)(\mathrm{d}^n/\mathrm{d}\zeta^n)\mathcal{T}_{2,i}(\zeta, Q^2/\mu^2, \alpha_s(\mu^2))|_{\zeta=0}, \qquad (14.97\text{b})$$

$$\begin{aligned}(p^+)^n v_{n,f}(\epsilon, \alpha_s(\mu^2)) & \\ &\equiv (p^+)^n \int_{-1}^{1} \mathrm{d}\xi\, \xi^n \psi_{f/h}(\xi, \epsilon, \alpha_s(\mu^2)) \\ &= \frac{1}{p^+}\int_{-\infty}^{\infty} \mathrm{d}^4 k (k^+)^n \operatorname{tr}[\tfrac{1}{2}\gamma^+ \chi_{f/h}(k^\mu, \epsilon, \alpha_s(\mu^2))], \\ &= \frac{1}{p^+}\tfrac{1}{2}\sum_\sigma \langle h(p, \sigma)|\bar{q}_f(0)\tfrac{1}{2}\gamma^+(\mathrm{i}\partial^+)^n q_f(0)|h(p, \sigma)\rangle_{A^+=0}, \qquad (14.97\text{c})\end{aligned}$$

$$(p^+)^n v_{n,g}(\epsilon, \alpha_s(\mu^2))$$
$$\equiv \frac{1}{2(p^+)^2} \sum_\sigma \langle h(p, \sigma)| F^+{}_\alpha(0)(\mathrm{i}\partial^+)^n F^{\alpha+}(0)|h(p, \sigma)\rangle_{A^+=0}. \quad (14.97\text{d})$$

(We specialize for now to the $A^+ = 0$ gauge.) In the second form for $v_{n,f}$ we have changed variables from ξ to k^+, and have extended the k^+ integral (see eq. (14.46)) to infinity, because the 'jet' function $\chi_{f/h}(\xi)$ vanishes for $|\xi| > 1$, as shown in Section 14.3. In the third form, we recognize that $v_{n,f}$ is the matrix element of a *local* operator, once it is separated from its convolution with the hard part by the expansion. Similar reasoning holds for the gluon matrix element. The $E_{n,i}$, like the short-distance functions in factorization theorems, are called coefficient functions.

In summary, we have expanded a structure function of the nonlocal operator

$$\int \mathrm{d}^4x\, \mathrm{e}^{\mathrm{i}q\cdot x} \langle h(p, \sigma)| T[j_\mu(x) j_\nu(0)]|h(p, \sigma)\rangle,$$

eq. (14.9), as a power series in ω. The terms in this expansion are short-distance coefficient functions times the matrix elements of *local* operators. Any such series is called an *operator product expansion.*

Fourier transform

The presence of local operators in eqs. (14.97) suggests that we reformulate the calculation in configuration space directly in terms of the original product of currents. To simplify matters, let us consider the trace of the product of electromagnetic currents. The same reasoning as above then gives (still in momentum space),

$$\int \mathrm{d}^4x\, \mathrm{e}^{\mathrm{i}q\cdot x} \langle h(p, \sigma)| T[j^\beta(x) j_\beta(0)]|h(p, \sigma)\rangle$$
$$= \sum_{n=0}^{\infty} \sum_{i=f,g} \mathcal{H}_{n,i}(q, \mu, \alpha_s(\mu^2))(p^+)^n v_{n,i}(\epsilon, \alpha_s(\mu^2)) + \bar{R}^{(s)}(\omega, Q) \quad (14.98)$$

where the $\mathcal{H}_{n,i}(q, \mu, \alpha_s(\mu^2))$ are specified by analogy with eq. (14.97b). Just as the $E_{n,i}$, they are of the form $(2q^-/Q^2)^n$ times a function of Q^2 only. We now define a new function of Q^2 only, $\bar{H}_{n,i}(Q^2, \mu^2, \alpha_s(\mu^2))$, by the relation

$$\mathcal{H}_{n,i}(q, \mu, \alpha_s(\mu^2)) = (-\mathrm{i}\partial/\partial q^+)^n \bar{H}_{n,i}(Q^2, \mu^2, \alpha_s(\mu^2)). \quad (14.99)$$

This makes it easier to take the Fourier transform of eq. (14.98) and put the matrix element back into configuration space. The result is

$$\langle h(p,\sigma)|T[j^\beta(x)j_\beta(0)]|h(p,\sigma)\rangle$$
$$= \sum_{n=0}^{\infty}\sum_{i=f,g} H_{n,i}(x^2,\mu^2,\alpha_s(\mu^2))(x_{\mu_1}\ldots x_{\mu_n})\langle h(p,\sigma)|O_i{}^{\mu_1\ldots\mu_n}(0)|h(p,\sigma)\rangle$$
$$+ R^{(s)}(x,p), \tag{14.100}$$

where we have returned to covariant notation, and where

$$H_{n,i}(x^2,\mu^2,\alpha_s(\mu^2)) \equiv \int \mathrm{d}^4q\,(2\pi)^{-4}\mathrm{e}^{-\mathrm{i}q\cdot x}\,\bar{H}_{n,i}(Q^2,\mu^2,\alpha_s(\mu^2)), \tag{14.101a}$$

$$R^{(s)}(x,p) \equiv \int \mathrm{d}^4q(2\pi)^{-4}\mathrm{e}^{-\mathrm{i}q\cdot x}\,\bar{R}^{(s)}(\omega,Q). \tag{14.101b}$$

In eq. (14.100), the tensor operators may be taken as

$$O_f{}^{\mu_1\ldots\mu_n}(0) = \bar{q}_f(0)\tfrac{1}{2}\gamma^{\mu_1}[\mathrm{iD}(A)^{\mu_2}]\ldots[\mathrm{iD}(A)^{\mu_n}]q_f(0),$$
$$O_g{}^{\mu_1\ldots\mu_n}(0) = F^{\mu_1}{}_{\alpha,a}(0)[\mathrm{iD}(A)^{\mu_2}]\ldots[\mathrm{iD}(A)^{\mu_{n-1}}]F^{\alpha\mu_n}{}_a(0), \tag{14.102}$$

in which we have generalized the matrix elements $v_{n,i}$ in eqs. (14.97c, d) from the $A^+=0$ gauge, by replacing each factor $\mathrm{i}\partial^\mu$ by the covariant derivative $\mathrm{iD}^\mu(A) = (\mathrm{i}\partial - g_sA)^\mu$. This makes the matrix elements, and therefore the coefficient functions $H_{n,i}$, gauge invariant. The matrix elements in eq. (14.100) can only be proportional to $p^\mu = p^+\delta_{\mu+}$, so that the covariant notation is exact.

It is possible to generalize the result given in eq. (14.100) both to arbitrary matrix elements *and* beyond the leading power in Q^2 (Zimmerman r1970, 1973, Brandt & Preparata 1971, Frishman 1970, Yndurain r1983). As a result, we may promote eq. (14.100) from an equality between matrix elements to the operator relation

$$T[j^\beta(x)j_\beta(0)] = \sum_{n=0}^{\infty}\sum_{I} H_{n,I}(x^2,\mu^2,\alpha_s(\mu^2))(x_{\mu_1}\ldots x_{\mu_n})O^{(I,n)\mu_1\ldots\mu_n}(0). \tag{14.103}$$

Here the sum over I extends to some independent set of local operators $O^{(I,n)}$ having the quantum numbers of the product of the currents and n (symmetric) tensor indices. n is often referred to as the *spin* of the operator (exercise 5.3). Even without specifying in detail what such a set of operators actually is, we can see what distinguishes the operators that give leading behavior to the structure functions from those that contribute to $R^{(s)}$ in eq. (14.100).

Light-cone singularities and twist

By comparing dimensions on either side of eq. (14.103), we conclude that for the operators in eqs. (14.102)

$$H_{n,I}(x^2, \mu^2, \alpha_s(\mu^2)) = (x^2)^{-2} h_{n,I}(x^2\mu^2, \alpha_s(\mu^2)), \qquad (14.104)$$

where $h_{n,I}(x^2\mu^2, \alpha_s(\mu^2))$ is a dimensionless function. (Minimal subtraction methods (Duncan & Furmanski 1983, Tkachov 1983, Llewellyn Smith & de Vries 1988) have been developed to define these short-distance coefficient functions.) As this example shows, the coefficient functions $H_{n,I}$ are generally singular on the light-cone. In fact, an x^{-4} singularity is necessary to produce approximate scaling behavior in momentum space, since the Fourier transform of $(x^2)^{-m}$ behaves as $(Q^2)^{m-2}$ for large Q. Remainder terms like $R^{(s)}(x, p)$, associated with corrections that decrease at least as fast as Q^{-2}, are less singular than x^{-4} on the light-cone; for instance,

$$(2\pi)^{-4}\int \mathrm{d}^4q\, \mathrm{e}^{-\mathrm{i}q\cdot x}(Q^2 + \mathrm{i}\epsilon)^{-1} = \mathrm{i}(2\pi)^{-2}(x^2 - \mathrm{i}\epsilon)^{-1}. \qquad (14.105)$$

Equation (14.103) is thus often referred to as a *light-cone expansion* for the operator product of currents. At truly short distances, however, where each x^μ is of the order of $\sqrt{x^2}$, the size of the terms in eq. (14.103) decreases with n. In this *short-distance* limit, it is the overall dimension of the combination $H_{n,I}\,(\prod_i x^{\mu_i})$ in eq. (14.103) that determines the importance of a given term, while in the light-cone expansion it is the dimension of the coefficient function $H_{n,I}$ alone that matters. The relevant quantities are thus in the short-distance limit

$$\dim(H_{n,I}\textstyle\prod_{i=1}^{n} x^{\mu_i}) = 6 - \dim[O^{(I,n)}], \qquad (14.106a)$$

and in the light-cone expansion

$$\dim(H_{n,I}) = 6 - \{\dim[O^{(I,n)}] - n\}. \qquad (14.106b)$$

The combination $\dim[O^{(I,n)}] - n$, which equals dimension $-$ spin, is usually referred to as the *twist* of the operator $O^{(I,n)}$ (Gross & Trieman 1971). In deeply inelastic scattering it is, as in eqs. (14.97), the operators of twist 2 that determine the (approximately scaling) leading behavior of the structure functions. *Higher-twist* operators govern nonleading powers in Q^2; they often involve more than two fields: for example, $(\bar\psi\gamma^+\psi)(\bar\psi\gamma^+\psi)$ has twist 4. This suggests that a factorization picture, based on the interactions of multiple partons, might describe Q^{-2} behavior. This is true in unpolarized deeply inelastic scattering (Ellis, Furmanski & Petronzio 1983, Jaffe 1983), as well as in the Drell–Yan process (Qiu & Sterman 1991).

Moments and the operator product expansion

The operator product expansion has a surprising relationship to the anomalous dimensions γ_n associated with moments of structure functions (Symanzik 1971, Christ, Hasslacher & Mueller 1972). To see how this comes about, we return to the analytic structure of $T_{\mu\nu}$, as exhibited in Fig. 14.2. Derivatives of $T_{\mu\nu}$ at the origin in the Taylor expansion, eq. (14.96) can be re-expressed as contour integrals in the ω-plane,

$$T_2{}^{(n)}(Q^2) \equiv \frac{1}{n!}\left(\frac{\mathrm{d}^n T_2}{\mathrm{d}\omega^n}\right)\bigg|_{\omega=0} = \frac{1}{2\pi\mathrm{i}}\int_{C_0} \mathrm{d}\omega\, \omega^{-n-1} T_2(\omega, Q^2), \tag{14.107}$$

where C_0 is any contour about the origin, as in Fig. 14.2(b). Applying Cauchy's theorem, we can expand C_0 into C_1, which encloses the branch cuts. The discontinuity across the branch cuts is exactly the imaginary part of $T_{\mu\nu}$ (see the reasoning for the vacuum expectation value of the product of currents, eq. (13.20)). Then, using the symmetry, eq. (14.11), between negative and positive ω, and the relation given in eq. (14.9) between W_2 and the imaginary part of T_2, we derive

$$\frac{1}{n!}\, T_2{}^{(n)}(Q^2) = \frac{1}{\pi}\int_0^1 \mathrm{d}x\, x^{n-1} W_2(x, Q^2) \quad (n \text{ even}), \tag{14.108}$$

where we have re-expressed the integral in terms of the variable $x = 1/\omega$. Now consider again the expansion of T_2 in ω, eq. (14.97a). By identifying the coefficient of ω^n in the latter with the expression for $T_2{}^{(n)}$ in eq. (14.108), we get a new expression for the even moments of W_2:

$$\frac{1}{\pi}\int_0^1 \mathrm{d}x\, x^{n-1} W_2(x, Q^2) = \omega^{-n} \sum_{i=f,g} E_{n,i}(q, \mu, \alpha_s(\mu^2))(p^+)^n v_{n,i}(\epsilon, \alpha_s(\mu^2)). \tag{14.109}$$

Referring to the definitions of $v_{n,i}(\epsilon, \alpha_s(\mu^2))$ in eqs. (14.97), we see that moments of W_2 with even n are proportional to the matrix elements of local operators in the operator produce expansion. As in eq. (14.77b), the renormalization-independence of the physical (in this case, singlet) moment leads to the (matrix) renormalization group equations:

$$[\mu\partial/\partial\mu + \beta\partial/\partial g_s]E_{n,i}(q, \mu, \alpha_s(\mu^2)) = [\gamma_n{}^{(s)}]_{ij} E_{n,j}(q, \mu, \alpha_s(\mu^2)) \tag{14.110}$$
$$[\mu\partial/\partial\mu + \beta\partial/\partial g_s]v_{n,i}(\epsilon, \alpha_s(\mu^2)) = -[\gamma_n{}^{(s)}]_{ij} v_{n,j}(\epsilon, \alpha_s(\mu^2)).$$

From the second of these equations, we see that the $[\gamma_n{}^{(s)}]_{ij}$ form an anomalous-dimension *matrix* for the set of twist 2 operators in eqs. (14.102), which mix under renormalization as discussed in Section 11.2:

$$\begin{gathered}(Z_n)_{ji}[O_i{}^{\mu_1\ldots\mu_n}(x)]_{\mathrm{R}} = [O_j{}^{\mu_1\ldots\mu_n}(x)]_0,\\ [\gamma_n{}^{(s)}]_{ij} = [Z_n{}^{-1}(\mu\partial/\partial\mu)Z_n]_{ij}.\end{gathered} \tag{14.111}$$

Note that the direct relation between moments and operators holds only for even n. Similar relations hold for the moments of the nonsinglet distributions, where the operators of eq. (14.102) are replaced by

$$O_{\rm ns}{}^{\mu_1\ldots\mu_n}(0) = \sum_f \bar{q}_f(0)\tfrac{1}{2}\gamma^{\mu_1}[\mathrm{i}D(A)^{\mu_2}]\ldots[\mathrm{i}D(A)^{\mu_n}]t_f q_f(0), \quad (14.112)$$

where t_f is any diagonal generator in the quark representation of the flavor symmetry. (In $F_2{}^{(\mathrm{p-n})}$, for instance, the relevant t_f would be the third component of iosopin.) In this case, of course, there is no operator mixing.

Vacuum expectation of the operator product

Another important application of the operator product expansion is to the total $\mathrm{e^+e^-}$ annihilation cross section. By eqs. (13.16) and (13.21), it is determined by the vacuum expectation value of the time-ordered product of electromagnetic currents. Again, specializing for simplicity to the scalar product, we can apply eq. (14.103) to obtain

$$\begin{aligned}\langle 0|T[j^\beta(x)j_\beta(0)]|0\rangle \\ &= \sum_I H_{0,I}(x^2, \mu^2, \alpha_s(\mu^2))\langle 0|O^{(I,0)}(0)|0\rangle \\ &= H_{0,1}(x^2, \mu^2, \alpha_s(\mu^2))\langle 0|\mathcal{I}|0\rangle + H_{0,q}(x^2, \mu^2, \alpha_s(\mu^2))\langle 0|\bar{q}(0)q(0)|0\rangle \\ &\quad + H_{0,F}(x^2, \mu^2, \alpha_s(\mu^2))\langle 0|F^{\mu\nu}{}_a(0)F_{\mu\nu,a}(0)|0\rangle + \ldots, \quad (14.113)\end{aligned}$$

where $\mathcal{I}$ is the unit operator, and where only spin-zero ($n=0$) operators contribute, because the momentum of the vacuum is zero. This operator is present in the deeply-inelastic-scattering operator product expansion, but does not connect the hadronic states to the currents in a nontrivial way. Thus is did not appear in our derivation of the expansion for matrix elements, eq. (14.100).

The first corrections to the (short-distance) expansion in eq. (14.113) are given by the three-dimensional operator $\bar{q}q$ and the four-dimensional operator F^2. The leading behavior in momentum space, proportional to q^2, is entirely determined by $H_{0,1}$, which behaves as $(x^2)^{-3}$ on the light cone. This is the coefficient that we compute in perturbation theory (Section 12.4). The vacuum expectation values of $\bar{q}q$ and F^2 determine power corrections to this result. $H_{0,q}$, however, vanishes in the massless limit because of chiral symmetry in QCD. These quantities are nonperturbative, since our perturbation theory is, from the outset, an expansion about zero field.

Exercises

14.1 Show that diagrams that contribute to $T_{\mu\nu}(\omega, q^2)$, eqs. (14.9), are analytic in ω for $q^2 < 0$ except for the normal thresholds of eq. (14.10). Assume that the state $|h(p, \sigma)\rangle$ is stable.

14.2 Use eqs. (14.9) to prove that the structure functions $T_a(\omega, Q^2)$ satisfy the symmetry property, eq. (14.11).

14.3 (a) Using the Landau equations, (13.5), show that the reduced diagrams of pinch surfaces for $T_{\mu\nu}$ with $p^2 = 0$, $q^2 < 0$ are of the type shown in Fig. 14.4 for physical gauges. (b) Carry out power counting for these pinch surfaces as in Section 13.4, and show that in a physical gauge, divergences are at worst logarithmic, and that, for leading pinch surfaces, no soft lines attach to the hard scattering subdiagram.

14.4 Prove that for $|\xi| > 1$ in Fig. 14.4 there is always a loop whose minus integral vanishes.

14.5 (a) Show that in the $A^+ = 0$ gauge the analogue of eq. (14.50b) for gluons may be represented as

$$\psi_{g/h}{}^{(\chi_f)}(\xi) \equiv \int_{n(\chi_f)} \mathrm{d}^4 k\, \delta(\xi - k^+/p^+)\, \mathrm{tr}\, [\mathrm{d}_{\mu\nu}\chi^{\mu\nu}{}_{g/h}(k, p, \epsilon)],$$

where $d_{11} = d_{22} = 1$, $d_{\mu\nu} = 0$ otherwise. (b) Show that the above expression is a contribution to the matrix element (14.57) that defines the MS gluon distribution in $A^+ = 0$ gauge.

14.6 The definition (14.53) for the quark density is a natural choice that is particularly useful for formal manipulations, but it is not the only choice. (a) Show that any other set of parton densities, related to $\phi_{i/h}(x, \mu^2)$ by

$$\phi'_{i/h}(x, \epsilon, \alpha_s(\mu^2)) = \sum_j \int_x^1 (\mathrm{d}\zeta/\zeta) D_{ij}(x/\zeta, \alpha_s(\mu^2)) \phi_{j/h}(\zeta, \epsilon, \alpha_s(\mu^2))$$

contains the same physical information as does the set $\phi_{j/h}$, so long as $D_{ij}(x/\zeta, \alpha_s(\mu^2))$ is short-distance dominated. (b) Determine a choice of $D_{ij}(x/\zeta, \alpha_s(q^2))$ that results in the following direct relation between F_2 and quark structure functions to all orders in perturbation theory:

$$(1/x) F_2{}^{(h)}(x, Q^2) = \sum_f Q_f{}^2 \phi'_{f/h}(x, \epsilon, \alpha_s(Q^2))$$

(Altarelli, Ellis & Martinelli 1979).

14.7 (a) Define an MS parton distribution appropriate to the theory $\phi^3{}_6$, and calculate it to one loop. (b) Calculate the anomalous dimensions associated with the moments of this distribution.

14.8 The path-ordered integral in the quark distribution, eq. (14.53b), may be defined as follows. Consider two points y^μ and Y^μ and a path $z^\mu(t)$ between these two points, where t is a parameter, defined such that

$$z(0) = y, \quad z(1) = Y.$$

The path-ordered exponential is defined for every such pair of points and path by the expression (see also exercise 2.9)

$$
\begin{aligned}
U(Y, y) & \\
&= P\exp\left\{ig\int_0^1 dt[dz^\mu(t)/dt]A_\mu(z(t))\right\} \\
&= 1 + ig\int_0^1 dt[dz^\mu(t)/dt]A_\mu(z(t)) \\
&\quad + (ig)^2\int_0^1 dt_2[dz^\mu(t_2)/dt_2]A_\mu(z(t_2))\int_0^{t_2} dt_1[dz^\mu(t_1)/dt_1]A_\mu(z(t_1)) + O(g^3),
\end{aligned}
$$

in which the nth term is the product of n integrals that divide up the path $z^\mu(t)$, with increasing t to the left. (a) show that under a gauge transformation $A'_\mu(x) = -\partial_\mu\alpha(x) + ig[\alpha(x), A_\mu(x)]$, the ordered exponential behaves as

$$U'(Y, y) = ig[\alpha(Y)U(Y, y) - U(Y, y)\alpha(y)],$$

and hence that (14.53b) is gauge invariant. (b) Show that as a matrix, $U(Y, y)$ is unitary.

14.9 Starting from eq. (14.68), derive the real-gluon contribution in eq. (14.70).

14.10 Compute the Born structure functions for neutrino–quark scattering, and show that in the parton model, valence-quark distributions may be determined from neutrino and antineutrino scattering (see Close r1979).

14.11 Show that the corrections to the leading scale breaking behavior of eqs. (14.82) are suppressed in the way described after eq. (14.82b).

15

Epilogue: Bound states and the limitations of perturbation theory

Factorization enables us to derive inclusive high-energy cross sections directly from the perturbative expansion in terms of elementary fields. Yet, the ubiquity of bound states in hadronic physics demands that we further bridge the gap between them and elementary fields. Similarly, to describe bound states such as positronium in QED fully, it is necessary to include field-theoretic corrections.

Under some circumstances, bound state masses and matrix elements are well understood in terms of infinite sums of Feynman diagrams, or of solutions to integral equations based on perturbation theory. The perturbative expansion can suggest expressions for bound state matrix elements, even when the corresponding wave functions are unknown. Certain quantities, however, do not seem accessible to perturbation theory, even summed to all orders. Among these are the matrix elements that appear in the operator product expansion in QCD, and the masses of light hadrons.

The Bethe–Salpeter equation and wave functions

The derivations that led to the reduction formulas eqs. (2.97) and (7.88) are not limited to elementary particles. It is easy to check that any asymptotic single-particle state produces a pole in any Green function with fields of the appropriate quantum numbers. Let us see, however, how such a pole can be generated in the language of perturbation theory. Consider, for example, the complex scalar three-point function,

$$G_3(K, k) = \int d^4x_1 d^4x_2 \exp[i(\tfrac{1}{2}K + k)\cdot x_1] \exp[i(\tfrac{1}{2}K - k)\cdot x_2] \times \langle 0|T[\phi^\dagger(x_2)\phi(x_1)|\phi(0)|^2]|0\rangle. \tag{15.1}$$

K^μ is the momentum flowing into the vertex associated with the composite field $|\phi(0)|^2 = \phi^\dagger(0)\phi(0)$ and k^μ is the relative momentum of the two external scalar lines. Our aim is to show how a two-particle bound state can appear in the K^μ-channel, a pole at $K^2 = M_B{}^2$, $M_B < 2m$. For K^2 and

$(\frac{1}{2}K \pm k)^2 > 0$, $G_3(K, k)$ has singularities only at normal thresholds, starting at $K^2 = 4m^2$, at any finite order in perturbation theory (Section 13.4). Thus, a bound state can only appear nonperturbatively.

To begin with, we observe that $G_3(K, k)$ satisfies a *Dyson–Schwinger equation* (Dyson 1949, Schwinger 1951a), illustrated in Fig. 15.1(*a*), of the form

$$G_3(K, k) = \mathrm{i}[(\tfrac{1}{2}K + k)^2 - m^2 + \mathrm{i}\epsilon]^{-1}\, \mathrm{i}[(\tfrac{1}{2}K - k)^2 - m^2 + \mathrm{i}\epsilon]^{-1} \times \left[1 + (2\pi)^{-4} \int \mathrm{d}^4 q\, \bar{\Gamma}_4(K, k, q) G_3(K, q)\right], \tag{15.2}$$

where $\bar{\Gamma}_4(K, k, q)$ is the sum of all truncated four-point diagrams that are two-particle irreducible in the K^μ-channel. That is, $\bar{\Gamma}_4$ cannot be disconnected by cutting two internal lines whose total momenta is K^μ. Examples of contributions to $\bar{\Gamma}_4$ are shown in Fig. 15.1(*b*). Equation (15.2) is an integral equation that describes the evolution of the two scalar particles produced at the composite vertex as they interact repeatedly through the 'potential' $\bar{\Gamma}_4$. Thus far, the equation is not fully defined, as $\bar{\Gamma}_4$ is itself an infinite sum in perturbation theory. If, however, we approximate $\bar{\Gamma}_4$ by some definite function, the equation becomes well defined. In this form, eq. (15.2) is known as a *Bethe–Salpeter equation* (Bethe & Salpeter 1951). To be specific, we may consider the Bethe–Salpeter equation for a theory in the *ladder approximation*, in which $\bar{\Gamma}_4$ is approximated by a single exchanged line (the first diagram in Fig. 15.1(*b*)). With QED in mind, we may take the exchanged propagator to be an 'instantaneous Coulomb

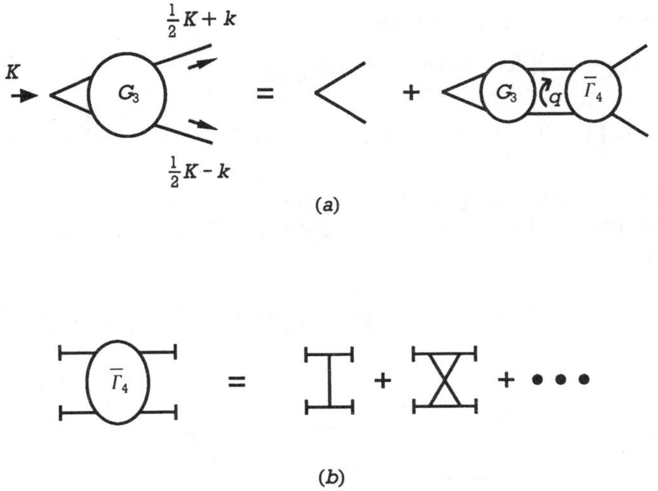

(*a*)

(*b*)

Fig. 15.1 Dyson–Schwinger equation for three-point function.

interaction' (exercise 7.8),

$$G_3(K, k) = \mathrm{i}[(\tfrac{1}{2}K + k)^2 - m^2 + \mathrm{i}\epsilon]^{-1}\, \mathrm{i}[(\tfrac{1}{2}K - k)^2 - m^2 + \mathrm{i}\epsilon]^{-1} \times \left[1 + (2\pi)^{-4} \int \mathrm{d}^4 q\, (\mathrm{i}g^2)|\mathbf{k} - \mathbf{q}|^{-2}\, G_3(K, q)\right], \qquad (15.3)$$

where g is the coupling constant. This corresponds to the infinite sum of diagrams shown in Fig. 15.2, in which the interaction is respresented by a dashed line.

In the following, we simplify even further by specializing to the nonrelativistic regime, in which $K_0 - 2m \ll m$ and $|\mathbf{k}| \ll m$. The integral over the relative energy k_0 of G_3 is of interest, because it measures the correlation between the two ϕ-fields at equal 'relative' time, $x_{1,0} = x_{2,0}$ in eq. (15.1). It is therefore suggestive of a nonrelativistic wave function. In the K^μ rest frame, $K^\mu \equiv (K_0, \mathbf{0})$, this integral is

$$\begin{aligned}\Phi(K_0, \mathbf{k}) &\equiv (2\pi)^{-1} \int_{-\infty}^{\infty} \mathrm{d}k_0\, G_3(K, k) \\ &= -(2\pi)^{-1} \int_{-\infty}^{\infty} \mathrm{d}k_0 [(\tfrac{1}{4}K_0{}^2 - m^2) + K_0 k_0 + k^2 + \mathrm{i}\epsilon]^{-1} \\ &\quad \times [(\tfrac{1}{4}K_0{}^2 - m^2) - K_0 k_0 + k^2 + \mathrm{i}\epsilon]^{-1} \\ &\quad \times \left[1 + (2\pi)^{-3} \int \mathrm{d}^3\mathbf{q}\, (\mathrm{i}g^2)|\mathbf{k} - \mathbf{q}|^{-2} \Phi(K_0, \mathbf{q})\right]. \qquad (15.4)\end{aligned}$$

There are four poles in the k_0 integral, two near $\pm (1/K_0)(\tfrac{1}{4}K_0{}^2 - m^2 - |\mathbf{k}|^2 + \mathrm{i}\epsilon)$ and two near $\pm (K_0 - \mathrm{i}\epsilon)$. The first pair traps the k_0 contour in the 'threshold' region, $k^2 \approx K \cdot k$, discussed at the end of Section 13.3. Closing the k_0 contour at infinity, and picking up one of the poles near the origin, gives

$$\Phi(K_0, \mathbf{k}) = (\mathrm{i}/4m^2)[K_0 - 2m - \mathbf{k}^2/m]^{-1} \times \left[1 + (2\pi)^{-3} \int \mathrm{d}^3\mathbf{q}\, (\mathrm{i}g^2)|\mathbf{k} - \mathbf{q}|^{-2}\, \Phi(K_0, \mathbf{q})\right] + \cdots, \qquad (15.5)$$

where corrections due to the pole at $|k_0| = |K_0|$ in the same half-plane are finite at the threshold $K_0 = 2m$. Neglecting these corrections, we rewrite eq. (15.5) as

$$[K_0 - 2m - \mathbf{k}^2/m]\, \Phi(K_0, \mathbf{k}) + (g^2/4m^2)(2\pi)^{-3} \int \mathrm{d}^3\mathbf{q}\, |\mathbf{k} - \mathbf{q}|^{-2}]\, \Phi(K_0, \mathbf{q}) = \mathrm{i}/4m^2. \qquad (15.6)$$

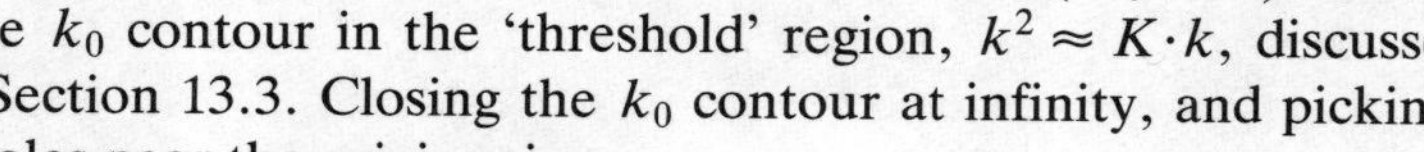

Fig. 15.2 Ladder approximation.

Now consider a solution $\psi(\mathbf{q})$ to the following momentum-space Schrödinger equation:

$$(\mathbf{k}^2/m)\psi(\mathbf{k}) - (g^2/4m^2)(2\pi)^{-3}\int d^3\mathbf{q}\,|\mathbf{k}-\mathbf{q}|^{-2}\psi(\mathbf{q}) = \mathscr{E}\,\psi(\mathbf{k}). \quad (15.7)$$

Here the energy $\mathscr{E}$ is negative. (This is the Schrödinger equation for a Coulomb force. Thus, we can be sure there are solutions, with the usual quantum numbers.) From $\psi(\mathbf{k})$ we can construct an approximate bound-state solution, accurate up to terms that are finite at the pole at $K_0 = 2m + \mathscr{E}$, where $\mathscr{E}$ is the binding energy:

$$\Phi(K_0, \mathbf{k}) = (\mathrm{i}/4m^2)\,[\psi(\mathbf{k})/[K_0 - M_\mathrm{B})], \quad M_\mathrm{B} \equiv 2m + \mathscr{E}. \quad (15.8)$$

By eq. (15.3), this pole also appears in the Green function $G_3(K, k)$:

$$G_3(K, k) = \mathrm{i}\Psi(K, k)/(K_0 - M_\mathrm{B}),$$
$$\Psi(K, k) = \mathrm{i}[(\tfrac{1}{2}K + k)^2 - m^2 + \mathrm{i}\epsilon]^{-1}\,\mathrm{i}[(\tfrac{1}{2}K - k)^2 - m^2 + \mathrm{i}\epsilon]^{-1} \quad (15.9)$$
$$\times [K_0 - M_\mathrm{B} + (\mathscr{E} - \mathbf{k}^2/m)\,\psi(\mathbf{k})].$$

In summary, nonperturbative poles may be generated in solutions to integral equations that originate in perturbation theory, such as the Bethe–Salpeter or, more generally, Dyson–Schwinger equation. The resulting poles come from an infinite series of diagrams; in our case, the ladder exchanges of Fig. 15.2. In addition, the residue at the pole, $\Psi(K, k)$, which gives the amplitude for the relative momenta of the constituents, is intimately related to the Schrodinger wave function $\psi(\mathbf{k})$. $\psi(\mathbf{k})$, and hence $\Psi(K, k)$, may be characterized by angular momentum, and, generalizing the scalar case, by spin. This approach to bound states has been extensively applied with great success to positronium in QED (Bodwin & Yennie 1978) and to the bound states of heavy quarks in QCD (Appelquist & Politzer 1975, Quigg & Rosner 1979, Kuhn & Zerwas r1988). Beyond the nonrelativistic approximation, of course, we must include wave functions for three or more constituents, but the principles remain the same.

Amplitudes for bound states

These observations enable us to formulate processes involving bound states in perturbation theory (Mandelstam 1955). A simple but nontrivial example is the vector form factor of a scalar bound state. For simplicity, we consider a real field. The form factor can be derived from the Green function

$$G_\mu{}^{(3)}(K_2, K_1)$$
$$= \int d^4y \exp(\mathrm{i}K_2\cdot y)\int d^4x \exp(-\mathrm{i}K_1\cdot x)\,\langle 0|T[\phi^2(y) j_\mu(0)\phi^2(x)]|0\rangle$$
$$= \prod_{i=1}^{2}\int d^4k_i\, G_3(K_i, k_i) T_\mu(\tfrac{1}{2}K_1 \pm k_1; \tfrac{1}{2}K_2 \pm k), \quad (15.10)$$

where $j_\mu(0)$ is the vector current, which is here supposed to act directly on the elementary fields with unit coupling. $G_\mu{}^{(3)}$ is given in perturbation theory by the diagrams in Fig. 15.3(*a*), in which the subdiagram T_μ is irreducible in both of the $K_i{}^2$ channels. As shown in Fig. 15.3(*b*) for a ϕ^3 theory, the perturbative expansion for T_μ begins at zeroth order in the coupling, where it may be written as

$$T_\mu{}^{(0)} = -\mathrm{i}j_\mu{}^{(0)}[(\tfrac{1}{2}K_1 - k_1)^2 - m^2] \times (2\pi)^4\delta^4[\tfrac{1}{2}K_1 - k_1 - (\tfrac{1}{2}K_2 - k_2)], \tag{15.11}$$

that is, as the product of one inverse propagator times a delta function, which reduces the integrals to a single loop.

The reduction process for $G_\mu{}^{(3)}(K_2, K_1)$ is carried out by repeating the steps that led to eq. (2.97) for elementary scalar fields. To proceed, we only need the analogue of eq. (4.15) for the single-particle pole in the ϕ^2 two-point function at $K^2 = M_\mathrm{B}{}^2$, which we can derive from the analogue of eq. (15.2) for a real scalar:

$$\begin{aligned} G_2(K) &= (2\pi)^{-4}\int \mathrm{d}^4k \int \mathrm{d}^4x_1\mathrm{d}^4x_2 \exp[\mathrm{i}(\tfrac{1}{2}K + k)\cdot x_1]\exp[\mathrm{i}(\tfrac{1}{2}K - k)\cdot x_2] \\ &\qquad \times \langle 0|T[\phi(x_2)\,\phi(x_1)\,\phi^2(0)]|0\rangle \\ &= \int \mathrm{d}^4u \exp(-\mathrm{i}K\cdot u)\langle 0|T[\phi^2(u)\,\phi^2(0)]|0\rangle \\ &\equiv \mathrm{i}R_\mathrm{B}\,(K^2 - M_\mathrm{B}{}^2 + \mathrm{i}\epsilon)^{-1} + \mathrm{O}(1). \end{aligned} \tag{15.12a}$$

Then, adopting the general form for the pole in G_3 given by the first equality in eqs. (15.9), we find

$$R_\mathrm{B} = 2M_\mathrm{B}\,(2\pi)^{-4}\int \mathrm{d}^4k\,\Psi(K, k). \tag{15.12b}$$

The bound state *S*-matrix element is now derived by summing to all orders in the G_3-subdiagrams in Fig. 15.3(*a*), and by applying the reduction formula eq. (2.97), with the value of R_B just derived:

$$\begin{aligned} S_\mu{}^{(3)}(K_2\,\text{out};\,K_1\,\text{in}) &= 2M_\mathrm{B}\left[(2\pi)^{-1}\int \mathrm{d}^4k\,\Psi(K, k)\right]^{-1} \\ &\quad \times \prod_{i=1}^{2}\int \mathrm{d}^4k_i\,\Psi(K_i, k_i)\,T_\mu(\tfrac{1}{2}K_1 \pm k_1;\,\tfrac{1}{2}K_2 \pm k_2). \end{aligned} \tag{15.13}$$

We shall not pursue bound-state scattering beyond this formula. To it must be added contributions from 'higher Fock states' with more than two constituents. We shall only note here that at high enough energies one may

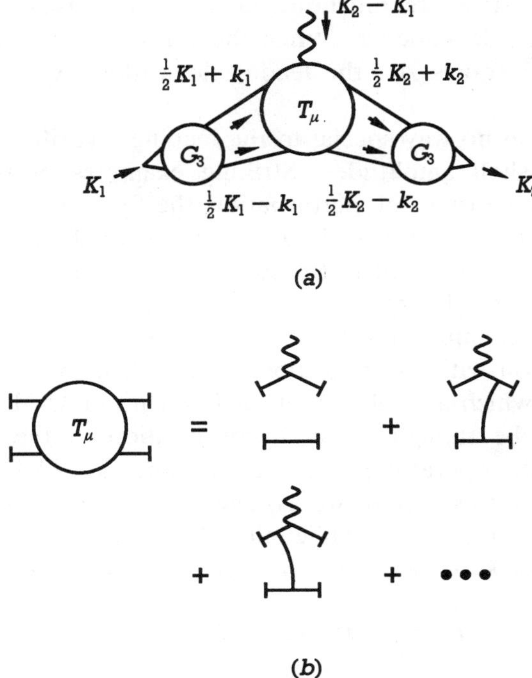

Fig. 15.3 Form factor for bound states.

apply reasoning in the style of the parton model to elastic scattering (Brodsky & Farrar 1973, Matveev, Muradyan & Tavkhelidze 1973). In this picture, only 'valence' constituents take part in a hard scattering, in which the analogue of T_μ in eq. (15.13) is treated in the Born approximation. In addition, it is possible to improve upon the parton picture by a factorization process, and to derive the evolution in momentum transfer of various elastic scattering processes (Farrar & Jackson 1979, Effremov & Radyushkin 1980a, Lepage & Brodsky 1980, Duncan & Mueller 1980, Brodsky & Lepage r1989). Other models of elastic scattering in QCD, at nonasymptotic energies, include contributions to T_μ at zeroth order (Nestorenko & Radyushkin 1982, 1983, Isgur & Lewellyn Smith 1989).

Convergence of the perturbative series

Bound state poles are, unfortunately, not the only examples of nonconvergence in the perturbative expansion. We already know that in QCD the perturbative effective coupling diverges at some value of the renormalization scale, as in eq. (12.94) at $\mu^2 = \Lambda^2$. Even when we take μ to be large, this phenomenon comes back to haunt us in any perturbative quantity that is sensitive to long distance effects. The leading logarithm expression for moments of the structure function, eq. (14.95), illustrates the failure of perturbative mass-dependence. When the transverse momentum cut-off is reached and m^2 equals Λ^2, the leading logarithm expression develops a singularity.

This effect is in no way special to the leading logarithm approximation, or even to on-shell amplitudes. Striking examples of the role of soft momenta at high orders of perturbation theory can be found even for Euclidean Green functions with arbitrary external momenta ('t Hooft 1977), which, as we showed in Section 13.3, are infrared finite order-by-order in perturbation theory.

Consider, for example, the pinch surface (Section 13.1) of a Euclidean two-point function with external momentum q^μ, $q^2 = -Q^2$, illustrated in Fig. 15.4(a), in which a single gluon, of momentum k^μ, has zero momentum. This may be thought of as a contribution to the matrix element $\langle 0|F^2(0)|0\rangle$ in the operator product expansion, eq. (14.103). To estimate the importance of this region, we assume that $0 < |k^\mu| < \epsilon Q$, for all components of k^μ, with $\epsilon < 1$, while all other lines in the diagram carry momenta of order $k^\mu = \mathrm{O}(Q)$. In this region, the integral is of the form

$$\begin{aligned} I_0 &= \int_0^{\epsilon Q} \mathrm{d}^4 k\, \Gamma_{\mu\nu}(q, k)(\delta_{\mu\nu}/k^2)k^2 \\ &= \pi^2 \Gamma_{\mu\mu}(q, 0) \int_0^{(\epsilon Q)^2} \mathrm{d}k^2\, k^2 + \cdots \\ &= \tfrac{1}{2}\pi^2 \Gamma_{\mu\mu}(q, 0)(\epsilon Q)^4 + \cdots, \end{aligned} \tag{15.14}$$

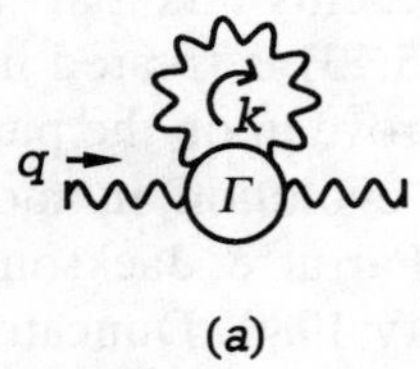

(a)

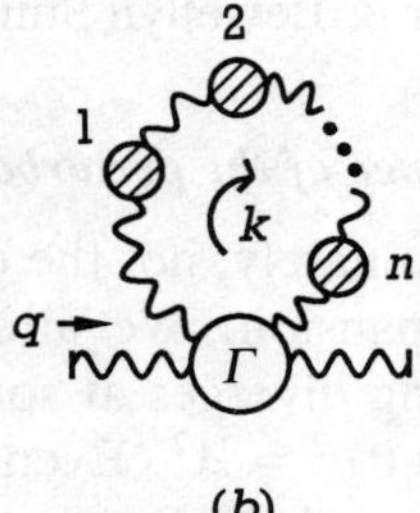

(b)

Fig. 15.4 Soft gluon corrections to infrared-finite process.

In the second equality, we have performed the three-dimensional Euclidean angular integrals, using eq. (9.37). $\Gamma_{\mu\nu}(q, k)$ represents the rest of the diagram, which, at this pinch surface, can be expanded in k^μ/Q. I_0 is a finite integral, and this region of momentum space is normally included in perturbative calculations of infrared safe quantities such as the photon self-energy. For example, a region of this type is included in eq. (12.86), the one-loop correction to the e^+e^- annihilation cross section.

Now consider the diagram of Fig. 15.4(*b*), in which the gluon has n one-loop self-energies. After renormalization, each of these self-energies has a logarithmic momentum-dependence, which for $SU(N)$ is, in Minkowski space,

$$\pi^{\mu\nu}(k) = \mathrm{i}(\alpha_s/3\pi)(5N/4 - T(\mathrm{F})n_f)(g^{\mu\nu}k^2 - k^\mu k^\nu)\ln(\mu^2/k^2) + \cdots, \tag{15.15}$$

with nonlogarithmic corrections. This result may be inferred easily from dimensional analysis and from the presence of a factor $\mu^{2\epsilon}$ in the gluon self-energy, whose ultraviolet pole is given in eq. (11.38). Assuming that $\Gamma_{\mu\nu}$ is itself transverse, the $k^\mu k^\nu$ terms vanish. Then, making the choice $\mu = \epsilon Q$, the diagram with n of these self-energy insertions behaves in Euclidean space as

$$\begin{aligned} I_n &= \pi^2 \Gamma_\mu{}^\mu(q, 0)[(\alpha_s/3\pi)(5N/4 - T(\mathrm{F})n_f)]^n \int_0^{(\epsilon Q)^2} \mathrm{d}k^2\, k^2 \ln^n[(\epsilon Q)^2/k^2] \\ &= 2^{-n-1}\pi^2 \Gamma_\mu{}^\mu(q, 0)[(\alpha_s/3\pi)(5N/4 - T(F)n_f)]^n(\epsilon Q)^4 \int_0^\infty \mathrm{d}x\, \mathrm{e}^{-x}x^n. \end{aligned} \tag{15.16}$$

In the second form, we have changed variables to $x = \ln[(\epsilon Q)^2/k^2]$. This is exactly the integral representation of the gamma function, eq. (9.31), so that

$$I_n = 2^{-n-1}\pi^2 \Gamma_\nu{}^\nu(q, 0)(\epsilon Q)^4[(\alpha_s/3\pi)(5N/4 - T(\mathrm{F})n_f)]^n \Gamma(n + 1), \tag{15.17}$$

with corrections that grow less rapidly with n. Thus, the single diagram in Fig. 15.4(*b*) is proportional to $n!$. Clearly, no matter how small the running coupling $\alpha_s((\epsilon Q)^2)$, the summation of these diagrams cannot converge unless we introduce an infrared cut-off. This is so, even though the diagrams are order-by-order infrared finite.

It is the conventional wisdom that in the absence of momentum space cut-offs all perturbative expansions in QCD diverge. This does not mean that they are useless, only that they are 'asymptotic expansions', whose higher orders can be expected to give approximations that improve up to a certain order for a fixed coupling, and then behave poorly. This problem is not unique to QCD, but seems to be characteristic of essentially all four-dimensional field theories (Dyson 1952, Lipatov 1977, Brézin, LeGuillou & Zinn-Justin 1977, Parisi 1979). Perhaps the most famous example can be found in QED. Consider the same set of diagrams as in Fig. 15.4(*b*), but this time in the abelian theory. Summing over n at fixed momentum, the effective photon propagator becomes

$$G^{\mu\nu}(p) = (-g^{\mu\nu}/p^2)\,[1 - (\alpha/3\pi)\ln(-p^2/m_e{}^2)]^{-1}. \qquad (15.18)$$

The propagator, including this set of diagrams, thus has a pole at the (fantastically large) Euclidean momentum $p^2 = -m_e{}^2 \exp(3\pi/\alpha)$ (Landau & Pomeranchuk 1955). It should be noted, however, that this and related observations all depend on special approximations and sets of diagrams, and that the true orders at which the putative asymptotic expansions begin to fail are unknown.

The essential assumption of the factorization program in QCD is that the short-distance behavior of the theory is perturbative. If this is so, calculations based on factorization can make sense, since factorization separates short-distance from long-distance regions in momentum space (Mueller 1985, Appell, Mackenzie & Sterman 1987). At any fixed order, soft gluons give power-suppressed finite corrections to infrared-safe quantities at high energy. If we sum to all orders, however, the perturbative series for any such power correction diverges, unless we include an explicit cut-off. Presumably, all this happens because, in the complete theory, infrared interactions produce new mass scales that are not present in perturbation theory. Examples of such 'dynamically generated' scales in QCD are thought to be the masses of the pion and proton (Appendix E). The failure of perturbation theory in the soft region goes hand in hand with the existence of these additional scales. Put differently, the failure of soft gluons to describe low-energy QCD is a sign that the vacuum state is itself essentially nonperturbative (Shifman, Vainshtein & Zakharov 1979, Rajaraman r1982, Shuryak r1988).

A few comments on dynamically generated masses will show why they cannot be computed in perturbation theory (Gross r1981). A mass M that is physically observable must obey the standard renormalization group equation

$$(\mu\,\mathrm{d}/\mathrm{d}\mu)M(\mu/\Lambda,\, g(\mu)) = 0, \qquad (15.19)$$

where we explicitly assume that M is independent of any mass scales apart from μ and Λ. Solving eq. (15.19), we find

$$M(\mu/\Lambda,\, g(\mu)) = (\mu/\mu_0)M(\mu_0/\Lambda,\, g(\mu_0))\exp\left[-\int_{g(\mu_0)}^{g(\mu)}\mathrm{d}g/\beta(g)\right]. \qquad (15.20)$$

Then the lowest-order beta function gives

$$M(\mu/\Lambda,\ g(\mu)) = (\mu/\mu_0)M(\mu_0/\Lambda,\ g(\mu_0))$$
$$\times \exp\{[8\pi^2/\beta^{(1)}][1/g^2(\mu_0) - 1/g^2(\mu)]\}. \qquad (15.21)$$

This expression has an essential singularity at $\bar{g}(\mu) = 0$, so that a perturbative expansion in powers of $\bar{g}$ is inappropriate. The issues of power corrections to high-energy cross sections, and of dynamically generated masses, thus signal a frontier for the applicability of perturbation theory.

Appendix A

Time evolution and the interaction picture

Time evolution

The time development, or dynamics, of a quantum mechanical theory is determined by its Hamiltonian $H(Q, P)$, itself an operator function of coordinates and momenta. We can summarize the time dependence of the matrix elements of any operator $A(Q, P)$ by the equation

$$-i\hbar(\mathrm{d}/\mathrm{d}t)\langle\psi|A(Q,P)|\phi\rangle = \langle\psi|[H(Q,P), A(Q,P)]|\phi\rangle. \quad \text{(A.1)}$$

The commutator is carried out at time t, and may be evaluated using the canonical relations, eqs. (2.9). In a sense, eq. (A.1) is sufficient, since all the physical information is in the matrix elements. For convenience, however, we would like to have separate equations for the time development of states and operators. To this end we write

$$-\mathrm{i}\hbar\,\mathrm{d}A/\mathrm{d}t = [M, A], \quad \mathrm{i}\hbar(\mathrm{d}/\mathrm{d}t)|\psi\rangle = N|\psi\rangle, \quad \text{(A.2)}$$

where $M + N = H$, and M and N are hermitian operators. A choice of M and N is said to define a *picture* of time development. The most common choices are the Heisenberg picture, $M = H$, $N = 0$, and the Schrödinger picture, $M = 0$, $N = H$. If H and A are not explicit functions of time themselves, eq. (A.2) can be solved trivially in these pictures. In the Heisenberg picture operators carry all the time dependence and states are constant. Then we have

$$-\mathrm{i}\hbar\,\mathrm{d}A(t)/\mathrm{d}t = [H, A(t)], \quad (\mathrm{d}/\mathrm{d}t)|\psi\rangle = 0,$$

$$A(t) = \exp[(\mathrm{i}/\hbar)H(t-t_0)]A(t_0)\exp[(-\mathrm{i}/\hbar)H(t-t_0)], \quad |\psi(t)\rangle = |\psi(t_0)\rangle, \quad \text{(A.3)}$$

for any fixed t_0. In the Schrödinger picture it is the other way around,

$$(\mathrm{d}/\mathrm{d}t)A(t) = 0, \quad \mathrm{i}\hbar(\mathrm{d}/\mathrm{d}t)|\psi(t)\rangle = H|\psi(t)\rangle,$$

$$A(t) = A(t_0), \quad |\psi(t)\rangle = \exp[(-\mathrm{i}/\hbar)H(t-t_0)]|\psi(t_0)\rangle. \quad \text{(A.4)}$$

Interaction picture

The interaction picture of time evolution led historically to the development of perturbation theory for fields (Tomonaga 1946, Schwinger 1948, 1949, Dyson 1949). It remains an important tool. Before going on, however, we should note that the interaction picture is not the only alternative to path integrals; a method inspired by the path integral, but which is in some sense intermediate between the two, was used by Feynmann (1949). Yet another formal approach may be found in Bogoliubov & Shirkov (r1980, Section 20). Nevertheless, the interaction picture and the path integral are sufficient to understand most of the literature.

In the interaction picture we use the freedom implicit in eq. (A.2) to allow operators to keep the time dependence of the Heisenberg representation of the *free* theory, enabling the states to take on a time dependence induced solely by the interaction Hamiltonian. This program is realized, in terms of eq. (A.2), when we choose $M = H_0$ and $N = H_{\mathrm{I}}$. The time development equations for operators $A(t)$ and states $|\psi(t)\rangle$ are given by

$$-\mathrm{i}\,\mathrm{d}A_{\mathrm{I}}(t)/\mathrm{d}t = [H_0, A_{\mathrm{I}}(t)],$$
$$\mathrm{i}(\mathrm{d}/\mathrm{d}t)|\psi(t)\rangle_{\mathrm{I}} = H_{\mathrm{I}}|n(t)\rangle_{\mathrm{I}}. \tag{A.5}$$

The subscript 'I' refers to the interaction picture. Here and below, we rescale to natural units (Section 2.3).

In eq. (A.5), H_0 and H_{I} are expressed in terms of the interaction-picture fields $\phi_{\mathrm{I}}(x)$ and $\pi_{\mathrm{I}}(x)$, which themselves satisfy eq. (A.5). Then the time development of $\phi_{\mathrm{I}}(x)$ is indistinguishable from that of the Heisenberg fields in the free-field theory, Chapter 2. If we Fourier-transform the fields, we can define interaction-picture creation and annihilation operators, which satisfy the same time development equations as in the free-field theory, eq. (2.39). As a result, H_0 is time-independent in the interaction picture. Then we can solve eq. (A.5) to get the time dependence of any operator $O_{\mathrm{I}}(t)$,

$$O_{\mathrm{I}}(t) = \exp[\mathrm{i}H_0(t - t')]O_{\mathrm{I}}(t')\exp[-\mathrm{i}H_0(t - t')], \tag{A.6}$$

which is also the same as in the Heisenberg picture for the free theory. Next, we solve eq. (A.5) for the time dependence of the interaction-picture states.

States and time ordering

Suppose we know a state of our system at time t. We can find a solution for all times in the form

$$|n(t')\rangle_{\mathrm{I}} = U(t', t)|n(t)\rangle_{\mathrm{I}}, \tag{A.7}$$

where $U(t', t)$ is an operator with boundary condition

$$U(t, t) = 1. \tag{A.8}$$

By definition, the products of two operators that represent consecutive time developments satisfy

$$U(t, t')U(t', t_0) = U(t, t_0), \tag{A.9}$$

the 'group property'.

Substituting eq. (A.7) into eq. (A.5) we find an equation for $U(t', t)$:

$$(\mathrm{d}/\mathrm{d}t)U(t', t) = -\mathrm{i}H_{\mathrm{I}}(t')U(t', t). \tag{A.10}$$

This looks like the equation for a simple exponential, and it would be if H_{I} were time-independent in the interaction picture. It has time dependence, however, and the solution is different, although related. It is the *time-ordered exponential* defined in exercise 2.9,

$$\begin{aligned} U(t', t) &= T\left\{\exp\int_t^{t'} \mathrm{d}\tau[-\mathrm{i}H_{\mathrm{I}}(\tau)]\right\} \\ &= 1 - i\int_t^{t'} \mathrm{d}\tau\, H_{\mathrm{I}}(\tau) + (-\mathrm{i})^2\int_t^{t'} \mathrm{d}\tau_2\, H_{\mathrm{I}}(\tau_2)\int_t^{\tau_2} \mathrm{d}\tau_1\, H_{\mathrm{I}}(\tau_1) + \ldots . \end{aligned} \tag{A.11}$$

$U(t, t')$ is unitary (exercise 2.9), which ensures that the normalization of an interaction-picture state is constant.

These results enable us to transform from the Heisenberg to the interaction picture for operators as well as states. Since matrix elements are independent of the picture used, we have

$${}_{\mathrm{H}}\langle m|O_{\mathrm{H}}(t)|n\rangle_{\mathrm{H}} = {}_{\mathrm{I}}\langle m(t)|O_{\mathrm{I}}(t)|n(t)\rangle_{\mathrm{I}}. \tag{A.12}$$

Equations (A.7) and (A.12) show that if we take as a boundary condition

$$|n(t_0)\rangle_{\mathrm{I}} = |n\rangle_{\mathrm{H}}, \tag{A.13}$$

then

$$O_{\mathrm{I}}(t) = U(t, t_0)O_{\mathrm{H}}(t)U(t_0, t). \tag{A.14}$$

Up to this point we have solved for the time dependence of both operators and states in the interaction picture. The operators, as promised, have free-field behavior, while the states are given as a power series in H_{I}. What remains is to find a natural choice of boundary conditions for the operators and states, so that we have a place to start the time evolution. For each such choice, we have a different interaction picture. These considerations naturally lead us back to the in- and out-states introduced in Section 2.5. We assume that the Heisenberg-picture Hilbert space is spanned by in- or out-states, referred to collectively as *asymptotic states*, which according to eq. (2.89), are individually complete.

In- and out-fields

It is the in-states that we would like to identify as the beginning of our time development in the interaction picture. The in-states are *Heisenberg* states $\{|\{p_i\}\,\text{in}\rangle_{\text{H}}\}$, for which the expectation values of all Heisenberg-picture operators $O_{\text{H}}(t)$ are appropriate to a set of isolated particles of momentum p_i as $t \to -\infty$. We then define interaction-picture states via eqs. (A.7) and (A.13) as

$$|\{p_i\}(t)\,\text{in}\rangle_{\text{I}} = U(t, -\infty)|\{p_i\}\,\text{in}\rangle_{\text{H}}, \tag{A.15}$$

and related interaction-picture fields, labelled ϕ_{in}, by

$$\phi_{\text{in}}(t) = U(t, -\infty)\phi_{\text{H}}(t)U(-\infty, t). \tag{A.16}$$

Another choice of starting points for the interaction picture is the Heisenberg-picture out-states. The corresponding interaction picture is found by developing backwards in time from $t = +\infty$, and the states and operators are defined by

$$\begin{aligned} |p_i(t)\,\text{out}\rangle_{\text{I}} &= U(t, +\infty)|p_i\,\text{out}\rangle_{\text{H}}, \\ \phi_{\text{out}}(t) &= U(t, +\infty)\phi_{\text{H}}(t)U(+\infty, t). \end{aligned} \tag{A.17}$$

Equations (A.16, A.17) apply to any interaction-picture operators, and relate the conjugate momenta for the in- and out-fields to those of the Heisenberg picture in the same way. We then verify by direct substitution, using the unitarity of the time evolution operator U, that the in- and out-fields satisfy the canonical commutation relations:

$$\begin{aligned} [\pi_{\text{as}}(\mathbf{x}, t), \phi_{\text{as}}(\mathbf{x}', t)] &= -\mathrm{i}\delta^3(\mathbf{x} - \mathbf{x}'), \\ [\phi_{\text{as}}(\mathbf{x}, t), \phi_{\text{as}}(\mathbf{x}', t)] &= [\pi_{\text{as}}(\mathbf{x}, t), \pi_{\text{as}}(\mathbf{x}', t)] = 0, \end{aligned} \tag{A.18}$$

where 'as' (asymptotic) refers to either in- or out-fields; commutators between in- and out-fields are not canonical. Now, the canonical commutation relations eq. (2.9) (compare eq. (A.18)), the time development eq. (2.5) (eq. (A.6)) and the explicit form of the free Hamiltonian H_0 were the only ingredients necessary to determine the time dependence of operators in the free-field theory. As noted above, all these ingredients are present for the asymptotic fields in the interaction picture, so that corresponding to eq. (2.41) we have

$$\phi_{\text{as}}(\mathbf{x}, t) = \int \frac{\mathrm{d}^3\mathbf{k}}{(2\pi)^{3/2}2\omega_k}\,[a_{\text{as}}(\mathbf{k})\mathrm{e}^{-\mathrm{i}\bar{k}\cdot x} + a^\dagger{}_{\text{as}}(\mathbf{k})\mathrm{e}^{\mathrm{i}\bar{k}\cdot x}], \tag{A.19}$$

and similarly for $\pi_{\text{as}}(x)$. Time-independent creation operators are defined again by $a_{\text{as}}(\mathbf{k}) = a_{\text{as}}(\mathbf{k}, t)\exp(\mathrm{i}\omega_k t)$. The asymptotic creation and annihilation operators also obey the same commutation relations as those for free fields:

$$[a_{\text{as}}(\mathbf{k}),\, a^\dagger{}_{\text{as}}(\mathbf{k})] = 2\omega_k \delta^3(\mathbf{k} - \mathbf{k}'),$$
$$[a_{\text{as}}(\mathbf{k}),\, a_{\text{as}}(\mathbf{k})] = [a^\dagger{}_{\text{as}}(\mathbf{k}),\, a^\dagger{}_{\text{as}}(\mathbf{k})] = 0. \tag{A.20}$$

Finally, we assume (Section 2.5), that at $t = -\infty$ the states of the system are effectively free, so that the in-field annihilation operators give zero when acting on the in-vacuum:

$$a_{\text{in}}(\mathbf{k})|0(-\infty)\ \text{in}\rangle_{\text{I}} = 0. \tag{A.21}$$

With these results in hand, we can derive a formal expression for Heisenberg-picture Green functions as a perturbation expansion in interaction-picture operators.

Vacuum states

For Green functions, the only states we need consider explicitly are the vacuum states. In the spirit of our comments above, if we identify Heisenberg-picture and interaction-picture states as $t_0 = -\infty$, by eq. (A.13) we have

$$|0(-\infty)\ \text{in}\rangle_{\text{I}} = |0\ \text{in}\rangle_{\text{H}} \equiv |0\ \text{in}\rangle, \tag{A.22}$$

which, as the notation suggests, we refer to as simply the in-vacuum. Perhaps surprisingly, we need not assume that the vacuum state $|0(+\infty)\ \text{out}\rangle_{\text{I}}$ is the same as the in-vacuum evolved from $t = -\infty$ to $t = +\infty$. From the spectral assumptions of Section 2.5 and momentum conservation, however, the in-vacuum state cannot evolve into any state but the vacuum state. Therefore, any of the vacuum states $|0\ \text{in}\rangle$, $|0(+\infty)\ \text{in}\rangle$ and $|0(+\infty)\ \text{out}\rangle$ differ by at most a phase, and in particular,

$${}_{\text{I}}\langle 0(+\infty)\ \text{in}\,|n(-\infty)\ \text{in}\rangle_{\text{I}} \equiv \delta_{n0}\mathrm{e}^{-\mathrm{i}\Phi} = \delta_{n0}\langle 0\ \text{in}|U(+\infty, -\infty)|0\ \text{in}\rangle, \tag{A.23}$$

where δ_{n0} picks the vacuum out of any sum over states. Since it is a phase, this vacuum-to-vacuum amplitude satisfies the relation

$$\langle 0\ \text{in}|U(+\infty, -\infty)|0\ \text{in}\rangle = [\langle 0\ \text{in}|U(-\infty, +\infty)|0\ \text{in}\rangle]^{-1}, \tag{A.24}$$

which we shall use below.

Green functions

Now consider a Green function in the Heisenberg picture, defined as the vacuum expectation value with respect to $|0\text{ in}\rangle$:

$$
\begin{aligned}
{}_{\mathrm{H}}\langle 0\text{ in}|T[\textstyle\prod_{i=1}^{n}\phi_{\mathrm{H}}(x_i)]|0\text{ in}\rangle_{\mathrm{H}}
&= \sum_{\text{perms}} \theta(t_n - t_{n-1}) \cdots \theta(t_2 - t_1) \\
&\quad \times {}_{\mathrm{H}}\langle 0\text{ in}|\phi_{\mathrm{H}}(x_n)\phi_{\mathrm{H}}(x_{n-1}) \cdots \phi_{\mathrm{H}}(x_1)|0\text{ in}\rangle_{\mathrm{H}}. \qquad \text{(A.25)}
\end{aligned}
$$

It is a simple exercise using eqs. (A.9), (A.16) and (A.23), to put this expression into the interaction picture in terms of in-fields:

$$
\begin{aligned}
&{}_{\mathrm{H}}\langle 0\text{ in}\,|T[\textstyle\prod_{i=1}^{n}\phi_{\mathrm{H}}(x_i)]|0\text{ in}\rangle_{\mathrm{H}} \\
&= \sum_{\text{perms}} \theta(t_n - t_{n-1}) \cdots \theta(t_2 - t_1)\, \langle 0\text{ in}|U(-\infty, +\infty)U(+\infty, t_n) \\
&\quad \times \phi_{\text{in}}(x_n)U(t_n, t_{n-1})\phi_{\text{in}}(x_{n-1}) \cdots \phi_{\text{in}}(x_1)U(\mathrm{t}_1, -\infty)|0\text{ in}\rangle. \qquad \text{(A.26)}
\end{aligned}
$$

Next, we insert a complete set of in-states between the factors $U(-\infty, +\infty)$ and $U(+\infty, t_n)$ in the matrix element, and use eqs. (A.23) and (A.24), which gives

$$
\begin{aligned}
&{}_{\mathrm{H}}\langle 0\text{ in}|T[\textstyle\prod_{i=1}^{n}\phi_{\mathrm{H}}(x_i)]|0\text{ in}\rangle_{\mathrm{H}} \\
&= \frac{\langle 0\text{ in}|T(\prod_{i=1}^{n}\phi_{\text{in}}(x_i)\exp\{\int_{-\infty}^{\infty}\mathrm{d}\tau[-\mathrm{i}H_{\mathrm{I}}(\tau)]\})|0\text{ in}\rangle}{\langle 0\text{ in}|\mathrm{T}(\exp\{\int_{-\infty}^{\infty}\mathrm{d}\tau[-\mathrm{i}H_{\mathrm{I}}(\tau)]\})|0\text{ in}\rangle}, \qquad \text{(A.27)}
\end{aligned}
$$

where we have used the explicit form of $U(t', t)$, eq (A.11), and the definition of time ordering. The fields in H_{I} are in-fields, so that eq. (A.27) is evaluated as in the free theory. Equation (A.27) is therefore a 'Gell-Mann–Low' formula, similar to eq. (3.89) derived from the path integral, and it may be used to derive Feynman rules, starting with Wick's theorem, eq. (3.86). The two formulas are identical when the interaction Hamiltonian equals the interaction Lagrangian, which is the case whenever the Lagrange density is free of derivative couplings. When derivative couplings are present, the difference between the operators in eqs. (3.89) and (A.27) is made up by the difference in the T^* product introduced in Section 4.4 and the standard time-ordered product in eq. (A.25). The Feynman rules derived from eq. (A.27) are thus the same as those from the path integral, with vertices based on the Lorentz invariant interaction Lagrangian. This result (described by Itzykson & Zuber r1980, Section 6-1-4) is often referred to as *Matthew's theorem* (Matthews 1949, Nishijima 1950, Rohrlich 1950).

Finally, we should note that the assumptions that go into the interaction picture in relativistic field theory are known to suffer from difficulties. Briefly, *Haag's theorem* (Hagg 1955) states that the unitary transformation (A.16) between the effectively free in-fields and the interaction Heisenberg fields is not strictly consistent with Poincaré invariance. More discussion of this fascinating point, which however, has not been shown to affect practical results, may be found in Barton (r1963), Streater & Wightman (r1964) and Bogoliubov, Logunov & Todorov (r1975).

Appendix B
Symmetry factors and generating functionals

Symmetry factors

A proof of the rule (iv′) for symmetry factors, eq. (3.95b, c) can be found from eq. (3.46), which gives interacting-field Green functions in terms of variations of the free-field generating functional. For definiteness, consider a ϕ^p-theory. Neglecting the external fields, a typical term that contributes after J is set to zero is proportional to

$$(1/m!)\left[\int d^4 y_j (1/p!)(\delta_J(y_j))^p\right]^m \times (1/n!)(1/2^n)\left[\int d^4 w\, d^4 z\, J(w)\, \Delta_F(w-z) J(z)\right]^n \Bigg|_{J=0}, \tag{B.1}$$

where $mp = 2n$. For this term, the overall numerical factor is $(1/m!)[(1/p!)^m](1/n!)(1/2^n)$. On taking the variations, however, the factors $1/n!$ and $1/2^n$ are always cancelled by permutations of the roles of the factors $\int d^4 w\, d^4 z\, J(w) \Delta_F(w-z) J(z)$, and by exchanges of sources $J(w)$ with $J(z)$ within each factor, respectively. That is, given any term in the expansion of eq. (B.1), each such permutation always gives another, identical, term in the expansion.

This leaves only the m-dependent part of the numerical weight. In general, $1/m!$ and $1/(p!)^m$ are cancelled in their turn by permutations of variations, $1/m!$ by permutations of the factors $\int d^4 y_j (1/p!)[\delta_J(y_j)]^p$, and $1/(p!)^m$ by permutations of the individual variations within each such factor. This, however, is not always the case. Symmetry factors occur whenever an exchange of the variations is equivalent to one of the permutations of sources above.

First consider the factors $1/(p!)^m$. They are associated with s_1 in eq. (3.95b). For any diagram with two or more lines attached to the same pair of vertices, as in Fig. B.1(a), any permutation of the $m_{ij} < p$ lines on vertex i, followed by the same permutation on vertex j, gives a pairing that has already been counted in the permutations of the m_{ij} factors $\int J(w) \Delta J(z)$,

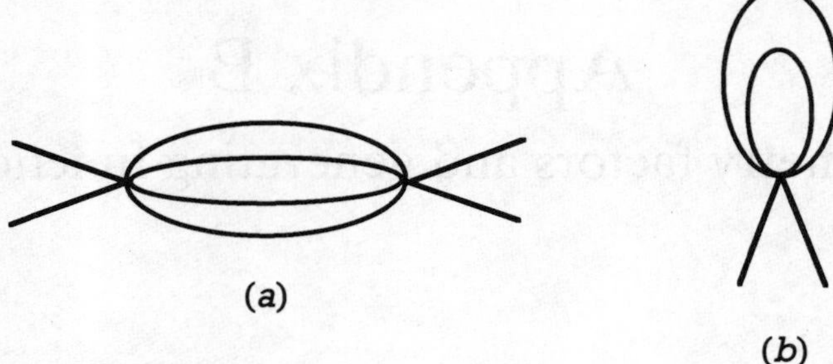

Fig. B.1 Diagrams illustrating symmetry factors.

where $w = y_i$ and $z = y_j$. These $m_{ij}!$ exchanges are therefore not counted again, and a symmetry factor $1/m_{ij}!$ remains. Similarly, in Fig. B.1(b), any exchange of the variation that act at the ends of a single propagator is equivalent to an exchange in the sources that originally sat at the ends of this propagator in eq. (B.1), so that these exchanges cannot be counted, and a factor $1/2^{m_k}$ is left over, where m_k is the number of lines connected at both ends to vertex k.

The remaining factor s_2 is associated with the factor of $1/m!$ in eq. (B.1). To understand this factor, it is useful to think in terms of the time-ordered diagrams introduced in Fig. 3.5. Suppose we fix the times in each of the m integrations y_j. Integrating over the $y_{j,0}$ gives $m!$ permutations of the roles of vertices, which cancels $1/m!$ providing that these permutations have not already been counted. Suppose, for a given time-ordered diagram, that permuting the vertices, while keeping their labels fixed, is the same as relabelling the vertices without otherwise changing the time ordering. Now a change in labels may be thought of as a permutation of lines, which is the same as a permutation of source factors, all of which have already been counted. Hence we find a symmetry factor $1/s_2$ whenever s_2 permutations of vertices in a time-ordered diagram gives an equivalent diagram. Referring, for instance, to Fig. 3.6, an exchange of vertices a and b always gives an equivalent time-ordered diagram, while an exchange of any other pair of vertices does not. Thus, as stated in Section 3.4, $s_2 = 2$ for this diagram.

Generating functional of connected diagrams

The proof that eq. (3.96) is the generating functional for connected diagrams is now very easy. Consider any contribution to $Z_{\mathscr{L}}[J]$. In general, it consists of a polynomial of the form $C_{\{i\}}\prod_i(\int G_i J_1 \cdots J_{n_i})^{a_i}$, where a_i is the power of connected diagram G_i, which has n_i external vertices (we have suppressed arguments and integration variables here), and $\{i\}$ represents the set $\{G_i\}$. $C_{\{i\}}$ is a constant factor, which we now determine. Suppose we define each diagram, G_j, to include all the symmetry factors that occur when it is the only factor in the expansion ($a_i = \delta_{ij}$). Then we only need consider symmetry factors due to the exchange of vertices between different G_i. These give new time-ordered versions of the overall diagram whenever the connected diagrams are different. When $a_j > 1$, however, the permutations of all the vertices in each factor of diagram G_j with every other identical factor leave the diagram unchanged. The symmetry factor is thus given by the number of such permutations, $a_i!$ for every i so that

$$C_{\{i\}} = \prod_i (1/a_i!). \tag{B.2}$$

Summing over all such terms, we see that $Z_{\mathscr{L}}[J] = \exp \sum_i (\int G_i J_1 \cdots J_{n_i})$, where the sum goes over every possible diagram, and eq. (3.96) follows immediately.

One-particle irreducible diagrams

Equation (3.100), which states that $\Gamma[\phi_0]$ is the generator of 1PI diagrams, may be proved from a nonlinear, iterative relation satisfied by the source-dependent classical field $\phi_0(x)$, eq. (3.97). The relation is

$$\phi_0(x) = -\mathrm{i}\int \mathrm{d}^4 y\, G_2(x-y)J(y) + \sum_{n=2}^{\infty} (1/n!) \int \mathrm{d}^4 y\, G_2(x-y) \int \gamma_{n+1}(y, x_1, \ldots, x_n) \prod_{j=1}^{n} \mathrm{d}^4 x_j\, \phi_0(x_j). \tag{B.3}$$

It is illustrated in Fig. B.2. Taking the variation $\mathrm{i}\delta_{J(x)}$ of any diagrammatic contribution to $G_{\mathrm{C}}[J]$ always results in a diagram with a propagator that begins at point x. Following this propagator, we encounter a series of 1PI

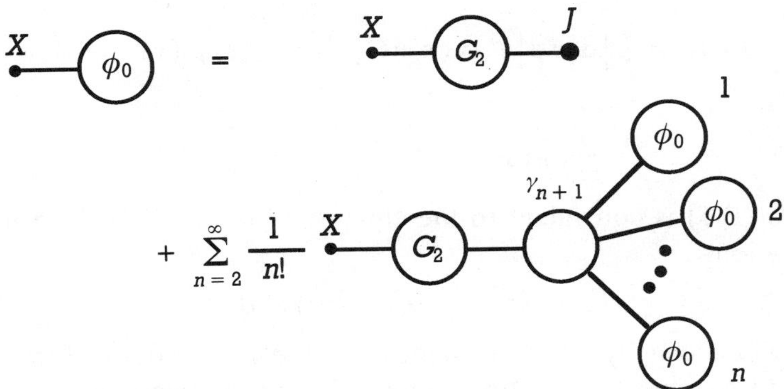

Fig. B.2 Graphical equivalent of eq. (B.3).

subdiagrams. If the first such diagrams are two-point functions, we include them in $G_2(x-y)$; the point y labels the vertex at which the line attached to x first enters a 1PI subdiagram, labelled γ_{n+1}, which is three-point or higher. The remaining external points of this diagram connect to n new propagators, each of which make the overall diagram one-particle reducible. Following each of these propagators leads to subdiagrams that are themselves of the form of the classical field ϕ_0. The factor $1/n!$ with n additional classical fields arises because γ_{n+1} is symmetric in each of its external points, so that any given diagram in which two classical fields $\phi_0(x)$ are exchanged is already included, assuming we use the complete subdiagram γ_{n+1}.

The proof of eq. (3.100) now follows from eq. (B.3), and eqs. (3.101) and (3.102), which fix the variations of $\Gamma[\phi_0]$ with respect to the source and the classical field. To proceed, we introduce the function $G_2{}^{-1}(y-x)$, which may be defined as the Fourier transform of the inverse of the complete momentum-space two-point Green function, normalized to give

$$\int \mathrm{d}^4 w\, G_2{}^{-1}(x-w) G_2(w-z) = \delta^4(x-z).$$

At zeroth order $G_2{}^{-1}(y-x) = \delta^4(y-x)[-(\partial_\mu \partial^\mu)^{(x)} - m^2]$, while at higher order it involves functions of $y-x$ resulting from diagrammatic corrections. Substituting eq. (3.102) into (B.3), and operating with $G_2{}^{-1}$ then gives

$$\int \mathrm{d}^4 x\, G_2{}^{-1}(y-x)\phi_0(x) = -\mathrm{i}\delta\Gamma[\phi_0]/\delta\phi_0(y)|_J + \sum_{n=2}^{\infty} \frac{1}{n!} \int \gamma_{n+1}(y, x_1, \ldots, x_n) \prod_{j=1}^{n} \mathrm{d}^4 x_j \phi_0(x_j). \tag{B.4}$$

This equation may be integrated to give $\Gamma[\phi_0]$ in the form

$$\Gamma[\phi_0] = \frac{1}{2}\int \mathrm{d}^4 x\, \mathrm{d}^4 x'\, \phi_0(x')\mathrm{i}G_2{}^{-1}(x'-x)\phi_0(x) + \sum_{m=3}^{\infty} \frac{1}{n!} \int [-\mathrm{i}\gamma_m(x_1, \ldots, x_m)] \prod_{j=1}^{m} \mathrm{d}^4 x_j \phi_0(x_j). \tag{B.5}$$

Equation (B.5) is equivalent to the equality of eq. (3.100) if we make the identifications

$$\Gamma_2 = \mathrm{i}G_2{}^{-1}, \quad \Gamma_n = -\mathrm{i}\gamma_n,\ n \geqslant 3. \tag{B.6}$$

The second is exactly what we want, since the Γ_n are defined to include a factor $-\mathrm{i}$ times the 1PI functions for $n \geqslant 3$. For the two-point function, we note that G_2 is the sum of diagrams shown in Fig. B.3; when the sum is

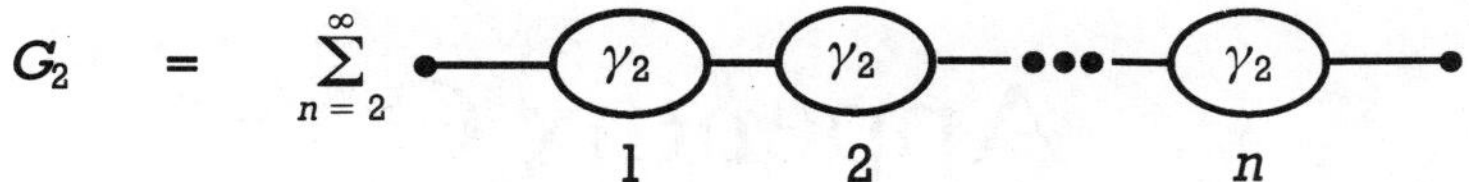

Fig. B.3 Relation of G_2 to γ_2.

evaluated we obtain in momentum space

$$G_2 = \mathrm{i}(p^2 - m^2 - \mathrm{i}\gamma_2)^{-1}. \tag{B.7}$$

The desired relation $\Gamma_2 = p^2 - m^2 - \mathrm{i}\gamma_2$ of eq. (3.98) then follows immediately, and eq. (3.100) is verified.

Appendix C

The standard model

In the following, we summarize the so-called standard model of fundamental interactions, from which we have drawn the gauge field theories discussed in the text. It is convenient to describe the degrees of freedom before including spontaneous symmetry breaking. The basic Lagrangian may be summarized as

$$\mathscr{L} = \mathscr{L}_{\text{vct}} + \mathscr{L}_{\text{Higgs}} + \mathscr{L}_{\text{lep}} + \mathscr{L}_{\text{quark}}, \tag{C.1}$$

corresponding to gauge bosons, Higgs scalar, leptonic fermions and hadronic fermions, respectively.

$SU(3) \times SU(2) \times U(1)$ vector Lagrange density (Sections 5.3, 5.4, 8.4)

For local gauge invariance, we must introduce three vector fields: for 'color' $SU(3)$ the gluon, denoted $G_{\bar{a}}{}^{\mu}$ here, and for $SU(2) \times U(1)$ the electroweak fields $B^{\mu}{}_{a}$ and C^{μ}, respectively:

$$\begin{aligned}\mathscr{L}_{\text{vct}} = &-\tfrac{1}{4}F_{\mu\nu,\bar{a}}(G)F^{\mu\nu}{}_{\bar{a}}(G) - \tfrac{1}{4}F_{\mu\nu,a}(B)F^{\mu\nu}{}_{a}(B)\\ &-\tfrac{1}{4}F_{\mu\nu}(C)F^{\mu\nu}(C),\end{aligned} \tag{C.2}$$

where

$$\begin{aligned}F_{\mu\nu,\bar{a}}(G) &= \partial_{\mu}G_{\nu,\bar{a}} - \partial_{\nu}G_{\mu,\bar{a}} - g_s f_{\bar{a}\bar{b}\bar{c}}G_{\mu,\bar{b}}G_{\mu,\bar{c}},\\ F_{\mu\nu,a}(B) &= \partial_{\mu}B_{\nu,a} - \partial_{\nu}B_{\mu,a} - g\epsilon_{abc}B_{\mu,b}B_{\mu,c},\\ F_{\mu\nu}(C) &= \partial_{\mu}C_{\nu} - \partial_{\nu}C_{\mu}.\end{aligned} \tag{C.3}$$

The $f_{\bar{a}\bar{b}\bar{c}}$ are the structure constants of $SU(3)$ and the ϵ_{abc} those of $SU(2)$. We denote the three independent couplings for the symmetries of $SU(3)$, $SU(2)$ and $U(1)$ as g_s, g, g', respectively.

Higgs density (Section 5.4)

We introduce a scalar field that is neutral under $SU(3)$ but that behaves as an $SU(2)$ doublet ($j = \frac{1}{2}$) and has a $U(1)$ charge, or *weak hypercharge*, (often denoted Y) of 1 in units of $\frac{1}{2}g'$. It thus couples to the $SU(2) \times U(1)$ gauge theory via the covariant derivative

$$\mathrm{D}_\mu(B, C) \equiv \partial_\mu + \mathrm{i}gB_{\mu,a}\tfrac{1}{2}\sigma_a + \tfrac{1}{2}\mathrm{i}g'C_\mu. \tag{C.4}$$

To this coupling, we add the very special potential term that gives the Higgs Lagrange density:

$$\mathscr{L}_{\text{Higgs}} = [D_\mu(B, C)\boldsymbol{\phi}]^* \cdot [\mathrm{D}^\mu(B, C)\boldsymbol{\phi}] + \mu^2\boldsymbol{\phi}^* \cdot \boldsymbol{\phi} - \lambda(\boldsymbol{\phi}^* \cdot \boldsymbol{\phi})^2. \tag{C.5}$$

Here and below, boldface denotes a two-dimensional weak $SU(2)$ spinor. The minimum of the purely Higgs potential is, as in eq. (5.111), at $\boldsymbol{\phi}_0^* \cdot \boldsymbol{\phi}_0 = \mu^2/2\lambda \equiv v^2/2$.

Leptonic density

We introduce three left-handed leptonic $SU(2)$ doublets and three right-handed singlets, assuming that neutrinos will remain massless even after spontaneous symmetry breaking:

$$\boldsymbol{\psi}_e^{(\mathrm{L})} = \begin{pmatrix} \nu_e^{(\mathrm{L})} \\ e^{(\mathrm{L})} \end{pmatrix}, \quad \boldsymbol{\psi}_\mu^{(\mathrm{L})} = \begin{pmatrix} \nu_\mu^{(\mathrm{L})} \\ \mu^{(\mathrm{L})} \end{pmatrix}, \quad \boldsymbol{\psi}_\tau^{(\mathrm{L})} = \begin{pmatrix} \nu_\tau^{(\mathrm{L})} \\ \tau^{(\mathrm{L})} \end{pmatrix}; \quad e^{(\mathrm{R})}, \mu^{(\mathrm{R})}, \tau^{(\mathrm{R})}. \tag{C.6}$$

In standard parlance, each pair ν_l, l^- constitutes a leptonic *generation*. The left-handed doublets are taken to have weak hypercharge $Y = -1$, and the right-handed singlets $Y = -2$. These values are chosen so that Y and I_3 are related to the electric charge Qe by

$$Y = 2(Q - I_3). \tag{C.7}$$

The full leptonic Lagrangian involves gauge and Yukawa couplings to the Higgs scalar:

$$\begin{aligned}\mathscr{L}_{\text{lep}} = &\sum_{i=e,\mu,\tau} \bar{\boldsymbol{\psi}}_i^{(\mathrm{L})}\left[\mathrm{i}\not{\partial} - g\sum_{a=1}^{3} \not{B}_a(\tfrac{1}{2}\sigma_a) + \tfrac{1}{2}g'\not{C}\right]\boldsymbol{\psi}_i^{(\mathrm{L})} \\ &+ \sum_{l=e,\mu,\tau} \bar{l}^{(\mathrm{R})}[\mathrm{i}\not{\partial} + g'\not{C}]l^{(\mathrm{R})} \\ &- \sum_{i=e,\mu,\tau} G_i\{\bar{l}_i^{(\mathrm{R})}[\boldsymbol{\phi}^\dagger \cdot \boldsymbol{\psi}_i^{(\mathrm{L})}] + [\bar{\boldsymbol{\psi}}_i^{(\mathrm{L})} \cdot \boldsymbol{\phi}]l_i^{(\mathrm{R})}\}.\end{aligned} \tag{C.8}$$

After reparameterizing the Higgs field according to spontaneous symmetry breaking, the Higgs couplings will give masses to the leptons e, μ and τ.

Quark density

As mentioned at the beginning of Section 8.1, the quarks fall into two classes: u, c, and t with $Q = \frac{2}{3}$ and d, s and b with $Q = -\frac{1}{3}$. We group these, like the leptons, into generations of left-handed doublets and right-handed singlets, according to

$$\psi_u{}^{(\mathrm{L})} = \begin{pmatrix} u^{(\mathrm{L})} \\ d'^{(\mathrm{L})} \end{pmatrix}, \quad \psi_c{}^{(\mathrm{L})} = \begin{pmatrix} c^{(\mathrm{L})} \\ s'^{(\mathrm{L})} \end{pmatrix}, \quad \psi_{\mathrm{t}}{}^{(\mathrm{L})} = \begin{pmatrix} t^{(\mathrm{L})} \\ b'^{(\mathrm{L})} \end{pmatrix}; \quad u^{(\mathrm{R})}, \ldots, b^{(\mathrm{R})}. \qquad \text{(C.9)}$$

Once again, each pair of fields $((u, d)$, etc.) constitutes a generation. The meaning of the primes will be discussed shortly. The weak hypercharges of the doublets and singlets are chosen to be consistent with eq. (C.7) $(Y_{\mathrm{L}} = \frac{1}{3}, Y_{\mathrm{R}} = 2Q)$. With these quantum numbers, all 'gauge anomalies' (Section 11.4) cancel in a theory with equal numbers of quark and lepton generations (Georgi & Glashow 1972, Bouchiat, Iliopoulos & Meyer 1972).

There are two important differences between the quark and leptonic cases. First, all the quarks possess right-handed components, which are expected to couple to mass terms as well as to the $U(1)$ charge. Related to this, the $I = -\frac{1}{2}$ entries d', s' and b' of the left-handed doublets are similar, but need not be identical to the fields d, s and b of definite mass that result from spontaneous symmetry breaking. Rather, the two sets of fields are related by a unitary transformation, which we denote V_{ij}, where i and j label the three $I = -\frac{1}{2}$ fields. Thus we have (in standard notation),

$$d' = V_{ud}d + V_{us}s + V_{ub}b, \qquad \text{(C.10)}$$

and so on. The matrix V_{ij} is known as the Kobayashi–Maskawa (1973) *mixing matrix*, or the KM matrix for short. In general, V is not far from being diagonal, with, for instance, $V_{us} \gg V_{ub}$ (Chau r1983, Paschos & Türke r1989). The 2×2 upper left-hand corner, involving only the d and s quark fields, is itself nearly a two-dimensional rotation:

$$d' \approx d \cos\theta_{\mathrm{C}} + s \sin\theta_{\mathrm{C}}, \qquad \text{(C.11)}$$

where θ_{C} is called the Cabbibo angle (Cabbibo 1963). The quark density, then, like the leptonic density, consists of gauge (now including gluon) and Yukawa interactions. To this end, the $I_3 = -\frac{1}{2}$ quarks may be coupled to the Higgs field just as the electron is. For the $I_3 = +\frac{1}{2}$ quarks, however, we must introduce the field

$$\bar{\boldsymbol{\phi}} = \mathrm{i}\sigma_2 \boldsymbol{\phi}^*, \qquad \text{(C.12)}$$

which, because of the factor σ_2, transforms according to the normal representation of weak $SU(2)$ (see eq. (5.7)). $\bar{\boldsymbol{\phi}}$, however, will develop a vacuum expectation value in its $I = +\frac{1}{2}$ component in contrast with $I = -\frac{1}{2}$ for $\boldsymbol{\phi}$.

We then have

$$\mathscr{L}_{\text{quark}} = \sum_{\text{all } q} \bar{q}\left[\mathrm{i}\partial\!\!\!/ - g_s \sum_{\bar{a}=1}^{8} G\!\!\!/_{\bar{a}}(\tfrac{1}{2}\lambda_{\bar{a}})\right]q$$

$$+ \sum_{i=u,c,t} \bar{\psi}_i^{(\mathrm{L})}\left[\mathrm{i}\partial\!\!\!/ - g\sum_{a=1}^{3} B\!\!\!/_a(\tfrac{1}{2}\sigma_a) - g'(Q - I_3)C\!\!\!/\right]\psi_i^{(\mathrm{L})}$$

$$+ \sum_{\text{all } q} \bar{q}^{(\mathrm{R})}(-g'QC\!\!\!/)q^{(\mathrm{R})}$$

$$- \sum_{d_i=d,s,b}\ \sum_{d'_j=d',s',b'} G_{d_i d'_j}[\bar{d}_i^{(\mathrm{R})}(\boldsymbol{\phi}^\dagger \cdot \psi_{d'_j}{}^{(\mathrm{L})}) + (\bar{\psi}_{d'_j}{}^{(\mathrm{L})} \cdot \boldsymbol{\phi})d_i^{(\mathrm{R})}]$$

$$- \sum_{u_i=u,c,t} G_q[\bar{u}_i^{(\mathrm{R})}(\bar{\boldsymbol{\phi}}^\dagger \cdot \psi_{u_i}{}^{(\mathrm{L})}) + (\bar{\psi}_{u_i}{}^{(\mathrm{L})} \cdot \bar{\boldsymbol{\phi}})u_i^{(\mathrm{R})}], \qquad \text{(C.13)}$$

where, for convenience of notation, we define $\psi_{d'_i}{}^{(\mathrm{L})} \equiv \psi_{u_i}{}^{(\mathrm{L})}$ for $d'_i = d'$, s', and b', and $u_i = u$, c and t, respectively. Here the assumption that d, s and b (no primes) are the fields of definite mass after spontaneous symmetry breaking means that the matrix product $G_{ij}V_{jk}$ is diagonal, with G_{ij} the matrix of $(I_3 = -\frac{1}{2})$-quark Yukawa couplings.

Let us compare eq. (C.13) to the leptonic case, eq. (C.8), where there is no mixing. A possible mixing matrix $V_{ll'}$ in the leptonic density could always be cancelled by a redefinition $\nu_i' = V_{ij}\nu_j$ of the neutrino fields since the latter have no diagonal Yukawa interactions (i.e. they will remain massless after spontaneous symmetry breaking). Thus, no mixing matrix is necessary for the leptons. In the hadronic sector this is no longer possible, precisely because of the purely diagonal Yukawa couplings for quarks with $I_3 = \frac{1}{2}$. Undoing the $I_3 = -\frac{1}{2}$ mixing would induce mixing in the $I_3 = \frac{1}{2}$ sector. Nevertheless, not all the parameters of the matrix V are relevant, since we are always free to redefine *any* of the fields $q^{(\mathrm{L})} + q^{(\mathrm{R})}$ by a global phase. A particularly important question is whether the relevant parameters of V include a complex phase, since if they do, the consequent interaction Lagrangian will no longer be real, and will generate time reversal or, equivalently, CP violation (see Appendix D).

In fact, this possibility is first realized for three generations of quarks, as is shown by the following simple counting argument. An $n \times n$ unitary matrix has, in general, n^2 free parameters, of which $2n - 1$ may be cancelled by phase transformations on the $2n$ quarks (one overall phase will not affect V). This leaves us with $n^2 - 2n + 1$ parameters. Now, an $n \times n$ real (that is orthogonal) unitary matrix has $\frac{1}{2}n(n-1)$ parameters. This leaves

$$n^2 - 2n + 1 - \tfrac{1}{2}n(n-1) = \tfrac{1}{2}(n-1)(n-2) \qquad \text{(C.14)}$$

parameters that cannot be absorbed into an orthogonal transformation, and which *may* produce complex coefficients in the Lagrangian, and break time reversal invariance. $n = 3$ is the smallest number of generations that allows time reversal invariance to be broken in this manner.

Spontaneous symmetry breaking (Section 5.4)

For the classical field theory, spontaneous symmetry breaking is just a change of variables in the fields. In Section 5.4, this was given in eq. (5.112), which automatically eliminates three of the Higgs boson degrees of freedom and defines the unitary gauge. The result is to replace the full field $\boldsymbol{\phi}(x)$ by

$$\boldsymbol{\phi}(x) = \begin{pmatrix} 0 \\ [v + \eta(x)]/2^{1/2} \end{pmatrix}. \tag{C.15}$$

In addition, exactly as in eq. (5.115), we introduce fields for the physical W and Z bosons and for the photon:

$$\begin{aligned} W^{\mu\pm} &= (B^\mu{}_1 \mp \mathrm{i}B^\mu{}_2)/2^{1/2}, \\ Z^\mu{}_0 &= (-C^\mu \sin\theta_\mathrm{W} + B^\mu{}_3 \cos\theta_\mathrm{W}), \\ A^\mu &= (C^\mu \cos\theta_\mathrm{W} + B^\mu{}_3 \sin\theta_\mathrm{W}), \end{aligned} \tag{C.16}$$

as well as the parameters

$$\begin{aligned} \sin\theta_\mathrm{W} = g'/(g'^2 + g^2)^{1/2}, \quad e \equiv gg'/(g'^2 + g^2)^{1/2} \\ M_\mathrm{W} = \tfrac{1}{2}vg, \quad M_\mathrm{Z} = \tfrac{1}{2}v(g'^2 + g^2)^{1/2} = M_\mathrm{W}/\cos\theta_\mathrm{W}. \end{aligned} \tag{C.17}$$

In these terms, the vector and Higgs densities, eqs. (C.2) and (C.5), may be combined and written as $((\mathrm{W}^\dagger)^* \equiv \mathrm{W}^-)$

$$\begin{aligned} \mathcal{L}_\mathrm{boson} = &-\tfrac{1}{4}F_{\mu\nu,\bar{a}}(G)F^{\mu\nu}{}_{\bar{a}}(G) - \tfrac{1}{2}\lambda(\partial\cdot G)^2 \\ &- \tfrac{1}{2}|\bar{D}_\mu W^+{}_\nu - \bar{D}_\nu W^+{}_\mu|^2 + M_\mathrm{W}{}^2(W^+{}_\mu)^* W^{+\mu} \\ &- \tfrac{1}{4}F_{\mu\nu}(Z_0)F^{\mu\nu}(Z_0) + \tfrac{1}{2}M_\mathrm{Z}{}^2 Z_0{}^2 - \tfrac{1}{4}F_{\mu\nu}(A)F^{\mu\nu}(A) \\ &- \tfrac{1}{2}\mathrm{i}[g\cos\theta_\mathrm{W} F_{\mu\nu}(Z_0) + eF_{\mu\nu}(A)](W^{+\mu}W^{-\nu} - W^{-\mu}W^{+\nu}) \\ &+ \tfrac{1}{4}g^2(W^+{}_\mu W^-{}_\nu - W^-{}_\mu W^+{}_\nu)^2 \\ &+ \tfrac{1}{8}[g^2(|W^+|^2 + |W^-|^2) + (g'^2 + g^2)Z_0{}^2](2v\eta + \eta^2) \\ &+ \tfrac{1}{2}[(\partial^\mu\eta)(\partial_\mu\eta) - 2\mu^2\eta^2] - v\lambda\eta^3 - \tfrac{1}{4}\lambda\eta^4, \end{aligned} \tag{C.18}$$

where for completeness we have added a covariant gauge-fixing term for the gluons, and where we define

$$\bar{\mathrm{D}}_\mu \equiv \partial_\mu + \mathrm{i}g(Z_{0,\mu}\cos\theta_\mathrm{W} + A_\mu \sin\theta_\mathrm{W}). \tag{C.19}$$

In this density we find the complex vector field W^+ (with antiparticle W^-), which is coupled to the other gauge bosons through the covariant derivative $\overline{\mathrm{D}}_\mu$ as well as through explicit (abelian) field strengths $F_{\mu\nu}(Z)$ and $F_{\mu\nu}(A)$. In addition, we have the neutral massive Z_0 and the photon, as well as the scalar Higgs particle, which has both three and four-point couplings to all four of the vector particles. Equation (C.18) exhibits manifest phase invariance for the charged W-field.

With the parameterization given in eq. (C.15) the $I = -\frac{1}{2}$ leptons and the quarks also acquire masses, through Yukawa terms of the form

$$-\bar{e}eG_e[v + \eta(x)]/2^{1/2}, \quad -\bar{u}uG_u[v + \eta(x)]/2^{1/2},$$
$$-\bar{d}_i d_i \sum_{d'_j} G_{d_i d'_j} V_{u_j d'_i}[v + \eta(x)]/2^{1/2}. \tag{C.20}$$

To write the resulting density, we use f to label quarks and leptons collectively, and $\psi_i^{(\mathrm{L})}$ to label the left-handed $SU(2)$ doublets of eqs. (C.6) and (C.9):

$$\begin{aligned}\mathscr{L}_{\text{lep}} + \mathscr{L}_{\text{quark}} &= \sum_{\text{all } f} \bar{f}[\mathrm{i}\not{\partial} - eQ_f\not{A} - m_f - (gm_f/2M_\mathrm{W})\eta]f + \sum_q \bar{q}(-g_s\gamma^\mu \tfrac{1}{2}\lambda_a)qG_{\mu,a} \\ &\quad - (g/2^{1/2}) \sum_{i=e,\mu,\tau,u,c,t} \bar{\psi}_i^{(\mathrm{L})}\gamma^\mu(\sigma^+ W^+{}_\mu + \sigma^- W^-{}_\mu)\psi_i^{(\mathrm{L})} \\ &\quad - (g/2\cos\theta_\mathrm{W}) \sum_{\text{all } f} \bar{f}\gamma^\mu(v_f - a_f\gamma_5)fZ_{0,\mu}.\end{aligned} \tag{C.21}$$

The sum over f covers all quark and lepton flavors, and v_f and a_f are vector and axial vector couplings to the Z_0 boson, given by

$$\begin{aligned} v_f &= I_{3,f} - 2Q_f \sin^2\theta_\mathrm{W}, \\ a_f &= I_{3,f}. \end{aligned} \tag{C.22}$$

The terms in this Lagrangian then describe electromagnetic, strong and weak interactions, the latter including a Yukawa coupling to the Higgs particle that is proportional to the ratio of the fermion mass to the W boson mass. Note that the mixing matrix V for the quarks disappears in the 'neutral' currents, which couple to the photon or the Z_0. The model thus has no tree-level processes in which the quark flavor is changed by a neutral current interaction (Glashow, Iliopoulos & Maiani 1970).

Parameters

The success of the standard model is its ability to incorporate a large set of experimental results. It requires, however, a large number of adjustable parameters to do so. (And, to the time of writing, the Higgs field and the t quark remain as yet unobserved.) In addition to three gauge couplings, we find nine independent Yukawa couplings, the Higgs mass and quartic coupling and four parameters (three real and one phase) in the KM matrix, for 18 free parameters. And this ignores the absence of any compelling reason why the three neutrinos should be exactly massless. If they *should* turn out to have nonzero masses, the model would have to incorporate another mixing matrix. This proliferation of parameters is evidently associated with our poor understanding of masses in the standard model. An up-to-date summary of values for the parameters may be found in the Particle Data Group's review of particle properties (r1990).

$\mathbf{R}_\xi$-gauges

The unitary gauge is convenient for exhibiting physical degrees of freedom, and for tree-level calculations. For other purposes, especially for the treatment of renormalization, however, it is useful to extend the formalism. To this end, we may define a less powerful, but more flexible change of variables:

$$\begin{gathered}\boldsymbol{\phi}(x) = \mathbf{v}/2^{1/2} + \mathbf{h}(x)/2^{1/2}, \\ \mathbf{v}^{\mathrm{t}} \equiv (0, v), \quad v^2 \equiv \mu^2/\lambda; \quad \mathbf{h}^{\mathrm{t}}(x) \equiv (2^{1/2} z^+(x), \quad \eta(x) + \mathrm{i}z(x)),\end{gathered} \tag{C.23}$$

where $z^+(x)$ is a complex field, $z(x)$ is real, and $\eta(x) \equiv (1/v)\,\mathrm{Re}\,\mathbf{v}\cdot\mathbf{h}(x)$ is the real part of the expansion of $\phi_2(x)$ about v. (The various factors of $2^{1/2}$ are chosen so that the quadratic part of the density takes a standard form given below.)

From all these changes, we derive

$$\begin{aligned}\mathcal{L}_{\text{Higgs}} = {} & \tfrac{1}{2}M_{\mathrm{W}}{}^2\{|W^+|^2 + |W^-|^2\} + \tfrac{1}{2}M_{\mathrm{Z}}{}^2 Z_0{}^2 \\ & + \tfrac{1}{2}[(\partial_\mu\eta)^2 - 2\mu^2\eta^2] + \tfrac{1}{2}(\partial_\mu z)^2 + |\partial_\mu z^+|^2 \\ & - \tfrac{1}{2}[g\partial_\mu B^\mu{}_a\,\mathrm{Im}\,(\mathbf{v}^{\mathrm{t}}\cdot\sigma_a\mathbf{h}) - \tfrac{1}{2}g'\partial_\mu C^\mu\,\mathrm{Im}\,(\mathbf{v}^{\mathrm{t}}\cdot\mathbf{h})] \\ & + \mathcal{L}_{\mathrm{vh}}[W^\pm, Z, A, \mathbf{h}],\end{aligned} \tag{C.24a}$$

where

$$\begin{aligned}& \mathcal{L}_{\mathrm{vh}}[W^\pm, Z, A, \mathbf{h}] \\ & = \mu^4/4\lambda - \lambda v\eta|\mathbf{h}|^2 - (\lambda/4)|\mathbf{h}|^4 \\ & \quad - 2^{-3/2}\mathrm{i}g\,[\mathbf{h}^+(W^+{}_\mu\sigma^+ + W^-{}_\mu\sigma^-)\partial^\mu\mathbf{h} - \partial^\mu\mathbf{h}^+(W^+{}_\mu\sigma^+ + W^-{}_\mu\sigma^-)\mathbf{h}] \\ & \quad - \mathrm{i}eA_\mu[(z^+)^*\partial^\mu z^+ - (\partial^\mu z^+)^* z^+]\end{aligned}$$

$$-\frac{g}{2\cos\theta_{\mathrm{W}}}\,Z_{0,\mu}\left\{\left(\frac{g^2-g'^2}{g^2+g'^2}\right)\mathrm{i}[(z^+)^*\partial^\mu z^+-(\partial^\mu z^+)^*z^+]\right.$$

$$\left.-\,[z\partial^\mu\eta-\partial^\mu z\,\eta]\right\}$$

$$+\left[\tfrac{1}{2}(\mathbf{h}+\mathbf{v})^{\mathrm{t}}\left\{2^{-1/2}g[W^+{}_\mu\sigma^++W^-{}_\mu\sigma^-]+eQA_\mu\right.\right.$$

$$\left.\left.+\left(\frac{g^2I_3-\frac{1}{2}g'^2}{(g^2+g'^2)^{1/2}}\right)Z_{0,\mu}\right\}^2\mathbf{h}+\mathrm{h.c.}\right] \qquad \text{(C.24b)}$$

Here we see, in turn, the usual vector boson mass terms, a Lagrange density for one massive real scalar field (η) and one real and one complex massless scalar field (z and z^+), which interact among themselves and with the vector bosons. From our discussion of the unitary gauge, we already know that the z and z^+ degrees of freedom are unphysical (they are 'would-be Goldstone bosons'), and this impression is strengthened by the terms in eq. (C.24a), in which they couple linearly to the equally unphysical divergences $\partial\cdot B_a$ and $\partial\cdot C$. In the following, we show how a gauge choice (the R_ξ gauge) can be used to give a more useful form to this density.

First let us see how a gauge choice can relate to the scalar degrees of freedom. As a preliminary step, consider a gauge density of the form of eq. (5.120), in which the gauge-fixing function $\mathcal{F}_a$ is built out of the Higgs field, rather than the gauge field,

$$\kappa\sum_{a=1}^{3}\mathcal{F}_a\mathcal{F}_a=\kappa\sum_{a=1}^{3}[\mathrm{Im}\,(\mathbf{v}^{\mathrm{t}}\cdot\sigma_a\mathbf{h})]^2=\kappa\{z^2+2|z^+|^2\}. \qquad \text{(C.25)}$$

In the limit $\kappa\to\infty$, such a gauge-fixing term enforces the condition $z=z^+=0$, because finite z or z^+ gives infinite action. Of course, we have only dared to make such a change in the Lagrange density because we know already that the z, z^+ degrees of freedom are unphysical. This procedure effectively reproduces the unitary gauge Lagrange density, eq. (5.117), since at $\zeta_i=0$ the gauge transformation eq. (5.113) is trivial, and (5.117) is found from the original density (5.109) by replacing ϕ_i with $(v+\eta)\delta_{i2}$.

The R_ξ gauges interpolate between the two forms of gauge fixing based on Higgs and gauge fields respectively. They are due to 't Hooft (1971b), Fujikawa, Lee & Sanda (1972) and Yao (1973). For $SU(2)\times U(1)$ they are conveniently defined by four terms, one for each generator:

$$\tfrac{1}{2}\sum_{a=0}^{3}\mathcal{F}_a\mathcal{F}_a=\tfrac{1}{2}\xi^{-1}(\partial_\mu C^\mu-\tfrac{1}{2}g'\xi\,\mathrm{Im}\,\mathbf{v}^{\mathrm{t}}\cdot\mathbf{h})^2$$

$$+\tfrac{1}{2}\xi^{-1}\sum_{a=1}^{3}(\partial_\mu B_a{}^\mu-\tfrac{1}{2}g\xi\,\mathrm{Im}\,\mathbf{v}^{\mathrm{t}}\cdot\sigma_a\mathbf{h})^2. \qquad \text{(C.26)}$$

The $\xi \to \infty$ limit reproduces, as above, the unitary gauge, while $\xi \to 0$ gives a covariant gauge in the $\lambda \to \infty$ limit (Lorentz gauge). For intermediate ξ, the unphysical degrees of freedom z, z^+ still appear, but with ξ-dependent mass terms. A big advantage of the R_ξ gauges is that linear interactions of the form $C^\mu \partial_\mu z$ cancel between off-diagonal terms in the original Lagrangian and the gauge-fixing terms, so that both physical and unphysical degrees of freedom appear with definite masses. The complete result for the bosonic fields, including the vector density from eq. (C.18) above, is of a still lengthy, but more standard, form:

$$\begin{aligned}
\mathcal{L}_{\text{boson}}(R_\xi) = &-\tfrac{1}{4}F_{\mu\nu,\bar{a}}(G)F^{\mu\nu}{}_{\bar{a}}(G) - \tfrac{1}{2}\lambda(\partial\cdot G)^2 \\
&- \tfrac{1}{2}|(\overline{\mathrm{D}}_\mu W^+{}_\nu - \overline{\mathrm{D}}_\nu W^+{}_\mu)|^2 \\
&+ M_{\mathrm{W}}{}^2(W^+{}_\mu)^*(W^{+\mu}) - \xi^{-1}|\partial\cdot W^+|^2 \\
&- \tfrac{1}{4}F_{\mu\nu}(Z_0)F^{\mu\nu}(Z_0) + \tfrac{1}{2}M_{\mathrm{Z}}{}^2 Z_0{}^2 - \tfrac{1}{2}\xi^{-1}(\partial\cdot Z_0)^2 \\
&- \tfrac{1}{4}F_{\mu\nu}(A)F^{\mu\nu}(A) - \tfrac{1}{2}\xi^{-1}(\partial\cdot A)^2 \\
&- \tfrac{1}{2}\mathrm{i}[g\cos\theta_{\mathrm{W}}\, F_{\mu\nu}(Z_0) + eF_{\mu\nu}(A)][W^{+\mu}W^{-\nu} - W^{-\mu}W^{+\nu}] \\
&+ \tfrac{1}{4}g^2(W^+{}_\mu W^-{}_\nu - W^-{}_\mu W^+{}_\nu)^2 \\
&+ \tfrac{1}{2}[(\partial_\mu\eta)^2 - 2\mu^2\eta^2] + \tfrac{1}{2}[(\partial_\mu z)^2 - \xi M_{\mathrm{Z}}{}^2 z^2] \\
&+ |\partial_\mu z^+|^2 - \xi M_{\mathrm{W}}{}^2|z^+|^2 + \mathcal{L}_{\mathrm{vh}}[W^\pm, Z, A, \mathbf{h}],
\end{aligned} \tag{C.27}$$

where $\mathcal{L}_{\mathrm{vh}}$ is given by eq. (C.24b). Our main interest in this somewhat complicated result is in the quadratic terms, where we now see, in addition to the vector masses, explicit gauge-fixing. These terms result in vector propagators of the form eq. (7.85). For ξ finite, their ultraviolet behavior is milder than in the unitary gauge, $\xi \to \infty$. As such, renormalization is easier to study in the R_ξ gauges. The cost of this improvement is the explicit appearance of the neutral $z(x)$ and complex $z^+(x)$ fields, now with gauge-dependent masses. In the $\xi \to \infty$ limit, these masses recede to infinity, and the $z^+(x)$ and $z(x)$ degrees of freedom may be dropped. Of course, this density must be supplemented by a ghost Lagrangian, which may be generated by the procedure of Section 7.3. The resulting Feynman rules are not particularly simple, but are conveniently summarized in an appendix of Cheng & Li (r1984) (although with an opposite sign for g and g', compared with the discussion above).

Appendix D
T, C and CPT

Together with parity (P) (Sections 1.5 and 6.6), the time reversal transformation (T)

$$\mathrm{T}^{\alpha}{}_{\beta} = -g^{\alpha\beta} \tag{D.1}$$

completes the extended Lorentz group. In this appendix, we discuss T, and charge conjugation (C), which interchanges particles and antiparticles. The three discrete operations, P, C and T, are deeply connected through the CPT (or PCT or TCP!) theorem, also briefly described below. We shall concentrate on the behavior of the free Dirac field under T and C, since this illustrates most of the complications of discrete symmetries in the standard model. Many more details and applications may be found in Sakurai (r1964) and Sachs (r1987).

Time reversal

Consider a solution ψ to the free Dirac equation

$$[\mathrm{i}\not{\partial} - m]\psi(\mathbf{x}, x^0) = 0. \tag{D.2}$$

Given ψ, we can generate a corresponding solution to the Dirac equation in the time-reversed coordinate system, in which $x'^0 = -x^0$, $\mathbf{x}' = \mathbf{x}$, as follows. Suppose we construct a matrix $\mathscr{T}$ such that

$$[\mathrm{i}\not{\partial}' - m] = \mathscr{T}[\mathrm{i}\not{\partial} - m]^*\mathscr{T}^{-1}, \tag{D.3}$$

where we define $\partial'_\mu \equiv -\partial^\mu$ as the time-reversed gradient. Then, analogously to eq. (5.52) for P, the spinor

$$\psi'(\mathbf{x}, -x^0) = \mathscr{T}\psi^*(\mathbf{x}, x^0) \tag{D.4}$$

satisfies the time-reversed Dirac equation

$$[\mathrm{i}\not{\partial}' - m]\psi'(\mathbf{x}, -x^0) = 0. \tag{D.5}$$

Note the complex conjugate in the relation between solutions, which will have special consequences below. The precise nature of $\mathscr{T}$ depends on the representation, but in the Dirac, Weyl or any other representation where only γ_2 is imaginary, the choice

$$\mathcal{T} = \mathcal{T}^{-1} = i\gamma^1\gamma^3 = \mathcal{T}^\dagger \tag{D.6}$$

serves our purposes, since then

$$\mathcal{T}(\gamma^\mu)^*\mathcal{T}^{-1} = \gamma_\mu. \tag{D.7}$$

In quantum field theory, the operator that effects this transformation is represented as

$$\psi(\mathbf{x}, -x^0) = \mathcal{T}V_{\mathrm{T}}\psi(\mathbf{x}, x^0)V_{\mathrm{T}}^{-1}, \tag{D.8}$$

where $\mathcal{T}V_{\mathrm{T}}\psi(\mathbf{x}, x_0)V_{\mathrm{T}}^{-1}$ replaces $\mathcal{T}\psi^*(\mathbf{x}, x_0)$ in eq. (D.4). In correspondence with the classical transformation, the operator V_{T} takes the complex conjugate of any c-number to its right, as, for instance, $V_{\mathrm{T}}u(p, s)\mathrm{e}^{-ip\cdot x} = u^*(p, s)\mathrm{e}^{ip\cdot x}V_{\mathrm{T}}$. From this relation we can derive its action on operators. The result depends on the basis used for the spinors (Section 6.3). For definiteness we shall work in the spin basis, where (eq. (6.35)) the solutions at arbitrary momenta are given explicitly in terms of rest-frame solutions by

$$\begin{aligned} u(p, s) &= [(2m)(p_0 + m)]^{-1/2}(\not{p} + m)u(q, s), \\ v(p, s) &= [(2m)(p_0 + m)]^{-1/2}(-\not{p} + m)v(q, s), \end{aligned} \tag{D.9}$$

where $q = (m, \mathbf{0})$. Use of the defining equation (D.7), along with the explicit rest-frame forms eq. (6.25) of the spinors then gives

$$\mathcal{T}u^*(\mathbf{p}, s) = i(-1)^{s+1/2}u(-\mathbf{p}, -s), \quad \mathcal{T}v^*(\mathbf{p}, s) = i(-1)^{s-1/2}v_\lambda(-\mathbf{p}, -s). \tag{D.10}$$

The action of V_{T} that gives the field transformation eq. (D.8) may be found by considering the plane wave expansion of that equation, which gives

$$\begin{aligned} V_{\mathrm{T}}b(\mathbf{p}, s)V_{\mathrm{T}}^{-1} &= -i(-1)^{-s-1/2}b(-\mathbf{p}, -s), \\ V_{\mathrm{T}}d^\dagger(\mathbf{p}, s)V_{\mathrm{T}}^{-1} &= i(-1)^{-s-1/2}d^\dagger(-\mathbf{p}, -s). \end{aligned} \tag{D.11}$$

The phases in these relations are basis-dependent, and a more general treatment, including explicit forms for V_{T}, may be found, for instance, in Bjorken & Drell (r1965, Chapter 15). For our purposes, however, it is sufficient to use the spin basis as an illustrative example.

The action of time reversal on free-particle states is now clear. It reverses spatial momenta and spins, while leaving energies and charges unchanged. In an interacting-field theory, this result remains relevant, because in- and out-states are labelled in terms of their asymptotic free-particle quantum numbers. In addition, the roles of in- and out-states are exchanged by time reversal, along with the directions of time and momentum. We may thus write

$$V_{\mathrm{T}}|\{\mathbf{p}_i, s_i\}^{(-)}, \{\mathbf{p}_j, s_j\}^{(+)} \text{ in (out)}\rangle$$
$$= \mathrm{e}^{\mathrm{i}\chi}|\{-\mathbf{p}_i, -s_i\}^{(-)}, \{-\mathbf{p}_j, -s_j\}^{(+)} \text{ out}' \text{ (in}')\rangle, \quad \text{(D.12)}$$

where the primes on the labels 'out' and 'in' on the right-hand side emphasize that the change is due to a change in coordinates (otherwise V_{T} would be the S-matrix), and where the value of χ is dependent on the state.

Equations (D.11) and (D.12) suggest that V_{T} might be unitary, but its property of complex conjugating c-numbers prevents this, as we now show. Consider the action of V_{T} on a state $|\psi\rangle$, expanded in terms of an orthonormal basis $\{|\alpha\rangle\}$, $\langle\alpha'|\alpha\rangle = \delta_{\alpha'\alpha}$:

$$V_{\mathrm{T}}|\psi\rangle = \sum_\alpha V_{\mathrm{T}}\langle\alpha|\psi\rangle|\alpha\rangle = \sum_\alpha \langle\psi|\alpha\rangle V_{\mathrm{T}}|\alpha\rangle. \quad \text{(D.13)}$$

Now assuming that the states $|\alpha'\rangle \equiv V_{\mathrm{T}}|\alpha\rangle$ are also orthonormal, we find

$$\langle\alpha|V_{\mathrm{T}}{}^\dagger V_{\mathrm{T}}|\psi\rangle = \langle\psi|\alpha\rangle = \langle\alpha|\psi\rangle^*. \quad \text{(D.14)}$$

This result can be easily extended to arbitrary pairs of states, and is to be contrasted with the behavior of a unitary operator U, for which $\langle\phi|U^\dagger U|\psi\rangle = \langle\phi|\psi\rangle$:

$$\langle\phi'|\psi'\rangle = \langle\phi|V_T{}^\dagger V_T|\psi\rangle = \langle\phi|\psi\rangle^* = \langle\psi|\phi\rangle, \quad \text{(D.15)}$$

where again $|\psi'\rangle \equiv V_T|\psi\rangle$. Operators such as time reversal, which obey this relation for arbitrary states, are said to be *antiunitary*, rather than unitary (Wigner 1932).

Bosonic fields

Free classical scalar fields, or gauge fields in Feynman gauge, obey differential equations that are quadratic in each derivative. As a result, the application of T is generally simpler than for fermions, because there is no need to introduce a matrix such as $\mathcal{T}$ to reverse the signs of certain derivatives. The behavior of free bosonic fields under T is determined only up to a sign. For instance, we may take

$$\phi'(\mathbf{x}, -x_0) = \pm\phi^*(\mathbf{x}, x_0), \quad \mathcal{A}'_\mu(\mathbf{x}, -x_0) = \pm[\mathcal{A}^\mu(\mathbf{x}, x_0)]^* \text{ (free field)}, \quad \text{(D.16a)}$$

for a scalar field $\phi(x)$ and a gauge field $\mathcal{A}^\nu$. Either choice is consistent with T invariance.

In an interacting-field theory, the relevant signs must be determined by the nature of interactions. For instance, the behavior of classical gauge fields under time reversal is determined by fermionic currents. Under time reversal, spatial currents change sign, while time components (charge densities) remain the same. Classical electromagnetic and other gauge fields inherit this property (exactly the negative of the behavior of x^μ), since currents act as sources for these fields. In the quantum theory we shall see that both vector and axial vector currents transform in the same way, and the corresponding transformation for the gauge-field operators is

$$\mathcal{A}_\mu(\mathbf{x}, -x_0) = V_\mathrm{T}\mathcal{A}^\mu(\mathbf{x}, x_0)V_\mathrm{T}^{-1}, \tag{D.16b}$$

where V_T is an appropriate antiunitary operator. In the following, we shall assume that this operator exists for any such field.

T invariance and noninvariance

As with parity, T invariance may be expressed in terms of the action of V_T on the Lagrangian. A natural generalization of form invariance to the behavior of the Lagrange interaction density under time reversal is

$$V_\mathrm{T}\mathcal{L}_\mathrm{I}(\mathbf{x}, x^0)V_\mathrm{T}^{-1} = \mathcal{L}_\mathrm{I}(\mathbf{x}, -x^0). \tag{D.17}$$

If this relation is satisfied, and if the complete Hamiltonian is time-independent, V_T will itself be time-independent, and the same V_T will apply to in- and out-states. Then, in analogy with the parity relation, eq. (6.89), we use the antiunitary property eq. (D.15) of V_T to give

$$\begin{aligned} S_{BA} &= \langle B \text{ out}|A \text{ in}\rangle = \langle B \text{ out}|V_\mathrm{T}{}^\dagger V_\mathrm{T}|A \text{ in}\rangle^* \\ &= \langle A' \text{ out}|B' \text{ in}\rangle = S_{A'B'}, \end{aligned} \tag{D.18}$$

where $|A' \text{ out}\rangle \equiv V_\mathrm{T}|A \text{ in}\rangle$ is the time-reversed state, including phases, with opposite momenta and spins, as above. This is the expression of time-reversal invariance in the S-matrix.

It is a straightforward exercise to test the leptonic and quark standard model Lagrange density for invariance under T. The most interesting part is the third term in eq. (C.21), the quark charged-current interaction. It may be expanded in terms of fields with definite mass by using the quark mixing matrix,

$$\begin{aligned} \mathcal{L}_\mathrm{I}(x_0) = \frac{-g}{2^{3/2}} \sum_{u_i=u,c,t} \sum_{d_j=d,s,b} \{&V_{u_id_j}\bar{u}_i(x_0)\gamma^\mu(1-\gamma_5)d_j(x_0)W^+{}_\mu(x_0) \\ &+V^*{}_{u_id_j}\bar{d}_j(x_0)\gamma^\mu(1-\gamma_5)u_i(x_0)[W^+{}_\mu(x_0)]^*\}, \end{aligned} \tag{D.19}$$

where we note the complex nature of the three-by-three Kobayashi–Maskawa (KM) matrix $V_{u_i d_j}$. The action of $V_{\rm T}$ on this density follows from eqs. (D.7), (D.8) and (D.16b),

$$V_{\rm T}\mathscr{L}_{\rm I}(x_0)V_{\rm T}^{-1} = \frac{-g}{2^{3/2}}\sum_{u_i=u,c,t}\sum_{d_j=d,s,b}\{V^*{}_{u_id_j}\bar{u}_i(-x_0)\gamma^\mu(1-\gamma_5)d_j(-x_0)W^+{}_\mu(-x_0) + V_{u_id_j}\bar{d}_j(-x_0)\gamma^\mu(1-\gamma_5)u_i(-x_0)[W^+{}_\mu(-x_0)]^*\} \neq \mathscr{L}_{\rm I}(-x_0). \tag{D.20}$$

The effect of T is to reverse the signs of complex phases in the KM matrix, as a result of the antiunitary nature of $V_{\rm T}$. Thus, T is not in general a good symmetry in the standard model, because of quark mixing. Leptonic interactions with massless neutrinos lack such mixing, and hence preserve T, as do the hadronic neutral currents in eq. (C.21). As a matter of experience, T noninvariance is small and has not been directly observed (Particle Data Group r1990). Its presence, however, is indirectly but convincingly implied by experimental evidence for CP violation, the combination of P and charge conjugation, C, to which we now turn.

Charge conjugation

Charge conjugation is defined to exchange the particle and antiparticle states associated with positive-energy and negative-energy solutions to the classical equations of motion. We shall discuss it in the same manner as T, starting with solutions to the Dirac equation. For C, it will be most natural to work in momentum space, where the Dirac equation (D.2) becomes

$$[\delta_w \not{p} - m]w(p, s) = 0, \tag{D.21}$$

where $\delta_w = +1$ for $w = u$ (positive-energy solution) and -1 for $w = v$ (negative-energy solution). Given $w(p, s)$, we are looking for a unique solution $w^{(\rm C)}(p, s)$, which obeys the Dirac equation with an opposite sign for the energy:

$$[-\delta_w \not{p} - m]w^{(\rm C)}(p, s) = 0. \tag{D.22}$$

Such a relation is given by the following, rather unobvious, definition,

$$\bar{w}^{(\rm C)}{}_\alpha(p, s) \equiv -w_\beta(p, s)(\mathscr{C}^{\rm t})_{\beta\alpha}, \tag{D.23}$$

where $\mathscr{C}$ is a matrix that satisfies

$$\mathscr{C}\gamma_\mu\mathscr{C}^{-1} = -(\gamma_\mu)^{\rm t}. \tag{D.24}$$

That is, $\mathscr{C}$ transforms a Dirac matrix into its negative transpose. The proof of eq. (D.22) is then quite simple, starting with the action of $\mathscr{C}$ on the original Dirac equation (D.21):

$$0 = \mathscr{C}(\delta_w \not{p} - m)w(p, s) = (-\delta_w \not{p}^{\mathrm{t}} - m)\mathscr{C}w(p, s). \tag{D.25}$$

Taking the transpose and using the definition (D.23) of $w^{(\mathrm{C})}$, we find

$$\bar{w}^{(\mathrm{C})}(-\delta_w \not{p} - m) = 0 \to (-\delta_w \not{p} - m)w^{(\mathrm{C})} = 0. \tag{D.26}$$

The actual matrix that plays the role of $\mathscr{C}$ depends on the representation, just as for $\mathscr{T}$ above. In any representation where γ^0 and γ^2 are symmetric (for example the Dirac or Weyl representations), a standard choice is

$$\mathscr{C} = \mathrm{i}\gamma^2\gamma^0. \tag{D.27}$$

This choice has the easily verified properties

$$\mathscr{C}^{-1} = \mathscr{C}^\dagger = \mathscr{C}^{\mathrm{t}} = -\mathscr{C}. \tag{D.28}$$

The operator version of the charge conjugation relation, eq. (D.23), can now be written in conventional form as

$$U_{\mathrm{C}}\bar{\psi}_\alpha U_{\mathrm{C}}{}^{-1} = -\psi_\beta(\mathscr{C}^{-1})_{\beta\alpha}, \quad U_{\mathrm{C}}\psi_\gamma U_{\mathrm{C}}{}^{-1} = \mathscr{C}_{\gamma\lambda}\bar{\psi}_\lambda, \tag{D.29}$$

where we have used the specific properties eq. (D.28) of $\mathscr{C}$. This operator does not change c-numbers, and turns out to be unitary.

As above, we can use the action of $\mathscr{C}$ on solutions in our spin basis, eq. (6.35), to determine how U_{C} acts on creation and annihilation operators:

$$\begin{aligned} \bar{v}_\gamma(p, s) &= (-1)^{s+1/2}\mathscr{C}_{\gamma\lambda}u_\lambda(p, s), \\ \bar{u}_\gamma(p, s) &= (-1)^{s+1/2}\mathscr{C}_{\gamma\lambda}v_\lambda(p, s). \end{aligned} \tag{D.30}$$

Considering a plane-wave expansion of the Dirac field in eq. (D.29), we find

$$\begin{aligned} U_C d(p, s)U_C{}^{-1} &= -(-1)^{s+1/2}b(p, s), \\ U_C b^\dagger(p, s)U_C{}^{-1} &= -(-1)^{s+1/2}d^\dagger(p, s), \end{aligned} \tag{D.31}$$

which is analogous to eq. (D.11) for time reversal. Then for asymptotic states we find

$$U_{\mathrm{C}}|\{\mathbf{p}_i, s_i\}^{(-)}, \{\mathbf{p}_j, s_j\}^{(+)} \text{ in (out)}\rangle = \mathrm{e}^{\mathrm{i}\xi}|\{\mathbf{p}_i, s_i\}^{(+)}, \{\mathbf{p}_j, s_j\}^{(-)} \text{ in (out)}\rangle, \tag{D.32}$$

where $\mathrm{e}^{\mathrm{i}\xi}$ is a state-dependent phase. The operation of charge conjugation thus reverses particle and antiparticle states, while leaving spins and momenta unchanged. It is also not difficult to prove that, given the definitions of eq. (D.31), C changes the sign of the charge operator, eq. (6.73c).

We note that the phases in eqs. (D.30) and (D.31) have the disadvantage that the charge conjugation operator treats different spin states differently. It is often more convenient to modify the spin basis in eq. (6.35) by the *definitions* (as in Lee r1981, Sachs r1987),

$$\bar{v}_\gamma(p, s) = \mathscr{C}_{\gamma\lambda} u_\lambda(p, s), \quad \bar{u}_\gamma(p, s) = \mathscr{C}_{\gamma\lambda} v_\lambda(p, s). \tag{D.30'}$$

C for bosons

Bosons again need interactions to define their behavior under U_C. Classically, we expect them to follow the behavior of the currents that produce them. For the photon, coupled to the electric current, we thus expect a change in sign:

$$U_C \mathscr{A}^\mu(x) U_C{}^{-1} = -\mathscr{A}^\mu(x). \tag{D.33}$$

In the quantum field, we shall find, however, that a simple relation such as this is possible only for fields that couple to pure vector or pure axial vector currents – a sure sign that charge conjugation is not a good symmetry in the standard model.

C invariance and violation

When U_C commutes with the Hamiltonian we can derive an analogue of the S-matrix relations of eq. (6.89) and (D.18) for parity and time reversal, respectively. It is

$$\langle B \text{ out} | A \text{ in} \rangle = \langle B_C \text{ out} | A_C \text{ in} \rangle, \tag{D.34}$$

where $|A_C \text{ in}\rangle$ is the charge-conjugated state as in eq. (D.32), whose particle and antiparticle labels are reversed and which may have an extra phase, but which is otherwise kinematically identical to the original state. C invariance, however, fails in any theory with mixed vector and axial vector couplings, as we now show.

Again, the action of U_C in the standard model is well illustrated by the hadronic charged-current part of the Lagrangian in eq. (C.21). We apply the transformations (D.29) and (D.33), and the defining property (D.24) of the matrix $\mathscr{C}$, and find

$$\begin{aligned} U_C \mathscr{L}_I U_C{}^{-1} = \frac{-g}{2^{3/2}} \sum_{u_i=u,c,t} \sum_{d_j=d,s,b} [&V^*{}_{u_i d_j} \bar{u}_i \gamma^\mu (1 + \gamma_5) d_j W^+{}_\mu \\ &+ V_{u_i d_j} \bar{d}_j \gamma^\mu (1 + \gamma_5) u_i (W^+{}_\mu)^*] \\ \neq \mathscr{L}_I, \end{aligned} \tag{D.35}$$

where we have also employed the anticommuting nature of the fields u_i and d_j. (In the neutral current case, we may assume normal ordering to avoid a delta-function contribution.) With the sign in eq. (D.33) for the vector field W^+ (or its opposite), U_C changes the relative sign of the axial vector and vector couplings, *and* exchanges V and V^*. Thus, it is not a symmetry of the standard model.

CP and CPT

Comparing eq. (D.35) with the action of the parity operator, in eq. (6.91), we recognize that the combination $U_C U_P$ leaves invariant any term in the Lagrangian that has only real couplings, since if P changes the sign of the axial vector couplings, C changes them back. For example, in the leptonic weak interactions, eq. (6.94), the combined symmetry CP holds. For the hadronic charged current interactions, however, even CP is not quite good, for the same reason as for T, that is, the complex phase of the quark mixing matrix. In the combination CPT, however, the action of C on the axial vector currents is cancelled by P, while its left-over action on the mixing matrix is cancelled by T. CPT is thus a good symmetry for the entire standard model.

This result is a special case of a general (CPT) theorem: the combined action of C, P and T leaves invariant the Lagrangian of any relativistic local theory (Streater & Wightman r1964, Bogoliubov, Logunov & Todorov r1975).

It is instructive in practical cases to note how the CPT theorem is satisfied by monomials in momentum and spin that may occur in cross sections. The simplest example has already been found in muon decay, eq. (8.76). The term $\mathbf{p}_e\cdot(s\mathbf{n})$, is not invariant under parity, since $\mathbf{p}_e$ changes sign, while $s\mathbf{n}$ does not. On the other hand, under C, the coefficient of this term changes sign, and the effect of *CP* is nil.

More generally, we may consider monomials of the form

$$\begin{aligned} &\epsilon_{\mu\nu\lambda\sigma} p_1{}^\mu p_2{}^\nu p_3{}^\lambda p_4{}^\sigma, \\ &\epsilon_{\mu\nu\lambda\sigma} p_1{}^\mu p_2{}^\nu p_3{}^\lambda s^\sigma, \end{aligned} \tag{D.36}$$

and so on, where the $p_i{}^\mu$ are momenta and s^σ is a spin vector. The first term violates both P and T, but preserves C and PT. By considering a frame where $s_0 = 0$, we see that the second violates T, but preserves P. The CPT theorem requires that it appear with a coefficient that changes sign under C, so that CT is conserved. Many of the experimental tests of discrete symmetry conservation and violation are based on this kind of analysis.

In the standard model, CP violation through quark mixing is of special interest, because it has been observed experimentally in the decay of neutral kaons (for reviews, see Wolfenstein r1986, Sachs r1987, Paschos & Türke r1989). Thus, as indicated above, the CPT theorem requires T violation as well.

Perhaps the most fundamental consequence of the CPT theorem is the equality of particle and antiparticle masses, even in the absence of C invariance. This follows by identifying the particle mass with the expectation value of the Hamiltonian, assumed to be hermitian:

$$m^{(-)} = \langle (q, s)^{(-)} | H | (q, s)^{(-)} \rangle = \langle (q, s)^{(-)} | V_{\mathrm{T}}{}^{\dagger} V_{\mathrm{T}} H | (q, s)^{(-)} \rangle, \quad \text{(D.37)}$$

where $|(q, s)^{(-)}\rangle$ is a particle state of (arbitrary) spin projection s. In the second equality we have used the hermiticity of H, and eq. (D.15) for the (antiunitary) action of V_{T}. We now insert $(U_{\mathrm{C}} U_{\mathrm{P}})^{\dagger}(U_{\mathrm{C}} U_{\mathrm{P}}) = I$ between $V_{\mathrm{T}}{}^{\dagger}$ and V_{T}, and assume that the combination $U_{\mathrm{C}} U_{\mathrm{P}} V_{\mathrm{T}}$ commutes with H:

$$(U_{\mathrm{C}} U_{\mathrm{P}} V_{\mathrm{T}}) H = H (U_{\mathrm{C}} U_{\mathrm{P}} V_{\mathrm{T}}). \quad \text{(D.38)}$$

Acting on the states with the resulting operators, we find from eq. (D.12) for the action of time reversal, and (6.87) and (D.32) for parity and charge conjugation respectively,

$$m^{(-)} = \langle (q, -s)^{(+)} | H | (q, -s)^{(+)} \rangle = m^{(+)}. \quad \text{(D.39)}$$

This is the desired result, since the antiparticle's mass does not depend on its spin direction.

Appendix E
The Goldstone theorem and $\pi^0 \to 2\gamma$

In this appendix, we discuss three related topics: the Goldstone theorem, chiral symmetry breaking in the strong interactions, and the extraordinary application of the axial anomaly to the decay of the neutral pion. These topics draw heavily on nonperturbative reasoning, and therefore stand somewhat outside our main line of discussion. At the same time, they are so central to our understanding of the standard model as to bear at least brief description.

The Goldstone theorem

As defined in Section 5.4, a symmetry is spontaneously broken when it does not leave the vacuum state invariant. According to the Goldstone theorem, a spontaneously broken symmetry implies the existence of zero-mass scalar particles. The proof of the Goldstone theorem given by Goldstone, Salam & Weinberg (1962) is very accessible, and the reader is referred to their paper for further details. We give a variant of their argument here.

Consider a set of generators Q_a, which commute with the Hamiltonian, $[Q_a, H] = 0$, but for which $Q_a|0\rangle = |a\rangle \neq |0\rangle$. That is, each Q_a acting on the vacuum gives a new, nonzero state. Evidently the vacuum is not unique. We are familiar with this situation from the standard-model Higgs Lagrange density with a wrong-sign mass term, eq. (5.110). In this case, the Higgs field has vacuum expectation value $2^{-1/2}v$ so that the vacuum state is not invariant under $SU(2)$ rotations. Here the Q_a are the generators of weak isospin.

In the general case, $[Q_a, H] = 0$ implies multiplets of degenerate states (Section 2.2). Spontaneous symmetry breaking, however, changes the meaning of this result. Consider a state $|s\rangle = \prod_i a_i{}^\dagger|0\rangle$, where $\{a_i{}^\dagger\}$ is some set of creation operators and $|0\rangle$ is the vacuum state. We may act on $|s\rangle$ with a unitary transformation $U \equiv \exp(\mathrm{i}\sum_a \beta_a Q_a)$ in the spontaneously broken symmetry. Using $[Q_a, H] = 0$ we easily verify that $|s_U\rangle = U|s\rangle$ is degenerate with $|s\rangle$. It is precisely these states that make up the degenerate multiplets referred to above. On the other hand, $|s_U\rangle = \prod_i (U a_i^{\dagger} U^{-1}\}|0_U\rangle$, where $|0_U\rangle \equiv U|0\rangle$. By assumption, $|0_U\rangle$, which is connected to the vacuum by a unitary transformation, is a different vacuum state. Thus, the states $|s_U\rangle$, although still degenerate with $|s\rangle$, are excitations about a different vacuum. They are not part of the normal space of states, since they cannot be reached without rotating the vacuum throughout all space. In this sense, the symmetry is 'hidden', rather than explicit, even though the Lagrangian may remain fully invariant under the relevant transformation. How then does a hidden symmetry manifest itself? It is precisely through the appearance of massless Goldstone bosons, one for each symmetry generator that 'rotates' the vacuum, as we now show.

Assuming that the Lagrangian is invariant under a global symmetry of the fields, Noether's theorem leads to the existence of a conserved current, $\partial^{\mu} j_{a,\mu} = 0$, given by eq. (1.61) for each independent parameter of the symmetry. Each current, in turn, gives rise to the conserved charge, $Q_a = \int \mathrm{d}^3\mathbf{x}\, j_a{}^0(t, \mathbf{x})$, which may or may not annihilate the vacuum, depending on the details of the system. Here we suppose Q_a does not:

$$Q_a|0\rangle = |a\rangle \neq 0. \tag{E.1}$$

This requires that the current $j_a{}^{\mu}(x)$ also has matrix elements between state $|a\rangle$ and the vacuum. Translation invariance (Section 2.2) implies that all states may be expanded in terms of momentum eigenstates, so that for each a there exists *some* state of definite momentum $|p, a\rangle$ such that

$$\langle p, a|j_a{}^{\mu}(x)|0\rangle = p^{\mu} f_a(p^2)\mathrm{e}^{\mathrm{i}p\cdot x} \neq 0, \tag{E.2}$$

where $f_a(p^2)$ is a function that transforms as a scalar, by Lorentz invariance. On the other hand, current conservation, $\partial_{\mu} j_a{}^{\mu} = 0$, may be interpreted as an operator statement between on-shell states (see the discussion in connection with eq. (11.97)). This fact, along with translation invariance, shows that current conservation implies

$$p_{\mu}\langle p, a|j_a{}^{\mu}(0)|0\rangle = p^2 f_a(p^2) = 0. \tag{E.3}$$

There are two ways of satisfying this relation: either $p^2 = 0$ and $f_a(p^2) = F_a \delta(p^2)$, where F_a is some normalization, or $f_a(p^2) = 0$ for all p^2. But if $f_a(p^2) \equiv 0$, then

$$\langle p, a|Q_a|0\rangle = \int \mathrm{d}^3\mathbf{x} \langle p, a|j_a{}^0(0, \mathbf{x})|0\rangle = p^0 (2\pi)^3 \delta^3(\mathbf{p}) f_a(p^2) = 0. \tag{E.4}$$

That is, $f_a(0) \equiv 0$, implies $\langle p, a|Q_a|0\rangle = 0$ as well. Then, if momentum eigenstates are complete, we must conclude $Q_a|0\rangle = 0$, a contradiction. Thus, we may assume that if $Q_a|0\rangle = |a\rangle \neq 0$, then $f_a(0)$ is *not* zero, and there exist states with $p^2 = 0$. (Note that $p^0 \delta^3(\mathbf{p})\delta(p^2) = \frac{1}{2}\delta^4(p)$, so we have actually only demonstrated the existence of states with $p^\mu = 0$.) These are the Goldstone bosons, and there is one for each charge Q_a that transforms the vacuum. We have seen in Section 5.4 that the Higgs phenomenon uses these degrees of freedom to convert massless to massive vector bosons. This is not the only possibility, however, and it depends upon the details of the underlying Lagrangian.

We emphasize that spontaneous symmetry breaking requires degenerate vacuum states. For theories with infinite numbers of degrees of freedom, the vacuum need not always be unique. On the other hand, spontaneous symmetry breaking does not occur in systems with finite number of degrees of freedom, or even in two-dimensional field theories (Mermin & Wagner 1966, Coleman 1973), because such systems always have unique ground states.

Partially conserved axial vector currents (PCAC) and chiral symmetry

Now let us apply this reasoning to the chiral transformation, eq. (11.98), and its associated currents $j^\mu{}_5 = \bar{\psi}\gamma^\mu\gamma_5\psi$. Suppose we consider a world with only strong interactions. The masses of quarks in the Lagrangian (which are entirely due to the Higgs phenomenon) vanish in this limit. Then by eq. (11.101), the axial current would be conserved (remember, we assume no electromagnetic interactions). Does this limit have anything to do with the real world? Important evidence that it does is given by the approximate operator relation (Gell-Mann & Levy 1960, Adler & Dashen r1968)

$$m_\pi{}^2 f_\pi \pi_a(x) = \partial_\mu j_5{}^\mu{}_a(x), \tag{E.5}$$

which identifies the pion field $\pi_a(x)$ as the divergence of the ath component of the axial isospin current. The normalization here is $\langle 0|\pi_i(0)|\pi_j(p)\rangle = \delta_{ij}$. Now, the decay of a charged pion is determined at lowest order in weak interactions by the matrix element $\langle 0|j_5{}^\mu{}_a(0)|\pi^\pm\rangle$. Equation (E.5) thus allows one to determine $f_\pi \approx 93$ MeV, once the pion decay rate is known, and f_π is called the *pion decay constant* (see, for example, Itzykson & Zuber r1980, Chapter 11).

Equation (E.5) goes by the name of *partially conserved axial vector current* (PCAC). Since m_π is much less than the nucleon mass, PCAC suggests that chiral symmetry is closer to being conserved than would be suggested by the scale of nucleon masses. Comparing the PCAC relation to eq. (11.101), $\partial_\mu j_5{}^\mu = 2mi\bar{\psi}\gamma_5\psi$ (where m is a quark mass), it is entirely likely that quark masses in the Lagrangian ('current masses') are much smaller than observed nucleon masses (Gasser & Leutwyler 1975, r1982, Langacker & Pagels 1979), at least for u and d quarks. We therefore assume that nucleon masses are generated indirectly through quark interactions, rather than directly from quark masses, and that nucleons in a world without weak interactions would have a spectrum much like those in our world. Let us discuss what this point of view means for chiral symmetry, specializing for simplicity to QCD with a single doublet of quarks (u, d).

With no mass terms, this Lagrangian is chiral invariant, and both the $SU(2)$ axial vector currents $j_a{}^\mu = \bar{\psi}\gamma^\mu\gamma_5(\frac{1}{2}\sigma_a)\psi$ and the vector currents $j^\mu = \bar{\psi}\gamma^\mu(\frac{1}{2}\sigma_a)\psi$ are exactly conserved. If the vacuum state were invariant under chiral transformations, physical states would be organized into degenerate multiplets (see Section 2.2). In this case, they would exhibit the full symmetry generated by axial vector charges $Q_a{}^5 = \int d^3\mathbf{x}\, j_{a,5}{}^0$ and vector charges $Q_a = \int d^3\mathbf{x}\, j_a{}^0$, whose joint algebra is

$$[Q_a, Q_b] = i\epsilon_{abc}Q_c, \quad [Q_a{}^5, Q_b{}^5] = i\epsilon_{abc}Q_c, \quad [Q_a, Q_b{}^5] = i\epsilon_{abc}Q_c{}^5. \tag{E.6}$$

These relations follow easily from the canonical anticommutation relations $[\psi_{\alpha,i}{}^\dagger(0, \mathbf{x}), \psi_{\beta,j}(0, \mathbf{y})]_+ = \delta_{\alpha\beta}\delta_{ij}\delta^3(\mathbf{x} - \mathbf{y})$, where α and β are Dirac indices, and i and j are $SU(2)$ indices; see eq. (6.74). (A simple change of variables to $(Q_a \pm Q_a{}^5)$ shows that the algebra (E.6) is $SU(2) \times SU(2)$.) From eq. (E.6), we might expect 'parity doubling', in which states of opposite parity are degenerate, since such states are connected by chiral transformations. According to our comments above, however, this is *not* the case, and, even with massless quarks, hadrons would have much the same masses as in the true standard model. Thus, we must conclude that the QCD vacuum is not invariant under axial transformations, and hence that chiral symmetry is spontaneously broken. There is therefore a Goldstone boson for each of the $Q_a{}^5$, which we identify as the same pion fields that appears in PCAC, eq. (E.5). Their potential partners in the multiplets of $SU(2) \times SU(2)$, a triplet of scalar particles, are expected to be much heavier, as indeed they are. (A pre-QCD review of this viewpoint is given in Weinberg (r1970).)

Chiral symmetry is also 'explicitly' broken in the standard model by the Higgs phenomenon. So, the pions are not massless, but they remain much lighter than other hadrons, a reflection of their exalted role as Goldstone bosons in an imaginary world with no electroweak interactions.

The process $\pi^0 \to 2\gamma$

Now let us re-introduce electroweak interactions. The arguments of Section 11.4 imply that *any* neutral axial current, in this case $j_5{}^\mu{}_3 = \bar{\psi}\gamma^\mu\gamma_5(\frac{1}{2}\sigma_3)\psi$, has an anomalous divergence, due to electromagnetic interactions, whenever it couples to charged elementary Dirac fermions. (Neutral gauge anomalies, however, cancel in the standard model.) So, to satisfy the anomaly equation (11.115) in the zero-mass limit, we must modify eq.

(E.5) to

$$m_\pi{}^2 f_\pi \pi_i(x) = \partial^\mu j_{5\mu,i}(x) - \delta_{i3}(\textstyle\sum_a I_{3,a}Q_a{}^2)(\alpha/2\pi)^* F_{\mu\nu}(A)F^{\mu\nu}(A), \quad \text{(E.7)}$$

where Q_a is the charge and $I_{3,a}$ is the third component of isospin of the ath fermion. Consider the matrix element of eq. (E.7) between a two-photon state and the vacuum. On the left-hand side, Bose statistics, current conservation and the pseudoscalar nature of the pion field give

$$\langle k_1, \epsilon_1; k_2\epsilon_2|\pi_0(0)|0\rangle = KH(q^2)/(q^2 - m_\pi{}^2), \qquad \text{(E.8)}$$

where we have explicitly exhibited the pion pole at $q^2 = (k_1 + k_2)^2 = m_\pi{}^2$. $H(q^2)$ is a scalar function, analytic at $q^2 = m_\pi{}^2$, and

$$K = \epsilon_1{}^{*\mu}\epsilon_2{}^{*\nu}\epsilon_{\alpha\beta\mu\nu}k_1{}^\alpha k_2{}^\beta. \qquad \text{(E.9)}$$

$H(m_\pi{}^2)K$ is the invariant matrix element from which we may calculate the $\pi^0 \to 2\gamma$ decay rate. We assume that $H(q^2)$ is nearly constant for $0 \leqslant q^2 \leqslant m_\pi{}^2$:

$$H(m_\pi{}^2) \approx H(0). \qquad \text{(E.10)}$$

That is, the matrix element, eq. (E.8) is dominated by the single-pion pole, and behaves as if the pion were an elementary particle, at least in this mass range. This makes sense, because there are no other hadronic states nearby in energy. Were (E.10) to fail, it would bring into question the very idea of a 'pion field'. Thus, if we can evaluate $H(0)$, we can estimate the pion decay rate.

When we take the two-photon matrix element, the tensor structure on the right-hand side of eq. (E.7) is the same as on the left, but only the axial current has a pole at the pion mass. Therefore, we may define

$$H(q^2) = (m_\pi{}^2 f_\pi)^{-1}[F(q^2) + (q^2 - m_\pi{}^2)G(q^2)], \qquad \text{(E.11)}$$

where

$$KF(q^2) \equiv (q^2 - m_\pi{}^2)\langle k_1, \epsilon_1; k_2, \epsilon_2|\partial_\mu j^\mu{}_{5,3}(0)|0\rangle, \qquad \text{(E.12a)}$$

$$KG(q^2) \equiv -\left(\sum_a I_{3,a}Q_a{}^2\right)(\alpha/2\pi)\langle k_1\epsilon_1; k_2\epsilon_2|^* F_{\mu\nu}F^{\mu\nu}(0)|0\rangle. \quad \text{(E.12b)}$$

$G(q^2)$, due directly to the anomaly, is given at lowest order in e by eq. (11.114):

$$G(q^2) = \left(\sum_a I_{3,a}Q_a{}^2\right)(2\alpha/\pi). \qquad \text{(E.13)}$$

At this order, it is q^2-independent. Higher orders may be shown to give small corrections (Adler r1970).

The evaluation of $F(0)$ would seem more challenging, since eq. (E.12a) involves an exact matrix element of the axial current, which includes the anomaly and a lot more. Surprisingly, it is not that bad. First, consider the triangle diagram, which is just $\epsilon^{*\sigma}(k_1)\epsilon^{*\rho}(k_2)$ times the tensor $T_{\sigma\rho}{}^{\mu}$ of eq. (11.109). We have from eq. (E.12a)

$$(q^2 - m_\pi{}^2)^{-1} KF(q^2)_{\text{triangle}} = \mathrm{i}q_\mu \langle 2\gamma | j^\mu{}_{5,3}(0)|0\rangle_{\text{triangle}}. \qquad \text{(E.14a)}$$

Now apply eq. (11.109) directly to the matrix element on the right, identifying $q^\mu = (k_1 + k_2)^\mu$, to find

$$(q^2 - m_\pi{}^2)^{-1} KF(q^2)_{\text{triangle}} = \mathrm{i}K(\tfrac{1}{2}q^2)(A_3 - A_6), \qquad \text{(E.14b)}$$

where A_3 and A_6 are invariant functions from the tensor expansion of the triangle diagram, eq. (11.106) (now times I_3). But the reasoning that led to eq. (11.109) depended only on symmetry and invariance. Therefore, we can take away the subscript 'triangle' in eq. (E.14), and replace A_3 and A_6 by two 'exact' functions $\bar{A}_3$, $\bar{A}_6$ that take higher-order strong-interaction contributions into account. Since there are no truly massless particles aside from the photon (although the pion is secretly a Goldstone boson, it still has nonzero mass), there are no poles at zero momentum, i.e. at $q^2 = 0$, in the $\bar{A}$'s. Then (E.14b) requires that $F(q^2)$ *vanishes* at $q^2 = 0$ (Sutherland 1967, Veltman 1967):

$$F(0) = 0. \qquad \text{(E.15)}$$

Combining this result with eq. (E.13) for $G(0)$ in eq. (E.11), we find

$$H(m_\pi{}^2) \approx H(0) = -\Big(\sum_a I_{3,a} Q_a{}^2\Big)(2\alpha/\pi f_\pi). \qquad \text{(E.16)}$$

This determines the matrix element eq. (E.8). Reducing out the pion field, we find an approximate expression for the decay matrix element:

$$\langle k_1, \epsilon_1; k_2 \epsilon_2 | \pi_0 \rangle = -K\Big(\sum_a I_{3,a} Q_a{}^2\Big)(2\alpha/\pi f_\pi). \qquad \text{(E.17)}$$

The decay width for $\pi_0 \to 2\gamma$ is then given by the general expression eq. (4.64),

$$\Gamma = \frac{1}{2(2m_\pi)} \int \frac{\mathrm{d}^3\mathbf{k}_1 \mathrm{d}^3\mathbf{k}_2}{(2\pi)^6 4\omega_1\omega_2} \sum_{\text{spins}} \Big| \sum_a (\tfrac{1}{2} I_{3,a} Q_a{}^2)(2\alpha/\pi f_\pi) K(k_i, \epsilon_i) \Big|^2$$

$$\times (2\pi)^4 \delta^4(q - k_1 - k_2)$$

$$= \Big(\sum_a I_{3,a} Q_a{}^2\Big)^2 \frac{\alpha^2 m_\pi{}^3}{16\pi^3 f_\pi{}^2}. \qquad \text{(E.18)}$$

Taking a single quark doublet of three colors, and using $f_\pi \approx 93$ MeV, we find a predicted decay width of 7.6 eV, close to the experimental value (Particle Data Group r1990). This is striking evidence for the existence of the color quantum number, since with fractionally charged quarks the answer is small by a factor of $3^2 = 9$ without color. Also, without the anomaly, the vanishing of $F(q^2)$ in eq. (E.11) would imply that the $\pi^0 \to 2\gamma$ cross section is strongly suppressed. As a result, this calculation is one of the triumphs of field theory, connecting as it does the arcana of ultraviolet behavior with experiment.

Appendix F

Groups, algebras and Dirac matrices

Metric

We use, throughout,

$$g^{\alpha\beta} = g_{\alpha\beta} = \mathrm{diag}\,(1, -1, -1, -1). \tag{F.1}$$

Lorentz group (Sections 1.5 and 5.1)

Generators in the defining representation:

$$(J_i)^\alpha{}_\beta = -\mathrm{i}\epsilon_{0i\alpha\beta}, \quad (K_i)^\alpha{}_\beta = -\mathrm{i}(\delta_{0\alpha}\delta_{i\beta} + \delta_{i\alpha}\delta_{0\beta}). \tag{F.2}$$

Standard form for proper Lorentz transformation:

$$\Lambda_{\boldsymbol{\omega},\boldsymbol{\theta}} = \exp\,(\mathrm{i}\boldsymbol{\omega}\cdot\mathbf{K})\exp\,(-\mathrm{i}\boldsymbol{\theta}\cdot\mathbf{J}). \tag{F.3}$$

$\Lambda_{\boldsymbol{\omega},\boldsymbol{\theta}}$ is a rotation by θ about unit vector $\boldsymbol{\theta}/\theta$ according to the right-hand rule, followed by a boost in direction $\hat{\boldsymbol{\omega}}$ with 'velocity' $v = \tanh\omega$.

Lie algebra:

$$[J_i, J_j] = \mathrm{i}\epsilon_{ijk}J_k, \quad [J_i, K_j] = \mathrm{i}\epsilon_{ijk}K_k, \quad [K_i, K_j] = -\mathrm{i}\epsilon_{ijk}J_k. \tag{F.4}$$

Second form for generators:

$$\begin{gathered} m_{0i} = -m_{i0} = K_i, \quad m_{ij} = -m_{ji} = \epsilon_{ijk}J_k, \\ (m_{\mu\nu})^\alpha{}_\beta = \mathrm{i}(g_\mu{}^\alpha g_{\nu\beta} - g_{\mu\beta}g_\nu{}^\alpha). \end{gathered} \tag{F.5}$$

Second form of Lie algebra:

$$[m_{\mu\nu}, m_{\lambda\sigma}] = -\mathrm{i}g_{\mu\lambda}m_{\nu\sigma} + \mathrm{i}g_{\mu\sigma}m_{\nu\lambda} + \mathrm{i}g_{\nu\lambda}m_{\mu\sigma} - \mathrm{i}g_{\nu\sigma}m_{\mu\lambda}. \tag{F.6}$$

Third form for generators:

$$M_i = \tfrac{1}{2}(J_i + \mathrm{i}K_i), \quad N_i = \tfrac{1}{2}(J_i - \mathrm{i}K_i). \tag{F.7}$$

Third form of Lie algebra:

$$[M_i, M_j] = \mathrm{i}\epsilon_{ijk}M_k, \quad [N_i, N_j] = \mathrm{i}\epsilon_{ijk}N_k, [M_i, N_j] = 0. \tag{F.8}$$

Extended Lorentz group in the defining representation:

$$\Lambda_{\boldsymbol{\omega},\boldsymbol{\theta}}{}^{(m,n)} = T^m P^n \Lambda_{\boldsymbol{\omega},\boldsymbol{\theta}} \quad (m,\, n = 0 \text{ or } 1), \qquad \text{(F.9)}$$
$$P^{\alpha}{}_{\beta} = g_{\alpha\beta} = -T^{\alpha}{}_{\beta}.$$

$\Lambda_{\boldsymbol{\omega},\boldsymbol{\theta}}{}^{(m,n)}$ is a general proper transformation, followed by a parity transformation and/or time reversal.

Poincaré group (Section 1.5)

Generators as they act on classical fields $\{\phi_a(x)\}$:

$$(\hat{p}_\mu)_{ab} = -\mathrm{i}\partial_\mu \delta_{ab}, \qquad \text{(F.10)}$$
$$(\hat{m}_{\mu\nu})_{ab} = \mathrm{i}(x_\mu \partial_\nu - x_\nu \partial_\mu)\delta_{ab} - (\Sigma_{\mu\nu})_{ab},$$

where the $\Sigma_{\mu\nu}$ act on field indices, and satisfy the algebra of the m's above.

Lie algebra:

$$[\hat{p}_\mu, \hat{p}_\nu] = 0,$$
$$[\hat{p}_\mu, \hat{m}_{\lambda\sigma}] = -\mathrm{i}g_{\mu\sigma}\hat{p}_\lambda + \mathrm{i}g_{\mu\lambda}\hat{p}_\sigma,$$
$$[\hat{m}_{\mu\nu}, \hat{m}_{\lambda\sigma}] = -\mathrm{i}g_{\mu\lambda}\hat{m}_{\nu\sigma} + \mathrm{i}g_{\mu\sigma}\hat{m}_{\nu\lambda} + \mathrm{i}g_{\nu\lambda}\hat{m}_{\mu\sigma} - \mathrm{i}g_{\nu\sigma}\hat{m}_{\mu\lambda}. \qquad \text{(F.11)}$$

***SU*(2)** (Section 1.4)

Generators in the defining representation:

$$t_a{}^{(\mathrm{F})} = \tfrac{1}{2}\sigma_a. \qquad \text{(F.12)}$$

Pauli matrices:

$$\sigma_1 = \begin{pmatrix} 0 & 1 \\ 1 & 0 \end{pmatrix}, \quad \sigma_2 = \begin{pmatrix} 0 & -\mathrm{i} \\ \mathrm{i} & 0 \end{pmatrix}, \quad \sigma_3 = \begin{pmatrix} 1 & 0 \\ 0 & -1 \end{pmatrix}. \qquad \text{(F.13)}$$

Lie algebra:

$$[\tfrac{1}{2}\sigma_a, \tfrac{1}{2}\sigma_b] = \mathrm{i}\epsilon_{abc}\tfrac{1}{2}\sigma_c. \qquad \text{(F.14)}$$

***SU*(3)** (Section 8.4)

Generators in the defining representation:

$$t_a{}^{(\mathrm{F})} = \tfrac{1}{2}\lambda_a. \qquad \text{(F.15)}$$

The Gell-Mann matrices λ_1–λ_8:

(i) a set that generalizes σ_1,

$$\lambda_1 = \begin{pmatrix} 0 & 1 & 0 \\ 1 & 0 & 0 \\ 0 & 0 & 0 \end{pmatrix}, \quad \lambda_4 = \begin{pmatrix} 0 & 0 & 1 \\ 0 & 0 & 0 \\ 1 & 0 & 0 \end{pmatrix}, \quad \lambda_6 = \begin{pmatrix} 0 & 0 & 0 \\ 0 & 0 & 1 \\ 0 & 1 & 0 \end{pmatrix}; \tag{F.16a}$$

(ii) a set that generalizes σ_2,

$$\lambda_2 = \begin{pmatrix} 0 & -\mathrm{i} & 0 \\ \mathrm{i} & 0 & 0 \\ 0 & 0 & 0 \end{pmatrix}, \quad \lambda_5 = \begin{pmatrix} 0 & 0 & -\mathrm{i} \\ 0 & 0 & 0 \\ \mathrm{i} & 0 & 0 \end{pmatrix}, \quad \lambda_7 = \begin{pmatrix} 0 & 0 & 0 \\ 0 & 0 & -\mathrm{i} \\ 0 & \mathrm{i} & 0 \end{pmatrix}; \tag{F.16b}$$

(iii) two traceless diagonal matrices,

$$\lambda_3 = \begin{pmatrix} 1 & 0 & 0 \\ 0 & -1 & 0 \\ 0 & 0 & 0 \end{pmatrix}, \quad \lambda_8 = \begin{pmatrix} 3^{-1/2} & 0 & 0 \\ 0 & 3^{-1/2} & 0 \\ 0 & 0 & -2/3^{1/2} \end{pmatrix}. \tag{F.16c}$$

Normalization for *SU*(*N*) (Section 8.4)

For $SU(2)$ and $SU(3)$, and in general for $SU(N)$, we define $T(R)$ and $C_2(R)$ for an arbitrary representation R by

$$\operatorname{tr} t_a{}^{(R)} t_b{}^{(R)} = T(R)\delta_{ab}, \tag{F.17}$$

$$\sum_a [t_a{}^{(R)}]^2 = C_2(R)I. \tag{F.18}$$

For $SU(N)$ the standard normalization is determined by the defining representation. We have, for $N \geqslant 2$,

$$T(\mathrm{F}) = \tfrac{1}{2},\ T(\mathrm{A}) = N; \quad C_2(\mathrm{F}) = (N^2-1)/2N,\ C_2(\mathrm{A}) = N, \tag{F.19}$$

where A is the adjoint representation defined from the structure constants C_{abc} by

$$[t_a{}^{(\mathrm{A})}]_{bc} = -\mathrm{i}C_{abc}. \tag{F.20a}$$

It is also useful to parameterize the simple products of matrices in the defining representation (Macfarlane, Sudbery & Weisz 1968),

$$t_a{}^{(\mathrm{F})} t_b{}^{(\mathrm{F})} = \tfrac{1}{2}(\mathrm{i}C_{abc} + d_{abc})t_c{}^{(\mathrm{F})} + \frac{1}{N}\delta_{ab}I \tag{F.20b}$$

where d_{abc} is a completely symmetric matrix that satisfies

$$\sum_{b,c} d_{abc} d_{ebc} = \frac{N^2 - 4}{N} \delta_{ae}. \tag{F.20c}$$

Dirac algebra (Section 5.2)

Defining relation for Dirac matrices:

$$\{\gamma_\mu, \gamma_\nu\}_+ = 2g_{\mu\nu}. \tag{F.21}$$

Weyl representation:

$$\gamma^i = \begin{pmatrix} 0 & \sigma_i \\ -\sigma_i & 0 \end{pmatrix}, \ \gamma^0 = \begin{pmatrix} 0 & \sigma_0 \\ \sigma_0 & 0 \end{pmatrix}, \tag{F.22}$$

where $\sigma_0 \equiv I_{2\times 2}$.

Dirac representation:

$$\gamma^0 = \begin{pmatrix} \sigma_0 & 0 \\ 0 & -\sigma_0 \end{pmatrix}, \ \gamma^i = \begin{pmatrix} 0 & \sigma_i \\ -\sigma_i & 0 \end{pmatrix}. \tag{F.23}$$

Commutators and γ_5:

$$\begin{gathered} \sigma_{\mu\nu} = (\mathrm{i}/2)[\gamma_\mu, \gamma_\nu], \\ \gamma_5 \equiv \mathrm{i}\gamma^0\gamma^1\gamma^2\gamma^3 = (\mathrm{i}/4!)\epsilon_{\mu\nu\lambda\sigma}\gamma^\mu\gamma^\nu\gamma^\lambda\gamma^\sigma, \end{gathered} \tag{F.24}$$

$$\gamma_5 = \begin{pmatrix} -\sigma_0 & 0 \\ 0 & \sigma_0 \end{pmatrix} \text{(Weyl)}, \ \ \gamma_5 = \begin{pmatrix} 0 & \sigma_0 \\ \sigma_0 & 0 \end{pmatrix} \text{(Dirac)}. \tag{F.25}$$

The sixteen elements of the Dirac algebra in four dimensions:

$$I_{4\times 4}, \ \ \{\gamma_\mu\}, \ \ \{\sigma_{\mu\nu}\}, \ \ \{\gamma_\mu\gamma_5\}, \ \ \gamma_5. \tag{F.26}$$

Trace and summation theorems (Sections 8.1 and 9.3)

n odd: $\quad \mathrm{tr}\,(\gamma_{\alpha_1} \times \cdots \times \gamma_{\alpha_n}) = 0,$

$n = 2$: $\quad \mathrm{tr}\,(\not{a}\not{b}) = \frac{1}{2}(2a\cdot b)\,\mathrm{tr}\,I = 4a\cdot b.$

$$\begin{aligned} n = 4: \ \ &\mathrm{tr}\,(\not{a}_1\not{a}_2\not{a}_3\not{a}_4) \\ &= 4[(a_1\cdot a_2)(a_3\cdot a_4) - (a_1\cdot a_3)(a_2\cdot a_4) + (a_1\cdot a_4)(a_2\cdot a_3)]. \end{aligned} \tag{F.27}$$

General even n:

$$\mathrm{tr}\,(\not{a}_1\not{a}_2 \times \cdots \times \not{a}_n) = \sum_{i=1}^{n-1} (a_n\cdot a_i)(-1)^{i-1}\,\mathrm{tr}\,(\not{a}_1\not{a}_2 \times \cdots \times \widehat{\not{a}}_i \times \cdots \times \not{a}_{n-1}), \tag{F.28}$$

where $\widehat{\not{a}}_i$ indicates that $\not{a}_i$ is omitted.

With γ_5:

$$\operatorname{tr}(\gamma_5\gamma_\mu\gamma_\nu\gamma_\lambda\gamma_\sigma) = 4\mathrm{i}\epsilon_{\mu\nu\lambda\sigma}. \tag{F.29}$$

Other identities:

$$\{\not{a}, \not{b}\}_+ = 2a\cdot b, \quad \not{a}^2 = a^2. \tag{F.30}$$

Summation identities in n dimensions:

$$\begin{aligned} \gamma_\mu\not{a}\gamma^\mu &= (2-n)\not{a}, \\ \gamma_\mu\not{a}\not{b}\gamma^\mu &= (n-4)\not{a}\not{b} + 4a\cdot b \\ \gamma_\mu\not{a}\not{b}\not{c}\gamma^\mu &= -(n-4)\not{a}\not{b}\not{c} - 2\not{c}\not{b}\not{a}. \end{aligned} \tag{F.31}$$

Solutions to the massive Dirac equation (Section 6.3 and exercise 6.2)

Defining equations:

$$\begin{aligned} (\not{p} - m)u(p, \lambda) &= \bar{u}(p, \lambda)(\not{p} - m) = 0, \\ (\not{p} + m)v(p, \lambda) &= \bar{v}(p, \lambda)(\not{p} + m) = 0. \end{aligned} \tag{F.32}$$

Normalization:

$$\begin{aligned} \bar{u}(p, \lambda)u(p, \lambda') &= |c|^2\delta_{\lambda\lambda'}, \; u^\dagger(p, \lambda)u(p, \lambda) = |c|^2(\omega_p/m), \\ \bar{v}(p, \lambda)v(p, \lambda') &= -|c|^2\delta_{\lambda\lambda'}, \; v^\dagger(p, \lambda)v(p, \lambda) = |c|^2(\omega_p/m), \\ \bar{u}(p, \lambda)v(p, \lambda') &= \bar{v}(p, \lambda)u(p, \lambda') = 0, \\ u^\dagger(p, \lambda)v(\bar{p}, \lambda) &= v^\dagger(p, \lambda)u(\bar{p}, \lambda) = 0. \end{aligned} \tag{F.33}$$

c is a normalization factor, chosen as $(2m)^{1/2}$ in the text; $\bar{p}^\mu \equiv (p_0, -\mathbf{p})$.

Projection operators (spin basis, $s^\mu = \lambda n^\mu$, $\lambda = \pm\frac{1}{2}$):

$$\begin{aligned} \sum_\lambda u(p, \lambda)_a\bar{u}(p, \lambda)_b &= (|c|^2/2m)(\not{p} + m)_{ab}, \\ \sum_\lambda v(p, \lambda)_a\bar{v}(p, \lambda)_b &= (|c|^2/2m)(\not{p} - m)_{ab}, \\ u(p, \lambda)_a\bar{u}(p, \lambda)_b &= (|c|^2/2m)[(\not{p} + m)(\tfrac{1}{2} + \gamma_5\not{s})]_{ab}, \\ v(p, \lambda)_a\bar{v}(p, \lambda)_b &= (|c|^2/2m)[(\not{p} - m)(\tfrac{1}{2} + \gamma_5\not{s})]_{ab}. \end{aligned} \tag{F.34}$$

Completeness:

$$\sum_\lambda u(p, \lambda)_a\bar{u}(p, \lambda)_b - \sum_\lambda v(q, \lambda)_a\bar{v}(q, \lambda)_b = \delta_{ab}. \tag{F.35}$$

Solutions to massless Dirac equation (Section 6.4)

Normalization:

$$\bar{u}(q,\lambda)u(q,\sigma) = \bar{v}(q,\lambda)v(q,\sigma) = \bar{u}(q,\lambda)v(q,\sigma) = \bar{v}(q,\lambda)u(q,\sigma) = 0. \tag{F.36}$$

Projection operators:

$$\begin{aligned} u_\alpha(q,\tfrac{1}{2})\bar{u}_\beta(q,\tfrac{1}{2}) &= \tfrac{1}{2}[(1+\gamma_5)\not{q}]_{\alpha\beta} = v_\alpha(q,-\tfrac{1}{2})\bar{v}_\beta(q,-\tfrac{1}{2}), \\ u_\alpha(q,-\tfrac{1}{2})\bar{u}_\beta(q,-\tfrac{1}{2}) &= \tfrac{1}{2}[(1-\gamma_5)\not{q}]_{\alpha\beta} = v_\alpha(q,\tfrac{1}{2})\bar{v}_\beta(q,\tfrac{1}{2}). \end{aligned} \tag{F.37}$$

Completeness:

$$\sum_\lambda u_\alpha(q,\lambda)\bar{u}_\beta(q,\lambda) = \not{q}_{\alpha\beta} = \sum_\lambda v_\alpha(q,\lambda)\bar{v}_\beta(q,\lambda). \tag{F.38}$$

Appendix G
Cross sections and Feynman rules

Cross sections (Section 4.3)

The fully differential cross section for the production of n identical particles is given by

$$\frac{\mathrm{d}^{3m}\sigma}{\prod_{i=1}^{n}\mathrm{d}^3\mathbf{l}_i/[(2\pi)^3 2\omega(\mathbf{l}_i)]} = \tfrac{1}{4}[(p_1\cdot p_2)^2 - m_1{}^2 m_2{}^2]^{-1/2} \times (1/n!)|M(p_1, p_2, \{l_i\})|^2 \times (2\pi)^4\delta^4(k_1 + k_2 - \textstyle\sum l_i). \tag{G.1}$$

For inclusive cross sections, we sum over final-state spins and average over initial-state spins and other unobserved quantum numbers. In general, we include a factor of $1/a!$ for any subset of a identical particles. The relation of M to the scattering matrix $S = 1 + \mathrm{i}T$ is

$$T(p_1, p_2, \{l_i\}) = (2\pi)^{-3(n+2)/2}(2\pi)^4\delta^4(k_1 + k_2 - \textstyle\sum l_i)M(p_1, p_2, \{l_i\}). \tag{G.2}$$

The elastic scattering cross sections for $p_1 + p_2 \rightarrow l_1 + l_2$ are

$$\begin{aligned}\frac{\mathrm{d}\sigma}{\mathrm{d}q^2} &= \frac{1}{64\pi|\mathbf{p}_{\mathrm{cm}}|^2 s}|M(s, t)|^2,\\ \frac{\mathrm{d}\sigma}{\mathrm{d}\Omega^*} &= \frac{|\mathbf{k}_{\mathrm{cm}}|}{64\pi^2 s|\mathbf{p}_{\mathrm{cm}}|}|M(s, t)|^2.\end{aligned} \tag{G.3}$$

In the case of identical particles, these cross sections count both particles. Ω^* refers to the center-of-mass system, and

$$\begin{aligned}|\mathbf{p}_{\mathrm{cm}}|^2 s &= (p_1\cdot p_2)^2 - m_1{}^2 m_2{}^2,\\ |\mathbf{k}_{\mathrm{cm}}|^2 s &= (l_1\cdot l_2)^2 - \mu_1{}^2\mu_2{}^2.\end{aligned} \tag{G.4}$$

Reduction formulas

For the scalar field (Section 2.5):

$$\mathrm{i}T(\{\mathbf{p}'_b\};\{\mathbf{p}_a\}) = \prod_b [R_\phi{}^{1/2}(2\pi)^{3/2}]^{-m-n}\bar{G}^{(\mathrm{T})}(\{p'_b{}^\mu\},\{p_a{}^\nu\}), \quad (\mathrm{G}.5)$$

where $\mathrm{i}R_\phi$ is the residue of the scalar two-point function (eq. (4.15)), and $\bar{G}^{(\mathrm{T})}$ is the truncated Green function with external propagators (and factors of i) removed, evaluated at $p_a{}^2 = p_b{}^2 = m^2$.

For Dirac fields (Section 7.4):

$$\mathrm{i}T(\{(\mathbf{p}'_k,\lambda'_k)^{(-)}\},\{(\mathbf{q}'_l,\sigma'_l)^{(+)}\};\{(\mathbf{p}_i,\lambda_i)^{(-)}\},\{(\mathbf{q}_j,\sigma_j)^{(+)}\})$$

$$= \prod_k [R_\psi{}^{-1/2}(2\pi)^{-3/2}\bar{u}_{a'_k}(p'_k,\lambda'_k)]\prod_j[-(2\pi)^{-3/2}R_\psi{}^{-1/2}\bar{v}_{b_j}(q_j,\sigma_j)]$$

$$\times\ \bar{G}^{(\mathrm{T})}{}_{\{a'_k;b'_l;a_i;b_j\}}(\{p'_k\},\{q'_l\};\{p_i\},\{q_j\})$$

$$\times \prod_l[-(2\pi)^{-3/2}R_\psi{}^{-1/2}v_{b'_l}(q'_l,\sigma'_l)]\prod_i[(2\pi)^{-3/2}R_\psi{}^{-1/2}u_{a_i}(p_i,\lambda_i)]. \quad (\mathrm{G}.6)$$

For vector fields (Section 7.4):

$$\mathrm{i}T(\{(\mathbf{q}'_j,\lambda'_j)\};\{(\mathbf{q}_i,\lambda_i)\})$$

$$= \prod_j[(2\pi)^{-3/2}R_A{}^{-1/2}\epsilon^{*\nu_j}(\mathbf{q}'_j,\lambda'_j)]\bar{G}^{(\mathrm{T})}{}_{\{\nu_j;\mu_i\}}(\{q'_j\};q_i\})$$

$$\times \prod_i[(2\pi)^{-3/2}R^{-1/2}\epsilon^{\mu_i}(\mathbf{q}_i,\lambda_i)]. \quad (\mathrm{G}.7)$$

One-loop integrals in dimensional regularization (Sections 9.3 and 9.4)

Minkowski-space forms:

$$\int \mathrm{d}^n k(k^2 + 2p\cdot k - M^2 + \mathrm{i}\epsilon)^{-s}$$

$$= (-1)^s \mathrm{i}\pi^{n/2}\frac{\Gamma(s-\frac{1}{2}n)}{\Gamma(s)}(p^2 + M^2 - \mathrm{i}\epsilon)^{n/2-s}, \quad (\mathrm{G}.8)$$

$$\int \mathrm{d}^n k\, k_\mu(k^2 + 2p\cdot k - M^2 + \mathrm{i}\epsilon)^{-s}$$

$$= -\mathrm{i}(-1)^s p_\mu \pi^{n/2}\frac{\Gamma(s-\frac{1}{2}n)}{\Gamma(s)}(p^2 + M^2 - \mathrm{i}\epsilon)^{n/2-s}, \quad (\mathrm{G}.9)$$

$$\int d^n k\, k_\mu k_\nu (k^2 + 2p\cdot k - M^2 + i\epsilon)^{-s}$$
$$= i(-1)^s \pi^{n/2}[\Gamma(s)]^{-1}(p^2 + M^2 - i\epsilon)^{n/2-s}$$
$$\times [p_\mu p_\nu \Gamma(s - \tfrac{1}{2}n) - \tfrac{1}{2}g_{\mu\nu}\Gamma(s - \tfrac{1}{2}n - 1)(p^2 + M^2)]. \qquad \text{(G.10)}$$

Angular integral in m dimensions:

$$\int d\Omega_m = [2\pi^{m/2}/\Gamma(\tfrac{1}{2}m)] \int_0^\pi d\theta_m \sin^{m-1}\theta_m$$
$$= 2\pi^{(m+1)/2}/\Gamma(\tfrac{1}{2}(m+1))$$
$$= 2^m \pi^{m/2}\Gamma(\tfrac{1}{2}m)/\Gamma(m). \qquad \text{(G.11)}$$

Feynman rules for scalars, QED and QCD (Sections 3.4, 8.1, 8.4 and 8.5)

(i) A Green function is the sum of all distinguishable diagrams with the correct number of external lines.

(ii) Associate with each internal line a factor i times the Feynman propagator:

$$i(2\pi)^{-4} \int d^4k\, (k^2 - m_\phi{}^2 + i\epsilon)^{-1}\delta_{ba},$$

$$i(2\pi)^{-4} \int d^4k [(\not{k} - m_f + i\epsilon)^{-1}]_{\beta\alpha}\delta_{ba}, \qquad \text{(G.12)}$$

$$i(2\pi)^{-4} \int d^4k (k^2 + i\epsilon)^{-1}[-g^{\mu\nu} + (1 - \lambda^{-1})k^\mu k^\nu/(k^2 + i\epsilon)]\delta_{ba},$$

for scalar ϕ, fermion f and massless vector lines in covariant gauge. The diagrammatic arrow of the fermion or charged scalar propagator points in the same direction as the flow of momentum k, and from a vertex with index α to one with index β, for fermions. δ_{ba} denotes that the propagators are all diagonal in the group indices.

(iii) Associate with each interaction vertex a factor $\kappa_{\mu\ldots}(p_i)(2\pi)^4\delta^4(\sum_i p_i{}^\mu)$, where the sum is over all momenta $p_i{}^\mu$ flowing into the vertex, and where $\kappa_{\mu\ldots}(p_i)$ is a tensor, possibly in vector, spinor and group indices, specified for a number of theories below. If the vertex is external, associate with it a factor $(2\pi)^4\delta^4(p + k)$, where p^μ and k^μ are the external and internal momenta flowing into the external vertex.

(iv) Associate a relative minus sign between graphs that differ by the exchange of two external fermion fields of the same flavor. The absolute sign of the Green function involving fermions, if needed, may be computed according to the convention introduced after eq. (7.43).

(v) Associate a factor (-1) with every fermion loop.

(vi) Symmetry factors are summarized for bosons in eqs. (3.95).

Lagrange density and vertices

In dimensional regularization, with $\epsilon = 2 - \frac{1}{2}n$,

(i) Real scalar Lagrange density:

$$\mathscr{L} = \tfrac{1}{2}[(\partial_\mu \phi)^2 - m^2\phi^2] - \tfrac{1}{6}g\mu^\epsilon \phi^3 - (\lambda/4!)\mu^{2\epsilon}\phi^4.$$

Figure G.1 illustrates three-point coupling: $\kappa = -ig$. Figure G.2 illustrates four-point coupling: $\kappa = -i\lambda$.

(ii) Scalar Lagrange density for quantum electrodynamics (covariant gauge fixing):

$$\mathscr{L} = |(\partial_\mu + ie\mu^\epsilon A_\mu)\phi|^2 - m^2|\phi|^2 - \tfrac{1}{4}g\mu^{2\epsilon}|\phi^2|^2$$
$$- \tfrac{1}{4}F_{\mu\nu}(A)F^{\mu\nu}(A) - \tfrac{1}{2}\lambda(\partial\cdot A)^2.$$

Figure G.3 shows a three-point vector–scalar vertex:

$$\kappa_\mu = -ie\mu^\epsilon(p_1 + p_2)_\mu.$$

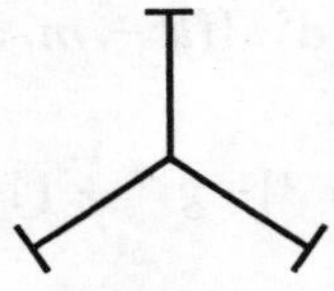

Fig. G.1

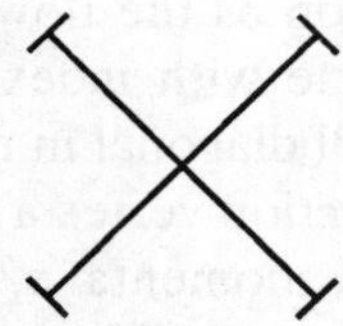

Fig. G.2

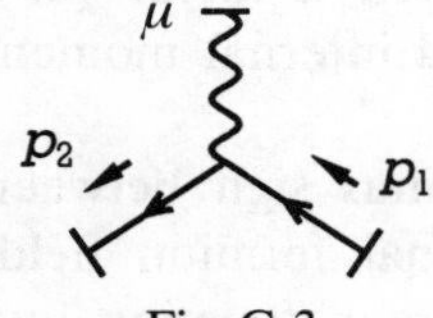

Fig. G.3

Figure G.4 shows a four-point vector–scalar vertex (seagull):

$$\kappa_{\mu\nu} = 2ie^2\mu^{2\epsilon}g_{\mu\nu}$$

Figure G.5 illustrates four-scalar coupling: $\kappa = -ig$.

(iii) Lagrange density for Dirac fermion quantum electrodynamics:

$$\mathcal{L} = \bar{\psi}(i\not{\partial} - e\mu^{\epsilon}\not{A} - m)\psi - \tfrac{1}{4}F_{\mu\nu}(A)F^{\mu\nu}(A) - \tfrac{1}{2}\lambda(\partial\cdot A)^2.$$

Figure G.6 shows a fermion–vector vertex: $\kappa_{\mu,\beta\alpha} = -ie(\gamma_\mu)_{\beta\alpha}$.

(iv) Lagrange density for $SU(N)$ gauge theory:

$$\mathcal{L} = \bar{\psi}(i\not{\partial} - g\mu^{\epsilon}\not{A}_a t_a{}^{(F)} - m)\psi - \tfrac{1}{4}F_{\mu\nu,a}(A, g)F^{\mu\nu}{}_a(A, g)$$
$$- \tfrac{1}{2}\lambda(\partial\cdot A_a)^2 - \bar{\eta}_a[(\partial_\mu)^2\delta_{ac} - gC_{abc}\partial_\mu A^\mu{}_b]\eta_c.$$

Figure G.7 shows a fermion–gluon vertex:

$$\kappa_{\mu,c;\beta\alpha,ji} = -ig[t_c{}^{(F)}]_{ji}(\gamma_\mu)_{\beta\alpha}.$$

Figure G.8 shows a three-gluon vertex:

$$\kappa^{\nu_1\nu_2\nu_3}{}_{a_1a_2a_3} = -g\mu^{\epsilon}C_{a_1a_2a_3}[g^{\nu_1\nu_2}(p_1 - p_2)^{\nu_3} + g^{\nu_2\nu_3}(p_2 - p_3)^{\nu_1}$$
$$+ g^{\nu_3\nu_1}(p_3 - p_1)^{\nu_2}]$$

Here $p_i{}^\mu$ is flowing into the vertex.

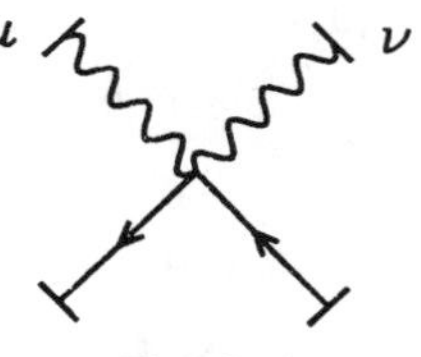

Fig. G.4

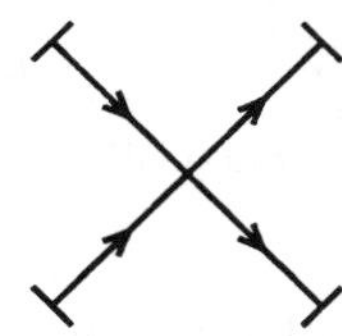

Fig. G.5

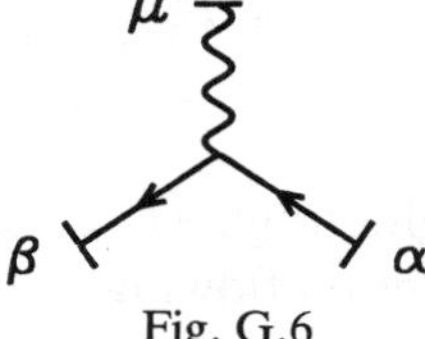

Fig. G.6

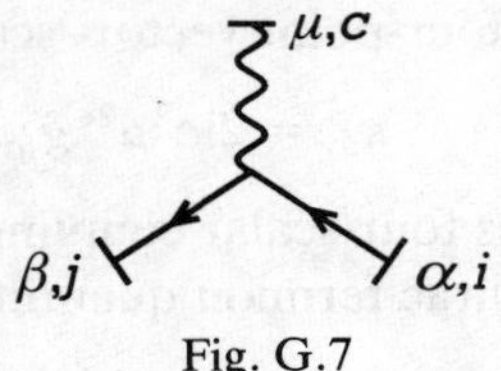

Fig. G.7

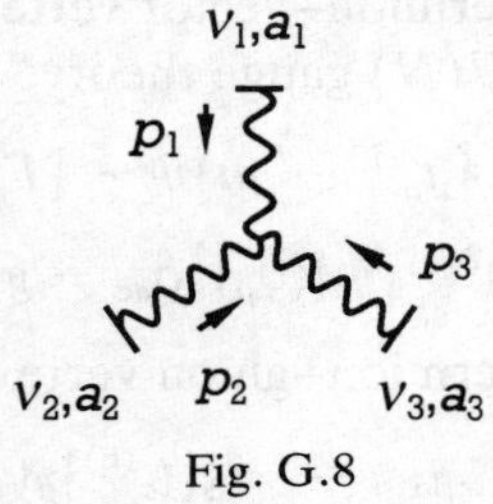

Fig. G.8

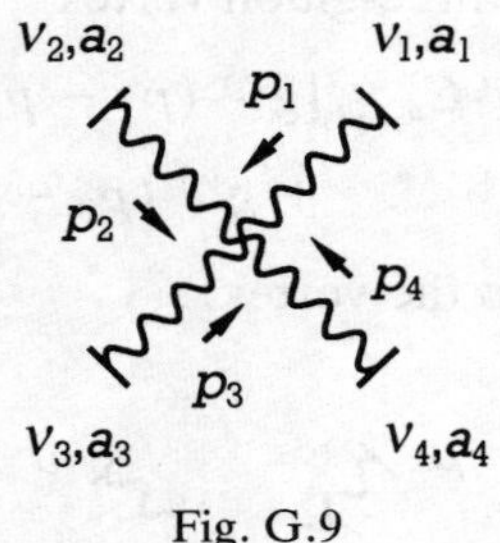

Fig. G.9

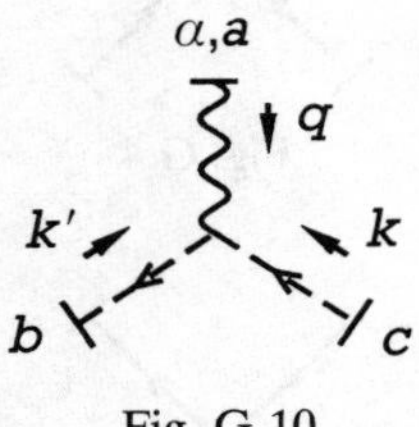

Fig. G.10

Figure G.9 shows a four-gluon vertex:

$$\begin{aligned}\kappa^{\nu_1\nu_2\nu_3\nu_4}{}_{a_1a_2a_3a_4} = -\mathrm{i}g^2\mu^{2\epsilon}[&C_{ea_1a_2}C_{ea_3a_4}(g^{\nu_1\nu_3}g^{\nu_2\nu_4} - g^{\nu_1\nu_4}g^{\nu_2\nu_3})\\ &+ C_{ea_1a_3}C_{ea_4a_2}(g^{\nu_1\nu_4}g^{\nu_3\nu_2} - g^{\nu_1\nu_2}g^{\nu_3\nu_4})\\ &+ C_{ea_1a_4}C_{ea_2a_3}(g^{\nu_1\nu_2}g^{\nu_4\nu_3} - g^{\nu_1\nu_3}g^{\nu_4\nu_2})].\end{aligned}$$

Figure G.10 shows a gluon–ghost vertex: $\kappa_{\alpha,abc} = g\mu^\epsilon C_{abc}k'_{\alpha}$. Here k' is the antighost momentum flowing into the vertex.

References

Section numbers are indicated in parentheses; 'i' refers to a chapter introduction and 'e' refers to chapter exercises.

Abarbanel, H.D.I. & Itzykson, C. (1969), *Phys. Rev. Lett.* **23**, 53 (12.3).
Adler, S.L. (1969), *Phys. Rev.* **177**, 2426 (11.4).
Adler, S.L. (1970), in *Lectures on Elementary Particles and Quantum Field Theory, Vol. 1*, eds S. Deser, M. Grisaru & H. Pendleton, MIT Press, Cambridge, MA (11.4, E).
Adler, S.L. (1972), *Phys. Rev.* **D5**, 3021 (12.3).
Adler, S.L. & Bardeen, W.A. (1969), *Phys. Rev.* **182**, 1517 (11.4).
Adler, S.L. & Dashen, R.F. (1968), *Current Algebras*, Benjamin, New York (E).
Altarelli, G. (1982), *Phys. Rep.* **81**, 1 (14.2, 14.4).
Altarelli, G. (1989), *Ann. Rev. Nucl. Part. Sci.* **39**, 357 (14.3).
Altarelli, G. & Parisi, G. (1977), *Nucl. Phys.* **B126**, 298 (14.2, 14.3, 14.4).
Altarelli, G., Ellis, R.K. & Martinelli, G. (1979), *Nucl. Phys.* **B157**, 461 (14.2, 14.3, 14e).
Amati, D., Petronzio, R. & Veneziano, G. (1978), *Nucl. Phys.* **B140**, 54; **B146**, 29 (14.3).
Anderson, P.W. (1962), *Phys. Rev.* **130**, 439 (5.4).
Antonelli, F., Consoli, M. & Corbo, G. (1980), *Phys. Lett.* **91B**, 90 (8.3).
Appell, D. Mackenzie, P. & Sterman, G. (1988), *Nucl. Phys.* **B309**, 259 (15).
Appelquist, T. & Carazzone, J. (1975), *Phys. Rev.* **D11**, 2856 (8.3).
Appelquist, T. & Georgi, H. (1973), *Phys. Rev.* **D8**, 4000 (12.4).
Appelquist, T. & Politzer, H.D. (1975), *Phys. Rev. Lett.* **34**, 43; *Phys. Rev.* **D12**, 1404 (15).
Ashmore, J.F. (1972), *Lettere al Nuovo Cim.* **4**, 289 (9.3).

Bardeen, W.A., Buras, A.J., Duke, D.W. & Muta, T. (1978), *Phys. Rev.* **D18**, 3998 (12.4)
Bargmann, V. & Wigner, E.P. (1946), *Proc, Nat. Acad. Sci. USA* **34**, 211 (6.2).
Barton, G. (1963), *Introduction to Advanced Field Theory*, Interscience Tracts on Physics and Astronomy No. 22, Interscience Publishers, New York (A).
Becchi, C., Rouet, A. & Stora, R. (1975), *Commun. Math. Phys.* **42**, 127 (11.3).
Bell, J.S. & Jackiw, R. (1969), *Nuovo Cim. Ser. 10* **60A**, 47 (11.4).
Bell, J.S. & van Royan, R. (1969), *Nuovo Cim. Ser. 10* **60A**, 62 (12.e).
Berezin, F.A. (1966), *The Method of Second Quantization*, Academic Press, New York (3.2, 7.1).

Bethe, H.A. (1947), *Phys. Rev.* **72**, 339 (12e).
Bethe, H.A. & Salpeter, E.E. (1951), *Phys. Rev.* **82**, 309; **84**, 1232 (15).
Bethke, J., Kunszt, Z., Soper, D.E. & Stirling, W.J. (1992), *Nucl. Phys.* **B370**, 310 (12.5).
Bhabha, H.J. (1936), *Proc. Roy. Soc. (London)* **A154**, 195 (8.1).
Bjorken, J.D. (1969), *Phys. Rev.* **179**, 1547 (8.6).
Bjorken, J.D. & Drell, S.D. (1964), *Relativistic Quantum Mechanics*, McGraw-Hill, New York (12e).
Bjorken, J.D. & Drell, S.D. (1965), *Relativistic Quantum Fields*, McGraw-Hill, New York (2.5, 3.4, 4e, 6.5, 6.6, D).
Bjorken, J.D. & Paschos, E.A. (1969), *Phys. Rev.* **185**, 1975 (8.6).
Bleuler, K. (1950), *Helv, Phys. Acta* **23**, 567 (6.5).
Bloch, F. & Nordsieck, A. (1937), *Phys. Rev.* **52**, 54 (12.2).
Bodwin, G.T. (1985), *Phys. Rev.* **D31**, 2616 (14.3).
Bodwin, G.T. & Yennie, D.R. (1978), *Phys. Rep.* **C43**, 267 (15).
Bogoliubov, N.N. & Parasiuk, O.S. (1957), *Acta Math.* **97**, 227 (10.4).
Bogoliubov, N.N & Shirkov, D.V. (1980), *Introduction to the Theory of Quantized Fields* (3rd edition), Wiley-Interscience, New York (6.5, 10.4, A).
Bogoliubov, N.N., Logunov, A.A. & Todorov, I.T. (1975), *Introduction to Axiomatic Field Theory*, Benjamin/Cummings, Reading, MA (2.5, 5.1, 6.5, 6.6, A, D).
Bohr, N. & Rosenfeld, L. (1933), *Kgl. Danske Vid. Sels. Mat.-fys. Medd.* **12**, 2 (2.2).
Bohr, N. & Rosenfeld, L. (1950), *Phys. Rev.* **78**, 794 (2.2).
Bollini, C.G. & Giambiagi, J.J. (1972), *Nuovo Cim. Ser. 11* **12B**, 20 (9.3).
Bouchiat, C., Iliopoulis, J., Meyer, P. (1972), *Phys. Lett.* **38B**, 519 (11.4, C).
Bovincini, G., ed. (1990), *QED Structure Functions*, Conference Proceedings No. 201, American Institute of Physics, New York (14.3).
Brandt, R.A. (1976), *Nucl. Phys.* **B116**, 413 (11.3).
Brandt, R.A. & Preparata, G. (1971), *Nucl. Phys.* **B27**, 541 (14.5).
Brauer, P. & Weyl, H. (1935), *Am. J. Math.* **57**, 425 (5.1).
Brézin, E., Le Guillou, J.C. & Zinn-Justin, J. (1977), *Phys. Rev.* **D15**, 1544; 1558 (15).
Brodsky, S.J. & Lepage, G.P. (1989), in *Perturbative Quantum Chromodynamics*, ed. A.H. Mueller, World Scientific, Singapore (15).
Brodsky, S.J. & Farrar, G.R. (1973), *Phys. Rev. Lett.* **31**, 1153 (15).
Burnett, T.H. & Kroll, N.M. (1968), *Phys. Rev. Lett.* **20**, 86 (12e).

Cabbibo, N. (1963), *Phys. Rev. Lett.* **10**, 531 (C).
Callan, C.G. (1981), in *Methods in Field Theory*, Les Houches 1975, Session XXVIII, eds R. Balian & J. Zinn-Justin, North-Holland/World Scientific, Singapore (10.5).
Callan, C.G. & Gross, D.J. (1969), *Phys. Rev. Lett.* **22**, 156 (14.3).
Callan, C.G. & Gross, D.J. (1975), *Phys. Rev.* **D11**, 2905 (13.4).
Candlin, D.J. (1956), *Nuovo Cim. Ser. 10* **4**, 231 (7.1).
Caswell, W.E. (1974), *Phys. Rev. Lett.* **33**, 244 (12.4).
Catani, S. & Trentadue, L. (1989), *Nucl. Phys.* **B327**, 323 (14.3).
Chau, L.L. (1983), *Phys. Rep.* **95**, 1 (C).
Cheng. T.-P. & Li, L.-F. (1984), *Gauge Theories of Elementary Particle Physics*, Oxford University Press, New York (C).
Christ, N.H. & Lee, T.D. (1980), *Phys. Rev.* **D22**, 939 (7.3).
Christ, N., Hasslacher, B. & Mueller, A.H. (1972), *Phys. Rev.* **D6**, 3543 (14.5).
Chung, V. (1965), *Phys. Rev.* **140**, B1110 (12.2).
Ciafaloni, M. (1989), in *Perturbative Quantum Chromodynamics*, ed. A.H. Mueller, World Scientific, Singapore (14.3).

Close, F.E. (1979), *An Introduction to Quarks and Partons*, Academic Press, London (8.6, 14.1, 14.4, 14e).
Cohen, E.R. & Taylor, B.N. (1987), *Rev. Mod. Phys.* **59**, 1121 (8.2).
Coleman, S. (1973), *Comm. Math. Phys.* **31**, 259 (E).
Coleman, S. (1985), 1974 Erice lecture, reprinted in *Aspects of Symmetry*, Cambridge University Press, Cambridge, (5.4).
Coleman, S. & Norton, R.E. (1965), *Nuovo Cim. Ser. 10* **38**, 438 (13.1).
Collins, J.C. (1980), *Phys. Rev.* **D22**, 1478 (12.3, 13.4).
Collins, J.C. (1984), *Renormalization*, Cambridge University Press, Cambridge (8.3, 9.3, 9.4, 10.3, 10.4, 11.2, 11.3).
Collins, J.C. (1989), in *Perturbative Quantum Chromodynamics*, ed. A.H. Mueller, World Scientific, Singapore (12.3, 13.4).
Collins, J.C. & Macfarlane, A.J. (1974), *Phys. Rev.* **D10**, 1201 (10.4, 10.5).
Collins, J.C. & Sterman, G. (1981), *Nucl. Phys.* **B185**, 172 (14.3).
Collins, J.C. & Soper, D.E. (1981), *Nucl. Phys.* **B193**, 381 (13.4).
Collins, J.C. & Soper, D.E. (1982), *Nucl. Phys.* **B194**, 445 (14.3).
Collins, J.C., Soper, D.E. & Sterman, G. (1988), *Nucl. Phys.* **B308**, 833 (13.4, 14.3).
Collins, J.C., Soper, D.E. & Sterman, G. (1989), in *Perturbative Quantum Chromodynamics*, ed. A.H. Mueller, World Scientific, Singapore (13.4, 14.3).
Combridge, B.L., Kripfganz, J. & Ranft, J. (1977), *Phys. Lett.* **70B**, 234 (8.4, 8.5).
Contopanagos, H.F. & Einhorn, M.B. (1992), *Phys. Rev.* **D45**, 1291; 1322 (11.2, 12.2, 13.5).
Corinaldesi, E. (1953), *Suppl. al Nuovo Cim. Ser 9* **10**, 83 (2.2).
Corson, E.M. (1981), *Introduction to Tensors, Spinors and Relativistic Wave-Equations* (2nd edition), Chelsea Publishing, New York (5.1).
Craig, K.H. & Llewellyn Smith, C.H. (1978), *Phys. Lett.* **72B**, 349 (14.3).
Creutz, M. (1983), *Quarks, Gluons and Lattices*, Cambridge University Press, Cambridge (12.4).
Creutz, M. & Wang, L.L. (1974), *Phys. Rev.* **D11**, 3749 (13.4).
Curci, G., Furmanski, W. & Petronzio, R. (1980), *Nucl. Phys.* **B175**, 27 (14.3).
Cutler, R. & Sivers, D. (1977), *Phys. Rev.* **D16**, 679 (8.4, 8.5).
Cutler, R. & Sivers, D. (1978), *Phys. Rev.* **D17**, 196 (8.5).

Davies, H. (1963), *Proc. Camb. Phil. Soc.* **59**, 147 (3.1).
Del Duca, V. (1990), *Nucl. Phys.* **B345**, 369 (12e).
de Wit, B.S. & Smith, J. (1986), *Field Theory in Particle Physics*, North Holland, Amsterdam (1.5).
DeWitt, B.S. (1967), *Phys. Rev.* **162**, 1195 (7.3).
DeWitt-Morette, C., Maheshwari, A. & Nelson, B. (1979), *Phys. Rep.* **50**, 255 (3i).
Di'Lieto, C., Gendron, S., Halliday, I.G. & Sachrajda, C.T. (1981), *Nucl. Phys.* **B183**, 223 (13.5).
Dirac, P.A.M. (1927), *Proc. Roy. Soc. (London)* **A114**, 243 (2.3).
Dirac, P.A.M. (1928), *Proc. Roy. Soc. (London)* **A117**, 610 (5.2, 6.5).
Dirac, P.A.M. (1951), *Proc. Roy. Soc. (London)* **A209**, 291 (7e).
Dokshitzer, Yu L., Dyakonov, D.I. & Troyan, S.I. (1980), *Phys. Rep.* **58**, 270 (14.4).
Dokshitzer, Yu L., Khoze, V.A. & Troyan, S.I. (1989), in *Perturbative Quantum Chromodynamics*, ed. A.H. Mueller, World Scientific, Singapore (14.3).
Doria, R., Frenkel, J. & Taylor, J.C. (1980), *Nucl. Phys.* **B168**, 93 (13.5).
Drell, S.D. & Yan, T.-M. (1971), *Ann. Phys. (New York)* **66**, 578 (8.6).
Duke, D.W. & Roberts, R.G. (1985), *Phys. Rep.* **120**, 275 (12.4).
Duncan, A. & Furmanski, W. (1983), *Nucl. Phys.* **B226**, 339 (14.5).

Duncan, A. & Mueller, A.H. (1980), *Phys. Rev.* **D21**, 1636 (15).
Dyson, F.J. (1949), *Phys. Rev.* **75**, 486; 1736 (15, A).
Dyson, F.J. (1952), *Phys. Rev.* **85**, 631 (15).

Eden, R.J., Landshoff, P.V., Olive, D.I. & Polkinghorne, J.C. (1966), *The Analytic S-Matrix*, Cambridge University Press, Cambridge (13.1, 13.4).
Edmonds A.R. (1960), *Angular Momentum in Quantum Mechanics*, Princeton University Press, Princeton (1.5).
Edwards, S.F. & Peierls, R.E. (1954), *Proc. Roy. Soc. (London)* **224**, 24 (3.1).
Efremov, A.V. & Radyushkin, A.V. (1980a), *Teor. Mat. Fiz.* **42**, 147 (*Theor. Math. Phys.* **42**, 97) (15).
Efremov, A.V. & Radyushkin, A.V. (1980b), *Teor. Mat. Fiz.* **44**, 17; 157 (*Theor. Math. Phys.* **44**, 573; 664) (14.3).
Eichten, E., Hinchliffe, I., Lane, K. & Quigg, C. (1984), *Rev. Mod. Phys.* **56**, 579 (14.4).
Ellis, R.K., Furmanski, W. & Petronzio, R. (1983), *Nucl. Phys.* **B212**, 29 (14.5).
Ellis, R.K., Georgi, H., Machacek, M., Politzer, H.D. & Ross, G.G. (1979), *Nucl. Phys.* **B152**, 285 (13.4, 14.3).
Englert, F. & Brout, R. (1964), *Phys. Rev. Lett.* **13**, 321 (5.4).

Faddeev, L.D. (1969), *Teor. Mat. Fiz.* **1**, 3 (*Theor. Math. Phys.* **1**, 1) (6.5).
Faddeev, L.D. (1981), in *Methods in Field Theory*, Les Houches 1975, Sessions XXVIII, eds R. Balian & J. Zinn-Justin, North-Holland/World Scientific, Singapore (3.2, 7.1).
Faddeev, L.D. & Popov, V.N. (1967), *Phys. Lett.* **25B**, 29 (7.3).
Faddeev, L.D. & Shatashvili, S.L. (1986), *Phys. Lett.* **167B**, 225 (11.4).
Faddeev, L.D. & Slavnov, A.A. (1980) *Gauge Fields, Introduction to Quantum Theory*, Benjamin/Cummings, Reading, MA (3.2).
Fadin, V.S. & Kuraev, E.A. (1985), *Yad. Fiz.* **41**, 733 (*Sov. J. Nucl. Phys.* **41**, 466) (14.3).
Farrar, G.R. & Jackson, D.R. (1979), *Phys. Rev. Lett.* **43**, 246 (15).
Fermi, E. (1932), *Rev. Mod. Phys.* **4**, 87 (6.5).
Fermi, E. (1934), *Zeit. für Physik* **88**, 161 (5.4).
Feynman, R.P. (1948a), *Phys. Rev.* **74**, 939 (2.4).
Feynman, R.P. (1948b), *Rev. Mod. Phys.* **20**, 367 (3i).
Feynman, R.P. (1949), *Phys. Rev.* **76**, 749; 769 (8.2, 9.1, A).
Feynman, R.P. (1951), *Phys. Rev.* **84**, 108 (3.1).
Feynman, R.P. (1963), *Acta Physica Polonica* **24**, 697 (7.3, 13.2).
Feynman, R.P. & Hibbs, A.R. (1965), *Quantum Mechanics and Path Integrals*, McGraw-Hill, New York (3i, 3e).
Feynman, R.P. (1969a), *Phys. Rev. Lett.* **23**, 1415 (8.6).
Feynman, R.P. (1969b), in *High Energy Collisions*, eds C.N. Yang, J.A. Cole, M. Good, R. Hwa & J. Lee-Franzini, Gordon and Breach, New York (8.6).
Feynman, R.P. (1972), *Photon–Hadron Interactions*, Benjamin, Reading, MA (8.6).
Fradkin, E.S. (1954), *Dokl. Akad. Nauk. SSSR* **98**, 47 (3.1).
Fradkin, E.S. (1955a), *Dokl. Akad. Nauk. SSSR*, **100**, 897 (3.1).
Fradkin, E.S. (1955b), *J. Exp. Theor. Phys. USSR (Sov. Phys. JETP* **1**, 604) (11.2).
Frampton, P.H. (1986), *Gauge Field Theories*, Benjamin/Cummings, Menlo Park, CA (11.3).
Friedman, J.I. & Kendall, H.W. (1972), *Ann. Rev. Nucl. Sci.* **22**, 203 (8.6, 14.2).
Frishman, Y. (1970), *Phys. Rev. Lett.* **25**, 966 (14.5).

Fritzsch, H., Gell-Mann, M. & Leutwyler, H. (1973), *Phys. Lett.* **47B**, 365 (8.4).
Fujikawa, K. (1979), *Phys. Rev. Lett.* **42**, 1195 (11.4).
Fujikawa, K. (1980), *Phys. Rev.* **D21**, 2848 (11.4).
Fujikawa, K., Lee, B.W. & Sanda, A.I. (1972), *Phys. Rev.* **D6**, 2923 (11.3, C).
Furry, W.H. (1937), *Phys. Rev.* **51**, 125 (11.1).

Garrod, C. (1966), *Rev. Mod. Phys.* **38**, 483 (3.1).
Gasser, J. & Leutwyler, H. (1975), *Nuc. Phys.* **B94**, 269 (12.4, E).
Gasser, J. & Leutwyler, H. (1982), *Phys. Rep.* **87**, 77 (12.4, E).
Gell-Mann, M. (1962), *Phys. Rev.* **125**, 1067 (8.4).
Gell-Mann, M. & Levy, M. (1960), *Nuovo Cim. Ser. 10* **16**, 705 (E).
Gell-Mann, M. & Low, F. (1951), *Phys. Rev.* **84**, 350 (3.4).
Georgi, H. (1982), *Lie Algebras in Particle Physics*, Benjamin/Cummings, Menlo Park, CA (1.4).
Georgi, H. & Glashow, S.L. (1972), *Phys. Rev.* **D6**, 429 (11.4, C).
Georgi, H. & Politzer, H.D. (1974), *Phys. Rev.* **D9**, 416 (14.4).
Glashow, S.L. (1961), *Nucl. Phys.* **22**, 579 (5.4).
Glashow, S.L., Iliopoulos, J. & Maiani, L. (1970), *Phys. Rev.* **D2**, 1285 (C).
Glimm, J. & Jaffe, A. (1981), *Quantum Physics, a Functional Integral Point of View*, Springer-Verlag, New York (3i, 3.1).
Goldstein, H. (1980), *Classical Mechanics* (2nd edition), Addison-Wesley, Reading, MA (1e).
Goldstone, J. (1961), *Nuovo Cim. Ser. 10* **19**, 154 (5.4).
Goldstone, J., Salam, A. & Weinberg, S. (1962), *Phys. Rev.* **127**, 965 (5.4, E).
Gordon, W. (1927), *Zeit. für Physik* **40**, 117 (1.2).
Gordon, W. (1928), *Zeit. für Physik* **50**, 630 (12.1).
Gorishny, S.G., Kataev, A.L. & Larin, S.A. (1991), Phys. Lett. **259B**, 144 (12.4).
Grammer, G., Jr & Yennie, D.R. (1973), *Phys. Rev.* **D8**, 4332 (12.2, 12.3).
Green, M.B., Schwarz, J.H. & Witten, E. (1987), *Superstring Theory*, Cambridge University Press, Cambridge (10i).
Greenberg, O.W. & Messiah, A.M.L. (1965), *Phys. Rev.* **138**, B1155 (6.5).
Gribov, V.N. (1978), *Nucl. Phys.* **B139**, 1 (7.3).
Gribov, V.N. & Lipatov, L.N. (1972a), *Yad. Fiz.* **15**, 781 (*Sov. J. Nucl. Phys.* **15**, 438) (14.2, 14.3).
Gribov, V.N. & Lipatov, L.N. (1972b), *Yad. Fiz.* **15**, 1218 (*Sov. J. Nucl. Phys.* **15**, 675) (14.2, 14.3).
Gribov, L.V., Levin, E.M. & Ryskin, M.G. (1983), *Phys. Rep.* **100**, 1 (14.3).
Gross, D.J. (1981), in *Methods in Field Theory*, Les Houches 1975, Session XXVIII, eds R. Balian & J. Zinn-Justin, North-Holland/World Scientific, Singapore (10.5, 11.1, 15).
Gross, D.J. & Jackiw, R. (1972), *Phys. Rev.* **D6**, 477 (11.4).
Gross, D.J. & Treiman, S.B. (1971), *Phys. Rev.* **D4**, 1059 (14.5).
Gross, D.J. & Wilczek, F. (1973), *Phys. Rev. Lett.* **30**, 1343; *Phys. Rev.* **D8**, 3633 (8.4, 11.1).
Gross, D.J. & Wilczek, F. (1974), *Phys. Rev.* **D9**, 980 (14.4).
Grosso-Pilcher, C. & Shochet, M.J. (1986), *Ann. Rev. Nucl. Part. Sci.* **36**, 1 (8.6).
Gupta, S.N. (1950), *Proc. Phys. Soc. (London)* **A68**, 681 (6.5).
Guralnik, G.S., Hagen, C.R. & Kibble, T.W.B. (1964), *Phys. Rev. Lett.* **13**, 585 (5.4).

Hagg, R. (1955), *Kgl. Danske Videnskab. Selskab. Mat.-Fys. Medd.* **29**, 12 (A).
Hahn, Y. & Zimmermann, W. (1968), *Commun. Math. Phys.* **10**, 330 (9.4).

Halpern, M.B., Jevicki, A. & Senjanovic, P. (1977), *Phys. Rev.* **D16**, 2476 (7.1).
Hamermesh, M. (1962), *Group Theory*, Addison-Wesley, Reading, MA (1.4, 14.3, 7.3).
Harada, K., Kaneko, T. & Sakai, N. (1979), *Nucl. Phys.* **B155**, 169; erratum, *Nucl. Phys.* **B165**, 545 (1980) (14.3).
Heisenberg, W. & Pauli, W. (1929), *Zeit. für Physik* **56**, 1 (2.1).
Hepp, K. (1966), *Comm. Math. Phys.* **2**, 301 (10.4).
Higgs, P.W. (1964), *Phys. Lett.* **12**, 132 (5.4).
Higgs, P.W. (1966), *Phys. Rev.* **145**, 1156 (5.4).
Hill, E.L. (1951), *Rev. Mod. Phys.* **23**, 253 (1.3).
Horgan, R. & Jacob, M. (1981), *Nucl. Phys.* **B179**, 441 (8.6).
Huang, K. (1982), *Quarks, Leptons and Gauge Fields*, World Scientific, Singapore (5.4).
Humpert, B. & van Neerven, W.L. (1979), *Phys. Lett.* **84B**, 327 (14.3).

Isgur, N. & Llewellyn Smith, C.H. (1989), *Nucl. Phys.* **B317**, 526 (15).
Itzykson, C. & Zuber, J.-B. (1980), *Quantum Field Theory*, McGraw-Hill, New York (2.5, 3.1, 3.4, 9.3, 9.4, 10.5, 11.3, A, E).

Jackiw, R. & Rajaraman, R. (1985), *Phys. Rev. Lett.* **54**, 1219 (11.4).
Jackson, J.D. (1975), *Classical Electrodynamics* (2nd edition), John Wiley and Sons, New York (1e, 12.2).
Jacob, M. & Wick, G.C. (1959), *Ann. Phys. (New York)* **7**, 404 (6.3, 6e).
Jaffe, R.L. (1983), *Nucl. Phys.* **B229**, 205 (14.3, 14.5).
Jauch, J.M. & Rohrlich, F. (1954), *Helv. Phys. Acta.* **27**, 613 (12.1, 12.2, 12.3).
Jauch, J.M. & Rohrlich, F. (1955), *The Theory of Photons and Electrons*, Addison-Wesley, Cambridge, MA (12.1, 12.3).
Johnson, K. & Baker, M. (1973), *Phys. Rev.* **D8**, 1110 (12.3).
Jones, D.R.T. (1974), *Nucl. Phys.* **B75**, 531 (12.4).
Jordan, P. & Klein, O. (1927), *Zeit. für Physik* **45**, 751 (2.1).
Jordan, P. & Wigner, E. (1928), *Zeit. für Physik* **47**, 631 (6.5).
Jost, R. (1965), *The General Theory of Quantized Fields*, Lectures in Applied Mathematics, Vol. IV, American Mathematical Society, Providence, RI (2.5, 5.1).

Källén, G. (1952), *Helv. Phys. Acta* **25**, 417 (4e).
Keller, J.B. & McLaughlin, D.W. (1975), *Amer. Math. Monthly* **82**, 451 (3i).
Kim, Y.S. & Noz, M.E. (1986), *Theory and Applications of the Poincaré Group*, Reidel, Boston, MA (5.1).
Kim, Y.S. & Wigner, E.P. (1987), *J. Math. Phys.* **28**, 1175 (6e).
Kinoshita, T. (1950), *Progr. Theor. Phys. (Kyoto)* **5**, 1045 (8.1).
Kinoshita, T. (1962), *J. Math. Phys.* **3**, 650 (13.5).
Kinoshita, T. & Lindquist, W.B. (1983), *Phys. Rev.* **D27**, 867 (11.1, 12.1).
Kinoshita, T. & Sirlin, A. (1958), *Phys. Rev.* **113**, 1652 (12.5).
Klein, O. (1927), *Zeit. für Physik* **41**, 407 (1.2).
Klein, O. & Nishina, Y. (1929), *Zeit. für Physik* **52**, 853 (8.2).
Kobayashi, M. & Maskawa, T. (1973), *Progr. Theor. Phys. (Kyoto)* **49**, 652 (C).
Komamiya, S., Le Diberder, F. *et al.* (1990), *Phys. Rev. Lett.* **64**, 987 (12.4).
Korchemsky, G.P. & Radyushkin, A.V. (1986), *Phys. Lett.* **B171**, 459 (13.4).
Kramer, G. & Lampe, B. (1989), *Fortschr. Phys.* **37**, 161 (12.5).
Kubar-André, J. & Paige, F.E. (1979), *Phys. Rev.* **D19**, 221 (14.3).
Kugo, T. & Ojima, I. (1979), *Suppl. Progr. Theor. Phys.* **66**, 1 (6.5).

Kuhn, J.H. & Zerwas, P.M. (1988), *Phys. Rep.* **167**, 321 (15).

Kulish, P.P. & Faddeev, L.D. (1970), *Teor. Mat. Fiz.* **4**, 153 (*Theor. Math. Phys.* **4**, 745) (12.2).

Labastida, J.M.F. (1984), *Nucl. Phys.* **B239**, 583 (13.5).

Lam, Y.-M. P. (1973), *Phys. Rev.* **D7**, 2943 (11.2, 11.3).

Lamb, W.E., Jr and Retherford, R.C. (1947), *Phys. Rev.* **72**, 241 (12e).

Landau, L.D. (1959), *Nuc. Phys.* **13**, 181 (13.1).

Landau, L.D. & Pomeranchuk, I. (1955), *Doklady Academii Nauk.* **102**, 489 (15).

Langacker, P. & Pagels, H. (1979), *Phys. Rev.* **D19**, 2070 (12.4, E).

Lee, B.W. (1981), in *Methods in Field Theory*, Les Houches 1975, Session XXVIII, eds R. Balian & J. Zinn-Justin, North-Holland/World Scientific, Singapore (11.3).

Lee, B.W. & Zinn-Justin, J. (1972), *Phys. Rev.* **D5**, 3121; 3137; 3155 (11.3).

Lee, B.W. & Zinn-Justin, J. (1973), *Phys. Rev.* **D7**, 1049 (11.3).

Lee, T.D. (1981), *Particle Physics and Introduction to Field Theory*, Harwood Academic Publishers, New York (D).

Lee, T.D. & Nauenberg, M. (1964), *Phys. Rev.* **133**, B1549 (13.5).

Lee, T.D. & Yang, C.N. (1956), *Phys. Rev.* **104**, 254 (6.6).

Lee, T.D. & Yang, C.N. (1962), *Phys. Rev.* **128**, 885 (5.4).

Lehmann, H. (1954), *Nuovo Cim. Ser. 9* **11**, 342 (4e).

Lehmann, H., Symanzik, K. & Zimmermann, W. (1955), *Nuovo Cim. Ser. 10* **1**, 205 (2.5).

Leibbrandt, G. (1975), *Rev. Mod. Phys.* **47**, 849 (11.1).

Leibbrandt, G. (1987), *Rev. Mod. Phys.* **59**, 1067 (7.3, 14.3).

Lepage, G.P. & Brodsky, S.J. (1980), *Phys. Rev.* **D22**, 2157 (15).

Levin, E.M. & Ryskin, M.G. (1990), *Phys. Rep.* **189**, 267 (14.3).

Levy, M. & Sucher, J. (1969), *Phys. Rev.* **186**, 1656 (12.3).

Libby, S.B. & Sterman, G. (1978), *Phys. Rev.* **D18**, 3252 (14.3).

Lipatov, L.N. (1974), *Yad. Fiz.* **20**, 181 (*Sov. J. Nucl. Phys.* **20**, 94 (1975)) (14.2).

Lipatov, L.N. (1977), *Pis'ma Zh. Eksp. Teor. Fiz.* **25**, 116 (*JETP Lett.* **25**, 104) (15).

Lipatov, L.N. (1989), in *Perturbative Quantum Chromodynamics*, ed. A.H. Mueller, World Scientific, Singapore (14.3).

Llewellyn Smith, C.H. & de Vries, J.P. (1988), *Nucl. Phys.* **B296**, 991 (14.5).

Low, F.E. (1955), *Phys. Rev.* **97**, 1392 (2.5).

Low, F.E. (1958), *Phys. Rev.* **110**, 974 (12e).

Lubański, J.K. (1942), *Physica* **9**, 310 (6.1).

Lynn, BW. & Wheater, J.F., eds (1984), *Workshop on Radiative Corrections in* $SU(2)_L \times U(1)$, World Scientific, Singapore (8.3).

Lyubarskii, G. Ya. (1960), *The Application of Group Theory in Physics*, trans. S. Dedijer, Pergamon, New York (5.1).

Macfarlane, A.J., Sudbery, A. & Weisz, R.H. (1968), *Commun. Math. Phys.* **11**, 77 (8.4).

Macfarlane, A.J. & Woo, G. (1974), *Nucl. Phys.* **B77**, 91; erratum, *Nucl. Phys.* **B86**, 548 (10.4, 10e, F).

Magnea, L. (1991), *Nucl. Phys.* **B 349**, 703 (14.3).

Majorana, E. (1937), *Nuovo Cim.* **14**, 171 (6.5).

Mandelstam, S. (1955), *Proc. Roy. Soc.* **A233**, 248 (15).

Mandelstam, S. (1958), *Phys. Rev.* **112**, 1344 (4.2).

Marciano, W.J. & Parsa, Z. (1986), *Ann. Rev. Nucl. Part. Sci.* **36**, 171 (8.3).

Marciano, W.J. & Sirlin, A. (1980), *Phys. Rev.* **D22**, 2695 (8.3).

Marinov, M.S. (1980), *Phys. Rep.* **60**, 1 (3i).
Marques, G.C. (1974), *Phys. Rev.* **D9**, 386 (13.4).
Matthews, P.T. (1949), *Phys. Rev.* **76**, 684; erratum, *ibid.*, 1489 (4.4, A).
Matthews, P.T. & Salam, A. (1955), *Nuovo Cim. Ser. 10* **2**, 120 (7.1).
Matsuura, T., van der Marck, S.C. & van Neerven, W.L. (1989), *Nucl. Phys.* **B319**, 570 (14.3).
Matveev, V.A., Muradyan, R.M. & Tavkhelidze, A.V. (1973), *Lettere al Nuovo Cim.* **7**, 719 (15).
Mermin, N.D. & Wagner, H. (1966), *Phys. Rev. Lett.* **17**, 1133 (E).
Mishra, S.R. & Sciulli, F. (1989), *Ann. Rev. Part. Nucl. Sci.* **39**, 259 (14.4).
Møller, C. (1932), *Ann. Phys.* **14**, 531 (8.1).
Mueller, A.H. (1974), *Phys. Rev.* **D9**, 963 (14.3).
Mueller, A.H. (1978), *Phys. Rev.* **D18**, 3705 (14.3).
Mueller, A.H. (1979), *Phys. Rev.* **D20**, 2037 (13.4).
Mueller, A.H. (1981), *Phys. Rep.* **73**, 237 (12.5).
Mueller, A.H. (1985), *Nucl. Phys.* **B250**, 327 (15).
Muhki, S. & Sterman, G. (1982), *Nucl. Phys.* **B206**, 221 (12.5).

Nakanishi, N. (1958), *Progr. Theor. Phys. (Kyoto)* **19**, 159 (8.1, 13.5)
Nambu, Y. & Jona-Lasinio, G. (1961), *Phys. Rev.* **122**, 345 (5.4).
Nelson, C.A. (1981), *Nucl. Phys.* **B186**, 187 (13.5).
Nesterenko, V.A. & Radyushkin, A.V. (1982), *Phys. Lett.* **115B** 410 (15).
Nesterenko, V.A. & Radyushkin, A.V. (1983), *Phys. Lett.* **128B**, 439 (15).
Newton, T.D. & Wigner, E. (1949), *Rev. Mod. Phys.* **21**, 400 (2.4).
Nishijima, K. (1950), *Progr. Theor. Phys.* **5**, 405 (4.4, A).
Noether, E. (1918), *Nachr. kgl. Ges. Wiss. Gottingen*, 235 (1.3).

OPAL Collaboration (1990), *Phys. Lett.* **235**, 389 (12.4).
Ore, F.R., Jr & Sterman, G. (1980), *Nucl. Phys.* **B165**, 93 (13.5).

Panofsky, W.K.H. & Phillips, M. (1962), *Classical Electricity and Magnetism* (2nd edition), Addison-Wesley, Reading, MA (5.3, 8.2, 8.6).
Parisi, G. (1979), *Phys. Rep.* **49**, 215 (15).
Particle Data Group (1990), Review of particle properties, *Phys. Lett.* **B 239**, 1 (4.3, 8.2, 8.6, 12.4, C, D, E).
Paschos, E.A. & Türke, U. (1989), *Phys. Rep.* **178**, 145 (C, D).
Pauli, W. (1940), *Phys. Rev.* **58**, 716 (6.5).
Pauli, W. & Villars, F. (1949), *Rev. Mod. Phys.* **21**, 434 (9.3).
Perl, M.L. (1974), *High Energy Hadron Physics*, John Wiley and Sons, New York (8.6).
Poggio, E.C. & Quinn, H.R. (1976), *Phys. Rev.* **D14**, 578 (13.3).
Polchinski, J. (1984), *Nucl. Phys.* **B231**, 269 (10.5).
Politzer, H.D. (1973), *Phys. Rev. Lett.* **30**, 1346 (11.1).
Politzer, H.D. (1977), *Nucl. Phys.* **B129**, 301 (14.3).
Proca, A. (1936), *J. Phys. Rad.* **7**, 347 (5.3).

Qiu, J. & Sterman, G. (1991), *Nucl. Phys.* **B353**, 137 (14.5).
Quigg, C. & Rosner, J.L. (1979), *Phys. Rep.* **56**, 167 (15).

Rajaraman, R. (1982), *Solitons and Instantons*, North Holland, New York (3.1, 3e, 15).
Rarita, W. & Schwinger, J. (1941), *Phys. Rev.* **60**, 61 (5e).

Rivier, D. (1949), *Helv. Phys. Acta* **22**, 265 (2.4).
Rohrlich, F. (1950), *Phys. Rev.* **80**, 666 (4.4, A).
Rose, M.E. (1957), *Elementary Theory of Angular Momentum*, Wiley, New York (1.5).
Roy, P. (1975), *Theory of Lepton–Hadron Processes at High Energies*, Oxford University Press, Oxford (8.6, 14.1).

Sachrajda, C.T. (1978), *Phys. Lett.* **73B**, 185 (14.3).
Sachs, R.G. (1987), *The Physics of Time Reversal*, The University of Chicago Press, Chicago (6.6, D).
Sakita, B. (1985), *Quantum Theory of Many-variable Systems and Fields*, World Scientific, Singapore (3.2).
Sakurai, J.J. (1964) *Invariance Principles and Elementary Particles*, Princeton University Press, Princeton (6.6, D).
Salam, A. (1968), in *Elementary Particle Physics (Nobel Symposium No. 8)*, ed. N. Svartholm, Almqvist and Wilsell, Stockholm (5.4).
Schulman, L.S. (1981), *Techniques and Applications of Path Integration*, John Wiley, New York (3i, 3e).
Schrödinger, E. (1926), *Ann. Physik* **81**, 109 (1.2).
Schweber, S.S. (1961), *An Introduction to Relativistic Quantum Field Theory*, Row, Peterson; Evanston, IL (2.4, 6.5).
Schwinger, J. (1948), *Phys. Rev.* **73**, 416 (12.1, A).
Schwinger, J. (1949), *Phys. Rev.* **74**, 1439; *ibid.*, **75**, 651 (A).
Schwinger, J. (1951a), *Proc. Nat. Acad. Sci.* **37**, 452; 455 (3.1, 15).
Schwinger, J. (1951b), *Phys. Rev.* **82**, 664 (11.4).
Sen, A. (1981), *Phys. Rev.* **D24**, 3281 (13.4).
Shifman, M.A., Vainshtein, A.I. & Zakharov, V.I. (1979), *Nucl. Phys.* **B147**, 385; 448; 519 (15).
Shuryak, E.V. (1988), *The QCD Vacuum, Hadrons and the Superdense Matter*, World Scientific, Singapore (15).
Sirlin, A. (1980), *Phys. Rev. D* **22**, 971 (8.3).
Slavnov, A.A. (1972), *Teor. Mat. Fiz.* **10**, 153 (*Theor. and Math. Phys.* **10**, 99) (11.1).
Slavnov, A.A. (1975), *Teor. Mat. Fiz.* **22**, 177 (*Theor. and Math. Phys.* **22**, 123 (1975)) (3.1).
Smith, P.F. (1989), *Ann. Rev. Nucl. Part. Sci.* **39**, 73 (8.6, 12.4).
Soper, D.E. (1976), *Classical Field Theory*, Wiley, New York (1e).
Soper, D.E. (1981), *Phys. Rev.* **D18**, 4590 (7.1).
Speer, E.R. (1968), *J. Math. Phys.* **9**, 1404 (9.3).
Speer, E.R. (1969), *Generalized Feynman Amplitudes*, Princeton University Press, Princeton (9.3).
Speer, E.R. (1974), *J. Math. Phys. 15*, 1 (9.4).
Sterman, G. (1976), *Phys. Rev.* **D14**, 2123 (13.3).
Sterman, G. (1978), *Phys. Rev.* **D17**, 2773; 2789 (13.4, 13.5).
Sterman, G. & Weinberg, S. (1977), *Phys. Rev. Lett.* **39**, 1436 (12.5).
Stevenson, P.M. (1978), *Phys. Lett. 78***B**, 451 (12.5).
Streater, R.F. & Wightman, A.S. (1964), *PCT, Spin and Statistics, and All That*, Benjamin, New York (2.5, 5.1, 6.5, 6.6, A, D).
Stueckelberg, E.C.G. (1942), *Helv. Phys. Acta* **15**, 23 (2.4).
Sudakov, V.V. (1956), *Zh. Eksp. Teor. Fiz.* **30**, 87 (*Sov. Phys. JETP* **3**, 65 (1956)) (12.3, 13.5).
Surguladze, L.R. & Samuel, M.A. (1991), *Phys. Rev. Lett.* **66**, 560 (12.4).
Sutherland, D.G. (1967), *Nucl. Phys.* **B2**, 433 (E).

Symanzik, K. (1971), *Commun. Math. Phys.* **23**, 49 (14.5).
Symanzik, K. (1973), *Commun. Math. Phys.* **34**, 7 (8.3).

Takahashi, Y. (1957), *Nuovo Cim. Ser. 10* **6**, 370 (11.2).
Taylor, J.C. (1971), *Nucl. Phys.* **B33**, 436 (11.1).
Taylor, J.C. (1976), *Gauge Theories of Weak Interactions*, Cambridge University Press, Cambridge (9.6).
Thacker, H.B. & Weisberger, W.I. (1976), *Phys. Rev.* **D14**, 2658 (13.4).
't Hooft, G. (1971a), *Nucl. Phys.* **B33**, 173 (8.5, 11.2, 11.3).
't Hooft, G. (1971b), *Nucl. Phys.* **B35**, 167 (11.3, C).
't Hooft, G. (1973), *Nucl. Phys.* **B61**, 455 (10.5).
't Hooft, G. (1977), in *The Whys of Subnuclear Physics*, Erice 1977, ed. A. Zichichi, Plenum, New York (15).
't Hooft, G. & Veltman, M. (1972a), *Nucl. Phys.* **B44**, 189 (9.3, 11.3).
't Hooft, G. & Veltman, M. (1972b), *Nucl. Phys.* **B50**, 318 (11.3).
't Hooft, G. and Veltman, M. (1973), *Diagrammar*, CERN 73–9, reprinted in *Particle Interactions at Very High Energies*, eds D. Speiser, F. Halzen & J. Weyers, Plenum Press, New York (1974) (9.4, 9.6, 11.1, 11.2, 11.3).
Tkachov, F.V. (1983), *Phys. Lett.* **124B**, 212 (14.5).
Tobocman, W. (1956), *Nuovo Cim. Ser. 10* **3**, 1213 (3.1).
Tomonaga, S. (1946), *Progr. Theor. Phys. (Kyoto)* **1**, 27 (A).
Tucci, R. (1985), *Phys. Rev.* **D32**, 945 (12.3).

van Nieuwenhuizen, P. (1981), *Phys. Rep.* **68**, 189 (7.1, 11.3).
Veltman, M. (1963), *Physica* **29**, 186 (8.1, 9.6).
Veltman, M. (1967), *Proc. Roy. Soc. (London)* **A301**, 107 (E).
Veltman, M. (1977), *Nucl. Phys.* **B123**, 89 (8.3)

Ward, J.C. (1950), *Phys. Rev.* **78**, 182 (11.1).
Warr, B.J. (1988), *Ann. Phys.* **183**, 1, 59 (10.5).
Weeks, B.G. (1979), *Phys. Lett.* **81B**, 377 (12.5).
Weinberg, S. (1960), *Phys. Rev.* **118**, 838 (9.4).
Weinberg, S. (1967), *Phys. Rev. Lett.* **19**, 1264 (5.4).
Weinberg, S. (1970), in *Lectures on Elementary Particles and Quantum Field Theory*, Vol. 1, eds S. Deser, M. Grisaru & H. Pendleton, MIT Press, Cambridge, MA (E).
Weinberg, S. (1973a), *Phys. Rev.* **D7**, 1068 (5.4).
Weinberg, S. (1973b), *Phys. Rev.* **D8**, 3497 (10.5).
Weinberg, S. (1973c), *Phys. Rev. Lett.* **31**, 494 (8.4).
Weyl, H. (1929), *Zeit. für Physik* **56**, 330 (5.2).
Wick, G.C. (1950), *Phys. Rev.* **80**, 268 (3.1, 3.4).
Wick, G.C. (1954), *Phys. Rev.* **96**, 1124 (3.1, 9.1).
Wigner, E. (1932), *Nach. Gessell. Wissen. Gottingen* **31**, 546 (2.2, D).
Wigner, E. (1939), *Ann. Math.* **40**, 149 (2.2, 6.2, 6.4).
Wilson, K. (1969), *Phys. Rev.* **179**, 1499 (14.5).
Wolfenstein, L. (1986), *Ann. Rev. Nucl. Part. Sci.* **36**, 137 (D).
Wu, S.L. (1984), *Phys. Rep.* **107**, 59 (8.6, 12.4, 12.5).
Wybourne, B.G. (1974), *Classical Groups for Physicists*, John Wiley, New York (1.4, 7.3).

Yang, C.N. & Mills, R.L. (1954), *Phys. Rev.* **96**, 191 (5.4).
Yang, C.N. & Feldman, D. (1950), *Phys. Rev.* **79**, 972 (2.5).

Yao, Y.-P. (1973), *Phys. Rev.* **D7** 1647 (C).

Yennie, D.R., Frautschi, S.C. & Suura, H. (1961), *Ann. Phys. (NY)* **13**, 379 (12.2).

Yndurain, F.J. (1983), *Quantum Chromodynamics*, Springer-Verlag, New York (14.4, 14.5).

Yukawa, H. (1935), *Proc. Phys. Math. Soc. Japan* **17**, 48 (5.4).

Zee, A. (1973), *Phys. Rev.* **D8**, 4038 (12.4).

Zimmermann, W. (1970), in *Lectures on Elementary Particles and Quantum Field Theory, Vol. 1*, eds S. Deser M. Grisaru & H. Pendleton, MIT Press, Cambridge, MA (10.4, 11.2, 14.5).

Zimmermann, W. (1973), *Ann. Phys. (New York)* **77**, 536; 570 (14.5).

Zinn-Justin, J. (1974), in *Trends in Elementary Particle Theory*, International Summer Institute on Theoretical Physics, Bonn 1974, eds H. Rollink & K. Dietz, Lecture Notes in Physics 37, Springer-Verlag, Berlin (3.1).

Index